Hole's
essentials of
Human Anatomy & Physiology

tenth edition

David Shier
Washtenaw Community College

Jackie Butler
Grayson County College

Ricki Lewis
Genetic Counselor
CareNet Medical Group

McGraw-Hill
Higher Education

Boston   Burr Ridge, IL   Dubuque, IA   New York   San Francisco   St. Louis
Bangkok   Bogotá   Caracas   Kuala Lumpur   Lisbon   London   Madrid   Mexico City
Milan   Montreal   New Delhi   Santiago   Seoul   Singapore   Sydney   Taipei   Toronto

McGraw-Hill Higher Education

HOLE'S ESSENTIALS OF HUMAN ANATOMY & PHYSIOLOGY, TENTH EDITION

Published by McGraw-Hill, a business unit of The McGraw-Hill Companies, Inc., 1221 Avenue of the Americas, New York, NY 10020. Copyright © 2009 by The McGraw-Hill Companies, Inc. All rights reserved. Previous editions © 2006, 2003, 2000, 1998, 1995, 1992, 1989, 1986 and 1983. No part of this publication may be reproduced or distributed in any form or by any means, or stored in a database or retrieval system, without the prior written consent of The McGraw-Hill Companies, Inc., including, but not limited to, in any network or other electronic storage or transmission, or broadcast for distance learning.

Some ancillaries, including electronic and print components, may not be available to customers outside the United States.

 This book is printed on recycled, acid-free paper containing 10% postconsumer waste.

1 2 3 4 5 6 7 8 9 0 QPD/QPD 0 9 8

ISBN 978-0-07-296563-6
MHID 0-07-296563-0

Publisher: *Michelle Watnick*
Senior Sponsoring Editor: *James F. Connely*
Developmental Editor: *Fran Schreiber*
Marketing Manager: *Lynn M. Breithaupt*
Senior Project Manager: *Jayne Klein*
Lead Production Supervisor: *Sandy Ludovissy*
Senior Media Project Manager: *Judi David*
Senior Coordinator of Freelance Design: *Michelle D. Whitaker*
Cover/Interior Designer: *Christopher Reese*
Senior Photo Research Coordinator: *John C. Leland*
Photo Research: *Toni Michaels/PhotoFind, LLC*
Supplement Producer: *Mary Jane Lampe*
Compositor: *Precision Graphics*
Typeface: 10.5/12 *ITC Garamond Book*
Printer: *Quebecor World Dubuque, IA*

(USE) Cover Image: © *Greg Epperson/Getty images*
Cover description: A climber making a tyrolean traverse, Joshua Tree National Park, CA. A tyrolean traverse is a mountaineering maneuver that involves traversing between two high points by traveling across two ropes strung between them.

The credits section for this book begins on page 586 and is considered an extension of the copyright page.

Library of Congress Cataloging-in-Publication Data

Shier, David.
    Hole's essentials of human anatomy & physiology / David Shier, Jackie Butler, Ricki Lewis. -- 10th ed.
      p. cm.
    Includes index.
    ISBN 978-0-07-296563-6 — ISBN 0-07-296563-0 (hard copy : alk. paper) 1. Human physiology.
2. Human anatomy. I. Butler, Jackie. II. Lewis, Ricki. III. Title. IV. Title: Hole's essentials of human anatomy and physiology. V. Title: Essentials of human anatomy and physiology.

QP34.5.S49 2009
612--dc22

                                                    2007037363

Brief Contents

Contents

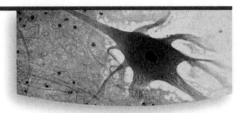
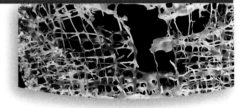

Unit 3
Integration and Coordination

Unit 4
Transport

Unit 5
Absorption and Excretion

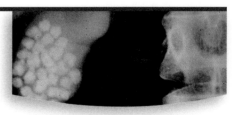

Unit 6
The Human Life Cycle

David Shier

David Shier has twenty-eight years of experience teaching anatomy and physiology, primarily to premedical, nursing, dental, and allied health students. He has effectively incorporated his extensive teaching experience into another student-friendly revision of *Hole's Essentials of Human Anatomy and Physiology* and *Hole's Human Anatomy and Physiology.* David has published in the areas of renal and cardiovascular physiology, the endocrinology of fluid and electrolyte balance, and hypertension. A faculty member in the Life Science Department at Washtenaw Community College, he is actively involved in a number of projects dealing with assessment, articulation, and the incorporation of technology into instructional design. David holds a Ph.D. in physiology from the University of Michigan.

Jackie Butler

Jackie Butler's professional background includes work at the University of Texas Health Science Center conducting research about the genetics of bilateral retinoblastoma. She later worked at Houston's M. D. Anderson Hospital investigating remission in leukemia patients. A popular educator for over twenty-five years at Grayson County College, Jackie teaches microbiology and human anatomy and physiology for health science majors. Her experience and work with students of various educational backgrounds have contributed significantly to another revision of *Hole's Essentials of Human Anatomy and Physiology* and *Hole's Human Anatomy and Physiology.* Jackie Butler received her B.S. and M.S. degrees from Texas A&M University, focusing on microbiology, including courses in immunology and epidemiology.

Ricki Lewis

Ricki Lewis is a science writer who earned her Ph.D. in genetics from Indiana University in 1980, where she worked on homeotic mutations in Drosophila. She is author of the McGraw-Hill Higher Education textbook *Human Genetics: Concepts and Applications,* now in its eighth edition, and an accompanying casebook; of the essay collection published by Blackwell Science; and of a soon-to-be published novel about stem cells. She has published thousands of articles and reports in a variety of magazines and journals, including *The Scientist* and *Nature.* Ricki is a fellow of the Alden March Bioethics Institute of Albany Medical College, a co-investigator of the Hidden Cancer Identification and Eradication Project at Weill Medical College, and is a course designer and instructor for the Northeast Regional Forensics Institute. Ricki has been a genetic counselor for CareNet Medical Group in Schenectady, NY and is a hospice volunteer. She is a member of the Undergraduate Education committee of the American Society of Human Genetics, and is a frequent public speaker.

A Note from John W. Hole, Jr.

When I prepared the first edition of *Essentials of Human Anatomy and Physiology,* a little more than a quarter of a century ago, I had been teaching the subject in a one semester course for about a dozen years, primarily to students in the nursing program of Rio Hondo College in Whittier, California. I had used several of the available textbooks and had found them in various ways to be unsatisfactory for our needs. As most teachers know, the best way to stimulate student learning is to somehow capture their interest in the subject matter. So when the opportunity arose to prepare a textbook to support a single semester program, I was pleased to give it a try and create a book that would meet the needs of my particular students, who as a group had minimal backgrounds in the physical and biological sciences. With the help and encouragement of members of the staff of the nursing department at Rio Hondo College, some of my former students, and a group of teaching consultants, each of whom provided many valuable suggestions during the preparation of the manuscript, I tried to design a textbook that was limited in scope to the essential anatomy and physiology material that my students needed in an accurate, interesting, and readable manner.

I am personally gratified that the textbook has found its place in many other classrooms over the years, and I am pleased to congratulate the present authors, who have continued to maintain the essential flavor of the book and have kept its contents up-to-date as well as adding many new, exciting, and attractive features to the presentation.

John W. Hole, Jr.

A Note from the Current Authors

Welcome to the tenth edition of *Hole's Essentials of Human Anatomy and Physiology.* In this new edition we continue our commitment to introducing the structure and function of the human body in an interesting and accessible manner.

We are authors, but first and foremost we are teachers. In much the way the first edition of this text evolved in John Hole's hands, this new edition continues to reflect what we and our reviewers do in class. Students have always come first in our approach to teaching and textbook authoring, but we now feel more excited than ever about the student-oriented, teacher-friendly quality of this text.

Along with updated versions of the extra resources that students and teachers alike have found so helpful over the years (Anatomy and Physiology Revealed®, text websites, etc.) we are especially pleased to present this tenth edition of *Hole's Essentials of Human Anatomy and Physiology* with a brand new feature. Each chapter is presented in a way that allows students (and teachers) to identify the specific learning outcomes associated with that chapter and to assess whether the student has mastered the related content. Teachers are likely familiar with outcomes assessment because of their importance for institutional accreditation. We are confident that students will benefit equally as they use this new feature to focus their study efforts and take an active role in monitoring their own progress. All of these resources are described in more detail in the Chapter Preview: Foundations for Success beginning on page xx.

Many of you are planning careers in health care, athletics, science, or education. We understand that you face the challenge of balancing family, work, and academics. Always remember that your course is not so much a hurdle on your journey as it is a stepping stone, even more so a foundation. We have written this book to help prepare you to travel that path.

David Shier
Jackie Butler
Ricki Lewis

Check out the Chapter Preview on page xx

Chapter Preview

was specifically designed to help students LEARN how to study at the collegiate level and efficiently use the tools available to them. It provides helpful study tips.

Chapter Preview

Foundations for Success

The Chapter Preview not only provides great study tips to offer a foundation for success, but it also offers tips on how to utilize this particular text.

OPENING VIGNETTE
Beginning each chapter is a vignette that discusses current events or research news relating to the subject matter in the chapter. These demonstrate applications of the concepts learned in the study of anatomy and physiology.

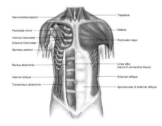

A photo on the opening page for each chapter generates interest.

It is a beautiful day. You can't help but stare wistfully out the window, the scent of spring blooms and sound of birds making it impossible to concentrate on what the instructor is saying. Gradually, the lecture fades as you become aware of your own breathing, the beating of your heart, and of a sheen of sweat that breaks out on your forehead in response to the radiant heat from the glorious day. Suddenly your reverie is cut short—the instructor has dropped a human anatomy and physiology textbook on your desk. You jump. Yelp. Your

heart hammers and a flash of fear grips your chest—but you soon realize what has happened and recover.

The message is clear: pay attention. So you do, tuning out the great outdoors and focusing on the lecture. In this course, you will learn all about the events that you have just experienced, including your response to the sudden stimulation of the instructor's wake-up call. This is a good reason to learn about how to stay focused in the course.

Learning Outcomes   *After you have studied this chapter, you should be able to:*

Each chapter begins with a list of outcomes you should be familiar with after studying the chapter. These are intended to help you master outcomes set by your instructor. Learning Outcomes will be tied to specific assessments (arrow) found at the end of each chapter.

P.1 Introduction
1. Explain the importance of an individualized approach to learning.

P.2 Strategies for Success
2. Summarize what you should do before attending class.
3. Identify student activities that enhance classroom experience.

4. List and describe several study techniques that can facilitate learning new material.

Topic of Interest

Burns

Slightly burned skin, as from a minor sunburn, may become warm and reddened (erythema) as dermal blood vessels dilate. This response may be accompanied by mild edema, and in time, the surface layer of skin may be shed. A burn injuring the epidermis alone is called a *superficial partial-thickness* (first degree) *burn*. Healing usually occurs within a few days to two weeks, with no scarring.

A burn that destroys some epidermis as well as some underlying dermis is a *deep partial-thickness* (second degree) *burn*. Fluid escapes from damaged dermal capillaries, and as it accumulates beneath the outer layer of epidermal cells, blisters appear. The injured region becomes moist and firm and may vary from dark red to waxy white. Such a burn most commonly occurs from exposure to hot objects, hot liquids, flames, or burning clothing.

Healing of a deep partial-thickness burn depends upon accessory structures of the skin that survive the injury because

they are deep in the dermis. These structures, which include hair follicles, sweat glands, and sebaceous glands, contain epithelial cells that divide and extend onto the surface of the injured dermis, spread over it, and form new epidermis.

A burn that destroys the epidermis, the dermis, and the accessory structures of the skin is a *full-thickness* (third degree) *burn*. The injured skin becomes dry and leathery, and may vary from red to black to white. A full-thickness burn usually is caused by immersion in hot liquids or prolonged exposure to hot objects, flames, or corrosive chemicals. Typically, most of the epithelial cells in the affected region are destroyed, and healing occurs only if epithelial cells divide and grow inward from the margin of the burn. If the injured area is extensive, it may require a transplant, using a thin layer of skin from an unburned region of the body (an autograft), cadaveric skin (a homograft), or a skin substitute (amniotic membrane from a human fetus, artificial membranes, or a tissue-engineered skin).

Remember when you were very young and encountered a substantial book for the first time? You were likely intimidated by its length, but were reassured that there were "a lot of pictures." This book has many illustrations too, all designed to help you master the material and become that person who you would want treating you.

Photographs and Line Art

Sometimes subdivisions have so many parts that the book goes to a third level, the "C-head." This information is presented in a slightly smaller font that identifies a specific section with an example.

Because line art can present different positions, layers, or perspectives, it can provide a unique view.

Photographs provide a realistic view of anatomy.

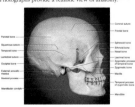

Macroscopic to Microscopic

Many figures show anatomical structures in a manner that is macroscopic to microscopic (or vice versa).

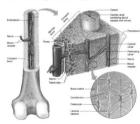

Flow Charts

Flow charts depict sequences of related events, steps of pathways, and complex concepts, easing comprehension. Other figures may show physiological processes.

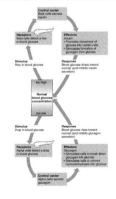

Anatomical Structures

Some figures illustrate the locations of anatomical structures.

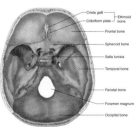

Other figures illustrate the functional relationships of anatomical structures.

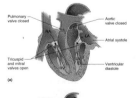

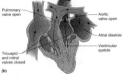

It also walks the student through the chapter explaining the features of the text.

Be sure to check it out!

Guided Tour Through a Chapter

Learning Outcomes & Assessments

This new feature highlights specific outcomes & then has specific assessments for each outcome.

8
Muscular System

DOUBLE THE MUSCLE. The newborn had an astonishing appearance—his prominent arm and thigh muscles looked as if he'd been weight-lifting in the womb. By five years of age, his muscles twice normal size, he could lift heavier weights than could many adults. He also had half the normal amount of body fat.

The boy's muscle cells cannot produce a protein called myostatin, which normally stops stem cells from making a muscle too large. In this boy a mutation turned off this genetic brake, and as a result his muscles bulge, their cells both larger and more numerous than those in the muscles of a normal child. The boy is healthy so far, but because myostatin is also made in cardiac muscle, he may develop heart problems.

Other species with myostatin mutations are well known. Naturally "double-muscled" cattle and sheep are valued for their high weights early in life. Chicken breeders lower myostatin production to yield meatier birds, and "mighty mice" with silenced myostatin genes are used in basic research to study muscle overgrowth. In clinical applications, researchers are investigating ways to block myostatin activity to stimulate muscle growth to reverse muscle-wasting from AIDS, cancer, and muscular dystrophy. Myostatin is also of interest in athletics. Theoretically, infants could be tested to identify those with myostatin gene variants that predict athletic prowess, given the right training. Myostatin could also be abused to enhance athletic performance.

For those of us not endowed with genetically doubled muscles, regular resistance training (weight lifting) can tone muscles and trim fat.

For those of us not endowed with double-muscle mutations, resistance (weight) training can increase the ratio of muscle to fat in our bodies, which offers several benefits. Because muscle burns calories at three times the rate of fat, a lean body is more energetically efficient. Weight-lifting increases muscle tone and bone density; lowers blood pressure; decreases the risks of developing arthritis, osteoporosis, and diabetes mellitus; and is even associated with improved self-esteem and fewer sick days.

Learning Outcomes   *After studying this chapter, you should be able to do the following:*

8.1 Introduction
1. List various outcomes of muscular actions. (p. 177)

8.2 Structure of a Skeletal Muscle
2. Describe how connective tissue is part of a skeletal muscle. (p. 177)
3. Name the major parts of a skeletal muscle fiber, and describe the function of each. (p. 177)
4. Discuss nervous stimulation of a skeletal muscle. (p. 180)

8.3 Skeletal Muscle Contraction
5. Identify the major events of skeletal muscle fiber contraction. (p. 181)

6. Describe the energy sources for muscle fiber contraction. (p. 183)
7. Describe how oxygen debt develops and how a muscle may become fatigued. (p. 184)

8.4 Muscular Responses
8. Distinguish between a twitch and a sustained contraction. (p. 187)
9. Explain how muscular contractions move body parts and help maintain posture. (p. 188)

8.5 Smooth Muscle
10. Distinguish between the structures and functions of multiunit smooth muscle and visceral smooth muscle. (p. 189)

11. Compare the contraction mechanisms of skeletal and smooth muscle fibers. (p. 189)

8.6 Cardiac Muscle
12. Compare the contraction mechanisms of cardiac and skeletal muscle fibers. (p. 190)

8.7 Skeletal Muscle Actions
13. Explain how the attachments, locations, and interactions of skeletal muscles make possible certain movements. (p. 192)

8.8 Major Skeletal Muscles
14. Describe the locations and actions of the major skeletal muscles of each body region. (p. 193)

176

Vignettes

lead into chapter content. They connect you to many areas of health care including technology, physiology, medical conditions, historical perspectives, and careers.

Learning Outcomes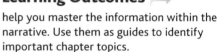

help you master the information within the narrative. Use them as guides to identify important chapter topics.

8.6 Cardiac Muscle (p. 190)
1. Like skeletal muscle cells, cardiac muscle cells have many actin and myosin filaments.
2. Cardiac muscle twitches last longer than skeletal muscle twitches.
3. Intercalated discs connect cardiac muscle cells.
4. A network of fibers contracts as a unit.
5. Cardiac muscle is self-exciting and rhythmic.

8.7 Skeletal Muscle Actions (p. 190)
The type of movement a skeletal muscle produces depends on the way the muscle attaches on either side of a joint.
1. Origin and insertion
   a. The immovable end is its origin, and the movable end of a skeletal muscle is its insertion.
   b. Some muscles have more than one origin.
2. Interaction of skeletal muscles
   a. Skeletal muscles function in groups.
   b. A prime mover is responsible for most of a movement. Synergists aid prime movers. Antagonists can resist the action of a prime mover.
   c. Smooth movements depend on antagonists giving way to the actions of prime movers.

8.8 Major Skeletal Muscles (p. 193)
1. Muscles of facial expression
   a. These muscles lie beneath the skin of the face and scalp and are used to communicate feelings through facial expression.
   b. They include the epicranius, orbicularis oculi, orbicularis oris, buccinator, zygomaticus, and platysma.
2. Muscles of mastication
   a. These muscles attach to the mandible and are used in chewing.
   b. They include the masseter and temporalis.

b. They include the external obli___, transversus abdominis, and re___
9. Muscles of the pelvic outlet
   a. These muscles form the floor ___ space within the pubic arch.
   b. They include the levator ani, s___, bulbospongiosus, and ischioca___
10. Muscles that move the thigh
   a. These muscles attach to the f___ pelvic girdle.
   b. They include the psoas major, ___, gluteus medius, gluteus minim___ adductor longus, adductor ma___
11. Muscles that move the leg
   a. These muscles connect the tib___ pelvic girdle.
   b. They include the biceps femori___, semimembranosus, sartorius, and the quadriceps femoris group.
12. Muscles that move the foot
   a. These muscles attach the femur, tibia, and fibula to bones of the foot.
   b. They include the tibialis anterior, fibularis tertius, extensor digitorum longus, gastrocnemius, soleus, flexor digitorum longus, tibialis posterior, and fibularis longus.

CHAPTER ASSESSMENTS

8.1 Introduction
1. The three types of muscle tissue are _____, _____, and _____. (p. 177)

8.2 Structure of a Skeletal Muscle
2. Describe the difference between a tendon and an aponeurosis. (p. 177)
3. Describe how connective tissue associates with skeletal muscle. (p. 177)
4. List the major parts of a skeletal muscle fiber, and describe the function of each part. (p. 177)
5. Describe a neuromuscular junction. (p. 180)
6. A neurotransmitter _____. (p. 180)
   a. binds actin filaments, causing them to slide
   b. travels across a synapse from a neuron to a muscle cell
   c. ferries ATP across a synapse
   d. travels across a synapse from a muscle cell to a neuron.
   e. is a contractile protein that is part of a skeletal muscle fiber.
7. Define *motor unit*. (p. 180)

8.3 Skeletal Muscle Contraction
8. List the major events of muscle fiber contraction and relaxation. (p. 181)
9. Describe how ATP and creatine phosphate interact. (p. 183)
10. Describe how muscles obtain oxygen. (p. 184)
11. Describe how an oxygen debt may develop. (p. 184)
12. Explain how muscles may become fatigued. (p. 185)
13. Explain how skeletal muscle function affects the maintenance of body temperature. (p. 186)

8.4 Muscular Responses
14. Define *threshold stimulus*. (p. 187)
15. Sketch a myogram of a single muscular twitch, and identify the latent period, period of contraction, and period of relaxation. (p. 187)
16. Explain motor unit *recruitment*. (p. 188)
17. Explain how skeletal muscle stimulation produces a sustained contraction. (p. 188)

Chapter Assessments <

check your understanding of the chapter's learning outcomes.

INTEGRATIVE ASSESSMENTS/ CRITICAL THINKING

OUTCOME 8.2
1. Discuss how connective tissue is part of the muscular system.

OUTCOME 8.3
2. As lactic acid and other substances accumulate in an active muscle they stimulate pain receptors, and the muscle may feel sore. How might the application of heat or substances that dilate blood vessels relieve such soreness?

OUTCOMES 8.3, 8.4
3. A woman takes her daughter to a sports medicine specialist and requests that she determine the percentage of fast- and slow-twitch fibers in the girl's leg muscles. The parent wants to know if the healthy girl should try out for soccer or cross-country running. Do you think this is a valid reason to test muscle tissue? Why or why not?

OUTCOMES 8.3, 8.4
4. Following an injury to a nerve, the muscle it supplies with motor nerve fibers may become paralyzed. How would you explain to a patient the importance of moving the disabled muscles passively or contracting them using electrical stimulation?

OUTCOMES 8.5, 8.8
5. What steps might be taken to minimize atrophy of the skeletal muscles in patients confined to bed for prolonged times?

Integrative Assessments/ Critical Thinking

questions involve relating information from various learning outcomes within a chapter and applying that information.

Aids to Understanding Words

help you remember scientific word meanings. Examine root words, stems, prefixes, suffices, pronunciations, and build a solid anatomy and physiology vocabulary.

Sample page 177:

Aids to Understanding Words

calat- [something inserted] intercalated disc: Membranous band that connects cardiac muscle cells.

erg- [work] synergist: Muscle that works with a prime mover to produce a movement.

hyper- [over, more] muscular *hypertrophy*: Enlargement of muscle fibers.

inter- [between] *intercalated* disc: Membranous band that connects cardiac muscle cells.

laten- [hidden] *latent* period: Time between application of a stimulus and the beginning of a muscle contraction.

myo- [muscle] *myofibril*: Contractile structure within a muscle cell.

sarco- [flesh] *sarcoplasm*: Material (cytoplasm) within a muscle fiber.

syn- [together] *synergist*: Muscle that works with a prime mover to produce a movement.

tetan- [stiff] *tetanic* contraction: Sustained muscular contraction.

-troph [well fed] muscular *hypertrophy*: Enlargement of muscle fibers.

8.1 INTRODUCTION

Talking and walking, breathing and sneezing—in fact, all movements—require muscles. Muscles are organs composed of specialized cells that use the chemical energy stored in nutrients to contract. Muscular actions also provide muscle tone, propel body fluids and food, generate the heartbeat, and distribute heat.

Muscles are of three types—skeletal muscle, smooth muscle, and cardiac muscle. This chapter focuses mostly on skeletal muscle, which attaches to bones and is under conscious control. Smooth muscle and cardiac muscle are discussed briefly.

8.2 STRUCTURE OF A SKELETAL MUSCLE

A skeletal muscle is an organ of the muscular system. It is composed of skeletal muscle tissue, nervous tissue, blood, and other connective tissues.

Connective Tissue Coverings

Layers of fibrous connective tissue called **fascia** (fash'e-ah) separate an individual skeletal muscle from adjacent muscles and hold it in position (fig. 8.1). This connective tissue surrounds each muscle and may project beyond its end to form a cordlike tendon. Fibers in a tendon may intertwine with those in a bone's periosteum, attaching the muscle to the bone. In other cases, the connective tissue forms broad fibrous sheets called **aponeuroses** (ap″o-nu-ro′sez), which may attach to bone or to the coverings of adjacent muscles (see figs. 8.17 and 8.19).

The layer of connective tissue that closely surrounds a skeletal muscle is called *epimysium* (fig. 8.1). Other layers of connective tissue, called *perimysium*, extend inward from the epimysium and separate the muscle tissue into small compartments. These compartments contain bundles of skeletal muscle fibers called *fascicles* (fasciculi). Each muscle fiber within a fascicle (fasciculus) lies within a layer of connective tissue in the form of a thin covering called *endomysium*. Layers of connective tissue, therefore, enclose and separate all parts of a skeletal muscle. This organization allows the parts to move somewhat independently. Many blood vessels and nerves pass through these layers.

A tendon, the attachment of a muscle to a bone, may become painfully inflamed and swollen following injury or the repeated stress of athletic activity, a condition called *tendinitis*. When the inflammation occurs in the connective tissue sheath of the tendon (the tenosynovium), it is called *tenosynovitis*. The tendons most commonly affected are those associated with the joint capsules of the shoulder, elbow, and hip and those that move the hand, thigh, and foot.

Skeletal Muscle Fibers

A skeletal muscle fiber is a single cell that contracts in response to stimulation and then relaxes when the stimulation ends. Each skeletal muscle fiber is a thin, elongated cylinder with rounded ends, and it may extend the full length of the muscle. Just beneath its cell membrane (or *sarcolemma*), the cytoplasm (or *sarcoplasm*) of the fiber has many small, oval nuclei and mitochondria (fig. 8.1). The sarcoplasm also contains many threadlike **myofibrils** (mi″o-fi′brilz) that lie parallel to one another.

Myofibrils play a fundamental role in muscle contraction. They consist of two kinds of protein filaments—thick ones composed of the protein **myosin** (mi′o-sin) and thin ones mainly composed of the protein **actin** (ak'tin) (figs. 8.2 and 8.3). (Two other thin filament proteins, troponin and tropomyosin, are discussed later on page 181.) The organization of these filaments produces the characteristic alternating light and dark *striations*, or bands, of a skeletal muscle fiber.

The striations form a repeating pattern of units called **sarcomeres** (sar′ko-mērz) along each muscle fiber. The myofibrils may be thought of as sarcomeres joined end-to-end (fig. 8.2). Muscle fibers, and in a way muscles themselves, may be considered a collection of sarcomeres. Sarcomeres are discussed later as the functional units of muscle contraction (p. 181).

Check Your Recall

by testing your understanding of key concepts.

Sample page 187:

Check Your Recall

11. Which biochemicals provide the energy to regenerate ATP?
12. What are the sources of oxygen for aerobic respiration?
13. How are lactic acid, oxygen debt, and muscle fatigue related?
14. What is the relationship between cellular respiration and heat production?

8.4 MUSCULAR RESPONSES

One way to observe muscle contraction is to remove a single muscle fiber from a skeletal muscle and connect it to a device that records changes in the fiber's length. Such experiments usually require an electrical device that can produce stimuli of varying strengths and frequencies.

Threshold Stimulus

When an isolated muscle fiber is exposed to a series of stimuli of increasing strength, the fiber remains unresponsive until a certain strength of stimulation called the **threshold stimulus** (thresh'old stim'u-lus) is applied. Once threshold is reached, an action potential is generated, resulting in a muscle impulse that spreads throughout the muscle fiber, releasing enough calcium ions from the sarcoplasmic reticulum to activate cross-bridge binding and contract that fiber. A single nerve impulse in a motor neuron normally releases enough ACh to bring the muscle fibers in its motor unit to threshold, generating a muscle impulse in each muscle fiber.

Recording of a Muscle Contraction

The contractile response of a single muscle fiber to a muscle impulse is called a **twitch.** A twitch consists of a period of contraction, during which the fiber pulls at its attachments, followed by a period of relaxation, during which the pulling force declines. These events can be recorded in a pattern called a myogram (fig. 8.12). Note that a twitch has a brief delay between the time of stimulation and the beginning of contraction. This is the **latent period**, which in human muscle may be less than 2 milliseconds.

When a muscle fiber is brought to threshold under a given set of conditions, it tends to contract completely, such that each twitch generates the same force. This has been referred to as an *all-or-none* response. This is misleading, however, because in normal use of muscles, the force generated by muscle fibers and by whole muscles must vary.

Figure 8.12
A myogram of a single muscle twitch.

Understanding the contraction of individual muscle fibers is important for understanding how muscles work, but such contractions by themselves are of little significance in day-to-day activities. Rather, the actions we need to perform usually require the contraction of multiple muscle fibers simultaneously. To record how a whole muscle responds to stimulation, a skeletal muscle can be removed from a frog or other small animal and mounted on a special device. The lever is then stimulated electrically; and when it contracts, it pulls on a lever. The lever's movement is recorded as a myogram. Because the myogram results from the combined twitches of muscle fibers taking part in the contraction, it looks essentially the same as the twitch contraction depicted in figure 8.12. The Topic of Interest on page 188 describes two types of twitches—the fatigue-resistant slow twitch and the fatigable fast twitch. Muscle fibers are either slow twitch or fast twitch.

The skeletal muscles of an average person have about half fast twitch and half slow twitch muscle fibers. In contrast, the muscles of an Olympic sprinter typically have more than 80% fast twitch muscle fibers, and those of an Olympic marathoner, more than 90% slow twitch muscle fibers.

Contractions of whole muscles enable us to perform everyday activities, but the force generated by those contractions must be controlled. For example, holding a styrofoam cup of coffee firmly enough that it does not slip through our fingers, but not so forcefully as to crush it, requires precise control of contractile force. In the whole muscle, the degree of tension developed reflects (1) the frequency at which individual muscle fibers are stimulated and (2) how many fibers take part in the overall contraction of the muscle.

Facts of Life

provide interesting bits of anatomy and physiology information, adding a touch of wonder to chapter topics.

Clinical Applications

are found in colored boxes and apply ideas and facts in the narrative to clinical situations.

Guided Tour Through a Chapter

Pull concepts together with real-life information

Topics of Interest

present disorders, physiological responses to environmental factors, and other topics of general interest.

Genetics Connections

explore the molecular underpinnings of familiar as well as not so familiar illnesses. Read about such topics as ion channel disorders, muscular dystrophy, and cystic fibrosis.

Clinical Terms

help you understand medical terminology. Lists of related terms often used in clinical situations are found at the end of several chapters.

Clinical Connections

are short discussions at the end of the chapter that give you a real-life connection to the material covered.

Chapter Summary Outlines

help you review the chapter's main ideas.

Organization **Illustrations**

found at the end of selected chapters conceptually link the highlighted body system to every other system and reinforce the dynamic interplay between systems. These illustrations help you review chapter concepts and reinforce the big picture in learning and applying the principles of anatomy and physiology.

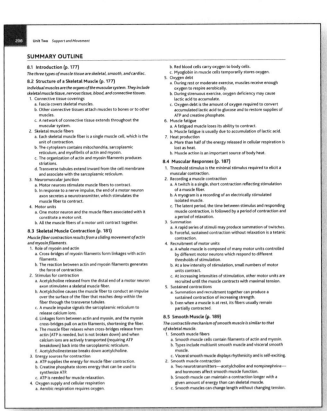

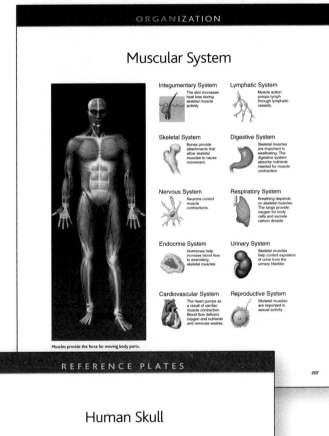

Reference Plates

offer vibrant detail of body structures.

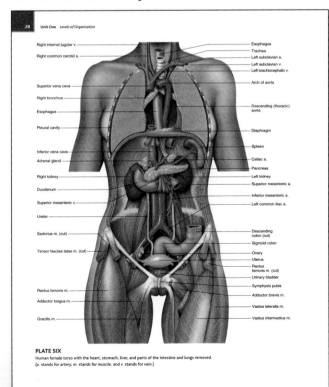

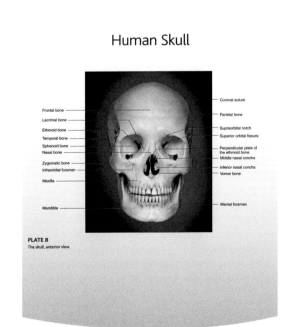

New to This Edition

Overview

- Learning Outcomes ➡ open the chapters, and are closely linked to Chapter Assessments and Integrated Assessments/Critical Thinking questions at each chapter's end ⬅. Many new questions.
- A new Chapter Preview—*Foundations for Success*—not only provides valuable study tips for students but also outlines how the students and instructors can utilize the pedagogical features of the text. **Be sure to check it out on page xx.**
- Many legends have been rewritten to more completely reflect text content, improving the pedagogical content of the figures.
- Added many new vignettes.

Chapter 1

- Improved figure 1.4 on internal environment.
- New figure 1.13 on directional terms.
- Improved figure 1.14 on anatomical planes.
- Greater emphasis on positive feedback.

Chapter 2

- New vignette tells the medical detective story of tracing a copper deficiency to zinc excess.
- Clarification that ionic bonds form crystals that dissociate in water.
- Explanation of buffers.
- Significant modification of figures 2.17 and 2.18, on amino acid and protein structure.

Chapter 3

- New vignette, "Watching Cells Specialize," explains derivation of human embryonic stem cells from 8-celled embryos.
- The *Genetics Connection* presents extreme and absent pain syndromes.

Chapter 4

- The rewritten vignette focuses on solving the arsenic problem in India and Bangladesh.
- A new *Topic of Interest,* "Beyond the Human Genome Project: Personalizing Medicine," stresses the many variations on "normal."

Chapter 5

- New vignette, "Donating Tissue for Research," offers compelling examples of people whose biological donations were used for profit.
- New *Clinical Connection* on tissue engineering tells how to build a bladder.

Chapter 6

- New small box describes tattoos.

Chapter 7

- New vignette, "Preventing Fragility Fractures," stresses the importance of attaining maximal bone density by age 30.
- Added bone classification at the start of the chapter.
- New figure 7.6 is a radiograph of epiphyseal plates, supplementing the illustration in figure 7.5.
- Coverage of factors affecting bone development, growth, and repair.
- New figure 7.37 illustrates all six types of synovial joints in place of detailed figures of shoulder, hip, and elbow.
- The *Topic of Interest* on repair of bone fractures now includes a list and figure showing types of fractures.

Chapter 8

- New vignette, "Double the Muscle," describes a mutation that made a newborn look like a weight-lifter—and is also found in chickens and cattle.
- Clearer explanation of events at the neuromuscular junction, including the muscle impulse.
- Improved explanation of the sliding filament model and the sarcomere as the functional unit of muscle contraction.
- Improved Table 8.1 summarizes the steps of muscle contraction.
- Updated discussion of control of contractile force of a whole muscle.
- Improved explanation of how the relationship of synergistic versus antagonistic muscles may change depending on the action.

Chapter 9

- New vignette, "Islands of Awareness in the Vegetative Brain," describes the surprising results of functional MRI on a woman thought to be unresponsive.
- Discussion of synapse earlier in chapter.
- New figure 9.12 of the resting potential and 9.13 of the action potential.

- Updated discussion of resting potential, synaptic potentials, threshold potentials, and the role of neurotransmitter.
- New figure 9.18 illustrates the components of a reflex arc.

Chapter 10

- Chapter renamed "The Senses."
- New vignette explores how the brain processes music.
- Updated figure 9.28 and discussion of Broca's area and Wernicke's area.
- Updated discussion of the five primary taste sensations.

Chapter 11

- Simplified figure 11.3 on the mechanism of non-steroid hormone action.
- Clarification of function of the intermediate lobe of the pituitary.
- The *Topic of Interest* on diabetes mellitus updates setback in islet transplants.

Chapter 12

- New vignette, "Universal Precautions" in a Marburg fever outbreak.
- Bands and segs distinction added to neutrophil discussion.
- New micrographs of white blood cells.
- New *Topic of Interest* on leukemia chronicles a young magazine editor whose life was saved with a new drug.
- Explanation for dissolving clots and box on tPA added.
- Immune thrombocytopenic purpura added to *Genetics Connection*.

Chapter 13

- Small box on aneurysms added.
- Figure on changes in the left ventricle during cardiac cycle moved to Appendix C.
- A schematic of a general reflex arc has been added to figure 13.16 for clarification.

Chapter 14

- Complement and natural killer cells added to the discussion of innate defenses.

Chapter 15

- New Vignette, "Microbes Are Us," describes the astonishing fact that most of the cells in the body (counting the gastrointestinal tract) are not our own.
- Segmentation and mixing movements added to discussion of small intestine.
- *Topic of Interest* on Inflammatory Bowel Disease replaces constipation.
- Figure 15.33 is the 2005 FDA food pyramid.
- Figure 15.34 is a new easy-to-use BMI table.
- New *Topic of Interest* on eating extremes discusses the paradox of eating disorders and obesity rampant in the same society.
- A new small box introduces stem cells harvested from teeth.

Chapter 16

- New vignette on "The Dangers of Secondhand Smoke."
- Updated figure 16.17 and description of respiratory control.

Chapter 17

- Vignette on *E. coli* food poisoning updated with recent spinach cases.
- Box on gout rewritten to discuss post-mortem diagnosis from little finger of King Charles I of Spain.
- Urethral differences in males and females added.

Chapter 18

- Treatment for heatstroke added to opening vignette.

Chapter 19

- New Vignette on "Selling Eggs."
- New breast cancer tests added to *Topic of Interest*.

Chapter 20

- New vignette, "Do Premature Babies Feel Pain?"
- Update on egg freezing added to *Topic of Interest*.
- Small box on ectopic pregnancy added.
- Teratogens added to text to complement *Topic of Interest*.
- New small box on morning sickness.
- Human milk is contrasted with that of cows and seals.
- Small box on premature birth survival statistics added.
- Using preimplantation genetic diagnosis to select against future cancer added to *Genetics Connection*.
- Biochemical excess and deficiency associated with preeclampsia added to *Clinical Connection*.

McGraw-Hill offers various tools and technology products to support the tenth edition of *Hole's Essentials of Human Anatomy and Physiology.* Students can order supplemental study materials by contacting their campus bookstore. Instructors can obtain teaching aides by calling the McGraw-Hill Customer Service Department at 1-800-338-3987, by visiting our Human Anatomy and Physiology catalog at www.mhhe.com, or by contacting their local McGraw-Hill sales representative.

Anatomy & Physiology Revealed® 2.0

This amazing multimedia tool is designed to help students learn and review human anatomy using cadaver specimens. Detailed cadaver photographs blended together with a state-of-the-art layering technique provide a uniquely interactive dissection experience. This easy-to-use program features the following sections:

- **Dissection**—Peel away layers of the human body to reveal structures beneath the surface. Structures can be pinned and labeled, just like in a real dissection lab. Each labeled structure is accompanied by detailed information and an audio pronunciation. Dissection images can be captured and saved.
- **Animation**—Compelling animations demonstrate muscle actions, clarify anatomical relationships, or explain difficult concepts.
- **Histology**—Study microscopic anatomy using labeled micrographs.
- **Imaging**—Labeled X-ray, MRI, and CT images familiarize students with the appearance of key anatomical structures as seen through different medical imaging techniques.
- **Self-Test**—Challenging exercises let students test their ability to identify anatomical structures in a timed practical exam format or traditional multiple choice. A results page provides analysis of test scores plus links back to all incorrectly identified structures for review.

Workbook to Accompany Anatomy & Physiology| REVEALED 2.0® by Robert B. Broyles Jr.

This workbook/study guide helps students utilize **Anatomy & Physiology|REVEALED 2.0®** for maximum study potential. The workbook uses the same art as **Anatomy & Physiology|REVEALED 2.0®** for easy correlation and includes review questions, tables, terminology questions, coloring activities, and important reminders on key content. New exercises on the Integumentary system, disarticulated bones, origins and insertions, and histology will help students get the most out of **Anatomy & Physiology|REVEALED 2.0®**.

Laboratory Manual for Hole's Essentials of Human Anatomy and Physiology, **Tenth Edition,**

by Terry R. Martin, Kishwaukee College, is designed to accompany the tenth edition of *Hole's Essentials of Human Anatomy & Physiology.*

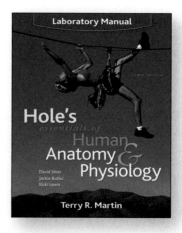

Anatomy & Physiology Laboratory Manual—Fetal Pig by Terry R. Martin, Kishwaukee College, provides excellent full-color photos of the dissected fetal pig with corresponding labeled art.

ARIS Course Management website (aris.mhhe.com)

McGraw-Hill's ARIS—Assessment, Review, and Instruction System—for *Hole's Essentials of Human Anatomy and Physiology*, tenth edition, is a complete electronic homework and course management system. Instructors can create and share course materials and assignments with colleagues with a few clicks of the mouse. Instructors can edit questions, import their own content, and create announcements and due dates for assignments. ARIS has automatic grading and reporting of easy-to-assign generated homework, quizzing, and testing. Once a student is registered in the course, all student activity within McGraw-Hill's ARIS website is automatically recorded and available to the instructor through a fully integrated grade book that can be downloaded to Excel. Instructors: To access ARIS, request registration information from your McGraw-Hill sales representative.

Text Website—aris.mhhe.com

The ARIS website that accompanies this text offers an extensive array of learning and teaching tools.

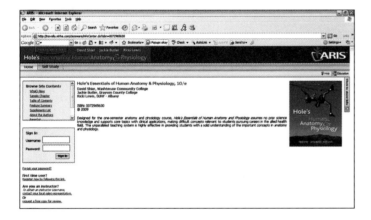

- Interactive Activities—Fun and exciting learning experiences await the student at the *Hole's Essentials of Human Anatomy and Physiology* website. Chapters offer a series of interactive activities like art labeling, animations, vocabulary flashcards, and more!
- Practice Quizzes at the *Hole's Essentials of Human Anatomy and Physiology* text website gauge student mastery of chapter content. Each chapter quiz is specially constructed to test student comprehension of key concepts. Immediate feedback to student responses explains why an answer is correct or incorrect.

Presentation Center—Complete set of electronic book images and assets for instructors

Build instructional materials wherever, whenever, and however you want!

Accessed from your textbook's ARIS website, **Presentation Center** is an online digital library containing photos, artwork, animations, and other media types that can be used to create customized lectures, visually enhanced tests and quizzes, compelling course websites, or attractive printed support materials. All assets are copyrighted by McGraw-Hill Higher Education, but can be used by instructors for classroom purposes. The visual resources in this collection include:

- **Art**   Full-color digital files of all illustrations in the book can be readily incorporated into lecture presentations, exams, or custom-made classroom materials. In addition, all files are pre-inserted into PowerPoint slides for ease of lecture preparation.
- **Photos**   The photos collection contains digital files of photographs from the text, which can be reproduced for multiple classroom uses.
- **Tables**   Every table that appears in the text has been saved in electronic form for use in classroom presentations and/or quizzes.
- **Animations**   Numerous full-color animations illustrating important processes are also provided. Harness the visual impact of concepts in motion by importing these files into classroom presentations or online course materials.

Also residing on your textbook's ARIS website are:

- **PowerPoint Lecture Outlines**   Ready-made presentations that combine art, and lecture notes are provided for each chapter of the text.
- **PowerPoint Slides**   For instructors who prefer to create their lectures from scratch, all illustrations, photos, and tables are pre-inserted by chapter into blank PowerPoint slides.

Test Bank—A computerized test bank that uses testing software to quickly create customized exams is available. The user-friendly program allows instructors to search for questions by topic or format, edit existing questions or add new ones; and scramble questions for multiple versions of the same test. Word files of the test bank questions are provided for those instructors who prefer to work outside the test-generator software.

Transparencies—Over 500 overhead transparencies includes all line art from the textbook.

Student Study Guide, by Nancy A. Sickels Corbett, offers chapter overviews, chapter outcomes, focus questions, mastery tests, study activities, and mastery test answers.

Course Delivery Systems—With help from our partners WebCT, Blackboard, Top-Class, eCollege, and other course management systems, professors can take complete control over their course content. Course cartridges containing text website content, online testing, and powerful student tracking features are readily available for use within these platforms.

eInstruction—This classroom performance system (CPS) utilizes wireless technology to bring interactivity into the classroom or lecture hall. Instructors and students receive immediate feedback through wireless response pads that are easy to use and engage students. eInstruction can assist instructors with:

- Taking attendance
- Administering quizzes and tests
- Creating a lecture with intermittent questions
- Using the CPS grade book to manage lectures and student comprehension
- Integrating interactivity into PowerPoint presentations.

Physiology Interactive Lab Simulations (Ph.I.L.S)—Ph.I.L.S. 3.0 contains *37 lab simulations* that allow students to perform experiments without using expensive lab equipment or live animals. This easy-

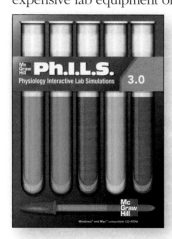

to-use software offers students the flexibility to change the parameters of every lab experiment, with no limit to the amount of times a student can repeat experiments or modify variables. This power to manipulate each experiment reinforces key physiology concepts by helping students to view outcomes, make predictions, and draw conclusions.

MediaPhys 3.0 CD-ROM—This interactive software tool offers detailed explanations, high quality illustrations and animations to provide students with a thorough introduction to the world of physiology—giving them a virtual tour of physiological processes. MediaPhys 3.0 is filled with interactive activities and quizzes to help reinforce physiology concepts that are often difficult to understand.

Electronic Books—If you or your students are ready for an alternative version of the traditional textbook, McGraw-Hill and VitalSource have partnered to bring you innovative and inexpensive electronic textbooks. By purchasing E-books from McGraw-Hill & VitalSource, students can save as much as 50% on selected titles delivered on the most advanced E-book platform available, VitalSource Bookshelf.

E-books from McGraw-Hill & VitalSource are smart, interactive, searchable and portable. VitalSource Bookshelf comes with a powerful suite of built-in tools that allow detailed searching, highlighting, note taking, and student-to-student or instructor-to-student note sharing. E-books from McGraw-Hill and VitalSource will help students study smarter and quickly find the information they need. And they will save money. Contact your McGraw-Hill sales representative to discuss E-book packaging options.

Acknowledgments

We would like to acknowledge the valuable contributions of the reviewers for the tenth edition who read either portions or all of the text, and who provided detailed criticisms, and ideas for improving the narrative and the illustrations. They include the following:

Diana Alagna, *Branford Hall Career Institute*
Michael J. Angilletta Jr., *Indiana State University*
Tammy Atchison, *Pitt Community College*
Ramona F. Atiles, *Career Institute of Health and Technology*
Patty Bostwick, *Florence Darlington Technical College*
Benjamin T.Y. Chan, *Space, University of Hong Kong*
Samuel Chen, *Moraine Valley Community College*
Bob DeLorme, *Gwinnett Technical College*
Eric Forman, *Sauk Valley Community College*
Yetao Gao, *Macau University of Science and Technology*
Amy Goode, *Illinois Central College*
Judy Y. Harris, *Kennebec Valley Community College*
Maria L. Hays, *Kennebec Valley Community College and Central Maine Community College*
Kevin A. Krown, *San Diego State University*
James McCaughern-Carucci, *Briarwood College*
Janice McClure, *Gwinnett Technical College*
Karen McCort, *Eastern New Mexico University Ruidoso*
Valerie Lin, *Nanyang Technological University*
Sue Passini, *Branford Hall Career Institute*
Babu P. Patlolla, *Alcorn State University*
Kristyn M. Powell-Holmes, *Utah Career College*
Christina Rauberts-Conklin, *Florida Metropolitan University*
Lawrence F. Roberge, *Goodwin College*
Dee Ann S. Sato, *Golden West College*
Danielle Schortzmann Wilken, *Goodwin College*
Kevin C. Zorda, *Branford Hall Career Institute*

And thanks to those who prepared ancillary materials:

- **Debbie Ambert,** *Pitt Community College,* revised the test bank questions.
- **Sheri Boyce,** *Messiah College,* revised the text website quiz and essay questions.
- **Melissa Eisenhauer,** *Trevecca Nazarene University,* revised the PowerPoint lecture outlines.
- **Jason Lapres,** *North Harris College,* prepared the new multi-layer quizzing found on the text website along with the APR Correlation Guide.
- **Michael Peters,** *Baker College,* once again revised the Instructor's Manual.
- **John Wakeman,** *Louisiana Technical University,* revised the study outlines found on the text website.

Any textbook is the result of hard work by a large team. Although we directed the revision, many "behind-the-scenes" people at McGraw-Hill were indispensable to the project. We would like to thank our editorial team of Jim Connely, Michelle Watnick, and Fran Schreiber; our production team, which included Jayne Klein, Sandy Ludovissy, Michelle Whitaker, John Leland, and Judi David; copyeditor Kennie Harris and freelance photo researcher Toni Michaels; Joanne Brummett of Precision Graphics; and most of all, John Hole, for giving us the opportunity and freedom to continue his classic work. We also thank our wonderfully patient families for their support.

David Shier
Jackie Butler
Ricki Lewis

Chapter Preview

The Chapter Preview not only provides great study tips to offer a foundation for success, but it also offers tips on how to utilize this particular text.

Foundations for Success

OPENING VIGNETTE
Beginning each chapter is a vignette that discusses current events or research news relating to the subject matter in the chapter. These demonstrate applications of the concepts learned in the study of anatomy and physiology.

A photo on the opening page for each chapter generates interest.

It is a beautiful day. You can't help but stare wistfully out the window, the scent of spring blooms and sound of birds making it impossible to concentrate on what the instructor is saying. Gradually, the lecture fades as you become aware of your own breathing, the beating of your heart, and of a sheen of sweat that breaks out on your forehead in response to the radiant heat from the glorious day. Suddenly your reverie is cut short—the instructor has dropped a human anatomy and physiology textbook on your desk. You jump. Yelp. Your heart hammers and a flash of fear grips your chest—but you soon realize what has happened and recover.

The message is clear: pay attention. So you do, tuning out the great outdoors and focusing on the lecture. In this course, you will learn all about the events that you have just experienced, including your response to the sudden stimulation of the instructor's wake-up call. This is a good reason to learn about how to stay focused in the course.

Learning Outcomes
After you have studied this chapter, you should be able to:

Each chapter begins with a list of outcomes you should be familiar with after studying the chapter. They are indicated with an arrow (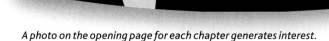). These are intended to help you master outcomes set by your instructor. Learning Outcomes will be tied to specific assessments found at the end of each chapter.

P.1 Introduction
1. Explain the importance of an individualized approach to learning.

P.2 Strategies for Success
2. Summarize what you should do before attending class.

3. Identify student activities that enhance classroom experience.

4. List and describe several study techniques that can facilitate learning new material.

Aids to Understanding Words *(Appendix A on page 567 has a complete list of Aids to Understanding Words.)*

> This section introduces building blocks of words that your instructor may assign. They are good investments of your time, since they can be used over and over and apply to many of the terms you will use in your career. Appendix A (p. 567) has a comprehensive list of these prefixes, suffixes, and root words.

ana- [up] *ana*tomy: the study of breaking up the body into its parts.

multi- [many] *multi*tasking: performing several tasks simultaneously.

physio- [relationship to nature] *physio*logy: the study of how body parts function.

P.1 INTRODUCTION

> Each chapter begins with an overview that tells you what to expect and why it is important.

Studying the human body can be overwhelming at times. The new terminology, used to describe body parts and how they work, can make it seem as if you are studying a foreign language. Learning all the parts of the body, along with the composition of each part, and how each part fits with the other parts to make the whole requires memorization. Understanding the way each body part works individually, as well as body parts working together, requires a higher level of knowledge, comprehension, and application. Identifying underlying structural similarities, from the macroscopic to the microscopic levels of body organization, taps more subtle critical thinking skills. This chapter will catalyze success in this active process of learning. (Remember that while the skills and tips discussed in this chapter relate to learning anatomy and physiology, they can be applied to other subjects.)

Students learn in different ways. Some students need to see the written word to remember it and the concept it describes or to actually write the words; others must hear the information or explain it to someone else. For some learners, true understanding remains elusive until a principle is revealed in a laboratory or clinical setting that provides a memorable context and engages all the senses.

> After each major section, a question or series of questions tests your understanding of the material. If you cannot answer the question(s), reread that section, being particularly on the lookout for the answer(s).

Check Your Recall

1. List some difficulties a student may experience when studying the human body.

P.2 STRATEGIES FOR SUCCESS

> Major divisions within a chapter are called "A-heads." They are numbered sequentially in very large, dark red type and identify major content areas within a chapter.

Many strategies for academic success are common sense, but it might help to review them. You may encounter new and helpful methods of learning.

Before Class

> The major divisions are divided into no-less-important subdivisions called "B-heads," identified by large, black type. These will help you organize the concepts that support the major divisions.

Before attending class, prepare by reading and outlining or taking notes on the assigned pages of the text. If outlining, leave adequate space between entries to allow room for note-taking during lectures. Or, fold each page of notes taken before class in half so that class notes can be written on the blank side of the paper across from the reading notes on the same topic. This introduces the topics of the next class lecture, as well as new terms. Some students team a vocabulary list with each chapter's notes. The outline or notes from the reading can be taken to class and expanded during the lecture.

As you read, you may feel the need for a "study break" or to "chill out." Other times, you may just need to shift gears. Try the following. Throughout the book are shaded boxes that present sidelights to the main focus of the text. Indeed, some of these may cover topics that your instructor chooses to highlight. Read them! They are interesting, informative, and a change of pace.

In a *hiatal hernia,* a portion of the stomach protrudes through a weakened area of the diaphragm, through the esophageal hiatus and into the thorax. As a result of a hiatal hernia, regurgitation (reflux) of gastric juice into the esophagus may inflame the esophageal mucosa, causing heartburn, difficulty in swallowing, or ulceration and blood loss. In response to the destructive action of gastric juice, columnar epithelium may replace the squamous epithelium that normally lines the esophagus. This condition, called *Barrett's esophagus,* increases the risk of developing esophageal cancer.

Approximately 40% of the body is skeletal muscle, and almost another 10% is smooth or cardiac muscle.

Genetics Connection

Coagulation Disorders

Hemophilia

Abnormalities of different clotting factors cause different forms of the bleeding disorder hemophilia, but hemophilia A is the most common. Factor VIII is deficient or absent. Symptoms of the hemophilias include severe hemorrhage following minor injuries, frequent nosebleeds, large intramuscular hematomas, and blood in the urine. The pattern of inheritance of hemophilia A is such that most affected individuals are male.

Hemophilia has left its mark on history. One of the earliest descriptions is in the Talmud, a second-century B.C. Jewish document, which reads, "If she circumcised her first child and he died, and a second one also died, she must not circumcise her third child." England's Queen Victoria (1819–1901) passed the hemophilia gene to several of her children, eventually spreading the condition to the royal families of Russia, Germany, and Spain. Hemophilia achieved notoriety when factor VIII pooled from blood donations was discovered to transmit HIV in 1985. Ninety percent of people with severe hemophilia who used such pooled factor VIII in the few years prior to that time developed AIDS.

von Willebrand Disease

The tendency to bleed and bruise easily may be a sign of *von Willebrand disease,* an inherited clotting disorder that is usually less severe, but much more common, than hemophilia. Affected persons lack a plasma protein, von Willebrand factor, that is secreted by the endothelial cells lining blood vessels. Von Willebrand factor enables platelets to adhere to damaged blood vessel walls, a key step preceding actual clotting. Sometimes, the condition can cause spontaneous bleeding from the mucous membranes of the gastrointestinal and urinary tracts. A person might not become aware of symptoms until excessive bleeding follows an injury. Von Willebrand disease is equally likely to affect males and females.

Immune Thrombocytopenic Purpura (ITP)

In ITP the immune system attacks platelets, dropping the count from 130,000 to 360,000 per microliter of blood to fewer than 10,000 to 30,000 per microliter. This autoimmune disorder affects three times as many females as males.

Symptoms of ITP include small purple marks on the skin where tiny blood vessels have broken, bleeding gums, nosebleeds, and bruising. Some people with ITP may never experience symptoms and may discover the condition only when a routine blood count reveals the platelet deficiency. Rarely, ITP causes a fatal brain bleed. Muscle aches, depression, and fatigue are also associated with the disorder. A form of ITP called gestational thrombocytopenia affects about 1 in 20 women late in pregnancy but clears up afterward.

Most people with ITP can live a normal life but must be careful to avoid activities that might cause injury, such as contact sports. Precautions must be taken during dental procedures and surgery. Various drugs are used to control ITP.

Topic of Interest

Burns

Slightly burned skin, as from a minor sunburn, may become warm and reddened (erythema) as dermal blood vessels dilate. This response may be accompanied by mild edema, and in time, the surface layer of skin may be shed. A burn injuring the epidermis alone is called a *superficial partial-thickness* (first degree) *burn*. Healing usually occurs within a few days to two weeks, with no scarring.

A burn that destroys some epidermis as well as some underlying dermis is a *deep partial-thickness* (second degree) *burn*. Fluid escapes from damaged dermal capillaries, and as it accumulates beneath the outer layer of epidermal cells, blisters appear. The injured region becomes moist and firm and may vary from dark red to waxy white. Such a burn most commonly occurs from exposure to hot objects, hot liquids, flames, or burning clothing.

Healing of a deep partial-thickness burn depends upon accessory structures of the skin that survive the injury because they are deep in the dermis. These structures, which include hair follicles, sweat glands, and sebaceous glands, include epithelial cells that divide and extend onto the surface of the injured dermis, spread over it, and form new epidermis.

A burn that destroys the epidermis, the dermis, and the accessory structures of the skin is a *full-thickness* (third degree) *burn*. The injured skin becomes dry and leathery, and may vary from red to black to white. A full-thickness burn usually is caused by immersion in hot liquids or prolonged exposure to hot objects, flames, or corrosive chemicals. Typically, most of the epithelial cells in the affected region are destroyed, and healing occurs only if epithelial cells divide and grow inward from the margin of the burn. If the injured area is extensive, it may require a transplant, using a thin layer of skin from an unburned region of the body (an autograft), cadaveric skin (a homograft), or a skin substitute (amniotic membrane from a human fetus, artificial membranes, or a tissue-engineered skin).

Remember when you were very young and encountered a substantial book for the first time? You were likely intimidated by its length, but were reassured that there were "a lot of pictures." This book has many illustrations too, all designed to help you master the material and become that person who you would want treating you.

Photographs and Line Art

Sometimes subdivisions have so many parts that the book goes to a third level, the "C-head." This information is presented in a slightly smaller font that identifies a specific section with an example.

Because line art can present different positions, layers, or perspectives, it can provide a unique view.

Photographs provide a realistic view of anatomy.

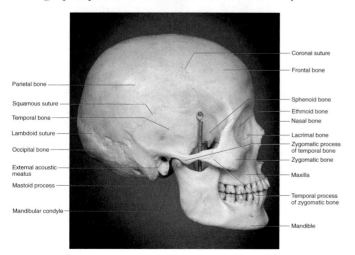

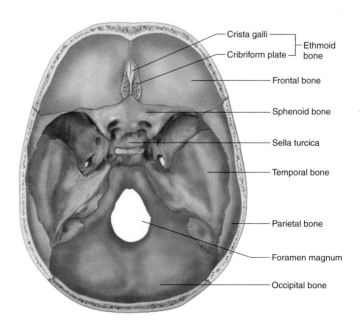

Macroscopic to Microscopic

Many figures show anatomical structures in a manner that is macroscopic to microscopic (or vice versa).

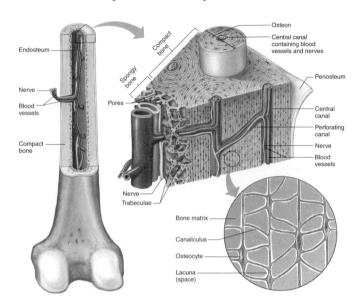

Anatomical Structures

Some figures illustrate the locations of anatomical structures.

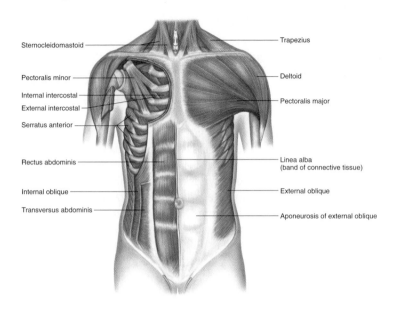

Flow Charts

Flow charts depict sequences of related events, steps of pathways, and complex concepts, easing comprehension. Other figures may show physiological processes.

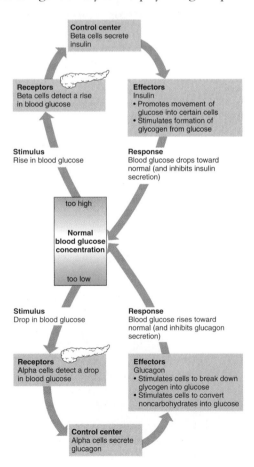

Other figures illustrate the functional relationships of anatomical structures.

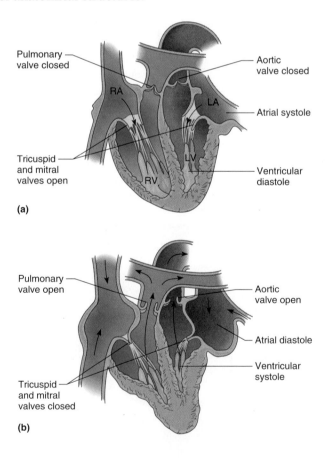

Organizational Tables

Organizational tables can help "put it all together," but are not a substitute for reading the text or having good lecture notes.

Table 5.6	Muscle and Nervous Tissues	
Type	**Function**	**Location**
Skeletal muscle tissue (striated)	Voluntary movements of skeletal parts	Muscles usually attached to bones
Smooth muscle tissue (lacks striations)	Involuntary movements of internal organs	Walls of hollow internal organs
Cardiac muscle tissue (striated)	Heart movements	Heart muscle
Nervous tissue	Sensory reception and conduction of nerve impulses	Brain, spinal cord, and peripheral nerves

As many resources as your text provides, it is critical that you attend class regularly, and be on time—even if the instructor's notes are posted on the Web. For many learners, hearing and writing new information is a better way to retain facts than just scanning notes on a computer screen. Attending lectures and discussion sections also provides more detailed and applied analysis of the subject matter, as well as a chance to ask questions.

During Class

Be alert and attentive in class. Take notes by adding to either the outline or notes taken while reading. Auditory learners benefit from recording the lectures and listening to them while driving or doing chores. This is called **multitasking**—doing more than one activity at a time.

Participate in class discussions, asking questions of the instructor and answering questions he or she poses. All of the students are in the class to learn, and many will be glad someone asked a question others would not be comfortable asking. Such student response can alert the instructor to topics that are misunderstood or not understood at all. However, respect class policy. Due to time constraints and class size, asking questions may be more appropriate after a large lecture class or during tutorial (small group) sessions.

After Class

In learning complex material, expediency is critical. Organize, edit, and review notes as soon after class as possible, fleshing out sections where the lecturer got ahead of the listener. Highlighting or underlining (in color, for visual learners) the key terms, lists, important points and major topics make them stand out, which eases both daily reviews and studying for exams.

Lists

Organizing information into lists or categories can minimize information overload, breaking it into manageable chunks. For example, when studying the muscles of the thigh it is easier to learn the insertion, origin, action, and nerve supply of the four muscles making up the *quadriceps femoris* as a group, because they all have the same insertion, action, and nerve supply . . . they differ only in their origins.

Mnemonic Devices

Another method for remembering information is the **mnemonic device.** One type of mnemonic device is a list of words, forming a phrase, in which the first letter of each word corresponds to the first letter of each word that must be remembered. For example, *F*requent *p*arade *o*ften *t*ests *s*oldiers' *e*ndurance stands for the skull bones *f*rontal, *p*arietal, *o*ccipital, *t*emporal, *s*phenoid, and *e*thmoid. Another type of mnemonic device is a word formed by the first letters of the items to be remembered. For example, *ipmat* represents the stages in the cell cycle: *i*nterphase, *p*rophase, *m*etaphase, *a*naphase, and *t*elophase.

Study Groups

Forming small study groups helps some students. Together the students review course material and compare notes. Working as a team and alternating leaders allows students to verbalize the information. Individual students can study and master one part of the assigned material, and then explain it to the others in the group, which incorporates the information into the memory of the speaker. Hearing the material spoken aloud also helps the auditory learner. Be sure to use anatomical and physiological terms, in explanations and everyday conversation, until they become part of your working vocabulary, rather than intimidating jargon. Most important of all—the group must stay on task, and not become a vehicle for social interaction. Your instructor may have suggestions or guidelines for setting up study groups.

Flash Cards

Flash cards may seem archaic in this computer age, but they are still a great way to organize and master complex and abundant information. The act of writing or drawing on a note card helps the tactile learner. Master a few new cards each day, and review cards from previous days, and use them all again at the end of the semester to prepare for the comprehensive final exam. They may even come in handy later, such as in studying for exams for admission to medical school or graduate school. Divide your deck in half and flip half of the cards so that the answer rather than the question is showing. Mix and shuffle them. Get used to identifying a structure or process from a description as well as giving a description when provided with a process or structure. This is more like what will be expected of you in the real world of the health-care professional.

Manage Your Time

Many of you have important obligations outside of class, such as jobs and family responsibilities. As important as these are, you still need to master this material on your path to becoming a health-care professional. Good time management skills are therefore essential in your study of human anatomy and physiology. In addition to class, lab, and study time, multitask. Spend time waiting for a ride, in a doctor's office, or on line reviewing notes or reading the text.

Daily repetition is helpful, so scheduling several short study periods each day can replace an end-of-semester crunch to cram for an exam. This does not take the place of time to prepare for the next class. Thinking about these suggestions for learning now can maximize study time throughout the semester, and, hopefully, lead to academic success. A working knowledge of the structure and function of the human body provides the foundation for all careers in the health sciences.

Check Your Recall

2. Why is it important to prepare before attending class?
3. Name two ways to participate in class discussions.
4. List several aids for remembering information.

SUMMARY OUTLINE

A summary of the chapter provides an outline to review major ideas and is a tool for organizing thoughts.

P.1 Introduction (page xxi)

Try a variety of methods to study the human body.

P.2 Strategies for Success (page xxi)

While strategies for academic success seem to be common sense, you might benefit from reminders of study methods.

1. Before class
   Read the assigned text material prior to the corresponding class meeting.
   a. Photographs give a realistic view and line art shows different perspectives.
   b. Macroscopic to microscopic show increase in detail.
   c. Flow charts depict sequences and steps.
   d. Figures of anatomical structures show locations.
   e. Organizational charts/tables summarize text.
2. During class
   Take notes and participate in class discussions.
3. After class
   a. Organize, edit, and review class notes.
   b. Mnemonic devices aid learning.
      (1) The first letters of the words to remember begin words of an easily recalled phrase.
      (2) The first letters of the items to be remembered form a word.
   c. Small study groups reviewing and vocalizing material can divide and conquer the learning task.
   d. Making flash cards helps the tactile learner.
   e. Time management skills encourage scheduled studying.
      (1) Repetition each day replaces cramming for exams.

CHAPTER ASSESSMENTS

Chapter assessments check mastery of learning outcomes. Assessments are tied to specific outcomes.

P.1 Introduction

1. Explain why the study of the human body can be overwhelming. (p. xxi)

P.2 Strategies for Success

2. Methods to prepare for class include: (p. xxi)
   a. reading the chapter.
   b. outlining the chapter.
   c. taking notes on the assigned reading.
   d. making a vocabulary list.
   e. all of the above.
3. Describe how you can participate in class discussions. (p. xxv)
4. Forming the phrase "*I passed my anatomy test*" to remember the cell cycle (interphase, prophase, metaphase, anaphase, telophase) is a _____ device. (p. xxv)
5. Name a benefit and a drawback of small study groups. (p. xxv)
6. Explain the value of repetition in learning and preparation for exams. (p. xxvi)

INTEGRATIVE ASSESSMENTS/
CRITICAL THINKING

> Integrative Assessments/Critical Thinking questions take the student beyond memorization to understanding and application.

⟹ **OUTCOME P.1**

1. Which study methods are most successful for you?

⟹ **OUTCOMES P.1, P.2**

2. Design a personalized study schedule.

> The student is directed to the text website at **aris.mhhe. com** for additional study tools. The student is also given information about the applicable Anatomy & Physiology Revealed® CD-ROM.

WEB CONNECTIONS

Visit the text website at **aris.mhhe.com** for additional quizzes, interactive learning exercises, and more.

Anatomy & Physiology | REVEALED® includes cadaver photos that allow you to peel away layers of the human body to reveal structures beneath the surface. This program also includes animations, radiologic imaging, audio pronunciations, and practice quizzing. To learn more visit **www.aprevealed.com**.

1

Introduction to Human Anatomy and Physiology

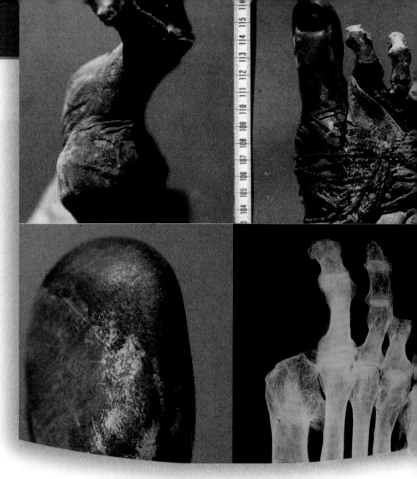

THE MUMMY'S TOE. No one knows her name, but she lived sometime between 1550 and 1300 B.C. in Thebes, a city in ancient Egypt. All that remains are pieces of her skeleton, bound with linen to preserve the general shape of her body in life. Yet, the telltale bones reveal a little of what her life was like.

The shape of the pelvic bones indicates that the person immortalized as the mummy was female. She was 50 to 60 years old when she died, according to the way the bony plates of her skull fit together and the lines of mineral deposition of a particularly well-preserved tooth. Among the preserved bones from the skull, pelvis, upper limbs, and right lower limbs, the part that easily commands the most attention is the right big toe, for it ends in a prosthesis, a manufactured replacement for a skeletal part. It was crucial for her balance and locomotion.

The mummy's toe tip is made of wood and painted a dark brown, perhaps to blend in with her skin color. It consists of a long part and two smaller parts that anchor the structure to the rest of the digit. Seven leather strings once attached it to the foot, and it even bears a fake nail. The fact that connective tissue and skin grew over the prosthesis reveals that her body had accepted the unnatural replacement part. Most amazing, however, is the shape of the prosthesis, which is remarkably like the body part it was intended to replace. Signs of wear indicate that it served its owner well. Although some prostheses found with mummies were placed after death to provide a complete skeleton for burial, this one was clearly used during the person's lifetime.

A wooden toe on an ancient Egyptian mummy reveals sophisticated knowledge of human anatomy and physiology from long ago.

The old woman with the replacement toe is evidence of sophisticated medical technology. Modern-day medical sleuths from the departments of pathology and diagnostic radiology at Ludwig-Maximillians University in Munich evaluated the ancient evidence using computerized tomography (CT) scans of the remnants of the natural toe. The researchers detected poor mineral content in the toe, plus calcium deposits in the largest blood vessel, the aorta, suggesting impaired circulation to the feet. Perhaps the mummy in life suffered from type 2 diabetes mellitus, which can cause poor circulation to the toes. If gangrene had set in, long-ago healers might have amputated the affected portion of the toe, replacing it with a very reasonable facsimile.

Learning Outcomes *After studying this chapter, you should be able to do the following:*

1.1 Introduction

1. Describe the early studies into the workings of the body. (p. 2)

1.2 Anatomy and Physiology

2. Define *anatomy* and *physiology*, and explain how they are related. (p. 3)

1.3 Levels of Organization

3. List the biological levels of organization and the characteristics of each. (p. 3)

1.4 Characteristics of Life

4. List and describe the major characteristics of life. (p. 4)

5. Define and give examples of metabolism. (p. 4)

1.5 Maintenance of Life

6. List and describe the major requirements of organisms. (p. 5)

7. Define *homeostasis,* and explain its importance to survival. (p. 5)

8. Describe the parts of a homeostatic mechanism and explain how they function together. (p. 6)

1.6 Organization of the Human Body

9. Describe the locations of the major body cavities. (p. 8)

10. List the organs located in each major body cavity. (p. 8)

11. Name the membranes associated with the thoracic and abdominopelvic cavities. (p. 10)

12. Name the major organ systems, and list the organs associated with each. (p. 12)

13. Describe the general functions of each organ system. (p. 12)

1.7 Anatomical Terminology

14. Properly use the terms that describe relative positions, body sections, and body regions. (p. 14)

Aids to Understanding Words   *(Appendix A on page 567 has a complete list of Aids to Understanding Words.)*

append- [to hang something] *append*icular: Pertaining to the limbs.

cardi- [heart] peri*cardi*um: Membrane that surrounds the heart.

cran- [helmet] *cran*ial: Pertaining to the portion of the skull that surrounds the brain.

dors- [back] *dors*al: Position toward the back.

homeo- [same] *homeo*stasis: Maintenance of a stable internal environment.

-logy [study of] physio*logy*: Study of body functions.

meta- [change] *meta*bolism: Chemical changes that occur within the body.

pariet- [wall] *pariet*al membrane: Membrane that lines the wall of a cavity.

pelv- [basin] *pelv*ic cavity: Basin-shaped cavity enclosed by the pelvic bones.

peri- [around] *peri*cardial membrane: Membrane that surrounds the heart.

pleur- [rib] *pleur*al membrane: Membrane that encloses the lungs and lines the thoracic cavity.

-stasis [standing still] homeo*stasis*: Maintenance of a stable internal environment.

-tomy [cutting] ana*tomy*: Study of structure, which often involves cutting or removing body parts.

1.1 INTRODUCTION

Modern medicine began with long-ago observations on the function, and malfunction, of the human body. The study of the human body probably began with our earliest ancestors, who must have been curious about how their bodies worked, as we are today. At first, their interests most likely concerned injuries and illnesses, because healthy bodies demand little attention from their owners. Early healers relied heavily on superstitions and notions about magic. However, as healers tried to help the sick, they began to discover useful ways of examining and treating the human body. They observed the effects of injuries, noticed how wounds healed, and examined cadavers to determine causes of death. They also found that certain herbs and potions could sometimes be used to treat coughs, headaches, fevers, and other common signs of illness.

Over time, people began to believe that humans could understand forces that caused natural events. They began observing the world around them more closely, asking questions and seeking answers. This set the stage for the development of modern medical science.

As techniques for making accurate observations and performing careful experiments evolved, knowledge of the human body expanded rapidly (fig. 1.1). At the same time, early medical providers coined many new terms to name body parts, describe their locations, and explain their functions and interactions. These terms, most of which originated from Greek and Latin words, formed the basis for the language of anatomy and physiology that persists today. (The names of some modern medical and applied sciences are listed on pages 17–19.)

Check Your Recall

1. What factors probably stimulated an early interest in the human body?

2. What kinds of activities helped promote the development of modern medical science?

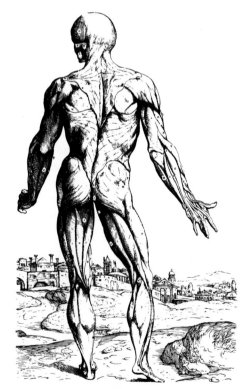

Figure 1.1

The study of the human body has a long history, as evidenced by this illustration from the second book of *De Humani Corporis Fabrica* by Andreas Vesalius, issued in 1543. (Note the similarity to the anatomical position, described later in this chapter.)

1.2 ANATOMY AND PHYSIOLOGY

Anatomy (ah-nat′o-me) is the branch of science that deals with the structure (morphology) of body parts—their forms and how they are organized. **Physiology** (fiz″e-ol′o-je), on the other hand, concerns the functions of body parts—what they do and how they do it.

The topics of anatomy and physiology are difficult to separate because the structures of body parts are so closely associated with their functions. Body parts form a well-organized unit—the human organism—and each part functions in the unit's operation. A particular body part's function depends on the way the part is constructed—that is, how its subparts are organized. For example, the organization of the parts in the human hand with its long, jointed fingers makes it easy to grasp objects; the hollow chambers of the heart are adapted to pump blood through tubular blood vessels; the shape of the mouth enables it to receive food; and teeth are shaped to break solid foods into small pieces (fig. 1.2).

Anatomy and physiology are ongoing as well as ancient fields. Researchers frequently discover new information about physiology, particularly at the molecular level. New parts of human anatomy are still discovered today, although less frequently. Recently, for example, researchers identified a small piece of connective tissue between the upper part of the spinal cord and a muscle at the back of the head. This connective tissue bridge may be the trigger for pain impulses in certain types of tension headaches.

In 2003, researchers completed deciphering the human genome—that is, the biochemical instructions that run the human body. Discovering the activities of our 20,500 or so genes, especially how they interact with each other as well as with environmental factors, is revealing many new details about physiology.

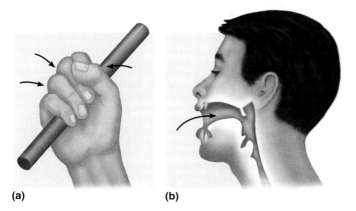

(a)   **(b)**

Figure 1.2

The structures of body parts make possible their functions:
(*a*) The hand is adapted for grasping, (*b*) the mouth for receiving food. (Arrows indicate movements associated with these functions.)

Check Your Recall

3. Why is it difficult to separate the topics of anatomy and physiology?

4. List several examples that illustrate how the structure of a body part makes possible its function.

1.3 LEVELS OF ORGANIZATION

Until the invention of magnifying lenses and microscopes about 400 years ago, anatomists were limited in their studies to what they could see with the unaided eye—large parts. But with these new tools, investigators discovered that larger body structures were made up of smaller parts, which in turn were composed of even smaller ones.

Figure 1.3 shows the levels of organization that modern-day scientists recognize. All materials, including those that make up the human body, are composed of chemicals. Chemicals consist of microscopic particles called **atoms,** which join to form **molecules.** Small molecules can combine in complex ways to form larger **macromolecules.**

In the human and other organisms, the basic unit of structure and function is a **cell,** which is microscopic. Although cells vary in size, shape, and specialized functions, all share certain characteristics. For instance, all cells of humans and other complex organisms contain structures called **organelles** (or″gah-nelz′) that carry out specific activities. Organelles are composed of aggregates of macromolecules, such as proteins, carbohydrates, lipids, and nucleic acids.

Cells may be organized into layers or other structures that have common functions. Such a group of cells forms a **tissue.** Groups of different tissues that interact form **organs**—complex structures with specialized functions—and groups of organs that function closely together comprise **organ systems.** Organ systems make up an **organism** (or′gah-nizm), which is a living thing.

Body parts can be described in terms of different levels of organization, such as the *atomic level,* the *molecular level,* or the *cellular level.* Furthermore, body parts differ in complexity from one level to the next. That is, atoms are less complex than molecules, molecules are less complex than organelles, tissues are less complex than organs, and so forth.

Chapters 2–6 discuss these levels of organization in more detail. Chapter 2 (pp. 31–36) describes the atomic and molecular levels. Chapter 3 (pp. 51–60) deals with organelles and cellular structures and functions, and chapter 4 explores cellular metabolism. Chapter 5 describes tissues and presents membranes (linings) as examples of organs, and chapter 6 considers the skin

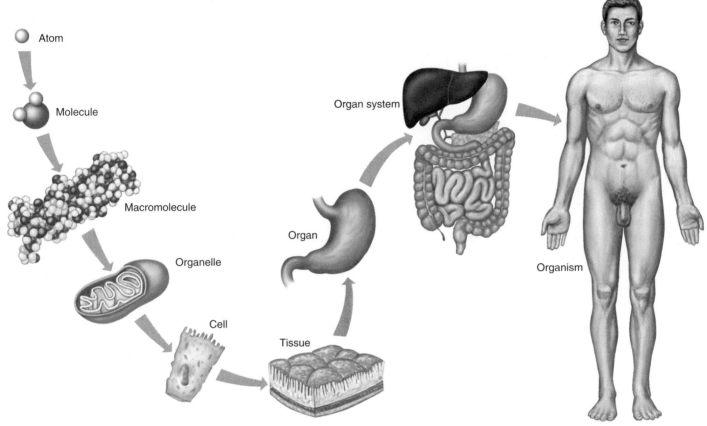

Figure 1.3

A human body is composed of parts within parts, with increasing complexity.

and its accessory organs as an example of an organ system. In the remaining chapters, the structures and functions of each of the other organ systems are described in detail.

> ### Check Your Recall
>
> 5. How does the human body illustrate levels of organization?
> 6. What is an organism?
> 7. How do body parts at different levels of organization vary in complexity?

1.4 CHARACTERISTICS OF LIFE

Before beginning a more detailed study of anatomy and physiology, it is helpful to consider some of the traits humans share with other organisms, particularly with other animals. As living organisms, we can move and respond to our surroundings. We start out as small individuals and then grow, eventually becoming able to reproduce. We gain energy by taking in or ingesting food, by breaking it down or digesting it, and by absorbing and assimilating it. The absorbed substances circulate throughout the internal environment of our bodies. We can then, by the process of respiration, use the energy in these nutrients for such vital functions as growth and repair of body parts. Finally, we excrete wastes from the body. All of these processes involve **metabolism** (mĕ-tab′o-lizm), the sum total of all of the chemical reactions in the body that break substances down and build them up. The reactions of metabolism enable us to acquire and use energy to fuel life processes. Table 1.1 summarizes the characteristics of life.

> ### Check Your Recall
>
> 8. What are the characteristics of life?
> 9. How are the characteristics of life dependent on metabolism?

Table 1.1	Characteristics of Life
Process	**Examples**
Movement	Change in position of the body or of a body part; motion of an internal organ
Responsiveness	Reaction to a change inside or outside the body
Growth	Increase in body size without change in shape
Reproduction	Production of new organisms and new cells
Respiration	Obtaining oxygen, removing carbon dioxide, and releasing energy from foods (Some forms of life do not use oxygen in respiration.)
Digestion	Breakdown of food substances into simpler forms that can be absorbed and used
Absorption	Passage of substances through membranes and into body fluids
Circulation	Movement of substances in body fluids
Assimilation	Changing absorbed substances into chemically different forms
Excretion	Removal of wastes produced by metabolic reactions

1.5 MAINTENANCE OF LIFE

The structures and functions of almost all body parts help maintain life. Even an organism's reproductive structures, whose primary function is to ensure that its species will continue into the future, may contribute to survival. For example, sex hormones help to strengthen bones.

Requirements of Organisms

Being alive requires certain environmental factors, including the following:

1. **Water** is the most abundant chemical in the body. It is required for many metabolic processes and provides the environment in which most of them take place. Water also transports substances within the organism and is important in regulating body temperature.
2. **Foods** are substances that provide the body with necessary chemicals (nutrients) in addition to water. Some of these chemicals are used as energy sources, others supply raw materials for building new living matter, and still others help regulate vital chemical reactions.
3. **Oxygen** is a gas that makes up about one-fifth of ordinary air. It is used to release energy from food substances. This energy, in turn, drives metabolic processes.
4. **Heat** is a form of energy. It is a product of metabolic reactions, and the degree of heat present partly determines the rate at which these reactions occur. Generally, the more heat, the more rapidly chemical reactions take place. (*Temperature* is a measure of the degree of heat.)
5. **Pressure** is an application of force to something. For example, the force on the outside of the body due to the weight of air above it is called *atmospheric pressure*. In humans, this pressure is important in breathing. Similarly, organisms living under water are subjected to *hydrostatic pressure*—a pressure a liquid exerts—due to the weight of water above them. In humans, heart action produces blood pressure (another form of hydrostatic pressure), which forces blood through blood vessels.

Health-care workers repeatedly monitor patients' *vital signs*—observable body functions that reflect essential metabolic activities. Vital signs indicate that a person is alive. Assessment of vital signs includes measuring body temperature and blood pressure and monitoring rates and types of pulse and breathing movements. Absence of vital signs signifies death. A person who has died displays no spontaneous muscular movements, including those of the breathing muscles and beating heart. A dead person does not respond to stimuli, and has no reflexes, such as the knee-jerk reflex and the pupillary reflexes of the eye. Brain waves cease with death, as demonstrated by a flat electroencephalogram (EEG), which signifies a lack of metabolic activity in the brain.

Although organisms require water, food, oxygen, heat, and pressure, these factors alone are not enough to ensure survival. Both the quantities and the qualities of such factors are also important. For example, the volume of water entering and leaving an organism must be regulated, as must the concentration of oxygen in body fluids. Similarly, survival depends on the quality as well as the quantity of food available—that is, food must supply the correct nutrients in adequate amounts.

Homeostasis

Factors in the external environment may change. If an organism is to survive, however, conditions within the fluid surrounding its body cells, which compose its **internal environment,** must remain relatively stable (fig. 1.4). In other words, body parts function only when the concentrations of water, nutrients, and oxygen and the conditions of heat and pressure remain within certain narrow limits. This condition of a stable internal environment is called **homeostasis** (ho″me-ō-sta′sis).

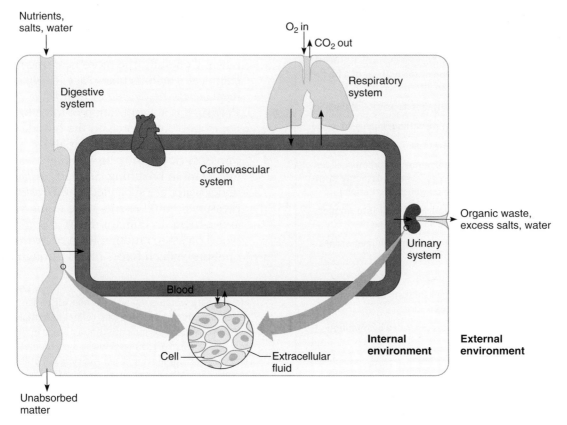

Figure 1.4

Our cells lie within an internal fluid environment (extracellular fluid). Concentrations of water, nutrients, and oxygen in the internal environment must be maintained within certain ranges to sustain life.

The body maintains homeostasis through a number of self-regulating control systems, or **homeostatic mechanisms,** that share the following three components (fig. 1.5):

- **Receptors** provide information about specific conditions (stimuli) in the internal environment.
- A **set point** tells what a particular value should be (such as body temperature at 37°C or, 98.6°F).
- **Effectors** cause responses that alter conditions in the internal environment.

A homeostatic mechanism works as follows. If the receptors measure deviations from the set point, effectors are activated that can return conditions toward normal. As conditions return toward normal, the deviation from the set point progressively lessens, and the effectors are gradually shut down. Such a response is called a **negative feedback** (neg′ah-tiv fēd′bak) mechanism, both because the deviation from the set point is corrected (moves in the opposite or negative direction) and because the correction reduces the action of the effectors. This latter aspect is important because it prevents a correction from going too far.

To better understand this idea of negative feedback, imagine a room equipped with a furnace and an air con-

ditioner (fig. 1.6). Suppose the room temperature is to remain near 20°C (68°F), so the thermostat is adjusted to an operating level, or set point, of 20°C. Because a thermostat senses temperature changes, it will signal the furnace to start and the air conditioner to stop whenever the room temperature drops below the set point. If the temperature rises above the set point, the thermostat will stop the furnace and start the air conditioner. As a result, the room will maintain a relatively constant temperature.

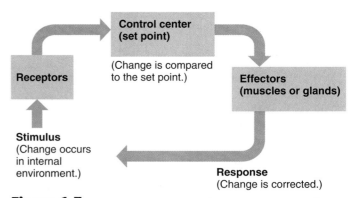

Figure 1.5

A homeostatic mechanism monitors a particular aspect of the internal environment and corrects any changes back to the value indicated by the set point.

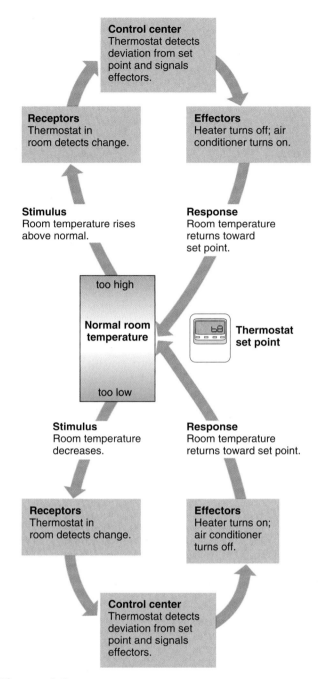

Figure 1.6

A thermostat signals an air conditioner and a furnace to turn on or off to maintain a relatively stable room temperature. This system is an example of a homeostatic mechanism.

A similar homeostatic mechanism regulates body temperature. Temperature receptors are scattered throughout the body. The "thermostat" is a temperature-sensitive region in a temperature control center of the brain. In healthy persons, the set point of the brain's thermostat is at or near 37°C (98.6°F).

If a person is exposed to cold and body temperature begins to drop, the temperature receptors sense this change and the temperature control center triggers heat-generating and heat-conserving activities. For example, small groups of muscles are stimulated to contract involuntarily, an action called *shivering*. Such muscular contractions produce heat, which helps warm the body. At the same time, blood vessels in the skin are signaled to constrict so that less warm blood flows through them. In this way, deeper tissues retain heat that might otherwise be lost.

If a person is becoming overheated, the brain's temperature control center triggers a series of changes that promote loss of body heat. Sweat glands in the skin secrete perspiration, and as this fluid evaporates from the surface, heat is carried away and the skin is cooled. At the same time, the brain center dilates blood vessels in the skin. This action allows the blood carrying heat from deeper tissues to reach the surface, where heat is lost to the outside (fig. 1.7). The brain stimulates an increase in heart rate, which sends a greater volume of blood into surface vessels, and an increase in breathing rate, which allows the lungs to expel more heat-carrying air. Body temperature regulation is discussed further in chapter 6 (pp. 124–125).

Another homeostatic mechanism regulates the blood pressure in the blood vessels (arteries) leading away from the heart. In this instance, pressure-sensitive receptors in the walls of these vessels sense changes in blood pressure and signal a pressure control center in the brain. If blood pressure is above the set point, the brain signals the heart chambers to contract more slowly and with less force. This decreased heart action sends less blood into the blood vessels, decreasing the pressure inside them. If blood pressure falls below the set point, the brain center signals the heart to contract more rapidly and with greater force. As a result, the pressure in the vessels increases. Chapter 13 (pp. 359–360) discusses regulation of blood pressure in more detail.

Human physiology offers many other examples of homeostatic mechanisms. All work by the same general process as the two preceding examples. Just as anatomical terms are used repeatedly throughout this book, so can the basic principles of a homeostatic mechanism be applied to the different organ systems. Homeostatic mechanisms maintain a relatively constant internal environment, yet physiological values may vary slightly in a person from time to time or from one individual to the next. Therefore, both normal values for an individual and the *normal range* for the general population are clinically important.

Most feedback mechanisms in the body are negative. However, sometimes change stimulates further change. A process that moves conditions away from the normal state is called a *positive feedback mechanism*. In blood clotting, for example, the chemicals that carry out clotting stimulate more clotting, minimizing bleeding (see chapter 12, p. 330). Another positive feedback mechanism increases the strength

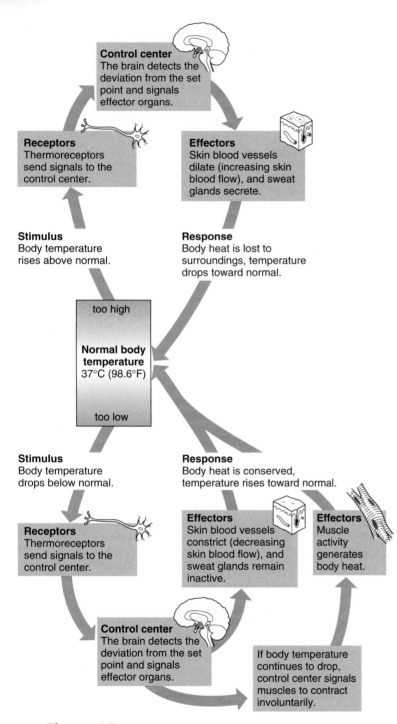

Control center
The brain detects the deviation from the set point and signals effector organs.

Receptors
Thermoreceptors send signals to the control center.

Effectors
Skin blood vessels dilate (increasing skin blood flow), and sweat glands secrete.

Stimulus
Body temperature rises above normal.

Response
Body heat is lost to surroundings, temperature drops toward normal.

too high

Normal body temperature
37°C (98.6°F)

too low

Stimulus
Body temperature drops below normal.

Response
Body heat is conserved, temperature rises toward normal.

Receptors
Thermoreceptors send signals to the control center.

Effectors
Skin blood vessels constrict (decreasing skin blood flow), and sweat glands remain inactive.

Effectors
Muscle activity generates body heat.

Control center
The brain detects the deviation from the set point and signals effector organs.

If body temperature continues to drop, control center signals muscles to contract involuntarily.

Figure 1.7
A homeostatic mechanism regulates body temperature.

of uterine contractions during childbirth, helping to bring the new individual into the world.

Positive feedback mechanisms usually produce unstable conditions, which might seem incompatible with homeostasis. However, the examples of positive feedback associated with normal health have very specific functions and are short-lived.

Check Your Recall

10. What requirements of organisms does the external environment provide?
11. Why is homeostasis important to survival?
12. Describe two homeostatic mechanisms.

1.6 ORGANIZATION OF THE HUMAN BODY

The human organism is a complex structure composed of many parts. Its major features include several body cavities, layers of membranes within these cavities, and a variety of organ systems.

Body Cavities

The human organism can be divided into an **axial** (ak'se-al) portion, which includes the head, neck, and trunk, and an **appendicular** (ap"en-dik'u-lar) portion, which includes the upper and lower limbs. Within the axial portion are the **cranial cavity,** which houses the brain; the **vertebral canal,** which contains the spinal cord within the sections of the backbone (vertebrae); the **thoracic** (tho-ras'ik) **cavity;** and the **abdomino-pelvic** (ab-dom"ĭ-no-pel'vik) **cavity.** The organs within these last two cavities are called **viscera** (vis'er-ah) (fig. 1.8*a*).

A broad, thin muscle called the **diaphragm** separates the thoracic cavity from the lower abdominopelvic cavity. The thoracic cavity wall is composed of skin, skeletal (voluntary) muscles, and various bones.

A region called the **mediastinum** (me"de-as-ti'num) separates the thoracic cavity into two compartments, which contain the right and left lungs. The remaining thoracic viscera—heart, esophagus, trachea, and thymus—are located within the mediastinum (fig. 1.8*b*).

The abdominopelvic cavity, which includes an upper abdominal portion and a lower pelvic portion, extends from the diaphragm to the floor of the pelvis. Its wall consists primarily of skin, skeletal muscles, and bones. The viscera within the **abdominal cavity** include the stomach, liver, spleen, gallbladder, kidneys, and most of the small and large intestines.

The **pelvic cavity** is the portion of the abdomino-pelvic cavity enclosed by the hip bones (see chapter 7, p. 157). It contains the terminal portion of the large intestine, the urinary bladder, and the internal reproductive organs.

Smaller cavities within the head include (fig. 1.9):

1. **Oral cavity,** containing the teeth and tongue.
2. **Nasal cavity,** located within the nose and divided into right and left portions by a nasal septum.

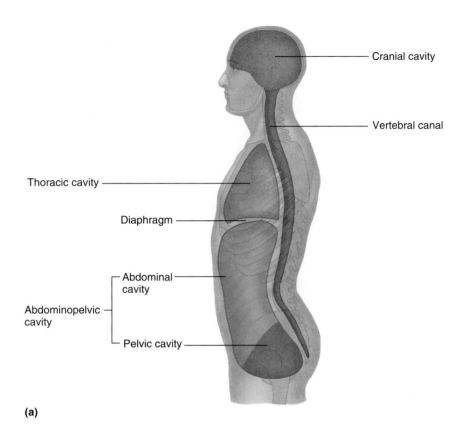

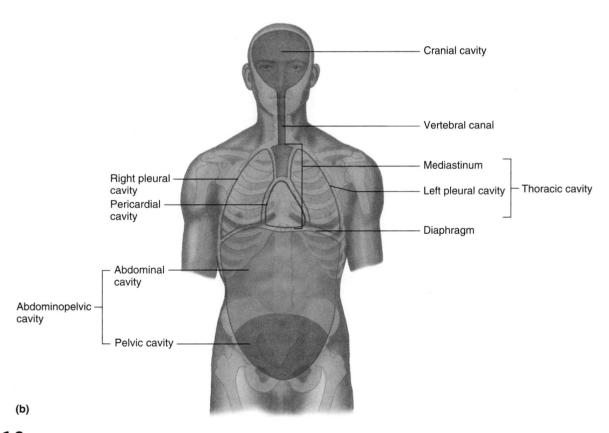

(a)

(b)

Figure 1.8

Major body cavities. (*a*) Lateral view. (*b*) Anterior view.

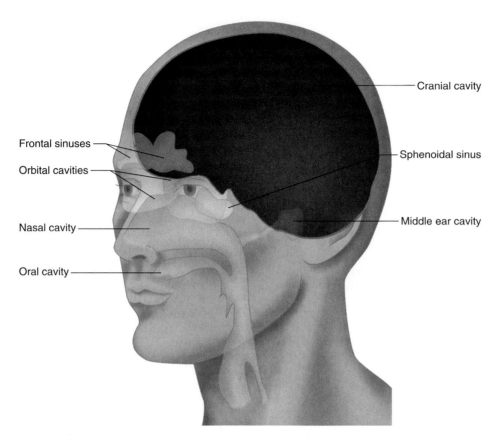

Cranial cavity

Frontal sinuses

Orbital cavities

Sphenoidal sinus

Nasal cavity

Middle ear cavity

Oral cavity

Figure 1.9

The cavities within the head include the cranial, oral, nasal, orbital, and middle ear cavities, as well as several sinuses. (Not all of the sinuses are shown.)

Several air-filled *sinuses* connect to the nasal cavity (see chapter 7, p. 142). These include the frontal and sphenoidal sinuses shown in figure 1.9.

3. **Orbital cavities,** containing the eyes and associated skeletal muscles and nerves.

4. **Middle ear cavities,** containing the middle ear bones.

Thoracic and Abdominopelvic Membranes

The walls of the right and left thoracic compartments, which contain the lungs, are lined with a membrane called the *parietal pleura* (fig. 1.10). A similar membrane, called the *visceral pleura,* covers the lungs themselves. (Note: **Parietal** (pah-ri′ĕ-tal) refers to the membrane attached to the wall of a cavity; **visceral** (vis′er-al) refers to the membrane that is deeper—toward the interior—and covers an internal organ, such as a lung.)

The parietal and visceral **pleural membranes** (ploo′ral mem′brānz) are separated by a thin film of watery fluid (serous fluid), which they secrete. While no actual space normally exists between these membranes, the potential space between them is called the *pleural cavity* (see figs. 1.8*b* and 1.10).

The heart, which is located in the broadest portion of the mediastinum, is surrounded by **pericardial** (per″ĭ-kar′de-al) **membranes.** A thin *visceral pericardium* covers the heart's surface and is separated from a thicker *parietal pericardium* by a small volume of fluid. The *pericardial cavity* (see figs. 1.8*b* and 1.10) is the potential space between these membranes.

In the abdominopelvic cavity, the lining membranes are called **peritoneal** (per″ĭ-to-ne′al) **membranes.** A *parietal peritoneum* lines the wall, and a *visceral peritoneum* covers each organ in the abdominal cavity (fig. 1.11). The *peritoneal cavity* is the potential space between these membranes.

Check Your Recall

13. What does *viscera* mean?

14. Which organ occupies the cranial cavity? The vertebral canal?

15. Name the cavities of the head.

16. Describe the membranes associated with the thoracic and abdominopelvic cavities.

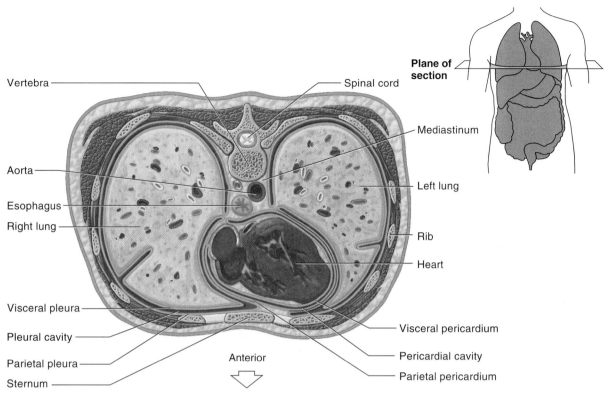

Figure 1.10

A transverse section through the thorax reveals the serous membranes associated with the heart and lungs (superior view).

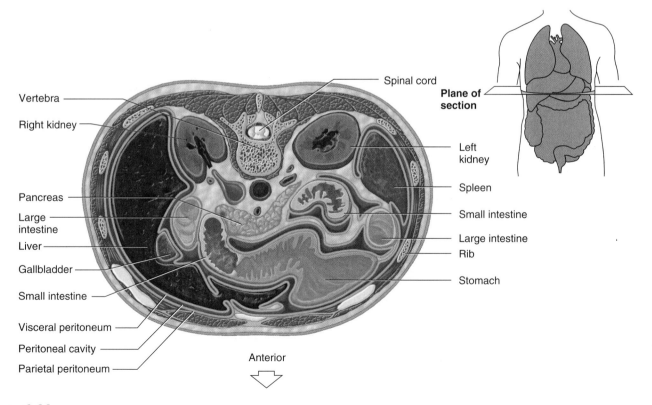

Figure 1.11

Transverse section through the abdomen (superior view).

Organ Systems

The human organism consists of several organ systems. Each system includes a set of interrelated organs that work together allowing each system to provide specialized functions that contribute to homeostasis (fig. 1.12). As you read about each system, you may want to consult the illustrations of the human torso in the Reference Plates (see pp. 23–29) and locate some of the organs described.

Body Covering

Organs of the **integumentary** (in-teg-u-men'tar-e) **system** (see chapter 6) include the skin and various accessory organs, such as the hair, nails, sweat glands, and sebaceous glands. These parts protect underlying tissues, help regulate body temperature, house a variety of sensory receptors, and synthesize certain products.

Support and Movement

The organs of the skeletal and muscular systems (see chapters 7 and 8) support and move body parts. The **skeletal** (skel'ĕ-tal) **system** consists of bones, as well as ligaments and cartilages that bind bones together. These parts provide frameworks and protective shields for softer tissues, are attachments for muscles, and act with muscles when body parts move. Tissues within bones also produce blood cells and store inorganic salts.

Muscles are the organs of the **muscular** (mus'ku-lar) **system.** By contracting and pulling their ends closer together, muscles provide forces that move body parts. They also maintain posture and are the main source of body heat.

Integration and Coordination

For the body to act as a unit, its parts must be integrated and coordinated. The nervous and endocrine systems control and adjust various organ functions, thus helping to maintain homeostasis.

The **nervous** (ner'vus) **system** (see chapter 9) consists of the brain, spinal cord, nerves, and sense organs (see chapter 10). Nerve cells within these organs communicate with each other and with muscles and glands using electrochemical signals called *nerve impulses.* Each impulse exerts a relatively short-term effect on its target. Some nerve cells are specialized sensory receptors that detect changes inside and outside the body. Other nerve cells receive the impulses transmitted from these sensory receptors and interpret and respond to the information received. Still other nerve cells carry impulses from the brain or spinal cord to muscles or glands, stimulating them to contract or to secrete products.

The **endocrine** (en'do-krin) **system** (see chapter 11) includes all the glands that secrete chemical messengers called *hormones.* The hormones, in turn, move away from the glands in body fluids, such as blood or tissue fluid (fluid from the spaces within tissues). A particular hormone affects only a particular group of cells, called its *target cells.* A hormone alters the metabolism of its target cells. Compared to nerve impulses, hormonal effects occur over a relatively long time period. Organs of the endocrine system include the hypothalamus of the brain; the pituitary, thyroid, parathyroid, and adrenal glands; and the pancreas, ovaries, testes, pineal gland, and thymus.

Transport

Two organ systems transport substances throughout the internal environment. The **cardiovascular** (kahr"de-o-vas'ku-lur) **system** (see chapters 12 and 13) includes the heart, arteries, veins, capillaries, and blood. The heart is a muscular pump that helps force blood through the blood vessels. Blood transports gases, nutrients, hormones, and wastes. It carries oxygen from the lungs and nutrients from the digestive organs to all body cells, where these biochemicals are used in metabolic processes. Blood also transports hormones and carries wastes from body cells to the excretory organs, where the wastes are removed from the blood and released to the outside.

The **lymphatic** (lim-fat'ik) **system** (see chapter 14) is closely related to the cardiovascular system. It is composed of the lymphatic vessels, lymph fluid, lymph nodes, thymus, and spleen. This system transports some of the tissue fluid back to the bloodstream and carries certain fatty substances away from the digestive organs and into the bloodstream. Cells of the lymphatic system are called lymphocytes, and they defend the body against infections by removing disease-causing microorganisms and viruses from tissue fluid.

Absorption and Excretion

Organs in several systems absorb nutrients and oxygen and excrete various wastes. For example, the organs of the **digestive** (di-jest'iv) **system** (see chapter 15) receive foods from the outside. Then they break down food molecules into simpler forms that can pass through cell membranes and be absorbed. Materials that are not absorbed are transported back to the outside and eliminated. Certain digestive organs also produce hormones and thus function as parts of the endocrine system. The digestive system includes the mouth, tongue, teeth, salivary glands, pharynx, esophagus, stomach, liver, gallbladder, pancreas, small intestine, and large intestine. Chapter 15 (pp. 429–437) also discusses nutrition.

The organs of the **respiratory** (re-spi'rah-to"re) **system** (see chapter 16) move air in and out and exchange gases between the blood and the air. More specifically, oxygen passes from the air within the lungs into the blood, and carbon dioxide leaves the blood and enters the air. The nasal cavity, pharynx, larynx, trachea, bronchi, and lungs are parts of this system.

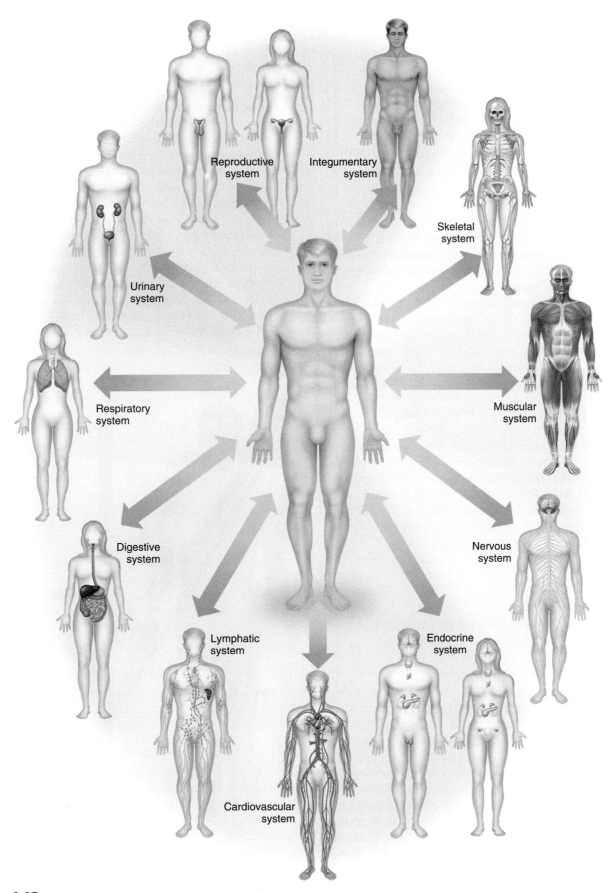

Figure 1.12

The organ systems in humans interact, maintaining homeostasis.

The **urinary** (u′rĭ-ner″e) **system** (see chapter 17) consists of the kidneys, ureters, urinary bladder, and urethra. The kidneys remove wastes from blood and help maintain the body's water and salt (electrolyte) concentrations. The product of these activities is urine. Other portions of the urinary system store urine and transport it outside the body. Chapter 18 discusses the urinary system's role in maintaining water and electrolyte concentrations and the acidity of the internal environment.

Reproduction

Reproduction is the process of producing offspring (progeny). Cells reproduce when they divide and give rise to new cells. However, the **reproductive** (re″pro-duk′tiv) **system** of an organism produces whole new organisms like itself (see chapter 19).

The male reproductive system includes the scrotum, testes, epididymides, ductus deferentia, seminal vesicles, prostate gland, bulbourethral glands, penis, and urethra. These parts produce and maintain sperm cells (spermatozoa). Components of the male reproductive system also transfer sperm cells into the female reproductive tract.

The female reproductive system consists of the ovaries, uterine tubes, uterus, vagina, clitoris, and vulva. These organs produce and maintain the female sex cells (egg cells, or oocytes), transport the female sex cells within the female reproductive system, and can receive the male sex cells (sperm cells) for the possibility of fertilizing an egg. The female reproductive system also supports development of embryos, carries fetuses to term, and functions in the birth process.

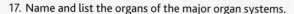

Check Your Recall

17. Name and list the organs of the major organ systems.
18. Describe the general functions of each organ system.

1.7 ANATOMICAL TERMINOLOGY

To communicate effectively with one another, researchers and clinicians have developed a set of precise terms to describe anatomy. These terms concern the relative positions of body parts, relate to imaginary planes along which cuts may be made, and describe body regions.

Use of such terms assumes that the body is in the **anatomical position.** This means that the body is standing erect, face forward, with the upper limbs at the sides and the palms forward. Note that the terms right and left refer to the "right" and "left" of the body in anatomical position.

Relative Positions

Terms of relative position describe the location of one body part with respect to another. They include the following (many of these terms are illustrated in figure 1.13):

1. **Superior** means that a body part is above another part or is closer to the head. (The thoracic cavity is superior to the abdominopelvic cavity.)
2. **Inferior** means that a body part is below another body part or is toward the feet. (The neck is inferior to the head.)
3. **Anterior** (or *ventral*) means toward the front. (The eyes are anterior to the brain.)
4. **Posterior** (or *dorsal*) is the opposite of anterior; it means toward the back. (The pharynx is posterior to the oral cavity.)
5. **Medial** refers to an imaginary midline dividing the body into equal right and left halves. A body part is medial if it is closer to this line than another part. (The nose is medial to the eyes.)

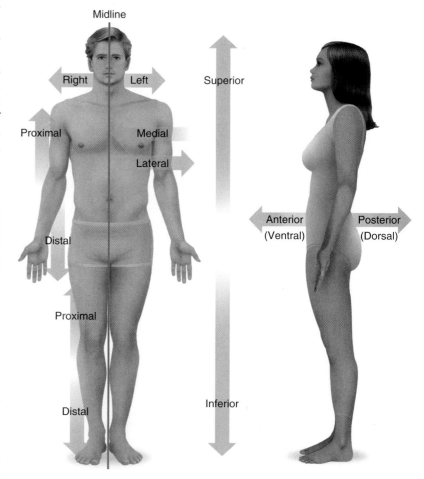

Figure 1.13

Relative positional terms describe a body part's location with respect to other body parts.

6. **Lateral** means toward the side with respect to the imaginary midline. (The ears are lateral to the eyes.)
7. **Bilateral** refers to paired structures, one of which is on each side. (The lungs are bilateral.)
8. **Ipsilateral** refers to structures on the same side. (The right lung and the right kidney are ipsilateral.)
9. **Contralateral** refers to structures on the opposite side. (A patient with a fractured right leg would have to bear weight on the contralateral—in this case, left—lower limb.)
10. **Proximal** describes a body part that is closer to a point of attachment to the trunk than another body part. (The elbow is proximal to the wrist.) Proximal may also refer to another reference point such as the proximal tubule, which is closer to the filtering structure in the kidney.
11. **Distal** is the opposite of proximal. It means that a particular body part is farther from a point of attachment to the trunk than another body part. (The fingers are distal to the wrist.) Distal may also refer to another reference point, such as decreased blood flow distal to occlusion of a coronary artery.
12. **Superficial** means situated near the surface. (The epidermis is the superficial layer of the skin.)

Peripheral also means outward or near the surface. It describes the location of certain blood vessels and nerves. (The nerves that branch from the brain and spinal cord are peripheral nerves.)
13. **Deep** describes parts that are more internal than superficial parts. (The dermis is the deep layer of the skin.)

Body Sections

Observing the relative locations and organization of internal body parts requires cutting or sectioning the body along various planes (fig. 1.14). The following terms describe such planes and the sections that result:

1. **Sagittal** refers to a lengthwise plane that divides the body into right and left portions. If a sagittal plane passes along the midline and divides the body into equal parts, it is called *median* (midsagittal). A sagittal section lateral to midline is called *parasagittal*.
2. **Transverse** (or *horizontal*) refers to a plane that divides the body into superior and inferior portions.
3. **Coronal** (or *frontal*) refers to a plane that divides the body into anterior and posterior portions.

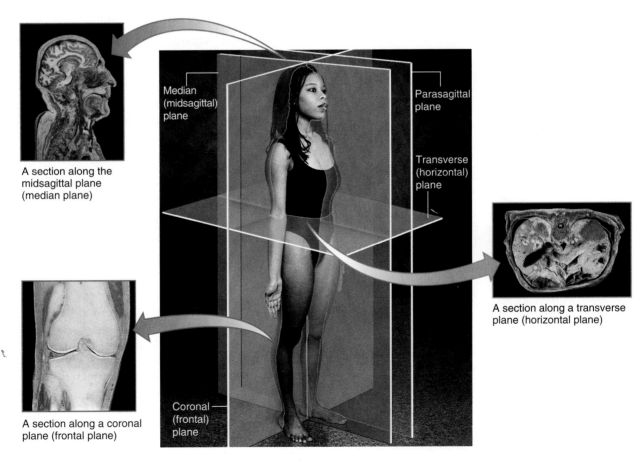

A section along the midsagittal plane (median plane)

Median (midsagittal) plane

Parasagittal plane

Transverse (horizontal) plane

A section along a transverse plane (horizontal plane)

Coronal (frontal) plane

A section along a coronal plane (frontal plane)

Figure 1.14

Observation of internal parts requires sectioning the body along various planes.

Sometimes, a cylindrical organ such as a long bone is sectioned. In this case, a cut across the structure is called a *cross section,* an angular cut is an *oblique section,* and a lengthwise cut is a *longitudinal section* (fig. 1.15).

Radiologists use a procedure called *computerized tomography,* or CT scanning, to visualize internal organ sections (fig. 1A). In this procedure, an X-ray-emitting device moves around the body region being examined. At the same time, an X-ray detector moves in the opposite direction on the other side. As the devices move, an X-ray beam passes through the body from hundreds of different angles. Since tissues and organs of varying composition within the body absorb X rays differently, the amount of X ray reaching the detector varies from position to position. A computer records the measurements from the X-ray detector, and combines them mathematically to create a sectional image of the internal body parts that can be viewed on a monitor.

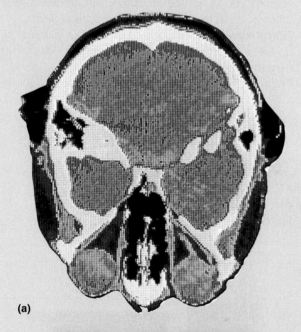

(a)

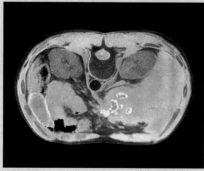

(b)

Figure 1A

Falsely colored CT (computerized tomography) scans of (*a*) the head and (*b*) the abdomen. *Note:* These are not shown in correct relative size.

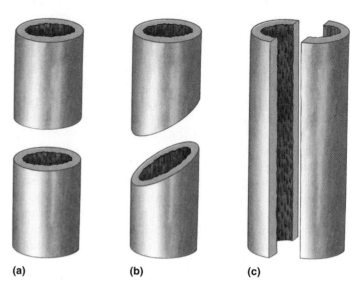

(a)   (b)   (c)

Figure 1.15

Cylindrical parts may be cut in (*a*) cross section, (*b*) oblique section, or (*c*) longitudinal section.

Body Regions

A number of terms designate body regions. The abdominal area, for example, is subdivided into the following nine regions, as figure 1.16*a* shows:

1. **Epigastric region** refers to the upper middle portion.
2. **Left** and **right hypochondriac regions** lie on each side of the epigastric region.
3. **Umbilical region** refers to the middle portion.
4. **Left** and **right lumbar regions** lie on each side of the umbilical region.
5. **Hypogastric region** refers to the lower middle portion.
6. **Left** and **right iliac regions** (left and right inguinal regions) lie on each side of the hypogastric region.

The abdominal area is also often subdivided into four quadrants, as figure 1.16*b* shows.

The following adjectives are commonly used to refer to various body regions, some of which are illustrated in figure 1.17:

abdominal (ab-dom′ĭ-nal) The region between the thorax and pelvis.
acromial (ah-kro′me-al) The point of the shoulder.
antebrachial (an″te-bra′ke-al) The forearm.
antecubital (an″te-ku′bĭ-tal) The space in front of the elbow.
axillary (ak′sĭ-ler″e) The armpit.
brachial (bra′ke-al) The arm.
buccal (buk′al) The cheek.
carpal (kar′pal) The wrist.
celiac (se′le-ak) The abdomen.
cephalic (sĕ-fal′ik) The head.

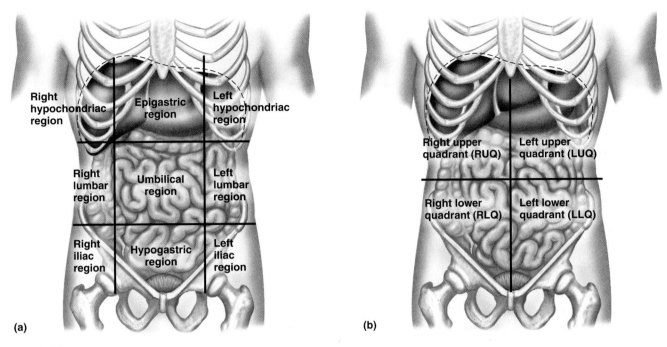

(a)   **(b)**

Figure 1.16

The abdominal area is commonly subdivided in two ways: (*a*) into nine regions and (*b*) into four quadrants.

cervical (ser′vĭ-kal) The neck.
costal (kos′tal) The ribs.
coxal (kok′sal) The hip.
crural (krōōr′al) The leg.
cubital (ku′bĭ-tal) The elbow.
digital (dij′ĭ-tal) The finger or toe.
dorsal (dor′sal) The back.
femoral (fem′or-al) The thigh.
frontal (frun′tal) The forehead.
genital (jen′ĭ-tal) The reproductive organs.
gluteal (gloo′te-al) The buttocks.
inguinal (ing′gwĭ-nal) The groin—the depressed area of the abdominal wall near the thigh.
lumbar (lum′bar) The loin—the region of the lower back between the ribs and the pelvis.
mammary (mam′er-e) The breast.
mental (men′tal) The chin.
nasal (na′zal) The nose.
occipital (ok-sip′ĭ-tal) The lower posterior region of the head.
oral (o′ral) The mouth.
orbital (or′bi-tal) The eye cavity.
otic (o′tik) The ear.
palmar (pahl′mar) The palm of the hand.
patellar (pah-tel′ar) The front of the knee.
pectoral (pek′tor-al) The chest.
pedal (ped′al) The foot.
pelvic (pel′vik) The pelvis.
perineal (per″ĭ-ne′al) The perineum—the region between the anus and the external reproductive organs.
plantar (plan′tar) The sole of the foot.

popliteal (pop″lĭ-te′al) The area behind the knee.
sacral (sa′kral) The posterior region between the hipbones.
sternal (ster′nal) The middle of the thorax, anteriorly.
sural (su′ral) The calf of the leg.
tarsal (tahr′sal) The instep of the foot.
umbilical (um-bil′ĭ-kal) The navel.
vertebral (ver′te-bral) The spinal column.

Check Your Recall

19. Describe the anatomical position.
20. Using the appropriate terms, describe the relative positions of several body parts.
21. Describe the three types of body sections.
22. Name the nine regions of the abdomen.

Some Medical and Applied Sciences

cardiology (kar″de-ol′o-je) Branch of medical science dealing with the heart and heart diseases.
cytology (si-tol′o-je) Study of the structure, function, and abnormalities of cells. Cytology and histology are subdivisions of microscopic anatomy.
dermatology (der″mah-tol′o-je) Study of the skin and its diseases.
endocrinology (en″do-krĭ-nol′o-je) Study of hormones, hormone-secreting glands, and their diseases.
epidemiology (ep″ĭ-de″me-ol′o-je) Study of the factors determining the distribution and frequency of health-related conditions in a defined human population.

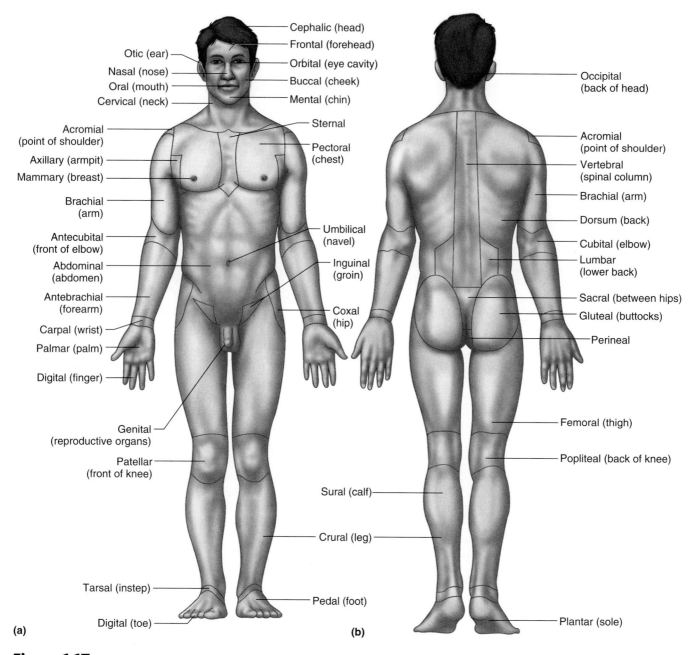

Cephalic (head)
Frontal (forehead)
Otic (ear)
Orbital (eye cavity)
Nasal (nose)
Buccal (cheek)
Oral (mouth)
Cervical (neck)
Mental (chin)

Occipital (back of head)

Acromial (point of shoulder)
Sternal
Pectoral (chest)
Acromial (point of shoulder)
Axillary (armpit)
Vertebral (spinal column)
Mammary (breast)
Brachial (arm)
Brachial (arm)
Dorsum (back)
Antecubital (front of elbow)
Umbilical (navel)
Cubital (elbow)
Abdominal (abdomen)
Inguinal (groin)
Lumbar (lower back)
Antebrachial (forearm)
Coxal (hip)
Sacral (between hips)
Gluteal (buttocks)
Carpal (wrist)
Palmar (palm)
Perineal
Digital (finger)

Genital (reproductive organs)
Femoral (thigh)
Patellar (front of knee)
Popliteal (back of knee)
Sural (calf)
Crural (leg)

Tarsal (instep)
Pedal (foot)
Plantar (sole)
Digital (toe)

(a)   (b)

Figure 1.17

Some terms used to describe body regions. (*a*) Anterior regions. (*b*) Posterior regions.

gastroenterology (gas″tro-en″ter-ol′o-je) Study of the stomach and intestines and their diseases.

geriatrics (jer″e-at′riks) Branch of medicine dealing with older individuals and their medical problems.

gerontology (jer″on-tol′o-je) Study of the aging process.

gynecology (gi″nĕ-kol′o-je) Study of the female reproductive system and its diseases.

hematology (hēm″ah-tol′o-je) Study of the blood and blood diseases.

histology (his-tol′o-je) Study of the structure and function of tissues. Histology and cytology are subdivisions of microscopic anatomy.

immunology (im″u-nol′o-je) Study of the body's resistance to infectious disease.

neonatology (ne″o-na-tol′o-je) Study of newborns and the treatment of their disorders.

nephrology (nĕ-frol′o-je) Study of the structure, function, and diseases of the kidneys.

neurology (nu-rol′o-je) Study of the nervous system and its disorders.

obstetrics (ob-stet′riks) Branch of medicine dealing with pregnancy and childbirth.

oncology (ong-kol′o-je) Study of cancers.

ophthalmology (of″thal-mol′o-je) Study of the eye and eye diseases.

orthopedics (or″tho-pe′diks) Branch of medicine dealing with the muscular and skeletal systems and their problems.

otolaryngology (o″to-lar″in-gol′o-je) Study of the ear, throat, and larynx, and their diseases.

pathology (pah-thol′o-je) Study of structural and functional changes that disease causes.

pediatrics (pe″de-at′riks) Branch of medicine dealing with children and their diseases.

pharmacology (fahr″mah-kol′o-je) Study of drugs and their uses in the treatment of disease.

podiatry (po-di′ah-tre) Study of the care and treatment of feet.

psychiatry (si-ki′ah-tre) Branch of medicine dealing with the mind and its disorders.

radiology (ra″de-ol′o-je) Study of X rays and radioactive substances and their uses in the diagnosis and treatment of diseases.

toxicology (tok″sĭ-kol′o-je) Study of poisonous substances and their effects upon body parts.

urology (u-rol′o-je) Branch of medicine dealing with the urinary system, apart from the kidneys (nephrology) and the male reproductive system, and their diseases.

SUMMARY OUTLINE

1.1 Introduction (p. 2)

1. Early interest in the human body probably developed as people became concerned about injuries and illnesses.
2. Primitive doctors began to learn how certain herbs and potions affected body functions.
3. The belief that humans could understand forces that caused natural events led to the development of modern science.
4. A set of terms originating from Greek and Latin words is the basis for the language of anatomy and physiology.

1.2 Anatomy and Physiology (p. 3)

1. Anatomy describes the form and organization of body parts.
2. Physiology considers the functions of anatomical parts.
3. The function of a body part depends on the way it is constructed.

1.3 Levels of Organization (p. 3)

The body is composed of parts with different levels of complexity.
1. Matter is composed of atoms.
2. Atoms join to form molecules.
3. Organelles are built of groups of large molecules (macromolecules).
4. Cells, which contain organelles, are the basic units of structure and function that form the body.
5. Cells are organized into tissues.
6. Tissues are organized into organs.
7. Organs that function closely together comprise organ systems.
8. Organ systems constitute the organism.
9. Beginning at the atomic level, these levels of organization differ in complexity from one level to the next.

1.4 Characteristics of Life (p. 4)

Characteristics of life are traits all organisms share.
1. These characteristics include:
   a. Movement—changing body position or moving internal parts.
   b. Responsiveness—sensing and reacting to internal or external changes.
   c. Growth—increasing size without changing shape.
   d. Reproduction—producing offspring.
   e. Respiration—obtaining oxygen, using oxygen to release energy from foods, and removing gaseous wastes.
   f. Digestion—breaking down food substances into component nutrients that the intestine can absorb.
   g. Absorption—moving substances through membranes and into body fluids.
   h. Circulation—moving substances through the body in body fluids.
   i. Assimilation—changing substances into chemically different forms.
   j. Excretion—removing body wastes.
2. Acquisition and use of energy constitute metabolism.

1.5 Maintenance of Life (p. 5)

The structures and functions of body parts maintain the life of the organism.
1. Requirements of organisms
   a. Water is used in many metabolic processes, provides the environment for metabolic reactions, and transports substances.
   b. Food supplies energy, raw materials for building new living matter, and chemicals necessary in vital reactions.
   c. Oxygen releases energy from food materials. This energy drives metabolic reactions.
   d. Heat is a product of metabolic reactions and helps govern the rates of these reactions.
   e. Pressure is an application of force to something. In humans, atmospheric and hydrostatic pressures help breathing and blood movements, respectively.
2. Homeostasis
   a. If an organism is to survive, the conditions within its body fluids must remain relatively stable.
   b. Maintenance of a stable internal environment is called *homeostasis*.
   c. Homeostatic mechanisms help regulate body temperature and blood pressure.
   d. Homeostatic mechanisms act through negative feedback.

1.6 Organization of the Human Body (p. 8)

1. Body cavities
   a. The axial portion of the body includes the cranial cavity, the vertebral canal, the thoracic cavity, and the abdominopelvic cavity.
   b. The diaphragm separates the thoracic and abdominopelvic cavities.
   c. The organs in a body cavity are called *viscera*.
   d. The mediastinum separates the thoracic cavity into right and left compartments.
   e. Body cavities in the head include the oral, nasal, orbital, and middle ear cavities.
2. Thoracic and abdominopelvic membranes
   a. Thoracic membranes
      (1) Pleural membranes line the thoracic cavity (parietal pleura) and cover the lungs (visceral pleura).
      (2) Pericardial membranes surround the heart (parietal pericardium) and cover its surface (visceral pericardium).
      (3) The pleural and pericardial cavities are the potential spaces between the respective parietal and visceral membranes.

b. Abdominopelvic membranes
   (1) Peritoneal membranes line the abdominopelvic cavity (parietal peritoneum) and cover the organs inside (visceral peritoneum).
   (2) The peritoneal cavity is the potential space between the parietal and visceral membranes.
3. Organ systems
   The human organism consists of several organ systems. Each system includes a set of interrelated organs.
   a. Body covering
      (1) The integumentary system includes the skin, hair, nails, sweat glands, and sebaceous glands.
      (2) It protects underlying tissues, regulates body temperature, houses sensory receptors, and synthesizes various substances.
   b. Support and movement
      (1) Skeletal system
         (a) The skeletal system is composed of bones, as well as cartilages and ligaments that bind bones together.
         (b) It provides a framework, protective shields, and attachments for muscles. It also produces blood cells and stores inorganic salts.
      (2) Muscular system
         (a) The muscular system includes the muscles of the body.
         (b) It moves body parts, maintains posture, and produces body heat.
   c. Integration and coordination
      (1) Nervous system
         (a) The nervous system consists of the brain, spinal cord, nerves, and sense organs.
         (b) It receives impulses from sensory parts, interprets these impulses, and acts on them by stimulating muscles or glands to respond.
      (2) Endocrine system
         (a) The endocrine system consists of glands that secrete hormones.
         (b) Hormones help regulate metabolism.
         (c) This system includes the hypothalamus of the brain and the pituitary, thyroid, parathyroid, and adrenal glands, as well as the pancreas, ovaries, testes, pineal gland, and thymus.
   d. Transport
      (1) Cardiovascular system
         (a) The cardiovascular system includes the heart, which pumps blood, and the blood vessels, which carry blood to and from body parts.
         (b) Blood transports oxygen, nutrients, hormones, and wastes.
      (2) Lymphatic system
         (a) The lymphatic system is composed of lymphatic vessels, lymph fluid, lymph nodes, thymus, and spleen.
         (b) It transports lymph fluid from tissues to the bloodstream, carries certain fatty substances away from the digestive organs, and aids in defending the body against disease-causing agents.
   e. Absorption and excretion
      (1) Digestive system
         (a) The digestive system receives foods, breaks down food molecules into nutrients that can pass through cell membranes, and eliminates materials that are not absorbed.
         (b) It includes the mouth, tongue, teeth, salivary glands, pharynx, esophagus, stomach, liver, gallbladder, pancreas, small intestine, and large intestine.
         (c) Some digestive organs produce hormones.
      (2) Respiratory system
         (a) The respiratory system takes in and sends out air and exchanges gases between the air and blood.
         (b) It includes the nasal cavity, pharynx, larynx, trachea, bronchi, and lungs.
      (3) Urinary system
         (a) The urinary system includes the kidneys, ureters, urinary bladder, and urethra.
         (b) It filters wastes from the blood and helps maintain water and electrolyte concentrations and the acidity of the internal environment.
   f. Reproduction
      (1) The reproductive systems produce new organisms.
      (2) The male reproductive system includes the scrotum, testes, epididymides, ductus deferentia, seminal vesicles, prostate gland, bulbourethral glands, urethra, and penis, which produce, maintain, and transport male sex cells (sperm cells).
      (3) The female reproductive system includes the ovaries, uterine tubes, uterus, vagina, clitoris, and vulva, which produce, maintain, and transport female sex cells (oocytes).

1.7 Anatomical Terminology (p. 14)

Terms with precise meanings help investigators communicate effectively.
1. Relative positions
   These terms describe the location of one part with respect to another part.
2. Body sections
   Body sections are planes along which the body may be cut to observe the relative locations and organization of internal parts.
3. Body regions
   Special terms designate various body regions.

CHAPTER ASSESSMENTS

◄ **1.1 Introduction**

1. Briefly describe the early development of knowledge about the human body. (p. 2)

◄ **1.2 Anatomy and Physiology**

2. Distinguish between anatomy and physiology. (p. 3)

3. Explain the relationship between the form and function of body parts. (p. 3)

◄ **1.3 Levels of Organization**

4. List the levels of organization within the human body and describe the characteristics of each. (p. 3)

◄ **1.4 Characteristics of Life**

5. List and describe ten characteristics of life. (p. 4)

6. Define *metabolism* and give examples. (p. 4)

◄ **1.5 Maintenance of Life**

7. List and describe five requirements of organisms. (p. 5)

8. Define *homeostasis,* and explain its importance. (p. 5)

9. Describe a general physiological control system. (p. 6)

10. Explain the control of body temperature. (p. 7)

11. Describe a homeostatic mechanism that helps regulate blood pressure. (p. 7)

1.6 Organization of the Human Body

12. Explain the difference between the axial and appendicular portions of the body. (p. 8)

13. Identify the cavities within the axial portion of the body. (p. 8)

14. Define *viscera*. (p. 8)

15. Describe the mediastinum and its contents. (p. 8)

16. List the cavities of the head and the contents of each cavity. (p. 8)

17. Name the body cavity that houses each of the following organs: (p. 8)
    a. Stomach   f. Rectum
    b. Heart   g. Spinal cord
    c. Brain   h. Esophagus
    d. Liver   i. Spleen
    e. Trachea   j. Urinary bladder

18. Distinguish between a parietal and a visceral membrane. (p. 10)

19. Name the major organ systems, and describe the general functions of each. (p. 12)

20. List the major organs that compose each organ system. (p. 12)

1.7 Anatomical Terminology

21. Write complete sentences using each of the following terms to correctly describe the relative locations of specific body parts: (p. 14)
    a. Superior   g. Proximal
    b. Inferior   h. Distal
    c. Anterior   i. Superficial
    d. Posterior   j. Peripheral
    e. Medial   k. Deep
    f. Lateral

22. Sketch the outline of a human body, and use lines to indicate each of the following sections: (p. 15)
    a. Sagittal
    b. Transverse
    c. Coronal

23. Sketch the abdominal area, and indicate the locations of the following regions: (p. 16)
    a. Epigastric   d. Hypochondriac
    b. Umbilical   e. Lumbar
    c. Hypogastric   f. Iliac

24. Provide the common name for the region to which each of the following terms refers: (p. 16)
    a. Acromial   m. Occipital
    b. Antebrachial n. Orbital
    c. Axillary   o. Otic
    d. Buccal   p. Palmar
    e. Celiac   q. Pectoral
    f. Coxal   r. Pedal
    g. Crural   s. Plantar
    h. Femoral   t. Popliteal
    i. Genital   u. Sacral
    j. Gluteal   v. Tarsal
    k. Inguinal   w. Umbilical
    l. Mental   x. Vertebral

INTEGRATIVE ASSESSMENTS/ CRITICAL THINKING

 OUTCOMES 1.2, 1.3, 1.4, 1.5

1. Which characteristics of life does an automobile have? Why is a car not alive?

OUTCOMES 1.4, 1.5

2. What environmental characteristics would be necessary for a human to survive on another planet?

OUTCOMES 1.2, 1.3, 1.4, 1.5, 1.6

3. Put the following in order, from smallest and simplest, to largest and most complex, and describe their individual roles in homeostasis: organ, molecule, organelle, atom, organ system, tissue, organism, cell, macromolecule.

OUTCOMES 1.5, 1.6

4. In health, body parts interact to maintain homeostasis. Illness can threaten the maintenance of homeostasis, requiring treatment. What treatments might be used to help control a patient's (*a*) body temperature, (*b*) blood oxygen concentration, and (*c*) water content?

OUTCOME 1.5

5. Suppose two individuals develop benign (noncancerous) tumors that produce symptoms because they occupy space and crowd adjacent organs. If one of these persons has the tumor in the thoracic cavity and the other has the tumor in the abdominopelvic cavity, which person would be likely to develop symptoms first? Why? Which might be more immediately serious? Why?

OUTCOMES 1.6, 1.7

6. If a patient complained of a "stomachache" and pointed to the umbilical region as the site of discomfort, which organs located in this region might be the source of the pain?

OUTCOMES 1.5, 1.6, 1.7

7. How might health-care professionals provide the basic requirements of life to an unconscious patient? Describe the body parts involved in the treatment, using correct directional and regional terms.

WEB CONNECTIONS

Visit the text website at **aris.mhhe.com** for additional quizzes, interactive learning exercises, and more.

Anatomy & Physiology | REVEALED includes cadaver photos that allow you to peel away layers of the human body to reveal structures beneath the surface. This program also includes animations, radiologic imaging, audio pronunciations, and practice quizzing. To learn more visit **www.aprevealed.com**.

The Human Organism

The series of illustrations that follows shows the major parts of the human torso. The first plate illustrates the anterior surface and reveals the superficial muscles on one side. Each subsequent plate exposes deeper organs, including those in the thoracic, abdominal, and pelvic cavities.

Chapters 6–19 of this textbook describe the organs and organ systems of the human organism in detail. As you read them, refer to these plates to visualize the locations of various organs and the three-dimensional relationships among them.

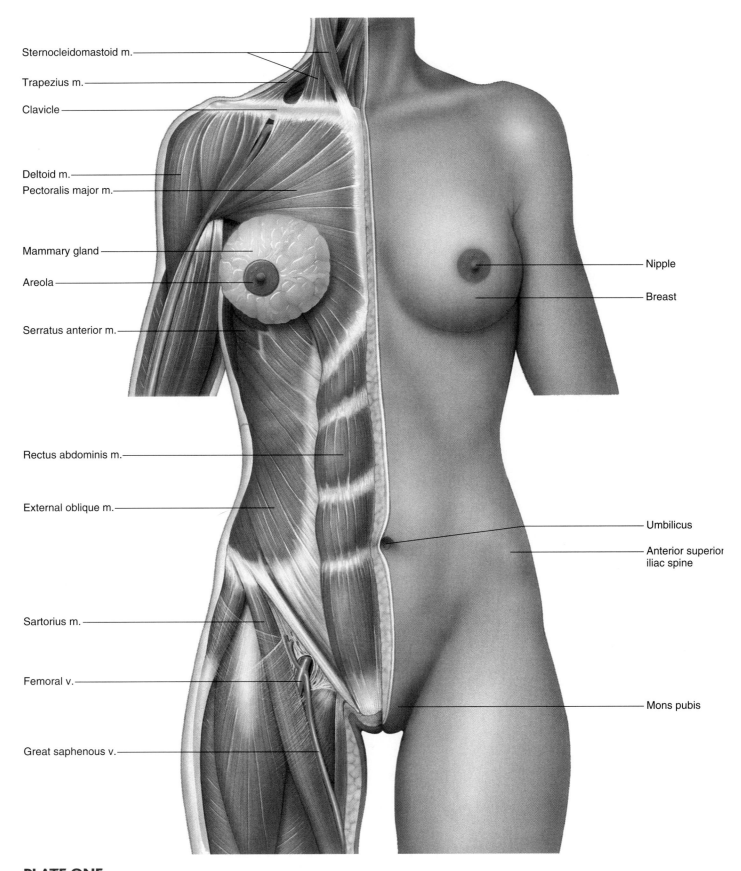

Sternocleidomastoid m.

Trapezius m.

Clavicle

Deltoid m.

Pectoralis major m.

Mammary gland

Areola

Serratus anterior m.

Rectus abdominis m.

External oblique m.

Sartorius m.

Femoral v.

Great saphenous v.

Nipple

Breast

Umbilicus

Anterior superior iliac spine

Mons pubis

PLATE ONE

Human female torso showing the anterior surface on one side and the superficial muscles exposed on the other side. (*m.* stands for *muscle,* and *v.* stands for *vein.*)

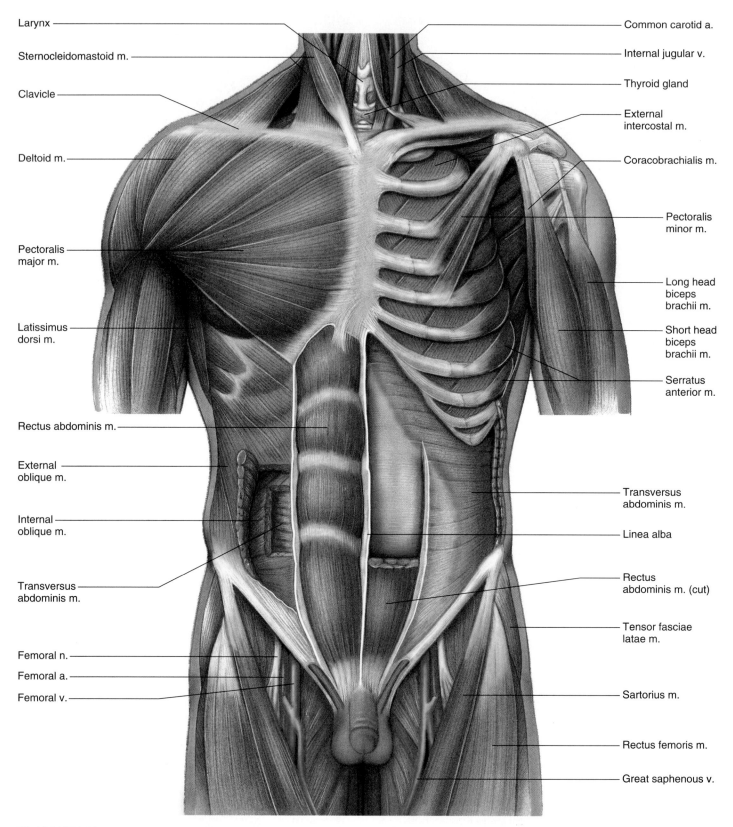

Larynx

Sternocleidomastoid m.

Clavicle

Deltoid m.

Pectoralis major m.

Latissimus dorsi m.

Rectus abdominis m.

External oblique m.

Internal oblique m.

Transversus abdominis m.

Femoral n.

Femoral a.

Femoral v.

Common carotid a.

Internal jugular v.

Thyroid gland

External intercostal m.

Coracobrachialis m.

Pectoralis minor m.

Long head biceps brachii m.

Short head biceps brachii m.

Serratus anterior m.

Transversus abdominis m.

Linea alba

Rectus abdominis m. (cut)

Tensor fasciae latae m.

Sartorius m.

Rectus femoris m.

Great saphenous v.

PLATE TWO

Human male torso with the deeper muscle layers exposed.

(*n.* stands for *nerve*, *a.* stands for *artery*, *m.* stands for *muscle*, and *v.* stands for *vein*.)

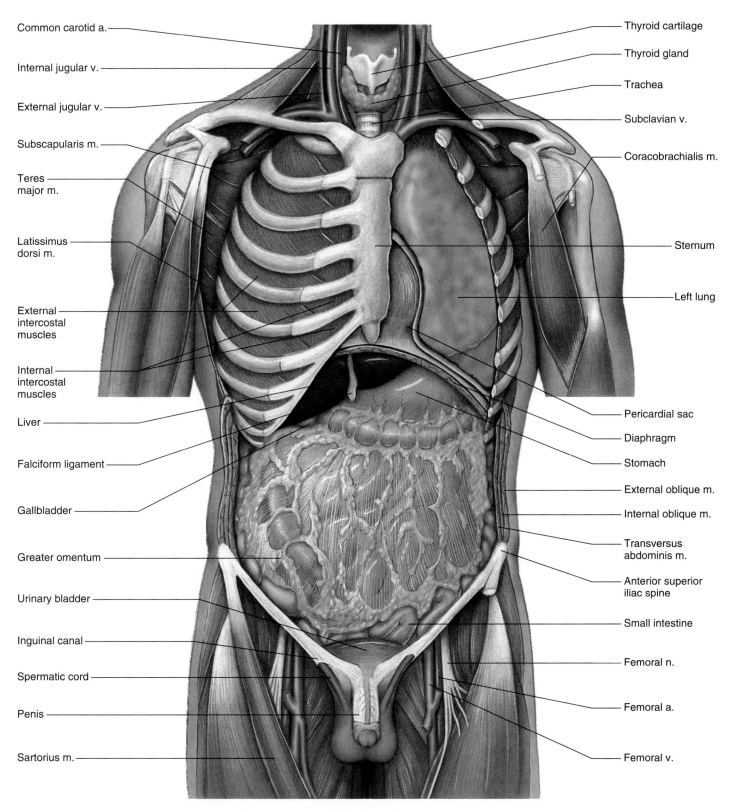

Common carotid a.

Internal jugular v.

External jugular v.

Subscapularis m.

Teres major m.

Latissimus dorsi m.

External intercostal muscles

Internal intercostal muscles

Liver

Falciform ligament

Gallbladder

Greater omentum

Urinary bladder

Inguinal canal

Spermatic cord

Penis

Sartorius m.

Thyroid cartilage

Thyroid gland

Trachea

Subclavian v.

Coracobrachialis m.

Sternum

Left lung

Pericardial sac

Diaphragm

Stomach

External oblique m.

Internal oblique m.

Transversus abdominis m.

Anterior superior iliac spine

Small intestine

Femoral n.

Femoral a.

Femoral v.

PLATE THREE

Human male torso with the deep muscles removed and the abdominal viscera exposed.

(*n.* stands for *nerve*, *a.* stands for *artery*, *m.* stands for *muscle*, and *v.* stands for *vein*.)

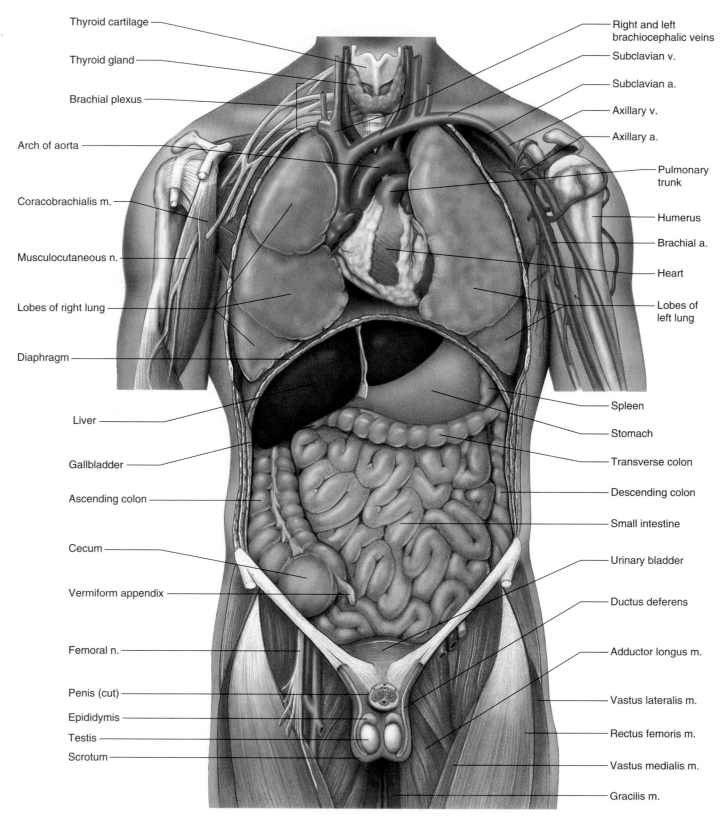

PLATE FOUR

Human male torso with the thoracic and abdominal viscera exposed.

(*n.* stands for *nerve, a.* stands for *artery, m.* stands for *muscle,* and *v.* stands for *vein.*)

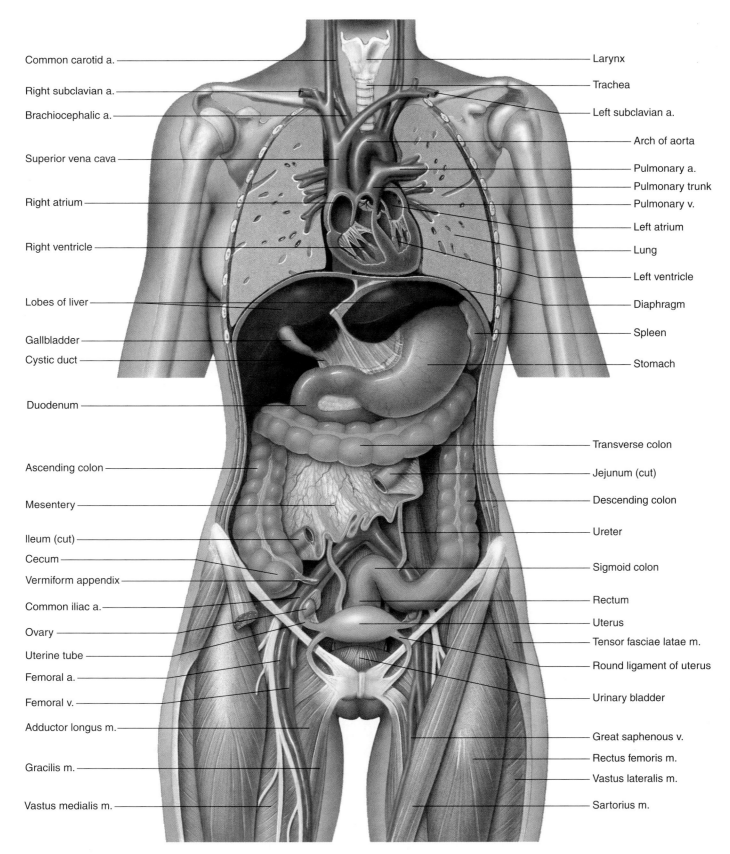

Common carotid a.

Right subclavian a.

Brachiocephalic a.

Superior vena cava

Right atrium

Right ventricle

Lobes of liver

Gallbladder

Cystic duct

Duodenum

Ascending colon

Mesentery

Ileum (cut)

Cecum

Vermiform appendix

Common iliac a.

Ovary

Uterine tube

Femoral a.

Femoral v.

Adductor longus m.

Gracilis m.

Vastus medialis m.

Larynx

Trachea

Left subclavian a.

Arch of aorta

Pulmonary a.

Pulmonary trunk

Pulmonary v.

Left atrium

Lung

Left ventricle

Diaphragm

Spleen

Stomach

Transverse colon

Jejunum (cut)

Descending colon

Ureter

Sigmoid colon

Rectum

Uterus

Tensor fasciae latae m.

Round ligament of uterus

Urinary bladder

Great saphenous v.

Rectus femoris m.

Vastus lateralis m.

Sartorius m.

PLATE FIVE

Human female torso with the lungs, heart, and small intestine sectioned and the liver reflected (lifted back).

(*a.* stands for *artery*, *m.* stands for *muscle*, and *v.* stands for *vein*.)

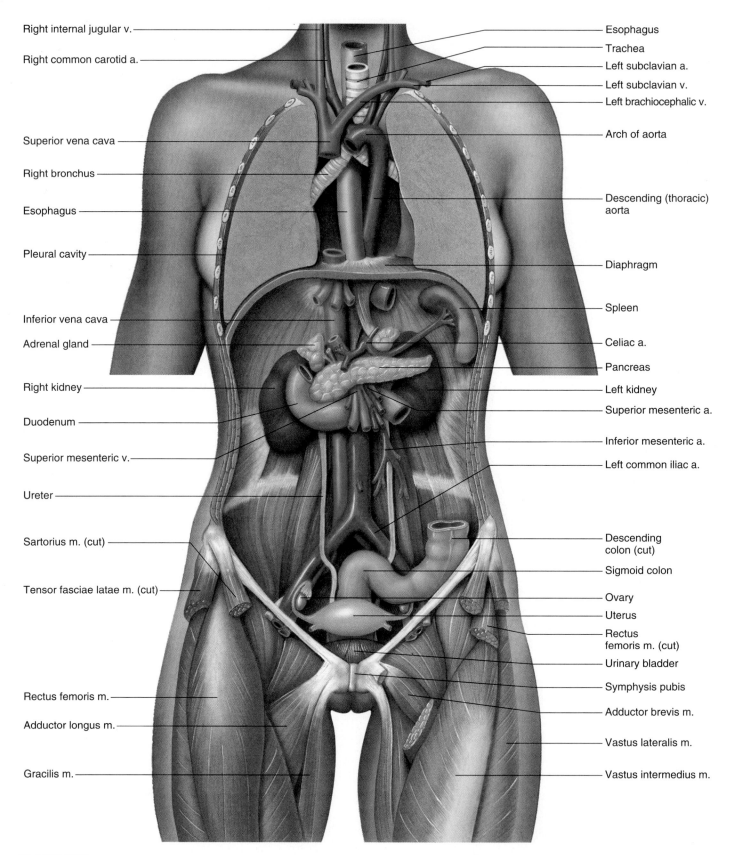

Right internal jugular v.

Right common carotid a.

Superior vena cava

Right bronchus

Esophagus

Pleural cavity

Inferior vena cava

Adrenal gland

Right kidney

Duodenum

Superior mesenteric v.

Ureter

Sartorius m. (cut)

Tensor fasciae latae m. (cut)

Rectus femoris m.

Adductor longus m.

Gracilis m.

Esophagus

Trachea

Left subclavian a.

Left subclavian v.

Left brachiocephalic v.

Arch of aorta

Descending (thoracic) aorta

Diaphragm

Spleen

Celiac a.

Pancreas

Left kidney

Superior mesenteric a.

Inferior mesenteric a.

Left common iliac a.

Descending colon (cut)

Sigmoid colon

Ovary

Uterus

Rectus femoris m. (cut)

Urinary bladder

Symphysis pubis

Adductor brevis m.

Vastus lateralis m.

Vastus intermedius m.

PLATE SIX

Human female torso with the heart, stomach, liver, and parts of the intestine and lungs removed.
(*a.* stands for *artery, m.* stands for *muscle,* and *v.* stands for *vein.*)

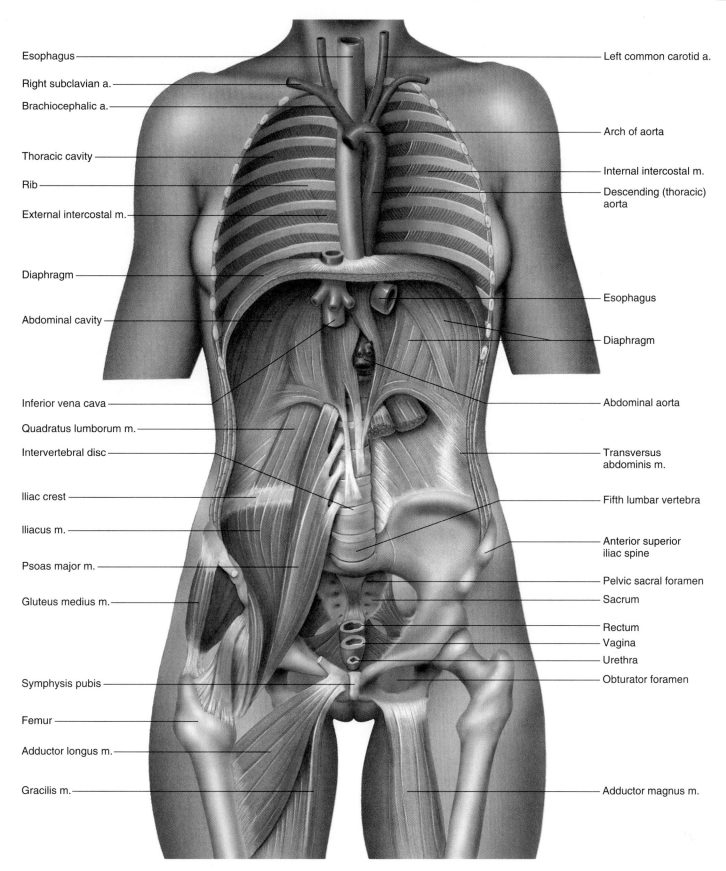

Esophagus

Right subclavian a.

Brachiocephalic a.

Thoracic cavity

Rib

External intercostal m.

Diaphragm

Abdominal cavity

Inferior vena cava

Quadratus lumborum m.

Intervertebral disc

Iliac crest

Iliacus m.

Psoas major m.

Gluteus medius m.

Symphysis pubis

Femur

Adductor longus m.

Gracilis m.

Left common carotid a.

Arch of aorta

Internal intercostal m.

Descending (thoracic) aorta

Esophagus

Diaphragm

Abdominal aorta

Transversus abdominis m.

Fifth lumbar vertebra

Anterior superior iliac spine

Pelvic sacral foramen

Sacrum

Rectum

Vagina

Urethra

Obturator foramen

Adductor magnus m.

PLATE SEVEN

Human female torso with the thoracic, abdominal, and pelvic viscera removed.

(*a.* stands for *artery,* and *m.* stands for *muscle.*)

2

Chemical Basis of Life

A TALE OF TWO TRACE ELEMENTS. Trace elements are minerals that make up only 0.005% of an adult's body. Because very small amounts are required, taking in too much of a trace element can be dangerous—but not only because of the excess of that mineral. Too much of one mineral can lead to a deficit of another. For instance, this happens with zinc and copper. These two elements are next to each other on the periodic table and are of similar size.

The average man requires 11 milligrams (mg) of zinc a day, and the average woman 9 mg. Zinc is available in several products: in multivitamins, supplements, throat lozenges, and other cold products, and as zinc oxide used to treat acne. Combining products without reading the labels to count the milligrams of zinc can lead to taking too much.

When zinc intake reaches 150 to 450 milligrams per day, one of the effects is to decrease the amount of copper in the body. The excess zinc triggers cells lining the small intestine to produce a protein called metallothionein, which transports both zinc and copper from the digestive tract into the bloodstream. The protein has a higher affinity for copper than for zinc, so once excess zinc triggers metallothionein production, the protein rapidly binds copper. Then the cell is sloughed off of the intestinal lining and leaves the body in a bowel movement before the copper can be absorbed. The result: copper deficiency. Copper deficiency affects the blood, causing anemia and neutropenia—too few red and white blood cells, respectively—which cause fatigue, fever, and increased susceptibility to infection.

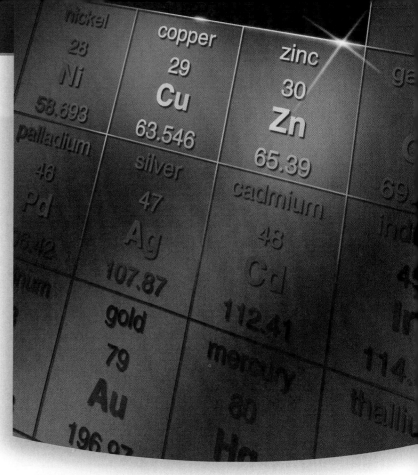

Too much zinc (Zn) causes too little copper (Cu), because atoms of these elements are so chemically similar that they compete for binding the same type of transport protein.

Tracing symptoms to zinc excess can require medical sleuthing. In one case, a teen was taking zinc supplements and using large amounts of zinc oxide cream. In another case a surgical patient had a huge wound that would not heal, and persistent fever. An alert young doctor noticed the patient's wife bringing in his nutritional supplements—the man had been taking 15 times the required daily amount of zinc for months! The excess zinc had led to copper deficiency, which was easily corrected. Yet another case was not as difficult to figure out—a man with schizophrenia had swallowed 461 coins, many of which were post-1981 pennies that are 97% zinc. He died of the trace mineral imbalance.

Learning Outcomes   *After studying this chapter, you should be able to do the following:*

2.1 Introduction

1. Give examples of how the study of living material depends on the study of chemistry. (p. 31)

2.2 Structure of Matter

2. Describe how atomic structure determines how atoms interact. (p. 32)

3. Describe the relationships between atoms and molecules. (p. 36)

4. Explain how molecular and structural formulas symbolize the composition of compounds. (p. 37)

5. Describe three types of chemical reactions. (p. 37)

6. Explain what acids, bases, and buffers are. (p. 38)

7. Define pH. (p. 39)

2.3 Chemical Constituents of Cells

8. List the major groups of inorganic chemicals common in cells. (p. 40)

9. Describe the functions of various types of organic chemicals in cells. (p. 41)

Aids to Understanding Words   *(Appendix A on page 567 has a complete list of Aids to Understanding Words.)*

di- [two] *di*saccharide: Compound whose molecules are composed of two joined saccharide units.

glyc- [sweet] *glyc*ogen: Complex carbohydrate composed of many joined sugar molecules.

lip- [fat] *lip*ids: Group of organic compounds that includes fats.

-lyt [dissolvable] electro*lyt*e: Substance that dissolves in water and releases ions.

mono- [one] *mono*saccharide: Compound whose molecules consist of a single saccharide unit.

poly- [many] *poly*unsaturated: Molecule with many double bonds between its carbon atoms.

sacchar- [sugar] mono*sacchar*ide: Sugar molecule composed of a single saccharide unit.

syn- [together] *syn*thesis: Process by which chemicals join to form new types of chemicals.

2.1 INTRODUCTION

At the cellular level of organization, chemistry, in a sense, becomes biology. A cell's working parts—its organelles—are intricate assemblies of macromolecules. Because the macromolecules that build the cells that build tissues and organs are themselves composed of atoms, the study of anatomy and physiology begins with chemistry.

Chemistry is the branch of science that considers the composition of matter and how this composition changes. Understanding chemistry is essential for understanding anatomy and physiology because body structures and functions result from chemical changes within cells. Indeed, the human body is composed of chemicals, including salts, water, proteins, carbohydrates, lipids, and nucleic acids. All of the food that we eat, liquids that we drink, and medications that we may take when we are sick, are chemicals.

2.2 STRUCTURE OF MATTER

Matter is anything that has weight and takes up space. This includes all the solids, liquids, and gases in our surroundings, as well as inside our bodies.

> Strictly speaking, matter has mass and takes up space. **Mass** refers to the amount of a substance, whereas **weight** refers to how heavy it is. If your weight on earth is 150 pounds, on the moon it would be only 25 pounds, but your mass (in kilograms) would be the same in both places. That is, you take up the same volume of space, but weigh less on the moon, because the force of gravity is lower on the moon. Since we are dealing with life on earth, and a constant gravity, we can consider mass and weight as roughly equivalent. Many of our students find it easier to think in terms of weight rather than mass.

Elements and Atoms

All matter is composed of fundamental substances called **elements** (el′ĕ-mentz). Examples include such common materials as iron, copper, silver, gold, aluminum, carbon, hydrogen, and oxygen. Although some elements exist in a pure form, they are found more frequently in combination with other elements.

Living organisms require about twenty elements. Of these, oxygen, carbon, hydrogen, and nitrogen make up more than 95% (by weight) of the human body (table 2.1). As the table shows, a one- or two-letter symbol represents each element.

Elements are composed of tiny particles called **atoms** (at′omz), which are the smallest complete units

Table 2.1	Elements in the Human Body	
Major Elements	**Symbol**	**Approximate Percentage of the Human Body (by weight)**
Oxygen	O	65.0%
Carbon	C	18.5
Hydrogen	H	9.5
Nitrogen	N	3.2
Calcium	Ca	1.5
Phosphorus	P	1.0
Potassium	K	0.4
Sulfur	S	0.3
Chlorine	Cl	0.2
Sodium	Na	0.2
Magnesium	Mg	0.1
		Total 99.9%
Trace Elements		
Chromium	Cr	Together less than 0.1%
Cobalt	Co	
Copper	Cu	
Fluorine	F	
Iodine	I	
Iron	Fe	
Manganese	Mn	
Zinc	Zn	

of elements. Atoms of an element are similar to each other, but they differ from the atoms that make up other elements. Atoms vary in size, weight, and the ways they interact with other atoms. Some atoms can combine with atoms like themselves or with other atoms by forming attractions called **chemical bonds,** whereas other atoms cannot form such bonds.

Atomic Structure

An atom consists of a central portion, called the **nucleus,** and one or more **electrons** (e-lek'tronz) that constantly move around it. The nucleus contains one or more relatively large particles called **protons** (pro'tonz). The nucleus also usually contains one or more **neutrons** (nu'tronz), which are similar in size to protons.

Electrons, which are extremely small, each carry a single, negative electrical charge (e$^-$), whereas protons each carry a single, positive electrical charge (p$^+$). Neutrons are uncharged and thus are electrically neutral (n^0) (fig. 2.1).

Because the nucleus contains the protons, it is always positively charged. However, the number of electrons outside the nucleus equals the number of protons. Therefore, a complete atom is electrically uncharged, or neutral.

The atoms of different elements have different numbers of protons. The number of protons in the atoms of a particular element is called the element's **atomic number.** Hydrogen, for example, whose atoms each have one proton, has the atomic number 1; carbon, whose atoms each have six protons, has the atomic number 6.

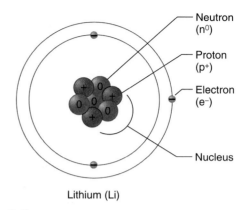

Lithium (Li)

Figure 2.1

An atom consists of subatomic particles. In an atom of the element lithium, three electrons move around a nucleus that consists of three protons and four neutrons.

The **atomic weight** of an atom of an element approximately equals the number of protons and neutrons in its nucleus; electrons have very little weight. Thus, the atomic weight of hydrogen, with one proton and no neutrons, is 1, whereas the atomic weight of carbon, with six protons and six neutrons, is 12 (table 2.2). In other words, an atom of carbon weighs about twelve times more than an atom of hydrogen.

All the atoms of a particular element have the same atomic number because they have the same number of protons and electrons. However, the atoms of an element vary in the number of neutrons in their nuclei; thus, they vary in atomic weight. For example, all oxygen atoms have eight protons in their nuclei, but these atoms may have eight, nine, or ten neutrons, corresponding to,

Table 2.2		Atomic Structure of Elements 1 Through 12						
Element	Symbol	Atomic Number	Atomic Weight	Protons	Neutrons	\multicolumn{3}{c}{Electrons in Shells}
						First	Second	Third
Hydrogen	H	1	1	1	0	1		
Helium	He	2	4	2	2	2 (inert)		
Lithium	Li	3	7	3	4	2	1	
Beryllium	Be	4	9	4	5	2	2	
Boron	B	5	11	5	6	2	3	
Carbon	C	6	12	6	6	2	4	
Nitrogen	N	7	14	7	7	2	5	
Oxygen	O	8	16	8	8	2	6	
Fluorine	F	9	19	9	10	2	7	
Neon	Ne	10	20	10	10	2	8 (inert)	
Sodium	Na	11	23	11	12	2	8	1
Magnesium	Mg	12	24	12	12	2	8	2

respectively, atomic weights of 16, 17, and 18. Atoms with the same atomic numbers but different atomic weights are called **isotopes** (i'so-tōps) of an element. Because a sample of an element is likely to include more than one isotope, the atomic weight of the element is often considered to be the average weight of the isotopes present. (See Appendix B, Periodic Table of the Elements, p. 568.)

How atoms interact reflects their number of electrons. Because the number of electrons in an atom is equal to its number of protons, all the isotopes of a particular element have the same number of electrons and react chemically in the same manner. Therefore, any of the isotopes of oxygen can play the same role in an organism's metabolic reactions.

Isotopes may be stable, or they may have unstable atomic nuclei that decompose, releasing energy or pieces of themselves. Unstable isotopes are called *radioactive* because they emit energetic particles, and the energy or atomic fragments they give off are called *radiation.*

Three common forms of radiation are alpha (α), beta (β), and gamma (γ). Alpha radiation consists of particles from atomic nuclei, each of which includes two protons and two neutrons, that travel slowly and can weakly penetrate matter. Beta radiation consists of much smaller particles (electrons) that travel more rapidly and penetrate matter more deeply. Gamma radiation is similar to X-ray radiation and is the most penetrating of these forms.

Each kind of radioactive isotope produces one or more forms of radiation, and each becomes less radioactive at a particular rate. The time required for an isotope to lose one-half of its radioactivity is called its *half-life.* Thus, the isotope of iodine called iodine-131, which emits one-half of its radiation in 8.1 days, has a half-life of 8.1 days. Half-lives vary greatly. The half-life of phosphorus-32 is 14.3 days; that of cobalt-60 is 5.26 years; and that of radium-226 is 1,620 years. The Topic of Interest on page 34 discusses some practical applications of radioactive isotopes.

Check Your Recall

1. What are elements?
2. Which elements are most common in the human body?
3. Where are electrons, protons, and neutrons located in an atom?
4. What is the difference between atomic number and atomic weight?

Bonding of Atoms

Atoms can attach to other atoms by forming chemical bonds. The chemical behavior of atoms results from interactions among their electrons. When atoms form chemical bonds they gain, lose, or share electrons. The electrons of an atom occupy one or more areas of space, called *shells,* around the nucleus (see table 2.2). For the elements up to atomic number 18, the maximum number of electrons that each of the first three inner shells can hold is as follows:

First shell (closest to the nucleus)	2 electrons
Second shell	8 electrons
Third shell	8 electrons

More complex atoms may have as many as eighteen electrons in the third shell. Simplified diagrams, such as those in figure 2.2, depict electron locations within the shells of atoms.

The electrons in the outermost shell of an atom determine its chemical behavior. Atoms such as helium, whose outermost electron shells are filled, have stable structures and are chemically inactive, or **inert** (see table 2.2). Atoms such as hydrogen or lithium, whose outermost electron shells are incompletely filled, tend to gain, lose, or share electrons in ways that empty or fill their outer shells. This enables them to achieve stable structures.

Atoms that gain or lose electrons become electrically charged and are called **ions** (i'onz). An atom of sodium, for example, has eleven electrons: two in the first shell, eight in the second shell, and one in the third shell (fig. 2.3). This atom tends to lose the electron from its outer shell, which leaves the second (now the outermost) shell filled and the new form stable (fig. 2.4*a*). In the process, sodium is left with eleven protons (11^+) in its nucleus and only ten electrons (10^-). As a result, the atom develops a net electrical charge of 1^+ and is called a sodium ion, symbolized Na^+.

A chlorine atom has seventeen electrons, with two in the first shell, eight in the second shell, and seven in the third shell. An atom of this type tends to accept a single electron, filling its outer shell and achieving stability (fig. 2.4*a*). In the process, the chlorine atom is left with seventeen protons (17^+) in its nucleus and eighteen electrons (18^-). The atom develops a net electrical charge of 1^- and is called a chloride ion, symbolized Cl^-.

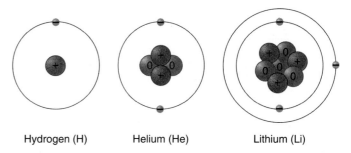

Hydrogen (H)   Helium (He)   Lithium (Li)

Figure 2.2

Electrons orbit the atomic nucleus. The single electron of a hydrogen atom is located in its first shell. The two electrons of a helium atom fill its first shell. Two of the three electrons of a lithium atom are in the first shell, and one is in the second shell.

Topic of Interest

Radioactive Isotopes

Because atomic radiation can be detected with special equipment, such as a scintillation counter, radioactive substances are useful in studying life processes (fig. 2A). A radioactive isotope can be introduced into an organism and then traced as it enters into metabolic activities. For example, the human thyroid gland is unique in using the element iodine in its metabolism. Therefore, radioactive iodine-131 is used to study thyroid functions and to evaluate thyroid disease (fig. 2B). Doctors use thallium-201, which has a half-life of 73.5 hours, to assess heart conditions, and gallium-67, with a half-life of 78 hours, to detect and monitor the progress of certain cancers and inflammatory diseases.

Atomic radiations also can change chemical structures, and in this way, alter vital cellular processes. For this reason, doctors sometimes use radioactive isotopes, such as cobalt-60, to treat cancers. The radiation from the cobalt preferentially kills the rapidly dividing cancer cells.

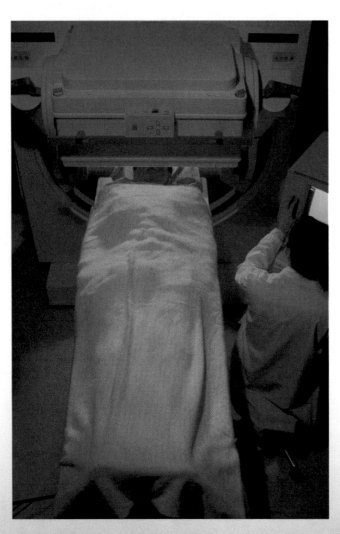

Figure 2A

Scintillation counters detect radioactive isotopes.

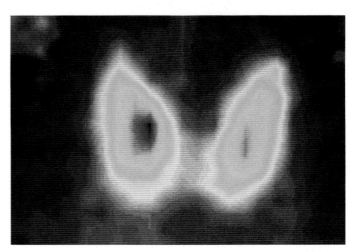

(a)

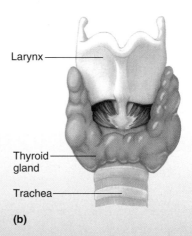

Larynx

Thyroid gland

Trachea

(b)

Figure 2B

(a) Scan of the thyroid gland 24 hours after the patient received radioactive iodine. Note how closely the scan in (a) resembles the shape of the thyroid gland, shown in (b).

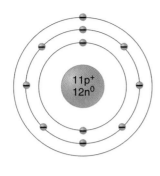

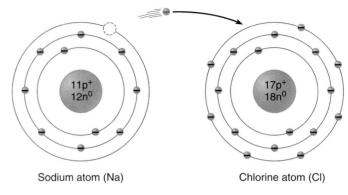

Sodium atom (Na)   Chlorine atom (Cl)

(a) Separate atoms

If a sodium atom loses an electron to a chlorine atom, the sodium atom becomes a sodium ion (Na^+), and the chlorine atom becomes a chloride ion (Cl^-).

Sodium atom contains
11 electrons (e^-)
11 protons (p^+)
12 neutrons (n^0)

Atomic number = 11
Atomic weight = 23

Figure 2.3

A sodium atom.

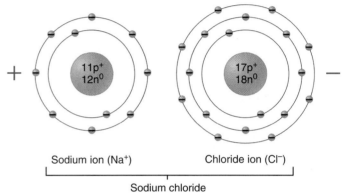

Sodium ion (Na^+)   Chloride ion (Cl^-)

Sodium chloride

(b) Bonded ions

These oppositely charged particles attract electrically and join by an ionic bond.

(c) Salt crystal

Ionically bonded substances form arrays such as a crystal of NaCl.

Figure 2.4

An ionic bond forms when one atom gains and another atom loses electrons (a) and then oppositely charged ions attract (b). Ionically bonded substances may form crystals (c).

Because oppositely charged ions attract, sodium and chloride ions react to form a type of chemical bond called an **ionic bond** (electrovalent bond). Sodium ions (Na^+) and chloride ions (Cl^-) uniting in this manner form the compound sodium chloride (NaCl), or table salt (fig. 2.4b). Some ions have an electrical charge greater than 1—for example, Ca^{+2} (or Ca^{++}).

Ionically bound substances do not form discreet molecules—instead, they form arrays, such as crystals of sodium chloride (fig. 2.4c). The molecular formulas for compounds like sodium chloride (NaCl) give the relative amounts of each element present. Atoms may also bond by sharing electrons, rather than by exchanging them. A hydrogen atom, for example, has one electron in its first shell but requires two electrons to achieve a stable structure (fig. 2.5). It may fill this shell by combining with another hydrogen atom in such a way that the two atoms share a pair of electrons. The two electrons then encircle the nuclei of both atoms, and each atom achieves a stable form. The chemical bond between the atoms that share electrons is called a **covalent bond.**

Carbon atoms, which have two electrons in their first shells and four electrons in their second shells, form covalent bonds when they unite with other atoms. In fact, carbon atoms (and certain other atoms) may bond in such a way that two atoms share one or more pairs of electrons. If one pair of electrons is shared, the resulting bond is called a *single covalent bond;* if two pairs of electrons are shared, the bond is called a *double covalent bond. Triple covalent bonds* are also possible between some atoms.

Different types of chemical bonds share electrons to different degrees. At one extreme is the ionic bond in which atoms gain or lose electrons. At the other extreme is the covalent bond in which the electrons

are shared equally. In between lies the covalent bond in which electrons are not shared equally, resulting in a molecule whose shape gives an uneven distribution of charges. Such a molecule is called **polar.** Unlike an ion, a polar molecule has an equal number of protons and electrons, but one end of the molecule has more than its share of electrons, becoming slightly negative, while the other end of the molecule has less than its share, becoming slightly positive. Typically, polar covalent bonds form where hydrogen atoms bond to oxygen

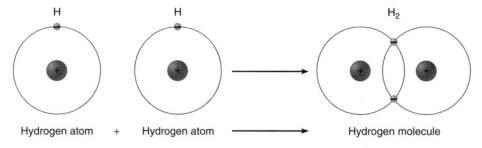

H   H   H₂

Hydrogen atom   +   Hydrogen atom   ⟶   Hydrogen molecule

Figure 2.5

A hydrogen molecule forms when two hydrogen atoms share a pair of electrons. A covalent bond forms between the atoms.

or nitrogen atoms. Water is an important polar molecule (fig. 2.6*a*).

The attraction of the positive hydrogen end of a polar molecule to the negative nitrogen or oxygen end of another polar molecule is called a **hydrogen bond.** Hydrogen bonds are weak, particularly at body temperature. For example, below 0°C, the hydrogen bonds between the water molecules shown in figure 2.6*b* are strong enough to form ice. As the temperature rises, increased molecular movement breaks the hydrogen bonds, and water becomes a liquid. Even at body temperature, hydrogen bonds are important in protein and nucleic acid structure. In these cases, hydrogen bonds form between polar regions of different parts of a single, very large molecule (see figs. 2.18 and 2.21). The contribution of hydrogen bonds to protein and nucleic acid structure is described in section 2.3, pages 44 and 46.

Molecules and Compounds

When two or more atoms bond, they form a new kind of particle called a **molecule** (mol'ĕ-kūl). If atoms of the same element bond, they produce molecules of that element. Gases of hydrogen, oxygen, and nitrogen consist of such molecules (see fig. 2.5).

When atoms of different elements bond, they form molecules called **compounds.** Two atoms of hydrogen, for example, can bond with one atom of oxygen to produce a molecule of the compound water (H_2O) (fig. 2.7). Table sugar (*sucrose*), baking soda, natural gas, beverage alcohol, and most drugs are compounds.

A molecule of a compound always consists of definite kinds and numbers of atoms. A molecule of water, for instance, always has two hydrogen atoms and one oxygen atom. If two hydrogen atoms bond with two oxygen atoms, the compound formed is not water, but hydrogen peroxide (H_2O_2). Table 2.3 summarizes the characteristics of the particles of matter discussed so far.

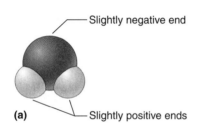

Slightly negative end

(a)   Slightly positive ends

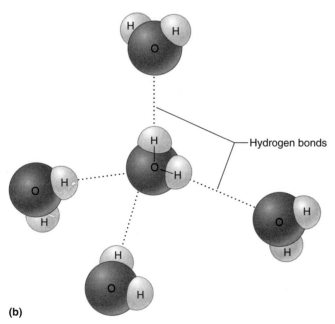

Hydrogen bonds

(b)

Figure 2.6

Water is a polar molecule. (*a*) Water molecules have equal numbers of electrons and protons but are polar because the electrons are shared unequally, creating slightly negative ends and slightly positive ends. (*b*) Hydrogen bonding connects water molecules.

Check Your Recall

5. What is an ion?

6. Describe two ways that atoms bond with other atoms.

7. Distinguish between a molecule and a compound.

8. Distinguish between an ion and a polar molecule.

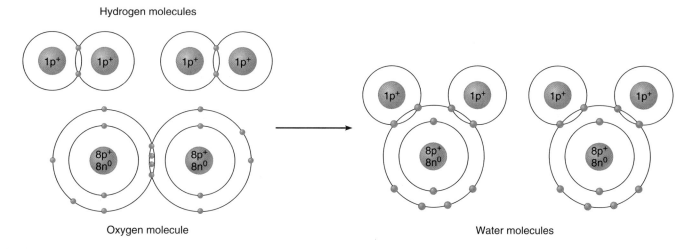

Hydrogen molecules

Oxygen molecule

Water molecules

Figure 2.7

Hydrogen molecules can combine with oxygen molecules, forming water molecules. The shared electrons represent covalent bonds.

Formulas

A **molecular formula** (mo-lek′u-lar for′mu-lah) represents the numbers and types of atoms in a molecule. Such a formula displays the symbols for the elements in the molecule and the number of atoms of each element. For example, the molecular formula for water is H_2O, which means that each water molecule consists of two atoms of hydrogen and one atom of oxygen (fig. 2.8). The molecular formula for the sugar *glucose* is $C_6H_{12}O_6$, indicating that each glucose molecule consists of six atoms of carbon, twelve atoms of hydrogen, and six atoms of oxygen.

Usually, the atoms of each element form a specific number of covalent bonds. Hydrogen atoms form single bonds, oxygen atoms form two bonds, nitrogen atoms form three bonds, and carbon atoms form four bonds. Symbols and lines can depict bonds as follows:

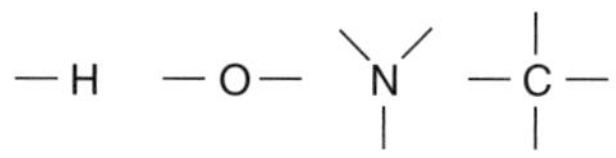

$$-H \quad -O- \quad \overset{\diagdown}{\underset{\diagup}{N}} \quad -\overset{|}{\underset{|}{C}}-$$

These representations show how atoms are joined and arranged in various molecules. Single lines represent single bonds, and double lines represent double bonds. Illustrations of this type are called **structural formulas** (struk′cher-al for′mu-lahz) (fig. 2.8). Three-dimensional models of structural formulas use different colors for the different kinds of atoms (fig. 2.9).

Chemical Reactions

Chemical reactions form or break bonds between atoms, ions, or molecules, generating new chemical combinations. For example, when two or more atoms (reactants) bond to form a more complex structure (product), the reaction is called **synthesis** (sin′thĕ-sis). Such a reaction is symbolized in this way:

$$A + B \rightarrow AB$$

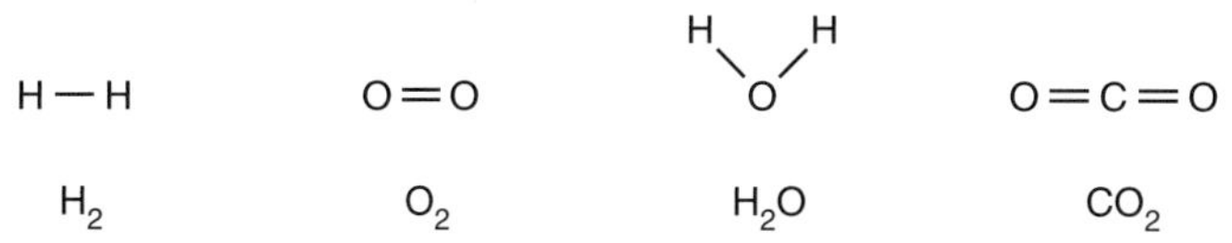

Table 2.3	Some Particles of Matter
Particle	**Characteristics**
Atom	Smallest particle of an element that has the properties of that element
Electron (e⁻)	Extremely small particle; carries a negative electrical charge and is in constant motion around a nucleus of an atom
Proton (p⁺)	Relatively large particle; carries a positive electrical charge and is found within a nucleus of an atom
Neutron (n⁰)	Relatively large particle; uncharged and thus electrically neutral; found within a nucleus of an atom
Molecule	Particle formed by the chemical union of two or more atoms
Ion	Atom or molecule that is electrically charged because it has gained or lost one or more electrons

$$H-H \qquad O=O \qquad \overset{\displaystyle H \diagdown \diagup H}{O} \qquad O=C=O$$

$$H_2 \qquad\qquad O_2 \qquad\qquad H_2O \qquad\qquad CO_2$$

Figure 2.8

Structural and molecular formulas for molecules of hydrogen, oxygen, water, and carbon dioxide. Note the double covalent bonds. (Triple covalent bonds are also possible between some atoms.)

(a)   A water molecule (H₂O), with the white parts depicting hydrogen atoms and the red part representing oxygen.

(b)   A glucose molecule ($C_6H_{12}O_6$), in which the black parts represent carbon atoms.

Figure 2.9

Three-dimensional molecular models depict spatial relationships of the constituent atoms.

If the bonds within a reactant molecule break so that simpler molecules, atoms, or ions form, the reaction is called **decomposition** (de″kom-po-zish′un). Decomposition is symbolized as follows:

$$AB \rightarrow A + B$$

Synthesis, which requires energy, is particularly important in the growth of body parts and the repair of worn or damaged tissues, which require buildup of larger molecules from smaller ones. In contrast, decomposition occurs when food molecules are digested into smaller ones that can be absorbed.

A third type of chemical reaction is an **exchange reaction.** In this reaction, parts of two different types of molecules trade positions as bonds are broken and new bonds are formed. The reaction is symbolized as follows:

$$AB + CD \rightarrow AD + CB$$

An example of an exchange reaction is when an acid reacts with a base, producing water and a salt. Acids and bases are described in the next section.

Many chemical reactions are reversible. This means that the product (or products) of the reaction can change back to the reactant (or reactants) that originally underwent the reaction. A **reversible reaction** is symbolized with a double arrow:

$$A + B \rightleftharpoons AB$$

Whether a reversible reaction proceeds in one direction or the other depends on such factors as the relative proportions of the reactant (or reactants) and product (or products), as well as the amount of available energy. Particular atoms or molecules that can change the rate (not the direction) of a reaction without being consumed in the process, called **catalysts,** speed many chemical reactions in the body so that they proceed fast enough to sustain the activities of life.

Acids and Bases

When ionically bound substances dissolve in water, the slightly negative and positive ends of the water molecules cause the ions to leave each other and interact with the water molecules instead. In this way, the polarity of water dissociates salts in the internal environment (fig. 2.10). For example, sodium chloride (NaCl) releases sodium ions (Na⁺) and chloride ions (Cl⁻) when it dissolves:

$$NaCl \rightarrow Na^+ + Cl^-$$

Since the resulting solution contains electrically charged particles (ions), it will conduct an electric current. Substances that release ions in water are, therefore, called **electrolytes** (e-lek′tro-lītz). **Acids** are electrolytes that release hydrogen ions (H⁺) in water. For example, in water, the compound hydrochloric acid (HCl) releases hydrogen ions (H⁺) and chloride ions (Cl⁻):

$$HCl \rightarrow H^+ + Cl^-$$

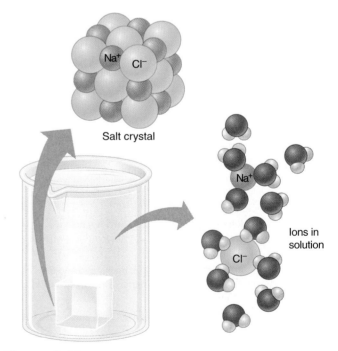

Salt crystal

Na⁺

Cl⁻

Ions in solution

Figure 2.10

The polar nature of water molecules dissociates sodium chloride (NaCl) in water, releasing sodium ions (Na⁺) and chloride ions (Cl⁻).

Electrolytes that release ions that bond with hydrogen ions are called **bases.** For example, the compound sodium hydroxide (NaOH) releases hydroxide ions (OH⁻) when placed in water:

$$NaOH \rightarrow Na^+ + OH^-$$

The hydroxide ions, in turn, can bond with hydrogen ions to form water; thus, sodium hydroxide is a base. Many bases are present in the body fluids, but because of the way they react in water, the concentration of hydroxide ions is a good estimate of the total base concentration. (Note: Some ions, such as OH⁻, consist of two or more atoms. However, such a group behaves like a single atom and usually remains unchanged during a chemical reaction.)

The concentrations of hydrogen ions (H⁺) and hydroxide ions (OH⁻) in body fluids greatly affect the chemical reactions that control certain physiological functions, such as blood pressure and breathing rate. Since their concentrations are inversely related (if one goes up, the other goes down), we need to keep track of only one of them. A value called **pH** measures hydrogen ion concentration.

The pH scale ranges from 0 to 14. A solution with a pH of 7.0, the midpoint of the scale, contains equal numbers of hydrogen and hydroxide ions and is said to be *neutral*. A solution that contains more hydrogen ions than hydroxide ions has a pH less than 7.0 and is *acidic*. A solution with fewer hydrogen ions than hydroxide ions has a pH greater than 7.0 and is *basic* (alkaline).

Figure 2.11 indicates the pH values of some common substances. Each whole number on the pH scale represents a tenfold difference in hydrogen ion concentration, and as the hydrogen ion concentration increases, the pH number gets smaller. Thus, a solution with a pH of 6 has ten times the hydrogen ion concentration of a solution with a pH of 7. This means that relatively small changes in pH can reflect large changes in hydrogen ion concentration.

Buffers are chemicals that resist pH change. They combine with hydrogen ions when these ions are in excess, or they donate hydrogen ions when these ions are depleted. Buffers and the regulation of the hydrogen ion concentration in body fluids are discussed further in chapter 18 (pp. 500–501).

The pH of human blood is about 7.4, and ranges from 7.35 to 7.45 (see fig. 2.11). If the pH drops below 7.35, the person has *acidosis;* if it rises above 7.45, the condition is *alkalosis*. Without medical intervention, a person usually cannot survive if blood pH drops to 6.9 or rises to 7.8 for more than a few hours. The general homeostatic mechanism described in chapter 1 (p. 6) may regulate pH.

Check Your Recall

9. What is a molecular formula? A structural formula?
10. Describe three kinds of chemical reactions.
11. Compare the characteristics of an acid with those of a base.
12. What does pH measure?
13. What is a buffer?

2.3 CHEMICAL CONSTITUENTS OF CELLS

Chemicals, including those that enter into metabolic reactions or are produced by them, can be divided into two large groups. Chemicals that include both carbon and hydrogen atoms are called **organic** (or-gan′ik). The rest are **inorganic** (in″or-gan′ik).

Inorganic substances usually dissociate in water to release ions; thus, they are *electrolytes*. Many organic compounds also dissolve in water, but they are more likely to dissolve in organic liquids, such as ether or

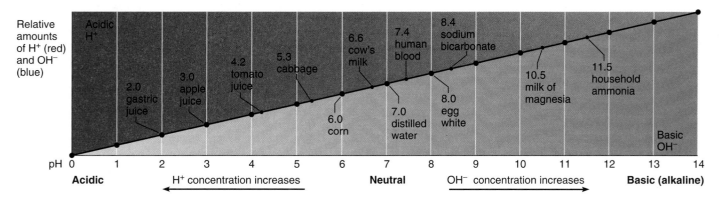

Figure 2.11

The pH scale measures hydrogen ion (H⁺) concentration. As the concentration of H⁺ increases, a solution becomes more acidic, and the pH value decreases. As the concentration of ions that bond with H⁺ (such as hydroxide ions) increases, a solution becomes more basic (alkaline), and the pH value increases. The pH values of some common substances are shown.

alcohol. Organic substances that dissolve in water usually do not release ions and are therefore called *nonelectrolytes*.

Inorganic Substances

Among the inorganic substances common in cells are water, oxygen, carbon dioxide, and a group of compounds called salts.

Water

Water is the most abundant compound in living material and accounts for about two-thirds of the weight of an adult human. It is the major component of blood and other body fluids, including those in cells.

Water is an important *solvent* because many substances readily dissolve in it. A substance dissolved in water, called a *solute*, is broken down into smaller and smaller pieces, eventually to molecular-sized particles, which may be ions. If two or more types of solutes are dissolved, they are much more likely to react with one another than were the original large pieces. Consequently, most metabolic reactions occur in water.

Water also plays an important role in moving chemicals in the body. For example, the aqueous (watery) portion of blood carries many vital substances, such as oxygen, sugars, salts, and vitamins, from the organs of digestion and respiration to the body cells.

Water can absorb and transport heat. Blood carries heat released from muscle cells during exercise from deeper parts of the body to the surface, where it may be lost to the outside.

Oxygen

Molecules of oxygen (O_2) enter the body through the respiratory organs and are transported throughout the body by the blood. The red blood cells bind and carry most of the oxygen. Cellular organelles use oxygen to release energy from the sugar glucose and other nutrients. The released energy drives the cell's metabolic activities.

Carbon Dioxide

Carbon dioxide (CO_2) is a simple, carbon-containing compound of the inorganic group. It is produced as a waste product when certain metabolic processes release energy, and it is exhaled from the lungs.

Salts

A salt is a compound composed of oppositely charged ions, such as sodium (Na^+) and chloride (Cl^-), which is the familiar table salt NaCl. Salts are abundant in tissues and fluids. They provide many necessary ions, including sodium (Na^+), chloride (Cl^-), potassium (K^+), calcium (Ca^{+2}), magnesium (Mg^{+2}), phosphate (PO_4^{-3}), carbonate (CO_3^{-2}), bicarbonate (HCO_3^-), and sulfate (SO_4^{-2}). These ions are important in metabolic processes, including transport of substances into and out of cells, muscle contraction, and nerve impulse conduction. Table 2.4 summarizes the functions of some of the inorganic substances in cells.

Table 2.4	Inorganic Substances Common in Cells	
Substance	**Symbol or Formula**	**Functions**
I. Inorganic molecules		
Water	H_2O	Major component of body fluids (chapter 12, p. 318); medium in which most biochemical reactions occur; transports chemicals (chapter 12, p. 325); helps regulate body temperature (chapter 6, pp. 124–125)
Oxygen	O_2	Used in energy release from glucose molecules (chapter 4, p. 82)
Carbon dioxide	CO_2	Waste product that results from metabolism (chapter 4, p. 82); reacts with water to form carbonic acid (chapter 16, p. 462)
II. Inorganic ions		
Bicarbonate ions	HCO_3^-	Helps maintain acid-base balance (chapter 18, p. 500)
Calcium ions	Ca^{+2}	Necessary for bone development (chapter 7, p. 139), muscle contraction (chapter 8, p. 182), and blood clotting (chapter 12, p. 330)
Carbonate ions	CO_3^{-2}	Component of bone tissue (chapter 7, p. 139)
Chloride ions	Cl^-	Helps maintain water balance (chapter 18, p. 493)
Magnesium ions	Mg^{+2}	Component of bone tissue (chapter 7, p. 139); required for certain metabolic processes (chapter 15, p. 435)
Phosphate ions	PO_4^{-3}	Required for synthesis of ATP, nucleic acids, and other vital substances (chapter 4, p. 80); component of bone tissue (chapter 7, p. 139); helps maintain polarization of cell membranes (chapter 9, p. 220)
Potassium ions	K^+	Required for polarization of cell membranes (chapter 9, p. 221)
Sodium ions	Na^+	Required for polarization of cell membranes (chapter 9, p. 221); helps maintain water balance (chapter 18, p. 493)
Sulfate ions	SO_4^{-2}	Helps maintain polarization of cell membranes (chapter 9, p. 220)

Check Your Recall

14. How do inorganic and organic molecules differ?
15. How do electrolytes and nonelectrolytes differ?
16. Name the inorganic substances common in body fluids.

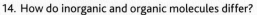

Organic Substances

Important groups of organic chemicals in cells include carbohydrates, lipids, proteins, and nucleic acids.

Carbohydrates

Carbohydrates (kar″bo-hi′drātz) provide much of the energy that cells require. They supply materials to build certain cell structures and often are stored as reserve energy supplies.

Carbohydrate molecules consist of atoms of carbon, hydrogen, and oxygen. These molecules usually have twice as many hydrogen as oxygen atoms—the same ratio of hydrogen to oxygen as in water molecules (H_2O). This ratio is easy to see in the molecular formulas of the carbohydrates glucose ($C_6H_{12}O_6$) and sucrose ($C_{12}H_{22}O_{11}$).

The carbon atoms of carbohydrate molecules join in chains whose lengths vary with the type of carbohydrate. For example, carbohydrates with shorter chains are called **sugars.**

Sugars with 6 carbon atoms (hexoses) are examples of *simple sugars,* or **monosaccharides** (mon″o-sak′ah-ridz). The simple sugars include glucose, fructose, and galactose, as well as the 5-carbon sugars ribose and deoxyribose. Figure 2.12 illustrates the structural formulas of glucose.

In *complex carbohydrates,* a number of simple sugar molecules link to form molecules of varying sizes (fig. 2.13). Some complex carbohydrates, such as sucrose (table sugar) and lactose (milk sugar), are *double sugars,* or **disaccharides** (di-sak′ah-ridz), whose molecules each consist of two simple sugar building blocks. Other complex carbohydrates are made up of many simple sugar units joined to form **polysaccharides** (pol″e-sak′ah-rīdz), such as plant starch. Animals, including humans, synthesize a polysaccharide similar to starch called *glycogen.*

Lipids

Lipids (lip′idz) are organic substances that are insoluble in water but soluble in certain organic solvents, such as ether and chloroform. Lipids include a variety of compounds—fats, phospholipids, and steroids—that have vital functions in cells. The most common lipids are fats.

Fats are used primarily to store energy for cellular activities. Fat molecules can supply more energy, gram for gram, than carbohydrate molecules.

Like carbohydrates, fat molecules are composed of carbon, hydrogen, and oxygen atoms. However, fats have a much smaller proportion of oxygen atoms than do carbohydrates. The formula for the fat tristearin, $C_{57}H_{110}O_6$, illustrates these characteristic proportions.

The building blocks of fat molecules are **fatty acids** and **glycerol** (glis′er-ol). Each glycerol molecule bonds with three fatty acid molecules to produce a single fat, or *triglyceride,* molecule (fig. 2.14).

The glycerol portions of all fat molecules are identical, but fats are diverse because there are many kinds of fatty acids. Fatty acid molecules differ in the lengths of their carbon atom chains, which usually have an even number of carbon atoms. The chains also vary in the way the carbon atoms bond. In some cases, the carbon atoms all join by single carbon–carbon bonds. This type of fatty acid is *saturated;* that is, each carbon atom is bound to as many hydrogen atoms as possible and is thus saturated with hydrogen atoms. Other fatty acid chains have not bound the maximum number of

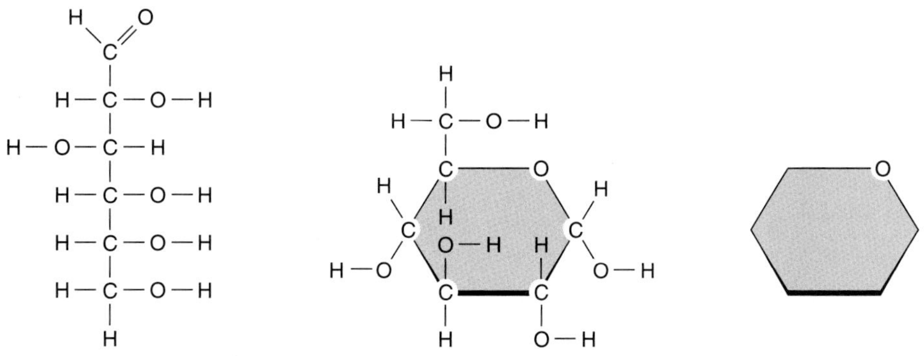

(a) Some glucose molecules ($C_6H_{12}O_6$) have a straight chain of carbon atoms.

(b) More commonly, glucose molecules form a ring structure.

(c) This shape symbolizes the ring structure of a glucose molecule.

Figure 2.12
Structural formulas depict a molecule of glucose.

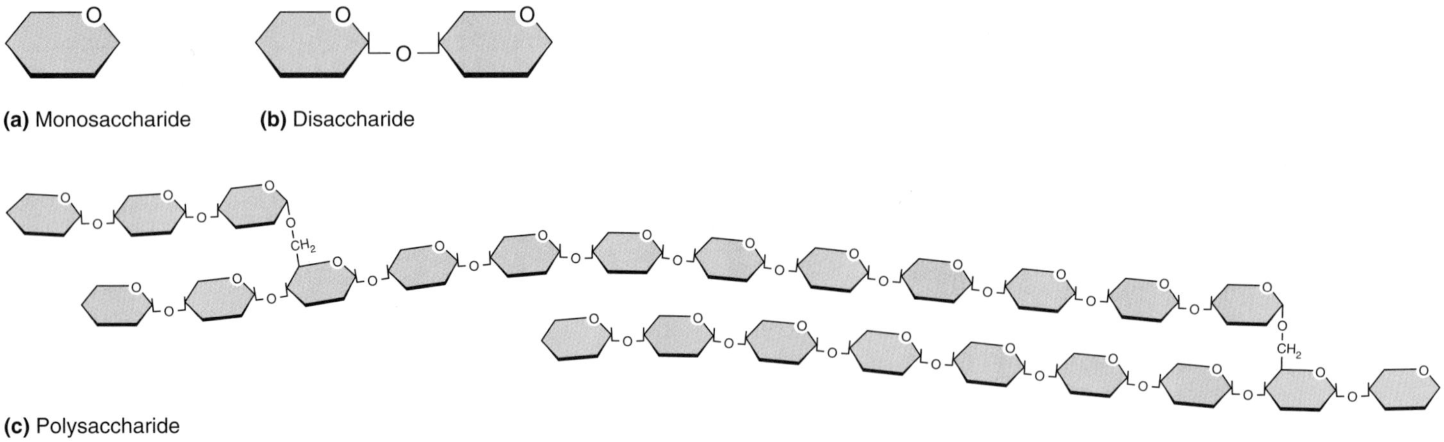

(a) Monosaccharide   **(b)** Disaccharide

(c) Polysaccharide

Figure 2.13

Carbohydrate molecules vary in size. (*a*) A monosaccharide molecule consists of one building block with 6 carbon atoms. (*b*) A disaccharide molecule consists of two of these building blocks. (*c*) A polysaccharide molecule consists of many such building blocks.

hydrogen atoms. These fatty acids, therefore, have one or more double bonds between carbon atoms, because a carbon atom must form four bonds to be stable. Fatty acid molecules with double bonds are *unsaturated,* and those with many double-bonded carbon atoms are *polyunsaturated.* Similarly, fat molecules that contain only saturated fatty acids are called *saturated fats,* and those that include unsaturated fatty acids are called *unsaturated fats.* The triglyceride molecule in figure 2.14 is an example of an unsaturated fat.

A **phospholipid** (fos″fo-lip′id) molecule is similar to a fat molecule in that it consists of a glycerol portion and fatty acid chains (fig. 2.15*a, b*). A phospholipid,

however, has only two fatty acid chains; in place of the third is a portion that includes a phosphate group. The phosphate portion is soluble in water (hydrophilic) and forms the "head" of the molecule, while the fatty acid portion is insoluble in water (hydrophobic) and forms a "tail" (fig. 2.15*c*). Phospholipids are important in cellular structures.

Steroid (ste′roid) molecules are complex structures that include four connected rings of carbon atoms (fig. 2.16). Among the more important steroids are cholesterol, which is in all body cells and is used to synthesize other steroids; sex hormones, such as estrogen, progesterone, and testosterone; and several hormones

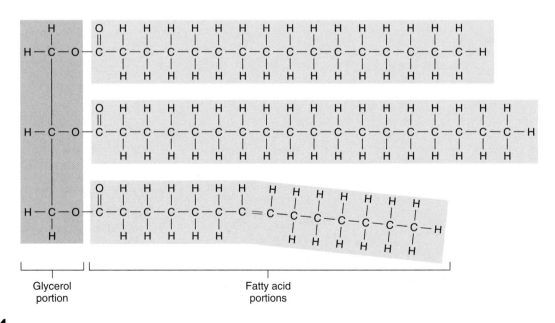

Glycerol portion

Fatty acid portions

Figure 2.14

A triglyceride molecule (fat) consists of a glycerol portion and three fatty acid portions. This is an example of an unsaturated fat. The double bond between carbon atoms in the unsaturated fatty acid is shown in red.

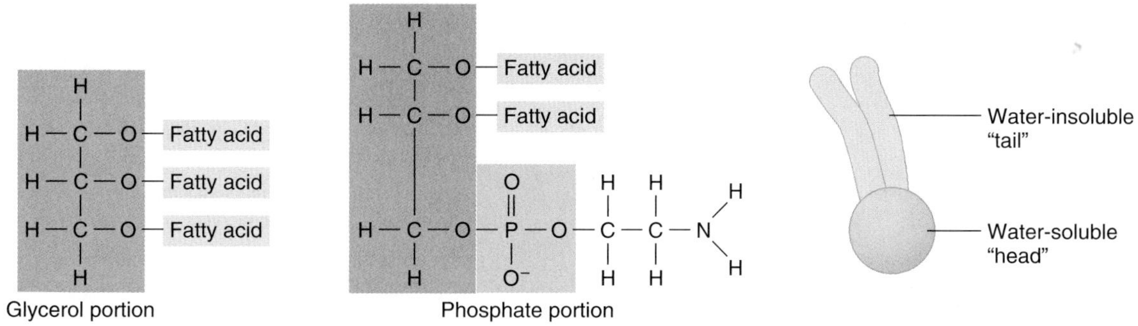

(a) A fat molecule

(b) A phospholipid molecule
(the unshaded portion may vary)

(c) Schematic representation
of a phospholipid molecule

Figure 2.15

Fats and phospholipids. (*a*) A fat molecule (triglyceride) consists of a glycerol and three fatty acids. (*b*) In a phospholipid molecule, a phosphate-containing group replaces one fatty acid. (*c*) Schematic representation of a phospholipid.

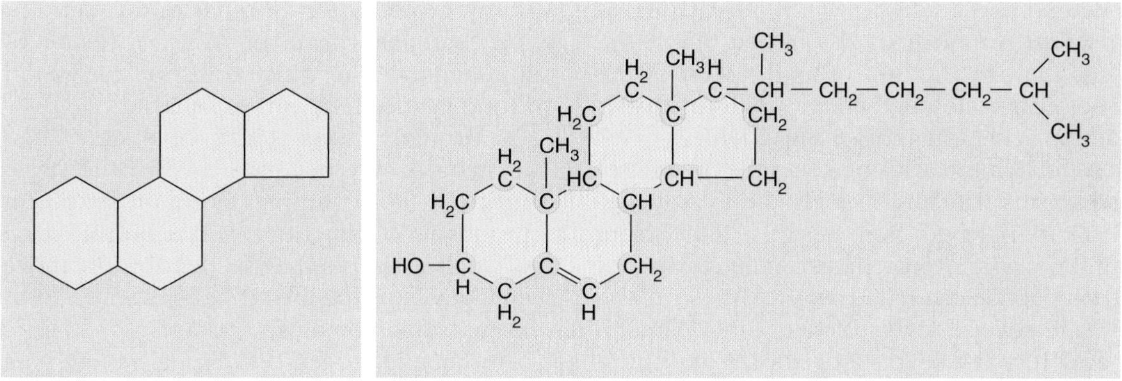

(a) General structure of a steroid   **(b)** Cholesterol

Figure 2.16

Steroid structure. (*a*) The general structure of a steroid. (*b*) The structural formula for cholesterol, a steroid widely distributed in the body.

from the adrenal glands. Chapters 11 and 19 discuss these steroids.

Table 2.5 lists the three important groups of lipids and their characteristics.

Saturated fats are more abundant in fatty foods that are solids at room temperature, such as butter, lard, and most animal fats. Unsaturated fats, on the other hand, are liquid at room temperature, such as soft margarine, seed oils (corn, sesame, soybean, sunflower, and peanut). Exceptions are coconut and palm kernel oils, which are high in saturated fats. The most heart-healthy fats are olive and canola oils, which are monounsaturated—that is, they have one carbon-carbon double bond.

Manufacturers of prepared foods sometimes harden vegetable oils by adding hydrogen atoms, producing a hydrogenated or "trans" fat. A diet high in saturated fats, trans fats, and cholesterol increases the risk of developing atherosclerosis, in which fatty deposits obstruct the inner linings of arteries. Some cities have banned the use of trans fats in restaurant food.

Proteins

Proteins (pro'tēnz) have a wide variety of functions in the body. Many serve as structural materials, energy sources, or hormones. Others combine with carbohydrates (glycoproteins) and function as receptors on cell surfaces, allowing cells to respond to particular kinds of molecules that bind to them. Proteins called *antibodies* detect and destroy foreign substances in the body. Metabolism could not occur fast enough to support life were it not for *enzymes*, which catalyze specific chemical reactions. (Enzymes are discussed in more detail in chapter 4, p. 79.)

A human body has more than 200,000 types of proteins, but only about 20,500 genes, which are the instructions for production of particular proteins. The information encoded in parts of different genes can combine to specify the many proteins. It is a little like assembling a large and diverse wardrobe from a few basic pieces of clothing.

Table 2.5	Important Groups of Lipids	
Group	**Basic Molecular Structure**	**Characteristics**
Triglycerides	Three fatty acid molecules bound to a glycerol molecule	Most common lipids in body; stored in fat tissue as an energy supply; fat tissue also provides insulation beneath the skin
Phospholipids	Two fatty acid molecules and a phosphate group bound to a glycerol molecule	Used as structural components in cell membranes; abundant in liver and parts of the nervous system
Steroids	Four connected rings of carbon atoms	Widely distributed in the body and have a variety of functions; include cholesterol, hormones of adrenal cortex, sex hormones, bile acids, and vitamin D

Like carbohydrates and lipids, proteins are composed of atoms of carbon, hydrogen, and oxygen. In addition, proteins always contain nitrogen atoms and, in some cases, sulfur atoms. The building blocks of proteins are **amino acids,** each of which has an *amino group* (—NH$_2$) at one end and a *carboxyl group* (—COOH) at the other (fig. 2.17*a*). Amino acids also have a *side chain,* or *R group* ("R" may be thought of as the "rest of the molecule"). The composition of the R group distinguishes one type of amino acid from another (fig. 2.17*b*).

Twenty different amino acids make up the proteins of most living organisms. The amino acids join in polypeptide chains that vary in length from less than 100 to more than 5,000 amino acids. A protein molecule consists of one or more polypeptide chains.

Proteins have several levels of structure: Primary, secondary, and tertiary levels are shown in figure 2.18*a–c*. Hydrogen bonding and even covalent bonding between atoms in different parts of the polypeptide give

the final protein a distinctive three-dimensional shape, or **conformation** (fig. 2.19). The conformation of a protein determines its function. Some proteins are long and fibrous, such as the keratin proteins that form hair, or fibrin, the protein whose threads form a blood clot. Many proteins are globular and function as enzymes, ion channels, carrier proteins, or receptors. Myoglobin and hemoglobin, which transport oxygen in muscle and blood, respectively, are globular.

In many cases, slight, reversible changes in conformation may occur as part of the protein's normal function. For example, some of the proteins involved in muscle contraction exert a pulling force as a result of such a shape change, leading to movement. Such changes in shape are reversible, so the protein can perform its function again and again.

When hydrogen bonds in a protein break as a result of exposure to excessive heat, radiation, electricity, pH changes, or various chemicals, a protein's unique shape

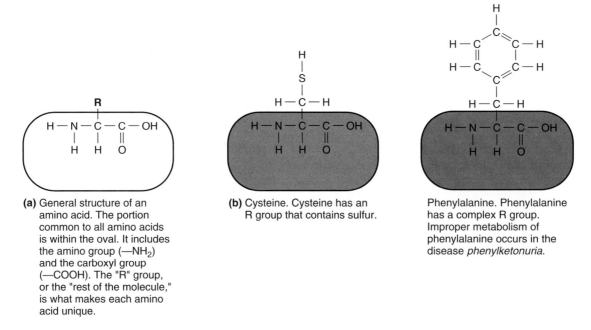

(a) General structure of an amino acid. The portion common to all amino acids is within the oval. It includes the amino group (—NH$_2$) and the carboxyl group (—COOH). The "R" group, or the "rest of the molecule," is what makes each amino acid unique.

(b) Cysteine. Cysteine has an R group that contains sulfur.

Phenylalanine. Phenylalanine has a complex R group. Improper metabolism of phenylalanine occurs in the disease *phenylketonuria*.

Figure 2.17

Amino acid structure. (*a*) An amino acid has an amino group, a carboxyl group, and a hydrogen atom that are common to all amino acid molecules, and a specific R group. (*b*) Some representative amino acids and their structural formulas. Each type of amino acid molecule has a particular shape due to its R group.

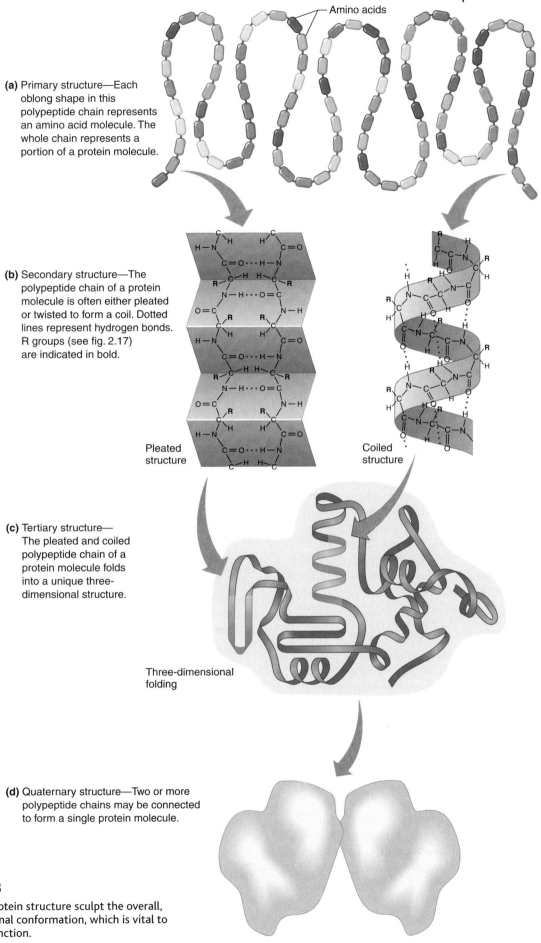

Amino acids

(a) Primary structure—Each oblong shape in this polypeptide chain represents an amino acid molecule. The whole chain represents a portion of a protein molecule.

(b) Secondary structure—The polypeptide chain of a protein molecule is often either pleated or twisted to form a coil. Dotted lines represent hydrogen bonds. R groups (see fig. 2.17) are indicated in bold.

Pleated structure

Coiled structure

(c) Tertiary structure— The pleated and coiled polypeptide chain of a protein molecule folds into a unique three-dimensional structure.

Three-dimensional folding

(d) Quaternary structure—Two or more polypeptide chains may be connected to form a single protein molecule.

Figure 2.18

The levels of protein structure sculpt the overall, three-dimensional conformation, which is vital to the protein's function.

Figure 2.19

A model of a portion of the protein collagen. The complex shape of a protein is characteristic of that protein and determines its functional properties.

may be changed dramatically, or *denatured.* Such proteins lose their special properties. For example, heat denatures the protein in egg white (albumin), changing it from a liquid to a solid. This is an irreversible change—a hard-boiled egg cannot return to its uncooked, runny state. Similarly, cellular proteins that are denatured may be permanently altered and lose their functions.

Proteins with more than one polypeptide chain have a fourth level of conformation, the *quaternary structure.* The constituent polypeptides are connected, often forming a very large protein (see fig. 2.18*d*). Hemoglobin is a quaternary protein made up of four separate polypeptide chains.

The conformation of a protein determines its function, and for most proteins, the conformation is always the same for a given amino acid sequence or primary structure (see Clinical Connection, pp. 46–47). Thus, it is the amino acid sequence that ultimately determines the role of a protein in the body. Genes, made of nucleic acid, contain the information for the amino acid sequences of all the body's proteins in a form that the cell can decode.

Nucleic Acids

Nucleic acids (nu-kle′ik as′idz) form genes and take part in protein synthesis. These molecules are generally very large and complex. They include atoms of carbon, hydrogen, oxygen, nitrogen, and phosphorus, which form building blocks called **nucleotides.** Each nucleotide consists of a 5-carbon *sugar* (ribose or deoxyribose), a *phosphate group,* and one of several *nitrogenous* (nitrogen-containing) *bases* (fig. 2.20). A nucleic acid molecule consists of a chain of many nucleotides (polynucleotide chain).

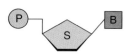

Figure 2.20

A nucleotide consists of a 5-carbon sugar (S = sugar), a phosphate group (P = phosphate), and a nitrogenous base (B = base).

Nucleic acids are of two types. One type—**RNA (ribonucleic acid)** (ri″bo-nu-kle′ik as′id)—is composed of molecules whose nucleotides have ribose. RNA usually is a single polynucleotide chain, but it can fold into various shapes that enable it to control when certain genes are accessed (fig. 2.21*a*). The second type—**DNA (deoxyribonucleic acid)** (de-ok′si-ri″bo-nu-kle″ik as′id)—has deoxyribose and forms a double polynucleotide chain. The two chains are held together by hydrogen bonds (fig. 2.21*b*).

DNA molecules store information in a type of molecular code created by the sequences of the four types of nitrogenous bases. Cells use this information to synthesize protein molecules. RNA molecules carry out protein synthesis. (Nucleic acids are discussed in more detail in chapter 4, pp. 83–89.) Certain nucleotides, such as adenosine triphosphate (ATP), have another role providing energy to certain chemical reactions (fig. 2.22). ATP is discussed further in chapter 4 (p. 80). Table 2.6 summarizes the four groups of organic compounds that build bodies.

Recall that water molecules are polar. Many larger molecules, including carbohydrates, proteins, and nucleic acids, have polar regions as well and dissolve easily in water as a result. Unlike electrolytes, however, they do not dissociate when they dissolve in water—they remain intact. Such water-soluble molecules are said to be hydrophilic (they "like" water).

Molecules that lack polar regions, such as triglycerides and steroids, do not dissolve in water ("oil and water don't mix"). Such molecules do dissolve in lipid and are said to be lipophilic (they "like" lipid).

Water solubility and lipid solubility are important factors in drug delivery and in the movement of substances throughout the body. Much of this will be discussed further in chapter 3 (pp. 61–62).

Check Your Recall

17. Compare the chemical composition of carbohydrates, lipids, proteins, and nucleic acids.
18. How does an enzyme affect a chemical reaction?
19. What is the chemical basis of the great diversity of proteins?
20. What are the functions of nucleic acids?

Clinical Connection

Many proteins undergo reversible changes in conformation as part of their normal function. In the case of a particular type of protein called a *prion,* some of the changes

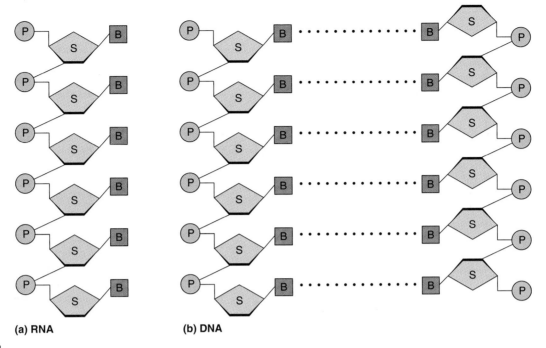

(a) RNA   **(b) DNA**

Figure 2.21

A schematic representation of nucleic acid structure. A nucleic acid molecule consists of (*a*) one (RNA) or (*b*) two (DNA) polynucleotide chains. DNA chains are held together by hydrogen bonds (dotted lines), and they entwine, forming a double helix.

in conformation cause disease. Worse, if a disease-causing prion bonds to a normal prion, it converts the normal prion to the disease-causing form.

Abnormal prions cause diseases called transmissible spongiform encephalopathies. More than eighty species of mammals are subject to these disorders. Familiar ones are bovine spongiform encephalopathy, better known as "mad cow disease," and in humans, variant Creutzfeldt-Jakob disease. The prion disorders turn parts of the brain into spongy masses, with steady loss of mental functions. Other diseases, including Alzheimer disease and Parkinson disease, are associated with misfolded proteins that are not infectious, but form gummy plaques in the brain. It isn't clear whether the plaques cause the symptoms or reflect the body's attempt to protect against the disease.

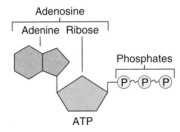

Figure 2.22

An ATP (adenosine triphosphate) molecule consists of an adenine, a ribose, and three phosphates. The wavy lines connecting the last two phosphates represent high-energy chemical bonds. When broken, these bonds release energy, which the cell uses for metabolic processes.

Table 2.6	Organic Compounds in Cells			
Compound	**Elements Present**	**Building Blocks**	**Functions**	**Examples**
Carbohydrates	C,H,O	Simple sugars	Provide energy, cell structure	Glucose, starch
Lipids	C,H,O (often P)	Glycerol, fatty acids, phosphate groups	Provide energy, cell structure	Triglycerides, phospholipids, steroids
Proteins	C,H,O,N (often S)	Amino acids	Provide cell structure, enzymes, energy	Albumins, hemoglobin
Nucleic acids	C,H,O,N,P	Nucleotides	Store information for protein synthesis; control cell activities	RNA, DNA

SUMMARY OUTLINE

2.1 Introduction (p. 31)

Chemistry describes the composition of substances and how chemicals react with each other. The human body is composed of chemicals.

2.2 Structure of Matter (p. 31)

1. Elements and atoms
   a. Matter is composed of elements.
   b. Some elements occur in pure form, but many are found combined with other elements.
   c. Elements are composed of atoms, which are the smallest complete units of elements.
   d. Atoms of different elements have characteristic size, weight, and ways of interacting.
2. Atomic structure
   a. An atom consists of one or more electrons surrounding a nucleus, which contains one or more protons and usually one or more neutrons.
   b. Electrons are negatively charged, protons are positively charged, and neutrons are uncharged.
   c. A complete atom is electrically neutral.
   d. An element's atomic number is equal to the number of protons in each atom. The atomic weight is equal to the number of protons plus the number of neutrons in each atom.
   e. Isotopes are atoms with the same atomic number but different atomic weights.
   f. Some isotopes are radioactive.
3. Bonding of atoms
   a. When atoms combine, they gain, lose, or share electrons.
   b. Electrons occupy shells around a nucleus.
   c. Atoms with completely filled outer shells are inert, but atoms with incompletely filled outer shells tend to gain, lose, or share electrons and thus achieve stable structures.
   d. Atoms that lose electrons become positively charged ions. Atoms that gain electrons become negatively charged ions.
   e. Ions with opposite electrical charges attract and form ionic bonds. Atoms that share electrons form covalent bonds.
   f. A polar covalently bonded molecule has an uneven distribution of charges.
   g. The attraction of positive to negative portions of polar covalent molecules is called a hydrogen bond.
4. Molecules and compounds
   a. Two or more atoms of the same element may bond to form a molecule of that element. Atoms of different elements may bond to form a molecule of a compound.
   b. Molecules consist of definite kinds and numbers of atoms.
5. Formulas
   a. A molecular formula represents the numbers and types of atoms in a molecule.
   b. A structural formula depicts the arrangement of atoms within a molecule.
6. Chemical reactions
   a. A chemical reaction breaks or forms bonds between atoms, ions, or molecules.
   b. Three types of chemical reactions are: synthesis, in which larger molecules form from smaller particles; decomposition, in which larger molecules are broken down into smaller

particles; and exchange reactions, in which the parts of two different molecules trade positions.
   c. Many reactions are reversible. The direction of a reaction depends on the proportions of reactants and end products, the energy available, and the presence of catalysts.
7. Acids and bases
   a. Compounds that release ions when they dissolve in water are electrolytes.
   b. Electrolytes that release hydrogen ions are acids, and those that release hydroxyl or other ions that react with hydrogen ions are bases.
   c. A value called pH represents a solution's concentration of hydrogen ions (H^+) and hydroxide ions (OH^-).
   d. A solution with equal numbers of H^+ and OH^- is neutral and has a pH of 7.0. A solution with more H^+ than OH^- is acidic and has a pH less than 7.0. A solution with fewer H^+ than OH^- is basic and has a pH greater than 7.0.
   e. Each whole number on the pH scale represents a tenfold difference in the hydrogen ion concentration.
   f. Buffers are chemicals that resist pH change.

2.3 Chemical Constituents of Cells (p. 39)

Molecules that have carbon and hydrogen atoms are organic and are usually nonelectrolytes. Other molecules are inorganic and are usually electrolytes.

1. Inorganic substances
   a. Water is the most abundant compound in cells and is a solvent in which chemical reactions occur. Water transports chemicals and heat.
   b. Oxygen releases energy from glucose and other nutrients. This energy drives metabolism.
   c. Carbon dioxide is produced when metabolism releases energy.
   d. Salts provide a variety of ions that metabolic processes require.
2. Organic substances
   a. Carbohydrates provide much of the energy that cells require and also contribute to cell structure. Their basic building blocks are simple sugar molecules.
   b. Lipids, such as fats, phospholipids, and steroids, supply energy and build cell parts. The basic building blocks of fats—the most common lipid—are molecules of glycerol and fatty acids.
   c. Proteins serve as structural materials, energy sources, hormones, cell surface receptors, and enzymes.
      (1) Enzymes speed chemical reactions without being consumed.
      (2) The building blocks of proteins are amino acids.
      (3) Proteins vary in the numbers, types, and sequences of their amino acids.
      (4) The amino acid chain of a protein molecule folds into a complex shape (conformation) that is maintained largely by hydrogen bonds.
      (5) Excessive heat, radiation, electricity, altered pH, or various chemicals can denature proteins.
   d. Nucleic acids are the genetic material and control cellular activities.
      (1) Nucleic acid molecules are composed of nucleotides.
      (2) The two types of nucleic acids are RNA and DNA.
      (3) DNA molecules store information that cell parts use to construct specific protein molecules. RNA molecules help synthesize proteins.

CHAPTER ASSESSMENTS

2.1 Introduction

1. Define *chemistry*. (p. 31)

2.2 Structure of Matter

2. Define *matter*. (p. 31)

3. Explain the relationship between elements and atoms. (p. 31)

4. List the four most abundant elements in the human body. (p. 31)

5. Describe the parts of an atom and where they are found within the atom. (p. 32)

6. Explain why a complete atom is electrically neutral. (p. 32)

7. Define *atomic number, atomic weight,* and *isotope.* (p. 32)

8. Explain how electrons are distributed within the electron shells of an atom. (p. 33)

9. An ionic bond forms when (p. 35)
    a. atoms share electrons.
    b. positively-charged and negatively-charged parts of polar covalent molecules attract.
    c. ions with opposite electrical charges attract.
    d. two atoms exchange protons.
    e. an element has two types of isotopes.

10. Explain the relationship between molecules and compounds. (p. 36)

11. Show the difference between a molecular formula and a structural formula. (p. 37)

12. The formula $C_6H_{12}O_6$ means _____. (p. 37)

13. Three major types of chemical reactions are _____, _____, and _____. (p. 37)

14. Explain what a reversible reaction is. (p. 38)

15. Define *catalyst.* (p. 38)

16. Define *acid* and *base.* (p. 38)

17. Explain what pH measures, and describe the pH scale. (p. 39)

18. Define *buffer.* (p. 39)

2.3 Chemical Constituents of Cells

19. Distinguish between electrolytes and nonelectrolytes. (p. 39)

20. Distinguish between inorganic and organic substances. (p. 39)

21. Describe the roles water and oxygen play in the human body. (p. 40)

22. List several ions in body fluids. (p. 40)

23. Describe the general characteristics of carbohydrates. (p. 41)

24. Distinguish between simple sugars and complex carbohydrates. (p. 41)

25. Describe the general characteristics of lipids, and list the three main kinds of lipids. (p. 41)

26. A triglyceride molecule consists of (p. 41)
    a. cholesterol and 3 fatty acids.
    b. 3 monosaccharides.
    c. 3 amino acids.
    d. 3 glycerols and 1 fatty acid.
    e. 3 fatty acids and 1 glycerol.

27. Explain the difference between saturated and unsaturated fats. (p. 42)

28. A hydrophilic molecule dissolves in (p. 42)
    a. lipid but not water.
    b. water but not lipid.
    c. neither lipid nor water.
    d. both lipid and water.
    e. alcohol and protein.

29. List at least three functions of proteins. (p. 43)

30. Describe four levels of protein structure. (p. 44)

31. Explain how protein molecules may denature. (p. 44)

32. Describe the structure of nucleic acids. (p. 46)

33. Explain the major functions of nucleic acids. (p. 46)

INTEGRATIVE ASSESSMENTS/ CRITICAL THINKING

OUTCOME 2.2

1. The thyroid gland metabolizes iodine, the most common form of which has a molecular weight of 127 (^{127}I). A physician wants to use a radioactive isotope of iodine (^{123}I) to test whether a patient's thyroid gland is metabolizing normally. Based on what you know about how atoms react, do you think this physician's plan makes sense or not?

OUTCOME 2.3

2. A topping for ice cream contains fructose, hydrogenated soybean oil, salt, and cellulose. What types of chemicals are in it?

3. At a restaurant, a waiter recommends a sparkling carbonated beverage, claiming that it contains no carbohydrates. The product label lists water and fructose as ingredients. Is the waiter correct?

4. A man on a very low-fat diet proclaims to his friend, "I'm going to get my cholesterol down to zero!" Is this desirable? Why or why not?

5. How would you explain the dietary importance of amino acids and proteins to a person who is following a diet composed primarily of carbohydrates?

6. A friend, while frying some eggs, points to the change in the egg white (which contains a protein called albumin) and explains that if the conformation of a protein changes, it will no longer have the same properties and will lose its ability to function. Do you agree or disagree with this statement?

OUTCOMES 2.2, 2.3

7. What acidic and basic substances do you encounter in your everyday activities? What acidic foods do you eat regularly? What basic foods do you eat?

WEB CONNECTIONS

Visit the text website at **aris.mhhe.com** for additional quizzes, interactive learning exercises, and more.

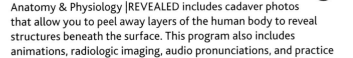

Anatomy & Physiology |REVEALED includes cadaver photos that allow you to peel away layers of the human body to reveal structures beneath the surface. This program also includes animations, radiologic imaging, audio pronunciations, and practice quizzing. To learn more visit **www.aprevealed.com**.

3

Cells

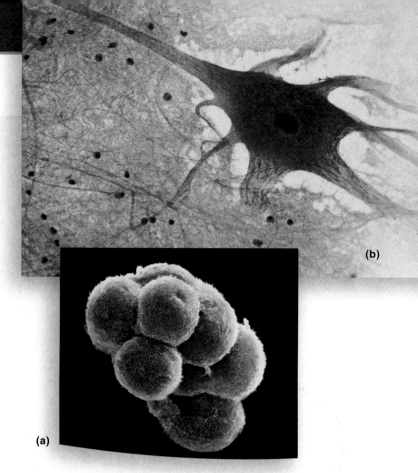

(b)

(a)

An embryonic stem (ES) cell can give rise to all other types of cells. A cell removed from an 8-celled embryo (a) can divide in culture, becoming an ES cell that can self-renew and give rise to specialized cells, such as a neuron (b).

WATCHING CELLS SPECIALIZE. The embryo consisted of eight identical-appearing cells. The cells were round and rather featureless—they looked nothing like the specialized cells that build most of a human body. Researchers took one cell from the eight, then froze the remaining seven-celled embryo. It would be implanted into the woman who had donated it, or perhaps a different woman, to continue development.

The experiment was a variation on a technique that has been used since the early 1990s. Called preimplantation genetic diagnosis, the technique samples a cell from an 8-celled embryo to test for a genetic disease in a family. The rest of the embryo is transferred to a woman's body to continue development if the disease hasn't been inherited. In the new twist on the protocol, researchers took the lone sampled cells from 8-celled embryos and cultured them in laboratory dishes filled with nutrients and growth factors.

Over the next few weeks, the cells, called blastomeres, divided to yield clumps. Many of the initial cells died, but they left behind cells that could both divide to make more of themselves and, if given an appropriate environment, specialize into differentiated cell types. These cells are human embryonic stem (hES) cells. In laboratory dishes, they gave rise to all other cell types, including blood-forming tissue, nervous tissue, liver, and intestinal and respiratory lining tissues. The hope is that such tissues grown in the laboratory could be used to treat certain illnesses and injuries. They could theoretically be grown to match individuals, so that a person's immune system would not reject them.

While researchers are developing ways to nurture hES cells into desired tissues without harming embryos, other sources of valuable stem cells are also being explored. For example, tissue removed from the brains of people undergoing surgery to treat intractable seizures—what would otherwise be medical waste—can be cultured to yield brain neurons. Even fat removed in liposuction procedures can be coaxed in culture to form smooth muscle, which can be used to repair damaged tissues. Stem cell biology may have many practical applications, but researchers still have much to learn about the basic biology.

Learning Outcomes   *After studying this chapter, you should be able to do the following:*

3.1 Introduction

1. Define what a cell is. (p. 51)
2. Explain how cells differ from one another. (p. 51)

3.2 Composite Cell

3. Explain how the structure of a cell membrane makes possible its functions. (p. 53)
4. Describe each type of organelle, and explain its function. (p. 55)

5. Describe the parts of the cell nucleus and their functions. (p. 59)

3.3 Movements Through Cell Membranes

6. Compare and contrast various ways that substances move through cell membranes. (p. 61)

3.4 The Cell Cycle

7. Describe the parts of the cell cycle and identify the major activities during each part. (p. 66)
8. Explain why regulation of the cell cycle is important to health. (p. 67)
9. Define differentiation. (p. 70)
10. Distinguish between a stem cell and a progenitor cell. (p. 70)

Aids to Understanding Words   *(Appendix A on page 567 has a complete list of Aids to Understanding Words.)*

cyt- [cell] *cyt*oplasm: Fluid (cytosol) and organelles that occupy the space between the cell membrane and the nuclear envelope.

endo- [within] *endo*plasmic reticulum: Complex of membranous structures within the cytoplasm.

hyper- [above] *hyper*tonic: Solution that has a greater osmotic pressure than body fluids.

hypo- [below] *hypo*tonic: Solution that has a lesser osmotic pressure than body fluids.

inter- [between] *inter*phase: Stage that occurs between mitotic divisions of a cell.

iso- [equal] *iso*tonic: Solution that has the same osmotic pressure as body fluids.

mit- [thread] *mit*osis: Process of cell division when threadlike chromosomes become visible within a cell.

phag- [to eat] *phag*ocytosis: Process by which a cell takes in solid particles.

pino- [to drink] *pino*cytosis: Process by which a cell takes in tiny droplets of liquid.

-som [body] ribo*som*e: Tiny, spherical structure that consists of protein and RNA and functions in protein synthesis.

3.1 INTRODUCTION

Recipe for a human being: cells, their products, and fluids. A cell, as the unit of life, is a world unto itself. To build a human, trillions of cells connect and interact, forming dynamic tissues, organs, and organ systems.

The estimated 75 trillion cells that make up an adult human body have much in common. Yet cells in different tissues vary considerably in size and shape, and typically, their three-dimensional forms make possible their functions, as illustrated in figure 3.1. For instance, nerve cells often have long, threadlike extensions that transmit electrical impulses from one part of the body to another.

Epithelial cells that line the inside of the mouth are thin, flattened, and tightly packed. This tile-like organization enables the cells to protect those beneath them. Muscle cells, which pull structures closer together, are slender and rodlike. The precise alignment of the fibers within muscle cells provides the strength to withstand the contraction that moves the structures to which they attach.

3.2 COMPOSITE CELL

Because cells vary greatly in size, shape, content, and function, describing a "typical" cell is impossible. The cell shown in figure 3.2 and described in this chapter is a

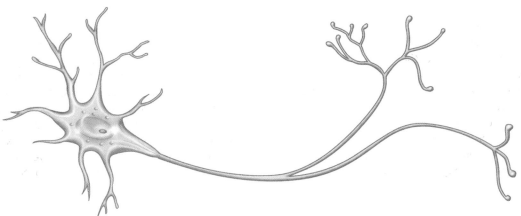

(a) A nerve cell's long extensions enable it to transmit impulses from one body part to another.

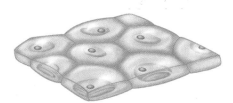

(b) The sheet-like organization of epithelial cells enables them to protect underlying cells.

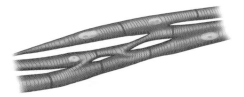

(c) The alignment of contractile proteins within muscle cells enables them to contract, pulling closer together the structures to which they attach.

Figure 3.1
Cells vary in structure and function.

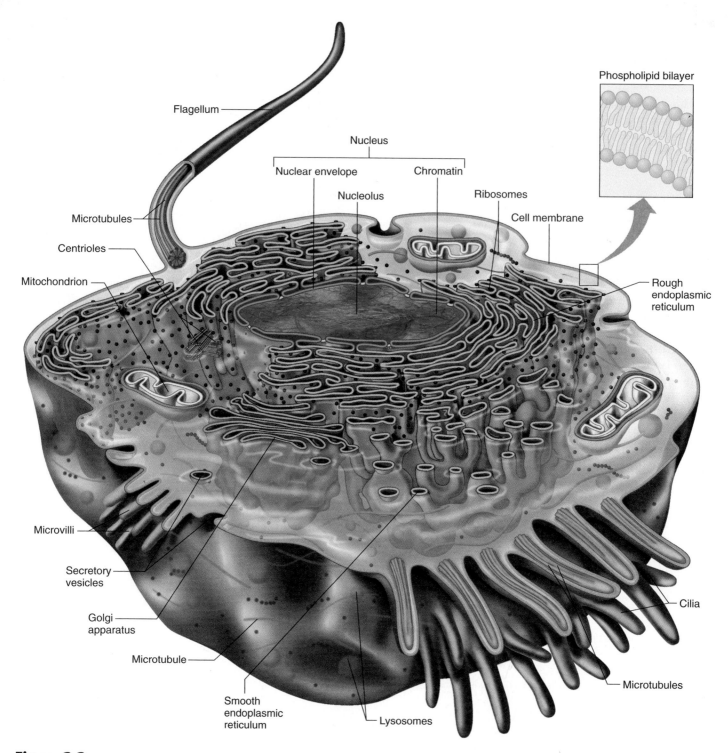

Figure 3.2

A composite cell illustrates the organelles and other structures found in cells. Specialized cells differ in the numbers and types of organelles, reflecting their functions.

composite cell that includes many known cell structures. In reality, any given cell has most, but not all, of these structures, and differing numbers of some of them.

Under the light microscope, a properly applied stain reveals three basic cell parts: the **cell membrane** (sel mem'-brān) that encloses the cell, the **nucleus** (nu'kle-

us) that houses the genetic material and controls cellular activities, and the **cytoplasm** (si'to-plazm) that fills out the cell.

Within the cytoplasm are specialized structures called **organelles** (or-gan-elz'), which can be seen clearly only under the higher magnification of electron microscopes. Organelles, suspended in a liquid called *cytosol,* perform specific functions, which divides the labor of the cell.

> ### ✓ Check Your Recall
>
> 1. Give three examples of how a cell's shape makes possible the cell's function.
> 2. Name the three major parts of a cell and their functions.
> 3. What are organelles and what are their general functions in a cell?

Cell Membrane

The cell membrane (also called the *plasma membrane*) is more than a simple boundary surrounding the cellular contents. It is an actively functioning part of the living material. The cell membrane regulates movement of substances in and out of the cell and is the site of much biological activity. Many of a cell's actions that enable it to survive and to interact with other cells use a molecular communication process called signal transduction. A series of molecules that are part of the cell membrane form pathways that detect signals from outside the cell and transmit them inward, where yet other molecules orchestrate the cell's response. The cell membrane also helps cells adhere to certain other cells, which is important in forming tissues.

General Characteristics

The cell membrane is extremely thin, flexible, and somewhat elastic. It typically has complex surface features with many outpouchings and infoldings that increase surface area (fig. 3.2). In addition to maintaining cell integrity, the cell membrane is **selectively permeable** (se-lek′tiv-le per′me-ah-bl) (also known as *semipermeable* or *differentially permeable*), which means that only certain substances can enter or leave the cell.

Cell Membrane Structure

A cell membrane is composed mainly of lipids and proteins, with fewer carbohydrates. Its basic framework is a double layer, or *bilayer,* of phospholipid molecules. Each phospholipid molecule includes a phosphate group and two fatty acids bound to a glycerol molecule (see chapter 2, p. 42). The water-soluble phosphate "heads" form the surfaces of the membrane, and the water-insoluble fatty acid "tails" make up the interior of the membrane. The lipid molecules can move sideways within the plane of the membrane. The two layers form a soft and flexible, but stable, fluid film.

The membrane's interior is oily because it consists largely of the fatty acid portions of the phospholipid molecules (fig. 3.3). Molecules such as oxygen and carbon dioxide, which are soluble in lipids, can easily pass through this bilayer. However, the bilayer is impermeable to water-soluble molecules, such as amino acids,

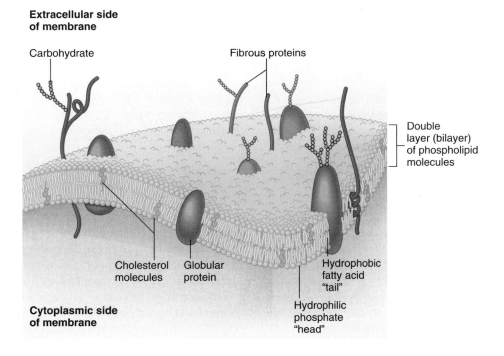

Extracellular side of membrane

Carbohydrate

Fibrous proteins

Double layer (bilayer) of phospholipid molecules

Cholesterol molecules

Globular protein

Hydrophobic fatty acid "tail"

Cytoplasmic side of membrane

Hydrophilic phosphate "head"

Figure 3.3

The cell membrane is composed primarily of phospholipids (and some cholesterol), with proteins scattered throughout the lipid bilayer and associated with its surfaces.

Faulty Ion Channels Cause Inherited Disease

What do abnormal pain intensity, irregular heartbeats, and cystic fibrosis have in common? All result from abnormal ion channels in cell membranes.

Ion channels are protein-lined tunnels in the phospholipid bilayer of a biological membrane. These passageways permit electrical signals to pass in and out of membranes as ions (charged particles). Many ion channels open or close like a gate in response to specific ions under specific conditions. These situations include a change in electrical forces across the membrane, binding of a molecule, or receiving biochemical messages from inside or outside the cell.

Ion channels are specific for calcium (Ca^{+2}), sodium (Na^+), potassium (K^+), or chloride (Cl^-). A cell membrane may have a few thousand ion channels specific for each of these ions. Ten million or more ions can pass through an ion channel in one second! Drugs may act by affecting ion channels, and abnormal ion channels cause certain disorders, including the following:

Absent or Excess Pain

The ten-year-old boy amazed the people on the streets of the small northern Pakistani town. He was completely unable to feel pain and had become a performer, stabbing knives through his arms and walking on hot coals to entertain crowds. Several other people in this community, where relatives often married relatives, were also unable to feel pain. Researchers studied the connected families and discovered a mutation that alters sodium channels on certain nerve cells. The mutation blocks the channels so that the message to feel pain cannot be sent. The boy died at age thirteen from jumping off a roof. His genes could protect him from pain, but pain protects against injury by providing a warning.

A different mutation affecting the same sodium channels causes drastically different symptoms. In "burning man syndrome," the channels become hypersensitive, opening and flooding the body with pain easily, in response to exercise, an increase in room temperature, or just putting on socks. In another condition, "paroxysmal extreme pain disorder," the sodium channels stay open too long, causing excruciating pain in the rectum, jaw, and eyes. Researchers are using the information from these genetic studies to develop new painkillers.

Long-QT Syndrome and Potassium Channels

Four children in a Norwegian family were born deaf, and three of them died at ages four, five, and nine. All of the children had inherited from unaffected "carrier" parents "long-QT syndrome associated with deafness." They had abnormal potassium channels in the heart muscle and in the inner ear. In the heart, the malfunctioning channels disrupted electrical activity, causing a fatal disturbance to the heart rhythm. In the inner ear, the abnormal channels caused an increase in the extracellular concentration of potassium ions, impairing hearing.

Cystic Fibrosis and Chloride Channels

A seventeenth-century English saying, "A child that is salty to taste will die shortly after birth," described the consequence of abnormal chloride channels in cystic fibrosis (CF), which is inherited from carrier parents. The major symptoms—difficulty breathing, frequent severe respiratory infections, and a clogged pancreas that disrupts digestion—all result from buildup of extremely thick mucous secretions.

Abnormal chloride channels in cells lining the lung passageways and ducts of the pancreas cause the symptoms of CF. The primary defect in the chloride channels also causes sodium channels to malfunction. The result: Salt trapped inside cells draws moisture in and thickens surrounding mucus. Gene therapy is attempting to supply patients' lung-lining cells with the instructions to produce normal chloride channels.

---

sugars, proteins, nucleic acids, and various ions. Cholesterol molecules embedded in the cell membrane's interior help make the membrane less permeable to water-soluble substances, while stabilizing the membrane with their rigid structure.

A cell membrane includes a few types of lipid molecules, but many kinds of proteins, which provide special functions. Membrane proteins are classified according to their positions. Membrane-spanning (transmembrane) proteins extend through the lipid bilayer and may protrude from one or both faces. Peripheral membrane proteins associate mostly with one side of the bilayer. Membrane proteins also vary in shape—they may be globular, rodlike, or fibrous. The cell membrane has been described as a "fluid mosaic" because its proteins are embedded in an oily background and therefore can move.

Membrane proteins have a variety of functions. Some form receptors on the cell surface that bind incoming hormones or growth factors, starting signal transduction. Other proteins transport ions or molecules across the cell membrane. The Genetics Connection above discusses how abnormal ion channels can affect health. Membrane proteins form selective channels that allow only particular ions to enter or leave. In nerve cells, for example, such selective channels control movement of sodium and potassium ions (see chapter 9, p. 220).

Proteins that extend inward from the inner face of the cell membrane anchor it to the protein rods and tubules that support the cell from within. Proteins that extend from the outer surface of the cell membrane mark the cell as part of a particular tissue or organ in a particular person. This identification as self is important for the functioning of the immune system (see chapter 14, p. 385). Many of these proteins are attached to carbohydrates, forming glycoproteins. Another type of protein on a cell's surface is a cellular adhesion molecule (CAM), which guides a cell's interactions with other cells. For example, a series of CAMs helps a white blood cell move to the site of an injury, such as a splinter in the skin.

Cytoplasm

The cytoplasm is the gel-like material in which organelles are suspended—it makes up most of a cell's volume. When viewed through a light microscope, cytoplasm usually appears as a clear jelly with specks scattered throughout. However, an electron microscope, which provides much greater magnification and the ability to distinguish fine detail (resolution), reveals that the cytoplasm contains networks of membranes and organelles suspended in the clear liquid *cytosol.* Cytoplasm also includes abundant protein rods and tubules that form a framework, or **cytoskeleton** (si″to-skel′e-ten), meaning "cell skeleton."

Cell activities occur mainly in the cytoplasm, where nutrients are received, processed, and used. The following organelles have specific functions in carrying out these activities:

1. **Endoplasmic reticulum** (en′do-plaz′mik rĕ-tik′u-lum) The endoplasmic reticulum (ER) is a complex organelle composed of membrane-bounded, flattened sacs, elongated canals, and fluid-filled, bubblelike sacs called *vesicles.* These membranous parts are interconnected and communicate with the cell membrane, the nuclear envelope, and other organelles. The ER provides a vast tubular network that transports molecules from one cell part to another.

    The endoplasmic reticulum participates in the synthesis of protein and lipid molecules. These molecules may leave the cell as secretions or be used within the cell for such functions as producing new ER or cell membrane as the cell grows.

    In many places, the ER's outer membrane is studded with many tiny, spherical structures called *ribosomes,* which give the ER a textured appearance when viewed with an electron microscope (fig. 3.4a, b). These parts of the ER are called *rough*

ER. The ribosomes are sites of protein synthesis and exist independently in the cytoplasm as well as associated with ER. Proteins being synthesized move through ER tubules to another organelle, the Golgi apparatus, for further processing.

ER that lacks ribosomes is called *smooth ER* (fig. 3.4c). Smooth ER contains enzymes important in lipid synthesis, absorption of fats from the digestive tract, and the metabolism of drugs. Cells that break down drugs and alcohol, such as liver cells, have extensive networks of smooth ER.

2. **Ribosomes** (ri′bo-sōmz) Ribosomes, where protein synthesis occurs, are attached to ER membranes or are scattered throughout the cytoplasm. Clusters of ribosomes in the cytoplasm, called *polysomes,* enable a cell to quickly manufacture proteins required in large amounts. All ribosomes are composed of protein and RNA molecules. Ribosomes provide enzymatic activity as well as a structural support for the RNA molecules that come together as the cell links amino acids to form proteins, discussed in Chapter 4 (p. 88).

3. **Golgi apparatus** (gol′je ap″ah-ra′tus) The Golgi apparatus is a stack of about six flattened, membranous sacs. This organelle refines, packages, and transports proteins synthesized on ribosomes associated with the ER. Proteins arrive at the Golgi apparatus enclosed in vesicles composed of the ER membrane. These vesicles fuse with the membrane at the innermost end of the Golgi apparatus, which is specialized to receive glycoproteins.

    As glycoproteins pass from layer to layer through the stacks of Golgi membrane, they are modified chemically. Sugar molecules may be added or removed. When the altered glycoproteins reach the outermost layer, they are packaged in bits of Golgi membrane, which bud off and form bubblelike structures called transport vesicles. Such a vesicle may then move to and fuse with the cell membrane, releasing its contents to the outside as a secretion (figs. 3.2 and 3.5). This process is called *exocytosis* (see page 65). Other vesicles, some of which bud off the cell membrane's inner face, may transport glycoproteins to organelles within the cell. The vesicles in a cell form a delivery service of sorts.

Check Your Recall

4. What is a selectively permeable membrane?

5. Describe the chemical structure of a cell membrane.

6. What are the functions of the endoplasmic reticulum?

7. Describe the functions of the Golgi apparatus.

8. How do organelles and other structures participate in secretion?

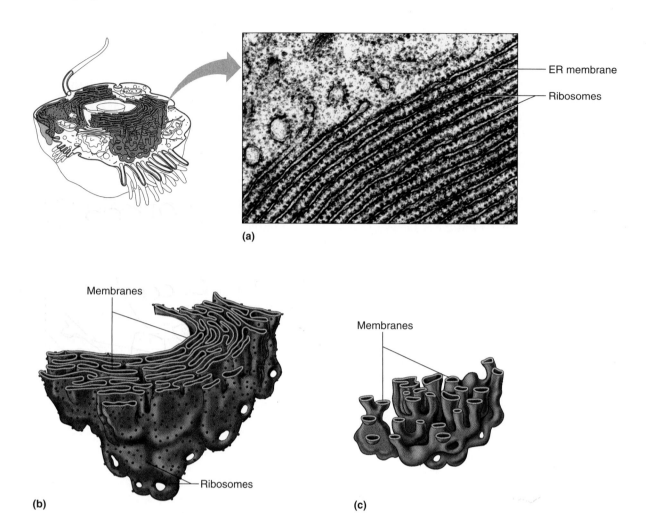

(a)

Membranes

Membranes

Ribosomes

(b)

(c)

Figure 3.4

The endoplasmic reticulum is the site of protein and lipid synthesis, and serves as a transport system. (*a*) A transmission electron micrograph of rough endoplasmic reticulum (ER) (28,500×). (*b*) Rough ER is dotted with ribosomes, whereas (*c*) smooth ER lacks ribosomes.

4. **Mitochondria** (mi″to-kon′dre-ah; *sing.* mi″to-kon′dre-on) Mitochondria are elongated, fluid-filled sacs that vary in size and shape. They can move slowly through the cytoplasm and reproduce by dividing. A mitochondrion has an outer and an inner layer (figs. 3.2 and 3.6). The inner layer is folded extensively into partitions called *cristae.* Connected to the cristae are enzymes that control some of the chemical reactions that release energy from certain nutrient molecules. Mitochondria are the major sites of chemical reactions that capture and store this energy within the chemical bonds of adenosine triphosphate (ATP), a chemical form that the cell can easily use. Very active cells, such as muscle cells, contain many thousands of mitochondria. (Chapter 4, p. 82, describes this energy-releasing function in more detail.) Mitochondria resemble bacterial cells and contain a small amount of their own DNA.

> Mitochondria are modern-day clues to the origin of life. According to the endosymbiont theory, mitochondria are the remnants of once-free-living bacterium-like cells that were engulfed by more complex primitive cells. In our cells today, mitochondria are crucial for extracting energy from nutrients.

5. **Lysosomes** (li′so-sōmz) Lysosomes, the "garbage disposals of the cell," are tiny membranous sacs (see fig. 3.2). They contain powerful enzymes that break down nutrient molecules or foreign particles. Certain white blood cells, for example, can engulf bacteria, which are then digested by the lysosomal enzymes. This is one way that white blood cells fight bacterial infections. Lysosomes also destroy worn cellular parts.

6. **Peroxisomes** (pĕ-roks′ĭ-sōmz) These membranous sacs are abundant in liver and kidney cells. They house enzymes that catalyze (speed) a variety of biochemical reactions, including synthesis of

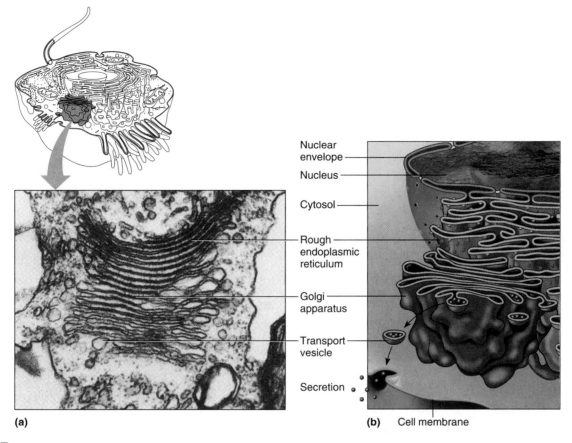

Nuclear envelope
Nucleus
Cytosol
Rough endoplasmic reticulum
Golgi apparatus
Transport vesicle
Secretion
Cell membrane

(a)

(b)

Figure 3.5

The Golgi apparatus processes secretions. (*a*) A transmission electron micrograph of a Golgi apparatus (48,500×). (*b*) The Golgi apparatus consists of membranous sacs that continually receive vesicles from the endoplasmic reticulum and produce vesicles that enclose secretions.

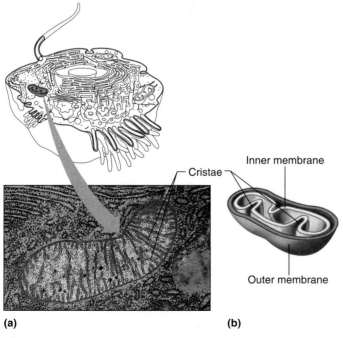

Cristae
Inner membrane
Outer membrane

(a)   (b)

Figure 3.6

A mitochondrion is a major site of energy reactions. (*a*) A transmission electron micrograph of a mitochondrion (28,000×). (*b*) Cristae partition this saclike organelle.

bile acids (used to digest fats); detoxification of hydrogen peroxide, a by-product of metabolism; breakdown of certain lipids and rare biochemicals; and detoxification of alcohol.

7. **Microfilaments and microtubules** Microfilaments and microtubules are two types of thin, threadlike strands within the cytoplasm. They form the cytoskeleton and are also part of certain structures that have specialized activities.

   Microfilaments are tiny rods of actin protein that form meshworks or bundles. They provide cell motility (movement). In muscle cells, for example, microfilaments aggregate to form *myofibrils,* which help these cells contract (see chapter 8, p. 177).

   Microtubules are long, slender tubes with diameters two or three times those of microfilaments (fig. 3.7). Microtubules are composed of molecules of a globular protein called tubulin, attached in a spiral to form a long tube.

8. **Centrosome** (sen′tro-sōm) The centrosome is a structure near the Golgi apparatus and nucleus. It is nonmembranous and consists of two hollow cylinders, called *centrioles,* which are composed of microtubules organized in nine groups of

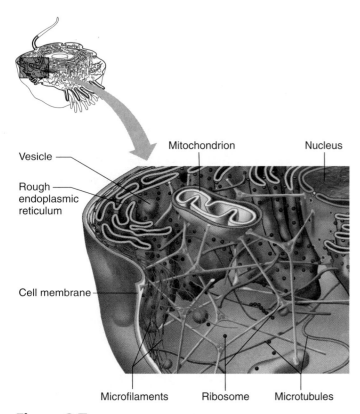

Vesicle

Rough endoplasmic reticulum

Cell membrane

Mitochondrion

Nucleus

Microfilaments   Ribosome   Microtubules

Figure 3.7

Components of the cytoskeleton. Microtubules built of tubulin and microfilaments built of actin help maintain the shape of a cell by forming a cytoskeleton within the cytoplasm. A cell's shape is critical to its function.

three (figs. 3.2 and 3.8). The centrioles lie at right angles to each other. During mitosis, the centrioles distribute chromosomes to newly forming cells.

9. **Cilia and flagella** Cilia and flagella are motile structures that extend from the surfaces of certain cells. They are composed of microtubules in a "9 + 2" array, similar to centrioles but with two additional microtubules in the center. Cilia and flagella are similar structures that differ mainly in length and abundance.

Cilia fringe the free surfaces of some epithelial (lining) cells. Each cilium is tiny and hairlike, and is attached beneath the cell membrane (see fig. 3.2). Cilia form in precise patterns. They move in a coordinated "to-and-fro" manner, so that rows of them beat in succession, producing a wave of motion that sweeps over the surface. This wave moves fluids, such as mucus, over the surface of certain tissues, including those that form the inner linings of the respiratory tubes (fig. 3.9*a*).

Flagella are considerably longer than cilia, and usually a cell has only a single flagellum. Flagella have an undulating wavelike motion, which begins at their base. The tail of a sperm cell is a flagellum that enables this motile cell to "swim" and is the only example of a flagellum in humans (fig. 3.9*b*).

10. **Vesicles** (ves′ĭ-k′lz) Vesicles (vacuoles) are membranous sacs formed by part of the cell membrane folding inward and pinching off. As a result, a tiny, bubblelike vesicle, containing some liquid or solid material formerly outside the cell, appears in the cytoplasm. As previously mentioned, the Golgi apparatus and ER also form vesicles that play a role in secretion (see fig. 3.2).

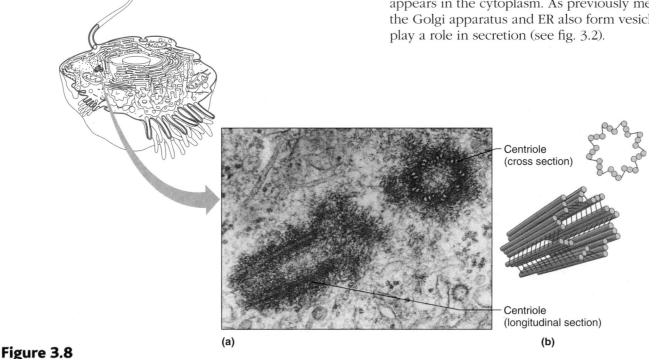

Centriole (cross section)

Centriole (longitudinal section)

(a)   (b)

Figure 3.8

Centrioles. (*a*) Transmission electron micrograph of the two centrioles in a centrosome (120,000×). (*b*) The centrioles lie at right angles to one another. These structures participate in apportioning the chromosomes of a dividing cell into two cells.

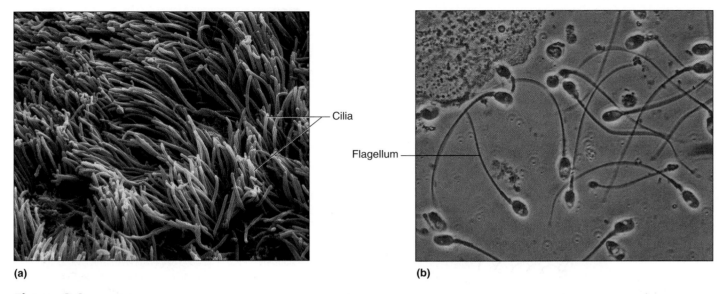

(a)   (b)

Figure 3.9

Cilia and flagella provide movement. (*a*) Cilia are common on the surfaces of certain cells, including those that form the inner lining of the respiratory tubes (5,800×). Cilia remove debris from the respiratory tract. One way that smoking harms health is by destroying cilia. (*b*) Flagella form the tails of these human sperm cells, enabling them to "swim" (840×).

Check Your Recall

9. Describe a mitochondrion.
10. What is the function of a lysosome?
11. How do microfilaments and microtubules differ?
12. What are some structures that consist of microtubules?
13. What is a centrosome and what does it do?
14. Where are cilia and flagella found, what are they composed of, and what do they do?

Cell Nucleus

The nucleus houses the genetic material (DNA), which directs all cell activities (figs. 3.2 and 3.10). It is a large, roughly spherical structure enclosed in a double-layered **nuclear envelope,** which consists of inner and outer lipid bilayer membranes. The nuclear envelope has protein-lined channels called *nuclear pores* that allow certain molecules to exit the nucleus. A nuclear pore is not just a hole, but a complex opening formed from 100 or so types of proteins.

The nucleus contains a fluid, called *nucleoplasm,* in which the following structures are suspended:

1. **Nucleolus** (nu-kle'o-lus) A nucleolus ("little nucleus") is a small, dense body composed largely of RNA and protein. It has no surrounding membrane and forms in specialized regions of certain chromosomes. Ribosomes form in the nucleolus, and then migrate through nuclear pores to the cytoplasm.

2. **Chromatin** Chromatin consists of loosely coiled fibers of DNA and protein that condense to form structures called **chromosomes** (kro'mo-sōmz). The DNA contains the information for protein synthesis. When the cell begins to divide, chromatin fibers coil tightly, and individual chromosomes become visible when stained and viewed under a light microscope.

Table 3.1 summarizes the structures and functions of organelles.

Check Your Recall

15. What structure separates the nuclear contents from the cytoplasm?
16. What is produced in the nucleolus?
17. What is chromatin?

3.3 MOVEMENTS THROUGH CELL MEMBRANES

The cell membrane is a selective barrier that controls which substances enter and leave the cell. Movements of substances into and out of cells include passive mechanisms that do not require cellular energy (diffusion, facilitated diffusion, osmosis, and filtration) and active mechanisms that use cellular energy (active transport, endocytosis, and exocytosis).

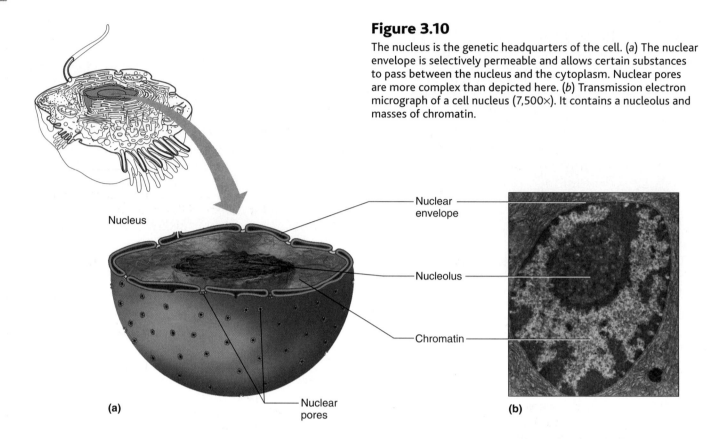

Figure 3.10

The nucleus is the genetic headquarters of the cell. (*a*) The nuclear envelope is selectively permeable and allows certain substances to pass between the nucleus and the cytoplasm. Nuclear pores are more complex than depicted here. (*b*) Transmission electron micrograph of a cell nucleus (7,500×). It contains a nucleolus and masses of chromatin.

Nucleus

Nuclear envelope

Nucleolus

Chromatin

Nuclear pores

(a)   (b)

Table 3.1	Structures and Functions of Cell Parts	
Cell Part(s)	**Structure**	**Function**
Cell membrane	Membrane composed of protein and lipid molecules	Maintains integrity of cell and controls passage of materials into and out of cell
Endoplasmic reticulum	Complex of interconnected membrane-bounded sacs and canals	Transports materials within cell, provides attachment for ribosomes, and synthesizes lipids
Ribosomes	Particles composed of protein and RNA molecules	Synthesize proteins
Golgi apparatus	Group of flattened, membranous sacs	Packages protein molecules for transport and secretion
Mitochondria	Membranous sacs with inner partitions	Release energy from nutrient molecules and change energy into a usable form
Lysosomes	Membranous sacs	Digest worn cellular parts or substances that enter cells
Peroxisomes	Membranous sacs	House enzymes that catalyze diverse reactions, including bile acid synthesis, lipid breakdown, and alcohol detoxification
Microfilaments and microtubules	Thin rods and tubules	Support the cytoplasm and help move substances and organelles within the cytoplasm
Centrosome	Nonmembranous structure composed of two rodlike centrioles	Helps distribute chromosomes to new cells during cell division
Cilia and flagella	Motile projections attached beneath the cell membrane	Cilia propel fluid over cellular surfaces, and a flagellum enables a sperm cell to move
Vesicles	Membranous sacs	Contain and transport various substances
Nuclear envelope	Double membrane that separates the nuclear contents from the cytoplasm	Maintains integrity of nucleus and controls passage of materials between nucleus and cytoplasm
Nucleolus	Dense, nonmembranous body composed of protein and RNA	Site of ribosome synthesis
Chromatin	Fibers composed of protein and DNA	Contains information for synthesizing proteins

Passive Mechanisms

Diffusion

Diffusion (dĭ-fu´zhun) (also called *simple diffusion*) is the process by which molecules or ions scatter or spread spontaneously from regions where they are in higher concentrations toward regions where they are in lower concentrations. This difference in concentration is called a *concentration gradient.* Atoms, molecules, and ions are said to diffuse down their concentration gradients.

Under natural conditions, molecules and ions constantly move at high speeds. Each particle travels in a separate path along a straight line until it collides and bounces off some other particle, changing direction, colliding again, and changing direction once more. Such random motion mixes molecules. At body temperature, small molecules such as water move more than a thousand miles per hour. However, the internal environment is crowded, from a molecule's point of view. A single molecule may collide with other molecules a million times each second.

Consider how molecules of sugar (a solute) in a sugar cube distribute, by diffusion, in a glass of water (a solvent). At first the sugar remains highly concentrated at the bottom of the glass (fig. 3.11). Diffusion causes the sugar molecules to move away from the area of high concentration and disperse into solution among the moving water molecules. Eventually, the sugar molecules become uniformly distributed in the water, a state called **equilibrium** (e″kwĭ-lib´re-um). Molecules continue to move after equilibrium occurs, but their concentrations no longer change.

Diffusion of a substance into or out of a cell can occur only if (1) the cell membrane is permeable to that substance, and (2) a concentration gradient exists such that the substance is at a higher concentration on one side of the membrane or the other (fig. 3.12). Consider oxygen and carbon dioxide, two substances to which cell membranes are permeable. In the body, diffusion is the process whereby oxygen enters cells and carbon dioxide leaves cells, but equilibrium is never reached. Intracellular oxygen is always low because oxygen is constantly used up in metabolic reactions. Extracellular oxygen is maintained at a high level by homeostatic mechanisms in the respiratory and cardiovascular systems. Thus, a concentration gradient always allows oxygen to diffuse into cells.

The level of carbon dioxide, which is a metabolic waste product, is always high inside cells. Homeostasis maintains a lower extracellular carbon dioxide level, so a concentration gradient always favors carbon dioxide diffusing out of cells (fig. 3.13).

Dialysis is a chemical technique that uses diffusion to separate small molecules from larger ones in a liquid. The artificial kidney uses a variant of this process—*hemodialysis*—to treat patients suffering from kidney damage or failure. An artificial kidney (dialyzer) passes blood from a patient through a long, coiled tubing composed of porous cellophane. The size of the pores allows smaller molecules carried in the blood, such as the waste material urea, to exit through the tubing, while larger molecules, such as those of blood proteins, remain inside the tubing. The tubing is submerged in a tank of dialyzing fluid (wash solution), which contains varying concentrations of different chemicals. The fluid has low concentrations of substances that should leave the blood and higher concentrations of those that should remain in the blood.

Altering the concentrations of molecules in the dialyzing fluid can control which molecules diffuse out of blood and which remain in it. For example, to remove blood urea, the dialyzing fluid must have a lower urea concentration than the blood; to maintain blood glucose concentration, the glucose concentration in the dialyzing fluid must be at least equal to that of the blood.

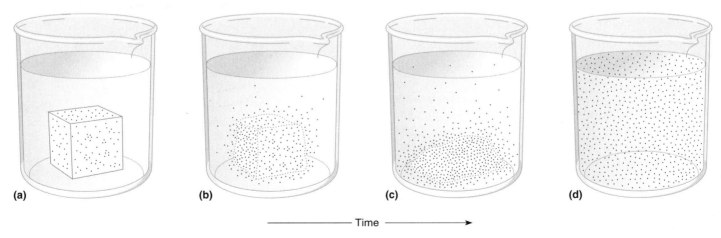

(a)   (b)   (c)   (d)

⸻ Time ⟶

Figure 3.11

A dissolving sugar cube illustrates diffusion. (*a–c*) A sugar cube placed in water slowly disappears as the sugar molecules dissolve and then diffuse from regions where they are more concentrated toward regions where they are less concentrated. (*d*) Eventually, the sugar molecules are distributed evenly throughout the water.

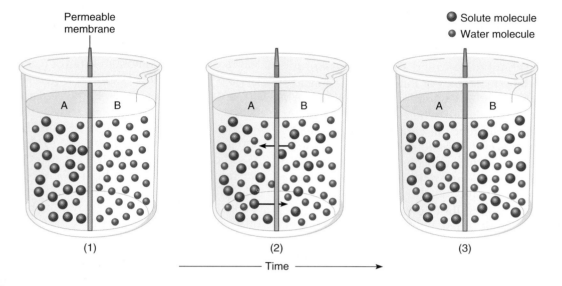

Permeable membrane

● Solute molecule
● Water molecule

A B   A B   A B

(1)   (2)   (3)

Time

Figure 3.12

Diffusion. (*1*) A membrane permeable to water and solute molecules separates a container into two compartments. Compartment *A* contains both types of molecules, while compartment *B* contains only water molecules. (*2*) As a result of molecular motions, solute molecules tend to diffuse from compartment *A* into compartment *B*. Water molecules tend to diffuse from compartment *B* into compartment *A*. (*3*) Eventually, equilibrium is reached.

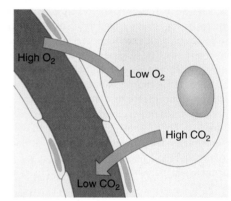

High O_2
Low O_2
High CO_2
Low CO_2

Figure 3.13

Diffusion enables oxygen to enter cells and carbon dioxide to leave.

Facilitated Diffusion

Substances that are not able to pass through the lipid bilayer need the help of membrane proteins to get across, a process known as **facilitated diffusion** (fah-sil"ĭ-tāt'ed dĭ-fu'zhun) (fig. 3.14). One form of facilitated diffusion uses the ion channels and pores described earlier. Molecules such as glucose and amino acids are not lipid soluble, but are too large to pass through membrane channels. They enter cells by another form of facilitated diffusion. In this process, which occurs in most cells, a glucose molecule combines with a special protein carrier molecule at the surface of the cell membrane. This union of the glucose and carrier molecules changes the shape of the carrier, enabling it to move glucose to the other side of the membrane. The carrier releases the glucose and then returns to its origi-

nal shape and picks up another glucose molecule. The hormone *insulin,* discussed in chapter 11 (p. 306), promotes facilitated diffusion of glucose through the membranes of certain cells.

Facilitated diffusion is similar to simple diffusion in that it only moves molecules from regions of higher

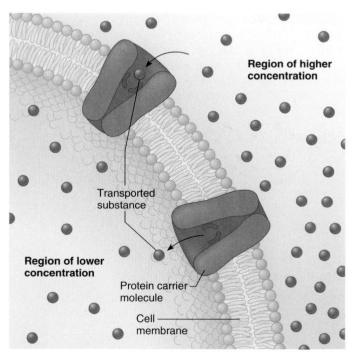

Region of higher concentration

Transported substance

Region of lower concentration

Protein carrier molecule

Cell membrane

Figure 3.14

Facilitated diffusion uses carrier molecules to transport some substances into or out of cells, from a region of higher concentration to one of lower concentration.

concentration toward regions of lower concentration. The number of carrier molecules in the cell membrane limits the rate of facilitated diffusion.

Osmosis

Osmosis (oz-mo′sis) is a special case of diffusion. It occurs whenever water molecules diffuse from a region of higher water concentration to a region of lower water concentration across a selectively permeable membrane, such as a cell membrane. In the example that follows, assume that the selectively permeable membrane is permeable to water molecules (the solvent), but impermeable to protein molecules (the solute).

In solutions, a higher concentration of solute (protein in this case) means a lower concentration of water; a lower concentration of solute means a higher concentration of water. This is because solute molecules take up space that water molecules would otherwise occupy.

Just like molecules of other substances, molecules of water diffuse from areas of higher concentration to areas of lower concentration. In figure 3.15, the presence of a greater concentration of protein in compartment *A* means that the water concentration there is less than the concentration of water in compartment *B*. Therefore, water diffuses from compartment *B* across the selectively permeable membrane and into compartment *A*. In other words, water moves from compartment

B into compartment *A* by osmosis. Protein, on the other hand, cannot diffuse out of compartment *A* because the selectively permeable membrane is impermeable to it.

Note in figure 3.15 that as osmosis occurs, the water level on side *A* rises. This ability of osmosis to generate enough pressure to lift a volume of water is called *osmotic pressure.* The greater the concentration of impermeant solute particles (protein in this case) in a solution, the *lower* the water concentration of that solution and the *greater* the osmotic pressure. Water always tends to diffuse toward solutions of greater osmotic pressure.

Since cell membranes are generally permeable to water, water equilibrates by osmosis throughout the body, and the concentration of water and solutes everywhere in the intracellular and extracellular fluids is essentially the same. Therefore, the osmotic pressure of the intracellular and extracellular fluids is the same. Any solution that has the same osmotic pressure as body fluids is called **isotonic** (fig. 3.16*a*).

Solutions with a higher osmotic pressure than body fluids are called **hypertonic.** If cells are put into a hypertonic solution, water moves by osmosis out of the cells into the surrounding solution, and the cells shrink (fig. 3.16*b*). Conversely, cells put into a **hypotonic** solution, which has a lower osmotic pressure than body fluids, tend to gain water by osmosis, and therefore they swell (fig. 3.16*c*).

The concentration of solute in solutions that are infused into body tissues or blood must be controlled. Otherwise, osmosis may cause cells to swell or shrink, impairing their function. For instance, if red blood cells are placed in distilled water (which is hypotonic to them), water diffuses into the cells, and they burst (hemolyze). On the other hand, red blood cells exposed to 0.9% NaCl solution (normal saline) do not change shape because this solution is isotonic to human cells. A red blood cell placed in a hypertonic solution shrinks.

Filtration

Molecules pass through membranes by diffusion or osmosis because of random movements. In other instances, the process of **filtration** (fil-tra′shun) forces molecules through membranes.

Filtration is commonly used to separate solids from water. One method is to pour a mixture of solids and water onto filter paper in a funnel. The paper is a porous membrane through which the small water molecules can pass, leaving behind the larger solid particles. *Hydrostatic pressure,* which is created by the weight of water on the paper due to gravity, forces the water molecules through to the other side. A familiar example of filtration is making coffee by the drip method.

In the body, tissue fluid forms when water and small dissolved substances are forced out through the

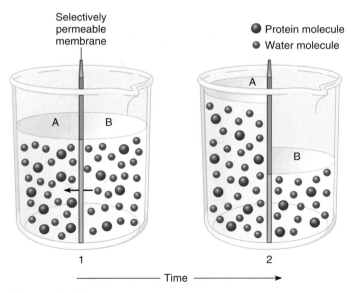

Figure 3.15

Osmosis. (*1*) A selectively permeable membrane separates the container into two compartments. At first, compartment *A* contains a higher concentration of protein (and a lower concentration of water) than compartment *B*. As a result of molecular motion, water diffuses by osmosis from compartment *B* into compartment *A*. (*2*) Because the membrane is impermeable to proteins, equilibrium can only be reached by diffusion of water. As water accumulates in compartment *A*, the water level on that side of the membrane rises.

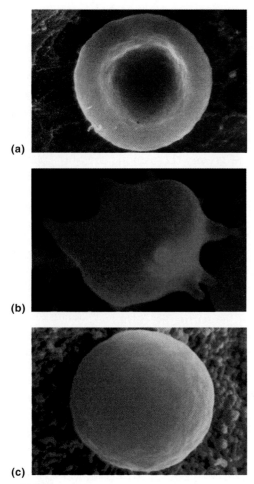

(a)

(b)

(c)

Figure 3.16

When red blood cells are placed (*a*) in an isotonic solution, equal volumes of water enter and leave cells, so the cells maintain their characteristic shapes. (*b*) In a hypertonic solution, more water leaves than enters, so cells shrink. (*c*) In a hypotonic solution, more water enters than leaves so cells swell and may burst (5,000×).

thin, porous walls of blood capillaries, but larger particles, such as blood protein molecules, are left inside (fig. 3.17). The force for this movement comes from blood pressure, generated mostly by heart action, which is greater within the vessel than outside it. However, the impermeable proteins tend to hold water in blood vessels by osmosis, thus preventing the formation of excess tissue fluid, a condition called **edema.** Filtration also occurs as the kidneys cleanse blood.

Check Your Recall

18. What kinds of substances diffuse most readily through a cell membrane?

19. Explain the differences between diffusion and osmosis.

20. Distinguish among hypertonic, hypotonic, and isotonic solutions.

21. Explain how filtration occurs in the body.

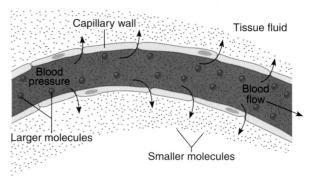

Figure 3.17

In filtration, blood pressure forces smaller molecules through tiny openings in the capillary wall. The larger molecules remain inside.

Active Mechanisms

When molecules or ions pass through cell membranes by diffusion, facilitated diffusion, or osmosis, their net movements are from regions of higher concentration toward regions of lower concentration. Sometimes, however, particles move from a region of lower concentration to one of higher concentration. This requires energy, which comes from cellular metabolism and, specifically, from a molecule called adenosine triphosphate (ATP).

Active Transport

Active transport (ak′tiv trans′port) is a process that moves particles through membranes from a region of lower concentration to a region of higher concentration. Sodium ions, for example, can diffuse passively into cells through protein channels in cell membranes, but their concentration typically remains much greater outside cells than inside. This occurs because active transport continually moves sodium ions through cell membranes from regions of lower concentration (inside) to regions of higher concentration (outside). Equilibrium is never reached.

Active transport is similar to facilitated diffusion in that it uses specific carrier molecules in cell membranes (fig. 3.18). It differs from facilitated diffusion in that particles move from areas of low concentration to areas of high concentration, and energy from ATP is required. Up to 40% of a cell's energy supply may be used to actively transport particles through cell membranes.

The carrier molecules in active transport are proteins with binding sites that combine with the particles being transported. Such a union triggers the release of energy, and this alters the shape of the carrier protein. As a result, the "passenger" particles move through the membrane. Once on the other side, the transported particles are released, and the carriers can accept other passenger molecules at that binding site. Because these carrier proteins transport substances from regions of low concentration to regions of high concentration, they are sometimes called "pumps."

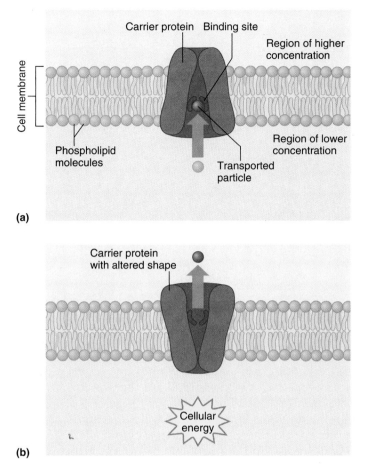

(a)

(b)

Figure 3.18

Active transport moves molecules against their concentration gradient. (*a*) During active transport, a molecule or an ion combines with a carrier protein, whose shape changes as a result. (*b*) This process, which requires cellular energy, transports the particle across the cell membrane.

Particles that are actively transported across cell membranes include sugars and amino acids as well as sodium, potassium, calcium, and hydrogen ions. Active transport also absorbs nutrient molecules into cells that form the intestinal lining.

Endocytosis and Exocytosis

Two processes use cellular energy to move substances into or out of a cell without actually crossing the cell membrane. In **endocytosis** (en″do-si-to′sis), molecules or other particles that are too large to enter a cell by diffusion, facilitated diffusion, or active transport are conveyed within a vesicle formed from a section of the cell membrane. In **exocytosis** (ek″so-si-to′sis), the reverse process secretes a substance stored in a vesicle from the cell. Nerve cells use exocytosis to release the neurotransmitter chemicals that signal other nerve cells, muscle cells, or glands.

Endocytosis happens in three ways: pinocytosis, phagocytosis, and receptor-mediated endocytosis. In **pinocytosis** (pi″no-si-to′sis), meaning "cell drinking," cells take in tiny droplets of liquid from their surroundings, as a small portion of the cell membrane indents. The open end of the tubelike part that forms seals off and produces a small vesicle, which detaches from the surface and moves into the cytoplasm. Eventually, the vesicular membrane breaks down, and the liquid inside becomes part of the cytoplasm. In this way, a cell can take in water and the particles dissolved in it, such as proteins, that otherwise might be too large to enter.

Phagocytosis (fag″o-si-to′sis), meaning "cell eating," is similar to pinocytosis, but the cell takes in solids rather than liquids. Certain kinds of white blood cells are called *phagocytes* because they can take in solid particles such as bacteria and cellular debris. When a phagocyte first encounters a particle, the particle attaches to the phagocyte's cell membrane. This stimulates a portion of the membrane to project outward, surround the particle, and slowly draw it inside the cell. The part of the membrane surrounding the particle detaches from the cell's surface, forming a vesicle that contains the particle (fig. 3.19).

Once a particle has been phagocytized inside a cell, a lysosome then combines with the newly formed vesicle, and the lysosomal digestive enzymes decompose the contents. The products of this decomposition may diffuse

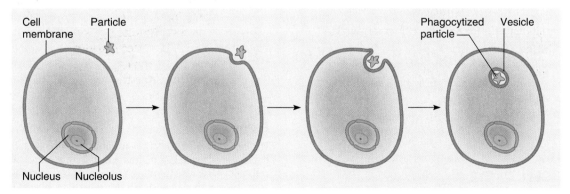

Figure 3.19

A cell may take in a solid particle from its surroundings by phagocytosis.

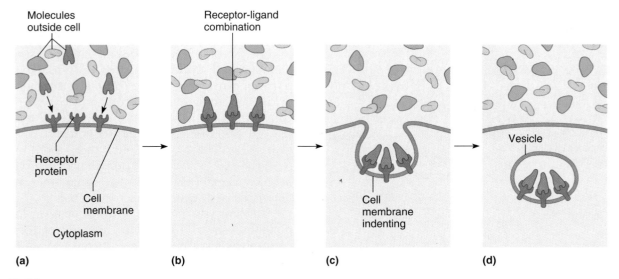

(a) (b) (c) (d)

Figure 3.20

Receptor-mediated endocytosis brings specific molecules into a cell. (*a,b*) A specific molecule binds to a receptor protein, forming a receptor-ligand combination. (*c*) The binding of the ligand to the receptor protein stimulates the cell membrane to indent. (*d*) Continued indentation forms a vesicle, which transports the molecule into the cytoplasm.

out of the lysosome and into the cytoplasm. Exocytosis usually expels any remaining residue from the cell.

Pinocytosis and phagocytosis engulf any molecules in the vicinity of the cell membrane. In contrast, **receptor-mediated endocytosis** moves very specific kinds of particles into the cell. In this process, protein molecules extend through a portion of the cell membrane to the outer surface, where they form receptors to which only specific molecules from outside the cell, called their *ligands,* can bind (fig. 3.20). Cholesterol molecules enter cells by receptor-mediated endocytosis.

A process called transcytosis combines endocytosis and exocytosis to transport a particle or structure across a cell. It is discussed in the Clinical Connection (p. 72) at the end of this chapter. Table 3.2 summarizes the types of movements into and out of cells.

Check Your Recall

22. What type of mechanism maintains unequal concentrations of ions on opposite sides of a cell membrane?

23. How are facilitated diffusion and active transport similar? How are they different?

24. What is the difference between endocytosis and exocytosis?

25. How is receptor-mediated endocytosis more specific than pinocytosis or phagocytosis?

26. What process combines endocytosis and exocytosis?

3.4 THE CELL CYCLE

The series of changes that a cell undergoes from the time it forms until it divides is called the *cell cycle* (fig. 3.21).

Superficially, this cycle seems rather simple: A newly formed cell grows for a time and then divides to form two new cells, which in turn may grow and divide. Yet the phases and timing of the cycle are quite complex, and include interphase, mitosis, cytoplasmic division (cytokinesis), and differentiation. Groups of special pro-

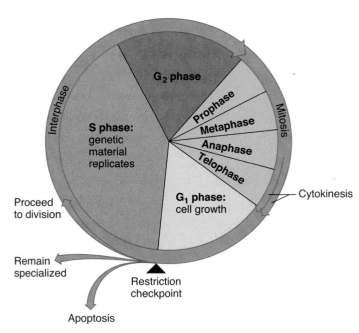

Figure 3.21

The cell cycle is divided into interphase, when cellular components duplicate, and cell division (mitosis and cytokinesis), when the cell splits in two, distributing its contents into two daughter cells. Interphase is divided into two gap, or growth, phases (G_1 and G_2) when specific molecules and structures duplicate, and a synthesis phase (S), when the genetic material replicates. Mitosis can be considered in stages—prophase, metaphase, anaphase, and telophase.

Table 3.2	Movements Through Cell Membranes		
Process	**Characteristics**	**Source of Energy**	**Example**
Passive mechanisms			
Diffusion	Molecules or ions move from regions of higher concentration toward regions of lower concentration.	Molecular motion	Exchange of oxygen and carbon dioxide in lungs
Facilitated diffusion	Carrier molecules move molecules through a membrane from a region of higher concentration to one of lower concentration.	Molecular motion	Movement of glucose through cell membrane
Osmosis	Water molecules move from regions of higher concentration toward regions of lower concentration through a selectively permeable membrane.	Molecular motion	Distilled water entering a cell
Filtration	Molecules are forced from regions of higher pressure to regions of lower pressure.	Hydrostatic pressure	Water molecules leaving blood capillaries
Active mechanisms			
Active transport	Carrier molecules move molecules or ions through membranes from regions of lower concentration toward regions of higher concentration.	Cellular energy (ATP)	Movement of various ions, sugars, and amino acids through membranes
Endocytosis			
Pinocytosis	Membrane engulfs droplets of liquid from surroundings.	Cellular energy	Membrane forming vesicles containing liquid and dissolved particles
Phagocytosis	Membrane engulfs particles from surroundings.	Cellular energy	White blood cell engulfing bacterial cell
Receptor-mediated endocytosis	Receptors bind specific ligands, and they are drawn into the cell.	Cellular energy	Cholesterol molecules entering cells
Exocytosis	Vesicle fuses with membrane to expel substances from cell.	Cellular energy	Secretion of certain hormones
Transcytosis	Combines endocytosis and exocytosis to ferry particles through cells.	Cellular energy	M cells in small intestine

teins interact at certain times in the cell cycle, called *checkpoints,* in ways that control the cell cycle. Of particular importance is the restriction checkpoint that determines a cell's fate—whether it will continue in the cell cycle and divide, move into a nondividing stage as a specialized cell, or die.

The cell cycle is very precisely regulated. Stimulation from a hormone or growth factor may trigger cell division. This occurs, for example, when the breasts develop into milk-producing glands during pregnancy. Disruption of the cell cycle can affect health: If cell division is too infrequent, a wound cannot heal; if too frequent, a cancer grows. The Topic of Interest on page 70 discusses cancer.

Most cells do not normally divide continually. If grown in the laboratory, most types of human cells divide only forty to sixty times. Presumably, such controls operate in the body too. Some cells may divide the maximum number of times, such as the cells that line the small intestine. Others normally do not divide, such

as nerve cells. A cell "knows" when to stop dividing because of a built-in "clock" in the form of the chromosome tips. These chromosome regions, called *telomeres,* shorten with each cell division. When the telomeres shorten to a certain length, the cell no longer divides.

Studies show that severe psychological or emotional stress can hasten telomere shortening. This may explain how stress can harm health.

Interphase

Before a cell actively divides, it must grow and duplicate much of its contents, so that two "daughter" cells can form from one. This period of preparedness is called **interphase.**

Once thought to be a time of rest, interphase is actually a time of great synthetic activity. During interphase, the cell obtains nutrients, utilizes them to manufacture new living material, and maintains routine "housekeeping" functions. The cell duplicates membranes, ribosomes, lysosomes, and mitochondria. Perhaps most importantly, the cell in interphase takes on the tremendous task of replicating its genetic material. This is important so that each of the two new cells will have a complete set of genetic instructions.

Interphase is considered in phases. DNA is replicated during the S (or synthesis) phase, which is bracketed by two gap (or growth) periods, called G_1 and G_2, when other structures are duplicated.

Cell Division

A cell can divide in two ways. One type of cell division is *meiosis,* which is part of *gametogenesis,* the formation of egg cells (in the female) and sperm cells (in the male). Because an egg fertilized by a sperm must have the normal complement of 46 chromosomes, both the egg and the sperm must first halve their normal chromosome number to 23 chromosomes. Meiosis, through a process called reduction division, accomplishes this. Only a few cells undergo meiosis, which is discussed in chapter 19 (pp. 509–511 and 519).

The other, much more common form of cell division increases cell number, which is necessary for growth and development and for wound healing. It consists of two separate processes: (1) division of the nucleus, called **mitosis** (mi-to′sis), and (2) division of the cytoplasm, called **cytokinesis** (si″to-ki-ne′sis). All cells can divide by mitosis, except sperm and egg cells, and red blood cells, which expel their nuclei.

Division of the nucleus must be very precise because it contains the DNA. Each new cell resulting from mitosis must have a complete and accurate copy of this information to survive. DNA replicates during interphase, but it is equally distributed into two cells in mitosis.

Although mitosis is often described in terms of stages, the process is continuous, without marked changes between one step and the next (fig. 3.22). Stages, however, indicate the sequence of major events. They are:

1. **Prophase** One of the first indications that a cell is going to divide is that the chromosomes become visible in the nucleus when stained. Because the cell has gone through S phase, each prophase chromosome is composed of two identical portions (chromatids), which are temporarily attached at a region on each called the *centromere.*

   The centrioles of the centrosome replicate just before mitosis begins. During prophase, the two newly formed centriole pairs move to opposite ends of the cell. Soon, the nuclear envelope and the nucleolus break up, disperse, and are no longer visible. Microtubules are assembled from tubulin proteins in the cytoplasm and associate with the centrioles and chromosomes. A spindle-shaped array of microtubules (spindle fibers) forms between the centrioles as they move apart.

2. **Metaphase** The chromosomes line up about midway between the centrioles, as a result of microtubule activity. Spindle fibers attach to the centromeres of each chromosome so that a fiber from one pair of centrioles contacts one centromere, and a fiber from the other pair of centrioles attaches to the other centromere.

3. **Anaphase** Soon the centromeres are pulled apart. As the chromatids separate, they become individual chromosomes. The separated chromosomes now move in opposite directions, once again guided by microtubule activity. The spindle fibers shorten and pull their attached chromosomes toward the centrioles at opposite ends of the cell.

4. **Telophase** The final stage of mitosis begins when the chromosomes complete their migration toward the centrioles. It is like prophase in reverse. As the chromosomes approach the centrioles, they begin to elongate and unwind from rodlike shapes into the threadlike fibers of chromatin. A nuclear envelope forms around each chromosome set, and nucleoli appear within the newly formed nuclei. Finally, the microtubules disassemble into free tubulin molecules.

Cytoplasmic Division

Cytoplasmic division (cytokinesis) begins during anaphase, when the cell membrane starts to constrict down the middle of the cell. This constriction, called a *cleavage furrow,* continues through telophase. Contraction of a ring of microfilaments, which assemble in the cytoplasm and attach to the inner surface of the cell membrane, divides the cytoplasm. The contractile ring lies at right angles to the microtubules that pulled the chromosomes to opposite sides of the cell. The ring pinches inward, separating the two newly formed nuclei and distributing about half of the organelles into each new cell. The new cells may differ slightly in size and number of organelles, but they contain identical genetic information.

Check Your Recall

27. Outline the cell cycle.
28. Explain regulation of the cell cycle.
29. Describe the events that occur during mitosis.
30. Why must division of DNA during mitosis be precise?

Cell Differentiation

Cells come from preexisting cells, by the processes of mitosis and cytokinesis. Cell division explains how a fertilized egg develops into an individual consisting of

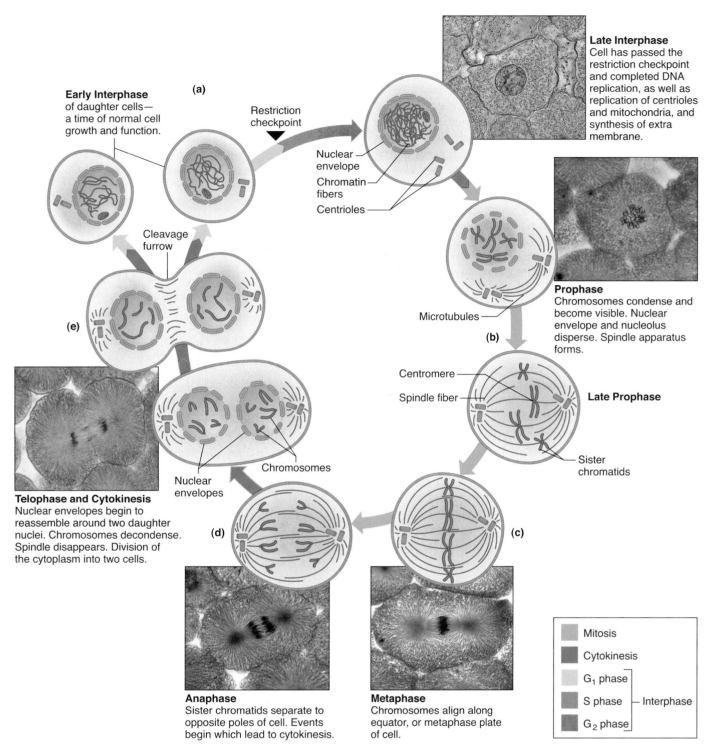

(a)

Early Interphase of daughter cells—a time of normal cell growth and function.

Restriction checkpoint

Late Interphase
Cell has passed the restriction checkpoint and completed DNA replication, as well as replication of centrioles and mitochondria, and synthesis of extra membrane.

Nuclear envelope
Chromatin fibers
Centrioles

Cleavage furrow

(e)

Microtubules

Prophase
Chromosomes condense and become visible. Nuclear envelope and nucleolus disperse. Spindle apparatus forms.

(b)

Centromere
Spindle fiber

Late Prophase

Sister chromatids

Telophase and Cytokinesis
Nuclear envelopes begin to reassemble around two daughter nuclei. Chromosomes decondense. Spindle disappears. Division of the cytoplasm into two cells.

Chromosomes
Nuclear envelopes

(d)

(c)

Anaphase
Sister chromatids separate to opposite poles of cell. Events begin which lead to cytokinesis.

Metaphase
Chromosomes align along equator, or metaphase plate of cell.

▮	Mitosis
▮	Cytokinesis
▮	G_1 phase
▮	S phase ⎤ Interphase
▮	G_2 phase ⎦

Figure 3.22

Mitosis and cytokinesis produce two cells from one. (*a*) During interphase, before mitosis, chromosomes are visible only as chromatin fibers. A single pair of centrioles is present, but not visible at this magnification. (*b*) In prophase, as mitosis begins, chromosomes have condensed and are easily visible when stained. The centrioles have replicated, and each pair moves to an opposite end of the cell. The nuclear envelope and nucleolus disappear, and spindle fibers associate with the centrioles and the chromosomes. (*c*) In metaphase, the chromosomes line up midway between the centrioles. (*d*) In anaphase, the centromeres are pulled apart by the spindle fibers, and the chromatids, now individual chromosomes, move in opposite directions. (*e*) In telophase, chromosomes complete their migration and unwind to become chromatin fibers, the nuclear envelope re-forms, and microtubules disassemble. Cytokinesis, which actually began during anaphase, continues during telophase. Not all chromosomes are shown in these drawings. (Micrographs approximately 360×)

Cancer

Cancer is a group of closely related diseases that can affect many different tissues. One in three of us will develop some form of cancer. These conditions result from changes in genes (mutations) that alter the cell cycle. Cancers share the following characteristics:

1. **Hyperplasia** Hyperplasia is uncontrolled cell division. Normal cells divide a set number of times, signaled by the shortening of chromosome tips. Cancer cells activate an enzyme, called *telomerase,* that continually rebuilds chromosomes, so that cells are not signaled to stop dividing.
2. **Dedifferentiation** Cancer cells lose many of the specialized structures and functions of the normal type of cell from which they descend, and are therefore said to be dedifferentiated (fig. 3A).
3. **Invasiveness** Cancer cells break through boundaries, called *basement membranes,* which separate cell layers within some organs.
4. **Angiogenesis** Cancer cells induce the extension of blood vessels, which nourish the cells and remove wastes, enabling the cancer to persist, grow, and spread.
5. **Metastasis** Cancer cells spread, or metastasize, to other tissues. Normal cells usually aggregate in groups of similar kinds. Small numbers of cancer cells can detach from their original mass and move from their place of origin, often into the bloodstream or lymphatic system. The cancer cells may establish new tumors elsewhere in the body.

Mutations in certain genes cause cancers. Such a mutation may activate a cancer-causing oncogene (a gene that normally controls mitotic rate) or inactivate a protective gene called a tumor suppressor. A person may inherit one abnormal

cancer-causing gene variant, present in all cells, and develop cancer when the second copy of that gene mutates in a cell of the organ that will be affected. This second mutation may occur in response to an environmental trigger. That is, cancer usually entails mutations in somatic (non-sex) cells.

Normal cells (with hairlike cilia)

Cancer cells

Figure 3A

The lack of cilia on these cancer cells, compared to the nearby cilia-fringed cells from which they arose, is one sign of their dedifferentiation (2,250×).

trillions of cells, of at least 260 specialized types. The process of specialization is called **differentiation.**

The ability to generate new cells is essential to the growth and repair of tissues. Cells that retain the ability to divide repeatedly without specializing, called **stem cells,** allow for this continual growth and renewal. A stem cell divides mitotically to yield either two daughter cells like itself (stem cells), or one daughter cell that is a stem cell and one that becomes partially specialized, termed a **progenitor cell.** The ability of a stem cell to divide and give rise to at least one other stem cell is called self-renewal. A progenitor cell's daughter cells can become any of a few cell types. For example, a neural stem cell divides to give rise to another stem cell and

a neural progenitor cell. The progenitor cell then can divide, and its daughter cells differentiate, becoming nervous tissue but not muscle or bone tissue. All of the differentiated cell types in a human body arise through such lineages of stem and progenitor cells. Figure 3.23 depicts a few of them.

Many organs have stem or progenitor cells that are stimulated to divide when injury or illness occurs. This action replaces cells, promoting healing. For example, one in 10,000 to 15,000 bone marrow cells is a stem cell that can give rise to blood as well as several other cell types. Stem cells in organs may have been set aside in the embryo or fetus as repositories of future healing. Alternatively, or perhaps also, stem or progenitor cells

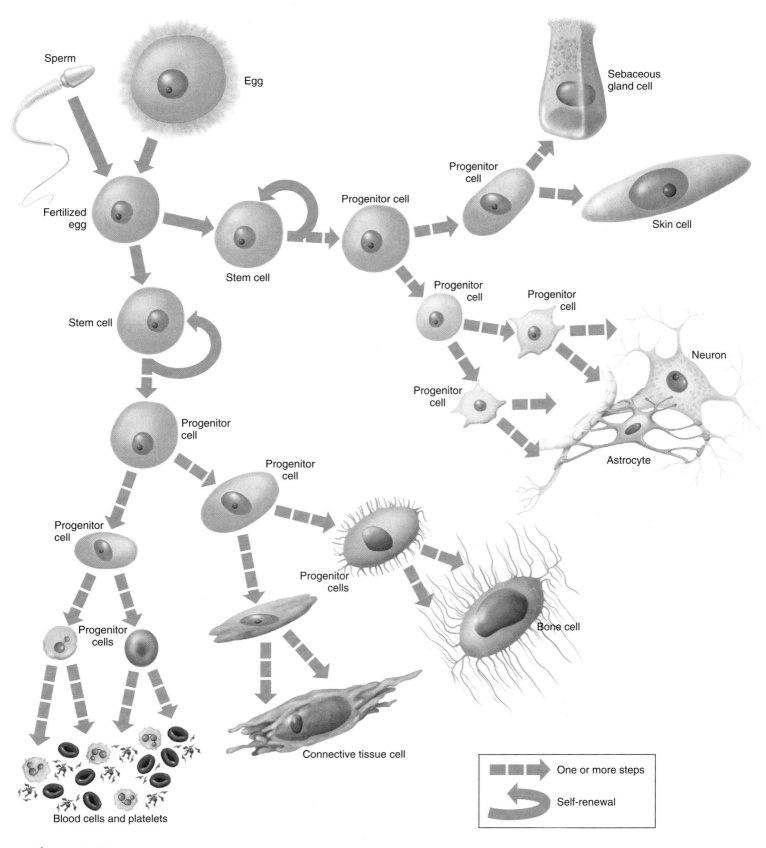

Sperm

Egg

Sebaceous gland cell

Fertilized egg

Progenitor cell

Stem cell

Progenitor cell

Progenitor cell

Skin cell

Stem cell

Progenitor cell

Progenitor cell

Neuron

Progenitor cell

Progenitor cell

Progenitor cell

Astrocyte

Progenitor cell

Progenitor cells

Progenitor cell

Bone cell

Progenitor cell

Progenitor cells

Connective tissue cell

Blood cells and platelets

One or more steps

Self-renewal

Figure 3.23

Cell lineages trace the paths of differentiation. All cells in the human body ultimately descend from stem cells, through the processes of cell division and differentiation. This simplified view depicts a few of the pathways that cells follow, grouping the cell types by the closeness of their lineages. Imagine the complexity of the lineages of the more than 260 human cell types!

may travel from the bone marrow to replace injured or dead cells in response to signals sent from the site of damage.

Throughout development, cells progressively specialize by utilizing different parts of the complete genetic instructions, or genome, that are present in each cell. That is, some genes are "turned on" in certain cells, and other genes are turned on in other cell types. This differential gene expression, for example, enables an immature bone cell to form from a progenitor cell by manufacturing proteins that bind bone mineral and an enzyme required for bone formation. An immature muscle cell, in contrast, forms from a muscle progenitor cell by accumulating contractile proteins. The bone progenitor does not produce contractile proteins, nor does the muscle progenitor produce mineral-binding proteins. However, each cell type also uses parts of the genome that enable it to synthesize molecules essential for basic survival. The final differentiated cell is like a library from which only some of the stored information is accessed.

Cell Death

A cell that does not divide or specialize has another option—it may die. **Apoptosis** (ap″o-to′sis) is a form of cell death that is a normal part of development, rather than the result of injury or disease. Apoptosis sculpts organs from naturally overgrown tissues. In the fetus, for example, apoptosis carves away webbing between developing fingers and toes, removes extra brain cells, and preserves only those immune system cells that recognize the body's cell surfaces. After birth, apoptosis occurs after a sunburn—it peels away damaged skin cells that might otherwise turn cancerous.

A cell in the throes of apoptosis goes through characteristic steps. It rounds up and bulges, the nuclear membrane breaks down, chromatin condenses, and enzymes cut the chromosomes into many equal-size pieces of DNA. Finally, the cell shatters into membrane-enclosed fragments, and a scavenger cell engulfs and destroys them. Apoptosis is a little like cleaning up a very messy room by placing garbage in many garbage bags, which are then removed.

Check Your Recall

31. Why must cells divide and specialize?
32. Distinguish between a stem cell and a progenitor cell.
33. How are new cells generated, and how do they specialize?
34. How is cell death a normal part of development?

Clinical Connection

In certain highly specialized cell types, endocytosis and exocytosis become structurally and functionally linked, in the process of transcytosis. In transcytosis, a particle enters by endocytosis, journeys through the cytoplasm, and then exits the cell by exocytosis from the other end. Transcytosis thereby escorts substances across single-cell layers. For example, it occurs in the lining of the small intestine, where rare cells called M cells sample bits of food, transporting the captured molecules through themselves, and then ejecting them by exocytosis to be met by a gathering of immune system cells inside the internal environment. From here, if the transported particles are recognized as foreign, they may stimulate an immune response, and other cells will then flood the small intestinal lining with specific antibodies against the potentially dangerous substance. Transcytosis through the M cells protects against some forms of food poisoning. Infectious prion proteins, discussed in the Clinical Connection in chapter 2 (pp. 46–47), can enter the body by evading the M cell barrier in the small intestine.

Other examples of transcytosis occur in the genital tract and in the rectum. HIV (the virus that causes AIDS) uses this mechanism to cross the epithelium and reach the bloodstream.

SUMMARY OUTLINE

3.1 Introduction (p. 51)
Cells vary considerably in size, shape, and function. The shapes of cells make possible their functions.

3.2 Composite Cell (p. 51)
A cell includes a cell membrane, cytoplasm, and nucleus. Organelles perform specific functions; the nucleus controls overall cell activities because it contains DNA.

 1. Cell membrane
    a. The cell membrane forms the outermost limit of the living material.
    b. It is a selectively permeable passageway that controls the entrance and exit of substances. Its molecules transmit signals.
    c. The cell membrane includes protein, lipid, and carbohydrate molecules.
    d. The cell membrane's framework is mainly a bilayer of phospholipid molecules.
    e. Molecules that are soluble in lipids pass through the cell membrane easily, but water-soluble molecules do not.
    f. Proteins function as receptors on membrane surfaces and form channels for the passage of ions and molecules.
    g. Patterns of surface carbohydrates associated with membrane proteins enable certain cells to recognize one another.
 2. Cytoplasm
    a. Cytoplasm contains membranes, organelles, and the rods and tubules of the cytoskeleton, suspended in cytosol.

b. The endoplasmic reticulum is a tubular communication system in the cytoplasm that transports lipids and proteins.

c. Ribosomes function in protein synthesis.

d. The Golgi apparatus packages glycoproteins for secretion.

e. Mitochondria contain enzymes that catalyze reactions that release energy from nutrient molecules.

f. Lysosomes contain digestive enzymes that decompose substances.

g. Peroxisomes house enzymes that catalyze bile acid synthesis, hydrogen peroxide degradation, lipid breakdown, and detoxification of alcohol.

h. Microfilaments (built of actin) and microtubules (built of tubulin) aid cellular movements and support and stabilize the cytoplasm and organelles. Together they form the cytoskeleton. Microtubules also form centrioles, cilia, and flagella.

i. The centrosome contains centrioles that aid in distributing chromosomes during cell division.

j. Cilia and flagella are motile extensions from cell surfaces.

k. Vesicles contain substances that recently entered the cell or that are to be secreted from the cell.

3. Cell nucleus

a. The nucleus is enclosed in a double-layered nuclear envelope.

b. It contains a nucleolus, which is the site of ribosome production.

c. It contains chromatin, which is composed of loosely coiled fibers of DNA and protein. As chromatin fibers condense, chromosomes become visible during cell division.

3.3 Movements Through Cell Membranes (p. 59)

The cell membrane is a barrier through which substances enter and leave a cell.

1. Passive mechanisms do not require cellular energy.

a. Diffusion

(1) Diffusion is the movement of molecules or ions from regions of higher concentration toward regions of lower concentration.

(2) In the body, diffusion exchanges oxygen and carbon dioxide.

b. Facilitated diffusion

(1) In facilitated diffusion, special carrier molecules move substances through the cell membrane.

(2) This process moves substances only from regions of higher concentration toward regions of lower concentration.

c. Osmosis

(1) Osmosis is diffusion of water molecules from regions of higher water concentration toward regions of lower water concentration through a selectively permeable membrane.

(2) Osmotic pressure increases as the number of impermeant particles dissolved in a solution increases.

(3) A solution is isotonic to a cell when it has the same osmotic pressure as the cell.

(4) Cells lose water when placed in hypertonic solutions and gain water when placed in hypotonic solutions.

d. Filtration

(1) Filtration is the movement of molecules from regions of higher hydrostatic pressure toward regions of lower hydrostatic pressure.

(2) Blood pressure causes filtration through porous capillary walls, forming tissue fluid.

2. Active mechanisms require cellular energy.

a. Active transport

(1) Active transport moves molecules or ions from regions of lower concentration toward regions of higher concentration.

(2) It requires cellular energy from ATP and carrier molecules in the cell membrane.

b. Endocytosis and exocytosis

(1) Endocytosis may convey relatively large particles into a cell. Exocytosis is the reverse of endocytosis.

(2) In pinocytosis, a cell membrane engulfs tiny droplets of liquid.

(3) In phagocytosis, a cell membrane engulfs solid particles.

(4) Receptor-mediated endocytosis moves specific types of particles into cells.

(5) Transcytosis combines endocytosis and exocytosis so that a particle can pass through a cell.

3.4 The Cell Cycle (p. 66)

The cell cycle includes interphase, mitosis, cytoplasmic division, and differentiation. It is highly regulated.

1. Interphase

a. During interphase, a cell duplicates membranes, ribosomes, organelles, and DNA.

b. Interphase terminates when mitosis begins.

2. Cell Division

a. Meiosis is a form of cell division that forms sex cells.

b. Mitosis is the division and distribution of genetic material to new cells.

c. The stages of mitosis are prophase, metaphase, anaphase, and telophase.

3. Cytoplasmic division distributes cytoplasm into two portions completing about the same time as mitosis.

4. Differentiation is the process of cell specialization. Stem cells provide new cells for growth and repair. They give rise to other stem cells and to progenitor cells that begin to specialize. Differentiation reflects the use of different portions of the genetic material.

5. A cell that does not divide or differentiate may undergo apoptosis, a form of cell death that is a normal part of development.

CHAPTER ASSESSMENTS

3.1 Introduction

1. An adult human body consists of about _____ cells. (p. 51)

a. 2 billion

b. 50 billion

c. 75 trillion

d. 8 quadrillion

2. Define *cell*. (p. 51)

3. Discuss how cells differ from one another. (p. 51)

3.2 Composite Cell

4. The three major parts of a cell are _____. (p. 52)

a. the nucleus, the nucleolus, and the nuclear envelope

b. the nucleus, cytoplasm, and the cell membrane

c. a nerve cell, an epithelial cell, and a muscle cell

d. endoplasmic reticulum, Golgi apparatus, and ribosomes

e. cytoplasm, organelles, and chromatin

5. Explain the general function of organelles. (p. 53)

6. Define *selectively permeable*. (p. 53)

7. Describe the structure of a cell membrane and explain how this structural organization provides the membrane's function. (p. 53)

8. List three functions of membrane proteins. (p. 54)

9. Match the following structures with their definitions: (pp. 55–58)

   (1) Golgi apparatus
   (2) mitochondria
   (3) peroxisomes
   (4) cilia
   (5) endoplasmic reticulum
   (6) cytoskeleton
   (7) vesicles
   (8) ribosomes

   A. Sacs that contain enzymes that catalyze a variety of specific biochemical reactions

   B. Structures on which protein synthesis occurs

   C. Structures that house the reactions that release energy from nutrients

   D. A network of microfilaments and micro-tubules that supports and shapes a cell

   E. A structure that modifies, packages, and exports glycoproteins

   F. Membrane-bounded sacs

   G. A network of membranous channels and sacs where lipids and proteins are synthesized

   H. Hairlike structures that extend from certain cell surfaces and wave about

10. List the parts of the nucleus and explain why each is important. (p. 59)

3.3 Movements Through Cell Membranes

11. Distinguish between active and passive mechanisms of movement across cell membranes. (p. 59)

12. Match the transport mechanisms on the left with their descriptions on the right. (pp. 61–66)

    (1) diffusion
    (2) facilitated diffusion
    (3) filtration
    (4) active transport
    (5) endocytosis
    (6) exocytosis

    A. The cell membrane engulfs a particle or substance, drawing it into the cell in a vesicle

    B. Movement down the concentration gradient with a carrier protein, without energy input

    C. Movement down the concentration gradient without a carrier protein or energy input

    D. A particle or substance leaves a cell in a vesicle that merges with the cell membrane

    E. Movement against the concentration gradient with energy input

    F. Hydrostatic pressure forces substances through membranes

13. Define *osmosis*. (p. 63)

14. Distinguish between hypertonic, hypotonic, and isotonic solutions. (p. 63)

15. Explain how phagocytosis differs from receptor-mediated endocytosis. (p. 65)

3.4 The Cell Cycle

16. Explain why it is important for the cell cycle to be highly regulated. (p. 67)

17. Distinguish between interphase and mitosis. (p. 67)

18. The period of the cell cycle when DNA replicates is _____. (p. 68)
    a. G_1 phase
    b. G_2 phase
    c. S phase
    d. prophase
    e. telophase

19. Explain how meiosis differs from mitosis. (p. 68)   *mitosis ↑ cell*
    meiosis ↓ cell

20. _____ occur simultaneously. (p. 68)
    a. G_1 phase and G_2 phase
    b. Interphase and mitosis
    c. Cytokinesis and telophase
    d. Prophase and metaphase
    e. Meiosis and mitotic metaphase

21. Describe the events of mitosis in sequence. (p. 68)

22. Define *differentiation*. (p. 70)

23. A stem cell _____. (p. 70)
    a. undergoes apoptosis
    b. self-renews
    c. is differentiated
    d. gives rise to only fully differentiated daughter cells
    e. forms from a progenitor cell

24. Describe the steps of apoptosis. (p. 72)

INTEGRATIVE ASSESSMENTS/ CRITICAL THINKING

OUTCOMES 3.1, 3.4

1. Why does a muscle cell contain many mitochondria, and a white blood cell contain many lysosomes?

OUTCOME 3.2

2. Organelles compartmentalize a cell, much as a department store displays related items together. What advantage does such compartmentalization offer a large cell? Cite two examples of organelles and the activities they compartmentalize.

3. In an inherited condition called glycogen cardiomyopathy, teenagers develop muscle weakness, which affects the heart as well as other muscles. Samples of the affected muscle cells contain huge lysosomes, swollen with the carbohydrate glycogen. How might this condition arise?

4. Exposure to tobacco smoke immobilizes cilia, and they eventually disappear. How might this effect explain why smokers have an increased incidence of coughing and respiratory infections?

OUTCOME 3.3

5. Liver cells are packed with glucose. What mechanism could be used to transport more glucose into a liver cell? Why would only this mode of transport work?

6. Which process—diffusion, osmosis, or filtration—is utilized in the following situations?
   a. Injection of a drug that is hypertonic to the tissues stimulates pain.
   b. The urea concentration in the dialyzing fluid of an artificial kidney is decreased.

7. What characteristic of cell membranes may explain why fat-soluble substances such as chloroform and ether rapidly affect cells?

OUTCOMES 3.2, 3.3

8. Many cancer drugs stop working because a large protein, called P-gp, in the cell membranes of some cancer cells pumps these types of drugs out of the cell before they can enter the cytoplasm. P-gp requires ATP to function, and it continually cycles from its site of synthesis in the cytoplasm, to the cell membrane, and then back inside the cell. What structures most likely transport P-gp?

OUTCOME 3.4

9. How would a drug that functions as an angiogenesis inhibitor be useful in treating cancer?

WEB CONNECTIONS

Visit the text website at **aris.mhhe.com** for additional quizzes, interactive learning exercises, and more.

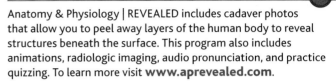

Anatomy & Physiology | REVEALED includes cadaver photos that allow you to peel away layers of the human body to reveal structures beneath the surface. This program also includes animations, radiologic imaging, audio pronunciation, and practice quizzing. To learn more visit **www.aprevealed.com**.

4

Cellular Metabolism

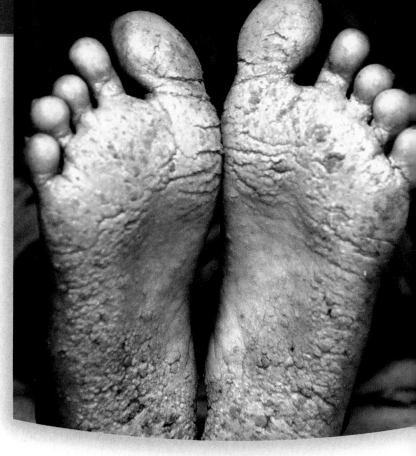

Chronic exposure to arsenic in drinking water causes the burning, colored lesions of arsenicosis.

ARSENIC POISONING. Disrupting the body's ability to extract energy from nutrients can have great effects on health. Arsenic is a chemical element that, if present in the body in excess, shuts down metabolism. It can do so suddenly or gradually.

Given in one large dose, arsenic causes chest pain, vomiting, diarrhea, shock, coma, and death. In contrast, a series of many small doses causes dark skin lesions that feel as if they are burning, along with numb hands and feet, and eventually skin cancer. Such gradual poisoning, called arsenicosis, may occur from contact with pesticides or environmental pollutants. The world's largest outbreak of arsenicosis, however, is due to a natural exposure.

When the World Bank and UNICEF began tapping into aquifers in India and Bangladesh in the late 1960s, they were trying to supply clean water to areas ravaged by sewage and industrial waste released from the many rivers by cycles of floods and droughts. Millions of people had already perished from diarrheal diseases due to the poor sanitation. But digging wells to provide clean water backfired when workers unwittingly penetrated a layer of sediment naturally rich in arsenic. The chemical has been leaching into the water in at least 2 million wells in Bangladesh alone ever since, reaching levels 50 times the safety limit set by the World Health Organization. Effects on health took several years to show up. When they did, the people thought arsenicosis was contagious. In addition to their physical pain, affected individuals bore the psychic pain of being shunned.

Arsenic damages the body by binding to bonds between sulfur atoms in proteins. The effects on metabolism largely stem from impairment of an enzyme that helps the breakdown products of glucose enter the mitochondria, where energy is extracted. Without the action of this enzyme, thanks to the arsenic exposure, cells run out of energy.

Today UNICEF is helping the people of India and Bangladesh avoid arsenic poisoning. They are diagnosing and treating arsenicosis, and providing tanks that the people are using to store rainwater. A vast education campaign has done much to quell the stigma of arsenicosis. Although cases will continue to appear for a few more decades, the use of alternate water sources has finally slowed the progression of this public health problem.

Learning Outcomes   *After studying this chapter, you should be able to do the following:*

4.1 ▶ Introduction

1. Briefly explain the role of metabolism. (p. 77)

4.2 ▶ Metabolic Reactions

2. Compare and contrast anabolism and catabolism. (p. 77)

4.3 ▶ Control of Metabolic Reactions

3. Describe how enzymes control metabolic reactions. (p. 79)

4. List the basic steps of an enzyme-catalyzed reaction. (p. 79)

4.4 ▶ Energy for Metabolic Reactions

5. Explain how cellular respiration releases chemical energy. (p. 80)

6. Describe how energy in the form of ATP becomes available for cellular activities. (p. 82)

4.5 ▶ Metabolic Pathways

7. Describe the general metabolic pathways of carbohydrates, lipids, and proteins. (p. 82)

4.6 ▶ DNA (Deoxyribonucleic Acid)

8. Explain how DNA carries genetic information. (p. 84)

9. Define gene and genome. (p. 84)

10. Describe how DNA molecules replicate. (p. 84)

4.7 ▶ Protein Synthesis

11. Define genetic code. (p. 85)

12. Describe the steps of protein synthesis. (p. 86)

Aids to Understanding Words   *(Appendix A on page 567 has a complete list of Aids to Understanding Words.)*

an- [without] *an*aerobic respiration: Respiratory process that proceeds without oxygen.

ana- [up] *ana*bolism: Cellular processes that use smaller molecules to build larger ones.

cata- [down] *cata*bolism: Cellular processes that break larger molecules into smaller ones.

mut- [change] *mut*ation: Change in a nucleic acid sequence.

-zym [causing to ferment] en*zym*e: Protein that initiates or speeds a chemical reaction without being consumed.

4.1 INTRODUCTION

Cells require energy and information to build bodies. In cells, the many chemical reactions of metabolism break down nutrients to release the energy in their chemical bonds and also build molecules to store energy. A cell uses some of that energy to copy the genetic material when it divides and to access genetic information to construct proteins from amino acids. Vital to all of these activities are special proteins called **enzymes** (en′zīmz) that control each of the interrelated reactions of metabolism, as well as the steps of DNA replication and protein synthesis.

4.2 METABOLIC REACTIONS

Metabolic reactions control a cell's use of energy and are of two major types. **Anabolism** (an″ah-bol′lizm) is the buildup of larger molecules from smaller ones, and requires energy. **Catabolism** (kă-tab′o-lizm) is the breakdown of larger molecules into smaller ones and

releases energy. The reactions of anabolism and catabolism store and release energy, respectively, and together constitute metabolism.

Anabolism

Anabolism provides the biochemicals required for cell growth and repair. For example, many simple sugar molecules (monosaccharides) are linked into a chain to form molecules of glycogen, a carbohydrate, using an anabolic process called **dehydration synthesis** (de″hi-dra′shun sin′thĕ-sis). As adjacent monosaccharide units join, an —OH (hydroxyl group) from one monosaccharide molecule and an —H (hydrogen atom) from an —OH group of another are removed. The —H and —OH react to produce a water molecule, and the monosaccharides are joined by a shared oxygen atom (fig. 4.1). As this process repeats, the molecular chain grows.

Dehydration synthesis also links glycerol and fatty acid molecules in fat (adipose) cells to form fat molecules (triglycerides). In this case, three hydrogen atoms are removed from a glycerol molecule, and an —OH group is removed from each of three fatty acid

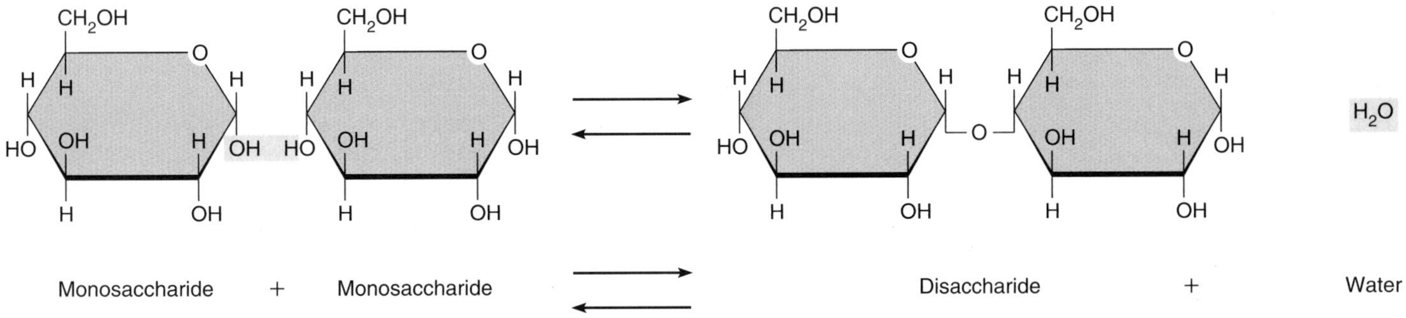

Monosaccharide   +   Monosaccharide ⇄ Disaccharide   +   Water

Figure 4.1

Building up and breaking down molecules. A disaccharide is formed from two monosaccharides in a dehydration synthesis reaction (arrows to the right). In the reverse reaction, hydrolysis, a disaccharide is broken down into two monosaccharides (arrows to the left).

molecules (fig. 4.2). The result is three water molecules and a single fat molecule. Shared oxygen atoms bind the glycerol and fatty acid portions.

Cells also use dehydration synthesis to join amino acid molecules, which eventually form protein molecules. When two amino acid molecules unite, an —OH from one and an —H from the —NH$_2$ group of another are removed. A water molecule forms, and the amino acid molecules are joined by a bond between a carbon atom and a nitrogen atom, called a *peptide bond* (fig. 4.3). Two bound amino acids form a *dipeptide,* and many linked in a chain form a *polypeptide.* Generally, a polypeptide that has a specific function and consists of 100 or more amino acid molecules is considered a *protein.* Some protein molecules consist of more than one polypeptide.

Catabolism

Physiological processes that break larger molecules into smaller ones constitute catabolism. An example of catabolism is **hydrolysis** (hi-drol'ĭ-sis), which decomposes carbohydrates, lipids, and proteins, and splits a water molecule in the process. For instance, hydrolysis of a disaccharide such as sucrose yields two monosaccharides (glucose and fructose) as a molecule of water splits:

$$C_{12}H_{22}O_{11} + H_2O \rightarrow C_6H_{12}O_6 + C_6H_{12}O_6$$
(Sucrose) (Water) (Glucose) (Fructose)

In this case, the bond between the simple sugars within the sucrose molecule breaks, and the water molecule supplies a hydrogen atom to one sugar molecule and a hydroxyl group to the other. Hydrolysis is the opposite of dehydration synthesis (see figs. 4.1, 4.2, and 4.3). Each of these reactions is reversible and can be summarized as follows:

Hydrolysis →

Disaccharide + Water ⇄ Monosaccharide + Monosaccharide

← Dehydration synthesis

Hydrolysis, which occurs during digestion, breaks down carbohydrates into monosaccharides, fats into glycerol and fatty acids, proteins into amino acids, and nucleic acids into nucleotides. (Chapter 15, pages 404–428, discusses digestion in more detail.)

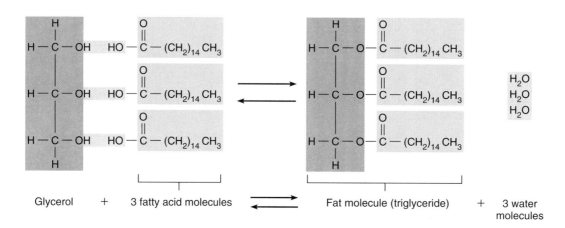

Figure 4.2

Forming a fat. A glycerol molecule and three fatty acid molecules react, yielding a fat molecule (triglyceride) in a dehydration synthesis reaction (arrows to the right). In the reverse reaction, hydrolysis, a triglyceride is broken down into three fatty acids and a glycerol (arrows to the left).

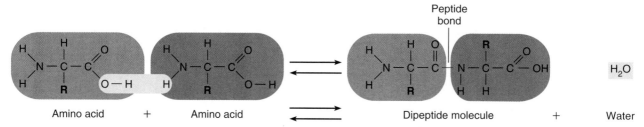

Figure 4.3

Peptide bonds link amino acids. When dehydration synthesis unites two amino acid molecules, a peptide bond forms between a carbon atom and a nitrogen atom, resulting in a dipeptide molecule (arrows to the right). In the reverse reaction, hydrolysis, a dipeptide molecule is broken down into two amino acids (arrows to the left).

Check Your Recall

1. What is the function of anabolism? Of catabolism?

2. What is the product of anabolism of monosaccharides? Of glycerol and fatty acids? Of amino acids?

3. Distinguish between dehydration synthesis and hydrolysis.

4.3 CONTROL OF METABOLIC REACTIONS

Specialized cells, such as nerve, muscle, or blood cells, carry out distinctive chemical reactions. However, all cells perform certain basic reactions, such as buildup and breakdown of carbohydrates, lipids, proteins, and nucleic acids. These reactions include hundreds of specific chemical changes that occur rapidly—yet in a coordinated fashion—thanks to enzymes.

Enzyme Action

Like other chemical reactions, metabolic reactions require energy to proceed. The temperature conditions in cells, however, are usually too mild to adequately promote the reactions that support life. Enzymes make these reactions possible.

Enzymes are complex molecules that promote chemical reactions by lowering the amount of energy, called the *activation energy,* required to start these reactions. In this way, enzymes speed the rates of metabolic reactions. This acceleration is called *catalysis,* and an enzyme is a catalyst. A small number of enzyme molecules go a long way, because they are not consumed in the reactions they catalyze and they are recycled.

The reaction that a particular enzyme catalyzes is very specific. Each enzyme acts only on a particular chemical, which is called its **substrate** (sub′strāt). Many enzymes are named after their substrates, with *-ase* as a suffix. A lipase, for example, catalyzes a reaction that breaks down a lipid. Another example of an enzyme is *catalase.* Its substrate is hydrogen peroxide, which is a toxic by-product of certain metabolic reactions. Catalase speeds breakdown of hydrogen peroxide into water and oxygen, preventing accumulation of hydrogen peroxide, which can damage cells.

Specific enzymes catalyze each of the hundreds of different chemical reactions that constitute cellular metabolism. Every cell contains hundreds of different enzymes, and each enzyme must recognize its specific substrate. This ability of an enzyme to identify its substrate arises from the three-dimensional shape, or conformation, of the enzyme molecule. Each enzyme's polypeptide chain twists and coils into a unique conformation that fits the specific shape of its substrate.

During an enzyme-catalyzed reaction, part of the enzyme molecule called the **active site** temporarily combines with portions of the substrate molecules, forming an enzyme-substrate complex (fig. 4.4). This interaction between the molecules distorts or strains certain chemical bonds within the substrates, altering their orientation so that they require less energy to react. The reaction occurs, product forms, and the enzyme is released in its original conformation.

An enzyme-catalyzed reaction can be summarized as follows:

$$\begin{array}{c}\text{Substrate} \\ \text{molecules}\end{array} + \begin{array}{c}\text{Enzyme} \\ \text{molecule}\end{array} \rightarrow \begin{array}{c}\text{Enzyme-} \\ \text{substrate} \\ \text{complex}\end{array} \rightarrow \begin{array}{c}\text{Product} \\ \text{(changed} \\ \text{substrates)}\end{array} + \begin{array}{c}\text{Enzyme} \\ \text{molecule}\end{array}$$

Many enzyme-catalyzed reactions are reversible, and in some cases the same enzyme catalyzes the reaction in both directions. The speed of the reaction depends partly on the number of enzyme and substrate molecules in the

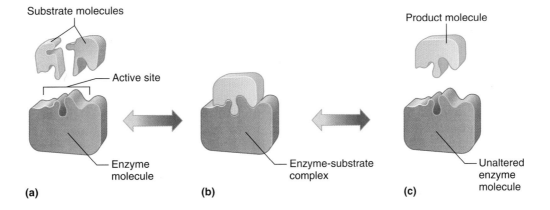

Figure 4.4

An enzyme-catalyzed reaction. (Many enzyme-catalyzed reactions, as depicted here, are reversible.) In the forward reaction (dark-shaded arrows), (*a*) the shapes of the substrate molecules fit the shape of the enzyme's active site. (*b*) When the substrate molecules temporarily combine with the enzyme, a chemical reaction occurs. (*c*) The result is a product molecule and an unaltered enzyme. The active site contorts somewhat as the substrate binds, such that formation of the enzyme-substrate complex is more like a hand fitting into a glove, which has some flexibility, than a key fitting into a lock.

cell, occuring more rapidly if the concentrations of the enzyme or the substrate increase. Also, the efficiency of different kinds of enzymes varies greatly. Some enzymes can process only a few substrate molecules per second, whereas others can catalyze many thousands of chemical reactions in a second.

Factors That Alter Enzymes

Almost all enzymes are proteins, and like other proteins, they can be denatured by exposure to heat, radiation, electricity, certain chemicals, or fluids with extreme pH values. For example, many enzymes become inactive at 45°C, and nearly all of them are denatured at 55°C. Some poisons work by denaturing enzymes. Potassium cyanide, for instance, interferes with respiratory enzymes, impairing a cell's ability to release energy from nutrient molecules.

Cofactors and Coenzymes

Some enzymes are inactive until they combine with a nonprotein component. Such a substance, called a **cofactor,** may be an ion of an element, such as copper, iron, or zinc, or a small organic molecule, called a **coenzyme** (ko-en′zīm). Many coenzymes are vitamin molecules. An example of a coenzyme is coenzyme A, which takes part in cellular respiration, discussed in section 4.4 (p. 82).

> **Check Your Recall**
> 4. What is an enzyme?
> 5. How does an enzyme recognize its substrate?
> 6. What factors affect the speed of an enzyme-controlled reaction?
> 7. What factors can denature enzymes?

4.4 ENERGY FOR METABOLIC REACTIONS

Energy is the capacity to change or move matter; that is, energy is the ability to do work. We recognize energy by what it causes to happen. Common forms of energy are heat, light, sound, electrical energy, mechanical energy, and chemical energy. Most metabolic processes use chemical energy.

Release of Chemical Energy

Chemical energy is held in the bonds between the atoms of molecules and is released when these bonds are broken, as in burning. Such a reaction usually starts by applying heat to activate the burning process. As the chemical burns, bonds break, and energy escapes as heat and light.

Cells "burn" glucose molecules in a process called **oxidation** (ok″sĭ-da′shun). The energy released by the oxidation of glucose powers the anabolic reactions of cells. However, the oxidation of biochemicals inside cells and the burning of substances outside cells differ in some ways.

Burning usually requires a relatively large input of energy, and most of the energy released escapes as heat or light. In cells, enzymes reduce the activation energy required for the oxidation that occurs in the reactions of *cellular respiration*. These reactions release the energy in the bonds of nutrient molecules. Cells can then capture about 40 percent of the energy released from breaking chemical bonds by transferring energy to special energy-carrying molecules. The rest of the liberated energy escapes as heat, which helps maintain body temperature.

> **Check Your Recall**
> 8. What is energy?
> 9. How does oxidation inside cells differ from the burning process outside cells?

Cellular Respiration

Cellular respiration occurs in three distinct, yet interconnected, series of reactions: **glycolysis,** the **citric acid cycle,** and the **electron transport chain,** shown schematically in figure 4.5. Glucose and oxygen are required for cellular respiration; the products of these reactions include CO_2, water, and energy. Although most of the energy is lost as heat, almost half is captured in the form of high-energy electrons that the cell can store through the synthesis of **ATP (adenosine triphosphate).**

ATP

Each ATP molecule includes a chain of three chemical groups called phosphates (fig. 4.6). As energy is released during cellular respiration, some of it is captured in the bond of the end phosphate. When energy is required for a metabolic reaction, this terminal phosphate bond breaks, releasing the stored energy. (The second phosphate bond is high energy too.) The cell uses ATP for a variety of functions, including active transport and synthesis of various compounds.

An ATP molecule that has lost its terminal phosphate becomes an ADP (adenosine diphosphate) molecule. The ADP can be converted back into ATP by the addition of energy and a third phosphate. Thus, as figure 4.7 shows, ATP and ADP molecules shuttle back and forth between the energy-releasing reactions of cellular respiration and the energy-utilizing reactions of the cell.

Glycolysis

1 The 6-carbon sugar glucose is broken down in the cytosol into two 3-carbon pyruvic acid molecules with a net gain of 2 ATP and the release of high-energy electrons.

Citric Acid Cycle

2 The 3-carbon pyruvic acids generated by glycolysis enter the mitochondria. Each loses a carbon (generating CO_2) and is combined with a coenzyme to form a 2-carbon acetyl coenzyme A (acetyl CoA). More high-energy electrons are released.

3 Each acetyl CoA combines with a 4-carbon oxaloacetic acid to form the 6-carbon citric acid, for which the cycle is named. For each citric acid, a series of reactions removes 2 carbons (generating 2 CO_2's), synthesizes 1 ATP, and releases more high-energy electrons. The figure shows 2 ATP, resulting directly from 2 turns of the cycle per glucose molecule that enters glycolysis.

Electron Transport Chain

4 The high-energy electrons still contain most of the chemical energy of the original glucose molecule. Special carrier molecules bring the high-energy electrons to a series of enzymes that store much of the remaining energy in more ATP molecules. The other products are heat and water. The function of oxygen as the final electron acceptor in this last step is why the overall process is called aerobic respiration.

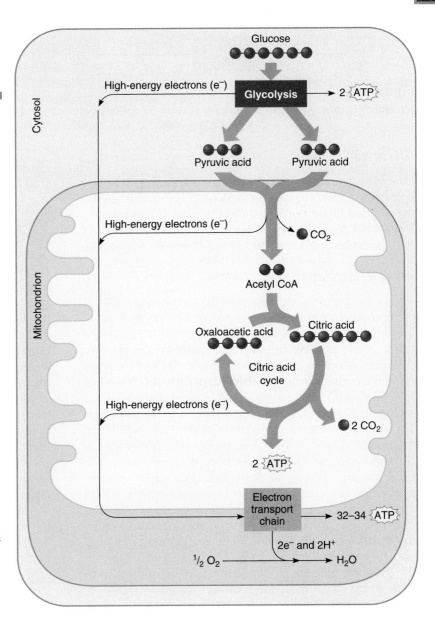

Figure 4.5

Glycolysis occurs in the cytosol and does not require oxygen. Aerobic respiration occurs in the mitochondria and only in the presence of oxygen. Their products include ATP, heat, CO_2, and water. Note that 2 ATP are generated by glycolysis, 2 result directly from the citric acid cycle, and 32–34 are generated by the electron transport chain. Thus, the total yield of ATP molecules per glucose molecule is 36–38, depending on the type of cell.

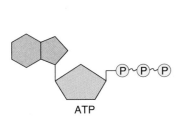

Figure 4.6

Phosphate bonds contain the energy stored in ATP.

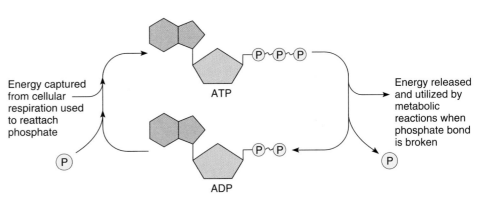

Figure 4.7

ATP provides energy for metabolic reactions. Cellular respiration generates ATP.

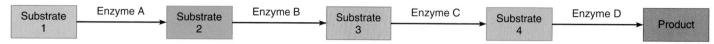

Figure 4.8

A metabolic pathway consists of a series of enzyme-controlled reactions leading to formation of a product.

Glycolysis

Cellular respiration begins with glycolysis, literally "the breaking of glucose" (see fig. 4.5). Glycolysis occurs in the cytosol (the liquid portion of the cytoplasm), and because it does not require oxygen, it is sometimes referred to as the **anaerobic** (an″a-er-o′bik) phase of cellular respiration.

Aerobic Respiration

The fate of pyruvic acid depends on oxygen availability. If oxygen is not present, then pyruvic acid enters an anaerobic pathway that yields lactic acid and taps limited energy. If oxygen is present in sufficient quantity, the pyruvic acid generated by glycolysis can enter the more energy efficient pathways of **aerobic respiration** (a″er-o′bik res″pĭ-ra″shun) in the mitochondria. After doing so, each molecule of pyruvic acid loses a carbon atom and binds a coenzyme to form a molecule of acetyl CoA, which can then combine with a 4-carbon compound to enter the citric acid cycle and then the electron transport chain. The final acceptor of electrons passed along the electron transport chain is oxygen. This is why the pathway is called "aerobic." (Because the reactions of the electron transport chain add phosphates to form ATP, they are also known as oxidative phosphorylation.) The aerobic reactions yield up to 36 ATP molecules per glucose (see fig. 4.5).

For each glucose molecule that is decomposed completely, up to 38 molecules of ATP can be produced. Two of these ATP molecules are the result of glycolysis, and the rest form during the aerobic phase. About half the energy released goes to ATP synthesis, while the rest ends up as heat.

In addition to releasing energy, the complete oxidation of glucose produces carbon dioxide and water. The carbon dioxide is eventually exhaled, and the water becomes part of the internal environment.

In humans, the volume of water produced by metabolism is far less than our daily water needs, so we must drink water to survive. In contrast, a small desert rodent, the kangaroo rat, can survive entirely on the water produced by aerobic respiration.

4.5 METABOLIC PATHWAYS

Like cellular respiration, anabolic and catabolic reactions in general have a number of steps that must occur in a particular sequence. The enzymes that control the rates of these reactions must act in a specific sequence. Such coordination suggests that the enzymes are positioned in exactly the same sequence as that of the reactions they control. For example, the enzymes responsible for aerobic respiration are located in tiny, stalked particles on the inner membranes (cristae) within the mitochondria, in the sequence in which they function. A sequence of enzyme-controlled reactions is called a **metabolic pathway** (fig. 4.8).

Recall that the rate of an enzyme-controlled reaction usually increases if either the number of substrate molecules or the number of enzyme molecules increases. However, the rate of a metabolic pathway is often determined by a regulatory enzyme responsible for one of its steps. The number of molecules of a regulatory enzyme is limited. Consequently, the enzyme supply can become saturated with substrate molecules whenever the substrate concentration exceeds a certain level. Once the enzyme is saturated, increasing the number of substrate molecules will no longer affect the reaction rate. In this way, a single enzyme can control the whole pathway.

As a rule, a *rate-limiting enzyme* is the first enzyme in a series. This position is important because if an enzyme at some point in the sequence were rate-limiting, an intermediate chemical in the pathway might accumulate.

This section has dealt with the metabolism of glucose. Fats and proteins can also be broken down to release energy for ATP synthesis. In all three cases, the final process is aerobic respiration, and the most common entry point is into the citric acid cycle as acetyl coenzyme A (acetyl CoA) (fig. 4.9). These metabolic pathways and their regulation are described further in chapter 15 (pp. 431–432).

Check Your Recall

10. What is the general function of ATP?
11. What happens during glycolysis?
12. What is the role of oxygen in cellular respiration?
13. What are the final products of cellular respiration?

Check Your Recall

14. What is a metabolic pathway?
15. What is a rate-limiting enzyme?

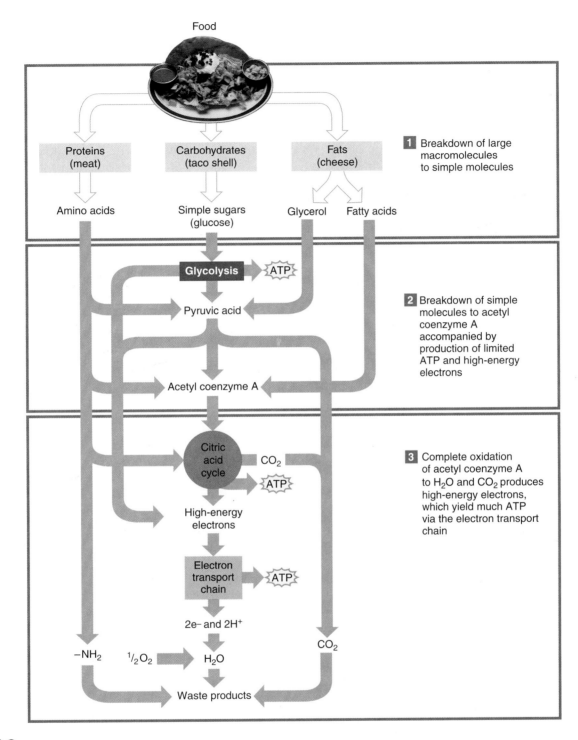

Figure 4.9
A summary of the breakdown (catabolism) of proteins, carbohydrates, and fats.

4.6 DNA (DEOXYRIBONUCLEIC ACID)

Because enzymes control the metabolic processes that enable cells to survive, cells must have instructions for producing these specialized proteins, as well as many other proteins. **DNA (deoxyribonucleic acid)** molecules hold such information in the form of a *genetic code*. This code "instructs" cells how to synthesize enzymes and other specific protein molecules.

Genetic Information

Children resemble their parents because of inherited traits. What actually passes from parents to child is genetic information in the form of DNA molecules from the parents' sex cells (egg and sperm). The portions of DNA molecules that contain the genetic information for making particular proteins are called **genes** (jēnz). Inherited traits are determined by the genes carried in the parents' sex cells. These cells fuse to form the first cell of an offspring's body. As an offspring develops, mitosis passes the information from cell to cell. Genes instruct cells to synthesize not only the enzymes that control metabolic pathways but also many other types of proteins.

All of the DNA in a cell constitutes the **genome** (je'nōm). Only about 2% of the human genome encodes protein; much of the rest controls when and where genes are used to guide protein synthesis. The human genome also includes many highly repetitive DNA sequences whose functions are not known.

DNA Molecules

As described in chapter 2 (p. 46), the building blocks of nucleic acids are nucleotides (see fig. 2.20, p. 46). They are joined so that the sugar and phosphate portions alternate, forming a long "backbone" to the polynucleotide chain (see fig. 2.21*b*, p. 47).

In a DNA molecule, the nitrogenous bases project from the backbone and bind weakly to the bases of the second strand (fig. 4.10). The resulting structure is like a ladder in which the uprights represent the alternating sugar and phosphate backbones of the two strands, and the rungs represent the organic bases. The organic base of a DNA nucleotide can be one of four types: *adenine* (A), *thymine* (T), *cytosine* (C), or *guanine* (G).

Both strands of a DNA molecule consist of nucleotides in a particular sequence. The sequence of bases along one of these strands may encode the genetic information that specifies a particular protein's amino acid sequence.

The nucleotide bases pair in specific ways because of their molecular shapes. An adenine will bond only to a thymine, and a cytosine will bond only to a guanine. As a consequence of this *complementary base pairing,* a DNA strand with the base sequence A, C, G, C joins a second strand with the complementary base sequence T, G, C, G (see the upper region of DNA in figure 4.10). The two strands run in opposite orientations, which is why the bases appear upside down in the illustrations.

The DNA molecule twists to form a double helix. A molecule of DNA is typically millions of base pairs long. The great length of DNA molecules may seem quite a challenge to copy, or replicate, when a cell divides, but a contingent of enzymes accurately and rapidly carries out this process.

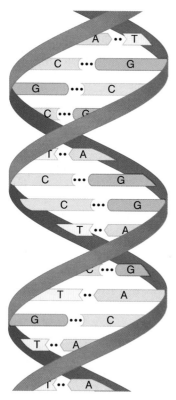

Figure 4.10

DNA structure. The molecular "ladder" of a double-stranded DNA molecule twists into a double helix. The ladder's "rungs" consist of complementary base pairs held together by hydrogen bonds—A with T (or T with A) and G with C (or C with G).

> Because the sequence of DNA nucleotides in each person's cells is unique (except for identical twins), a person can be matched to a sample of cells by analyzing the cells' DNA. This technique is called DNA profiling. DNA profiling compares thirteen sites in the human genome that vary greatly in populations, rather than entire sequenced genomes. It was used extensively to identify victims of the World Trade Center collapse on September 11, 2001. DNA from human remains was matched to DNA from blood relatives and to skin and hair cells from victims' toothbrushes and hairbrushes. DNA profiling was less useful in the Indian Ocean tsunami of 2004 and hurricane Katrina in 2005 because most human remains were washed away.

DNA Replication

DNA molecules undergo **replication** (rep"lĭ-ka'shun) during interphase of the cell cycle. In this way, each newly formed cell receives a copy of the pre-existing cell's genetic information so that the new cell can synthesize the proteins necessary to maintain life functions, build and repair cell parts, and perform other metabolic functions.

As replication begins, hydrogen bonds between complementary base pairs of the double strands in each DNA molecule break (fig. 4.11). Then the double-stranded structure pulls apart and unwinds, exposing the nitrogenous bases of its nucleotides. An enzyme called DNA polymerase brings in new DNA nucleotides, forming complementary pairs with the exposed bases. Other enzymes knit together the sugar-phosphate backbone.

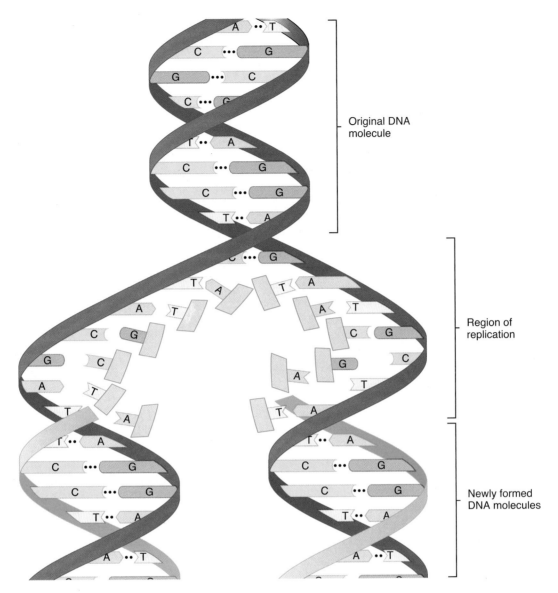

Figure 4.11

DNA replication forms two double helices from one. When a DNA molecule replicates, its original strands separate locally. A new strand of complementary nucleotides forms along each original strand.

In this way, a new strand of complementary nucleotides forms along each of the old strands. This replication produces two complete DNA molecules, each with one old strand of the original molecule and one new strand. These two DNA molecules become incorporated into replicated copies of a chromosome and separate during mitosis so that one passes to each of the newly forming cells. The Topic of Interest (p. 86) describes errors in DNA replication that cause sequence changes, or mutations.

Check Your Recall

16. Why must DNA molecules be replicated?

17. How is replication accomplished?

4.7 PROTEIN SYNTHESIS

DNA provides the genetic instructions that a cell needs to synthesize proteins. Manufacturing proteins is a multistep, enzyme-catalyzed process.

The Genetic Code—Instructions for Making Proteins

Cells can synthesize specific proteins because the sequence of nucleotide bases in the DNA of genes specifies a particular sequence of amino acid building blocks of a protein molecule. This correspondence of gene and protein building block sequence is called the **genetic code.**

Topic of Interest

Mutations

It is easy to make an error when typing a paragraph consisting of several hundred letters. DNA replication, which is similar to copying such a paragraph, is also error-prone. A newly replicated gene may have too many or too few bases, or an "A" where the complementary base should be a "C." Fortunately, cells have several mechanisms that scan newly replicated DNA, detect mutations and correct them. When this DNA damage response (repair) system fails, health may suffer. Mutations may occur spontaneously or may be induced by agents called mutagens, such as certain toxic chemicals or ionizing radiation.

Thousands of inherited illnesses result from mutations. Certain genetic tests can detect a particular mutation before symptoms of the associated illness begin. This is possible because the mutated gene is present in an individual from the time of conception. For example, a genetic test may identify the mutation that causes the neurological disorder Huntington disease in an eighteen-year-old, even though the symptoms—personality changes and uncontrollable, constant movements—probably will not appear for another two decades or longer. Such predictive tests are also available for inherited forms of breast cancer. Predictive testing is controversial, particularly when the illness is not treatable.

Not all mutations are harmful. About 1% of the individuals in European populations have a mutation that makes HIV infection impossible. The gene that is mutant normally encodes a protein to which the virus must bind in order to enter immune system cells. Without this protein, the virus cannot bind to and enter human cells. In Asian and African populations, this mutation is very rare.

Each of the twenty types of amino acids is represented in a DNA molecule by a particular sequence of three nucleotides. The DNA sequence G, G, T represents one type of amino acid; G, C, A represents another; and T, T, A another. Other nucleotide sequences encode the instructions for beginning or ending the synthesis of a protein molecule. Thus, the sequence of nucleotides in a DNA molecule denotes the order of amino acids of a protein molecule, as well as where to start or stop that protein's synthesis.

Transcription

Because DNA molecules are trapped within a cell's nucleus and protein synthesis occurs in the cytoplasm, genetic information must be carried from the nucleus into the cytoplasm. Copying of the information in DNA into RNA (ribonucleic acid), which can exit the nucleus, is called **transcription.** This process both allows the genetic information to be maintained in the nucleus for transmission when the cell divides and mobilizes that information to be accessed in the cytoplasm to synthesize proteins. **Messenger RNA (mRNA)** is the type of RNA that carries a gene's message out of the nucleus. Other types of RNA help to build proteins.

RNA (ribonucleic acid) molecules differ from DNA molecules in several ways (Table 4.1). RNA molecules are single-stranded, and their nucleotides include the sugar ribose rather than deoxyribose. Like DNA, each RNA nucleotide includes one of four nitrogenous bases. However, whereas adenine, cytosine, and guanine nucleotides are in both DNA and RNA, thymine nucleotides are only in DNA. In place of thymine nucleotides, RNA molecules have *uracil* (U) nucleotides.

The enzyme RNA polymerase synthesizes mRNA following the rules of complementary base pairing. For example, the DNA sequence A, T, G, C, G, specifies the complementary mRNA bases U, A, C, G, C (fig. 4.12). Specific DNA sequences outside the actual genes signal RNA as to which of the two DNA strands contains the information to build a protein. RNA polymerase also recognizes sequences in the DNA that indicate where the gene

Table 4.1	A Comparison of DNA and RNA Molecules	
	DNA	**RNA**
Main location	Part of chromosomes in nucleus	In the cytoplasm
5-carbon sugar	Deoxyribose	Ribose
Basic molecular structure	Double-stranded	Single-stranded
Nitrogenous bases included	Adenine, thymine, cytosine, guanine	Adenine, uracil, cytosine, guanine
Major functions	Contains genetic code for protein synthesis; replicates prior to cell division	mRNA carries transcribed DNA information to cytoplasm and acts as template for synthesis of protein molecules; tRNA carries amino acids to mRNA

begins, where it stops, and the correct direction to read the DNA, just like a sentence. When the RNA polymerase reaches the end of the gene, it releases the newly formed mRNA. Transcription is complete.

Translation

Each amino acid in a protein is specified by three contiguous bases in the DNA sequence. Those amino acids, in the proper order, are represented by a series of three-base sequences, called **codons** (ko′donz), in mRNA (table 4.2). In addition to the mRNA codons that specify amino acids, the sequence AUG represents the "start" of a gene, and three other mRNA base sequences indicate "stop." To guide protein synthesis, an mRNA molecule must leave the nucleus and associate with a ribosome in the cytoplasm. There the series of codons on mRNA are translated from the "language" of nucleic acids to the "language" of amino acids. This process of protein synthesis is appropriately called **translation.**

Building a protein molecule requires ample supplies of the correct amino acids in the cytoplasm and positioning them in the order specified along a strand of mRNA. A second kind of RNA molecule called **transfer RNA (tRNA)** correctly aligns amino acids, which are then linked by an enzyme to form proteins (fig. 4.13). Like mRNA, tRNA is synthesized in the nucleus and sent into the cytoplasm, where it assists in constructing a protein molecule.

Because twenty different types of amino acids form biological proteins, at least twenty different types of tRNA molecules must be available, one for each type

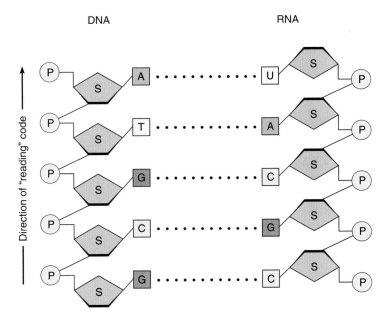

DNA   **RNA**

Direction of "reading" code

Figure 4.12

Transcription of RNA from DNA. When an RNA molecule is synthesized beside a strand of DNA, complementary nucleotides bond as in a double-stranded DNA molecule, with one exception: RNA contains uracil nucleotides (U) in place of thymine nucleotides (T).

of amino acid. Each type of tRNA has a region at one end that consists of three nucleotides that complementary base pair to a specific mRNA codon. The three nucleotides in the tRNA are called an **anticodon** (an″tĭ-ko′don). In this way, tRNA carries its amino acid to a correct position on an mRNA strand. This action occurs

Table 4.2	**Codons (mRNA Three-Base Sequences)**								
		SECOND LETTER							
		U		**C**		**A**		**G**	
U	UUU, UUC	phenylalanine (phe)	UCU, UCC, UCA, UCG	serine (ser)	UAU, UAC	tyrosine (tyr)	UGU, UGC	cysteine (cys)	U, C
	UUA, UUG	leucine (leu)			UAA	STOP	UGA	STOP	A
					UAG	STOP	UGG	tryptophan (trp)	G
C	CUU, CUC, CUA, CUG	leucine (leu)	CCU, CCC, CCA, CCG	proline (pro)	CAU, CAC	histidine (his)	CGU, CGC, CGA, CGG	arginine (arg)	U, C
					CAA, CAG	glutamine (gln)			A, G
A	AUU, AUC, AUA	isoleucine (ile)	ACU, ACC, ACA, ACG	threonine (thr)	AAU, AAC	asparagine (asn)	AGU, AGC	serine (ser)	U, C
	AUG	START methionine (met)			AAA, AAG	lysine (lys)	AGA, AGG	arginine (arg)	A, G
G	GUU, GUC, GUA, GUG	valine (val)	GCU, GCC, GCA, GCG	alanine (ala)	GAU, GAC	aspartic acid (asp)	GGU, GGC, GGA, GGG	glycine (gly)	U, C
					GAA, GAG	glutamic acid (glu)			A, G

FIRST LETTER   **THIRD LETTER**

on a ribosome (see fig. 4.13). (There are 64 possible types of tRNA anticodons, because there are this many possible triplets. Some amino acids may bind to more than one type of tRNA.)

Shortly after protein synthesis begins, a ribosome binds an mRNA molecule. A tRNA molecule with the complementary anticodon holds its amino acid and binds the first mRNA codon. A second tRNA then binds the next codon, bringing its amino acid to an adjacent site on the ribosome. Then a peptide bond forms between the two amino acids, beginning a chain. The first tRNA molecule is released from its amino acid and is recycled to the cytoplasm (fig. 4.14). This process repeats as the ribosome moves along the mRNA molecule, the amino acids delivered by the tRNA molecules added one at a time to the developing protein. Enzymes control protein synthesis.

As the protein molecule forms, it folds into its unique conformation and is then released to become a separate functional molecule. Correct protein folding

is essential to health. In cells, misfolded proteins are threaded through spool-shaped structures called *proteasomes*. Here, they are either refolded into the functional conformation, or destroyed if they are too abnormal. Table 4.3 summarizes protein synthesis.

A gene that is transcribed and translated into a protein is said to be *expressed*. The types and amounts of proteins in a cell, which can change with changing conditions, largely determine the function a cell performs in the body. Gene expression is the basis for cell differentiation, described in chapter 3 (pp. 70–72). The Topic of Interest (pp. 90–91) considers the roles that analyzing gene expression will play in health care.

Check Your Recall

18. What is the function of DNA?

19. How is information carried from the nucleus to the cytoplasm?

20. How are protein molecules synthesized?

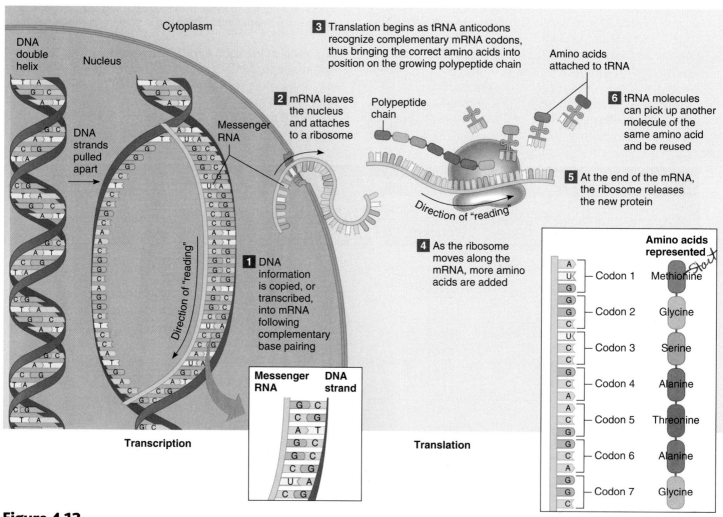

Figure 4.13

The steps of protein synthesis. DNA information is transcribed into mRNA, which in turn is translated into a sequence of amino acids. The inset shows some examples of the correspondence between mRNA codons and the specific amino acids that they encode.

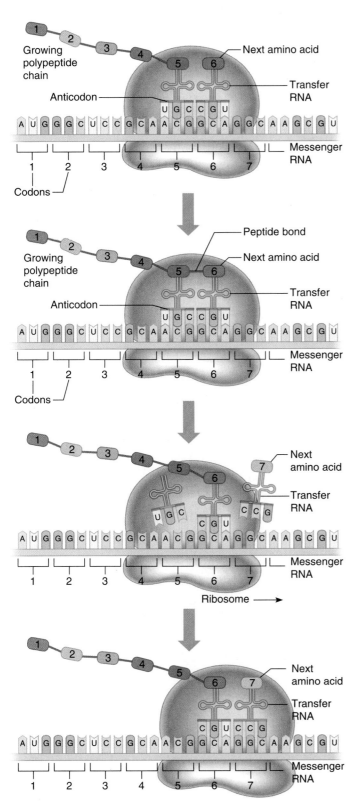

Figure 4.14

A closer look at protein synthesis. Molecules of transfer RNA (tRNA) attach to and carry specific amino acids, aligning them in the sequence determined by the codons of mRNA. These amino acids, connected by peptide bonds, form a polypeptide chain of a protein molecule.

Table 4.3	Protein Synthesis
Transcription (occurs in the nucleus)	**Translation (occurs in the cytoplasm)**
1. RNA polymerase associates with the base sequence of one strand of a gene.	1. A ribosome binds to the mRNA molecule near the codon at the beginning of the messenger strand.
2. Other enzymes unwind the DNA molecule, exposing a portion of the gene.	2. A tRNA molecule with the complementary anticodon associates with the ribosome, and the amino acid it carries becomes part of the chain.
3. RNA polymerase moves along the exposed gene and polymerizes an mRNA molecule, whose nucleotides are complementary to those of the gene strand.	3. This process repeats for each codon in the mRNA sequence as the ribosome moves along the mRNA's length.
4. When the RNA polymerase reaches the end of the gene, the newly formed mRNA molecule is released.	4. Enzymes associated with the ribosome catalyze peptide bonds, forming a chain of amino acids.
5. The mRNA molecule passes through a pore in the nuclear envelope and enters the cytoplasm.	5. As the chain of amino acids grows, it folds into the unique shape of a functional protein molecule.
	6. The completed protein molecule is released. The mRNA molecule, ribosome, and tRNA molecules can function repeatedly to synthesize other protein molecules.

Clinical Connection

Sperm cells that cannot swim cannot fertilize an egg cell. One cause of such nonmotile sperm is the presence of white blood cells in semen. The blood cells produce toxic compounds called reactive oxygen species, which bind to sperm cell membranes. Within the membranes overlying the sperm head are some of the enzymes that function in glycolysis and cellular respiration. When the reactive oxygen species destroy these enzymes, the sperm cell cannot manufacture enough ATP to enable it to move effectively. If many sperm cells cannot move sufficiently, fertility declines. Clinical trials are testing drug candidates that exert antioxidant effects that might be used to treat this form of infertility in the male.

Beyond the Human Genome Project: Personalizing Medicine

Sequencing of the human genome was completed in 2000, refined two years later, and many researchers are still discovering and describing the functions of individual genes. At the same time, investigators are teasing other sorts of information from our genetic instructions, described here.

Profiling Gene Expression

Proteomics is the analysis of gene expression—where and when genes are accessed to produce proteins. To analyze gene expression, researchers use small squares of glass or nylon called DNA microarrays, or "chips," to immobilize many genes of interest. A DNA chip for diabetes mellitus, for example, bears gene variants that reflect how the body handles glucose transport and uptake into cells. A chip for cardiovascular disease includes thousands of genes whose protein products control blood pressure, blood clotting, and the synthesis, transport, and metabolism of cholesterol and other lipids. In an analysis, the cell type affected in the condition is sampled, and its messenger RNA molecules are collected and copied,

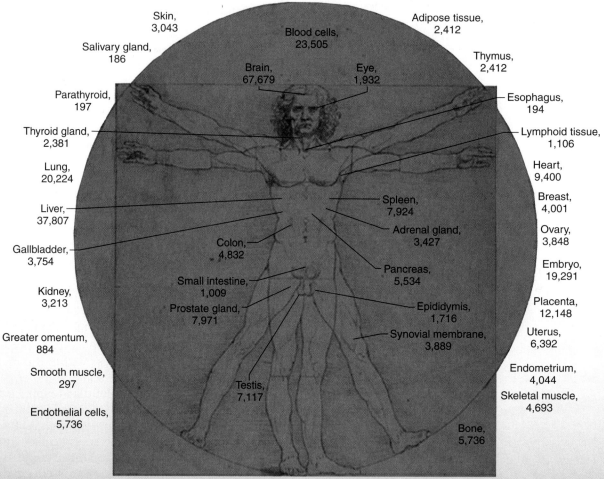

Figure 4A

One way to study the human genome is to determine which genes are expressed in different tissues. DNA microarrays (chips) are used to identify the genes that are expressed in particular tissues and organs. This illustration depicts the numbers of genes expressed in the designated body parts. Note that the liver, which carries out thousands of biochemical reactions, uses the information in 37,807 genes, compared to the 2,412 genes expressed in less active adipose (fat) tissue. Not surprisingly, the brain has the greatest number of expressed genes.

using an enzyme, into DNA molecules. The mRNAs from a specialized cell reflect the types of proteins manufactured there. The DNA copies are labeled and then added to the DNA chip. Where the sample sequence is complementary to the DNA on the chip, a label molecule lights up. The resulting pattern of fluorescent spots reveals, to a computer, which genes are expressed. This information enables researchers to view disease at a molecular level, which can suggest new types of treatments. Figure 4A summarizes information from many gene expression analyses.

DNA chips allow researchers to make compelling comparisons. A muscle cell from a person with diabetes mellitus expresses different genes than a muscle cell from a person who does not have this metabolic disorder. DNA chips are particularly valuable in diagnosing and monitoring cancer. For certain cancers, DNA chips can predict which drugs will most likely be effective, and how likely the cancer is to spread.

Identifying Our Genetic Differences

The human genome sequence indicates that we are 99.9% similar to each other, and researchers are focusing on the tiny percentage of difference. One way to do this is to identify "single nucleotide polymorphisms," or SNPs (pronounced "snips"). A SNP is a single base site in the DNA sequence that differs in more than 1% of a population (fig. 4B). ("Polymorphism" means "many forms.") A mutation may also affect a single DNA base, but it is much rarer than a SNP. The human genome is riddled with millions of SNPs, about one for every 1,000 DNA bases.

A SNP by itself can be helpful or harmful—or, in most instances, have no effect at all. But the patterns generated by many SNPs are valuable for their predictive power, because different SNP groupings on a section of chromosome, termed *haplotypes,* may correlate very strongly with different disorders and therefore be used to estimate risk.

To guide clinical decision making, an *association study* examines SNP combinations in populations to confirm if they are found almost exclusively in people with a particular disorder. Then, the correlations between certain SNP patterns and elevated disease risks can be applied to refine and speed diagnoses and prognoses and even predict an individual's likely response to a particular drug. SNP tests to identify haplotypes may become a standard part of medical practice, beginning with predicting disease susceptibilities at birth or even before.

Copy Number Variants

Mutations and SNPs are ways that people can differ in their DNA sequences. Another way is by the number of repeats of particular sequences, which is called a *copy number variant* (fig. 4C). A repeated sequence may range from a few bases to millions, and copies may lie next to each other on a chromosome, far apart on a chromosome, or part of different chromosomes. Sequencing of the human genome did not reveal copy number variants, but these may be very important to health.

Susceptibility to HIV infection is an example of a trait that is affected by the number of copies of a specific gene. The gene *CCL3L1* encodes a protein that binds to the receptor protein that HIV uses to enter a human cell. If the *CCL3L1* protein blocks the receptor, HIV can't get in and the person does not become infected. Population-level studies show that people who have the fastest-progressing HIV infections have the *fewest* copies of the *CCL3L1* gene, and people not infected with HIV tend to have the *most* copies of the *CCL3L1* gene.

We still have much to learn about the significance of SNPs, gene expression differences, and copy number variants. What is becoming clear, though, is that there is a range of what we can call a "normal" human genome.

A C C T C T A T C T C A A C G C G   96% of population

A C C T C T A T A T C A A C G C G   4% of population = SNP
(a)

A C C T C T A T C T C A A C G C G A T T C G C G T T A T A C G C T G A
= haplotype 1

A C C T C T A T A T C A A C G C T A T T C G C G T T A T A C G A T G A
= haplotype 2 (3 SNPs)
(b)

Figure 4B

SNPs distinguish individuals. A SNP is a site in the genome that differs in more than 1% of a population (a). A haplotype is a section of a chromosome that includes more than one SNP (b).

Moe   A C C T C T A T C T C A A C G C

Larry   A C C T C T A T C A T C A T C T C A A C G C

Curly   A C C T C T A T C A T C A T C A T C T C A A C G C

Figure 4C

Copy number differences are very common among human genomes. In this example, the 3-base sequence A T C varies in copy number on this hypothetical section of chromosome for these three individuals. In actuality, such repeated sequences may be millions of bases long. Their significance is not well understood, but they are one way to distinguish individuals.

SUMMARY OUTLINE

4.1 Introduction (p. 77)

A cell continuously carries on metabolic reactions.

4.2 Metabolic Reactions (p. 77)

1. Anabolism
   a. Anabolism builds large molecules from smaller molecules.
   b. In dehydration synthesis, water forms, and smaller molecules join by sharing atoms.
   c. Carbohydrates are synthesized from monosaccharides, fats are synthesized from glycerol and fatty acids, and proteins are synthesized from amino acids.
2. Catabolism
   a. Catabolism breaks down larger molecules into smaller ones.
   b. In hydrolysis, a water molecule is split as an enzyme breaks the bond between two portions of a molecule.
   c. Hydrolysis breaks down carbohydrates into monosaccharides, fats into glycerol and fatty acids, proteins into amino acids, and nucleic acids into nucleotides.

4.3 Control of Metabolic Reactions (p. 79)

Enzymes control metabolic reactions, which include many specific chemical changes.

1. Enzyme action
   a. Enzymes are molecules that promote metabolic reactions without being consumed.
   b. An enzyme acts upon a specific substrate.
   c. The shape of an enzyme molecule fits the shape of its substrate molecule.
   d. When an enzyme combines with its substrate, the substrate changes, lowering the energy necessary for a reaction to proceed. A product forms, and the enzyme is released in its original form.
   e. The speed of an enzyme-controlled reaction depends partly upon the number of enzyme and substrate molecules present and the enzyme's efficiency.
2. Factors that alter enzymes
   a. Almost all enzymes are proteins. Harsh conditions cause them to lose their shape, or denature.
   b. Factors that may denature enzymes include heat, radiation, electricity, certain chemicals, and extreme pH values.

4.4 Energy for Metabolic Reactions (p. 80)

Energy is the capacity to do work. Common forms of energy include heat, light, sound, electrical energy, mechanical energy, and chemical energy.

1. Release of chemical energy
   a. Most metabolic processes use chemical energy released when molecular bonds break.
   b. The energy released from glucose breakdown during cellular respiration drives the reactions of cellular metabolism.
2. Cellular respiration
   a. ATP
      (1) Energy is captured in the bond of the terminal phosphate of each ATP molecule.
      (2) When a cell requires energy, the terminal phosphate bond of an ATP molecule breaks, releasing stored energy.
      (3) An ATP molecule that loses its terminal phosphate becomes an ADP molecule.
      (4) An ADP molecule that captures energy and a phosphate becomes ATP.
   b. Glycolysis
      (1) The first phase of glucose decomposition does not require oxygen.
      (2) Some of the energy released is transferred to ATP molecules.
   c. Aerobic respiration
      (1) The second phase of glucose decomposition requires oxygen.
      (2) Considerably more ATP molecules form during this phase than during the anaerobic phase.
      (3) The final products of glucose breakdown are carbon dioxide, water, and energy (ATP and heat).

4.5 Metabolic Pathways (p. 82)

Metabolic processes consist of chemical reactions that occur in a certain sequence. A sequence of enzyme-controlled reactions constitutes a metabolic pathway.

1. Rate-limiting enzymes present in limited quantities determine the rates of metabolic pathways.
2. Rate-limiting enzymes become saturated when substrate concentrations increase above a certain level.

4.6 DNA (Deoxyribonucleic Acid) (p. 83)

DNA molecules contain information that instructs a cell how to synthesize enzymes and other proteins.

1. Genetic information
   a. Inherited traits result from DNA information passed from parents to offspring.
   b. A gene is a portion of a DNA molecule that contains the genetic information for making a particular protein.
2. DNA molecules
   a. A DNA molecule consists of two strands of nucleotides twisted into a double helix.
   b. The nucleotides of a DNA strand are in a particular sequence.
   c. The nucleotides of each strand pair with those of the other strand in a complementary fashion.
3. DNA replication
   a. When a cell divides, each new cell requires a copy of the older cell's genetic information.
   b. DNA molecules replicate during interphase of the cell cycle.
   c. Each new DNA molecule contains one old strand and one new strand.

4.7 Protein Synthesis (p. 85)

Genes provide instructions for making proteins, which are involved directly in many aspects of cell function.

1. The genetic code—Instructions for making proteins
   a. The sequence of nucleotides in a DNA molecule encodes the sequence of amino acids in a protein molecule.
   b. RNA molecules transfer genetic information from the nucleus to the cytoplasm.
   c. Transcription
      (1) RNA molecules are usually single-stranded; they contain ribose instead of deoxyribose and uracil nucleotides in place of thymine nucleotides.
      (2) Messenger RNA (mRNA) molecules consist of nucleotide sequences that are complementary to those of exposed strands of DNA.
      (3) Messenger RNA molecules associate with ribosomes and provide patterns for the synthesis of protein molecules.
   d. Translation
      (1) A ribosome binds to an mRNA molecule.

(2) Molecules of tRNA position amino acids along a strand of mRNA.

(3) Amino acids released from the tRNA molecules join and form a protein molecule that folds into a unique shape.

CHAPTER ASSESSMENTS

4.1 Introduction

1. Explain the relationship between genes and cellular metabolism. (p. 77)

2. Explain why enzymes are important in the body. (p. 77)

4.2 Metabolic Reactions

3. Distinguish between catabolism and anabolism. (p. 77)

4. Distinguish between dehydration synthesis and hydrolysis. (p. 77)

4.3 Control of Metabolic Reactions

5. Describe how an enzyme interacts with its substrate. (p. 79)

6. Define *active site*. (p. 79)

7. The process of changing the shape of an enzyme to the point where it loses function is called _____. (p. 80)
   a. active site
   b. substrate
   c. product
   d. denaturation
   e. conformation

8. Define *cofactor*. (p. 80)

4.4 Energy for Metabolic Reactions

9. Explain how the oxidation of molecules inside cells differs from the burning of substances outside cells. (p. 80)

10. Explain the importance of ATP, and the relationship of ATP to ADP. (p. 80)

11. Distinguish between anaerobic and aerobic phases of cellular respiration. (p. 82)

12. Match the parts of cellular respiration on the left to the associated activities on the right. (p. 82)
    (1) electron transport chain
    (2) glycolysis
    (3) citric acid cycle

    A. Glucose molecules are broken down into pyruvic acid

    B. Carrier molecules and enzymes extract energy and store it as ATP, releasing water and heat

    C. Pyruvic acid molecules enter mitochondria, where CO_2 and high-energy electrons are released

13. Identify the final acceptor of the electrons released in the reactions of cellular respiration. (p. 82)

4.5 Metabolic Pathways

14. Define *metabolic pathway*. (p. 82)

15. Explain how one enzyme can control the rate of a metabolic pathway. (p. 82)

16. Identify the cellular respiration pathway where glucose, fats, and proteins commonly enter. (p. 82)

4.6 DNA (Deoxyribonucleic Acid)

17. Identify the part of a DNA molecule that encodes information. (p. 84)

18. Distinguish between a gene and a genome. (p. 84)

19. DNA information provides instructions for the cell to _____. (p. 84)
    a. manufacture carbohydrate molecules
    b. extract energy
    c. manufacture RNA from amino acids
    d. synthesize protein molecules.

20. Explain why DNA replication is essential. (p. 84)

21. Describe the events of DNA replication. (p. 84)

4.7 Protein Synthesis

22. If a strand of DNA has the sequence A T G C G A T C C G C then the sequence of an mRNA molecule transcribed from it is _____. (p. 86)

23. Distinguish between transcription and translation. (p. 86)

24. Describe the function of a ribosome in protein synthesis. (p. 87)

25. Calculate the number of amino acids that a DNA sequence of 27 nucleotides encodes. (p. 87)

26. Define *gene expression*. (p. 88)

INTEGRATIVE ASSESSMENTS/ CRITICAL THINKING

OUTCOME 4.2
1. How can the same biochemical be both a reactant (starting material) and a product?

OUTCOME 4.4
2. After finishing a grueling marathon, a runner exclaims, "Whew, I think I've used up all my ATP!" Could this be possible?

OUTCOMES 4.3, 4.4
3. What effect might changes in the pH of body fluids or body temperature that accompany illness have on enzymes?

OUTCOME 4.5
4. Explain how proteins assist in DNA replication.

OUTCOME 4.6
5. A mutation that deletes one or two DNA nucleotides changes gene function more drastically than a substitution of one nucleotide for another type, or the removal or addition of three contiguous DNA nucleotides. Why?

OUTCOMES 4.4, 4.6
6. People who have mutations in genes that encode proteins that participate in ATP formation in mitochondria are never seen. Why might this be?

7. Several specific types of chemical bonds are discussed in this chapter. Describe each of the following, and explain why each is important.
   a. High-energy phosphate bond
   b. Peptide bond
   c. The bond between an mRNA codon and a tRNA anticodon

WEB CONNECTIONS

Visit the text website at **aris.mhhe.com** for additional quizzes, interactive learning exercises, and more.

AP R

Anatomy & Physiology | REVEALED includes cadaver photos that allow you to peel away layers of the human body to reveal structures beneath the surface. This program also includes animations, radiologic imaging, audio pronunciations, and practice quizzing. To learn more visit **www.aprevealed.com**.

5

Tissues

(a)

(b)

Henrietta Lacks (a) checked into Johns Hopkins Hospital in 1951, where she was treated for cervical cancer. A sample of her cells went on to become a standard cell line used in laboratories all over the world (b). Her husband was told, after her death later that year, only that the cells would be used in research. The family did not discover how widespread their relative's unwitting legacy was until 1976.

DONATING TISSUE FOR RESEARCH. If you donate cells or tissues to research, and the material is used to develop a commercially available test or treatment, can you profit from the donation? According to legal precedent in the United States, probably not.

Perhaps the most famous cell donor was Henrietta Lacks, whose cervical cancer cells, sampled shortly before her death in 1951 and dubbed HeLa in her honor, went on to become laboratory standards, without her consent. More recent were three cases in which tissue donors requested that they be compensated for use of their cells or tissues, but lost:

- In a 1990 case, John Moore sued the Regents of the University of California because part of his spleen, removed to treat leukemia years earlier, had been used to culture a line of cells that was patented by his physician and a researcher. The cultured cells were used to derive valuable substances. Moore lost the case.

- William Catalona was a researcher at Washington University in St. Louis. Here he developed the prostate-specific antigen test that assesses prostate cancer risk, with the help of thousands of patients who donated their tissue. When Dr. Catalona switched universities, he asked his patients to request that their tissues be transferred too. Washington University took identifying names off of the samples, which effectively negated the patients' rights to request transfer of their samples—which six thousand of them had done. In 2006 a judge ruled in Washington University's favor, fearing that allowing patients to move their samples would adversely affect cell and tissue banks.

- Debbie and Dan Greenberg organized parents of children with Canavan disease, a rare inherited disorder, to donate their children's blood, urine, and autopsy tissue to a researcher at Miami

Children's Hospital, to help identify the causative gene. Once the gene was discovered, a diagnostic and carrier test were eventually developed and patented, and soon the families were paying for tests that their children's tissues had helped develop. The case was settled, the genetic tests remain commercialized, and the gene itself is available for research use.

These cases have spurred tissue banks to take measures to ensure that tissue donors are aware of their rights. The Coriell Cell Repository, for example, is supported by the National Institutes of Health and includes the following wording in its informed consent form: "Submission of my sample to the Repository may give scientists valuable research material that can help them to develop new diagnostic tests, new treatments, and new ways to prevent disease."

Learning Outcomes
After studying this chapter, you should be able to do the following:

5.1 Introduction

1. List the four major tissue types, and tell where each is located in the body. (p. 95)

5.2 Epithelial Tissues

2. Describe the general characteristics and functions of epithelial tissues. (p. 95)

3. Name the types of epithelium, and for each type, identify an organ in which that type is found. (p. 96)

4. Explain how glands are classified. (p. 101)

5.3 Connective Tissues

5. Compare and contrast the general cellular components, structures, fibers, and extracellular matrix (where applicable) in each type of connective tissue. (p. 102)

6. Explain the major functions of each type of connective tissue. (p. 105)

5.4 Types of Membranes

7. Distinguish among the four major types of membranes. (p. 109)

5.5 Muscle Tissues

8. Distinguish among the three types of muscle tissues. (p. 110)

5.6 Nervous Tissues

9. Describe the general characteristics and functions of nervous tissues. (p. 111)

Aids to Understanding Words

(Appendix A on page 567 has a complete list of Aids to Understanding Words.)

adip- [fat] *adip*ose tissue: Tissue that stores fat.

chondr- [cartilage] *chondr*ocyte: Cartilage cell.

-cyt [cell] osteo*cyte*: Bone cell.

epi- [upon] *epi*thelial tissue: Tissue that covers all free body surfaces.

-glia [glue] neuro*glia*l cells: Cells that support neurons; part of nervous tissue.

inter- [between] *inter*calated disc: Band between adjacent cardiac muscle cells.

macr- [large] *macr*ophage: Large phagocytic cell.

os- [bone] *os*seous tissue: Bone tissue.

pseud- [false] *pseud*ostratified epithelium: Tissue with cells that appear to be in layers, but are not.

squam- [scale] *squam*ous epithelium: Tissue with flattened or scalelike cells.

strat- [layer] *strat*ified epithelium: Tissue with cells that are in layers.

5.1 INTRODUCTION

Cells, the basic units of structure and function within the human organism, are organized into groups and layers called **tissues** (tish'uz). Each type of tissue is composed of similar cells specialized to carry on a particular function.

The tissues of the human body are of four major types: *epithelial, connective, muscle,* and *nervous.* Epithelial tissues form protective coverings and function in secretion and absorption. Connective tissues support soft body parts and bind structures together. Muscle tissues produce body movements, and nervous tissues conduct impulses that help control and coordinate body activities.

Table 5.1 compares the four major tissue types. Throughout this chapter, simplified line drawings (for example, fig. 5.1*a*) are included with each micrograph (for example, fig. 5.1*b*) to emphasize the distinguishing characteristics of the specific tissue, as well as a locator icon (an example of where in the body that particular tissue may be found).

Check Your Recall

1. What is a tissue?
2. List the four major types of tissues.

5.2 EPITHELIAL TISSUES

General Characteristics

Epithelial (ep"ĭ-the'le-al) **tissues** are found throughout the body. Since epithelium covers organs, forms the inner lining of body cavities, and lines hollow organs, it always has a *free (apical) surface*—one that is exposed to the outside or to an open space internally. The underside of this tissue is anchored to connective tissue by a thin, nonliving layer, called the **basement membrane.**

As a rule, epithelial tissues lack blood vessels. However, nutrients diffuse to epithelium from underlying connective tissues, which have abundant blood vessels.

Epithelial cells readily divide. As a result, injuries heal rapidly as new cells replace lost or damaged ones. Skin cells and cells that line the stomach and intestines are continually damaged and replaced.

Epithelial cells are tightly packed. Consequently, these cells form effective protective barriers in such structures as the outer layer of the skin and the lining of the mouth. Other epithelial functions include secretion, absorption, and excretion.

Epithelial tissues are classified according to shape and number of layers of cells. Epithelial tissues that are composed of thin, flattened cells are *squamous;* those

Table 5.1	Types of Tissue		
Type	**Function**	**Location**	**Distinguishing Characteristics**
Epithelial	Protection, secretion, absorption, excretion	Cover body surfaces, cover and line internal organs, compose glands	Lack blood vessels, readily divide; cells are tightly packed
Connective	Bind, support, protect, fill spaces, store fat, produce blood cells	Widely distributed throughout body	Mostly have good blood supply; cells are farther apart than epithelial cells
Muscle	Movement	Attached to bones, in the walls of hollow internal organs, heart	Able to contract in response to specific stimuli
Nervous	Transmit impulses for coordination, regulation, integration, and sensory reception	Brain, spinal cord, nerves	Cells communicate with each other and with other body parts

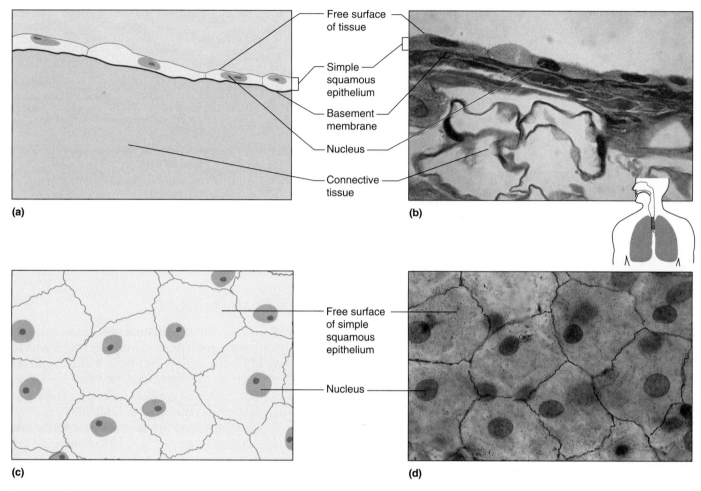

Figure 5.1
Simple squamous epithelium consists of a single layer of tightly packed, flattened cells (670×). (*a*) and (*b*) side view, (*c*) and (*d*) surface view. In one example, it lines the air sacs of the lungs.

with cube-shaped cells are *cuboidal;* and those with tall, elongated cells are *columnar;* those with single layers of cells are *simple;* those with two or more layers of cells are *stratified.* In the following descriptions, note that the free surfaces of epithelial cells are modified in ways that reflect their specialized functions.

Check Your Recall

3. List the general characteristics of epithelial tissues.
4. Describe the classification of epithelium in terms of shape and number of cell layers.

Simple Squamous Epithelium

Simple squamous (skwa′mus) **epithelium** consists of a single layer of thin, flattened cells. These cells fit tightly together, somewhat like floor tiles, and their nuclei are usually broad and thin (fig. 5.1).

Substances pass rather easily through simple squamous epithelium, which is common at sites of diffusion and filtration. For instance, simple squamous epithelium lines the air sacs (alveoli) of the lungs where oxygen and carbon dioxide are exchanged. It also forms the walls of capillaries, lines the insides of blood and lymph vessels, and covers the membranes that line body cavities. However, because it is so thin and delicate, simple squamous epithelium is easily damaged.

Simple Cuboidal Epithelium

Simple cuboidal epithelium consists of a single layer of cube-shaped cells. These cells usually have centrally located, spherical nuclei (fig. 5.2).

Simple cuboidal epithelium covers the ovaries and lines most of the kidney tubules and the ducts of certain glands, such as the salivary glands, thyroid gland, pancreas, and liver. In the kidneys, this tissue functions in secretion and absorption; in glands, it secretes glandular products.

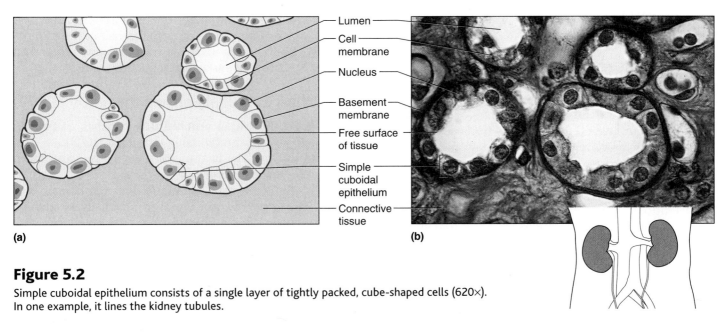

(a)   (b)

Figure 5.2

Simple cuboidal epithelium consists of a single layer of tightly packed, cube-shaped cells (620×). In one example, it lines the kidney tubules.

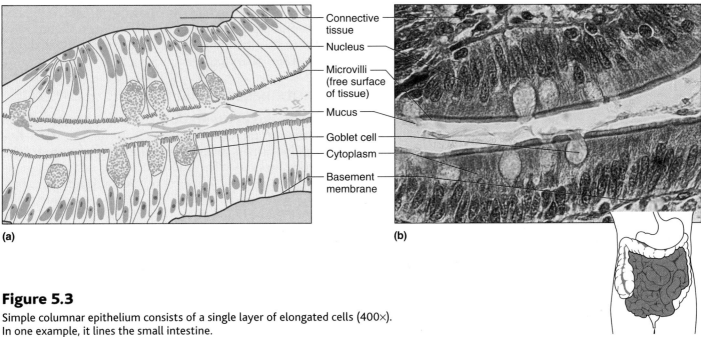

(a)   (b)

Figure 5.3

Simple columnar epithelium consists of a single layer of elongated cells (400×). In one example, it lines the small intestine.

Simple Columnar Epithelium

The cells of **simple columnar epithelium** are elongated; that is, they are longer than they are wide. This tissue is composed of a single layer of cells whose nuclei are elongated, like the shape of the cells themselves; this layer of cells is usually located at about the same level, near the basement membrane (fig. 5.3). The cells of this tissue can be ciliated or nonciliated. *Cilia* extend from the free surfaces of the cells and move constantly (see chapter 3, p. 58). In the female reproductive tubes, cilia aid in moving egg cells to the uterus.

Nonciliated simple columnar epithelium lines the uterus and most organs of the digestive tract, including the stomach and the small and large intestines. Because its cells are elongated, this tissue is thick, which enables it to protect underlying tissues. Simple columnar epithelium also secretes digestive fluids and absorbs nutrients from digested food.

Simple columnar cells, specialized for absorption, often have many tiny, cylindrical processes extending from their surfaces. These processes, called *microvilli,* increase the surface area of the cell membrane where it is exposed to substances being absorbed.

Typically, specialized, flask-shaped glandular cells are scattered among the columnar cells of simple columnar epithelium. These cells, called *goblet cells,* secrete a protective fluid, called *mucus,* onto the free surface of the tissue (see fig. 5.3).

Pseudostratified Columnar Epithelium

The cells of **pseudostratified** (soo″do-strat′ĭ-fīd) **columnar epithelium** appear stratified or layered, but they are not. A layered effect occurs because the nuclei are at two or more levels in the row of aligned cells. However, the cells, which vary in shape, all reach the basement membrane, even though some of them may not contact the free surface.

Pseudostratified columnar epithelial cells commonly have cilia, which extend from the free surfaces of the cells. Goblet cells scattered throughout this tissue secrete mucus, which the cilia sweep away (fig. 5.4).

Pseudostratified columnar epithelium lines the passages of the respiratory system. Here, the mucus-covered linings are sticky and trap dust and microorganisms that enter with the air. The cilia move the mucus and its captured particles upward and out of the airways.

Stratified Squamous Epithelium

The many cell layers of **stratified squamous epithelium** make this tissue relatively thick. Cells divide in the deeper layers, and newer cells push older ones farther outward, where they flatten (fig. 5.5). In naming stratified epithelial tissues based on shape of cells, the appearance of the top layer of cells is used.

Stratified squamous epithelium forms the outer layer of the skin (*epidermis*). As skin cells age, they accumulate a protein called *keratin* and then harden and die. This "keratinization" produces a covering of dry, tough, protective material that prevents water and other substances from escaping underlying tissues and blocks various chemicals and microorganisms from entering.

Stratified squamous epithelium also lines the oral cavity, esophagus, vagina, and anal canal. In these parts, the tissue is not keratinized; it stays soft and moist, and the cells on its free surfaces remain alive.

Stratified Cuboidal Epithelium

Stratified cuboidal epithelium consists of two or three layers of cuboidal cells that form the lining of a lumen (fig. 5.6). The layering of the cells provides more protection than the single layer affords.

Stratified cuboidal epithelium lines the larger ducts of the mammary glands, sweat glands, salivary glands, and pancreas. It also forms the lining of developing ovarian follicles and seminiferous tubules, which are parts of the female and male reproductive systems, respectively.

Stratified Columnar Epithelium

Stratified columnar epithelium consists of several layers of cells (fig. 5.7). The superficial cells are columnar, whereas the basal layers consist of cuboidal cells. Small amounts of stratified columnar epithelium are found in the male urethra and ductus deferens and in parts of the pharynx.

Transitional Epithelium

Transitional epithelium is specialized to change in response to increased tension. It forms the inner lining of the urinary bladder and lines the ureters and the superior urethra. When the wall of one of these organs contracts, the tissue consists of several layers of cuboidal cells; however, when the organ is distended, the tissue stretches, and the physical relationships among the cells change (fig. 5.8). In addition to providing an expandable lining, transitional epithelium forms a barrier that helps prevent the contents of the urinary tract from diffusing back into the internal environment.

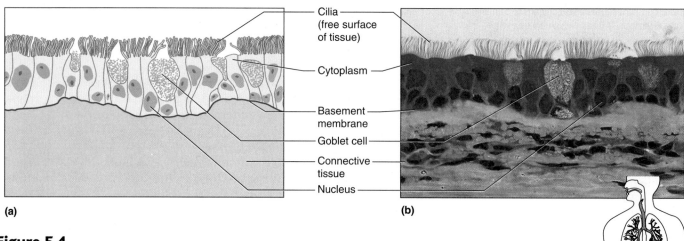

(a)   (b)

Cilia (free surface of tissue)
Cytoplasm
Basement membrane
Goblet cell
Connective tissue
Nucleus

Figure 5.4
Pseudostratified columnar epithelium appears stratified because the cell nuclei are located at different levels (255×). In one example, it lines the passages of the respiratory system.

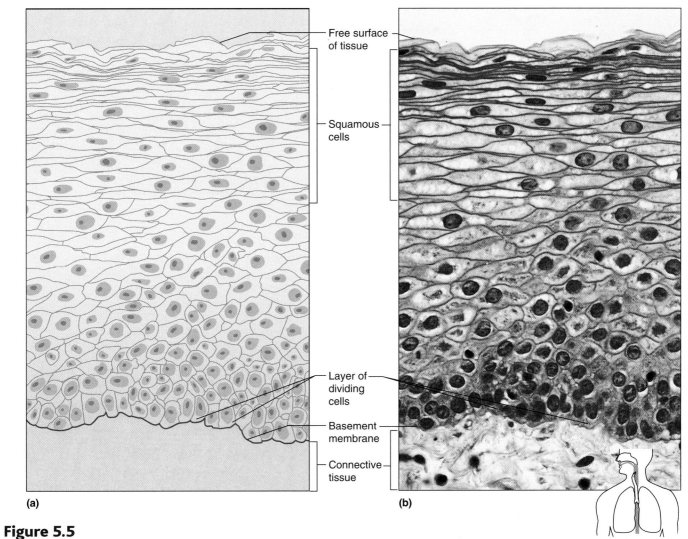

Free surface of tissue

Squamous cells

Layer of dividing cells

Basement membrane

Connective tissue

(a)

(b)

Figure 5.5

Stratified squamous epithelium consists of many layers of cells (385×).
In one example, it lines the oral cavity and esophagus.

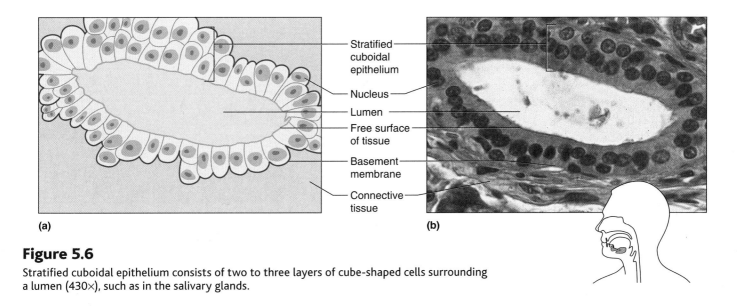

Stratified cuboidal epithelium

Nucleus

Lumen

Free surface of tissue

Basement membrane

Connective tissue

(a)

(b)

Figure 5.6

Stratified cuboidal epithelium consists of two to three layers of cube-shaped cells surrounding
a lumen (430×), such as in the salivary glands.

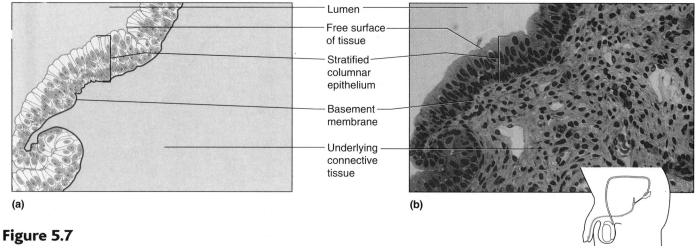

(a)   (b)

Figure 5.7

Stratified columnar epithelium consists of a superficial layer of columnar cells overlying several layers of cuboidal cells (220×). In one example, it is found in the male urethra.

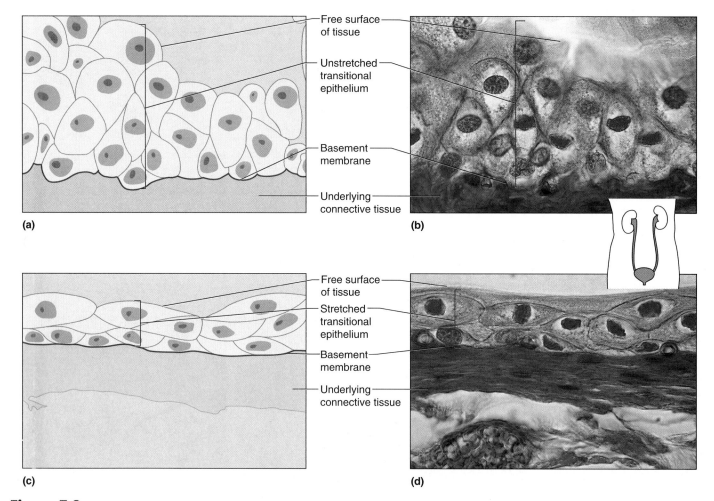

(a)   (b)

(c)   (d)

Figure 5.8

Transitional epithelium. (*a* and *b*) When the organ wall contracts, transitional epithelium is unstretched and consists of many layers (675×). (*c* and *d*) When the organ is distended, the tissue stretches and appears thinner (675×). Transitional epithelium lines the urinary bladder and the ureters and part of the urethra.

Up to 90% of all human cancers are *carcinomas,* which are growths that originate in epithelium. Most carcinomas begin on surfaces that contact the external environment, such as skin, linings of the airways in the respiratory tract, or linings of the stomach or intestine in the digestive tract. This observation suggests that the more common cancer-causing agents may not penetrate tissues very deeply.

Glandular Epithelium

Glandular epithelium is composed of cells that are specialized to produce and secrete substances into ducts or into body fluids. Such cells are usually found within columnar or cuboidal epithelium, and one or more of these cells constitute a *gland.* Glands that secrete their products into ducts that open onto surfaces, such as the skin or the lining of the digestive tract, are called **exocrine glands.** Glands that secrete their products into tissue fluid or blood are called **endocrine glands.** (Endocrine glands are discussed in chapter 11.)

Exocrine glands are classified according to the ways these glands secrete their products (fig. 5.9). Glands that release fluid by exocytosis are called **merocrine** (mer′o-krin) **glands.** Glands that lose small portions of their glandular cell bodies during secretion are called **apocrine** (ap′o-krin) **glands.** Glands that release entire cells that disintegrate to release cell secretions are called **holocrine** (ho′lo-krin) **glands.** Table 5.2 summarizes these glands and their secretions.

Most exocrine secretory cells are merocrine, and they can be further subclassified based on their secretion of serous fluid or mucus. *Serous fluid* is typically watery, and has a high concentration of enzymes. Serous cells secreting this fluid, that lubricates, are commonly associated with the visceral and parietal membranes of the thoracic and abdominopelvic cavities. The thicker fluid, *mucus,* is rich in the glycoprotein *mucin* and abundantly secreted by cells, for protection, in the inner linings of the digestive, respiratory, and reproductive systems. Mucous cells and goblet cells secrete mucus, but in different parts of the body. Table 5.3

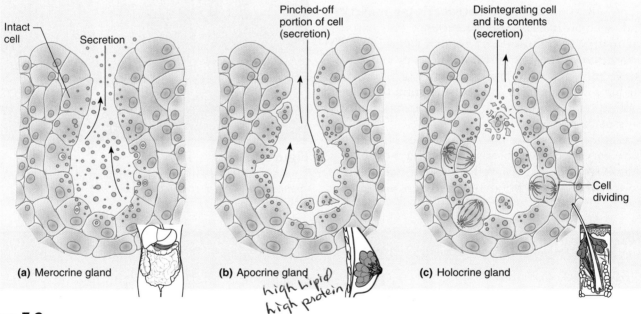

Figure 5.9

Types of exocrine glands. (*a*) Merocrine glands release secretions without losing cytoplasm. (*b*) Apocrine glands lose small portions of their cell bodies during secretion. (*c*) Holocrine glands release entire cells filled with secretory products.

Table 5.2	Types of Exocrine Glandular Secretions	
Type of Gland	**Description of Secretion**	**Example**
Merocrine glands	A fluid product released through the cell membrane by exocytosis	Salivary glands, pancreatic glands, sweat glands of the skin
Apocrine glands	Cellular product and portions of the free ends of glandular cells pinched off during secretion	Mammary glands, ceruminous glands lining the external ear canal
Holocrine glands	Disintegrated entire cells filled with secretory products	Sebaceous glands of the skin

summarizes the characteristics of the different types of epithelial tissues.

> ## Check Your Recall
>
> 5. Describe the special functions of each type of epithelium.
> 6. Distinguish between exocrine glands and endocrine glands.
> 7. Explain how exocrine glands are classified.
> 8. Distinguish between a serous cell and a mucous cell.

5.3 CONNECTIVE TISSUES

General Characteristics

Connective (kŏ-nek′tiv) **tissues** bind structures, provide support and protection, serve as frameworks, fill spaces, store fat, produce blood cells, protect against infections, and help repair tissue damage. Connective tissue cells are farther apart than epithelial cells, and they have an abundance of **extracellular matrix** (eks″trah-sel′u-lar ma′triks) between them. This extracellular matrix is composed of *protein fibers,* and a *ground substance* consisting of nonfibrous protein and other molecules, and fluid. The consistency of the extracellular matrix varies from fluid to semisolid to solid. The Topic of Interest on page 103 details the importance of this tissue component.

Most connective tissue cells can divide. These tissues have varying degrees of vascularity, but most have good blood supplies and are well nourished. Some connective tissues, such as bone and cartilage, are quite rigid. Loose connective tissue (areolar), adipose tissue, and dense connective tissue are more flexible.

Major Cell Types

Connective tissues contain a variety of cell types. Some cells are called *fixed cells* because they reside in the tissue for an extended period of time. These include fibroblasts and mast cells. Other cells, such as macrophages, are *wandering cells.* They move through and appear in tissues temporarily, usually in response to an injury or infection.

Fibroblasts (fi′bro-blastz) are the most common type of fixed cell in connective tissue. These large, star-shaped cells produce fibers by secreting proteins into the extracellular matrix of connective tissues (fig. 5.10).

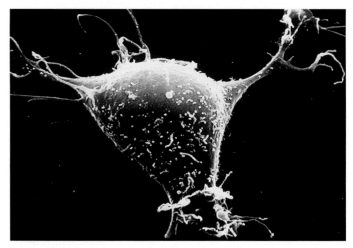

Figure 5.10
Scanning electron micrograph of a fibroblast (4,000×), the most abundant cell type of connective tissue.

Table 5.3	Epithelial Tissues	
Type	**Function**	**Location**
Simple squamous epithelium	Filtration, diffusion, osmosis; covers surface	Air sacs of the lungs, walls of capillaries, linings of blood and lymph vessels
Simple cuboidal epithelium	Secretion, absorption	Surface of ovaries, linings of kidney tubules, and linings of ducts of certain glands
Simple columnar epithelium	Absorption, secretion, protection	Linings of uterus, stomach, and intestines
Pseudostratified columnar epithelium	Protection, secretion, movement of mucus	Linings of respiratory passages
Stratified squamous epithelium	Protection	Outer layer of skin, linings of oral cavity, throat, vagina, and anal canal
Stratified cuboidal epithelium	Protection	Linings of larger ducts of mammary glands, sweat glands, salivary glands, and pancreas
Stratified columnar epithelium	Protection, secretion	Part of the male urethra and parts of the pharynx
Transitional epithelium	Distensibility, protection	Inner lining of urinary bladder and linings of ureters and part of urethra
Glandular epithelium	Secretion	Salivary glands, sweat glands, endocrine glands

A New View of the Body's Glue: The Extracellular Matrix

The traditional description of connective tissue matrix as "intercellular material" suggested that it merely fills the spaces between cells. However, when cell biologists looked beyond the abundant collagens that comprise much of the matrix, they discovered a complex and changing recipe of different molecules that modifies the tissue to suit different organs and conditions. Not only does this material outside cells—the extracellular matrix, or ECM—serve as a scaffolding to organize cells into tissues, but it relays the biochemical signals that control cell division, differentiation, repair, and migration.

The ECM has two basic components: the basement membrane that covers cell surfaces, and the rest of the material between cells, called the interstitial matrix. The basement membrane is mostly composed of tightly packed collagenous fibers from which large, cross-shaped glycoproteins called laminins extend. The laminins (and other glycoproteins such as fibronectin and tenascin) traverse the interstitial matrix and touch receptors, called integrins, on other cells. In this way, the ECM connects cells into tissues. The ECM is versatile, and includes at least twenty types of collagen and precursor versions of hormones, enzymes, growth factors, and immune system biochemicals (cytokines). These molecules are activated under certain conditions.

The components of the ECM are always changing, as its cells synthesize proteins while enzymes called proteases break down specific proteins. The balance of components is important to maintaining and repairing organ structure. Disrupt the balance, and disease can result. Here are three common examples:

Cancer

The spread of a cancerous growth uses the ability of fibroblasts to contract as they close a wound, where they are replaced with normal epithelium. Chemical signals from cancer cells cause fibroblasts to become more contractile (myofibroblasts), as well as to take on the characteristics of cancer cells. At the same time, alterations in laminins loosen the connections of the fibroblasts to surrounding cells. This abnormal flexibility enables the changed fibroblasts to migrate, and the cancer spreads. Normally, fibroblasts secrete abundant collagen (figure 5A).

Liver Fibrosis

In fibrosis, a part of all chronic liver diseases, collagen deposition increases so that the ECM exceeds its normal 3% of the organ. Healthy liver ECM sculpts a framework that supports the epithelial and vascular tissues of the organ. In response to a damaging agent such as a virus, alcohol, or a toxic drug,

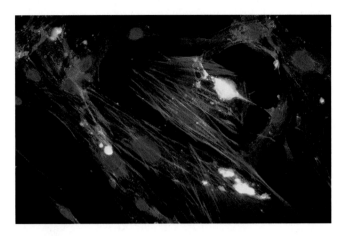

Figure 5A

The fibroblast connective tissue cells shown here have been taken from fetal skin. Fibroblasts form connective tissue by secreting extracellular matrix material such as collagen. (Immunofluorescent light micrograph, 225×.) Fibroblasts produce abundant collagens, of various types. Collagens make up more than half of the extracellular matrix in most parts of the body. The extracellular matrix is particularly important before birth, when organs form.

hepatic stellate cells secrete collagenous fibers in the areas where the epithelium and blood vessels meet. Such limited fibrosis seals off the affected area, preventing its spread. But if the process continues—if an infection is not treated or the noxious stimulus not removed—the ECM grows and eventually blocks the interaction between liver cells and the bloodstream. The liver tissue eventually hardens, a dangerous condition called *cirrhosis*.

Heart Failure and Atherosclerosis

The heart's ECM organizes cells into a three-dimensional network that coordinates their contractions into the rhythmic heartbeat necessary to pump blood. It consists of collagen, fibronectin, laminin, and elastin surrounding cardiac muscle cells and myofibroblasts, and is also in the walls of arteries. Heart failure and atherosclerosis reflect imbalances of collagen production and degradation. As in the liver, the natural response of ECM buildup is to wall off an area where circulation is blocked, but if it continues, the extra scaffolding stiffens the heart, which can ultimately lead to heart failure. In atherosclerosis, excess ECM accumulates on the interior linings of arteries, blocking blood flow. During a myocardial infarction (heart attack), collagen synthesis and deposition increase in affected and nonaffected heart parts, which is why damage can continue even after pain starts.

Macrophages (mak'ro-fājez), or histiocytes, originate as white blood cells (see chapter 14, p. 384) and are almost as numerous as fibroblasts in some connective tissues. They are specialized to carry on phagocytosis. Macrophages can move about and function as scavenger and defensive cells that clear foreign particles from tissues (fig. 5.11).

Mast cells are large and widely distributed in connective tissues. They are usually near blood vessels (fig. 5.12). Mast cells release *heparin,* which prevents blood clotting, and *histamine,* which promotes some of the

reactions associated with inflammation and allergies (see chapter 14, p. 393).

Connective Tissue Fibers

Fibroblasts produce three types of connective tissue fibers: collagenous fibers, elastic fibers, and reticular fibers. Of these, collagenous and elastic fibers are the most abundant.

Collagenous (kol-laj'ĕ-nus) **fibers** are thick threads of the protein *collagen*. They are grouped in long, parallel bundles, and are flexible but only slightly elastic. More importantly, they have great tensile strength—that is, they resist considerable pulling force. Thus, collagenous fibers are important components of body parts that hold structures together, such as **ligaments** (which connect bones to bones) and **tendons** (which connect muscles to bones).

Tissue containing abundant collagenous fibers is called *dense connective tissue*. It appears white, and for this reason, collagenous fibers are sometimes called *white fibers*.

Elastic fibers are composed of a protein called *elastin*. These thin fibers branch, forming complex networks. Elastic fibers are weaker than collagenous fibers, but they stretch easily and can resume their original lengths and shapes. Elastic fibers are common in body parts that are frequently stretched, such as the vocal cords. They are sometimes called *yellow fibers* because tissues well supplied with them appear yellowish.

Reticular fibers are very thin collagenous fibers. They are highly branched and form delicate supporting networks in a variety of tissues, including the spleen. Table 5.4 summarizes the major cells and tissue fibers of connective tissue, including their functions.

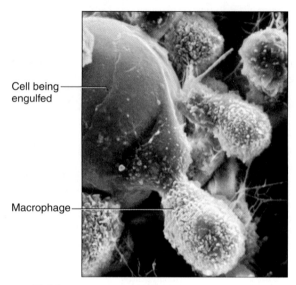

Cell being engulfed

Macrophage

Figure 5.11
Macrophages are scavenger cells common in connective tissues. This scanning electron micrograph shows a number of macrophages engulfing parts of a larger cell (3,300×).

When skin is exposed to prolonged and intense sunlight, connective tissue fibers lose elasticity, and the skin stiffens and becomes leathery. In time, the skin may sag and wrinkle. Collagen injections may temporarily smooth out wrinkles. However, collagen applied as a cream to the skin does not combat wrinkles because collagen molecules are far too large to actually penetrate the skin.

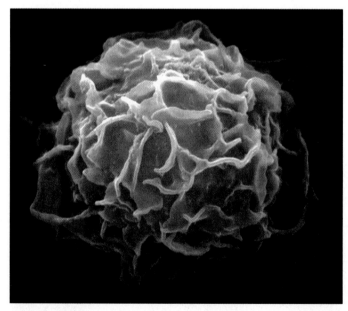

Figure 5.12
Scanning electron micrograph of a mast cell (6,600×), which releases heparin and histamine.

Table 5.4	Components of Connective Tissue
Cell Type	**Function**
Fibroblasts	Produce fibers
Macrophages	Carry on phagocytosis
Mast cells	Secrete heparin and histamine
Tissue Fibers	**Function**
Collagenous	Hold structures together with great tensile strength
Elastic	Stretch easily
Reticular	Lend delicate support

Categories of Connective Tissue

Connective tissue is classified into two categories. *Connective tissue proper* includes loose connective tissue and dense connective tissue. The *specialized connective tissues* include cartilage, bone, and blood.

Loose Connective Tissue

Loose connective tissue includes areolar tissue, adipose tissue, and reticular connective tissue. **Areolar** (ah-re′o-lar) **tissue** forms delicate, thin membranes throughout the body. The cells of this tissue, mainly fibroblasts, are located some distance apart and are separated by a gel-like extracellular matrix containing many collagenous and elastic fibers that fibroblasts secrete (fig. 5.13). Areolar tissue binds the skin to the underlying organs and fills spaces between muscles. It lies beneath most layers of epithelium, where its many blood vessels nourish nearby epithelial cells.

Adipose (ad′ĭ-pō s) **tissue,** or fat, develops when certain cells (adipocytes) store fat as droplets in their cytoplasm and enlarge (fig. 5.14). When such cells become so abundant that they crowd other cell types, they form adipose tissue. Adipose tissue lies beneath the skin, in spaces between muscles, around the kidneys, behind the eyeballs, in certain abdominal membranes, on the surface of the heart, and around certain joints. Adipose tissue cushions joints and some organs, such as the kidneys. It also insulates beneath the skin, and it stores energy in fat molecules.

The average adult has between 40 and 50 billion fat cells.

Overeating and lack of exercise can increase the size of adipose cells, leading to overweight or obesity. During periods of fasting, however, fat supplies energy, and adipocytes lose fat, shrink, and become more like fibroblasts.

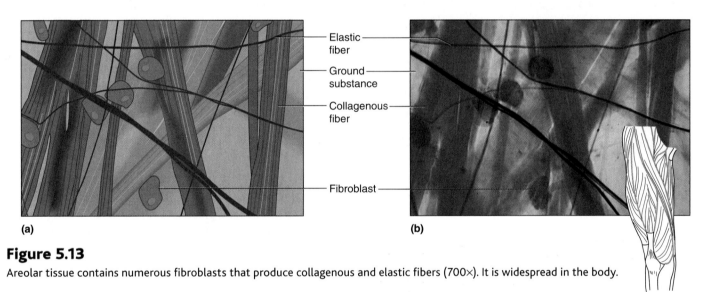

Elastic fiber
Ground substance
Collagenous fiber
Fibroblast

(a)   (b)

Figure 5.13
Areolar tissue contains numerous fibroblasts that produce collagenous and elastic fibers (700×). It is widespread in the body.

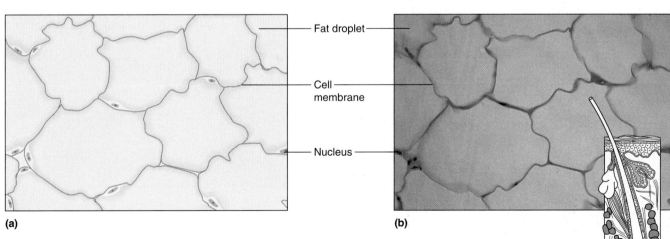

Fat droplet
Cell membrane
Nucleus

(a)   (b)

Figure 5.14
Adipose tissue cells contain large fat droplets that push the nuclei close to the cell membranes (450×). Adipose tissue beneath the skin provides insulation.

Reticular connective tissue is composed of thin, collagenous fibers in a three-dimensional network. It helps to provide the framework of certain internal organs, such as the liver and spleen.

Dense Connective Tissue

Dense connective tissue consists of many closely packed, thick, collagenous fibers and a fine network of elastic fibers. It has few cells, most of which are fibroblasts (fig. 5.15).

Collagenous fibers of dense connective tissue are very strong, enabling the tissue to withstand pulling forces. As parts of tendons and ligaments, dense connective tissue binds muscle to bone and bone to bone. This type of tissue is also in the protective white layer of the eyeball and in the deeper skin layers. The blood supply to dense connective tissue is poor, slowing tissue repair.

> ## Check Your Recall
>
> 9. What are the general characteristics of connective tissues?
> 10. What are the characteristics of collagen and elastin?
> 11. What feature distinguishes adipose tissue from other connective tissues?
> 12. Explain the difference between loose connective tissue and dense connective tissue.

Cartilage

Cartilage (kar′ti-lij) is a rigid connective tissue. It provides support, frameworks, and attachments, protects underlying tissues, and forms structural models for many developing bones.

Cartilage extracellular matrix is abundant and is largely composed of collagenous fibers embedded in a gel-like ground substance. Cartilage cells, or **chondrocytes** (kon′dro-sītz), occupy small chambers called *lacunae* and lie completely within the extracellular matrix (fig. 5.16).

A cartilaginous structure is enclosed in a covering of connective tissue called the *perichondrium*. The perichondrium contains blood vessels that provide cartilage cells with nutrients by diffusion. The lack of a direct blood supply to cartilage tissue is why torn cartilage heals slowly and why chondrocytes do not divide frequently.

Different types of extracellular matrix distinguish three types of cartilage. **Hyaline cartilage,** the most common type, has very fine collagenous fibers in its extracellular matrix and looks somewhat like white glass. Hyaline cartilage is the type shown in figure 5.16. It is found on the ends of bones in many joints, in the soft part of the nose, and in the supporting rings of the respiratory passages. Hyaline cartilage is also important in the development and growth of most bones (see chapter 7, p. 134).

Elastic cartilage has a dense network of elastic fibers and thus is more flexible than hyaline cartilage (fig. 5.17). It provides the framework for the external ears and for parts of the larynx.

Fibrocartilage, a very tough tissue, has many collagenous fibers (fig. 5.18). It is a shock absorber for structures that are subjected to pressure. For example, fibrocartilage forms pads (intervertebral discs) between the individual bones (vertebrae) of the spinal column. It also cushions bones in the knees and in the pelvic girdle.

> Between ages thirty and seventy, a nose may lengthen and widen by as much as half an inch, and the ears may lengthen by a quarter inch, because the cartilage in these areas continues to grow as we age.

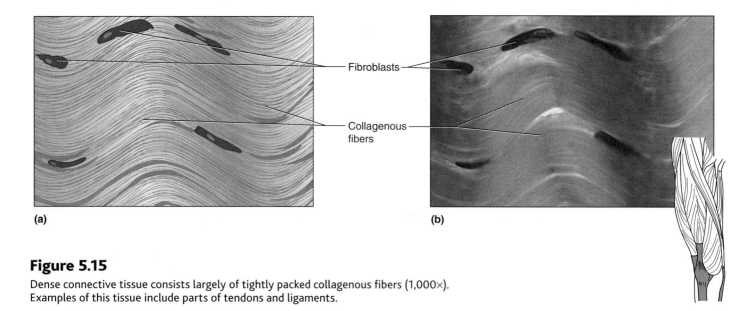

(a)   (b)

Fibroblasts

Collagenous fibers

Figure 5.15

Dense connective tissue consists largely of tightly packed collagenous fibers (1,000×). Examples of this tissue include parts of tendons and ligaments.

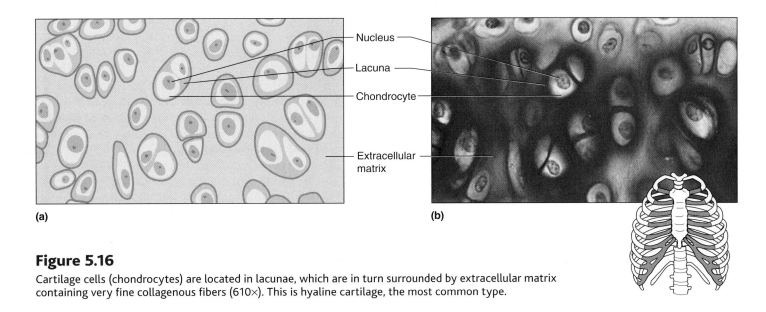

(a)   (b)

Figure 5.16

Cartilage cells (chondrocytes) are located in lacunae, which are in turn surrounded by extracellular matrix containing very fine collagenous fibers (610×). This is hyaline cartilage, the most common type.

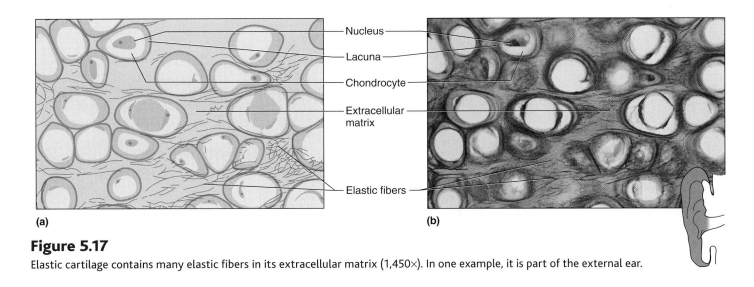

(a)   (b)

Figure 5.17

Elastic cartilage contains many elastic fibers in its extracellular matrix (1,450×). In one example, it is part of the external ear.

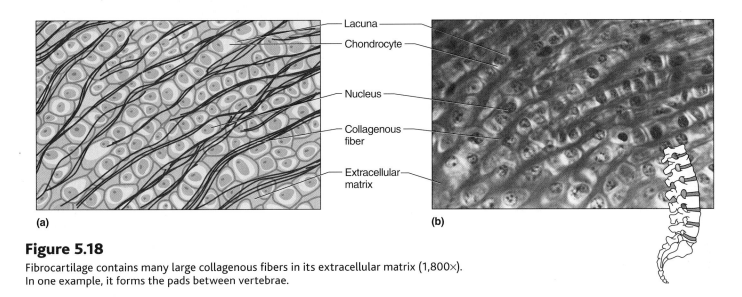

(a)   (b)

Figure 5.18

Fibrocartilage contains many large collagenous fibers in its extracellular matrix (1,800×). In one example, it forms the pads between vertebrae.

Bone

Bone is the most rigid connective tissue. Its hardness is largely due to mineral salts, such as calcium phosphate and calcium carbonate, between cells. This extracellular matrix also has many collagenous fibers, which are flexible and reinforce the mineral components of bone.

Bone internally supports body structures. It protects vital parts in the cranial and thoracic cavities, and is an attachment for muscles. Bone also contains red marrow, which forms blood cells, and it stores and releases inorganic chemicals such as calcium and phosphorus.

Bone matrix is deposited in thin layers called *lamellae,* which form concentric patterns around tiny longitudinal tubes called *central canals,* or Haversian canals

(fig. 5.19). Bone cells, or **osteocytes** (os′te-o-sītz), are located in lacunae, which are rather evenly spaced within the lamellae. Consequently, osteocytes also form concentric circles.

In a bone, the osteocytes and layers of extracellular matrix, which are concentrically clustered around a central canal, form a cylinder-shaped unit called an **osteon** (os′te-on), or Haversian system. Many osteons cemented together form the substance of bone.

Each central canal contains a blood vessel, which places every bone cell near a nutrient supply. In addition, bone cells have many cytoplasmic processes that extend outward and pass through very small tubes in the extracellular matrix called *canaliculi.* These cellular processes connect with the membranes of nearby cells.

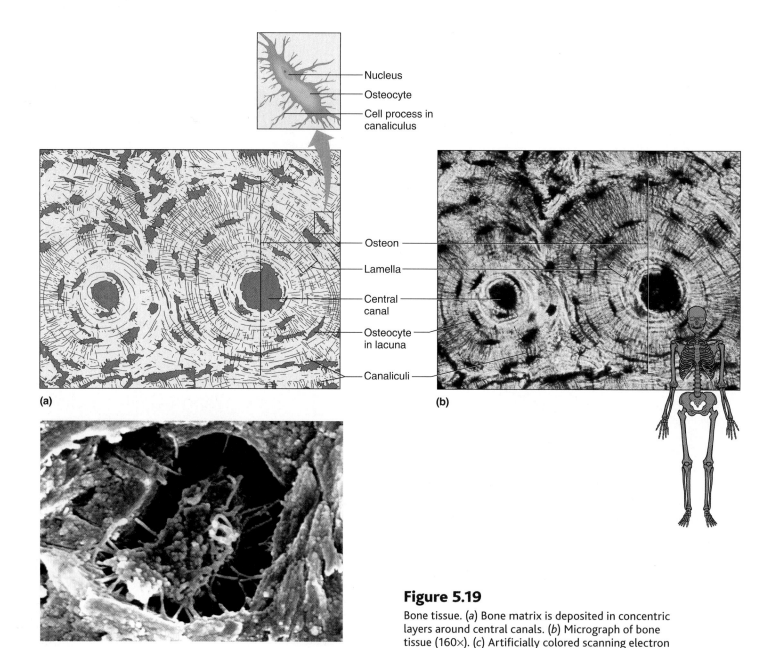

Nucleus
Osteocyte
Cell process in canaliculus

Osteon
Lamella
Central canal
Osteocyte in lacuna
Canaliculi

(a)

(b)

(c)

Figure 5.19

Bone tissue. (*a*) Bone matrix is deposited in concentric layers around central canals. (*b*) Micrograph of bone tissue (160×). (*c*) Artificially colored scanning electron micrograph of an osteocyte within a lacuna (6,000×).

As a result, materials can move rapidly between blood vessels and bone cells. Thus, in spite of its inert appearance, bone is a very active tissue that heals much more rapidly than does injured cartilage. (The microscopic structure of bone is described in more detail in chapter 7, p. 132.)

Blood

Blood transports a variety of materials between interior body cells and those that exchange substances with the external environment. In this way, blood helps maintain stable internal environmental conditions. Blood is composed of *formed elements* suspended in a fluid extracellular matrix called *blood plasma*. The formed elements include *red blood cells, white blood cells,* and cell fragments called *platelets* (fig. 5.20). Most blood cells form in red marrow within the hollow parts of certain long bones. Chapter 12 describes blood in detail. Table 5.5 lists the characteristics of the connective tissues.

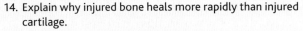

Check Your Recall

13. Describe the general characteristics of cartilage.
14. Explain why injured bone heals more rapidly than injured cartilage.
15. What are the major components of blood?

5.4 TYPES OF MEMBRANES

After discussing epithelial and connective tissues, membranes are better understood. **Epithelial membranes** are thin, sheetlike structures composed of epithelium and underlying connective tissue covering body surfaces and lining body cavities. The three major types of epithelial membranes are *serous, mucous,* and *cutaneous.*

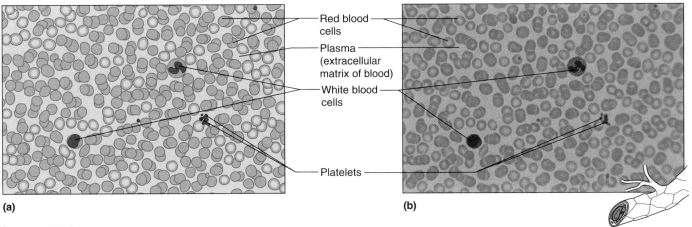

(a)   **(b)**

Figure 5.20
Blood tissue consists of red blood cells, white blood cells, and platelets suspended in a fluid extracellular matrix, the blood plasma (425×).

Table 5.5	Connective Tissues	
Type	**Function**	**Location**
Loose connective tissue		
Areolar tissue	Binds organs, holds tissue fluids	Beneath skin, between muscles, beneath epithelial tissues
Adipose tissue	Protects, insulates, stores fat	Beneath skin, around kidneys, behind eyeballs, on surface of heart
Reticular connective tissue	Supports	Walls of liver and spleen
Dense connective tissue	Binds organs	Tendons, ligaments, deeper layers of skin
Hyaline cartilage	Supports, protects, provides framework	Nose, ends of bones, rings in the walls of respiratory passages
Elastic cartilage	Supports, protects, provides flexible framework	Framework of external ear and parts of larynx
Fibrocartilage	Supports, protects, absorbs shock	Between bony parts of spinal column, parts of pelvic girdle and knee
Bone	Supports, protects, provides framework	Bones of skeleton
Blood	Transports substances, helps maintain stable internal environment	Throughout body within a closed system of blood vessels and heart chambers

Serous (se′rus) **membranes** line body cavities that lack openings to the outside. These membranes form the inner linings of the thorax (parietal pleura) and abdomen (parietal peritoneum), and they cover the organs within these cavities (visceral pleura and visceral peritoneum, respectively, as shown in figs. 1.10 and 1.11, p. 11). A serous membrane consists of a layer of simple squamous epithelium and a thin layer of loose connective tissue. The cells of a serous membrane secrete watery *serous fluid,* which lubricates membrane surfaces.

Mucous (mucus) **membranes** line cavities and tubes that open to the outside of the body, including the oral and nasal cavities and the tubes of the digestive, respiratory, urinary, and reproductive systems. A mucous membrane consists of epithelium overlying a layer of loose connective tissue. Goblet cells within a mucous membrane secret *mucus.*

The **cutaneous** (ku-ta′ne-us) **membrane** is more commonly called the skin. It is described in detail in chapter 6.

Another type of membrane is a **synovial** (si-nove-al) **membrane,** which lines joints and is discussed further in chapter 7 (p. 162). A synovial membrane is composed entirely of connective tissues.

> ### Check Your Recall
> 16. Name the four types of membranes, and explain how they differ.

5.5 MUSCLE TISSUES

General Characteristics

Muscle (mus′el) **tissues** are able to contract; that is, their elongated cells, or *muscle fibers,* can shorten. As they contract, muscle fibers pull at their attached ends, and this action moves body parts. The three types of muscle tissue—skeletal, smooth, and cardiac—are discussed in more detail in chapter 8.

> Approximately 40% of the body is skeletal muscle, and almost another 10% is smooth or cardiac muscle.

Skeletal Muscle Tissue

Skeletal muscle tissue is found in muscles that attach to bones and is controlled by conscious effort. For this reason, it is often called *voluntary* muscle tissue. The long, threadlike cells of skeletal muscle have alternating light and dark cross-markings called *striations.* Each cell has many nuclei (fig. 5.21). A nerve cell can stimulate a muscle cell to contract, and then the muscle cell relaxes when stimulation stops.

The muscles built of skeletal muscle tissue move the head, trunk, and limbs. They enable us to make facial expressions, write, talk, sing, chew, swallow, and breathe, essentially carrying out all voluntary movements.

Smooth Muscle Tissue

Smooth muscle tissue is so-called because its cells do not have striations. Smooth muscle cells are shorter than skeletal muscle cells and are spindle-shaped, each with a single, centrally located nucleus (fig. 5.22). This tissue comprises the walls of hollow internal organs, such as the stomach, intestine, urinary bladder, uterus, and blood vessels. Unlike skeletal muscle, smooth muscle usually cannot be stimulated to contract by conscious efforts. Thus, its actions are *involuntary.* For example, smooth muscle tissue moves food through

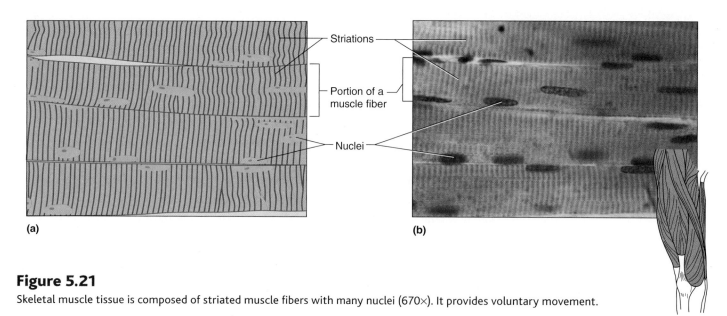

(a)   (b)

Striations

Portion of a muscle fiber

Nuclei

Figure 5.21
Skeletal muscle tissue is composed of striated muscle fibers with many nuclei (670×). It provides voluntary movement.

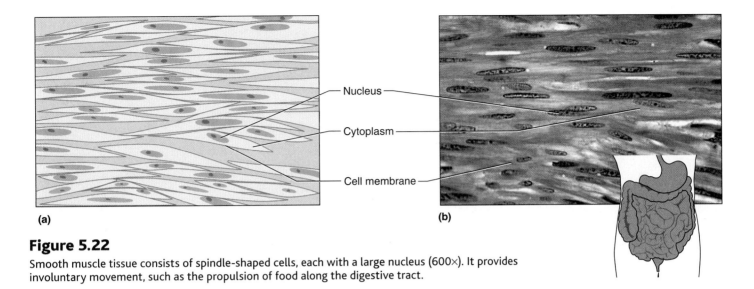

Figure 5.22
Smooth muscle tissue consists of spindle-shaped cells, each with a large nucleus (600×). It provides involuntary movement, such as the propulsion of food along the digestive tract.

the digestive tract, constricts blood vessels, and empties the urinary bladder.

Cardiac Muscle Tissue

Cardiac muscle tissue is only in the heart. Its cells, which are striated and branched, are joined end to end, forming intricate networks. Each cardiac muscle cell has a single nucleus (fig. 5.23). Where it touches another cell is a specialized intercellular junction called an *intercalated disc,* discussed further in chapter 8, p. 190.

Cardiac muscle, like smooth muscle, is controlled involuntarily. This tissue makes up the bulk of the heart and pumps blood through the heart chambers and into blood vessels.

Check Your Recall

17. List the general characteristics of muscle tissues.

18. Distinguish among skeletal, smooth, and cardiac muscle tissues.

The cells of different tissues vary greatly in their abilities to divide. Cells that divide continuously include the epithelial cells of the skin and inner lining of the digestive tract and the connective tissue cells that form blood cells in red bone marrow. However, skeletal and cardiac muscle cells and nerve cells do not usually divide at all after differentiating.

Fibroblasts respond rapidly to injuries by increasing in number and fiber production. They are often the principal agents of repair in tissues that have limited abilities to regenerate. For instance, fibroblasts form scar tissue after a heart attack occurs. Many organs include pockets of stem or progenitor cells that can divide and replace damaged, differentiated cells, under certain conditions.

5.6 NERVOUS TISSUES

Nervous (ner′vus) **tissues** are found in the brain, spinal cord, and peripheral nerves. The basic cells are called **neurons** (nu′ronz), or nerve cells (fig. 5.24). Neurons sense certain types of changes in their surroundings. They respond by transmitting nerve impulses along cellular processes called *axons* to other neurons or to muscles or glands. Because neurons communicate with each other and with muscle and gland cells, they can coordinate, regulate, and integrate many body functions.

In addition to neurons, nervous tissue includes **neuroglial cells** (nu-rog′le-ahl selz), shown in figure 5.24. Neuroglial cells divide and are crucial to the functioning of neurons. These cells support and bind the components of nervous tissue, carry on phagocytosis, and help supply nutrients to neurons by connecting them to blood vessels. They also play a role in cell-to-cell communication. Nervous tissue is discussed in more detail in chapter 9. Table 5.6 summarizes the general characteristics of muscle and nervous tissues.

Check Your Recall

19. Describe the general characteristics of nervous tissues.

20. Distinguish between neurons and neuroglial cells.

Clinical Connection

Each year in the United States, about ten thousand people need their bladders repaired or replaced. Typically a urologic surgeon replaces part of the bladder with part of the large intestine. However, the function of the intestine is to absorb, and the function of the bladder is to hold waste. Tissue engineering is providing a better way to build a bladder, which is a balloon-like organ

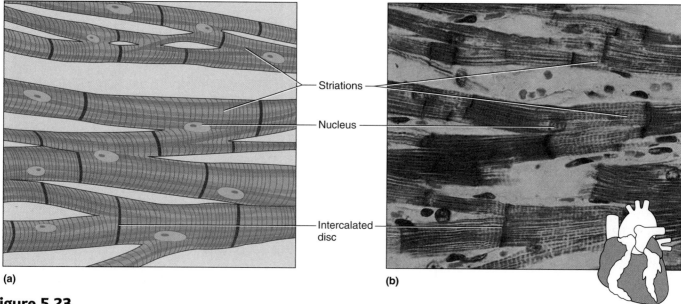

Figure 5.23

Cardiac muscle cells are branched and interconnected, with a single nucleus each (360×). Their contraction is involuntary.

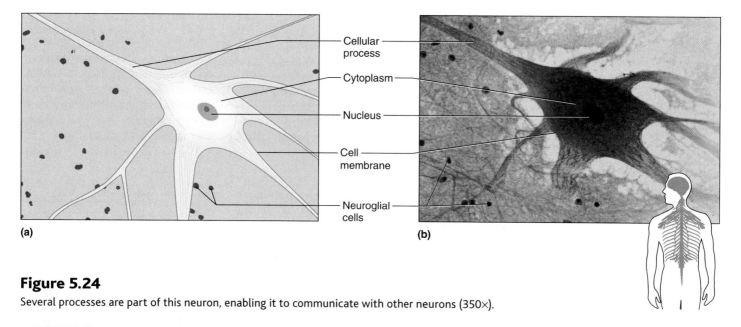

Figure 5.24

Several processes are part of this neuron, enabling it to communicate with other neurons (350×).

Table 5.6	Muscle and Nervous Tissues	
Type	**Function**	**Location**
Skeletal muscle tissue (striated)	Voluntary movements of skeletal parts	Muscles usually attached to bones
Smooth muscle tissue (lacks striations)	Involuntary movements of internal organs	Walls of hollow internal organs
Cardiac muscle tissue (striated)	Heart movements	Heart muscle
Nervous tissue	Sensory reception and conduction of nerve impulses	Brain, spinal cord, and peripheral nerves

that has smooth muscle on the outside and lining tissue (urothelium) and connective tissue on the inside.

In tissue engineering, a patient's cells are grown with a synthetic scaffold to form an implant that can bolster or replace a failing body part. Because the cells originate in the patient, the immune system does not reject them. Tissue engineering has provided bioengineered skin to help burn patients heal, replacement cartilage, and even blood vessels.

Researchers created replacements for part of the urinary bladders of seven children and teens who have spina bifida, a birth defect, in which the malfunctioning bladder can harm the kidneys. Each patient donated a postage-stamp-size sample of bladder tissue, which consisted of about a million cells. The researchers separated the cells and let them divide in culture, expanding to 1.5 billion cells within seven weeks. Then the cultured cells were seeded onto synthetic, three-dimensional domes. After the cells formed confluent layers, the domes were surgically attached to the lower portions of the patients' bladders, after removing the upper portions. The scaffolds degenerated over time, leaving new bladder built from the patients' own cells.

The bladder replacements were done several years ago but were reported only recently—time was needed to assess their functioning. While the patients still need to be catheterized, they no longer must wear diapers because their urine flow is more normal. The replacement bladders have also apparently halted the kidney damage. Further testing is ongoing, but replacing failing bladders with intestine may be on the way out, thanks to tissue engineering.

SUMMARY OUTLINE

5.1 Introduction (p. 95)
Tissues are groups of cells with specialized structural and functional roles. The four major types of human tissue are epithelial, connective, muscle, and nervous.

5.2 Epithelial Tissues (p. 95)
1. General characteristics
   a. Epithelial tissue covers organs, lines cavities and hollow organs, and is the major tissue of glands.
   b. Epithelium is anchored to connective tissue by a basement membrane, lacks blood vessels, consists of tightly packed cells, and is replaced continuously.
   c. It functions in protection, secretion, absorption, and excretion.
   d. Epithelial tissues are classified according to cell shape and number of layers of cells.
2. Simple squamous epithelium
   a. This tissue consists of a single layer of thin, flattened cells.
   b. It functions in gas exchange in the lungs and lines blood and lymph vessels and various body cavities.
3. Simple cuboidal epithelium
   a. This tissue consists of a single layer of cube-shaped cells.
   b. It carries on secretion and absorption in the kidneys and various glands.
4. Simple columnar epithelium
   a. This tissue is composed of elongated cells with nuclei near the basement membrane.
   b. It lines the uterus and digestive tract.
   c. Absorbing cells often have microvilli.
   d. This tissue has goblet cells that secrete mucus.
5. Pseudostratified columnar epithelium
   a. Nuclei located at two or more levels give this tissue a stratified appearance.
   b. Cilia that are part of this tissue move mucus over the surface.
   c. It lines passageways of the respiratory system.
6. Stratified squamous epithelium
   a. This tissue is composed of many layers of cells.
   b. It protects underlying cells.
   c. It forms the outer layer of the skin and lines the oral cavity, esophagus, vagina, and anal canal.
7. Stratified cuboidal epithelium
   a. This tissue is composed of two or three layers of cube-shaped cells.
   b. It lines the larger ducts of the mammary glands, sweat glands, salivary glands, and pancreas.
   c. It protects.
8. Stratified columnar epithelium
   a. The top layer of cells in this tissue has elongated columns. Cube-shaped cells make up the bottom layers.
   b. It is in the male urethra and ductus deferens and parts of the pharynx.
   c. This tissue protects and secretes.
9. Transitional epithelium
   a. This tissue is specialized to become distended.
   b. It is in the walls of various organs of the urinary tract.
10. Glandular epithelium
   a. Glandular epithelium is composed of cells that are specialized to secrete substances.
   b. A gland consists of one or more cells.
      (1) Exocrine glands secrete into ducts.
      (2) Endocrine glands secrete into tissue fluid or blood.
   c. Exocrine glands are classified according to the composition of their secretions.
      (1) Merocrine glands secrete fluid without loss of cytoplasm.
         (a) Serous cells secrete watery fluid with a high enzyme content.
         (b) Mucous cells secrete mucus.
      (2) Apocrine glands lose portions of their cells during secretion.
      (3) Holocrine glands release cells filled with secretory products.

5.3 Connective Tissues (p. 102)
1. General characteristics
   a. Connective tissue connects, supports, protects, provides frameworks, fills spaces, stores fat, produces blood cells, protects against infection, and helps repair damaged tissues.
   b. Connective tissue cells usually have considerable extracellular matrix between them.
   c. This extracellular matrix consists of fibers, a ground substance and fluid.
   d. Major cell types
      (1) Fibroblasts produce collagenous and elastic fibers.
      (2) Macrophages are phagocytes.
      (3) Mast cells may release heparin and histamine, and usually are near blood vessels.
   e. Connective tissue fibers
      (1) Collagenous fibers are composed of collagen and have great tensile strength.
      (2) Elastic fibers are composed of elastin and are very elastic.
      (3) Reticular fibers are very fine, collagenous fibers.
2. Categories of connective tissue
   Connective tissue proper includes loose connective tissue and dense connective tissue. Specialized connective tissue includes cartilage, bone, and blood.
   a. Loose connective tissue
      (1) Areolar tissue forms thin membranes between organs and binds them. It is beneath most layers of epithelium and between muscles.

(2) Adipose tissue stores fat. It is found beneath the skin, in certain abdominal membranes, and around the kidneys, heart, and various joints.

(3) Reticular connective tissue is composed of thin, collagenous fibers. It helps provide the framework of the liver and spleen.

b. Dense connective tissue

(1) This tissue is largely composed of strong, collagenous fibers.

(2) It is found in the tendons, ligaments, white portions of the eyes, and the deeper skin layers.

c. Cartilage

(1) Cartilage provides a supportive framework for various structures.

(2) Its extracellular matrix is composed of fibers and a gel-like ground substance.

(3) Cartilaginous structures are enclosed in a perichondrium.

(4) Cartilage lacks a direct blood supply and is slow to heal.

(5) Major types are hyaline cartilage, elastic cartilage, and fibrocartilage.

d. Bone

(1) The extracellular matrix of bone contains mineral salts and collagen.

(2) Its cells are usually organized in concentric circles around central canals. Canaliculi connect them.

(3) Bone is an active tissue that heals rapidly.

e. Blood

(1) Blood transports substances and helps maintain a stable internal environment.

(2) Blood is composed of red blood cells, white blood cells, and platelets suspended in plasma.

(3) Blood cells develop in red marrow in the hollow parts of long bones.

5.4 Types of Membranes (p. 109)

1. Epithelial membranes are thin, covering surfaces and lining cavities. Serous, mucous, and cutaneous membranes are epithelial membranes.

2. Serous membranes, composed of epithelium and loose connective tissue, are membranes that line body cavities lacking openings to the outside. The cells of a serous membrane secrete serous fluid to lubricate membrane surfaces.

3. Mucous membranes, composed of epithelium and loose connective tissue, are membranes that line body cavities opening to the outside. Goblet cells within these membranes secrete mucus.

4. The cutaneous membrane is the external body covering commonly called skin.

5. Synovial membranes, composed entirely of connective tissues, line joints.

5.5 Muscle Tissues (p. 110)

1. General characteristics

a. Muscle tissues contract, moving structures that are attached to them.

b. The three types are skeletal, smooth, and cardiac muscle tissues.

2. Skeletal muscle tissue

a. Muscles containing this tissue usually are attached to bones and controlled by conscious effort.

b. Cells, or muscle fibers, are long and threadlike.

c. Muscle cells contract when stimulated by nerve cells, and then relax when stimulation stops.

3. Smooth muscle tissue

a. This tissue is in the walls of hollow internal organs.

b. Usually it is involuntarily controlled.

4. Cardiac muscle tissue

a. This tissue is found only in the heart.

b. Cells are joined by intercalated discs and form branched networks.

5.6 Nervous Tissues (p. 111)

1. Nervous tissues are in the brain, spinal cord, and peripheral nerves.

2. Neurons (nerve cells)

a. Neurons sense changes and respond by transmitting nerve impulses to other neurons or to muscles or glands.

b. They coordinate, regulate, and integrate body activities.

3. Neuroglial cells

a. Some of these cells bind and support nervous tissue.

b. Others carry on phagocytosis.

c. Still others connect neurons to blood vessels.

d. They also play a role in cell-to-cell communication.

CHAPTER ASSESSMENTS

5.1 Introduction

1. Which of the following is a major tissue type in the body? (p. 95)

a. epithelial

b. nervous

c. muscle

d. connective

e. all of the above

2. Indicate where each major type of tissue can be found in the body. (p. 95)

5.2 Epithelial Tissues

3. A general characteristic of epithelial tissues is that _____. (p. 95)

a. numerous blood vessels are present

b. cells are spaced apart

c. cells readily divide

d. there is much extracellular matrix between cells

4. Explain how the structure of epithelial tissues provides for the functions of epithelial tissues. (p. 95)

5. Match the epithelial tissue on the left to an organ in which the tissue is found. (pp. 96–101)

(1) simple squamous epithelium	A. lining of intestines
(2) simple cuboidal epithelium	B. lining of ducts of mammary glands
(3) simple columnar epithelium	C. lining of urinary bladder
(4) pseudostratified columnar epithelium	D. salivary glands
(5) stratified squamous epithelium	E. air sacs of lungs
(6) stratified cuboidal epithelium	F. respiratory passages
(7) stratified columnar epithelium	G. ductus deferens
(8) transitional epithelium	H. lining of kidney tubules
(9) glandular epithelium	I. outer layer of skin

6. Distinguish between exocrine and endocrine glands. (p. 101)

7. A gland that secretes substances out of cells by exocytosis is a(n) _____. (p. 101)

a. merocrine gland

b. apocrine gland

c. holocrine gland

5.3 Connective Tissues

8. Define *extracellular matrix*. (p. 102)

9. Describe three major types of connective tissue cells. (p. 102)

10. Distinguish between collagen and elastin. (p. 104)

11. Compare and contrast the different types of loose connective tissue. (p. 105)

12. Define *dense connective tissue*. (p. 106)

13. Explain why injured dense connective tissue and cartilage are usually slow to heal. (p. 106)

14. Name the types of cartilages and describe their differences and similarities. (p. 106)

15. Describe how bone cells are organized in bone tissue. (p. 108)

16. The fluid extracellular matrix of blood is called _____. (p. 109)
    a. white blood cells
    b. red blood cells
    c. platelets
    d. plasma
    e. bone marrow

5.4 Types of Membranes

17. Identify the locations of four types of membranes in the body and indicate the types of tissues making up each membrane. (p. 110)

5.5 Muscle Tissues

18. Compare and contrast skeletal, smooth, and cardiac muscle tissues. (p. 110)

5.6 Nervous Tissues

19. Distinguish between neurons and neuroglial cells with respect to their functions. (p. 111)

INTEGRATIVE ASSESSMENTS/ CRITICAL THINKING

OUTCOME 5.3

1. Select a skin care product with added collagen and elastin, then connect this product's use to the tissue normally consisting of these fibers.

2. Joints such as the elbow, shoulder, and knee contain considerable amounts of cartilage and dense connective tissue. Explain why joint injuries are often very slow to heal.

3. Disorders of collagen are characterized by deterioration of connective tissues. Why would you expect such diseases to produce widely varying symptoms?

OUTCOMES 5.2, 5.3, 5.5, 5.6

4. Assess which of the four major tissue types carcinogens (cancer-causing agents) would most influence and least influence. (Hint: Carcinogens act on dividing cells.)

WEB CONNECTIONS

Visit the text website at **aris.mhhe.com** for additional quizzes, interactive learning exercises, and more.

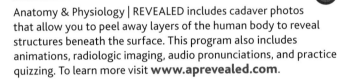

Anatomy & Physiology | REVEALED includes cadaver photos that allow you to peel away layers of the human body to reveal structures beneath the surface. This program also includes animations, radiologic imaging, audio pronunciations, and practice quizzing. To learn more visit **www.aprevealed.com**.

6

Integumentary System

THE ORIGIN OF HAIR. Hairs are extensions of specialized cells in the outer skin layer that some of us spend enormous amounts of time washing, drying, curling, straightening, styling, coloring, plucking, and shaving. Yet, compared to other mammals, we humans appear relatively hairless. Appearances are deceiving—we actually have as many hair follicles per square inch of skin as a chimpanzee. The chimp's hairs, however, are longer, thicker, and darker than ours.

All of the 5 million hair follicles of an adult human form during the fifth month of prenatal development, coating the fetus with a downy layer called lanugo. In most newborns, the lanugo has receded beneath the skin surface, perhaps leaving a bit of fuzziness on the ear tips or elsewhere. Persistence of this early hair accounts for much of the difference in hairiness between us and other primates. In a very rare inherited condition in humans called hypertrichosis, some of the lanugo remains and grows long. In less enlightened times, people with severe cases were exhibited in circuses as ape-men or werewolves.

In other mammals, hair provides warmth. It is absent in aquatic mammals such as whales and manatees, and reduced in their semi-aquatic cousins, such as hippos, presumably because a furry coat would impair swimming. What advantages might lighter coats have afforded our ancestors that can explain why this almost uniquely human trait has persisted?

One theory maintains that less hair enabled us to successfully conquer grasslands. Furry, four-footed animals can run fast for a short

We have about the same density of hair follicles as our nearest relatives, the chimps and other great apes, but many of our hair follicles remain beneath the surface of our skin after birth. Hence, this human and her gorilla friend differ greatly in hairiness.

time, and then slow down due to heating up. With hair dense only atop the head, protecting against sunburn, two-footed humans could run for longer times, enabling them to hunt. The lack of hair enabled our sweat glands to efficiently cool the body. Our hair has also persisted in places where our individual scents cling, which is essential for reproduction and offspring-parent bonding. Yet another explanation for our less hairy appearance is the "parasite-reduction hypothesis": Fur entraps fleas, lice, and ticks, which spread infectious disease. Shed the fur, and we shed the parasites.

Learning Outcomes   *After studying this chapter, you should be able to do the following:*

6.1 Introduction

1. Define organ, and name the large organ of the integumentary system. (p. 117)

6.2 Skin and Its Tissues

2. List the general functions of the skin. (p. 117)
3. Describe the structure of the layers of the skin. (p. 117)

4. Summarize the factors that determine skin color. (p. 119)

6.3 Accessory Structures of the Skin

5. Describe the anatomy and physiology of each accessory structure of the skin. (p. 122)

6.4 Regulation of Body Temperature

6. Explain how the skin helps regulate body temperature. (p. 124)

6.5 Healing of Wounds

7. Describe the events that are part of wound healing. (p. 125)

Aids to Understanding Words   *(Appendix A on page 567 has a complete list of Aids to Understanding Words.)*

cut- [skin] sub*cut*aneous: Beneath the skin.

derm- [skin] *derm*is: Inner layer of the skin.

epi- [upon] *epi*dermis: Outer layer of the skin.

follic- [small bag] hair *follic*le: Tubelike depression in which a hair develops.

kerat- [horn] *kerat*in: Protein produced as epidermal cells die and harden.

melan- [black] *melan*in: Dark pigment produced by certain cells.

seb- [grease] *seb*aceous gland: Gland that secretes an oily substance.

6.1 INTRODUCTION

Two or more kinds of tissues grouped together and performing specialized functions constitute an **organ.** The skin, the largest organ in the body by weight, and the various accessory structures associated with it (hair, finger nails, sensory receptors, and glands) make up the **integumentary** (in-teg-u-men′tar-e) **system.** The skin forms a barrier between ourselves and the outside, and as such is vitally important to the functioning of all other organs.

6.2 SKIN AND ITS TISSUES

One of the larger and more versatile organs of the body, the skin is vital in maintaining homeostasis. In addition to providing a protective covering, the skin helps regulate body temperature, retards water loss from deeper tissues, houses sensory receptors, synthesizes various biochemicals, and excretes small quantities of wastes.

Skin cells help produce vitamin D, which is necessary for normal bone and tooth development. This vitamin can be obtained in the diet or can form from a substance (dehydrocholesterol) that is synthesized by cells in the digestive system. When dehydrocholesterol reaches the skin by means of the blood and is exposed to ultraviolet light from the sun, it is converted to vitamin D.

Certain skin cells (keratinocytes) assist the immune system by producing hormonelike substances that stimulate development of certain white blood cells (T lymphocytes). These cells defend against infection by disease-causing bacteria and viruses (see chapter 14, pp. 386–388).

The skin includes two distinct layers (fig. 6.1). The outer layer, called the **epidermis** (ep″ĭ-der′mis), is composed of stratified squamous epithelium. The inner layer, or **dermis** (der′mis), is thicker than the epidermis, and it includes connective tissue consisting of collagenous and elastic fibers, epithelial tissue, smooth muscle tissue, nervous tissue, and blood. A *basement membrane* anchors the epidermis to the dermis and separates these two skin layers.

Beneath the dermis is loose connective tissue, predominantly adipose tissue, that binds the skin to the underlying organs forming the **subcutaneous** (sub″ku-ta′ne-us) **layer** (hypodermis). As its name indicates, this layer is beneath the skin and not a true layer of the skin. The collagenous and elastic fibers of this layer are continuous with those of the dermis. Most of these fibers run parallel to the surface of the skin, extending in all directions. As a result, no sharp boundary separates the dermis and the subcutaneous layer. The adipose tissue of the subcutaneous layer insulates, helping to conserve body heat and impeding the entrance of heat from the outside. The subcutaneous layer also contains the major blood vessels that supply the skin and underlying adipose tissue.

If the skin of a 150-pound person were spread out flat, it would cover approximately 20 square feet. People who have lost a large amount of weight often develop rashes and skin breakdown due to excess skin. "Body contouring" surgery removes excess skin.

Check Your Recall

1. List the general functions of the skin.
2. Name the tissue in the outer layer of the skin.
3. Name the tissues in the inner layer of the skin.
4. Name the tissues in the subcutaneous layer beneath the skin.
5. What are the functions of the subcutaneous layer?

Intradermal injections are injected into the skin. *Subcutaneous injections* are administered through a hollow needle into the subcutaneous layer beneath the skin. Subcutaneous injections and *intramuscular injections,* administered into muscles, are sometimes called hypodermic injections.

Some substances are introduced through the skin by means of an adhesive transdermal patch that includes a small reservoir containing a drug. The drug passes from the reservoir through a permeable membrane at a known rate. It then diffuses into the epidermis and enters the blood vessels of the dermis. Transdermal patches deliver drugs that protect against motion sickness, alleviate chest pain associated with heart disease, and lower blood pressure. A transdermal patch that delivers nicotine is used to help people stop smoking.

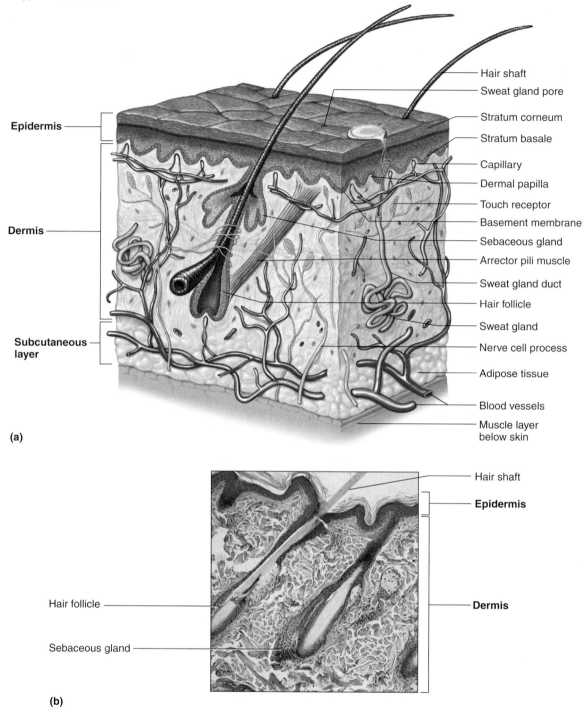

Epidermis

Dermis

Subcutaneous layer

- Hair shaft
- Sweat gland pore
- Stratum corneum
- Stratum basale
- Capillary
- Dermal papilla
- Touch receptor
- Basement membrane
- Sebaceous gland
- Arrector pili muscle
- Sweat gland duct
- Hair follicle
- Sweat gland
- Nerve cell process
- Adipose tissue
- Blood vessels
- Muscle layer below skin

(a)

Hair shaft

Epidermis

Hair follicle

Sebaceous gland

Dermis

(b)

Figure 6.1

Skin. (*a*) The skin is an organ that includes two layers, the epidermis and dermis, that lie atop a subcutaneous ("beneath the skin") layer. A section of skin. (*b*) This light micrograph depicts the layered structure of the skin (75×).

Epidermis

Because the epidermis is composed of stratified squamous epithelium, it lacks blood vessels. However, the deepest layer of epidermal cells, called the *stratum basale* (stra′tum ba′sal), or stratum germinativum, is close to the dermis and is nourished by dermal blood vessels (fig. 6.1*a*). As the cells of this layer divide and grow, the older epidermal cells are pushed away from the dermis toward the skin surface. The farther the cells move, the poorer their nutrient supply becomes, and in time, they die.

The older cells (keratinocytes) harden in a process called **keratinization** (ker″ah-tin″ĭ-za′shun). The cytoplasm fills with strands of a tough, fibrous, waterproof *keratin* protein. As a result, many layers of tough, tightly packed dead cells accumulate in the outermost epidermis, forming a layer called the *stratum corneum* (kor'ne-um). Dead cells that compose it are eventually shed.

The thickness of the epidermis varies from region to region. In most areas, only four layers can be distinguished: the *stratum basale, stratum spinosum* (spi-no'sum), *stratum granulosum* (gran"u-lo'sum), and *stratum corneum*. An additional layer, the *stratum lucidum* (loo'sid-um), is in the thickened skin of the palms and soles (fig. 6.2). The stratum lucidum may be missing where the epidermis is thin over the rest of the body.

In healthy skin, production of epidermal cells is closely balanced with loss of dead cells from the stratum corneum, so that the skin does not wear away completely. In fact, the rate of cell division increases where the skin is rubbed or pressed regularly, causing growth of thickened areas called *calluses* on the palms and soles, and keratinized conical masses on the toes called *corns.*

The epidermis has important protective functions. It shields the moist underlying tissues against excess water loss, mechanical injury, and the effects of harmful chemicals. When intact, the epidermis also keeps out disease-causing microorganisms (pathogens).

Specialized cells in the epidermis called *melanocytes* produce **melanin** (mel'ah-nin), a dark pigment that provides skin color (fig. 6.3*a*). Melanin absorbs ultraviolet radiation in sunlight, preventing mutations in the DNA of skin cells and other damaging effects. Melanocytes lie in the deepest portion of the epidermis. Although they are the only cells that can produce melanin, the pigment also may be present in other epidermal cells nearby. This happens because melanocytes have long, pigment-containing cellular extensions that pass upward between epidermal cells and the extensions transfer melanin granules into these other cells by a process called *cytocrine secretion*. Nearby epidermal cells may contain more melanin than the melanocytes (fig. 6.3*b*). The Topic of Interest on page 121 discusses skin cancer arising from melanocytes and other epidermal cells.

Skin Color

Skin color is due largely to melanin. All people have about the same number of melanocytes in their skin. Differences in skin color result from differences in the amount of melanin that melanocytes produce and in the distribution and size of the pigment granules. Skin color is mostly genetically determined. If genes instruct melanocytes to produce abundant melanin, the skin is dark.

Environmental and physiological factors also influence skin color. Sunlight, ultraviolet light from sunlamps, or X rays stimulate production of additional pigment. Blood in the dermal vessels may affect skin color as physiological changes occur. When blood is well oxygenated, the blood pigment (hemoglobin) is bright red, making the skin of light-complexioned people appear pinkish. On the other hand, when blood oxygen concentration is low, hemoglobin is dark red, and the skin

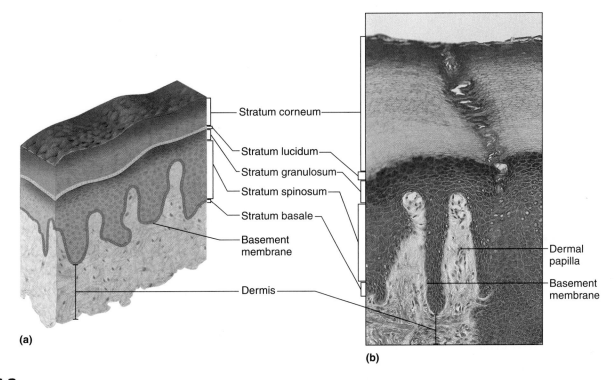

Stratum corneum

Stratum lucidum

Stratum granulosum

Stratum spinosum

Stratum basale

Basement membrane

Dermis

Dermal papilla

Basement membrane

(a)

(b)

Figure 6.2

Epidermis of thick skin. (*a*) The layers of the epidermis are distinguished by changes in cells as they are pushed toward the surface of the skin. (*b*) Light micrograph of skin (30×).

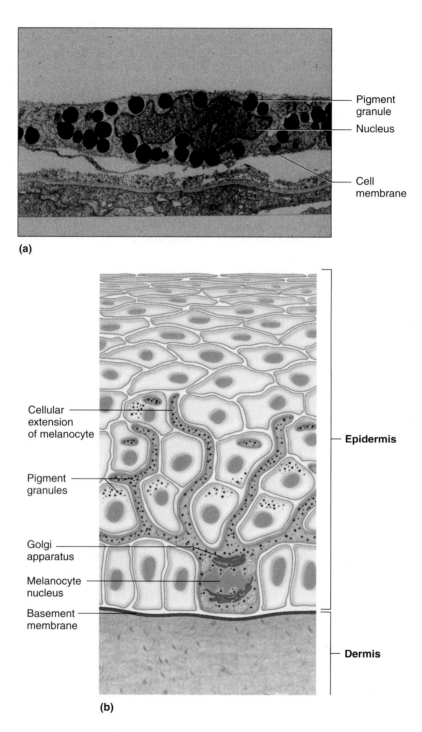

(a)

(b)

Figure 6.3

Melanocytes produce melanin. (*a*) This transmission electron micrograph shows a melanocyte with pigment-containing granules (10,600×). (*b*) A melanocyte may have pigment-containing extensions that pass between epidermal cells and transfer pigment into them. Note that much of the melanin is deposited above the nucleus, where the pigment can absorb UV radiation from outside before the DNA is damaged.

appears bluish—a condition called *cyanosis.* Other physiological factors affect skin color. For example, a diet high in yellow vegetables may turn skin orange-yellow, because these foods are rich in a pigment called β-carotene. Biochemical imbalances may also affect skin color. In newborns who have jaundice, for example, build-up of a substance called bilirubin turns the skin yellowish.

Check Your Recall

6. Explain how the epidermis is formed.

7. Distinguish between the stratum basale and the stratum corneum.

8. What is the function of melanin?

9. What factors influence skin color?

Topic of Interest

Skin Cancer

Skin cancer usually arises from nonpigmented epithelial cells within the deep layer of the epidermis or from melanocytes. Skin cancers originating from epithelial cells are called *cutaneous carcinomas* (squamous cell carcinoma or basal cell carcinoma); those arising from melanocytes are *cutaneous melanomas* (melanocarcinomas or malignant melanomas) (fig. 6A).

Cutaneous carcinomas are the most common type of skin cancer, occurring most frequently in light-skinned people over forty years of age. These cancers usually appear in individuals who are regularly exposed to sunlight, such as farmers, sailors, athletes, and sunbathers, and may result because normally protective apoptosis—peeling of sun-damaged cells—fails to occur.

Cutaneous carcinomas often develop from hard, dry, scaly growths (lesions) that have reddish bases. Such lesions may be either flat or raised and usually firmly adhere to the skin, appearing most often on the neck, face, or scalp. Cutaneous carcinomas are typically slow growing and can usually be cured completely by surgical removal or radiation treatment.

Cutaneous melanomas are pigmented with melanin, often with a variety of colored areas, such as variegated brown, black, gray, or blue. Melanomas usually have irregular rather than smooth outlines, and may feel bumpy.

People of any age may develop cutaneous melanomas. These cancers seem to be caused by short, intermittent exposure to high-intensity sunlight. Thus, risk of melanoma increases in persons who stay indoors but occasionally sustain blistering sunburns. Melanoma is not associated with sustained sun exposure, as are the other types of skin cancers.

A cutaneous melanoma, usually appearing on the back or limbs, may arise from normal-appearing skin or from a mole (nevus). The lesion spreads horizontally through the skin, but eventually may thicken and grow downward, invading deeper tissues. If the melanoma is surgically removed while it is in its horizontal growth phase, it may be arrested. Once it thickens and spreads into deeper tissues, it becomes difficult to treat, and survival rate is very low.

To reduce the chances of developing skin cancer, avoid exposing the skin to high-intensity sunlight, use sunscreens and sunblocks, and examine the skin regularly. Report any unusual lesions—particularly those that change in color, shape, or surface texture—to a physician at once.

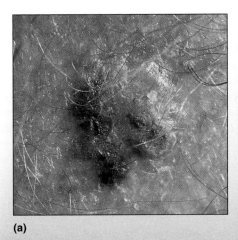
(a)

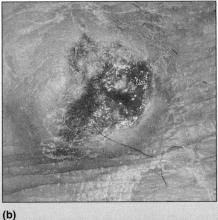
(b)

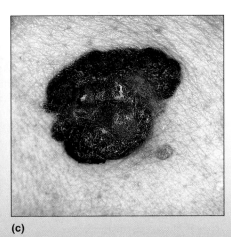
(c)

Figure 6A

Skin cancer. (*a*) Squamous cell carcinoma. (*b*) Basal cell carcinoma. (*c*) Melanoma.

Dermis

The boundary between the epidermis and dermis is uneven because epidermal ridges project inward and conical projections of dermis, called dermal papillae, extend into the spaces between the ridges (see fig. 6.1*a*). Dermal papillae can be found in skin all over the body, but they are most abundant in the hands and feet. Fingerprints form from these undulations of the skin at the distal end of the palmar surface of a finger. Genes determine fingerprint patterns, but the patterns can change slightly as a fetus moves and presses the forming ridges against the uterine wall. For this reason, even the fingerprints of identical twins are not exactly alike.

The dermis binds the epidermis to underlying tissues (see fig. 6.1*a*). It is largely composed of dense connective tissue that includes tough collagenous fibers and elastic fibers within a gel-like ground substance. Networks of these fibers give the skin toughness and elasticity.

Dermal blood vessels supply nutrients to all skin cells. These vessels also help regulate body temperature, as explained later in this chapter on pages 124–125.

> Because dermal blood vessels supply nutrients to the epidermis, interference with blood flow may kill epidermal cells. For example, when a person lies in one position for a prolonged period, the weight of the body pressing against the bed blocks the skin's blood supply. If cells die, the tissues begin to break down (necrosis), and a *pressure ulcer* (also called a decubitus ulcer or bedsore) may appear.
>
> Pressure ulcers usually form in the skin overlying bony projections, such as on the hip, heel, elbow, or shoulder. Frequently changing body position or massaging the skin to stimulate blood flow in regions associated with bony prominences can prevent pressure ulcers.

Nerve cell processes are scattered throughout the dermis. Motor processes carry impulses out from the brain or spinal cord to dermal muscles and glands. Sensory processes carry impulses away from specialized sensory receptors, such as touch receptors in the dermis, and into the brain or spinal cord. The dermis also contains accessory structures including hair follicles, sebaceous (oil-producing) glands, and sweat glands (see fig. 6.1*a*).

> To create a tattoo, very fine needles inject inks into the dermis. The color is permanent, because dermis cells are not shed, as are cells of the epidermis. To remove a tattoo, a laser shatters the ink molecules, and the immune system removes the resulting debris. Before laser removal became available in the late 1980s, unwanted tattoos were scraped, frozen, or cut away—all painful procedures.

Check Your Recall

10. What kinds of tissues make up the dermis?
11. What are the functions of these tissues?

6.3 ACCESSORY STRUCTURES OF THE SKIN

Nails

Nails are protective coverings on the ends of the fingers and toes. Each nail consists of a *nail plate* that overlies a surface of skin called the *nail bed*. Specialized epithelial cells that are continuous with the epithelium of the

skin produce the nail bed. The whitish, thickened, half-moon–shaped region (lunula) at the base of a nail plate covers the most actively growing region. The epithelial cells here divide, and the newly formed cells become keratinized. This gives rise to tiny, keratinized scales that become part of the nail plate, pushing it forward over the nail bed. In time, the plate extends beyond the end of the nail bed and with normal use gradually wears away (fig. 6.4).

> The thumbnail grows the slowest; the middle nail grows the fastest.

Hair Follicles

Hair is present on all skin surfaces except the palms, soles, lips, nipples, and parts of the external reproductive organs. Each hair develops from a group of epidermal cells at the base of a tubelike depression called a **hair follicle** (hār fol'i-kl) (figs. 6.1 and 6.5). This follicle extends from the surface into the dermis and contains the *hair root*. The epidermal cells at its base are nourished from dermal blood vessels in a projection of connective tissue at the deep end of the follicle. As these epidermal cells divide and grow, older cells are pushed toward the surface. The cells that move upward and away from their nutrient supply become keratinized and die. Their remains constitute the structure of a developing *hair shaft* that extends away from the skin surface (fig. 6.6). In other words, a hair is composed of dead epidermal cells.

Genes determine hair color by directing the type and amount of pigment that epidermal melanocytes produce. Dark hair has more of the brownish-black

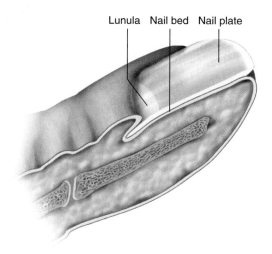

Lunula   Nail bed   Nail plate

Figure 6.4

Nails grow from epithelial cells that divide and become keratinized, forming the rest of the nail.

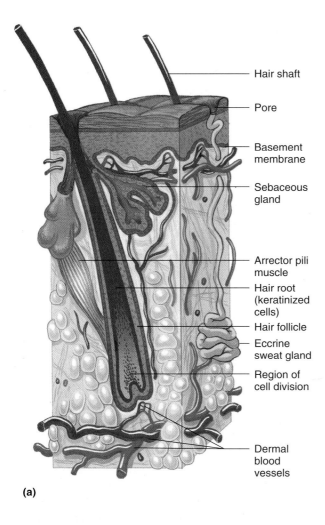

Hair shaft

Pore

Basement membrane

Sebaceous gland

Arrector pili muscle

Hair root (keratinized cells)

Hair follicle

Eccrine sweat gland

Region of cell division

Dermal blood vessels

(a)

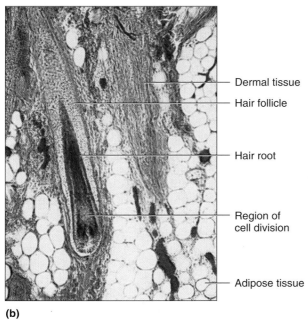

Dermal tissue

Hair follicle

Hair root

Region of cell division

Adipose tissue

(b)

Figure 6.5

Hair follicle. (*a*) A hair grows from the base of a hair follicle when epidermal cells divide and older cells move outward and become keratinized. Stem cells keep the hair growing. (*b*) Light micrograph of a hair follicle (160×).

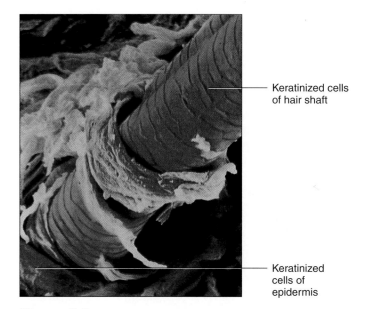

Keratinized cells of hair shaft

Keratinized cells of epidermis

Figure 6.6

This scanning electron micrograph shows a hair emerging from the epidermis (875×).

eumelanin (u-mel′ah-nin), while blonde hair and red hair have more of the reddish-yellow **pheomelanin** (fe″o-mel′ah-nin). The white hair of a person with the inherited condition *albinism* lacks melanin altogether. A mixture of pigmented and unpigmented hair usually appears gray.

A bundle of smooth muscle cells, forming the *arrector pili muscle*, attaches to each hair follicle (see figs. 6.1*a* and 6.5*a*). This muscle is positioned so that a short hair within the follicle stands on end when the muscle contracts. If a person is emotionally upset or very cold, nerve impulses may stimulate the arrector pili muscles to contract, causing gooseflesh, or goose bumps.

Just above the "bulge" region at the base of a hair follicle are stem cells that can give rise to hair as well as epidermal cells. The first clue to the existence of these "young transient amplifying cells" was that new skin in burn patients arises from hair follicles. Then, experiments in mice that mark stem cells and their descendants showed that the young transient amplifying cells give rise to both hair and skin. Manipulating these stem cells could someday treat baldness or extreme hairiness (hirsutism).

Sebaceous Glands

Sebaceous glands (se-ba′shus glandz) contain groups of specialized epithelial cells and are usually associated with hair follicles (figs. 6.5*a* and 6.7). They are holocrine glands (see chapter 5, p. 101) that secrete an oily mixture of fatty material and cellular debris called *sebum* through small ducts into the hair follicles. Sebum helps keep the hair and skin soft, pliable, and waterproof.

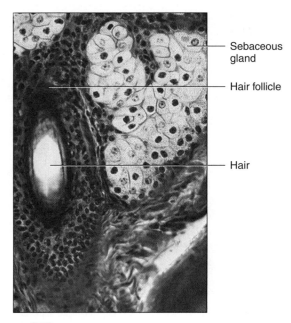

Sebaceous gland

Hair follicle

Hair

Figure 6.7

A sebaceous gland secretes sebum into a hair follicle, shown here in oblique section (300×).

Many teens are all too familiar with a disorder of the sebaceous glands called *acne* (acne vulgaris). Overactive and inflamed glands in some body regions become plugged and surrounded by small, red elevations containing blackheads (comedones) or pimples (pustules).

Sweat Glands

Sweat (swet) **glands,** or sudoriferous glands, are exocrine glands that are widespread in the skin. Each gland consists of a tiny tube that originates as a ball-shaped coil in the deeper dermis or superficial subcutaneous layer. The coiled portion of the gland is closed at its deep end and is lined with sweat-secreting epithelial cells.

The most abundant sweat glands, the **eccrine** (ek'rin) **glands,** respond throughout life to body temperature elevated by environmental heat or physical exercise (see fig. 6.5*a*). These glands are common on the forehead, neck, and back, where they produce profuse sweat on hot days or during intense physical activity.

The fluid (sweat) that eccrine glands secrete is carried away by a tube (duct) that opens at the surface as a *pore.* Sweat is mostly water, but it also contains small amounts of salt and wastes, such as urea and uric acid. Thus, sweating is also an excretory function.

Other sweat glands, known as **apocrine glands,** become active at puberty. Although they are currently called apocrine, these glands secrete by the same mechanism as eccrine glands, usually when a person is emotionally upset, frightened, in pain, or during sexual arousal. They are most numerous in the axillary regions

and groin. Ducts of these glands open into hair follicles. The secretions of these glands develop a scent as they are metabolized by skin bacteria.

Other sweat glands are structurally and functionally modified to secrete specific fluids, such as the *ceruminous glands* of the external ear canal that secrete earwax. The female *mammary glands* that secrete milk are another example of modified sweat glands.

The average square inch (6.45 square centimeters) of skin holds 650 sweat glands, 20 blood vessels, 60,000 melanocytes, and more than a thousand nerve endings.

Check Your Recall

12. Describe the structure of a nail bed.
13. Explain how a hair forms.
14. What is the function of the sebaceous glands?
15. Distinguish between the eccrine and apocrine sweat glands.

6.4 REGULATION OF BODY TEMPERATURE

Regulation of body temperature is vitally important because even slight shifts can disrupt the rates of metabolic reactions. Normally, the temperature of deeper body parts remains close to a set point of 37°C (98.6°F). Maintenance of a stable temperature requires that the amount of heat the body loses be balanced by the amount it produces. The skin plays a key role in the homeostatic mechanism that regulates body temperature.

Heat is a product of cellular metabolism; thus, the more active cells of the body are the major heat producers. These cells include skeletal and cardiac muscle cells and the cells of certain glands, such as the liver.

As body temperature rises, nerve impulses stimulate structures in the skin and other organs to release heat. For example, during physical exercise, active muscles release heat, which the blood carries away. The warmed blood reaches the part of the brain (the hypothalamus) that controls the body's temperature set point, which signals muscles in the walls of dermal blood vessels to relax. As these vessels dilate (vasodilation), more blood enters them, and some of the heat the blood carries escapes to the outside.

At the same time that the skin loses heat, the nervous system stimulates the eccrine sweat glands to become active and to release sweat onto the skin surface. As this fluid evaporates (changes from a liquid to a gas), it carries heat away from the surface, cooling the skin further.

If too much heat is lost and body temperature drops, as may occur in a very cold environment, the brain triggers different responses in the skin structures. Muscles in the walls of dermal blood vessels are stimulated to contract; this decreases the flow of heat-carrying blood through the skin and helps reduce heat loss. Also, the sweat glands remain inactive, decreasing heat loss by evaporation. If body temperature continues to drop, the nervous system may stimulate muscle fibers in the skeletal muscles throughout the body to contract slightly. This action requires an increase in the rate of cellular respiration and produces heat as a by-product. If this response does not raise body temperature to normal, small groups of muscles may contract rhythmically with greater force, and the person begins to shiver, generating more heat. Chapter 1 introduced this type of homeostatic mechanism (fig. 1.7, p. 8).

Most of the body's heat (80%) escapes through the head.

Deviation from the normal range for body temperature impairs health and may be lethal. People with severe spinal cord injuries can no longer control body temperature, which fluctuates depending upon the environment.

In hypothermia, core body temperature falls below 95°F. The body becomes so cold that it cannot maintain function. Symptoms of worsening hypothermia include a gradual loss of coordination, stiffening muscles, confusion, fatigue, and slow, shallow breathing. When core temperature falls to 87.8°F, the skin turns a bluish-gray, weakness intensifies, and consciousness ebbs away.

In hyperthermia, core body temperature exceeds 106°F. The skin becomes hot, dry, and flushed, and the person becomes weak, dizzy, and nauseous, with headache and a rapid, irregular pulse. The vignette that opens chapter 18 (p. 491) describes a fatal case of hyperthermia in an athlete.

Check Your Recall

16. Why is regulation of body temperature so important?
17. How does the body lose excess heat?
18. Which actions help the body conserve heat?

6.5 HEALING OF WOUNDS

A wound and the area surrounding it usually become red and painfully swollen. This is the result of *inflammation,* which is a normal response to injury or stress. Blood vessels in affected tissues dilate and become more permeable, allowing fluids to leak into the damaged tissues. Inflamed skin may become reddened, warm, swollen, and painful to touch (table 6.1). How-

Table 6.1	Inflammation
Symptom	**Cause**
Redness	Vasodilation, more blood in area
Heat	Large amount of blood accumulating in area and as a by-product of increased metabolic activity in tissue
Swelling	Increased permeability of blood vessels, fluids leaving blood go into tissue spaces (edema)
Pain	Injury to neurons and increased pressure from edema

ever, the dilated blood vessels provide the tissues with more nutrients and oxygen, which aids healing.

The specific events in healing depend on the nature and extent of the injury. If a break in the skin is shallow, epithelial cells along its margin are stimulated to divide more rapidly than usual, and the newly formed cells fill the gap.

If the injury extends into the dermis or subcutaneous layer, blood vessels break, and the escaping blood forms a clot in the wound. The blood clot and dried tissue fluids form a *scab* that covers and protects underlying tissues. Before long, fibroblasts migrate into the injured region and begin forming new collagenous fibers that bind the edges of the wound together. Suturing or otherwise closing a large break in the skin speeds this process.

As healing continues, blood vessels extend into the area beneath the scab. Phagocytic cells remove dead cells and other debris. Eventually, the damaged tissues are replaced, and the scab sloughs off. If the wound is extensive, the newly formed connective tissue may appear on the surface as a *scar.*

In large, open wounds, healing may be accompanied by formation of small, rounded masses called *granulations* that develop in the exposed tissues. A granulation consists of a new branch of a blood vessel and a cluster of collagen-secreting fibroblasts that the vessel nourishes. In time, some of the blood vessels are resorbed, and the fibroblasts move away, leaving a scar that is largely composed of collagenous fibers. The Topic of Interest on page 126 includes descriptions of healing that occurs following tissue damage resulting from a burn.

Check Your Recall

19. What is the tissue response to inflammation?
20. Distinguish between the activities necessary to heal a wound in the epidermis and those necessary to heal a wound in the dermis.
21. Explain the role of phagocytic cells in wound healing.
22. Define granulation.

Burns

Slightly burned skin, as from a minor sunburn, may become warm and reddened (erythema) as dermal blood vessels dilate. This response may be accompanied by mild edema, and in time, the surface layer of skin may be shed. A burn injuring the epidermis alone is called a *superficial partial-thickness* (first degree) *burn*. Healing usually occurs within a few days to two weeks, with no scarring.

A burn that destroys some epidermis as well as some underlying dermis is a *deep partial-thickness* (second degree) *burn*. Fluid escapes from damaged dermal capillaries, and as it accumulates beneath the outer layer of epidermal cells, blisters appear. The injured region becomes moist and firm and may vary from dark red to waxy white. Such a burn most commonly occurs from exposure to hot objects, hot liquids, flames, or burning clothing.

Healing of a deep partial-thickness burn depends upon accessory structures of the skin that survive the injury because they are deep in the dermis. These structures, which include hair follicles, sweat glands, and sebaceous glands, include epithelial cells that divide and extend onto the surface of the injured dermis, spread over it, and form new epidermis.

A burn that destroys the epidermis, the dermis, and the accessory structures of the skin is a *full-thickness* (third degree) *burn*. The injured skin becomes dry and leathery, and may vary from red to black to white. A full-thickness burn usually is caused by immersion in hot liquids or prolonged exposure to hot objects, flames, or corrosive chemicals. Typically, most of the epithelial cells in the affected region are destroyed, and healing occurs only if epithelial cells divide and grow inward from the margin of the burn. If the injured area is extensive, it may require a transplant, using a thin layer of skin from an unburned region of the body (an autograft), cadaveric skin (a homograft), or a skin substitute (amniotic membrane from a human fetus, artificial membranes, or a tissue-engineered skin).

Common Skin Disorders

acne (ak′ne) Disease of the sebaceous glands that produces blackheads and pimples.

alopecia (al″o-pe′she-ah) Hair loss, usually sudden.

athlete's foot (ath′-lētz foot) Fungus (*Tinea pedis*) infection usually in the skin of the toes and soles.

birthmark (berth′ mark) Congenital blemish or spot on the skin, visible at birth or soon after.

boil (boil) Bacterial infection (furuncle) of the skin, produced when bacteria enter a hair follicle.

carbuncle (kar′bung-kl) Bacterial infection, similar to a boil, that spreads into the subcutaneous tissues.

cyst (sist) Liquid-filled sac or capsule.

dermatitis (der″mah-ti′tis) Inflammation of the skin.

eczema (ek′zĕ-mah) Noncontagious skin rash that produces itching, blistering, and scaling.

erythema (er″ĭ-the′mah) Reddening of the skin due to dilation of dermal blood vessels in response to injury or inflammation.

herpes (her′pēz) Infectious disease of the skin, caused by the herpes simplex virus and characterized by recurring formations of small clusters of vesicles.

impetigo (im″pĕ-ti′go) Contagious disease of bacterial origin, characterized by pustules that rupture and become covered with loosely held crusts.

keloid (ke′loid) Elevated, enlarging fibrous scar usually initiated by an injury.

mole (mōl) Fleshy skin tumor (nevus) that is usually pigmented; colors range from brown to black.

pediculosis (pĕ-dik″u-lo′sis) Disease produced by an infestation of lice.

pruritus (proo-ri′tus) Itching of the skin.

psoriasis (so-ri′ah-sis) Chronic skin disease characterized with red patches covered with silvery scales.

pustule (pus′tūl) Elevated, pus-filled area on the skin.

scabies (ska′bēz) Disease resulting from an infestation of mites.

seborrhea (seb″o-re′ah) Hyperactivity of the sebaceous glands, causing greasy skin and dandruff.

ulcer (ul′ser) Open sore.

urticaria (ur″tĭ-ka′re-ah) Allergic reaction of the skin that produces reddish, elevated patches (hives).

vitiligo (vit′ĭ-li′go) Loss of melanocytes in parts of the epidermis, producing whitened areas of skin.

wart (wort) Flesh-colored, raised area caused by a viral infection.

Clinical Connection

When a mammal is hurt, it licks the wound—for good reason. Saliva is rich in epidermal growth factor (EFG), a protein that speeds healing. EGF stimulates fibroblasts in the skin to secrete collagen, which helps to fill in the damaged area. An experiment with people who have leg ulcers demonstrated another effect of EGF—proliferation of stem cells. Researchers compared the skin of patients with leg ulcers given EGF to patients who did not receive the treatment and to healthy individuals. The healthy people and untreated patients had very few skin stem cells, located near hair follicles and at the bottom of the basement membrane. The treated patients, however, had many more stem cells, grouped into "islands" that traversed more than one layer, particularly where the epidermis dips down into the dermis.

Integumentary System

Skeletal System

Vitamin D activated by the skin helps provide calcium needed for bone matrix.

Lymphatic System

The skin, acting as a barrier, provides an important first line of defense for the immune system.

Muscular System

Involuntary muscle contractions (shivering) work with the skin to control body temperature. Muscles act on facial skin to create expressions.

Digestive System

Excess calories may be stored as subcutaneous fat. Vitamin D activated by the skin stimulates dietary calcium absorption.

Nervous System

Sensory receptors provide information about the outside world to the nervous system. Nerves control the activity of sweat glands.

Respiratory System

Stimulation of skin receptors may alter respiratory rate.

Endocrine System

Hormones help to increase skin blood flow during exercise. Other hormones stimulate either the synthesis or the decomposition of subcutaneous fat.

Urinary System

The kidneys help compensate for water and electrolytes lost in sweat.

Cardiovascular System

Skin blood vessels play a role in regulating body temperature.

Reproductive System

Sensory receptors play an important role in sexual activity and in the suckling reflex.

The skin provides protection, contains sensory receptors, and helps control body temperature.

SUMMARY OUTLINE

6.1 Introduction (p. 117)

An organ is formed by two or more tissues grouped together and performing specialized functions. The skin, the largest organ in the body, is part of the integumentary system.

6.2 Skin and Its Tissues (p. 117)

Skin is a protective covering, helps regulate body temperature, retards water loss, houses sensory receptors, synthesizes various biochemicals, and excretes wastes. It is composed of an epidermis and a dermis separated by a basement membrane. Beneath the skin is the subcutaneous layer that binds the skin to underlying organs, stores fat, and contains blood vessels that supply the skin.

1. Epidermis
   a. The deepest layer of the epidermis, called the stratum basale, contains cells that divide.
   b. Epidermal cells undergo keratinization as they mature and are pushed toward the surface.
   c. The outermost layer, called the stratum corneum, is composed of dead epidermal cells.
   d. The epidermis protects underlying tissues against water loss, mechanical injury, and the effects of harmful chemicals.
   e. Melanin protects underlying cells from the effects of ultraviolet light.
   f. Melanocytes transfer melanin to nearby epidermal cells.
   g. Melanin provides skin color.
      (1) All people have about the same number of melanocytes.
      (2) Skin color is due largely to the amount of melanin and the distribution and size of pigment granules in the epidermis.
      (3) Environmental and physiological factors, as well as genes, influence skin color.
2. Dermis
   a. The dermis binds the epidermis to underlying tissues.
   b. Dermal blood vessels supply nutrients to all skin cells and help regulate body temperature.
   c. Nerve fibers are scattered throughout the dermis.
      (1) Some dermal nerve fibers carry impulses to muscles and glands of the skin.
      (2) Other dermal nerve fibers are associated with sensory receptors in the skin, and carry impulses to the brain and spinal cord.
   d. The dermis also has hair follicles, sebaceous glands, and sweat glands.

6.3 Accessory Structures of the Skin (p. 122)

1. Nails
   a. Nails are protective covers on the ends of fingers and toes.
   b. Specialized epidermal cells that are keratinized make up nails.
   c. The keratin of nails is harder than that produced by the skin's epidermal cells.
2. Hair follicles
   a. Each hair develops from epidermal cells at the base of a tubelike hair follicle.
   b. As newly formed cells develop and grow, older cells are pushed toward the surface and undergo keratinization.
   c. A bundle of smooth muscle cells is attached to each hair follicle.
   d. Hair color is determined by genes that direct the amount of eumelanin or pheomelanin that melanocytes associated with hair follicles produce.

3. Sebaceous glands
   a. Sebaceous glands are usually associated with hair follicles.
   b. Sebaceous glands secrete sebum, which helps keep the skin and hair soft and waterproof.
4. Sweat glands
   a. Each sweat gland is a coiled tube.
   b. Sweat is primarily water but also has salts and wastes.
   c. Eccrine sweat glands respond to elevated body temperature; apocrine glands respond to emotional stress.

6.4 Regulation of Body Temperature (p. 124)

Regulation of body temperature is vital because heat affects the rates of metabolic reactions. The normal temperature of deeper body parts is close to a set point of 37°C (98.6°F).

1. When body temperature rises above the normal set point, dermal blood vessels dilate, and sweat glands secrete sweat.
2. If body temperature drops below the normal set point, dermal blood vessels constrict, and sweat glands become inactive.
3. Excessive heat loss stimulates skeletal muscles to contract involuntarily.

6.5 Healing of Wounds (p. 125)

Skin injuries trigger inflammation. The affected area becomes red, warm, swollen, and tender.

1. Dividing epithelial cells fill in shallow cuts in the epidermis.
2. Clots close deeper cuts, sometimes leaving a scar where connective tissue replaces skin.
3. Granulations form in large, open wounds as part of the healing process.

CHAPTER ASSESSMENTS

◀ 6.1 Introduction

1. Two or more tissues grouped together and performing specialized functions defines a(n) _____. (p. 117)
   a. organelle   c. organ
   b. cell   d. organ system

2. The largest organ(s) in the body is (are) the _____. (p. 117)
   a. liver   c. lungs
   b. intestines   d. skin

◀ 6.2 Skin and Its Tissues

3. Functions of the skin include _____. (p. 117)
   a. retarding water loss   d. excretion
   b. body temperature regulation   e. all of the above
   c. sensory reception

4. List the remaining functions of skin not mentioned in question 3. (p. 117)

5. The epidermis is composed of layers of _____ tissue. (p. 118)

6. The _____ layer of epidermal cells contains older keratinized cells and dead cells. (p. 118)
   a. stratum corneum   d. stratum spinosum
   b. stratum lucidum   e. stratum basale
   c. stratum granulosum

7. Discuss the function of melanin, other than providing color to the skin. (p. 119)

8. List and describe the influence of each factor affecting skin color. (p. 119)

9. The dermis is composed primarily of what kind of tissue? (p. 121)

6.3 Accessory Structures of the Skin

10. Describe how nails are formed and relate the structure of nails to their function. (p. 122)

11. Distinguish between a hair and a hair follicle. (p. 122)

12. Sebaceous glands are _____ glands that secrete _____. (p. 123)

13. Compare and contrast eccrine and apocrine sweat glands. (p. 124)

6.4 Regulation of Body Temperature

14. Explain how body heat is produced. (p. 124)

15. Explain how sweat glands help regulate body temperature. (p. 124)

16. Describe the body's responses to decreasing body temperature. (p. 125)

6.5 Healing of Wounds

17. Explain how the healing of superficial breaks in the skin differs from the healing of deeper wounds. (p. 125)

INTEGRATIVE ASSESSMENTS/ CRITICAL THINKING

OUTCOME 6.2

1. Everyone's skin contains about the same number of melanocytes, even though people are of many different skin colors. How is this possible?

2. Which of the following would result in the more rapid absorption of a drug: a subcutaneous injection or an intradermal injection? Why?

3. How is it protective for skin to peel after a severe sunburn?

OUTCOMES 6.2, 6.4

4. A premature infant typically lacks subcutaneous adipose tissue and the small body has a relatively large surface area compared to its volume. How do these factors affect the ability of a premature infant to regulate its body temperature?

OUTCOMES 6.2, 6.5

5. As a rule, a superficial partial-thickness burn is more painful than one involving deeper tissues. How would you explain this observation?

OUTCOMES 6.2, 6.3, 6.4, 6.5

6. What special problems would result from the loss of 50% of a person's functional skin surface? How might this person's environment be modified to partially compensate for such a loss?

WEB CONNECTIONS

Visit the text website at **aris.mhhe.com** for additional quizzes, interactive learning exercises, and more.

AP R INTEGUMENTARY SYSTEM

Anatomy & Physiology | REVEALED includes cadaver photos that allow you to peel away layers of the human body to reveal structures beneath the surface. This program also includes animations, radiologic imaging, audio pronunciation, and practice quizzing. To learn more visit **www.aprevealed.com**.

7

Skeletal System

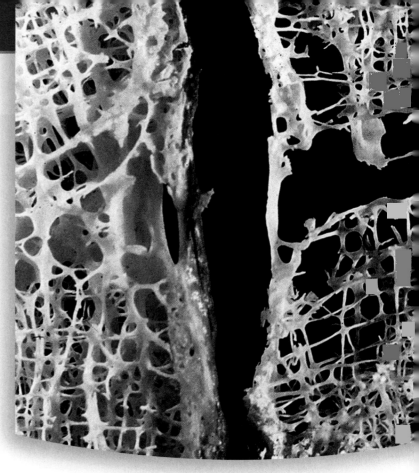

Spaces develop in bones when a person has osteoporosis. The vertebra on the left is normal; the one on the right has been weakened by osteoporosis.

PREVENTING "FRAGILITY FRACTURES." Skeletal health is a matter of balance. Before age 30, cells that form new bone tissue counter cells that degrade it, so that living bone is in a constant state of remodeling. Then the balance shifts so that bone is lost, especially in women past menopause, due to hormonal changes. This imbalance may progress to osteopenia or the more severe osteoporosis.

A "fragility fracture" is a telltale sign of dangerously low bone density. This is a fracture that happens after a fall from less than standing height, which a strong, healthy skeleton could resist. Fragility fractures occur in 1.5 million people in the United States each year, yet despite this warning sign, only one-fourth to one-third of them are followed up with bone scans and treatment to build new bone tissue. Since 1995, five new drugs have become available to treat osteoporosis. One class, the bisphosphonates, actually builds new bone.

Osteopenia and osteoporosis are common. The surgeon general estimates that half of all people over age 50 have one of these conditions, which amounts to 10 million with osteoporosis and another 35 million with osteopenia. Screening is advised for all individuals over age 65, as well as for those with risk factors. The most telling predictor is a previous fragility fracture. Other risk factors include a family history of osteoporosis, recent height loss, and older age.

Osteopenia and osteoporosis are assessed by measuring bone mineral density (BMD). This is most often done in the hip bone and lower spine with a technique called dual-energy X-ray absorptiometry. Osteopenia ("low bone mass") is defined as BMD at least 1 to 2.5 standard deviations below the mean. Osteoporosis is defined as bone mass at least 2.5 standard deviations below the mean for young adults. These measurements produce T values. Another measurement, a Z value, compares BMD to other individuals of a person's age and is used for individuals under age 65.

These two conditions are not just concerns of people approaching retirement age, because they can be prevented. Researchers think that what puts people at risk is failure to attain maximal possible bone density by age 30. To keep bones as strong as possible for as long as possible, it is essential to get at least 30 minutes of exercise daily (some of which should be weight-bearing), consume enough daily calcium (1,000–1,200 mg) and vitamin D (200 IU), and not smoke. There is much you can do to promote skeletal health—at any age.

Learning Outcomes   *After studying this chapter, you should be able to do the following:*

Aids to Understanding Words   (Appendix A on page 567 has a complete list of Aids to Understanding Words.)

acetabul- [vinegar cup] *acetabul*um: Depression of the hip bone that articulates with the head of the femur.

ax- [axis] *ax*ial skeleton: Upright portion of the skeleton that supports the head, neck, and trunk.

-blast [bud] osteo*blast*: Cell that will form bone tissue.

carp- [wrist] *carp*als: Wrist bones.

-clast [break] osteo*clast*: Cell that breaks down bone tissue.

condyl- [knob] *condyl*e: Rounded, bony process.

corac- [a crow's beak] *corac*oid process: Beaklike process of the scapula.

cribr- [sieve] *cribr*iform plate: Portion of the ethmoid bone with many small openings.

crist- [crest] *crist*a galli: Bony ridge that projects upward into the cranial cavity.

fov- [pit] *fov*ea capitis: Pit in the head of a femur.

glen- [joint socket] *glen*oid cavity: Depression in the scapula that articulates with the head of the humerus.

inter- [among, between] *inter*vertebral disc: Structure located between adjacent vertebrae.

intra- [inside] *intra*membranous bone: Bone that forms within sheetlike masses of connective tissue.

meat- [passage] auditory *meat*us: Canal of the temporal bone that leads inward to parts of the ear.

odont- [tooth] *odont*oid process: Toothlike process of the second cervical vertebra.

poie- [make, produce] hemato*poie*sis: Process that forms blood cells.

7.1 INTRODUCTION

Halloween skeletons and the skull-and-crossbones symbol for poison and pirates may make bones seem like lifeless objects. However, bone consists of a variety of very active, living tissues: bone tissue, cartilage, dense connective tissue, blood, and nervous tissue. Bones are not only very much alive but also multifunctional. Bones, the organs of the skeletal system, provide points of attachment for muscles, protect and support softer tissues, house blood-producing cells, store inorganic salts, and form passageways for blood vessels and nerves.

7.2 BONE STRUCTURE

The bones of the skeletal system differ greatly in size and shape. However, they are similar in structure, development, and functions.

Bone Classification

Bones are classified according to their shapes—long, short, flat, irregular, or sesamoid (round).

- **Long bones** have long longitudinal axes and expanded ends. Examples of long bones are the forearm and thigh bones.
- **Short bones** are somewhat cubelike, with their lengths and widths roughly equal. The bones of the wrists and ankles are this type.
- **Flat bones** are platelike structures with broad surfaces, such as the ribs, scapulae, and some bones of the skull.
- **Irregular bones** have a variety of shapes and are usually connected to several other bones. Irregular bones include the vertebrae that comprise the backbone and many facial bones.

- **Sesamoid (round) bones** are usually small and nodular and are embedded within tendons adjacent to joints. The kneecap (patella) is a sesamoid bone.

Parts of a Long Bone

The femur, the long bone in the thigh, illustrates the structure of bone (fig. 7.1). At each end of such a bone is an expanded portion called an **epiphysis** (e-pif′ĭ-sis) (plural, *epiphyses*), which articulates (forms a joint) with another bone. One epiphysis, called the proximal epiphysis, is nearest to the center of the body. The other, called the distal epiphysis, is farthest from the center of the body. On its outer surface, the articulating portion of the epiphysis is coated with a layer of hyaline cartilage called **articular cartilage** (ar-tik′u-lar kar′tĭ-lij). The shaft of the bone, between the epiphyses, is called the **diaphysis** (di-af′ĭ-sis).

A tough, vascular covering of fibrous tissue called the **periosteum** (per″e-os′te-um) completely encloses the bone, except for the articular cartilage on the bone's ends. The periosteum is firmly attached to the bone, and periosteal fibers are continuous with the connecting ligaments and tendons. The periosteum also helps form and repair bone tissue.

A bone's shape makes possible its functions. For example, bony projections called *processes* provide sites where ligaments and tendons attach; grooves and openings form passageways for blood vessels and nerves; and a depression of one bone may articulate with a process of another.

The wall of the diaphysis is mainly composed of tightly packed tissue called **compact bone** (kom′pakt bōn), or cortical bone. This type of bone has a continuous extracellular matrix with no spaces. The epiphyses, in contrast, are composed largely of **spongy bone** (spun′je bōn), or cancellous bone, with thin layers of compact bone on their surfaces. Spongy bone consists

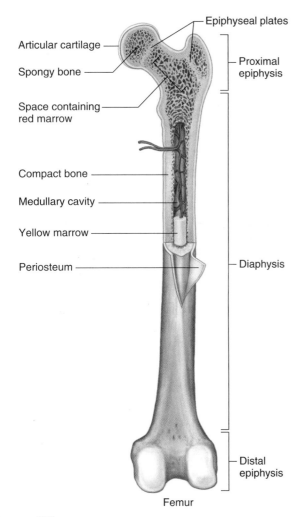

Figure 7.1

Major parts of a long bone. This is a femur, the long bone in the thigh.

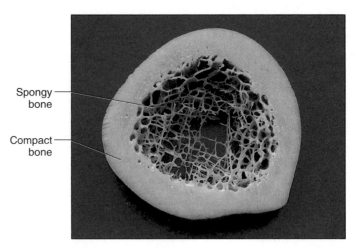

Figure 7.2

This cross section of a long bone reveals a layer of spongy bone beneath a layer of compact bone.

of numerous branching bony plates. Irregular connecting spaces between these plates help reduce the bone's weight (fig. 7.2). The bony plates are most highly developed in the regions of the epiphyses that are subjected to compressive forces. Both compact and spongy bone are strong and resist bending.

Compact bone in the diaphysis of a long bone forms a semirigid tube, which has a hollow chamber called the **medullary cavity** (med'u-lār"e kav'ĭ-te) that is continuous with the spaces of the spongy bone. A thin layer of cells called the **endosteum** (en-dos'te-um) lines these areas, and a specialized type of soft connective tissue called **marrow** (mar'o) fills them.

Microscopic Structure

Recall from chapter 5 (p. 108) that bone cells called *osteocytes* occupy very small, bony chambers called *lacunae,* which form concentric circles around *central canals* (Haversian canals) (fig. 7.3; see fig. 5.19, p. 108). Osteocytes communicate with nearby cells by means

of cellular processes passing through *canaliculi*. The extracellular matrix of bone tissue is largely composed of collagen and inorganic salts (calcium phosphate). Collagen gives bone its strength and resilience, and inorganic salts make it hard and resistant to crushing.

In compact bone, the osteocytes and layers of extracellular matrix concentrically clustered around a central canal form a cylinder-shaped unit called an *osteon* (Haversian system). Many of these units cemented together form the substance of compact bone.

Each central canal contains blood vessels (usually capillaries) and nerve fibers surrounded by loose connective tissue. The blood in these vessels nourishes bone cells associated with the central canal.

Central canals extend longitudinally through bone tissue, and transverse *perforating canals* (Volkmann's canals) connect them. Perforating canals contain larger blood vessels and nerves by which the smaller blood vessels and nerve fibers in central canals communicate with the surface of the bone and the medullary cavity (fig. 7.3).

Spongy bone is also composed of osteocytes and extracellular matrix, but the bone cells do not aggregate around central canals. Instead, the cells lie within the *trabeculae* and get nutrients from substances diffusing into canaliculi that lead to the surface of these thin, bony plates.

Check Your Recall

1. Explain how bones are classified.
2. List five major parts of a long bone.
3. How do compact and spongy bone differ in structure?
4. Describe the microscopic structure of compact bone.

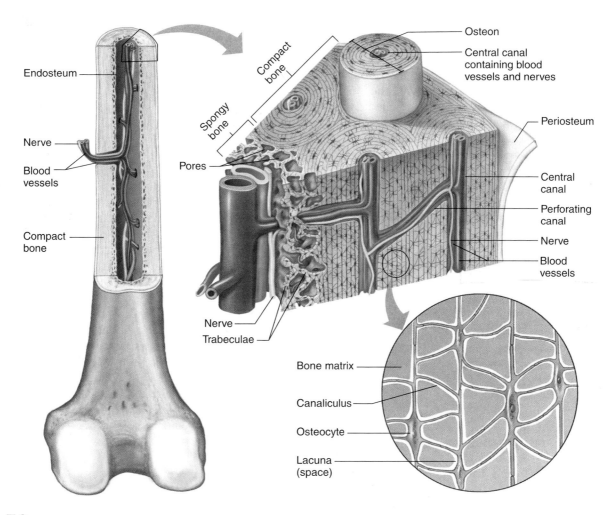

Figure 7.3

Compact bone is composed of osteons cemented by bone matrix. Drawing is not to scale. Extensions from osteocytes communicate through tunnel-like canaliculi.

7.3 BONE DEVELOPMENT AND GROWTH

Parts of the skeletal system begin to form during the first few weeks of prenatal development, and bony structures continue to develop and grow into adulthood. Bones form by replacing existing connective tissues in either of two ways: (1) Intramembranous bones originate between sheetlike layers of connective tissues. (2) Endochondral bones begin as masses of cartilage that are later replaced by bone tissue.

Intramembranous Bones

The broad, flat bones of the skull are **intramembranous bones** (in″trah-mem′brah-nus bōnz) (fig. 7.4). During their development, membranelike layers of unspecialized, or relatively undifferentiated, connective tissues appear at the sites of the future bones. Then, some of the partially differentiated progenitor cells enlarge and further differentiate into bone-forming

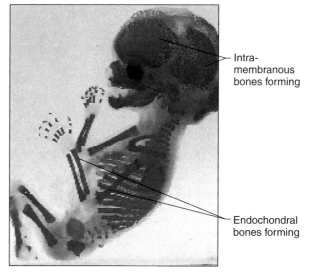

Figure 7.4

Intramembranous bones in the fetus form by replacing unspecialized connective tissue. Endochondral bones form from cartilage "models" that are gradually replaced with the harder tissue of bone. Note the stained, developing bones of this fourteen-week fetus.

cells called **osteoblasts** (os'te-o-blastz). The osteoblasts become active within the layers of connective tissue and deposit bony matrix around themselves. As a result, spongy bone tissue forms in all directions within the layers of connective tissues. When extracellular matrix completely surrounds osteoblasts, they are called **osteocytes.** Eventually, cells of the membranous tissues that persist outside the developing bone give rise to the periosteum. Osteoblasts on the inside of the periosteum form a layer of compact bone over the surface of the newly formed spongy bone. The formation of bone is called **ossification** (os"ĭ-fi-ka'shun).

Endochondral Bones

Most of the bones of the skeleton are **endochondral bones** (en"do-kon'dral bōnz). They develop in the fetus from masses of hyaline cartilage shaped like future bony structures (fig. 7.4). These cartilaginous models grow rapidly for a time and then begin to change extensively.

In a long bone, changes begin in the center of the diaphysis, where the cartilage slowly breaks down and disappears (fig. 7.5). At about the same time, a periosteum forms from connective tissue that encircles the developing diaphysis. Blood vessels and osteoblasts from the periosteum invade the disintegrating cartilage, and spongy bone forms in its place. This region of bone formation is called the *primary ossification center,* and bone tissue develops from it toward the ends of the cartilaginous structure. Meanwhile, osteoblasts from the periosteum deposit a thin layer of compact bone around the primary ossification center.

The epiphyses of the developing bone remain cartilaginous and continue to grow. Later, *secondary ossification centers* appear in the epiphyses, and spongy bone forms in all directions from them. As spongy bone is deposited in the diaphysis and in the epiphysis, a band of cartilage called the **epiphyseal plate** (ep"ĭ-fiz'e-al plāt), or metaphysis, remains between these two ossification centers.

The cartilaginous tissue of the epiphyseal plate includes layers of young cells that are undergoing mitosis and producing new cells. As these cells enlarge and extracellular matrix forms around them, the cartilaginous plate thickens, lengthening the bone. At the same time, calcium salts accumulate in the extracellular matrix adjacent to the oldest cartilaginous cells, and as the extracellular matrix calcifies, the cells begin to die.

In time, large, multinucleated cells called **osteoclasts** (os'te-o-klastz) break down the calcified extracellular matrix. These large cells originate in bone marrow when certain single-nucleated white blood cells (monocytes) fuse (see chapter 12, p. 323).

Osteoclasts secrete an acid that dissolves the inorganic component of the calcified matrix, and their lysosomal enzymes digest the organic components. After osteoclasts remove the extracellular matrix, bone-building osteoblasts invade the region and deposit new bone tissue in place of the calcified cartilage.

A long bone continues to lengthen while the cartilaginous cells of the epiphyseal plates are active (fig. 7.6). However, once the ossification centers of the diaphysis and epiphyses meet and the epiphyseal plates

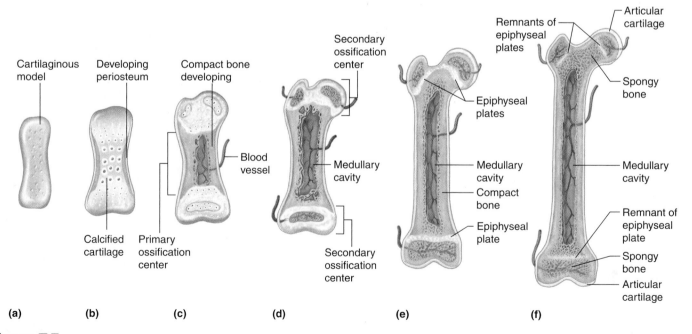

(a)   (b)   (c)   (d)   (e)   (f)

Figure 7.5

Major stages (*a–d* fetal, *e* child, *f* adult) in the development of an endochondral bone. (Relative bone sizes are not to scale.)

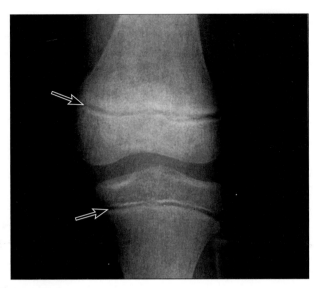

Figure 7.6

Radiograph showing epiphyseal plates (arrows) in a child's bones indicates that the bones are still lengthening.

ossify, lengthening is no longer possible in that end of the bone.

A developing long bone thickens as compact bone is deposited on the outside, just beneath the periosteum. As this compact bone forms on the surface, osteoclasts erode other bone tissue on the inside. The resulting space becomes the medullary cavity of the diaphysis, which later fills with marrow. The bone in the central regions of the epiphyses and diaphysis remains spongy, and hyaline cartilage on the ends of the epiphyses persists throughout life as articular cartilage.

If an epiphyseal plate is damaged before it ossifies, elongation of the long bone may cease prematurely, or growth may be uneven. For this reason, injuries to the epiphyses of a young person's bones are of special concern. An epiphysis is sometimes altered surgically in order to equalize the growth rate of bones developing at very different rates.

Homeostasis of Bone Tissue

After the intramembranous and endochondral bones form, the actions of osteoclasts and osteoblasts continually remodel them. Throughout life, osteoclasts resorb bone matrix, and osteoblasts replace it. Hormones that regulate blood calcium help control these opposing processes of *resorption* and *deposition* of matrix (see chapter 11, pp. 300–301). As a result, the total mass of bone tissue of an adult skeleton normally remains nearly constant, even though 3–5% of bone calcium is exchanged each year.

Factors Affecting Bone Development, Growth, and Repair

A number of factors influence bone development, growth, and repair. These include nutrition, hormonal secretions, and physical exercise. For example, vitamin D is necessary for proper absorption of calcium in the small intestine. In the absence of this vitamin, calcium (provided it is present through dietary consumption) is poorly absorbed, and the inorganic salt portion of bone matrix lacks calcium, softening and thereby deforming bones. Growth hormone secreted by the pituitary gland stimulates division of the cartilage cells in the epiphyseal plates. Sex hormones stimulate ossification of the epiphyseal plates. Physical exercise pulling on muscular attachments to bones stresses the bones, stimulating the bone tissue to thicken and strengthen. The Topic of Interest on pages 136–137 discusses bone repair.

In bone cancers, abnormally active osteoclasts destroy bone tissue. Interestingly, cancer of the prostate gland can have the opposite effect. If the cancer cells reach the bone marrow, as they do in most cases of advanced prostatic cancer, these cells stimulate osteoblast activity, which promotes formation of new bone on the surfaces of the bony plates.

Check Your Recall

5. Describe the development of an intramembranous bone.
6. Explain how an endochondral bone develops.
7. Explain how osteoclasts and osteoblasts remodel bone.
8. Explain how nutritional factors, hormones, and physical exercise affect bone development and growth.

7.4 BONE FUNCTION

Bones shape, support, and protect body structures. They also aid body movements, house tissues that produce blood cells, and store various inorganic salts.

Support and Protection

Bones give shape to structures such as the head, face, thorax, and limbs. They also provide support and protection. For example, the bones of the lower limbs, pelvis, and backbone support the body's weight. The bones of the skull protect the eyes, ears, and brain. Those of the rib cage and shoulder girdle protect the heart and lungs, whereas the bones of the pelvic girdle protect the lower abdominal and internal reproductive organs.

Topic of Interest

Bone Fractures

A *fracture* is a break in a bone. A fracture is classified by its cause as a traumatic, spontaneous, or pathologic fracture and by the nature of the break as a greenstick, fissured, comminuted, transverse, oblique, or spiral fracture (fig. 7A). A broken bone exposed to the outside by an opening in the skin is termed a compound (open) fracture.

Whenever a bone breaks, blood vessels within it rupture, and the periosteum is likely to tear. Blood escaping from the broken vessels spreads through the damaged area and soon forms a blood clot, or *hematoma*. Vessels in surrounding tissues dilate, swelling and inflaming the tissues.

Within days or weeks, developing blood vessels and large numbers of osteoblasts from the periosteum invade the hematoma. The osteoblasts rapidly divide in the regions close to the new blood vessels, building spongy bone nearby. Granulation tissue develops, and in regions farther from a blood supply, fibroblasts produce masses of fibrocartilage. Meanwhile, phagocytic cells begin to remove the blood clot, as well as any dead or damaged cells in the affected area. Osteoclasts also appear and resorb bone fragments, aiding in "cleaning up" debris.

In time, fibrocartilage fills the gap between the ends of the broken bone. This mass, termed a *cartilaginous callus,* is later replaced by bone tissue in much the same way that the hya-

A *greenstick* fracture is incomplete, and the break occurs on the convex surface of the bend in the bone.

A *fissured* fracture is an incomplete longitudinal break.

A *comminuted* fracture is complete and fragments the bone.

A *transverse* fracture is complete, and the break occurs at a right angle to the axis of the bone.

An *oblique* fracture occurs at an angle other than a right angle to the axis of the bone.

A *spiral* fracture is caused by excessive twisting of a bone.

Figure 7A
Various types of fractures.

line cartilage of a developing endochondral bone is replaced. That is, the cartilaginous callus breaks down, blood vessels and osteoblasts invade the area, and a *bony callus* fills the space.

Typically, more bone is produced at the site of a healing fracture than is required to replace the damaged tissues. Osteoclasts remove the excess, and the final result is a bone shaped very much like the original (fig. 7B).

Physicians can help the bone-healing process. The first casts to immobilize fractured bones were introduced in

Philadelphia in 1876, and soon after, doctors began using screws and plates internally to align healing bone parts. Today, orthopedic surgeons also use rods, wires, and nails. These devices have become lighter and smaller; many are built of titanium. A device called a hybrid fixator treats a broken leg using metal pins internally to align bone pieces. The pins are anchored to a metal ring device worn outside the leg.

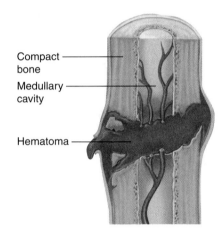

(a) Blood escapes from ruptured blood vessels and forms a hematoma.

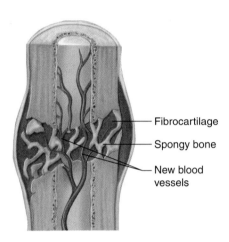

(b) Spongy bone forms in regions close to developing blood vessels, and fibrocartilage forms in more distant regions.

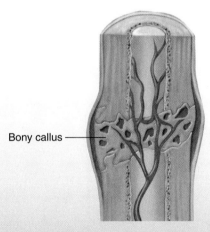

(c) A bony callus replaces fibrocartilage.

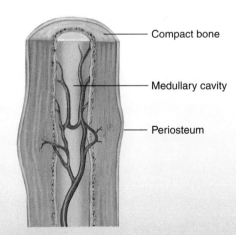

(d) Osteoclasts remove excess bony tissue, restoring new bone structure much like the original.

Figure 7B
Major steps (*a–d*) in repair of a fracture.

Body Movement

Whenever limbs or other body parts move, bones and muscles interact as simple mechanical devices called **levers** (lev′erz). A lever has four basic components: (1) a rigid bar or rod, (2) a fulcrum or pivot on which the bar turns, (3) an object that is moved against resistance, and (4) a force that supplies energy to move the bar.

The actions of bending and straightening the upper limb at the elbow illustrate bones and muscles functioning as levers. When the upper limb bends, the forearm bones represent the rigid bar, the elbow joint is the fulcrum, the hand is moved against the resistance provided by the weight, and the force is supplied by muscles on the anterior side of the arm (fig. 7.7*a*). One of these muscles, the *biceps brachii,* is attached by a tendon to a projection on a bone (radius) in the forearm, a short distance below the elbow.

When the upper limb straightens at the elbow, the forearm bones again serve as the rigid bar, the elbow joint serves as the fulcrum, and the hand moves against the resistance by pulling on the rope to raise the weight (fig. 7.7*b*). However, this time, the *triceps brachii,* a muscle located on the posterior side of the arm, supplies the force. A tendon of this muscle attaches to a projection on a forearm bone (ulna) at the point of the elbow.

Blood Cell Formation

The process of blood cell formation, called **hematopoiesis** (he″mă-to-poi-e′sis), begins in the *yolk sac,* which lies outside the human embryo (see chapter 20, p. 549). Later in development, blood cells are manufactured in the liver and spleen, and still later, they form in bone marrow.

Marrow is a soft, netlike mass of connective tissue within the medullary cavities of long bones, in the irregular spaces of spongy bone, and in the larger central canals of compact bone tissue. The two kinds of marrow are red marrow and yellow marrow. *Red marrow* functions in the formation of red blood cells (erythrocytes), white blood cells (leukocytes), and blood platelets. Red marrow's color comes from the oxygen-carrying pigment **hemoglobin** in the red blood cells.

In an infant, red marrow occupies the cavities of most bones. With increasing age, however, yellow marrow replaces much of it. *Yellow marrow* stores fat; it is not active in blood cell production. In an adult, red marrow is primarily found in the spongy bone of the skull, ribs, sternum, clavicles, vertebrae, and hip bones. However, if the body requires more blood, yellow marrow can be replaced by extensions of red bone marrow from elsewhere in the bone, which then reverts to yellow marrow when there is enough or a surplus of blood. Chapter 12 (pp. 320, 323, and 325) describes blood cell formation in more detail.

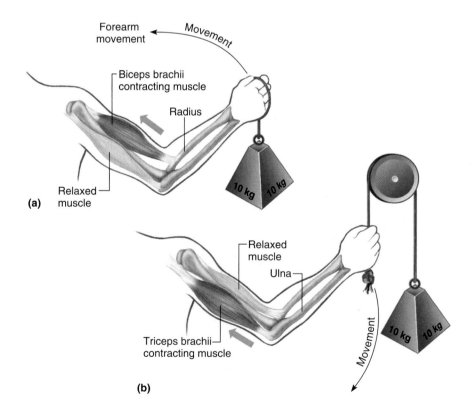

(a)

(b)

Figure 7.7

Bones and muscles form lever systems when they interact to move body parts. (*a*) When the forearm bends at the elbow or (*b*) when the forearm straightens at the elbow, the bones and muscles function as a lever.

In bone marrow transplant (BMT) a hollow needle and syringe are used to remove normal red marrow cells from the spongy bone of a donor, or stem cells (which can give rise to specialized blood cells) are separated out from the donor's bloodstream. Stem cells from the umbilical cord of a newborn can be used in place of bone marrow.

The donor is selected because the pattern of molecules on his or her cell surfaces closely matches that of the recipient. In 30% of BMTs, the donor is a blood relative. The cells are injected into the bloodstream of the recipient, whose own marrow has been intentionally destroyed with radiation or chemotherapy. If all goes well, the donor cells travel to the spaces within bones that red marrow normally occupies and replenish the blood supply with healthy cells. About 15% of the time, the patient dies from infection because the immune system rejects the transplant, or because the transplanted tissue attacks the recipient, a condition called graft-versus-host disease.

BMT is used to treat more than sixty types of illnesses, mostly blood disorders such as sickle cell disease and leukemias. In cancer treatment, BMTs enable a patient to withstand high doses of radiation or chemotherapy, which usually damages bone marrow. BMT is used when other cancer treatments have failed. Bone marrow may become a major part of "regenerative medicine," because it contains a variety of stem cells that can replenish many types of tissues. In one clinical trial, for example, a patient's own bone marrow progenitor cells are taken from the hip bone and injected into failing heart muscle.

Storage of Inorganic Salts

Bones store calcium. The extracellular matrix of bone tissue is rich in calcium salts, mostly in the form of calcium phosphate. Vital metabolic processes require calcium. When the blood is low in calcium, parathyroid hormone stimulates osteoclasts to break down bone tissue, which releases calcium salts from the extracellular matrix into the blood. A high blood calcium level inhibits osteoclast activity, and calcitonin from the thyroid gland stimulates osteoblasts to form bone tissue, storing excess calcium in the extracellular matrix (fig. 7.8). Chapter 11 (p. 300) describes the details of this homeostatic mechanism. Maintaining sufficient blood calcium levels is important in muscle contraction, nerve impulse conduction, blood clotting, and other physiological processes.

Bone tissue includes fewer magnesium, sodium, potassium, and carbonate ions than it does other constituents. Bones also accumulate certain harmful metallic elements, such as lead, radium, or strontium. These are not normally present in the body, but are sometimes ingested accidentally.

Check Your Recall

9. Name the major functions of bones.
10. Distinguish between the functions of red marrow and yellow marrow.
11. List the substances normally stored in bone tissue.

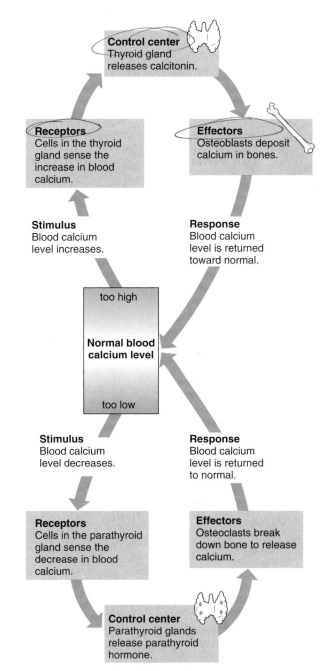

Figure 7.8
Hormones regulate deposition and resorption of bone calcium.

7.5 SKELETAL ORGANIZATION

For purposes of study, it is convenient to divide the skeleton into two major portions—an axial skeleton and an appendicular skeleton (fig. 7.9). The **axial skeleton** consists of the bony and cartilaginous parts that support and protect the organs of the head, neck, and trunk. These parts include:

1. **Skull** The skull is composed of the **cranium** (kra′ne-um), or brain case, and the *facial bones*.

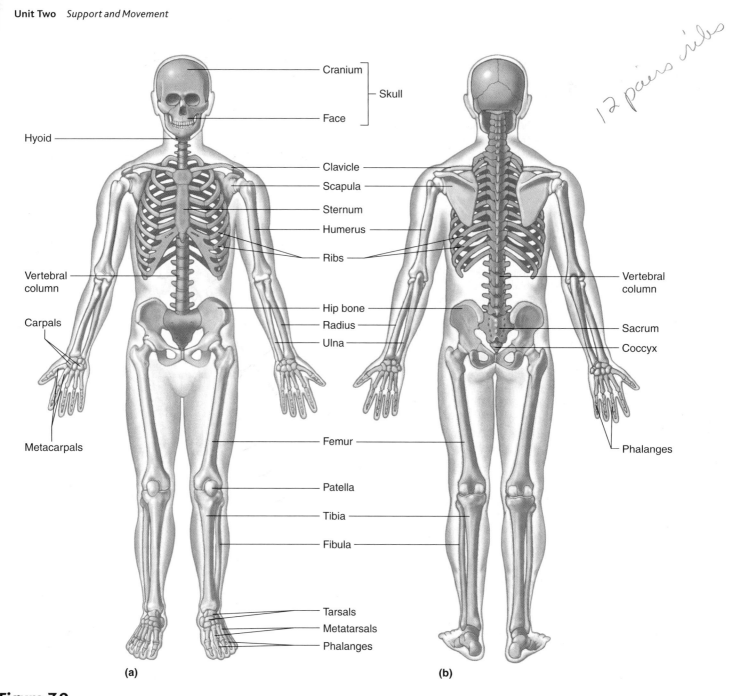

12 pairs ribs

(a) **(b)**

Figure 7.9

Major bones of the skeleton. (*a*) Anterior view. (*b*) Posterior view. The axial portion is shown in orange, and the appendicular portions are shown in yellow.

2. **Hyoid bone** The hyoid (hi′oid) bone is located in the neck between the lower jaw and the larynx. It supports the tongue and is an attachment for certain muscles that help move the tongue during swallowing.

3. **Vertebral column** The vertebral column (backbone) consists of many vertebrae separated by cartilaginous *intervertebral discs*. Near its distal end, several vertebrae fuse to form the **sacrum** (sa′krum), which is part of the pelvis. The **coccyx**

(kok′siks), a small, rudimentary tailbone composed of several fused vertebrae, is attached to the end of the sacrum.

4. **Thoracic cage** The thoracic cage protects the organs of the thoracic cavity and the upper abdominal cavity. It is composed of twelve pairs of **ribs,** which articulate posteriorly with thoracic vertebrae. The thoracic cage also includes the **sternum** (ster′num), or breastbone, to which most of the ribs attach anteriorly.

The **appendicular skeleton** consists of the bones of the upper and lower limbs and the bones that anchor the limbs to the axial skeleton. It includes:

1. **Pectoral** (pek′to-ral) **girdle** A **scapula** (scap′u-lah) and a **clavicle** (klav′ĭ-k′l) bone form the pectoral girdle on both sides of the body. The pectoral girdle connects the bones of the upper limbs to the axial skeleton and aids in upper limb movements.
2. **Upper limbs** Each upper limb consists of a **humerus** (hu′mer-us), or arm bone, two forearm bones—a **radius** (ra′de-us) and an **ulna** (ul′nah)—and a hand. The humerus, radius, and ulna articulate with each other at the elbow joint. At the distal end of the radius and ulna is the hand. There are eight **carpals** (kar′pals), or wrist bones. The five bones of the palm are called **metacarpals** (met″ah-kar′pals), and the fourteen finger bones are called **phalanges** (fah-lan′jēz); singular, *phalanx* (fa′lanks).
3. **Pelvic girdle** Two hip bones form the pelvic girdle and are attached to each other anteriorly and to the sacrum posteriorly. They connect the bones of the lower limbs to the axial skeleton and, with the sacrum and coccyx, form the **pelvis.**

4. **Lower limbs** Each lower limb consists of a **femur** (fe′mur), or thigh bone, two leg bones—a large **tibia** (tib′e-ah) and a slender **fibula** (fib′u-lah)—and a foot. The femur and tibia articulate with each other at the knee joint, where the **patella** (pah-tel′ah) covers the anterior surface. At the distal ends of the tibia and fibula is the foot. There are seven **tarsals** (tahr′sals), or ankle bones. The five bones of the instep are called **metatarsals** (met″ah-tahr′sals), and the fourteen bones of the toes (like the fingers) are called **phalanges.**

Table 7.1 lists the bones of the adult skeleton, and table 7.2 lists terms that describe skeletal structures.

> The skeleton of an average 160-pound body weighs about 29 pounds.

Check Your Recall

12. Distinguish between the axial and appendicular skeletons.
13. List the bones of the axial skeleton and of the appendicular skeleton.

Table 7.1	Bones of the Adult Skeleton

1. Axial Skeleton

a. Skull
  8 cranial bones
    frontal 1   temporal 2
    parietal 2   sphenoid 1
    occipital 1   ethmoid 1
  14 facial bones
    maxilla 2   lacrimal 2
    zygomatic 2   nasal 2
    palatine 2   vomer 1
    inferior nasal concha 2
    mandible 1
                                    22 bones

b. Middle ear bones
    malleus 2
    incus 2
    stapes 2
                                    6 bones

c. Hyoid
    hyoid bone 1
                                    1 bone

d. Vertebral column
    cervical vertebrae 7
    thoracic vertebrae 12
    lumbar vertebrae 5
    sacrum 1
    coccyx 1
                                    26 bones

e. Thoracic cage
    rib 24
    sternum 1
                                    25 bones

2. Appendicular Skeleton

a. Pectoral girdle
    scapula 2
    clavicle 2
                                    4 bones

b. Upper limbs
    humerus 2
    radius 2
    ulna 2
    carpal 16
    metacarpal 10
    phalanx 28
                                    60 bones

c. Pelvic girdle
    hip bone 2
                                    2 bones

d. Lower limbs
    femur 2
    tibia 2
    fibula 2
    patella 2
    tarsal 14
    metatarsal 10
    phalanx 28
                                    60 bones

    Total   206 bones

Table 7.2	Terms Used to Describe Skeletal Structures	
Term	**Definition**	**Examples**
Condyle (kon′dīl)	A rounded process that usually articulates with another bone	Occipital condyle of occipital bone (fig. 7.13)
Crest (krest)	A narrow, ridgelike projection	Iliac crest of ilium (fig. 7.28)
Epicondyle (ep″ĭ-kon′dīl)	A projection situated above a condyle	Medial epicondyle of humerus (fig. 7.24)
Facet (fas′et)	A small, nearly flat surface	Rib facet of thoracic vertebra (fig. 7.17)
Fontanel (fon″tah-nel′)	A soft spot in the skull where membranes cover the space between bones	Anterior fontanel between frontal and parietal bones (fig. 7.16)
Foramen (fo-ra′men)	An opening through a bone that usually is a passageway for blood vessels, nerves, or ligaments	Foramen magnum of occipital bone (fig. 7.13)
Fossa (fos′ah)	A relatively deep pit or depression	Olecranon fossa of humerus (fig. 7.24)
Fovea (fo′ve-ah)	A tiny pit or depression	Fovea capitis of femur (fig. 7.30)
Head (hed)	An enlargement on the end of a bone	Head of humerus (fig. 7.24)
Meatus (me-a′tus)	A tubelike passageway within a bone	External acoustic meatus of ear (fig. 7.12)
Process (pros′es)	A prominent projection on a bone	Mastoid process of temporal bone (fig. 7.12)
Sinus (si′nus)	A cavity within a bone	Frontal sinus of frontal bone (fig. 7.15)
Spine (spīn)	A thornlike projection	Spine of scapula (fig. 7.23)
Suture (soo′cher)	An interlocking line of union between bones	Lambdoid suture between occipital and parietal bones (fig. 7.12)
Trochanter (tro-kan′ter)	A relatively large process	Greater trochanter of femur (fig. 7.30)
Tubercle (tu′ber-kl)	A small, knoblike process	Greater tubercle of humerus (fig. 7.24)
Tuberosity (tu″bĕ-ros′ĭ-te)	A knoblike process usually larger than a tubercle	Radial tuberosity of radius (fig. 7.25)

7.6 SKULL

A human skull usually consists of twenty-two bones that, except for the lower jaw, are firmly interlocked along *sutures* (soo′cherz) (fig. 7.10). Eight of these interlocked bones make up the cranium, and fourteen form the facial skeleton. The **mandible** (man′dĭ-b′l), or lower jawbone, is a movable bone held to the cranium by ligaments. (Three other bones in each middle ear are discussed in chapter 10, p. 269.) Reference plates 8–11 on pages 173–175 show the human skull and its parts.

Cranium

The **cranium** encloses and protects the brain, and its surface provides attachments for muscles that make chewing and head movements possible. Some of the cranial bones contain air-filled cavities called *paranasal sinuses,* which are lined with mucous membranes and connected by passageways to the nasal cavity (fig. 7.11). Sinuses reduce the skull's weight and increase the intensity of the voice by serving as resonant sound chambers.

The eight bones of the cranium, shown in figures 7.10 and 7.12, are:

1. **Frontal bone** The frontal (frun′tal) bone forms the anterior portion of the skull above the eyes. On the upper margin of each orbit (the bony socket of the eye), the frontal bone is marked by a *supraorbital foramen* (or *supraorbital notch* in some skulls), through which blood vessels and nerves pass to the tissues of the forehead. Within the frontal bone are two *frontal sinuses,* one above each eye near the midline (see fig. 7.11).

2. **Parietal bones** One parietal (pah-ri′ĕ-tal) bone is located on each side of the skull just behind the frontal bone (fig. 7.12). Together, the parietal bones form the bulging sides and roof of the cranium. They are fused at the midline along the *sagittal suture,* and they meet the frontal bone along the *coronal suture.*

3. **Occipital bone** The occipital (ok-sip′ĭ-tal) bone joins the parietal bones along the *lambdoid* (lam′doid) *suture* (figs. 7.12 and 7.13). It forms the back of the skull and the base of the cranium. Through a large opening on its lower surface called the *foramen magnum* pass nerve fibers from the brain, which enter the vertebral canal to become part of the spinal cord. Rounded processes called *occipital condyles,* located on each side of the foramen magnum, articulate with the first vertebra (atlas) of the vertebral column.

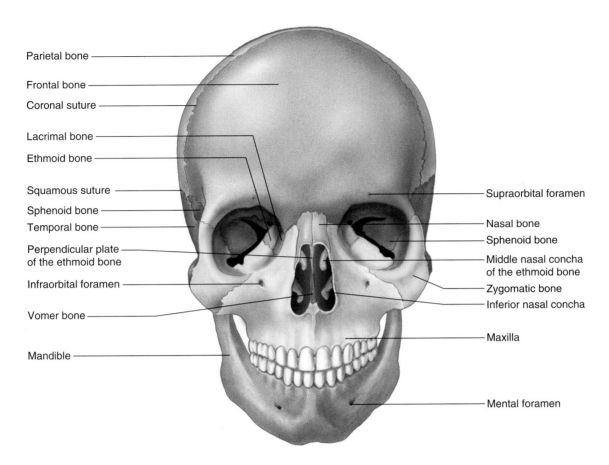

Parietal bone

Frontal bone

Coronal suture

Lacrimal bone

Ethmoid bone

Squamous suture

Sphenoid bone

Temporal bone

Perpendicular plate
of the ethmoid bone

Infraorbital foramen

Vomer bone

Mandible

Supraorbital foramen

Nasal bone

Sphenoid bone

Middle nasal concha
of the ethmoid bone

Zygomatic bone

Inferior nasal concha

Maxilla

Mental foramen

Figure 7.10
Anterior view of the skull.

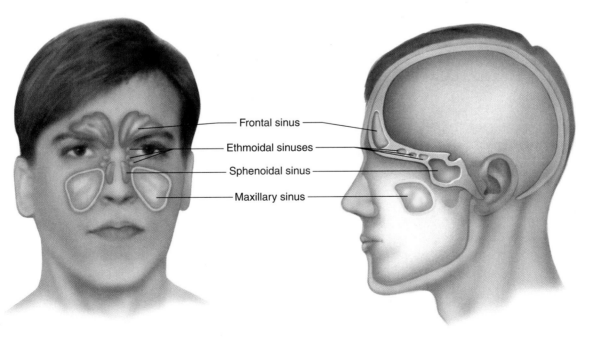

Frontal sinus

Ethmoidal sinuses

Sphenoidal sinus

Maxillary sinus

Figure 7.11
Locations of the paranasal sinuses.

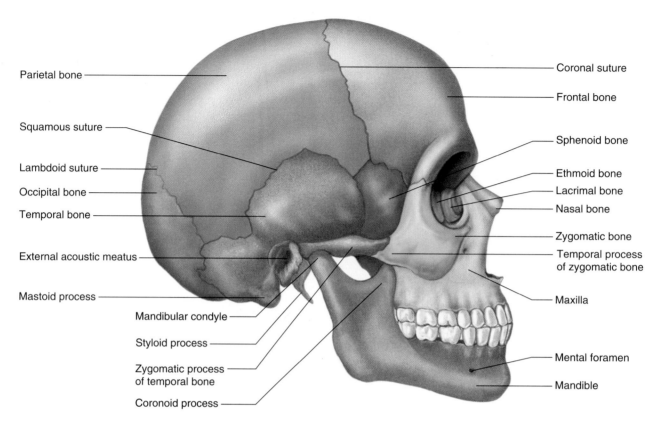

Figure 7.12

Right lateral view of the skull.

Parietal bone

Squamous suture

Lambdoid suture

Occipital bone

Temporal bone

External acoustic meatus

Mastoid process

Mandibular condyle

Styloid process

Zygomatic process
of temporal bone

Coronoid process

Coronal suture

Frontal bone

Sphenoid bone

Ethmoid bone

Lacrimal bone

Nasal bone

Zygomatic bone

Temporal process
of zygomatic bone

Maxilla

Mental foramen

Mandible

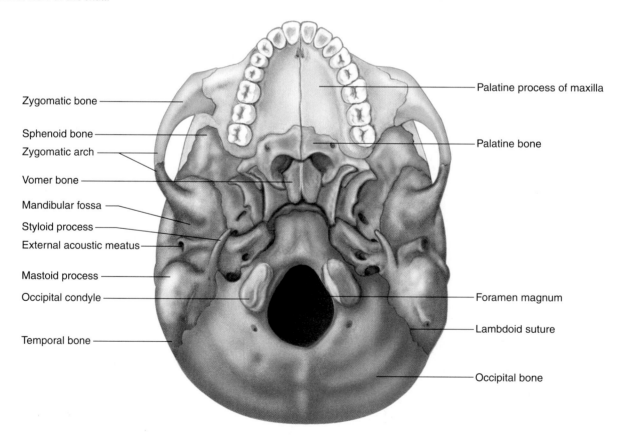

Figure 7.13

Inferior view of the skull.

Zygomatic bone

Sphenoid bone

Zygomatic arch

Vomer bone

Mandibular fossa

Styloid process

External acoustic meatus

Mastoid process

Occipital condyle

Temporal bone

Palatine process of maxilla

Palatine bone

Foramen magnum

Lambdoid suture

Occipital bone

4. **Temporal bones** A temporal (tem′po-ral) bone on each side of the skull joins the parietal bone along a *squamous suture* (see figs. 7.10 and 7.12). The temporal bones form parts of the sides and the base of the cranium. Located near the inferior margin is an opening, the *external acoustic meatus,* which leads inward to parts of the ear. The temporal bones have depressions called the *mandibular fossae* that articulate with condyles of the mandible. Below each external acoustic meatus are two projections—a rounded *mastoid process* and a long, pointed *styloid process.* The mastoid process provides an attachment for certain muscles of the neck, whereas the styloid process anchors muscles associated with the tongue and pharynx. A *zygomatic process* projects anteriorly from the temporal bone, joins the *zygomatic bone,* and helps form the prominence of the cheek.

5. **Sphenoid bone** The sphenoid (sfe′noid) bone is wedged between several other bones in the anterior portion of the cranium (figs. 7.12 and 7.13). This bone helps form the base of the cranium, the sides of the skull, and the floors and sides of the orbits. Along the midline within the cranial cavity, a portion of the sphenoid bone indents to form the saddle-shaped *sella turcica* (sel′ah tur′si-ka). The pituitary gland occupies this depression. The sphenoid bone also contains two *sphenoidal sinuses* (see fig. 7.11).

6. **Ethmoid bone** The ethmoid (eth′moid) bone is located in front of the sphenoid bone (figs. 7.12 and 7.14). It consists of two masses, one on each side of the nasal cavity, which are joined horizontally by thin *cribriform* (krib′rĭ-form) *plates.* These plates form part of the roof of the nasal cavity (fig. 7.14).

Projecting upward into the cranial cavity between the cribriform plates is a triangular process of the ethmoid bone called the *crista galli* (kris′tă gal′li) (cock's comb). Membranes that enclose the brain attach to this process (figs. 7.14 and 7.15). Portions of the ethmoid bone also form sections of the cranial floor, the orbital walls, and the nasal cavity walls. A *perpendicular plate* projects downward in the midline from the cribriform plates and forms most of the nasal septum (fig. 7.15).

Delicate scroll-shaped plates called the *superior nasal conchae* (kong′ke) and the *middle nasal conchae* project inward from the lateral portions of the ethmoid

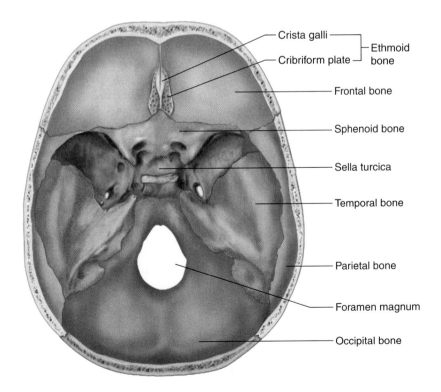

Figure 7.14

Floor of the cranial cavity, viewed from above.

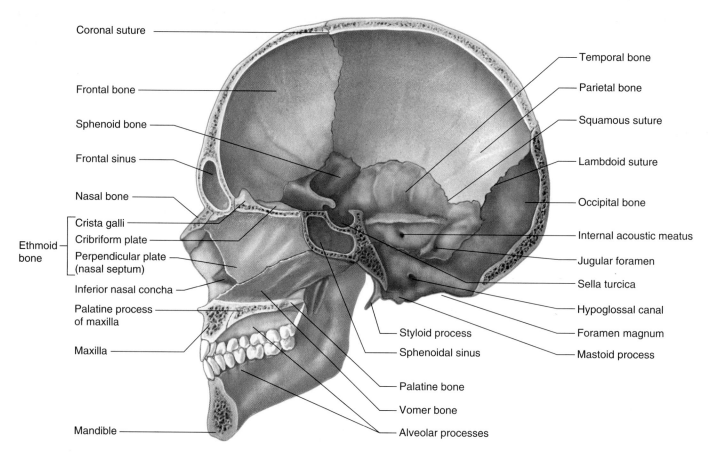

Figure 7.15
Sagittal section through the skull.

bone toward the perpendicular plate (see fig. 7.10). The lateral portions of the ethmoid bone contain many small air spaces, the *ethmoidal sinuses* (see fig. 7.11).

Facial Skeleton

The **facial skeleton** consists of thirteen immovable bones and a movable lower jawbone. These bones form the basic shape of the face and provide attachments for muscles that move the jaw and control facial expressions. The bones of the facial skeleton are:

1. **Maxillae** The maxillae (mak-sil′e; singular, *maxilla,* mak-sil′ah) form the upper jaw (see figs. 7.12 and 7.13). Portions of these bones comprise the anterior roof of the mouth (*hard palate*), the floors of the orbits, and the sides and floor of the nasal cavity. They also contain the sockets of the upper teeth. Inside the maxillae, lateral to the nasal cavity, are *maxillary sinuses,* the largest of the sinuses (see fig. 7.11).

During development, portions of the maxillae called *palatine processes* grow together and fuse along the midline to form the anterior section of the hard palate. The inferior border of each maxillary bone projects downward, forming an *alveolar* (al-ve′o-lar) *process* (fig. 7.15). Together, these processes form a horseshoe-shaped *alveolar arch* (dental arch). Teeth occupy cavities in this arch (dental alveoli). Dense connective tissue binds teeth to the bony sockets.

Sometimes, fusion of the palatine processes of the maxillae is incomplete at birth; the result is a *cleft palate*. Infants with a cleft palate may have trouble suckling because of the opening between the oral and nasal cavities. A temporary prosthetic device (artificial palate) may be inserted into the mouth, or a special type of nipple can be placed on bottles until surgery can be performed to correct the condition.

2. **Palatine bones** The L-shaped palatine (pal′ah-tīn) bones are located behind the maxillae (see figs. 7.13 and 7.15). The horizontal portions form the posterior section of the hard palate and the floor of the nasal cavity. The perpendicular portions help form the lateral walls of the nasal cavity.

3. **Zygomatic bones** The zygomatic (zi″go-mat′ik) bones form the prominences of the cheeks below and to the sides of the eyes (see figs. 7.12 and 7.13). These bones also help form the lateral walls and the floors of the orbits. Each bone has a *temporal process,* which extends posteriorly to join the zygomatic process of a temporal bone. Together, these processes form a *zygomatic arch.*

4. **Lacrimal bones** A lacrimal (lak′rĭ-mal) bone is a thin, scalelike structure located in the medial wall of each orbit between the ethmoid bone and the maxilla (see figs. 7.10 and 7.12).

5. **Nasal bones** The nasal (na′zal) bones are long, thin, and nearly rectangular (see figs. 7.10 and 7.12). They lie side by side and are fused at the midline, where they form the bridge of the nose.

6. **Vomer bone** The thin, flat vomer (vo′mer) bone is located along the midline within the nasal cavity (see figs. 7.10 and 7.15). Posteriorly, it joins the perpendicular plate of the ethmoid bone, and together they form the nasal septum.

7. **Inferior nasal conchae** The inferior nasal conchae are fragile, scroll-shaped bones attached to the lateral walls of the nasal cavity (see figs. 7.10 and 7.15). Like the superior and middle conchae, the inferior conchae support mucous membranes within the nasal cavity.

8. **Mandible** The mandible is a horizontal, horseshoe-shaped body with a flat portion projecting upward at each end (see figs. 7.10 and 7.12). This projection is divided into two processes—a posterior *mandibular condyle* and an anterior *coronoid process.* The mandibular condyles articulate with the mandibular fossae of the temporal bones (see fig. 7.13), whereas the coronoid processes provide attachments for muscles used in chewing. A curved bar of bone on the superior border of the mandible, the *alveolar arch,* contains the hollow sockets (dental alveoli) that bear the lower teeth.

Infantile Skull

At birth, the skull is incompletely developed, with fibrous membranes connecting the cranial bones. These membranous areas of incomplete intramembranous ossification are called **fontanels** (fon″tah-nelz′) or, more commonly, soft spots (fig. 7.16). They permit some movement between the bones, so that the developing skull is partially compressible and can slightly change shape. This enables an infant's skull to more easily pass through the birth canal. Eventually, the fontanels close as the cranial bones grow together.

Other characteristics of an infantile skull include a relatively small face with a prominent forehead and large orbits. The jaw and nasal cavity are small, the sinuses are incompletely formed, and the frontal bone is in two parts. The skull bones are thin, but they are also somewhat flexible and thus are less easily fractured than adult skull bones.

Check Your Recall

14. Locate and name each of the bones of the cranium.
15. Locate and name each of the facial bones.
16. Explain how an adult skull differs from that of an infant.

7.7 VERTEBRAL COLUMN

The **vertebral column** extends from the skull to the pelvis and forms the vertical axis of the skeleton. It is composed of many bony parts, called **vertebrae** (ver′tĕ-brā), that are separated by masses of fibrocartilage called *intervertebral discs* and are connected to one another by ligaments (fig. 7.17). The vertebral column supports the head and trunk of the body. It also protects the spinal cord, which passes through a *vertebral canal* formed by openings in the vertebrae.

A Typical Vertebra

Although vertebrae in different regions of the vertebral column have special characteristics, they also have features in common. A typical vertebra has a drum-shaped *body,* which forms the thick, anterior portion of the bone (fig. 7.18). A longitudinal row of these vertebral bodies supports the weight of the head and trunk. The intervertebral discs, which separate adjacent vertebral bodies, cushion and soften the forces from movements such as walking and jumping.

Projecting posteriorly from each vertebral body are two short stalks called *pedicles* (ped′ĭ-k'lz). Two plates called *laminae* (lam′i-ne) arise from the pedicles and fuse in the back to become a *spinous process.* The pedicles, laminae, and spinous process together complete a bony *vertebral arch* around the *vertebral foramen,* through which the spinal cord passes.

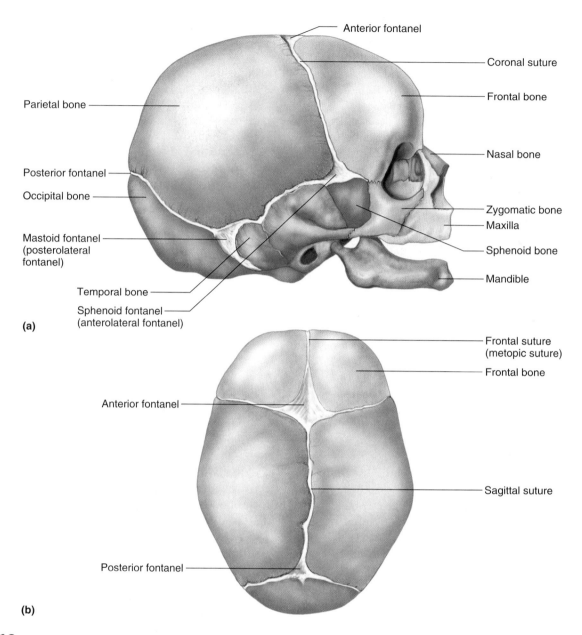

Figure 7.16

Fontanels. (*a*) Right lateral view and (*b*) superior view of the infantile skull.

If the laminae of the vertebrae fail to unite during development, the vertebral arch remains incomplete, causing a condition called *spina bifida*. The contents of the vertebral canal protrude outward. This problem occurs most frequently in the lumbosacral region. Spina bifida is associated with folic acid deficiency in certain genetically susceptible individuals.

Between the pedicles and laminae of a typical vertebra is a *transverse process*, which projects laterally and posteriorly. Ligaments and muscles are attached to the dorsal spinous process and the transverse processes. Projecting upward and downward from each vertebral arch are *superior* and *inferior articular processes*. These processes bear cartilage-covered facets by which each vertebra is joined to the one above and the one below it.

On the lower surfaces of the vertebral pedicles are notches that align with adjacent vertebrae to form openings called *intervertebral foramina* (in"ter-ver'tĕ-bral fo-ram'ĭ-nah) (see fig. 7.17). These openings provide passageways for spinal nerves.

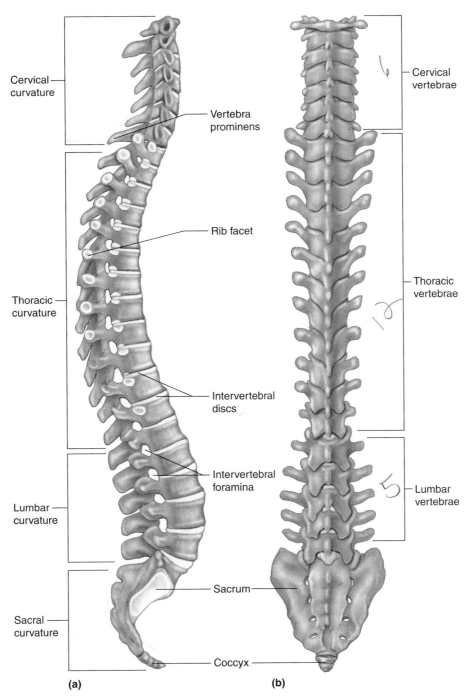

Cervical
curvature

Vertebra
prominens

Rib facet

Thoracic
curvature

Intervertebral
discs

Intervertebral
foramina

Lumbar
curvature

Sacral
curvature

Cervical
vertebrae

Thoracic
vertebrae

Lumbar
vertebrae

Sacrum

Coccyx

(a)   (b)

Figure 7.17

The curved vertebral column consists of many vertebrae separated by intervertebral discs. (*a*) Right lateral view. (*b*) Posterior view.

Cervical Vertebrae

Seven **cervical vertebrae** comprise the bony axis of the neck (see fig. 7.17). The transverse processes of these vertebrae are distinctive because they have *transverse foramina,* which are passageways for arteries leading to the brain (see fig. 7.18*a*). Also, the spinous processes of the second through the fifth cervical vertebrae are uniquely forked (bifid). These processes provide attachments for muscles.

Two of the cervical vertebrae are of special interest: the atlas and the axis (fig. 7.19). The first vertebra, or **atlas** (at′las), supports the head. On its superior surface are two kidney-shaped *facets* that articulate with the occipital condyles.

The second cervical vertebra, or **axis** (ak′sis), bears a toothlike *dens* (odontoid process) on its body. This process projects upward and lies in the ring of the atlas. As the head is turned from side to side, the atlas pivots around the dens.

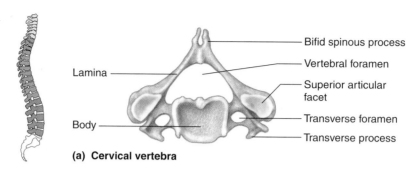

(a) Cervical vertebra

- Bifid spinous process
- Vertebral foramen
- Superior articular facet
- Transverse foramen
- Transverse process
- Lamina
- Body

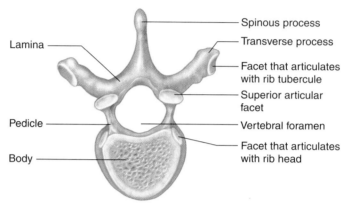

(b) Thoracic vertebra

- Spinous process
- Transverse process
- Facet that articulates with rib tubercule
- Superior articular facet
- Vertebral foramen
- Facet that articulates with rib head
- Lamina
- Pedicle
- Body

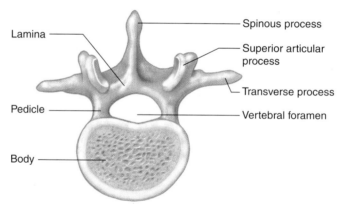

(c) Lumbar vertebra

- Spinous process
- Superior articular process
- Transverse process
- Vertebral foramen
- Lamina
- Pedicle
- Body

Figure 7.18

Superior view of (*a*) a cervical vertebra, (*b*) a thoracic vertebra, and (*c*) a lumbar vertebra.

Giraffes and humans have the same number of vertebrae in their necks . . . seven.

Thoracic Vertebrae

The twelve **thoracic vertebrae** are larger than the cervical vertebrae (see fig. 7.17). Each vertebra has a long, pointed spinous process, which slopes downward, and facets on the sides of its body, which articulate with a rib (see fig. 7.18*b*).

Beginning with the third thoracic vertebra and moving inferiorly, the bodies of these bones increase in size. Thus, they are adapted to bear increasing loads of body weight.

Lumbar Vertebrae

Five **lumbar vertebrae** are in the small of the back (loin) (see fig. 7.17). These vertebrae are adapted with larger and stronger bodies to support more weight than the vertebrae above them (see fig. 7.18*c*).

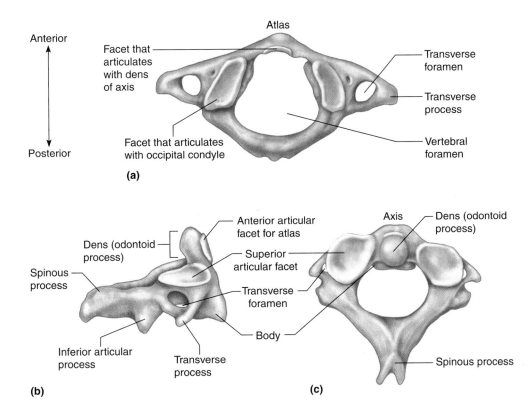

Figure 7.19

Atlas and axis. (a) Superior view of the atlas. (b) Right lateral view and (c) superior view of the axis.

Sacrum Base Vertebrae Column

The **sacrum** (sa′krum) is a triangular structure, composed of five fused vertebrae, that forms the base of the vertebral column (fig. 7.20). The spinous processes of these fused bones form a ridge of *tubercles*. To the sides of the tubercles are rows of openings, the *posterior sacral foramina,* through which nerves and blood vessels pass.

The vertebral foramina of the sacral vertebrae form the *sacral canal,* which continues through the sacrum to an opening of variable size at the tip, called the *sacral hiatus* (sa′kral hi-a′tus). On the ventral surface of the sacrum, four pairs of *anterior sacral foramina* provide passageways for nerves and blood vessels.

Coccyx

The **coccyx** (kok′siks), or tailbone, is the lowest part of the vertebral column and is usually composed of four fused vertebrae (fig. 7.20). Ligaments attach it to the margins of the sacral hiatus.

Changes in the intervertebral discs can cause back problems. Each disc is composed of a tough outer layer of fibrocartilage and an elastic central mass. With age, these discs degenerate—the central masses lose firmness, and the outer layers thin and weaken, developing cracks. Extra pressure, as when a person falls or lifts a heavy object, can break the outer layer of a disc, squeezing out the central mass. Such a rupture may press on the spinal cord or on a spinal nerve that branches from it. This condition—a ruptured or herniated disc—may cause back pain and numbness or the loss of muscular function in the parts innervated by the affected spinal nerve.

Check Your Recall

17. Describe the structure of the vertebral column.

18. Describe a typical vertebra.

19. Explain how the structures of cervical, thoracic, and lumbar vertebrae differ.

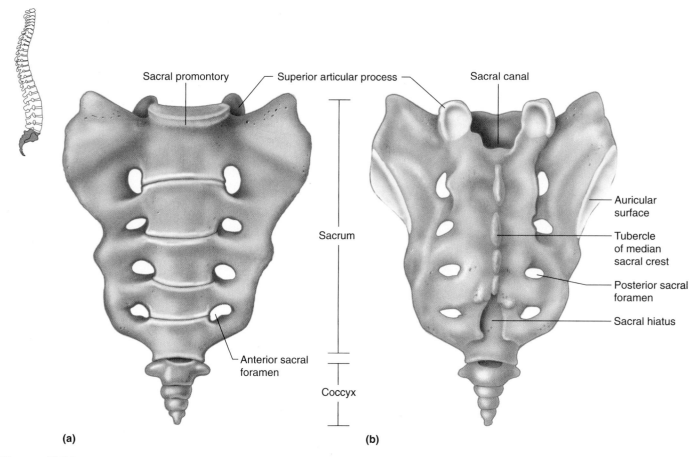

Sacral promontory — Superior articular process — Sacral canal

Sacrum

Anterior sacral foramen

Coccyx

Auricular surface

Tubercle of median sacral crest

Posterior sacral foramen

Sacral hiatus

(a) (b)

Figure 7.20

Sacrum and coccyx. (*a*) Anterior view and (*b*) posterior view.

7.8 THORACIC CAGE

The **thoracic cage** includes the ribs, the thoracic vertebrae, the sternum, and the costal cartilages that attach the ribs to the sternum (fig. 7.21). These bones support the pectoral girdle and upper limbs, protect the viscera in the thoracic and upper abdominal cavities, and play a role in breathing.

Ribs

The usual number of **ribs** is twenty-four—one pair attached to each of the twelve thoracic vertebrae. The first seven rib pairs, *true ribs* (vertebrosternal ribs), join the sternum directly by their costal cartilages. The remaining five pairs are called *false ribs,* because their cartilages do not reach the sternum directly. Instead, the cartilages of the upper three false ribs (vertebrochondral ribs) join the cartilages of the seventh rib. The last two (or sometimes three) rib pairs are called *floating ribs* (vertebral ribs) because they have no cartilaginous attachments to the sternum.

A typical rib has a long, slender shaft, which curves around the chest and slopes downward. On the posterior end is an enlarged *head* by which the rib articulates with a *facet* on the body of its own vertebra and with the body of the next higher vertebra. A *tubercle,* close to the head of the rib, articulates with the transverse process of the vertebra.

Sternum

The **sternum,** or breastbone, is located along the midline in the anterior portion of the thoracic cage (fig. 7.21). This flat, elongated bone develops in three parts—an upper *manubrium* (mah-nu′bre-um), a middle *body,* and a lower *xiphoid* (zīf′oid) *process* that projects downward. The manubrium articulates with the clavicles by facets on its superior border.

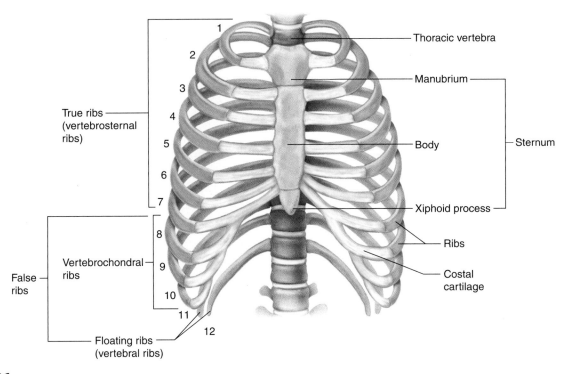

Figure 7.21
The thoracic cage includes the ribs, the thoracic vertebrae, the sternum, and the costal cartilages that attach the ribs to the sternum.

The clavicles brace the freely movable scapulae, helping to hold the shoulders in place. They also provide attachments for muscles of the upper limbs, chest, and back.

Scapulae

The **scapulae** (skap'u-le), or shoulder blades, are broad, somewhat triangular bones located on either side of the upper back (figs. 7.22 and 7.23). A *spine* divides the posterior surface of each scapula into unequal portions. This spine leads to two processes—an *acromion* (ah-kro'me-on) *process* that forms the tip of the shoulder and a *coracoid* (kor'ah-koid) *process* that curves anteriorly and inferiorly to the clavicle. The acromion process articulates with the clavicle and provides attachments for muscles of the upper limb and chest. The coracoid process also provides attachments for upper limb and chest muscles. Between the processes is a depression called the *glenoid cavity* (glenoid fossa of the scapula) that articulates with the head of the arm bone (humerus).

> **Check Your Recall**
> 20. Which bones compose the thoracic cage?
> 21. What are the differences among true, false, and floating ribs?
> 22. Name the three parts of the sternum.

7.9 PECTORAL GIRDLE

The **pectoral girdle,** or shoulder girdle, is composed of four parts—two clavicles and two scapulae (fig. 7.22). Although the word *girdle* suggests a ring-shaped structure, the pectoral girdle is an incomplete ring. It is open in the back between the scapulae, and the sternum separates its bones in front. The pectoral girdle supports the upper limbs and is an attachment for several muscles that move them.

Clavicles

The **clavicles,** or collarbones, are slender, rodlike bones with elongated S shapes (fig. 7.22). Located at the base of the neck, they run horizontally between the manubrium and the scapulae.

> **Check Your Recall**
> 23. Which bones form the pectoral girdle?
> 24. What is the function of the pectoral girdle?

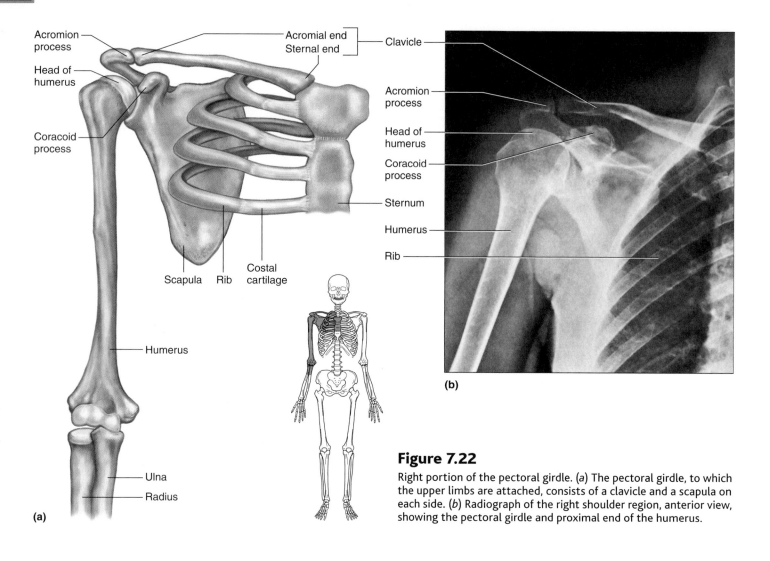

(a)

(b)

Figure 7.22

Right portion of the pectoral girdle. (*a*) The pectoral girdle, to which the upper limbs are attached, consists of a clavicle and a scapula on each side. (*b*) Radiograph of the right shoulder region, anterior view, showing the pectoral girdle and proximal end of the humerus.

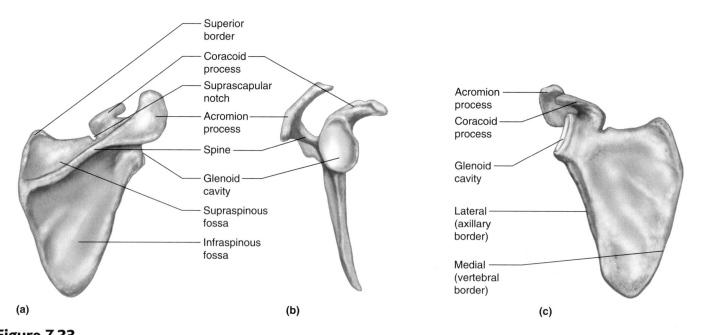

(a) **(b)** **(c)**

Figure 7.23

Right scapula. (*a*) Posterior surface. (*b*) Lateral view showing the glenoid cavity that articulates with the head of the humerus. (*c*) Anterior surface.

7.10 UPPER LIMB

The bones of the upper limb form the framework of the arm, forearm, and hand. They also provide attachments for muscles, and they function in levers that move limb parts. These bones include a humerus, a radius, an ulna, carpals, metacarpals, and phalanges (see fig. 7.9).

Humerus

The **humerus** is a long bone that extends from the scapula to the elbow (fig. 7.24). At its upper end is a smooth, rounded *head* that fits into the glenoid cavity of the scapula. Just below the head are two processes—a *greater tubercle* on the lateral side and a *lesser tubercle* on the anterior side. These tubercles provide attachments for muscles that move the upper limb at the shoulder. Between them is a narrow furrow, the *intertubercular groove.*

The narrow depression along the lower margin of the humerus head separates it from the tubercles and is called the *anatomical neck.* Just below the head and the tubercles is a tapering region called the *surgical neck,* so named because fractures commonly occur there. Near the middle of the bony shaft on the lateral side is a rough, V-shaped area called the *deltoid tuberosity.* It provides an attachment for the muscle (deltoid) that raises the upper limb horizontally to the side.

At the lower end of the humerus are two smooth *condyles* (a lateral *capitulum* and a medial *trochlea*) that articulate with the radius on the lateral side and the ulna on the medial side. Above the condyles on either side are *epicondyles,* which provide attachments for muscles and ligaments of the elbow. Between the epicondyles anteriorly is a depression, the *coronoid* (kor'o-noid) *fossa,* that receives a process of the ulna (coronoid process) when the elbow bends. Another depression on the posterior surface, the *olecranon* (o"lek'ra-non) *fossa,* receives an ulnar process (olecranon process) when the upper limb straightens at the elbow.

Radius

The **radius,** located on the thumb side of the forearm, extends from the elbow to the wrist and crosses over the ulna when the hand is turned so that the palm

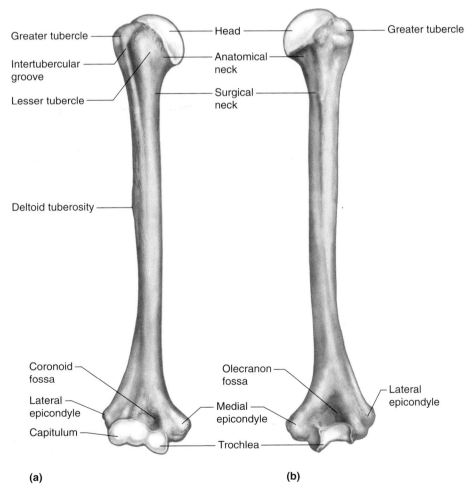

Greater tubercle — Head — Greater tubercle

Intertubercular groove — Anatomical neck

Lesser tubercle — Surgical neck

Deltoid tuberosity

Coronoid fossa — Olecranon fossa — Lateral epicondyle

Lateral epicondyle — Medial epicondyle

Capitulum — Trochlea

(a)   (b)

Figure 7.24

Right humerus. (*a*) Anterior surface. (*b*) Posterior surface.

faces backward (fig. 7.25). A thick, disclike *head* at the upper end of the radius articulates with the humerus and a notch of the ulna (radial notch). This arrangement allows the radius to rotate.

On the radial shaft just below the head is a process called the *radial tuberosity.* It is an attachment for a muscle (biceps brachii) that bends the upper limb at the elbow. At the distal end of the radius, a lateral *styloid* (sti'loid) *process* provides attachments for ligaments of the wrist.

Ulna

The **ulna** is longer than the radius and overlaps the end of the humerus posteriorly (fig. 7.25). At its proximal end, the ulna has a wrenchlike opening, the *trochlear* (trok'le-ar) *notch,* that articulates with the humerus. Two processes on either side of this notch, the *olecranon process* and the *coronoid process,* provide attachments for muscles.

At the distal end of the ulna, its knoblike *head* articulates laterally with a notch of the radius (ulnar notch) and with a disc of fibrocartilage inferiorly. This disc, in turn, joins a wrist bone (triquetrum). A medial *styloid process* at the distal end of the ulna provides attachments for wrist ligaments.

Hand

The hand is made up of the wrist, palm, and fingers. The skeleton of the wrist consists of eight small **carpal bones** that are firmly bound in two rows of four bones each. The resulting compact mass is called a *carpus* (kar'pus). The carpus articulates with the radius and with the fibrocartilaginous disc on the ulnar side. Its distal surface articulates with the metacarpal bones. Figure 7.26 names the individual bones of the carpus.

Five **metacarpal bones,** one in line with each finger, form the framework of the palm or *metacarpus* (met"ah-kar'pus) of the hand. These bones are cylindrical,

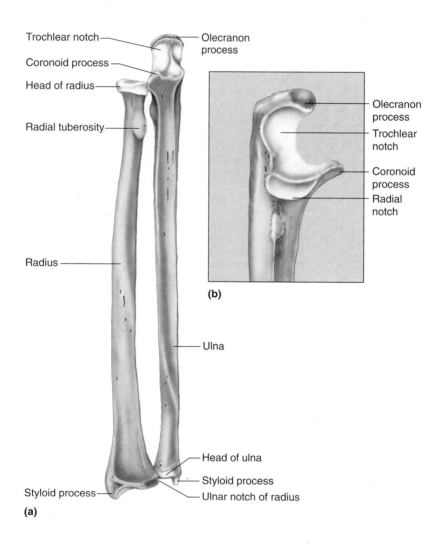

Trochlear notch

Coronoid process

Head of radius

Radial tuberosity

Olecranon process

Olecranon process

Trochlear notch

Coronoid process

Radial notch

Radius

(b)

Ulna

Head of ulna

Styloid process

Ulnar notch of radius

Styloid process

(a)

Figure 7.25

Right radius and ulna. (*a*) The head of the radius articulates with the radial notch of the ulna, and the head of the ulna articulates with the ulnar notch of the radius. (*b*) Lateral view of the proximal end of the ulna.

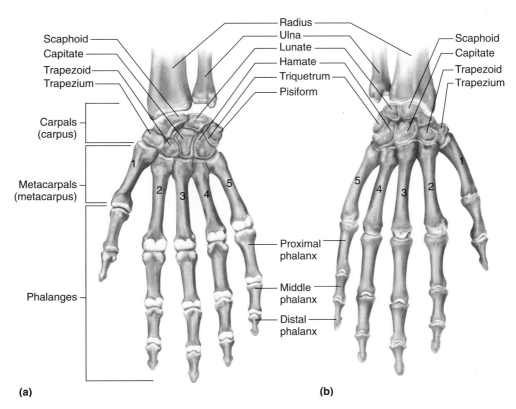

Figure 7.26

Right hand. (*a*) Anterior view. (*b*) Posterior view.

with rounded distal ends that form the knuckles of a clenched fist. They are numbered 1–5, beginning with the metacarpal of the thumb (fig. 7.26). The metacarpals articulate proximally with the carpals and distally with the phalanges.

The **phalanges** are the finger bones. Each finger has three phalanges—a proximal, a middle, and a distal phalanx—except the thumb, which has two (it lacks a middle phalanx).

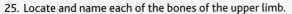

Check Your Recall

25. Locate and name each of the bones of the upper limb.
26. Explain how the bones of the upper limb articulate with one another.

7.11 PELVIC GIRDLE

The **pelvic girdle** consists of two hip bones (coxal bones, pelvic bones, or innominate bones) which articulate with each other anteriorly and with the sacrum posteriorly. The sacrum, coccyx, and pelvic girdle together form the bowl-shaped **pelvis** (fig. 7.27). The pelvic girdle supports the trunk of the body, provides attachments for the lower limbs, and protects the urinary bladder, the distal end of the large intestine, and the internal reproductive organs.

Each hip bone develops from three parts—an ilium, an ischium, and a pubis (fig. 7.28). These parts fuse in the region of a cup-shaped cavity called the *acetabulum* (as″ĕ-tab′u-lum). This depression, on the lateral surface of the hip bone, receives the rounded head of the femur (thigh bone).

The **ilium** (il′e-um), which is the largest and uppermost portion of the hip bone, flares outward, forming the prominence of the hip. The margin of this prominence is called the *iliac crest*.

Posteriorly, the ilium joins the sacrum at the *sacroiliac* (sa″kro-il′e-ak) *joint*. A projection of the ilium, the *anterior superior iliac spine,* can be felt lateral to the groin and provides attachments for ligaments and muscles.

The **ischium** (is′ke-um), which forms the lowest portion of the hip bone, is L-shaped, with its angle, the *ischial tuberosity,* pointing posteriorly and downward. This tuberosity has a rough surface that provides attachments for ligaments and lower limb muscles. It also supports the weight of the body during sitting. Above the ischial tuberosity, near the junction of the ilium and ischium, is a sharp projection called the *ischial spine.* The distance between the ischial spines is the shortest diameter of the pelvic outlet.

The **pubis** (pu′bis) constitutes the anterior portion of the hip bone. The two pubic bones join at the midline, forming a joint called the *symphysis pubis* (sin′fi-sis

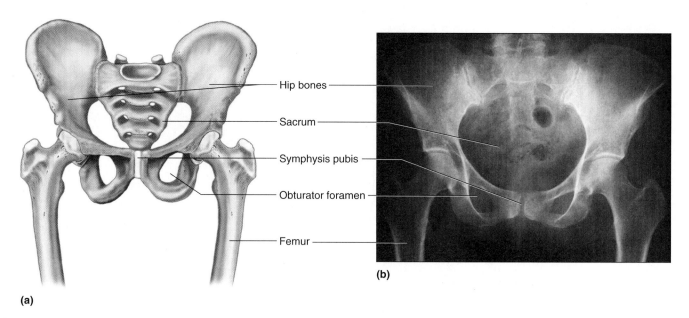

Figure 7.27

Pelvic girdle. (*a*) The pelvic girdle is formed by two hip bones. The pelvis includes the pelvic girdle as well as the sacrum and the coccyx. (*b*) Radiograph of the pelvic girdle showing the sacrum, coccyx, and proximal ends of the femurs.

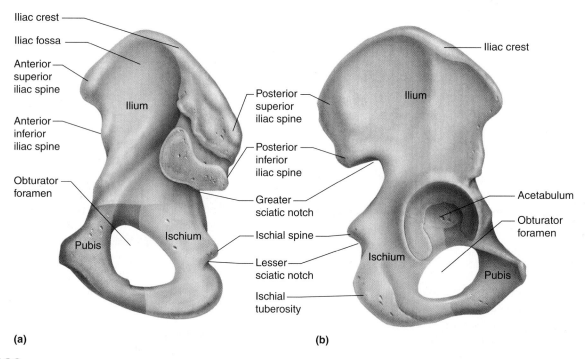

Figure 7.28

Right hip bone. (*a*) Medial surface. (*b*) Lateral view.

pu'bis). The angle these bones form below the symphysis is the *pubic arch* (fig. 7.29).

A portion of each pubis passes posteriorly and downward to join an ischium. Between the bodies of these bones on either side is a large opening, the *obturator foramen,* which is the largest foramen in the skeleton (see figs. 7.27 and 7.28).

If a line were drawn along each side of the pelvis from the sacral promontory downward and anteriorly to the upper margin of the symphysis pubis, it would mark the *pelvic brim* (linea terminalis) (fig. 7.29). This margin separates the lower, or lesser (true), pelvis from the upper, or greater (false), pelvis. Table 7.3 summarizes some differences in the female and male pelves and other skeletal structures.

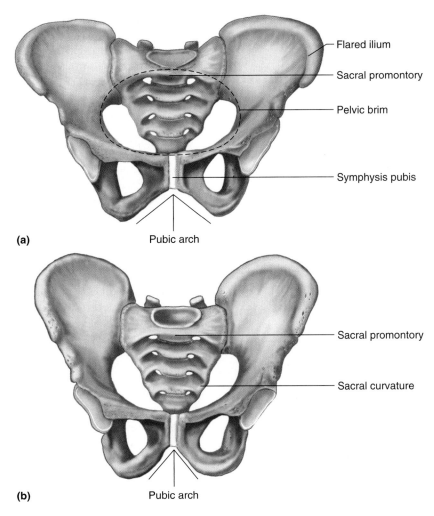

Flared ilium

Sacral promontory

Pelvic brim

Symphysis pubis

(a)   Pubic arch

Sacral promontory

Sacral curvature

(b)   Pubic arch

Figure 7.29

The female pelvis is usually wider in all diameters and roomier than that of the male. (*a*) Female pelvis. (*b*) Male pelvis.

Table 7.3	Differences Between the Female and Male Skeletons
Part	**Differences**
Skull	Female skull is smaller and lighter, with less conspicuous muscular attachments. Female facial area is rounder, jaw is smaller, and mastoid process is less prominent than those of a male.
Pelvic girdle	Female hip bones are lighter, thinner, and have less obvious muscular attachments. The obturator foramina and acetabula are smaller and farther apart than those of a male.
Pelvic cavity	Female pelvic cavity is wider in all diameters and is shorter, roomier, and less funnel-shaped. The distances between the ischial spines and ischial tuberosities are greater than in a male.
Sacrum	Female sacrum is wider, the first sacral vertebra projects forward to a lesser degree, and the sacral curvature is bent more sharply posteriorly than in a male.
Coccyx	Female coccyx is more movable than that of a male.

Check Your Recall

27. Locate and name each bone that forms the pelvis.

28. Name the bones that fuse to form a hip bone.

7.12 LOWER LIMB

Bones of the lower limb form the frameworks of the thigh, leg, and foot. They include a femur, a tibia, a fibula, tarsals, metatarsals, and phalanges (see fig. 7.9).

Femur

The **femur,** or thigh bone, is the longest bone in the body and extends from the hip to the knee (fig. 7.30). A large, rounded *head* at its proximal end projects medially into the acetabulum of the hip bone. On the head, a pit called the *fovea capitis* marks the attachment of a ligament (ligamentum capitis). Just below the head are a constriction, or *neck,* and two large processes—a superior, lateral *greater trochanter* and an inferior, medial *lesser trochanter*. These processes provide attachments for muscles of the lower limbs and buttocks.

The strongest bone in the body, the femur, is hollow. Ounce for ounce, it has greater pressure tolerance and bearing strength than a rod of equivalent size made of cast steel.

At the distal end of the femur, two rounded processes, the *lateral* and *medial condyles,* articulate with the tibia of the leg. A **patella,** or kneecap, also articulates with the femur on its distal anterior surface (see fig. 7.9). It is located in a tendon that passes anteriorly over the knee.

Hip fracture is one of the more serious causes of hospitalization among elderly persons. The site of hip fracture is most commonly the neck of a femur or the region between the trochanters of a femur. Often a hip fracture is a cause of a fall, rather than the result of a fall.

Tibia

The **tibia,** or shin bone, is the larger of the two leg bones and is located on the medial side (fig. 7.31). Its proximal end is expanded into *medial* and *lateral condyles,* which have concave surfaces and articulate with the condyles of

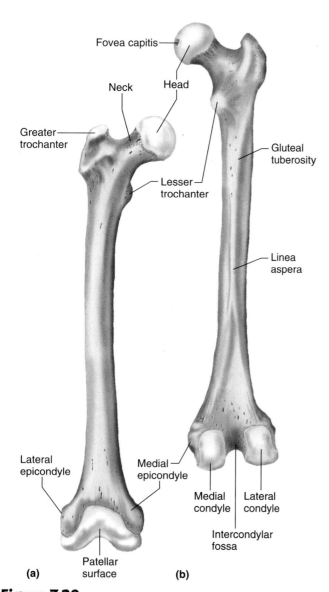

Figure 7.30

Right femur. (*a*) Anterior surface. (*b*) Posterior surface.

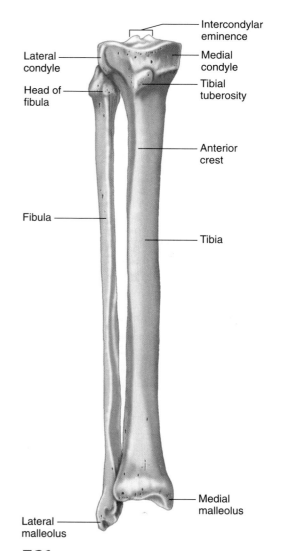

Figure 7.31

Right tibia and fibula, anterior view.

the femur. Below the condyles, on the anterior surface, is a process called the *tibial tuberosity,* which provides an attachment for the *patellar ligament*—a continuation of the patella-bearing tendon.

At its distal end, the tibia expands to form a prominence on the inner ankle called the *medial malleolus* (mah-le'o-lus), which is an attachment for ligaments. On its lateral side is a depression that articulates with the fibula. The inferior surface of the tibia's distal end articulates with a large bone (the talus) in the ankle.

Fibula

The **fibula** is a long, slender bone located on the lateral side of the tibia (fig. 7.31). Its ends are slightly enlarged into a proximal *head* and a distal *lateral malleolus.* The head articulates with the tibia just below the lateral condyle; however, it does not enter into the knee joint and does not bear any body weight. The lateral malleolus articulates with the ankle and protrudes on the lateral side.

Foot

The foot is made up of the ankle, the instep, and the toes. The ankle, or *tarsus* (tahr'sus), is composed of seven **tarsal bones** (figs. 7.32 and 7.33). These bones are arranged so that one of them, the **talus** (ta'lus), can move freely where it joins the tibia and fibula. The remaining tarsal bones are bound firmly together, forming a mass supporting the talus. Figure 7.33 names the individual bones of the tarsus.

The largest of the tarsals, the **calcaneus** (kal-ka'ne-us), or heel bone, is located below the talus, where it projects backward to form the base of the heel. The calcaneus helps support body weight and provides an attachment for the muscles that move the foot.

The instep, or *metatarsus* (met"ah-tar'sus), consists of five elongated **metatarsal bones** that articulate with the tarsus. They are numbered 1–5, beginning on the medial side (fig. 7.33). The heads at the distal ends of these bones form the ball of the foot. The tarsals and metatarsals are arranged and bound by ligaments to form the arches of the foot. A longitudinal arch extends from the heel to the toe, and a transverse arch stretches across the foot. These arches provide a stable, springy base for the body. Sometimes, however, the tissues that bind the metatarsals weaken, producing fallen arches, or flat feet.

The **phalanges** of the toes, which are similar to those of the fingers, align and articulate with the metatarsals. Each toe has three phalanges—a proximal, a middle, and a distal phalanx—except the great toe, which lacks a middle phalanx.

Check Your Recall

29. Locate and name each of the bones of the lower limb.
30. Explain how the bones of the lower limb articulate with one another.
31. Describe how the foot is adapted to support the body.

7.13 JOINTS

Joints (articulations) are functional junctions between bones. They bind parts of the skeletal system, make possible bone growth, permit parts of the skeleton to change shape during childbirth, and enable the body to move in response to skeletal muscle contractions. Joints vary considerably in structure and function. If classified according to the degree of movement they make possible,

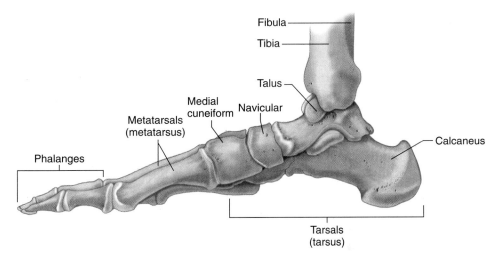

Figure 7.32

Right foot. The talus moves freely where it articulates with the tibia and fibula.

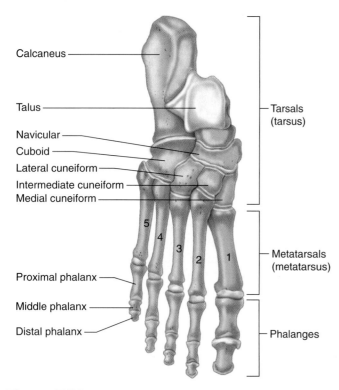

Figure 7.33
Right foot, viewed superiorly.

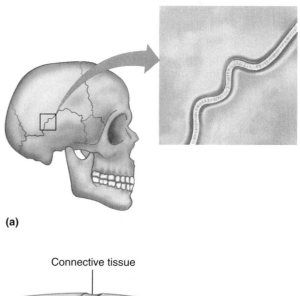

(a)

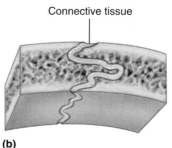

(b)

Figure 7.34
Fibrous joints. (*a*) The fibrous joints between the bones of the skull are immovable and are called sutures. (*b*) A thin layer of connective tissue connects the bones at the suture.

joints can be immovable (synarthrotic), slightly movable (amphiarthrotic), or freely movable (diarthrotic). Joints also can be grouped according to the type of tissue (fibrous, cartilaginous, or synovial) that binds the bones together at each junction. Currently, this structural classification by tissue type is more commonly used.

A human body has 230 joints.

Fibrous Joints

Fibrous (fi′brus) **joints** lie between bones that closely contact one another. A thin layer of dense connective tissue joins the bones at such joints, as in a *suture* between a pair of flat bones of the skull (fig. 7.34). Generally, no appreciable movement (synarthrotic) takes place at a fibrous joint. Some fibrous joints, such as the joint in the leg between the distal ends of the tibia and fibula, have limited movement (amphiarthrotic).

Cartilaginous Joints

Hyaline cartilage, or fibrocartilage, connects the bones of **cartilaginous** (kar″tĭ-lah′jin-us) **joints.** For example, joints of this type separate the vertebrae of the vertebral column. Each intervertebral disc is composed of a band of fibrocartilage (annulus fibrosus) surrounding a pulpy or gelatinous core (nucleus pulposus). The disc absorbs shocks and helps equalize pressure between adjacent vertebrae when the body moves (see fig. 7.17).

Due to the slight flexibility of the discs, cartilaginous joints allow limited movement (amphiarthrotic), as when the back is bent forward or to the side or is twisted. Other examples of cartilaginous joints include the symphysis pubis and the first rib with the sternum.

Synovial Joints *Most Joints*

Most joints within the skeletal system are **synovial** (sĭ-no′ve-al) **joints,** which allow free movement (diarthrotic). They are more complex structurally than fibrous or cartilaginous joints.

The articular ends of the bones in a synovial joint are covered with hyaline cartilage (articular cartilage), and a surrounding, tubular capsule of dense connective tissue holds them together (fig. 7.35). This *joint capsule* is composed of an outer layer of ligaments and an inner lining of *synovial membrane,* which secretes synovial

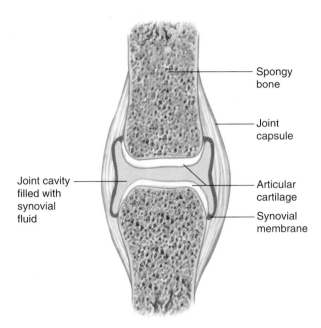

Figure 7.35

The generalized structure of a synovial joint.

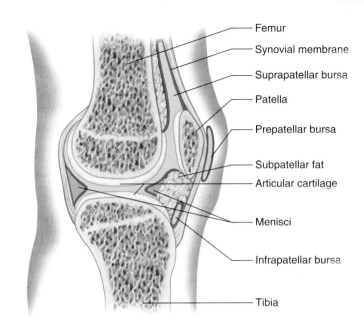

Figure 7.36

Menisci separate the articulating surfaces of the femur and tibia. Several bursae are associated with the knee joint.

fluid. Having a consistency similar to that of uncooked egg white, synovial fluid lubricates joints.

Some synovial joints have flattened, shock-absorbing pads of fibrocartilage called **menisci** (mĕ-nis′ke) (singular, *meniscus*) between the articulating surfaces of the bones (fig. 7.36). Such joints may also have fluid-filled sacs called **bursae** (ber′se) associated with them. Each bursa is lined with synovial membrane, which may be continuous with the synovial membrane of a nearby joint cavity. Bursae are commonly located between tendons and underlying bony prominences, as in the patella of the knee or the olecranon process of the elbow. They aid the movement of tendons that glide over these bony parts or over other tendons. Figure 7.36 shows and names some of the bursae associated with the knee.

Based on the shapes of their parts and the movements they permit, synovial joints are classified as follows:

1. A **ball-and-socket joint** consists of a bone with a globular or slightly egg-shaped head that articulates with the cup-shaped cavity of another bone. Such a joint allows a wider range of motion than does any other kind, permitting movements in all planes, as well as rotational movement around a central axis. The shoulder and hip have joints of this type (fig. 7.37).
2. In a **condyloid joint,** or **ellipsoidal joint,** an oval-shaped condyle of one bone fits into an elliptical cavity of another bone, such as in the joints

between the metacarpals and phalanges (fig. 7.37). This type of joint permits a variety of movements in different planes; rotational movement, however, is not possible.
3. The articulating surfaces of **gliding joints,** or **plane joints,** are nearly flat or slightly curved. Most of the joints within the wrist (fig. 7.37) and ankle, as well as those between the articular processes of adjacent vertebrae, belong to this group. They allow sliding and twisting movements. The sacroiliac joints and the joints formed by ribs 2–7 connecting with the sternum are also gliding joints.
4. In a **hinge joint,** the convex surface of one bone fits into the concave surface of another, as in the elbow (fig. 7.37) and the joints of the phalanges. Such a joint resembles the hinge of a door in that it permits movement in one plane only.
5. In a **pivot joint,** the cylindrical surface of one bone rotates within a ring formed of bone and ligament. Movement is limited to the rotation around a central axis. The joint between the proximal ends of the radius and the ulna is of this type (fig. 7.37).
6. A **saddle joint** forms between bones whose articulating surfaces have both concave and convex regions. The surface of one bone fits the complementary surface of the other. This physical relationship permits a variety of movements, as in the joint between the carpal (trapezium) and metacarpal bones of the thumb (fig. 7.37).

Table 7.4 summarizes the types of joints.

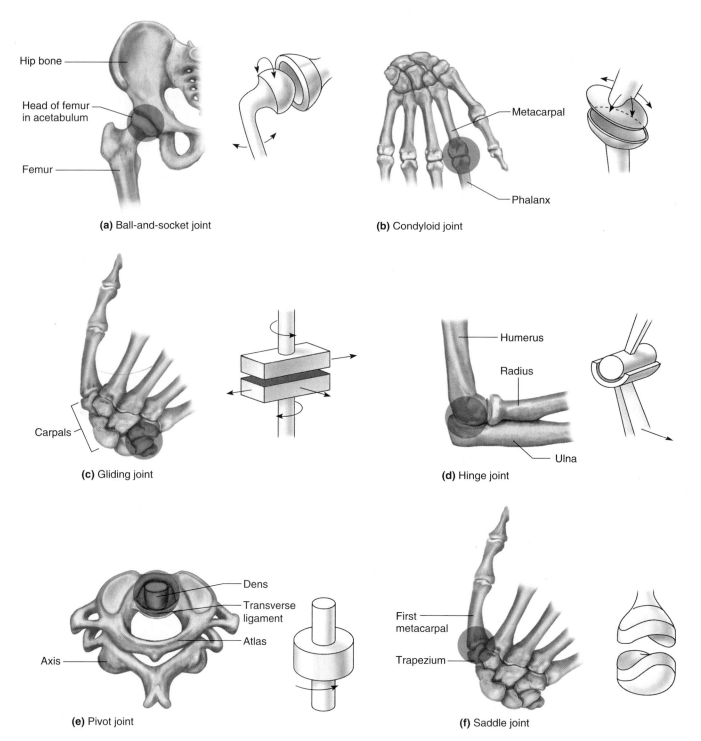

(a) Ball-and-socket joint

Hip bone

Head of femur in acetabulum

Femur

(b) Condyloid joint

Metacarpal

Phalanx

(c) Gliding joint

Carpals

(d) Hinge joint

Humerus

Radius

Ulna

(e) Pivot joint

Dens

Transverse ligament

Atlas

Axis

(f) Saddle joint

First metacarpal

Trapezium

Figure 7.37

Types and examples of synovial (freely movable) joints.

Table 7.4   Types of Joints

Type of Joint	Description	Possible Movements	Examples
Fibrous	Articulating bones are fastened together by a thin layer of dense connective tissue.	None	Suture between bones of skull, joint between the distal ends of tibia and fibula
Cartilaginous	Articulating bones are connected by hyaline cartilage or fibrocartilage.	Limited movement, as when back is bent or twisted	Joints between the bodies of vertebrae, symphysis pubis
Synovial	Articulating ends of bones are surrounded by a joint capsule of ligaments and synovial membranes; ends of articulating bones are covered by hyaline cartilage and separated by synovial fluid.	Allow free movement (see the following list)	
1. Ball-and-socket	Ball-shaped head of one bone articulates with cup-shaped cavity of another.	Movements in all planes and rotation	Shoulder, hip
2. Condyloid or Ellipsoidal	Oval-shaped condyle of one bone articulates with elliptical cavity of another.	Variety of movements in different planes, but no rotation	Joints between the metacarpals and phalanges
3. Gliding or Plane	Articulating surfaces are nearly flat or slightly curved.	Sliding or twisting	Joints between various bones of wrist and ankle, sacroiliac joints, joints between ribs 2–7 and sternum
4. Hinge	Convex surface of one bone articulates with concave surface of another.	Flexion and extension	Elbow, joints of phalanges
5. Pivot	Cylindrical surface of one bone articulates with ring of bone and ligament.	Rotation around a central axis	Joint between the proximal ends of radius and ulna
6. Saddle	Articulating surfaces have both concave and convex regions; the surface of one bone fits the complementary surface of another.	Variety of movements, mainly in two planes	Joint between the carpal and metacarpal of thumb

Arthritis is a group of disorders that cause inflamed, swollen, and painful joints. More than a hundred different types of arthritis affect millions of people worldwide. The most common forms are *rheumatoid arthritis* and *osteoarthritis*.

In rheumatoid arthritis, which is the most painful and debilitating of the arthritic diseases, the synovial membrane of a freely movable joint becomes inflamed and thickened. Then the articular cartilage is damaged, and fibrous tissue infiltrates, interfering with joint movement. In time, the joint may ossify, fusing the articulating bones. Rheumatoid arthritis is an autoimmune disorder in which the immune system attacks the body's healthy tissues.

Osteoarthritis is a degenerative disorder that occurs as a result of aging, but an inherited form may appear as early as one's thirties. In osteoarthritis, articular cartilage softens and disintegrates gradually, roughening the articular surfaces. Joints become painful, and movement is restricted. Osteoarthritis most often affects joints that are used the most over a lifetime, such as those of the fingers, hips, knees, and lower parts of the vertebral column.

Types of Joint Movements

Skeletal muscle action produces movements at synovial joints. Typically, one end of a muscle is attached to a relatively immovable or fixed part on one side of a joint, and the other end of the muscle is fastened to a movable part on the other side. When the muscle contracts, its fibers pull its movable end, the *insertion,* toward its fixed end, the *origin,* and a movement occurs at the joint.

The following terms describe movements at joints (figs. 7.38, 7.39, and 7.40).

flexion (flek'shun) Bending parts at a joint so that the angle between them decreases and the parts come closer together (bending the knee).

extension (ek-sten'shun) Straightening parts at a joint so that the angle between them increases and the parts move farther apart (straightening the knee).

dorsiflexion (dor″sĭ-flek'shun) Movement at the ankle that brings the foot closer to the shin (walking on heels).

plantar flexion (plan'tar flek'shun) Movement at the ankle that brings the foot farther from the shin (walking or standing on toes).

hyperextension (hi″per-ek-sten'shun) Extension of the parts at a joint beyond the anatomical position (bending the head back beyond the upright position); often used to describe an abnormal extension beyond the normal range of motion resulting in injury.

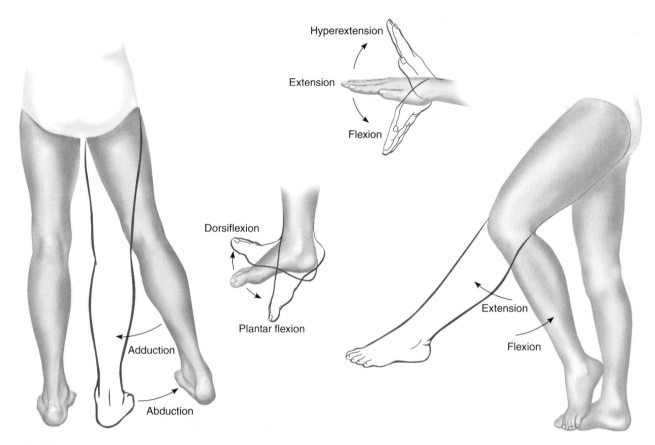

Figure 7.38

Joint movements: adduction, abduction, dorsiflexion, plantar flexion, hyperextension, extension, and flexion.

abduction (ab-duk′shun) Moving a part away from the midline (lifting the upper limb horizontally to form a right angle with the side of the body).

adduction (ah-duk′shun) Moving a part toward the midline (returning the upper limb from the horizontal position to the side of the body).

rotation (ro-ta′shun) Moving a part around an axis (twisting the head from side to side).

circumduction (ser″kum-duk′shun) Moving a part so that its end follows a circular path (moving the finger in a circular motion without moving the hand).

pronation (pro-na′shun) Turning the hand so that the palm is downward or facing posteriorly (in anatomical position).

supination (soo″pĭ-na′shun) Turning the hand so that the palm is upward or facing anteriorly (in anatomical position).

eversion (e-ver′zhun) Turning the foot so the plantar surface faces laterally.

inversion (in-ver′zhun) Turning the foot so the plantar surface faces medially.

retraction (re-trak′shun) Moving a part backward (pulling the head backward).

protraction (pro-trak′shun) Moving a part forward (thrusting the head forward).

elevation (el″ĕ-va′shun) Raising a part (shrugging the shoulders).

depression (de-presh′un) Lowering a part (drooping the shoulders).

Check Your Recall

32. Describe the characteristics of the three major types of joints.

33. Name six types of synovial joints.

34. What terms describe movements possible at synovial joints?

Injuries to the elbow, shoulder, and knee are commonly diagnosed and treated using a procedure called *arthroscopy*. Arthroscopy enables a surgeon to visualize the interior of a joint and perform diagnostic or therapeutic procedures, guided by the image on a video screen. An arthroscope is a thin, tubular instrument about 25 centimeters long containing optical fibers that transmit an image. The surgeon inserts the device through a small incision in the joint capsule. Arthroscopy is far less invasive than conventional surgery. Many runners have undergone uncomplicated arthroscopy and raced just weeks later.

Figure 7.39

Joint movements: rotation, circumduction, supination, and pronation.

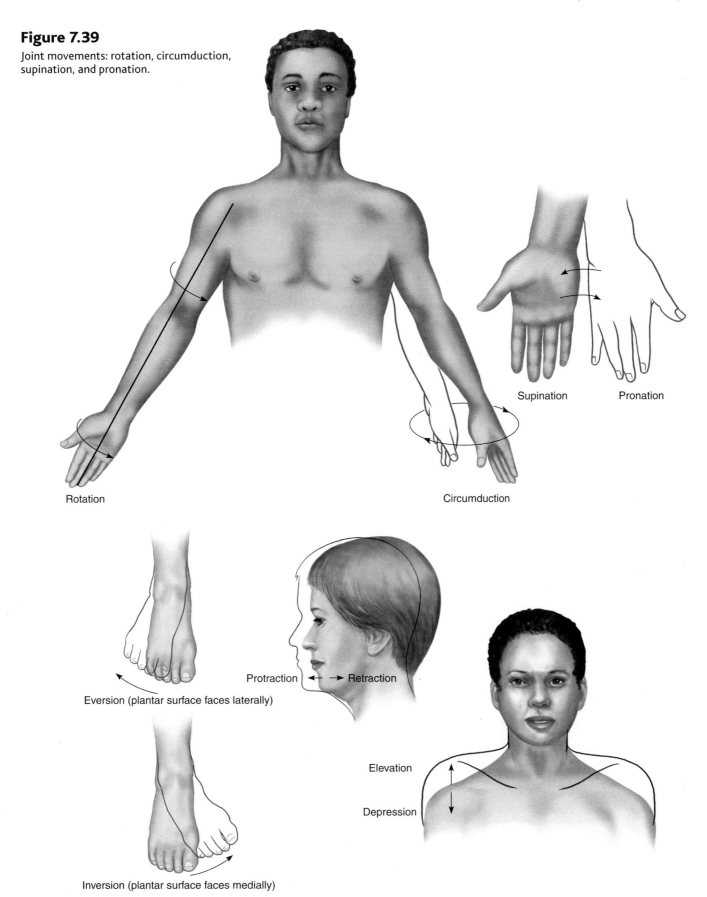

Rotation

Supination   Pronation

Circumduction

Eversion (plantar surface faces laterally)

Inversion (plantar surface faces medially)

Protraction ← → Retraction

Elevation

Depression

Figure 7.40

Joint movements: eversion, inversion, retraction, protraction, elevation, and depression.

Skeletal System

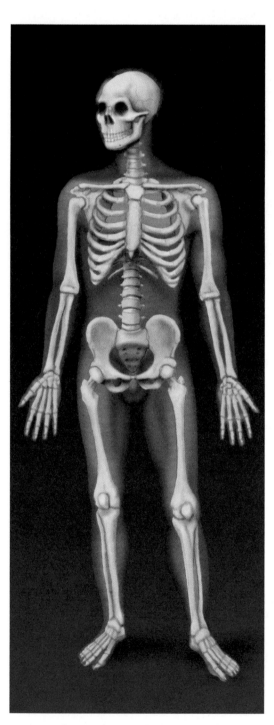

Integumentary System

Vitamin D, activated in the skin, plays a role in calcium absorption and availability for bone matrix.

Muscular System

Muscles pull on bones to cause movement.

Nervous System

Proprioceptors sense the position of body parts. Pain receptors warn of trauma to bone. Bones protect the brain and spinal cord.

Endocrine System

Some hormones act on bone to help regulate blood calcium levels.

Cardiovascular System

Blood transports nutrients to bone cells. Bone helps regulate plasma calcium levels, important to heart function.

Lymphatic System

Cells of the immune system originate in the bone marrow.

Digestive System

Absorption of dietary calcium provides material for bone matrix.

Respiratory System

Ribs and muscles work together in breathing.

Urinary System

The kidneys and bones work together to help regulate blood calcium levels.

Reproductive System

The pelvis helps support the uterus during pregnancy. Bones provide a source of calcium during lactation.

Bones provide support, protection, and movement and also play a role in calcium balance.

Clinical Terms Related to the Skeletal System

acromegaly (ak″ro-meg′ah-le) Abnormal enlargement of facial features, hands, and feet in adults as a result of overproduction of growth hormone.

ankylosis (ang″kĭ-lo′sis) Abnormal stiffness of a joint or fusion of bones at a joint, often due to damage to the joint membranes from chronic rheumatoid arthritis.

arthralgia (ar-thral′je-ah) Pain in a joint.

arthrocentesis (ar″thro-sen-te′sis) Puncture of and removal of fluid from a joint cavity.

arthrodesis (ar″thro-de′sis) Surgery to fuse the bones at a joint.

arthroplasty (ar′thro-plas″te) Surgery to make a joint more movable.

Colles fracture (kol′ēz frak′cher) Fracture at the distal end of the radius that displaces the smaller fragment posteriorly.

epiphysiolysis (ep″ĭ-fiz″e-ol′ĭ-sis) Separation or loosening of the epiphysis from the diaphysis of a bone.

hemarthrosis (hem″ar-thro′sis) Blood in a joint cavity.

laminectomy (lam″ĭ-nek′to-me) Surgical removal of the posterior arch of a vertebra, usually to relieve the symptoms of a ruptured intervertebral disc.

lumbago (lum-ba′go) Dull ache in the lumbar region of the back.

orthopedics (or″tho-pe′diks) Medical specialty that prevents, diagnoses, and treats diseases and abnormalities of the skeletal and muscular systems.

ostealgia (os″te-al′je-ah) Pain in a bone.

ostectomy (os-tek′to-me) Surgical removal of a bone.

osteitis (os″te-i′tis) Inflammation of bone tissue.

osteochondritis (os″te-o-kon-dri′tis) Inflammation of bone and cartilage tissues.

osteogenesis (os″te-o-jen′ĕ-sis) Bone development.

osteogenesis imperfecta (os″te-o-jen′ĕ-sis im-per-fek′tah) Inherited condition of deformed and abnormally brittle bones.

osteoma (os″te-o′mah) Tumor composed of bone tissue.

osteomalacia (os″te-o-mah-la′she-ah) Softening of adult bone due to a disorder in calcium and phosphorus metabolism, usually caused by vitamin D deficiency.

osteomyelitis (os″te-o-mi″ĕ-li′tis) Bone inflammation caused by the body's reaction to a bacterial or fungal infection.

osteonecrosis (os″te-o-ne-kro′sis) Death of bone tissue. This condition occurs most commonly in the femur head in elderly persons and may be due to obstructed arteries supplying the bone.

osteopathology (os″te-o-pah-thol′o-je) Study of bone diseases.

osteotomy (os″te-ot′o-me) Cutting a bone.

synovectomy (sin″o-vek′to-me) Surgical removal of the synovial membrane of a joint.

Clinical Connection

When the twenty-year-old professional soccer player jammed his left toe at high speed against the ball and howled in pain, he thought it would get better in a few days, as such injuries usually do. This time, the injured toe started to turn bluish-red immediately, as a hematoma formed beneath the nail. The pain continued—for weeks. Pus swelled from beneath the darkened nail. Finally, barely able to walk, let alone continue playing his sport, the athlete consulted a physician, who, assuming the wound was infected, prescribed antibiotics and an anti-inflammatory cream. But the unrelenting pain was not due to infection. The young man finally went to an emergency department, where a sample of the pus revealed no bacteria. Instead, X rays clearly indicated an osteochondroma, a spike of bone emerging 4 millimeters from the dorsal terminal phalanx of the left great toe, capped with cartilage. Usually an osteochondroma is a benign bone tumor that arises during fetal development. In this case, however, the physician in charge suspected that the soccer player's spike was a response to trauma—followed by failure to rest afterwards. Surgery removed the spike, and a month later, the athlete was back on the field.

SUMMARY OUTLINE

7.1 Introduction (p. 131)

Individual bones are the organs of the skeletal system. A bone contains very active tissues.

7.2 Bone Structure (p. 131)

Bone structure reflects its function.

1. Bones are classified according to their shapes including long, short, flat, irregular, and sesamoid (round).
2. Parts of a long bone
   a. Epiphyses at each end are covered with articular cartilage and articulate with other bones.
   b. The shaft of a bone is called the diaphysis.
   c. Except for the articular cartilage, a bone is covered by a periosteum.
   d. Compact bone has a continuous extracellular matrix with no gaps.
   e. Spongy bone has irregular interconnecting spaces between bony plates that reduce the weight of bone.
   f. Both compact and spongy bone are strong and resist bending.
   g. The diaphysis contains a medullary cavity filled with marrow.
3. Microscopic structure
   a. Compact bone contains osteons cemented together.
   b. Central canals contain blood vessels that nourish the cells of osteons.
   c. Diffusion from the surface of the thin, bony plates nourishes the cells of spongy bone.

7.3 Bone Development and Growth (p. 133)

1. Intramembranous bones
   a. Intramembranous bones develop from layers of unspecialized connective tissues.
   b. Osteoblasts within the membranous layers form bone tissue.
   c. Mature bone cells are called osteocytes.

2. Endochondral bones
   a. Endochondral bones develop as hyaline cartilage that is later replaced by bone tissue.
   b. The primary ossification center appears in the diaphysis, whereas secondary ossification centers appear in the epiphyses.
   c. An epiphyseal plate remains between the primary and secondary ossification centers.
   d. The epiphyseal plates are responsible for lengthening.
   e. Long bones continue to lengthen until the epiphyseal plates are ossified.
   f. Growth in thickness is due to ossification beneath the periosteum.
3. Homeostasis of bone tissue
   a. Osteoclasts break down bone matrix and osteoblasts deposit bone matrix to continually remodel bone.
   b. The total mass of bone remains nearly constant.
4. Factors affecting bone development, growth, and repair include nutrition, hormonal secretions, and physical exercise.

7.4 Bone Function (p. 135)

1. Support and protection
   a. Bones shape and form body structures.
   b. Bones support and protect softer underlying tissues.
2. Body movement
   a. Bones and muscles function together as levers.
   b. A lever consists of a bar, a pivot (fulcrum), a resistance, and a force that supplies energy.
3. Blood cell formation
   a. At different ages, hematopoiesis occurs in the yolk sac, liver and spleen, and red bone marrow.
   b. Red marrow houses developing red blood cells, white blood cells, and blood platelets. Yellow marrow stores fat.
4. Storage of inorganic salts
   a. Bones store calcium in the extracellular matrix of bone tissue, which contains large quantities of calcium phosphate.
   b. When blood calcium is low, osteoclasts break down bone, releasing calcium salts. When blood calcium is high, osteoblasts form bone tissue and store calcium salts.
   c. Bone stores small amounts of magnesium, sodium, potassium, and carbonate ions.

7.5 Skeletal Organization (p. 139)

1. The skeleton can be divided into axial and appendicular portions.
2. The axial skeleton consists of the skull, hyoid bone, vertebral column, and thoracic cage.
3. The appendicular skeleton consists of the pectoral girdle, upper limbs, pelvic girdle, and lower limbs.

7.6 Skull (p. 142)

The skull consists of twenty-two bones: eight cranial bones and fourteen facial bones.
1. Cranium
   a. The cranium encloses and protects the brain.
   b. Some cranial bones contain air-filled paranasal sinuses.
   c. Cranial bones include the frontal bone, parietal bones, occipital bone, temporal bones, sphenoid bone, and ethmoid bone.
2. Facial skeleton
   a. Facial bones form the basic shape of the face and provide attachments for muscles.
   b. Facial bones include the maxillae, palatine bones, zygomatic bones, lacrimal bones, nasal bones, vomer bone, inferior nasal conchae, and mandible.

3. Infantile skull
   a. Fontanels connect incompletely developed bones.
   b. The proportions of the infantile skull are different from those of an adult skull.

7.7 Vertebral Column (p. 147)

The vertebral column extends from the skull to the pelvis and protects the spinal cord. It is composed of vertebrae separated by intervertebral discs.
1. A typical vertebra
   a. A typical vertebra consists of a body and a bony vertebral arch, which surrounds the spinal cord.
   b. Notches on the upper and lower surfaces provide intervertebral foramina through which spinal nerves pass.
2. Cervical vertebrae
   a. Transverse processes bear transverse foramina.
   b. The atlas (first vertebra) supports and balances the head.
   c. The dens of the axis (second vertebra) provides a pivot for the atlas when the head is turned from side to side.
3. Thoracic vertebrae
   a. Thoracic vertebrae are larger than cervical vertebrae.
   b. Facets on the sides articulate with the ribs.
4. Lumbar vertebrae
   a. The vertebral bodies are large and strong.
   b. They support more body weight than other vertebrae.
5. Sacrum
   a. The sacrum is a triangular structure formed of five fused vertebrae.
   b. Vertebral foramina form the sacral canal.
6. Coccyx
   a. The coccyx, composed of four fused vertebrae, forms the lowest part of the vertebral column.
   b. It acts as a shock absorber when a person sits.

7.8 Thoracic Cage (p. 152)

The thoracic cage includes the ribs, thoracic vertebrae, sternum, and costal cartilages. It supports the pectoral girdle and upper limbs, protects viscera, and functions in breathing.
1. Ribs
   a. Twelve pairs of ribs attach to the twelve thoracic vertebrae.
   b. Costal cartilages of the true ribs join the sternum directly. Those of the false ribs join it indirectly or not at all.
   c. A typical rib has a shaft, a head, and tubercles that articulate with the vertebrae.
2. Sternum
   a. The sternum consists of a manubrium, body, and xiphoid process.
   b. It articulates with the clavicles.

7.9 Pectoral Girdle (p. 153)

The pectoral girdle is composed of two clavicles and two scapulae. It forms an incomplete ring that supports the upper limbs and provides attachments for muscles.
1. Clavicles
   a. Clavicles are rodlike bones located between the manubrium and the scapulae.
   b. They hold the shoulders in place and provide attachments for muscles.
2. Scapulae
   a. The scapulae are broad, triangular bones.
   b. They articulate with the humerus of each upper limb and provide attachments for muscles.

7.10 Upper Limb (p. 155)

Bones of the upper limb provide the frameworks and attachments of muscles, and function in levers that move the limb and its parts.

1. Humerus
   a. The humerus extends from the scapula to the elbow.
   b. It articulates with the radius and ulna at the elbow.
2. Radius
   a. The radius is located on the thumb side of the forearm between the elbow and the wrist.
   b. It articulates with the humerus, ulna, and wrist.
3. Ulna
   a. The ulna is longer than the radius and overlaps the humerus posteriorly.
   b. It articulates with the radius laterally and with a disc of fibrocartilage inferiorly.
4. Hand
   a. The wrist is composed of eight carpal bones that form a carpus.
   b. The palm or metacarpus includes five metacarpal bones and fourteen phalanges compose the fingers.

7.11 Pelvic Girdle (p. 157)

The pelvic girdle consists of two hip bones that articulate with each other anteriorly and with the sacrum posteriorly.

1. The sacrum, coccyx, and pelvic girdle form the bowl-shaped pelvis.
2. Each hip bone consists of an ilium, ischium, and pubis, which are fused in the region of the acetabulum.
   a. The ilium
      (1) The ilium is the largest portion of the hip bone.
      (2) It joins the sacrum at the sacroiliac joint.
   b. The ischium
      (1) The ischium is the lowest portion of the hip bone.
      (2) It supports the body weight when sitting.
   c. The pubis
      (1) The pubis is the anterior portion of the hip bone.
      (2) The pubic bones are fused anteriorly at the symphysis pubis.

7.12 Lower Limb (p. 159)

Bones of the lower limb provide frameworks for the thigh, leg, and foot.

1. Femur
   a. The femur extends from the hip to the knee.
   b. The patella articulates with the femur's anterior surface.
2. Tibia
   a. The tibia is located on the medial side of the leg.
   b. It articulates proximally with the femur and distally with the talus of the ankle.
3. Fibula
   a. The fibula is located on the lateral side of the tibia.
   b. It articulates with the ankle but does not bear body weight.
4. Foot
   a. The ankle consists of the tarsus formed by the talus and six other tarsal bones.
   b. The instep or metatarsus includes five metatarsals, and fourteen phalanges compose the toes.

7.13 Joints (p. 161)

Joints can be classified according to degree of movement as well as according to the type of tissue that binds the bones together.

1. Fibrous joints
   a. Bones at fibrous joints are tightly joined by a layer of dense connective tissue.
   b. Little (amphiarthrotic) or no movement (synarthrotic) occurs at a fibrous joint.

2. Cartilaginous joints
   a. A layer of cartilage joins the bones of cartilaginous joints.
   b. Such joints allow limited movement (amphiarthrotic).
3. Synovial joints
   a. The bones of a synovial joint are covered with hyaline cartilage and held together by a fibrous joint capsule.
   b. The joint capsule consists of an outer layer of ligaments and an inner lining of synovial membrane.
   c. Pads of fibrocartilage, menisci, act as shock absorbers in some synovial joints.
   d. Bursae are located between tendons and underlying bony prominences.
   e. Synovial joints that allow free movement (diarthrotic) include ball-and-socket, condyloid, gliding, hinge, pivot, and saddle.
4. Types of joint movements
   a. Muscles fastened on either side of a joint produce the movements of synovial joints.
   b. Joint movements include flexion, extension, dorsiflexion, plantar flexion, hyperextension, abduction, adduction, rotation, circumduction, pronation, supination, eversion, inversion, retraction, protraction, elevation, and depression.

CHAPTER ASSESSMENTS

7.1 Introduction

1. Active, living tissues found in bone include _____. (p. 131)
   a. blood
   b. nervous tissue
   c. dense connective tissue
   d. bone tissue
   e. all of the above

7.2 Bone Structure

2. Sketch a typical long bone, and label its epiphyses, diaphysis, medullary cavity, periosteum, and articular cartilages. On the sketch, designate the locations of compact and spongy bone. (p. 131)

3. Discuss the functions of the parts labeled in the sketch you made for question 2. (p. 131)

4. Differentiate between the microscopic structure of compact bone and spongy bone. (p. 132)

7.3 Bone Development and Growth

5. Explain how the development of intramembranous bone differs from that of endochondral bone. (p. 133)

6. _____ are mature bone cells, whereas _____ are bone-forming cells and _____ are bone-resorbing cells. (p. 134)

7. Explain the function of an epiphyseal plate. (p. 134)

8. Physical exercise pulling on muscular attachments to bones stimulates _____. (p. 135)

7.4 Bone Function

9. Give several examples of how bones support and protect body parts. (p. 135)

10. List and describe other functions of bones. (p. 138)

7.5 Skeletal Organization

11. Bones of the head, neck, and trunk compose the _____ skeleton; bones of the limbs and their attachments compose the _____ skeleton. (p. 139)

7.6–7.12 (Skull–Lower Limb)

12. Name the bones of the cranium and the facial skeleton. (pp. 142–147)

13. Describe a typical vertebra, and distinguish among the cervical, thoracic, and lumbar vertebrae. (pp. 147–150)

14. Name the bones that compose the thoracic cage. (p. 152)

15. The clavicle and scapula form the _____ girdle, whereas the hip bones and sacrum form the _____ girdle. (pp. 153 and 157)

16. Name the bones of the upper and lower limbs. (pp. 155–161)

17. Match the parts listed on the left with the bones listed on the right. (pp. 142–161)

    (1) Foramen magnum   A. Maxilla
    (2) Mastoid process   B. Occipital bone
    (3) Palatine process   C. Temporal bone
    (4) Sella turcica   D. Femur
    (5) Deltoid tuberosity   E. Fibula
    (6) Greater trochanter   F. Humerus
    (7) Lateral malleolus   G. Radius
    (8) Medial malleolus   H. Sternum
    (9) Radial tuberosity   I. Tibia
    (10) Xiphoid process   J. Sphenoid bone

7.13 Joints

18. Describe and give an example of a fibrous joint, a cartilaginous joint, and a synovial joint. (p. 162)

19. Name an example of each type of synovial joint, and describe the parts of the joint as they relate to the movement(s) allowed by that particular joint. (p. 163)

20. Joint movements occur when a muscle contracts and the muscle fibers pull the muscle's movable end of attachment to the bone, the _____, toward its fixed end, the _____. (p. 165)

21. Match the movement on the left with the appropriate description on the right. (pp. 165–166)

    (1) Rotation   A. turning palm upward
    (2) Supination   B. decreasing angle between parts
    (3) Extension   C. moving part forward
    (4) Eversion   D. moving part around axis
    (5) Protraction   E. moving part toward midline
    (6) Flexion   F. turning foot so plantar surface faces laterally
    (7) Pronation   G. increasing angle between parts
    (8) Abduction   H. lowering a part
    (9) Depression   I. turning palm downward
    (10) Adduction   J. moving part away from midline

INTEGRATIVE ASSESSMENTS/ CRITICAL THINKING

OUTCOME 7.2

1. How does the structure of a bone make it strong yet lightweight?

OUTCOME 7.3

2. When a child's bone is fractured, growth may be stimulated at the epiphyseal plate of that bone. What problems might this extra growth cause in an upper or lower limb before the growth of the other limb compensates for the difference in length?

OUTCOMES 7.3, 7.4, 7.11

3. Archaeologists discover skeletal remains of humanlike animals in Ethiopia. Examination of the bones suggests that the remains represent four types of individuals. Two of the skeletons have bone densities that are 30% less than those of the other two skeletons. The skeletons with the lower bone mass also have broader front pelvic bones. Within the two groups defined by bone mass, smaller skeletons have bones with evidence of epiphyseal plates, but larger bones have only a thin line where the epiphyseal plates should be. Give the age group and gender of the individuals in this find.

OUTCOME 7.13

4. Based upon knowledge of joint structures, which could be more satisfactorily replaced by a prosthetic device, a hip joint or a knee joint? Why?

WEB CONNECTIONS

Visit the text website at **aris.mhhe.com** for additional quizzes, interactive learning exercises, and more.

AP R SKELETAL SYSTEM

Anatomy & Physiology | REVEALED includes cadaver photos that allow you to peel away layers of the human body to reveal structures beneath the surface. This program also includes animations, radiologic imaging, audio pronunciation, and practice quizzing. To learn more visit **www.aprevealed.com**.

Human Skull

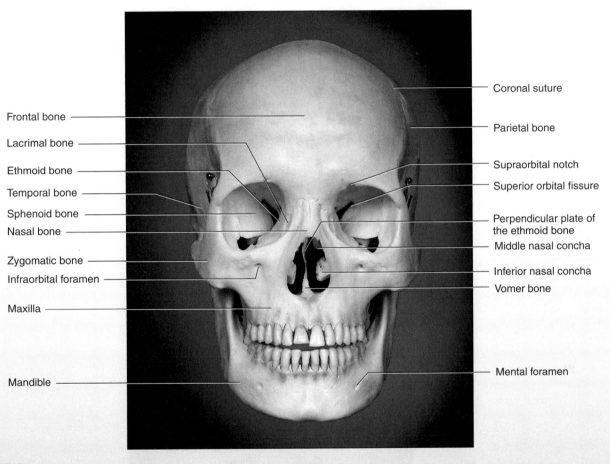

Frontal bone

Lacrimal bone

Ethmoid bone

Temporal bone

Sphenoid bone

Nasal bone

Zygomatic bone

Infraorbital foramen

Maxilla

Mandible

Coronal suture

Parietal bone

Supraorbital notch

Superior orbital fissure

Perpendicular plate of
the ethmoid bone

Middle nasal concha

Inferior nasal concha

Vomer bone

Mental foramen

PLATE 8
The skull, anterior view.

7

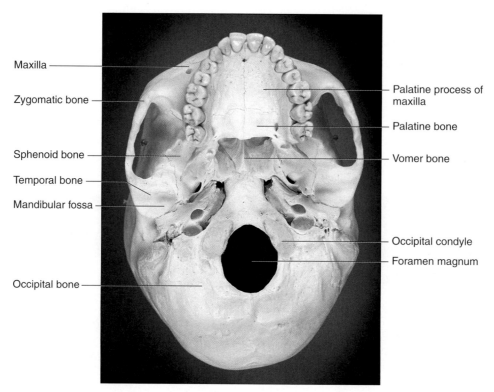

PLATE 9

The skull, inferior view.

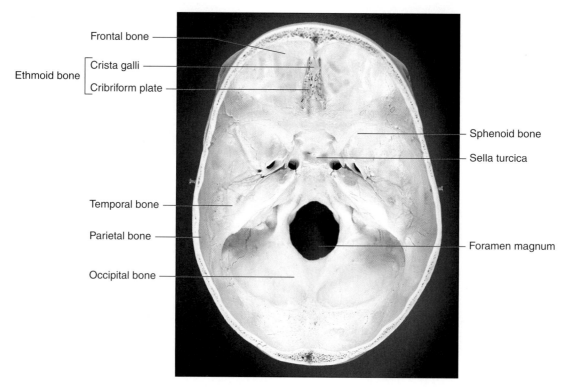

PLATE 10

The skull, floor of the cranial cavity.

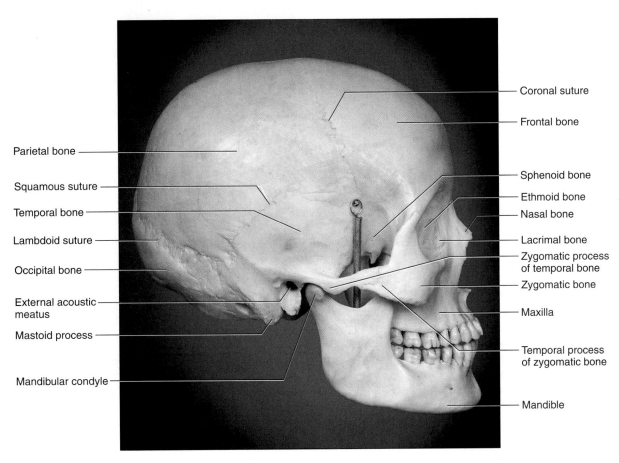

Coronal suture

Frontal bone

Parietal bone

Sphenoid bone

Squamous suture

Ethmoid bone

Temporal bone

Nasal bone

Lambdoid suture

Lacrimal bone

Occipital bone

Zygomatic process
of temporal bone

Zygomatic bone

External acoustic
meatus

Maxilla

Mastoid process

Temporal process
of zygomatic bone

Mandibular condyle

Mandible

PLATE 11

The skull, lateral view.

8

Muscular System

DOUBLE THE MUSCLE. The newborn had an astonishing appearance—his prominent arm and thigh muscles looked as if he'd been weight-lifting in the womb. By five years of age, his muscles twice normal size, he could lift heavier weights than could many adults. He also had half the normal amount of body fat.

The boy's muscle cells cannot produce a protein called myostatin, which normally stops stem cells from making a muscle too large. In this boy a mutation turned off this genetic brake, and as a result his muscles bulge, their cells both larger and more numerous than those in the muscles of a normal child. The boy is healthy so far, but because myostatin is also made in cardiac muscle, he may develop heart problems.

Other species with myostatin mutations are well known. Naturally "double-muscled" cattle and sheep are valued for their high weights early in life. Chicken breeders lower myostatin production to yield meatier birds, and "mighty mice" with silenced myostatin genes are used in basic research to study muscle overgrowth. In clinical applications, researchers are investigating ways to block myostatin activity to stimulate muscle growth to reverse muscle-wasting from AIDS, cancer, and muscular dystrophy. Myostatin is also of interest in athletics. Theoretically, infants could be tested to identify those with myostatin gene variants that predict athletic prowess, given the right training. Myostatin could also be abused to enhance athletic performance.

For those of us not endowed with genetically doubled muscles, regular resistance training (weight lifting) can tone muscles and trim fat.

For those of us not endowed with double-muscle mutations, resistance (weight) training can increase the ratio of muscle to fat in our bodies, which offers several benefits. Because muscle burns calories at three times the rate of fat, a lean body is more energetically efficient. Weight-lifting increases muscle tone and bone density; lowers blood pressure; decreases the risks of developing arthritis, osteoporosis, and diabetes mellitus; and is even associated with improved self-esteem and fewer sick days.

Learning Outcomes   *After studying this chapter, you should be able to do the following:*

8.1 Introduction

1. List various outcomes of muscular actions. (p. 177)

8.2 Structure of a Skeletal Muscle

2. Describe how connective tissue is part of a skeletal muscle. (p. 177)

3. Name the major parts of a skeletal muscle fiber, and describe the function of each. (p. 177)

4. Discuss nervous stimulation of a skeletal muscle. (p. 180)

8.3 Skeletal Muscle Contraction

5. Identify the major events of skeletal muscle fiber contraction. (p. 181)

6. Describe the energy sources for muscle fiber contraction. (p. 183)

7. Describe how oxygen debt develops and how a muscle may become fatigued. (p. 184)

8.4 Muscular Responses

8. Distinguish between a twitch and a sustained contraction. (p. 187)

9. Explain how muscular contractions move body parts and help maintain posture. (p. 188)

8.5 Smooth Muscle

10. Distinguish between the structures and functions of multiunit smooth muscle and visceral smooth muscle. (p. 189)

11. Compare the contraction mechanisms of skeletal and smooth muscle fibers. (p. 189)

8.6 Cardiac Muscle

12. Compare the contraction mechanisms of cardiac and skeletal muscle fibers. (p. 190)

8.7 Skeletal Muscle Actions

13. Explain how the attachments, locations, and interactions of skeletal muscles make possible certain movements. (p. 192)

8.8 Major Skeletal Muscles

14. Describe the locations and actions of the major skeletal muscles of each body region. (p. 193–205)

Aids to Understanding Words

(Appendix A on page 567 has a complete list of Aids to Understanding Words.)

calat- [something inserted] inter*calat*ed disc: Membranous band that connects cardiac muscle cells.

erg- [work] syn*erg*ist: Muscle that works with a prime mover to produce a movement.

hyper- [over, more] muscular *hyper*trophy: Enlargement of muscle fibers.

inter- [between] *inter*calated disc: Membranous band that connects cardiac muscle cells.

laten- [hidden] *laten*t period: Time between application of a stimulus and the beginning of a muscle contraction.

myo- [muscle] *myo*fibril: Contractile structure within a muscle cell.

sarco- [flesh] *sarco*plasm: Material (cytoplasm) within a muscle fiber.

syn- [together] *syn*ergist: Muscle that works with a prime mover to produce a movement.

tetan- [stiff] *tetan*ic contraction: Sustained muscular contraction.

-troph [well fed] muscular hyper*troph*y: Enlargement of muscle fibers.

8.1 INTRODUCTION

Talking and walking, breathing and sneezing—in fact, all movements—require muscles. Muscles are organs composed of specialized cells that use the chemical energy stored in nutrients to contract. Muscular actions also provide muscle tone, propel body fluids and food, generate the heartbeat, and distribute heat.

Muscles are of three types—skeletal muscle, smooth muscle, and cardiac muscle. This chapter focuses mostly on skeletal muscle, which attaches to bones and is under conscious control. Smooth muscle and cardiac muscle are discussed briefly.

8.2 STRUCTURE OF A SKELETAL MUSCLE

A skeletal muscle is an organ of the muscular system. It is composed of skeletal muscle tissue, nervous tissue, blood, and other connective tissues.

Connective Tissue Coverings

Layers of fibrous connective tissue called **fascia** (fash′e-ah) separate an individual skeletal muscle from adjacent muscles and hold it in position (fig. 8.1). This connective tissue surrounds each muscle and may project beyond its end to form a cordlike tendon. Fibers in a tendon may intertwine with those in a bone's periosteum, attaching the muscle to the bone. In other cases, the connective tissue forms broad fibrous sheets called **aponeuroses** (ap″o-nu-ro′sez), which may attach to bone or to the coverings of adjacent muscles (see figs. 8.17 and 8.19).

The layer of connective tissue that closely surrounds a skeletal muscle is called *epimysium* (fig. 8.1). Other layers of connective tissue, called *perimysium,* extend inward from the epimysium and separate the muscle tissue into small compartments. These compartments contain bundles of skeletal muscle fibers called *fascicles* (fasciculi). Each muscle fiber within a fascicle (fasciculus) lies within a layer of connective tissue in the form of a thin covering called *endomysium.* Layers of con-

nective tissue, therefore, enclose and separate all parts of a skeletal muscle. This organization allows the parts to move somewhat independently. Many blood vessels and nerves pass through these layers.

> A tendon, the attachment of a muscle to a bone, may become painfully inflamed and swollen following injury or the repeated stress of athletic activity, a condition called *tendinitis.* When the inflammation occurs in the connective tissue sheath of the tendon (the tenosynovium), it is called *tenosynovitis.* The tendons most commonly affected are those associated with the joint capsules of the shoulder, elbow, and hip and those that move the hand, thigh, and foot.

Skeletal Muscle Fibers

A skeletal muscle fiber is a single cell that contracts in response to stimulation and then relaxes when the stimulation ends. Each skeletal muscle fiber is a thin, elongated cylinder with rounded ends, and it may extend the full length of the muscle. Just beneath its cell membrane (or *sarcolemma*), the cytoplasm (or *sarcoplasm*) of the fiber has many small, oval nuclei and mitochondria (fig. 8.1). The sarcoplasm also contains many threadlike **myofibrils** (mi″o-fi′brilz) that lie parallel to one another.

Myofibrils play a fundamental role in muscle contraction. They consist of two kinds of protein filaments—thick ones composed of the protein **myosin** (mi′o-sin) and thin ones mainly composed of the protein **actin** (ak′tin) (figs. 8.2 and 8.3). (Two other thin filament proteins, troponin and tropomyosin, are discussed later on page 181.) The organization of these filaments produces the characteristic alternating light and dark *striations,* or bands, of a skeletal muscle fiber.

The striations form a repeating pattern of units called **sarcomeres** (sar′ko-mĕrz) along each muscle fiber. The myofibrils may be thought of as sarcomeres joined end-to-end (fig. 8.2). Muscle fibers, and in a way muscles themselves, may be considered a collection of sarcomeres. Sarcomeres are discussed later as the functional units of muscle contraction (p. 181).

Figure 8.1

A skeletal muscle is composed of a variety of tissues, including layers of connective tissue. Fascia covers the surface of the muscle, epimysium lies beneath the fascia, and perimysium extends into the structure of the muscle where it separates muscle cells into fascicles. Endomysium separates individual muscle fibers.

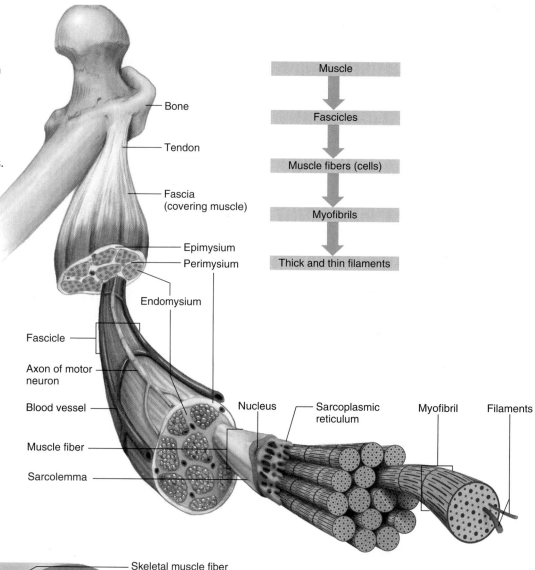

Muscle

↓

Fascicles

↓

Muscle fibers (cells)

↓

Myofibrils

↓

Thick and thin filaments

Bone

Tendon

Fascia (covering muscle)

Epimysium
Perimysium

Endomysium

Fascicle

Axon of motor neuron

Blood vessel

Muscle fiber

Sarcolemma

Nucleus

Sarcoplasmic reticulum

Myofibril

Filaments

Figure 8.2

Skeletal muscle fiber.
(*a*) A skeletal muscle fiber contains many myofibrils, each consisting of
(*b*) repeating units called sarcomeres. The characteristic striations of a sarcomere reflect the organization of actin and myosin filaments.

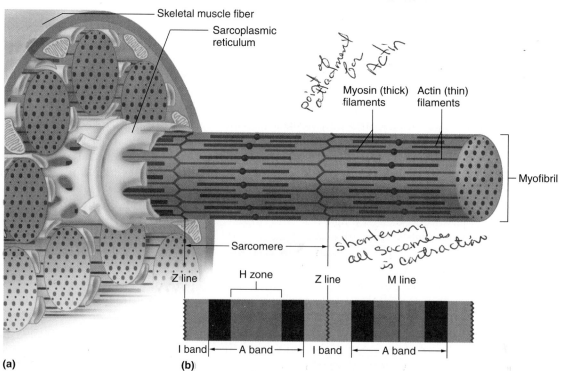

Skeletal muscle fiber

Sarcoplasmic reticulum

Point of attachment for Actin

Myosin (thick) filaments

Actin (thin) filaments

Myofibril

Sarcomere

Shortening all sarcomeres is contraction

Z line

H zone

Z line

M line

I band

A band

I band

A band

(a)

(b)

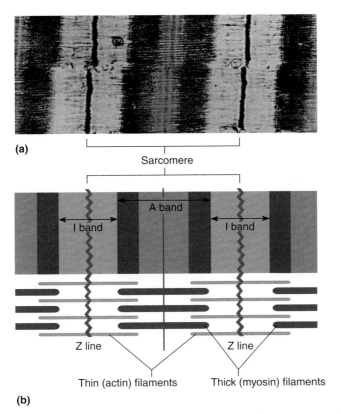

(a)

Sarcomere

A band

I band | I band

Z line Z line

Thin (actin) filaments Thick (myosin) filaments

(b)

Figure 8.3

A sarcomere is a functional unit of muscle contraction.
(a) Micrograph (16,000×). (b) The spatial relationship of thin and thick filaments in a sarcomere makes contraction possible.

Muscle cells are packed with actin and myosin, but several other types of protein are important in muscle function too. Discovery of the gene that causes the two most common forms of muscular dystrophy took many years because the protein that is absent or incomplete, called dystrophin, comprises only 0.002% of the protein in skeletal muscle. (see Genetics Connection on page 191.)

The striation pattern of skeletal muscle fibers has two main parts. The first, the *I bands* (the light bands), are composed of thin actin filaments directly attached to structures called *Z lines*.

The second part of the striation pattern consists of the *A bands* (the dark bands), which are composed of thick myosin filaments overlapping thin actin filaments. The A band consists of a region where the thick and thin filaments overlap, and a central region (*H zone*) consisting only of thick filaments, plus a thickening known as the *M line* (fig. 8.2). The M line consists of proteins that help hold the thick filaments in place. A sarcomere extends from one Z line to the next (figs. 8.2 and 8.3).

Within the sarcoplasm of a muscle fiber is a network of membranous channels that surrounds each myofibril and runs parallel to it (fig. 8.4). These membranes form the **sarcoplasmic reticulum,** which corresponds to the endoplasmic reticulum of other cells.

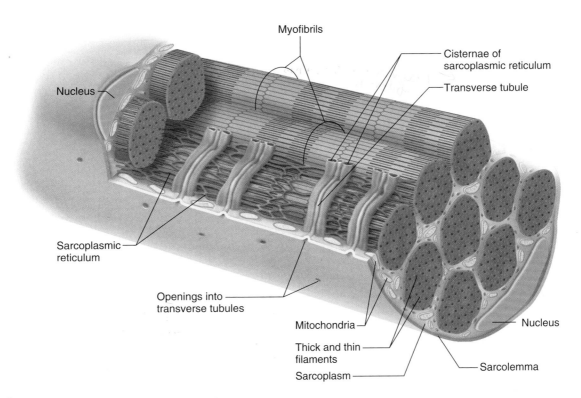

Myofibrils

Cisternae of sarcoplasmic reticulum

Transverse tubule

Nucleus

Sarcoplasmic reticulum

Openings into transverse tubules

Mitochondria

Thick and thin filaments

Sarcoplasm

Nucleus

Sarcolemma

Figure 8.4

Within the sarcoplasm of a skeletal muscle fiber are a network of sarcoplasmic reticulum and a system of transverse tubules.

Another set of membranous channels, called **transverse tubules** (T tubules), extends inward as invaginations from the fiber's membrane and passes all the way through the fiber. Thus, each tubule opens to the outside of the muscle fiber and contains extracellular fluid. Furthermore, each transverse tubule lies between two enlarged portions of the sarcoplasmic reticulum called *cisternae,* near the region where the actin and myosin filaments overlap. The sarcoplasmic reticulum and transverse tubules activate the muscle contraction mechanism when the fiber is stimulated.

Muscle fibers and their associated connective tissues are flexible but can tear if overstretched. This type of injury, common in athletes, is called *muscle strain.* The seriousness of the injury depends on the degree of damage the tissues sustain. If the strain is mild, only a few muscle fibers are injured, the fascia remains intact, and loss of function is minimal. In a severe strain, however, many muscle fibers as well as the fascia tear, and muscle function may be completely lost. Such a severe strain is painful and produces discoloration and swelling.

Check Your Recall

1. Describe how connective tissue is part of a skeletal muscle.
2. Describe the general structure of a skeletal muscle fiber.
3. Explain why skeletal muscle fibers appear striated.
4. Explain the relationship between the sarcoplasmic reticulum and the transverse tubules.

Neuromuscular Junction  *Site where motor neuron & muscle fiber meet is this*

Recall from chapter 5 (p. 111) that neurons (nerve cells) play a role in body communication by conducting nerve impulses. Neurons that control effectors, including skeletal muscle, are called **motor neurons.**

Each skeletal muscle fiber is functionally (but not physically) connected to the axon of a motor neuron that passes outward from the brain or the spinal cord, in much the same way that you can talk into a cell phone although your mouth is not in direct physical contact with it. This functional connection is called a **synapse** (sin'aps). Neurons communicate with the cells that they control by releasing chemicals, called **neurotransmitters** (nu″ro-trans′mit-erz), at synapses. Normally, a skeletal muscle fiber contracts only upon stimulation by a motor neuron.

The connection between the motor neuron and the muscle fiber is called a **neuromuscular junction.** Here, the muscle fiber membrane is specialized to form a **motor end plate.** In this region of the muscle fiber, nuclei and mitochondria are abundant, and the cell membrane (sarcolemma) is extensively folded (fig. 8.5).

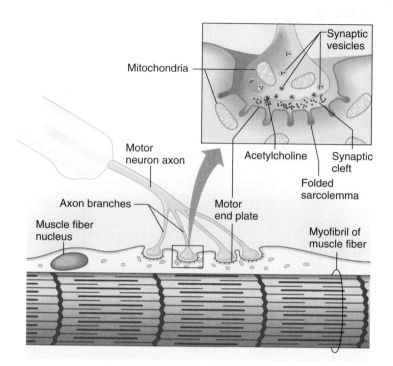

Figure 8.5
A neuromuscular junction includes the end of a motor neuron and the motor end plate of a muscle fiber.

The end of the motor neuron branches and projects into recesses of the muscle fiber membrane. The cytoplasm at the distal ends of these motor neuron axons is rich in mitochondria and contains many tiny vesicles (synaptic vesicles) that store neurotransmitters.

When a nerve impulse traveling from the brain or spinal cord reaches the end of a motor neuron axon, some of the vesicles release neurotransmitter molecules into the gap (synaptic cleft) between the neuron and the motor end plate of the muscle fiber. This action stimulates the muscle fiber to contract.

Motor Units

A muscle fiber usually has a single motor end plate. The axons of motor neurons, however, are densely branched. By means of these branches, one motor neuron may connect to many muscle fibers. When a motor neuron transmits an impulse, all of the muscle fibers it links to are stimulated to contract simultaneously. Together, a motor neuron and the muscle fibers that it controls constitute a **motor unit** (mo'tor u'nit) (fig. 8.6).

Check Your Recall

5. Which two structures approach each other at a neuromuscular junction?
6. Describe a motor end plate.
7. What is the function of a neurotransmitter?
8. What is a motor unit?

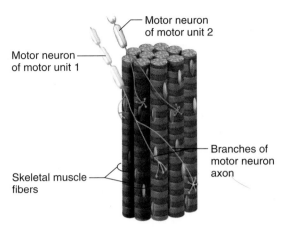

Figure 8.6

Two motor units. Muscle fibers within a motor unit are innervated by a single neuron and may be distributed throughout the muscle.

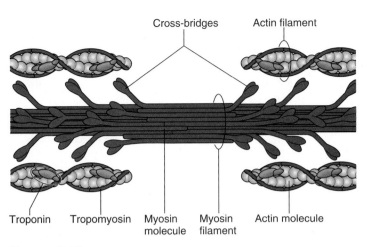

Figure 8.7

Thick filaments are composed of the protein myosin, and thin filaments are composed primarily of the protein actin. Myosin molecules have cross-bridges that extend toward nearby actin filaments.

8.3 SKELETAL MUSCLE CONTRACTION

A muscle fiber contraction is a complex interaction of organelles and molecules in which myosin binds to actin and exerts a pulling action. The result is a movement within the myofibrils in which the filaments of actin and myosin slide past one another. This action shortens the muscle fiber so that it pulls on its attachments.

Role of Myosin and Actin

A myosin molecule is composed of two twisted protein strands with globular parts called cross-bridges projecting outward along their lengths. Many of these molecules comprise a myosin (thick) filament (fig. 8.7). An actin molecule is a globular structure with a binding site to which the myosin cross-bridges can attach. Many actin molecules twist into a double strand (helix), forming an actin (thin) filament. The proteins **troponin** and **tropomyosin** are also part of the actin filament (fig. 8.7).

The sarcomere is considered the functional unit of skeletal muscles because the contraction of an entire skeletal muscle can be described in terms of the shortening of sarcomeres within its muscle fibers. The force that shortens the sarcomeres comes from the cross-bridges pulling on the thin filaments. A myosin cross-bridge can attach to an actin binding site and bend slightly, pulling on the actin filament. Then the head can release, straighten, combine with another binding site further down the actin filament, and pull again (fig. 8.8).

The **sliding filament model** includes all of these events and gets its name from the way the sarcomeres shorten. Thick and thin filaments do not change length. Rather, they slide past one another, with the thin filaments sliding toward the center of the sarcomere from both ends (fig. 8.9).

The globular portions of the myosin filaments contain an enzyme, **ATPase,** which catalyzes the breakdown of ATP to ADP and phosphate. The reaction releases energy that puts the myosin cross-bridge in a "cocked" position (see chapter 4, p. 80). When a cocked cross-bridge binds to actin, it pulls on the thin filament. After the cross-bridge pulls, another ATP binding to the cross-bridge causes it to be released from actin even before the ATP splits. This cycle repeats as long as ATP is available as an energy source and as long as the muscle fiber is stimulated to contract.

Stimulus for Contraction

A skeletal muscle fiber normally does not contract until a neurotransmitter stimulates it. The neurotransmitter in this case is **acetylcholine** (as″ĕ-til-ko′lēn). This neurotransmitter is synthesized in the cytoplasm of the motor neuron and stored in vesicles at the distal end of the motor neuron axons. When a nerve impulse (described in chapter 9, p. 224) reaches the end of a motor neuron axon, some of the vesicles release their acetylcholine into the space (synaptic cleft) between the motor neuron axon and the motor end plate (see fig. 8.5).

Acetylcholine diffuses rapidly across the synaptic cleft and binds to specific protein molecules (receptors) in the muscle fiber membrane, increasing membrane permeability to sodium ions. Entry of these charged particles into the muscle cell stimulates a **muscle impulse** (mus′el im′puls), which is very much like a nerve impulse. The impulse passes in all directions over the surface of the muscle fiber membrane and travels through the transverse tubules, deep into the fiber, until it reaches the sarcoplasmic reticulum (see fig. 8.4).

The sarcoplasmic reticulum contains a high concentration of calcium ions. In response to a muscle impulse,

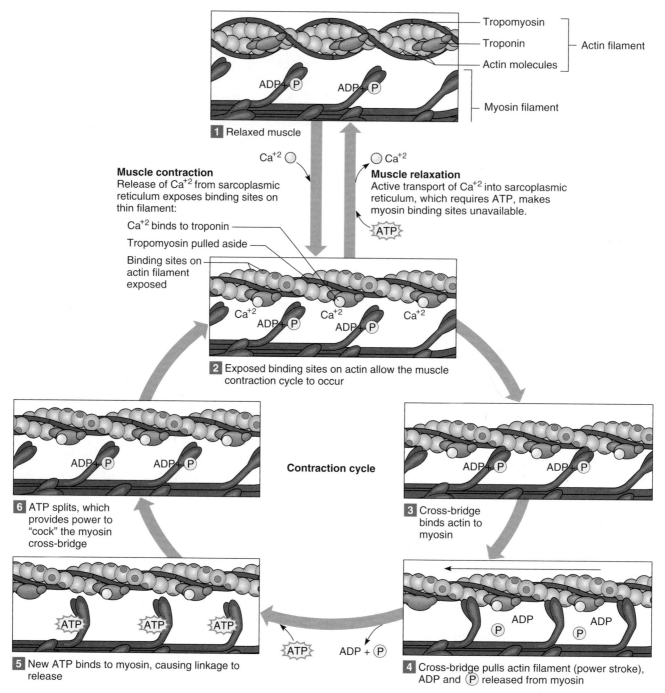

Figure 8.8

The sliding filament model. (*1*) and (*2*) When calcium ion concentration rises, binding sites on actin filaments open, and cross-bridges attach. (*3*) and (*4*) Upon binding to actin, cross-bridges spring from the cocked position and pull on actin filaments. (*5*) ATP binds to the cross-bridge (but is not yet broken down), causing the cross-bridge to release from the actin filament. (*6*) ATP breakdown provides energy to "cock" the unattached myosin cross-bridge. As long as ATP and calcium ions are present, the cycle continues. When calcium ion concentration is low, the muscle remains relaxed.

the membranes of the cisternae become more permeable to these ions, and the calcium ions diffuse into the sarcoplasm of the muscle fiber.

When a high concentration of calcium ions is in the sarcoplasm, troponin and tropomyosin interact in a way that exposes binding sites on actin. As a result, linkages form between the actin and myosin filaments, and the muscle fiber contracts (see figs. 8.8 and 8.9). The contraction, which also requires ATP, continues as long as nerve impulses release acetylcholine.

When the nerve impulses cease, two events lead to muscle relaxation. First, the acetylcholine that stimulated

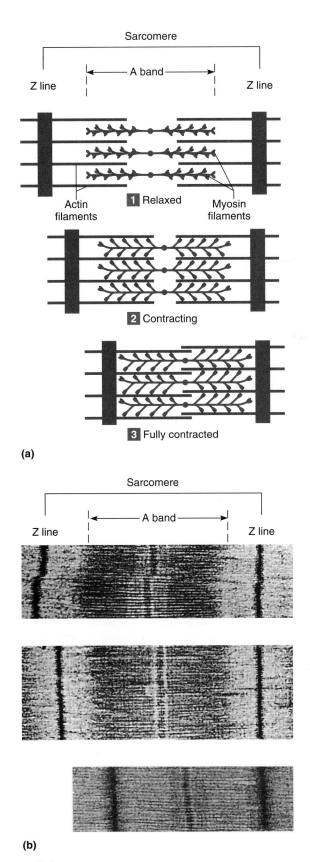

(a)

(b)

Figure 8.9

When a skeletal muscle contracts, (*a*) individual sarcomeres shorten as thick and thin filaments slide past one another. (*b*) This transmission electron micrograph shows a sarcomere shortening during muscle contraction (23,000×).

the muscle fiber is rapidly decomposed by the enzyme **acetylcholinesterase** (as′ĕ-til-ko″lin-es′ter-ās). This enzyme is present at the neuromuscular junction on the membranes of the motor end plate. Acetylcholinesterase prevents a single nerve impulse from continuously stimulating the muscle fiber.

The second event in relaxation occurs once acetylcholine is broken down and the stimulus to the muscle fiber ceases. Calcium ions are actively transported back into the sarcoplasmic reticulum, which decreases the calcium ion concentration of the sarcoplasm. The linkages between actin and myosin filaments break, and consequently, the muscle fiber relaxes. Table 8.1 summarizes the major events leading to muscle contraction and relaxation.

The bacterium *Clostridium botulinum* produces a poison, called botulinum toxin, that can prevent the release of acetylcholine from motor neuron axons at neuromuscular junctions, causing *botulism*, a very serious form of food poisoning. Most cases of botulism are caused by eating home-processed food that has not been heated enough to kill the bacteria in it or to inactivate the toxin.

Botulinum toxin blocks stimulation of muscle fibers, paralyzing muscles, including those responsible for breathing. Without prompt medical treatment, the fatality rate for botulism is high. Very small amounts of botulism toxin ("botox") are injected into facial skin to temporarily smooth wrinkles by preventing local muscles from contracting. If done to excess, botox can cause the face to appear frozen.

Check Your Recall

9. Explain how a motor nerve impulse can trigger a muscle contraction.

10. Explain how the filaments of a myofibril interact during muscle contraction.

Energy Sources for Contraction

ATP molecules supply the energy for muscle fiber contraction. However, a muscle fiber has only enough ATP to enable it to contract for a very short time, so when a fiber is active, ATP must be regenerated.

The initial source of energy available to a contracting muscle comes from existing ATP molecules in the cell. Almost immediately, however, cells must regenerate ATP from ADP and phosphate. The molecule that makes this possible is **creatine phosphate** (kre′ah-tin fos′fāt). Like ATP, creatine phosphate contains high-energy phosphate bonds, and it is four to six times more abundant in muscle fibers than ATP. Creatine

Table 8.1	Major Events of Muscle Contraction and Relaxation

Muscle Fiber Contraction

1. A nerve impulse travels down a motor neuron axon.

2. The motor neuron terminal releases the neurotransmitter acetylcholine (ACh).

3. ACh binds to ACh receptors.

4. The sarcolemma is stimulated, and a muscle impulse travels over the surface of the muscle fiber and deep into the fiber through the transverse tubules.

5. The muscle impulse reaches the sarcoplasmic reticulum and calcium channels open.

6. Calcium ions diffuse from the sarcoplasmic reticulum into the sarcoplasm and bind to troponin molecules.

7. Tropomyosin molecules move and expose specific sites on actin.

8. Actin and myosin form linkages.

9. Thin (actin) filaments are pulled toward the center of the sarcomere by myosin cross-bridges.

10. The muscle fiber shortens as a contraction occurs.

Muscle Fiber Relaxation

1. Acetylcholinesterase decomposes acetylcholine, and the muscle fiber membrane is no longer stimulated.

2. Calcium ions are actively transported into the sarcoplasmic reticulum.

3. ATP breaks linkages between actin and myosin filaments without breakdown of the ATP itself.

4. Breakdown of ATP "cocks" the cross-bridges

5. Troponin and tropomyosin molecules inhibit the interaction between myosin and actin filaments.

6. The muscle fiber remains relaxed, yet ready, until stimulated again.

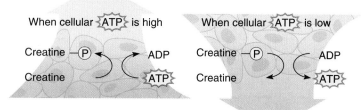

Figure 8.10

Creatine phosphate may be used to replenish ATP stores when ATP levels in a muscle cell are low.

to support this aerobic respiration from the lungs to body cells. Red blood cells carry the oxygen, loosely bound to molecules of **hemoglobin,** the pigment responsible for the red color of blood. Another pigment, **myoglobin,** is synthesized in muscle cells and imparts the reddish-brown color of skeletal muscle tissue. Like hemoglobin, myoglobin can combine loosely with oxygen. This ability to temporarily store oxygen reduces a muscle's requirement for a continuous blood supply during muscular contraction (fig. 8.11).

Oxygen Debt

When a person is resting or is moderately active, the respiratory and cardiovascular systems can usually supply sufficient oxygen to skeletal muscles to support aerobic respiration. However, this is not the case when skeletal muscles are used strenuously for even a minute or two. In this situation, muscle fibers must increasingly use anaerobic respiration to obtain energy.

In one form of anaerobic respiration, glucose molecules are broken down by glycolysis to yield *pyruvic acid* (see chapter 4, p. 82). Because the oxygen supply is low, however, the pyruvic acid reacts to produce *lactic acid,* which may accumulate in the muscles (fig. 8.11). Lactic acid diffuses into the bloodstream and eventually reaches the liver. In liver cells, reactions requiring ATP synthesize glucose from lactic acid.

During strenuous exercise, available oxygen is used primarily to synthesize the ATP the muscle fiber requires to contract, rather than to make ATP for synthesizing glucose from lactic acid. Consequently, as lactic acid accumulates, a person develops an **oxygen debt** (ok′sĭ-jen det) that must be repaid. Oxygen debt equals the amount of oxygen liver cells require to convert the accumulated lactic acid into glucose, plus the amount muscle cells require to restore ATP and creatine phosphate to their original concentrations.

The conversion of lactic acid back into glucose is slow. Repaying an oxygen debt following vigorous exercise may take several hours.

phosphate, however, cannot directly supply energy to a cell's energy-utilizing reactions. Instead, it stores excess energy released from the mitochondria. When ATP supply is sufficient, an enzyme in the mitochondria (creatine phosphokinase) catalyzes the synthesis of creatine phosphate, which stores excess energy in its phosphate bonds (fig. 8.10).

As ATP decomposes, the energy from creatine phosphate can be transferred to ADP molecules, converting them back into ATP. Active muscle, however, rapidly exhausts the supply of creatine phosphate. When this happens, the muscle fibers use cellular respiration of glucose as an energy source for synthesizing ATP.

Oxygen Supply and Cellular Respiration

As chapter 4 describes (p. 82), glycolysis can take place in the absence of oxygen. However, the more complete breakdown of glucose occurs in the mitochondria and requires oxygen. The blood carries the oxygen required

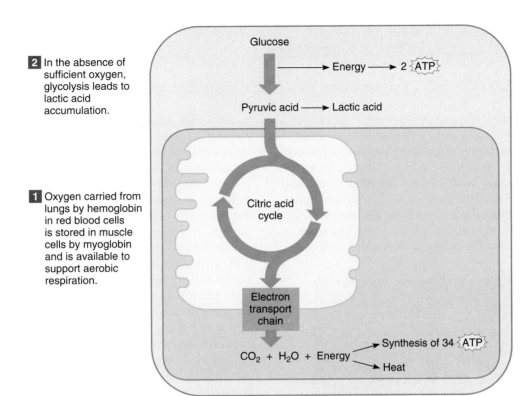

2 In the absence of sufficient oxygen, glycolysis leads to lactic acid accumulation.

1 Oxygen carried from lungs by hemoglobin in red blood cells is stored in muscle cells by myoglobin and is available to support aerobic respiration.

Glucose

Energy → 2 ATP

Pyruvic acid → Lactic acid

Citric acid cycle

Electron transport chain

CO_2 + H_2O + Energy → Synthesis of 34 ATP
→ Heat

Figure 8.11

The oxygen required to support aerobic respiration is carried in the blood and stored in myoglobin. In the absence of sufficient oxygen, pyruvic acid is converted to lactic acid by anaerobic respiration. The maximum number of ATPs generated per glucose molecule varies with cell type; in skeletal muscle, it is 36 (2 + 34).

The metabolic capacity of a muscle may change with training. With high-intensity exercise that depends more on glycolysis for ATP, a muscle synthesizes more glycolytic enzymes, and its capacity for glycolysis increases. With aerobic exercise, more capillaries and mitochondria form, and the muscle's capacity for aerobic respiration is greater. Table 8.2 summarizes muscle metabolism, and the Topic of Interest on page 186 discusses the dangerous use of steroid drugs to enhance muscle performance.

Muscle Fatigue

A muscle exercised strenuously for a prolonged period may lose its ability to contract, a condition called *fatigue*. Interruption in the muscle's blood supply or, rarely, lack of acetylcholine in motor neuron axons may cause fatigue. However, fatigue is most likely to arise from accumulation of lactic acid in the muscle as a result of anaerobic respiration. The lactic acid buildup lowers pH, and as a result, muscle fibers no longer respond to stimulation.

Occasionally, a muscle becomes fatigued and cramps at the same time. A cramp is a painful condition in which a muscle undergoes a sustained involuntary contraction. Cramps are thought to occur when changes in the extracellular fluid surrounding the muscle fibers and their motor neurons somehow trigger uncontrolled stimulation of the muscle. *Lack ATP required Return Ca ions back to Sarcoplasmic R so muscle fibers can relax.*

Table 8.2	Muscle Metabolism	
Type of Exercise	Low to moderate intensity: Blood flow provides sufficient oxygen for cellular requirements	High intensity: Oxygen supply is not sufficient for cellular requirement
Pathway Used	Glycolysis, leading to pyruvic acid formation and aerobic respiration	Glycolysis, leading to lactic acid formation
ATP Production	36 ATP per glucose for skeletal muscle	2 ATP per glucose
Waste Product	Carbon dioxide is exhaled	Lactic acid accumulates

Steroids and Athletes—An Unhealthy Combination

In the 1988 summer Olympics held in Seoul, South Korea, Canadian Ben Johnson flew past his competitors in the 100-meter run. But seventy-two hours later, officials rescinded the gold medal he won for his record-smashing time of 9.79 seconds, after a urine test revealed traces of the drug stanozolol, a synthetic stand-in for the steroid hormone testosterone (fig. 8A). Johnson's natural testosterone level was only 15% of normal—evidence of negative feedback acting because of an outside supply of the hormone. Steroid abuse among athletes, both amateur and professional, continues today, and among high school and college athletes is on the increase, according to the American College of Sports Medicine.

Athletes who abuse steroids do so to take advantage of the hormone's ability to increase muscular strength. But improved performance today may have consequences tomorrow. Steroids hasten adulthood, stunting height and causing early hair loss. In males, excess steroid hormones lead to breast development, and in females to a deepened voice, hairiness, and a male physique. The kidneys, liver, and heart may be damaged, and atherosclerosis may develop because steroids raise LDL and lower HDL—the opposite of a healthy cholesterol profile. In males, the body mistakes the synthetic steroids for the natural hormone and lowers its own production of testosterone—as Ben Johnson found out. Infertility may result. Steroids can also cause psychiatric symptoms, including delusions, depression, and violence.

Steroid abuse began in Nazi Germany, where Hitler used the drugs to fashion his "super race." Ironically, steroids were also used shortly after the war to build up the bodies of concentration camp survivors. In the 1950s, Soviet athletes began using steroids in the Olympics, and a decade later, U.S. athletes did the same. In 1976, the International Olympic Committee banned steroid use and required urine tests for detecting the drugs.

Anabolic steroids were created for medical purposes. They were first used clinically in the 1930s to treat underdevelopment of the testes and the resulting deficit in testosterone. In the 1950s, physicians used anabolic steroids to treat anemia and muscle-wasting disorders, and to bulk up patients whose muscles had atrophied due to extended bed rest. In the 1960s, anabolic steroids were used to treat some forms of short stature and dwarfism, a practice that was discontinued when pure preparations of human growth hormone became available through recombinant DNA technology. Today, anabolic steroids are used to treat wasting associated with AIDS.

Figure 8A

Canadian track star Ben Johnson ran away with the gold medal in the 100-meter race at the 1988 Summer Olympics—but then had to return the award when traces of a steroid drug showed up in his urine. Drug abuse continues to be a problem among amateur as well as professional athletes.

Several hours after death, the skeletal muscles undergo a partial contraction that fixes the joints. This condition, *rigor mortis,* may continue for 72 hours or more. It results from an increase in membrane permeability to calcium ions and a decrease in ATP in muscle fibers, which prevents relaxation. Thus, the actin and myosin filaments of the muscle fibers remain linked until the muscles begin to decompose.

Heat Production

Less than half of the energy released in cellular respiration is available for use in metabolic processes; the rest becomes heat. Although all active cells generate heat, muscle tissue is a major heat source because muscle is such a large proportion of the total body mass. Blood transports heat generated in muscle to other tissues, which helps maintain body temperature.

8.4 MUSCULAR RESPONSES

One way to observe muscle contraction is to remove a single muscle fiber from a skeletal muscle and connect it to a device that records changes in the fiber's length. Such experiments usually require an electrical device that can produce stimuli of varying strengths and frequencies.

Threshold Stimulus

When an isolated muscle fiber is exposed to a series of stimuli of increasing strength, the fiber remains unresponsive until a certain strength of stimulation called the **threshold stimulus** (thresh'old stim'u-lus) is applied. Once threshold is reached, an action potential is generated, resulting in a muscle impulse that spreads throughout the muscle fiber, releasing enough calcium ions from the sarcoplasmic reticulum to activate cross-bridge binding and contract that fiber. A single nerve impulse in a motor neuron normally releases enough ACh to bring the muscle fibers in its motor unit to threshold, generating a muscle impulse in each muscle fiber.

Recording of a Muscle Contraction

The contractile response of a single muscle fiber to a muscle impulse is called a **twitch.** A twitch consists of a period of contraction, during which the fiber pulls at its attachments, followed by a period of relaxation, during which the pulling force declines. These events can be recorded in a pattern called a myogram (fig. 8.12). Note that a twitch has a brief delay between the time of stimulation and the beginning of contraction. This is the **latent period,** which in human muscle may be less than 2 milliseconds.

When a muscle fiber is brought to threshold under a given set of conditions, it tends to contract completely, such that each twitch generates the same force. This has been referred to as an *all-or-none* response. This is misleading, however, because in normal use of muscles, the force generated by muscle fibers and by whole muscles must vary.

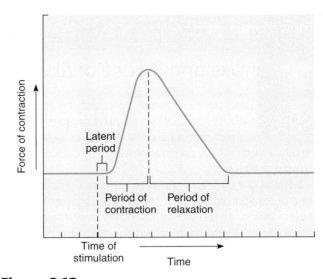

Figure 8.12

A myogram of a single muscle twitch.

Understanding the contraction of individual muscle fibers is important for understanding how muscles work, but such contractions by themselves are of little significance in day-to-day activities. Rather, the actions we need to perform usually require the contraction of multiple muscle fibers simultaneously. To record how a whole muscle responds to stimulation, a skeletal muscle can be removed from a frog or other small animal and mounted on a special device. The muscle is then stimulated electrically, and when it contracts, it pulls on a lever. The lever's movement is recorded as a myogram. Because the myogram results from the combined twitches of muscle fibers taking part in the contraction, it looks essentially the same as the twitch contraction depicted in figure 8.12. The Topic of Interest on page 188 describes two types of twitches—the fatigue-resistant slow twitch and the fatigable fast twitch. Muscle fibers are either slow twitch or fast twitch.

The skeletal muscles of an average person have about half fast twitch and half slow twitch muscle fibers. In contrast, the muscles of an Olympic sprinter typically have more than 80% fast twitch muscle fibers, and those of an Olympic marathoner, more than 90% slow twitch muscle fibers.

Contractions of whole muscles enable us to perform everyday activities, but the force generated by those contractions must be controlled. For example, holding a styrofoam cup of coffee firmly enough that it does not slip through our fingers, but not so forcefully as to crush it, requires precise control of contractile force. In the whole muscle, the degree of tension developed reflects (1) the frequency at which individual muscle fibers are stimulated and (2) how many fibers take part in the overall contraction of the muscle.

Use and Disuse of Skeletal Muscles

Skeletal muscles are very responsive to use and disuse. Forcefully exercised muscles enlarge, which is called *muscular hypertrophy.* Conversely, an unused muscle undergoes *atrophy,* decreasing in size and strength.

The way a muscle responds to use also depends on the type of exercise. A muscle contracting with lower intensity, during swimming or running, activates *slow-twitch fibers,* which are oxidative and thus fatigue-resistant. With use, these specialized muscle fibers develop more mitochondria, and more extensive capillary networks envelop them. Such changes increase the slow-twitch fibers' ability to resist fatigue during prolonged exercise, although their sizes and strengths may remain unchanged.

Forceful exercise, such as weight lifting, in which a muscle exerts more than 75% of its maximum tension, utilizes *fast-twitch fibers,* which may be glycolytic and thus fatigable. In response to strenuous exercise, these fibers produce new filaments of actin and myosin, the diameters of the muscle fibers increase, and the entire muscle enlarges. However, the muscular hypertrophy does not produce new muscle fibers.

The strength of a muscular contraction is directly proportional to the diameter of the activated muscle fibers. Consequently, an enlarged muscle can produce stronger contractions than before. Such a change, however, does not increase the muscle's ability to resist fatigue during activities like swimming or running.

If regular exercise stops, the capillary networks shrink, and the number of mitochondria within the muscle fibers drops. The number of actin and myosin filaments decreases, and the entire muscle atrophies. Such atrophy commonly occurs when accidents or diseases interfere with motor nerve impulses and prevent them from reaching muscle fibers. An unused muscle may shrink to less than half its usual size within a few months.

The fibers of muscles whose motor neurons are severed not only shrink, but also may fragment and, in time, be replaced by fat or fibrous connective tissue. However, reinnervation within the first few months following an injury may restore muscle function.

Summation

The force that a muscle fiber can generate is not limited to the maximum force of a single twitch. A muscle fiber exposed to a series of stimuli of increasing frequency reaches a point when it is unable to completely relax before the next stimulus in the series arrives. When this happens, the force of individual twitches combines by the process of **summation.** When the resulting forceful, sustained contraction lacks even partial relaxation, it is called a **tetanic** (tĕ-tan'ik) **contraction,** or tetanus (fig. 8.13).

Recruitment of Motor Units

Anatomically, the muscle fibers within a muscle are organized into motor units, each of which is controlled by a single motor neuron. Each motor unit is also a functional unit, because a nerve impulse in its motor neuron will cause all of the fibers in that motor unit to contract at the same time. A whole muscle is composed of many motor units controlled by different motor neurons, which respond to different thresholds of stimulation. If only the motor neurons with low thresholds are stimulated, few motor units contract. At higher intensities of stimulation, other motor neurons respond, and more motor units are activated. Such an increase in the

number of motor units being activated is called **recruitment** (re-krōōt'ment). As the intensity of stimulation increases, recruitment of motor units continues until, finally, all possible motor units in that muscle are activated and the muscle contracts with maximal tension.

Sustained Contractions

At the same time twitches combine to increase force, the strength of the contractions may increase due to the recruitment of motor units. The smaller motor units, which have finer fibers, are most easily stimulated and tend to respond earlier in the series of stimuli. The larger motor units, which have thicker fibers, respond later and more forcefully. Summation and recruitment together can produce a *sustained contraction* of increasing strength.

Sustained contractions of whole muscles enable us to perform everyday activities. Such contractions are responses to a rapid series of impulses transmitted from the brain and spinal cord on motor neuron axons.

Even when a muscle appears to be at rest, its fibers undergo some sustained contraction. This is called **muscle tone** (tonus). Muscle tone is a response to nerve impulses that originate repeatedly from the spinal cord and stimulate a few muscle fibers. Muscle tone is particularly important in maintaining posture. If muscle tone

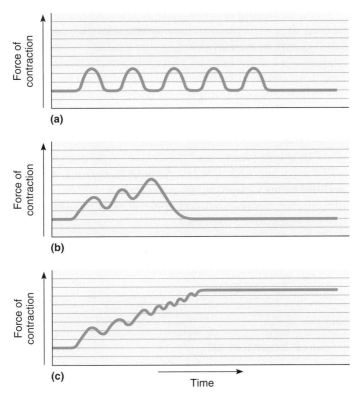

Figure 8.13

Myograms of (a) a series of twitches, (b) summation, and (c) a tetanic contraction. Note that stimulation frequency increases from one myogram to the next.

is suddenly lost, as happens when a person loses consciousness, the body collapses.

> When skeletal muscles contract very forcefully, they may generate up to 50 pounds of pull for each square inch of muscle cross section. Consequently, large muscles, such as those in the thigh, can pull with several hundred pounds of force. Occasionally, this force is so great that the tendons of muscles tear away from their attachments to the bones (*muscle pull*).

Check Your Recall

15. Define *threshold stimulus*.
16. What is recruitment?
17. Distinguish between a twitch and a sustained contraction.
18. How is muscle tone maintained?

8.5 SMOOTH MUSCLE

The contractile mechanism of smooth muscles is essentially the same as for skeletal muscles. The cells of smooth muscle, however, have some important structural and functional differences from the other types of muscle.

Smooth Muscle Fibers

Recall from chapter 5 (p. 110) that smooth muscle cells are elongated, with tapering ends. They contain filaments of actin and myosin in myofibrils that extend the lengths of the cells. However, these filaments are organized differently and more randomly than those in skeletal muscle. Consequently, smooth muscle cells lack striations. The sarcoplasmic reticulum in these cells is not well developed.

The two major types of smooth muscles are multi unit and visceral. In **multiunit smooth muscle,** the muscle fibers are separate rather than organized into sheets. Smooth muscle of this type is found in the irises of the eyes and in the walls of blood vessels. Typically, multiunit smooth muscle tissue contracts only in response to stimulation by motor nerve impulses or certain hormones.

Visceral smooth muscle is composed of sheets of spindle-shaped cells in close contact with one another (see fig. 5.22, p. 111). This more common type of smooth muscle is found in the walls of hollow organs, such as the stomach, intestines, urinary bladder, and uterus.

Fibers of visceral smooth muscles can stimulate each other. When one fiber is stimulated, the impulse moving over its surface may excite adjacent fibers, which in turn stimulate still others. Visceral smooth muscles also display *rhythmicity*, a pattern of repeated contractions. Rhythmicity is due to self-exciting fibers that deliver spontaneous impulses periodically into surrounding muscle tissue. These two features—transmission of impulses from cell to cell and rhythmicity—are largely responsible for the wavelike motion, called **peristalsis,** that occurs in certain tubular organs, such as the intestines, and helps force the contents of these organs along their lengths.

Smooth Muscle Contraction

Smooth muscle contraction resembles skeletal muscle contraction in a number of ways. Both mechanisms include reactions of actin and myosin, both are triggered by membrane impulses and an increase in intracellular calcium ions, and both use energy from ATP. However, these two types of muscle tissue also have significant differences.

Recall that acetylcholine is the neurotransmitter in skeletal muscle. Two neurotransmitters affect smooth muscle—acetylcholine and norepinephrine. Each of these neurotransmitters stimulates contractions in some smooth muscles and inhibits contractions in others (see chapter 9, p. 251). Also, a number of hormones affect smooth muscle, stimulating contractions in some cases and altering the degree of response to neurotransmitters in others.

Smooth muscle is slower to contract and to relax than skeletal muscle. On the other hand, smooth muscle can maintain a forceful contraction longer with a given amount of ATP. Also, unlike skeletal muscle, smooth muscle fibers can change length without changing tautness; therefore, smooth muscles in the stomach and intestinal walls can stretch as these organs fill, yet maintain the pressure inside these organs.

Check Your Recall

19. Describe two major types of smooth muscle.
20. What special characteristics of visceral smooth muscle make peristalsis possible?
21. How does smooth muscle contraction differ from that of skeletal muscle?

8.6 CARDIAC MUSCLE

Cardiac muscle is found only in the heart. Its mechanism of contraction is essentially the same as that of skeletal and smooth muscle, but with some important differences.

Cardiac muscle is composed of branching, striated cells interconnected in three-dimensional networks (see fig. 5.23, p. 112). Each cell has many filaments of actin and myosin, similar to those in skeletal muscle. A cardiac muscle cell also has a sarcoplasmic reticulum, many mitochondria, and a system of transverse tubules. However, the cisternae of cardiac muscle fibers are less well developed and store less calcium than those of skeletal muscle. On the other hand, the transverse tubules of cardiac muscle are larger, and they release many calcium ions into the sarcoplasm in response to muscle impulses. This extra calcium from the transverse tubules comes from the extracellular fluid and causes cardiac muscle twitches to be longer than skeletal muscle twitches.

The opposing ends of cardiac muscle cells are connected by crossbands called *intercalated discs*. These bands form from elaborate junctions between cell membranes. The discs help to join cells and to transmit the force of contraction from cell to cell. Intercalated discs also allow muscle impulses to pass freely so that they travel rapidly from cell to cell.

When one portion of the cardiac muscle network is stimulated, the resulting impulse passes to the other parts of the network, and the whole structure contracts as a functional unit. Cardiac muscle is also self-exciting and rhythmic. Consequently, a pattern of contraction and relaxation repeats again and again and causes the rhythmic contractions of the heart.

Table 8.3 summarizes the characteristics of the three types of muscle tissue. The Genetics Connection on page 191 considers several inherited diseases that affect the muscular system.

Check Your Recall

22. How is cardiac muscle similar to smooth muscle?
23. How is cardiac muscle similar to skeletal muscle?
24. What is the function of intercalated discs?
25. What characteristic of cardiac muscle contracts the heart as a unit?

8.7 SKELETAL MUSCLE ACTIONS

Skeletal muscles provide a variety of body movements, as described in chapter 7 (pp. 165–166). Each muscle's movement depends largely on the kind of joint it is associated with and the way the muscle attaches on either side of that joint.

Table 8.3	Types of Muscle Tissue		
	Skeletal	**Smooth**	**Cardiac**
Major Location	Skeletal muscles	Walls of hollow viscera, blood vessels	Wall of the heart
Major Function	Movement of bones at joints, maintenance of posture	Movement of viscera, peristalsis, vasoconstriction	Pumping action of the heart
Cellular Characteristics			
Striations	Present	Absent	Present
Nucleus	Many nuclei	Single nucleus	Single nucleus
Special features	Well-developed transverse tubule system	Lacks transverse tubules	Well-developed transverse tubule system; intercalated discs separating adjacent cells
Mode of Control	Voluntary	Involuntary	Involuntary
Contraction Characteristics	Contracts and relaxes rapidly	Contracts and relaxes slowly; self-exciting; rhythmic	Network of cells contracts as a unit; self-exciting; rhythmic

Genetics Connection

Inherited Diseases of Muscle

A variety of inherited conditions affect muscle tissue. These disorders differ in the nature of the genetic defect, the type of protein that is abnormal in form or function, and the particular muscles in the body that are impaired.

The Muscular Dystrophies—Missing Proteins

A muscle cell is packed with filaments of actin and myosin. Less abundant, but no less important, is a protein called *dystrophin*. It holds skeletal muscle cells together by linking actin in the cell to glycoproteins (called *dystrophin-associated glycoproteins,* or DAGs) that are part of the cell membrane. This helps attach the cell to the surrounding extracellular matrix. Missing or abnormal dystrophin or DAGs cause muscular dystrophies. These illnesses vary in severity and age of onset, but in all cases, muscles weaken and degenerate. Eventually, fat and connective tissue replace muscle.

Duchenne muscular dystrophy (DMD) is the most severe type of the illness (fig. 8B). Symptoms begin by age five and affect only boys. By age thirteen, the person cannot walk, and by early adulthood he usually dies from failure of the respira-

Figure 8B
This young man has Duchenne muscular dystrophy. The condition has not yet severely limited his activities, but he shows the hypertrophied (overdeveloped) calf muscles that result from his inability to rise from a sitting position the usual way—an early sign of the illness.

tory muscles. In DMD, dystrophin is often missing. In Becker muscular dystrophy, symptoms begin in early adulthood, are less severe, and result from underproduction of dystrophin. Limb-girdle muscular dystrophy causes weakness in the upper limbs, usually noticeable in a person's thirties. This form of muscular dystrophy is often the result of a missing or abnormal DAG, which causes the other DAGs to be deficient too.

Charcot-Marie-Tooth Disease— A Duplicate Gene

Charcot-Marie-Tooth disease causes a slowly progressing weakness in the muscles of the hands and feet and a decrease in tendon reflexes in these parts. In this illness, an extra gene impairs the insulating sheath around affected nerve cells, so that nerve cells cannot adequately stimulate the involved muscles. Physicians perform two tests—electromyography and nerve conduction velocities—to diagnose Charcot-Marie-Tooth disease. It is also possible to test for the gene mutation to establish a diagnosis.

Myotonic Dystrophy—An Expanding Gene

Myotonic dystrophy delays muscle relaxation following contraction (myotonia), which causes facial and limb weakness, cataracts, and an irregular heartbeat. It is caused by inheriting either of two "expanding genes" that actually grow with each generation. As the gene enlarges, symptoms increase in severity or begin at an earlier age. For example, a grandfather might experience only mild weakness in his forearms, but his daughter might have more noticeable arm and leg weakness. By the third generation, affected children might suffer severe muscle impairment. For many years, physicians attributed the worsening of symptoms over generations to psychological causes. We now know there is a physical basis for the phenomenon. The expanded genes are transcribed into messenger RNA molecules that are too large to leave the nucleus, so that the proteins they encode are not synthesized.

Hereditary Idiopathic Dilated Cardiomyopathy—A Tiny Glitch

This very rare inherited form of heart failure usually begins in a person's forties and is lethal in 50% of cases within five years of diagnosis, unless a heart transplant can be performed. The condition is caused by a tiny genetic error in a form of actin found only in cardiac muscle, where it is the predominant component of the thin filaments. A change in a single DNA building block (nucleotide base) disturbs actin's ability to anchor to the Z lines in heart muscle cells. The mutation prevents actin from effectively transmitting the force of contraction, which gradually causes the heart chambers to enlarge and eventually to fail.

Origin and Insertion

Recall that bones forming movable joints function as levers (see chapter 7, p. 138). One end of a skeletal muscle usually fastens to a relatively immovable or fixed part at a movable joint, and the other end connects to a movable part on the other side of that joint. The immovable end of the muscle is called its **origin** (or'ĭ-jin), and the movable end is its **insertion** (in-ser'shun). When a muscle contracts, its insertion is pulled toward its origin.

Some muscles have more than one origin or insertion. The *biceps brachii* in the arm, for example, has two origins. This is reflected in the name *biceps,* which means "two heads." (Note: The head of a muscle is the part nearest its origin.) One head of the muscle attaches to the coracoid process of the scapula, and the other head arises from a tubercle above the glenoid cavity of the scapula. The muscle extends along the front surface of the humerus and is inserted by means of a tendon on the radial tuberosity of the radius. When the biceps brachii contracts, its insertion is pulled toward its origin, and the forearm flexes at the elbow (fig. 8.14).

The movements termed *flexion* and *extension* describe changes in the angle between bones that meet at a joint. For example, flexion of the elbow refers to a movement of the forearm that bends the elbow, or decreases the angle. Alternatively, one could say that flexion of the elbow results from the action of the biceps brachii on the radius of the forearm.

Since students often find it helpful to think of movements in terms of the specific actions of the muscles involved, we may also describe flexion and extension in

these terms. Thus, the action of the biceps brachii may be described as "flexion of the forearm at the elbow," and the action of the quadriceps group as "extension of the leg at the knee." We believe this occasional departure from strict anatomical terminology eases understanding and learning.

Interaction of Skeletal Muscles

Skeletal muscles almost always function in groups. Consequently, for a particular body movement to occur, a person must do more than contract a single muscle; instead, after learning to make a particular movement, the person wills the movement to occur, and the nervous system stimulates the appropriate group of muscles.

Careful observation of body movements indicates the special roles of muscles. For instance, when the upper limb is lifted horizontally away from the side, a contracting *deltoid* muscle provides most of the movement and is said to be the **prime mover** (prīm mōōv'er), also referred to as an **agonist** (ag'o-nist). However, while a prime mover is acting, certain nearby muscles are also contracting. In the case of the contracting deltoid muscle, nearby muscles help hold the shoulder steady and in this way make the prime mover's action more effective. Muscles that contract and assist the prime mover are called **synergists** (sin'er-jistz).

Certain muscles act as **antagonists** (an-tag'o-nistz) to prime movers. These muscles can resist a prime mover's action and cause movement in the opposite direction. For example, the antagonist of the prime mover that raises the upper limb can lower the upper limb, or the antagonist of the prime mover that bends the upper limb can straighten it (see fig. 7.7, p. 138). If both a prime mover and its antagonist contract simultaneously, the part they act upon remains rigid. Consequently, smooth body movements depend on antagonists relaxing and, thus, giving way to the prime movers whenever the prime movers contract. Once again, the nervous system controls these complex actions.

Sometimes the relationship between two muscles changes. For example, the pectoralis major and latissimus dorsi are antagonistic for flexion and extension of the shoulder. However, they are synergistic for medial rotation of the shoulder. Thus, the role of a muscle must be learned in the context of a particular movement.

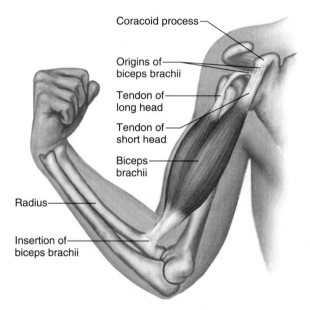

Coracoid process

Origins of biceps brachii

Tendon of long head

Tendon of short head

Biceps brachii

Radius

Insertion of biceps brachii

Figure 8.14

The biceps brachii has two heads that originate on the scapula. A tendon inserts this muscle on the radius.

Check Your Recall

26. Distinguish between the origin and the insertion of a muscle.

27. Define *prime mover.*

28. What is the function of a synergist? An antagonist?

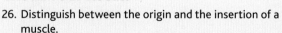

8.8 MAJOR SKELETAL MUSCLES

The section that follows discusses the locations, actions, and attachments of some of the major skeletal muscles. (Figures 8.15 and 8.16 and reference plates 1 and 2, pp. 23–24, show the locations of the superficial skeletal muscles—those near the surface.)

Note that the names of these muscles often describe them. A name may indicate a muscle's relative size, shape, location, action, number of attachments, or the direction of its fibers, as in the following examples:

pectoralis major Of large size (major) and located in the pectoral region (chest).
deltoid Shaped like a delta or triangle.

extensor digitorum Extends the digits (fingers or toes).
biceps brachii Having two heads (biceps) or points of origin and located in the brachium (arm).
sternocleidomastoid Attached to the sternum, clavicle, and mastoid process.
external oblique Located near the outside, with fibers that run obliquely (in a slanting direction).

Muscles of Facial Expression

A number of small muscles that lie beneath the skin of the face and scalp enable us to communicate feelings through facial expression (fig. 8.17*a*). Many of these muscles, located around the eyes and mouth, are responsible for such expressions as surprise, sadness,

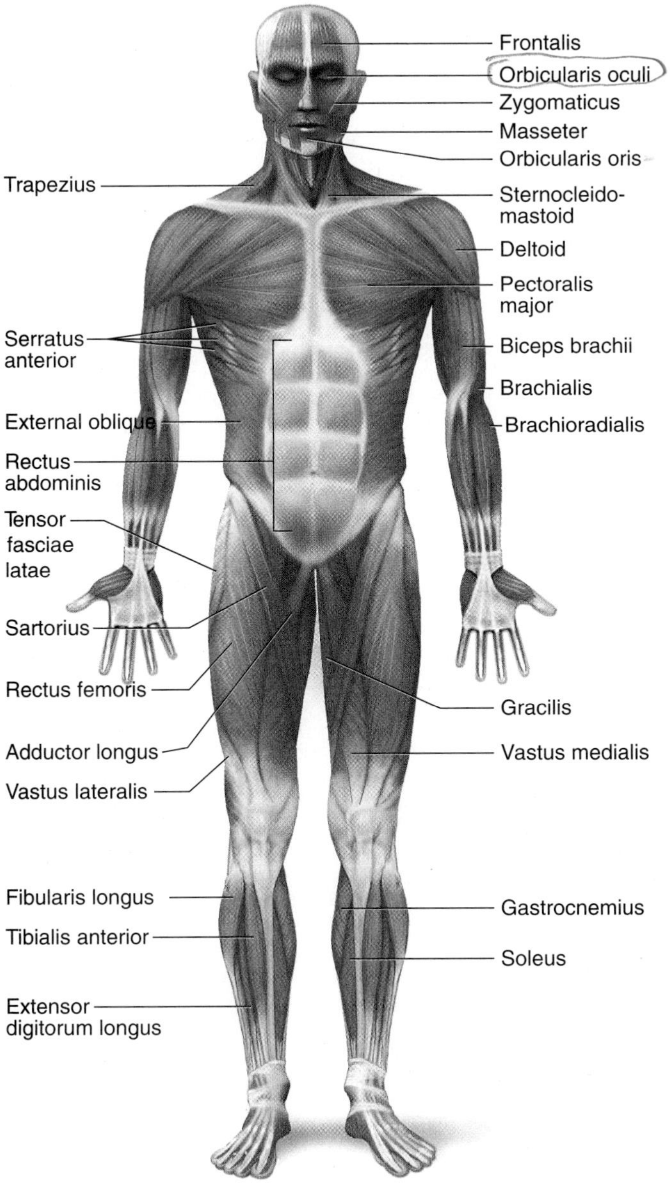

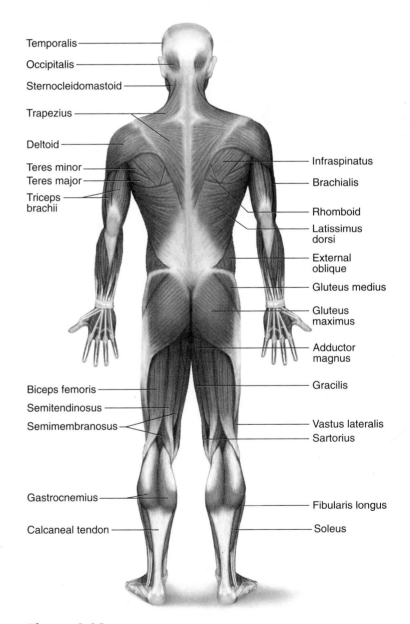

Figure 8.15
Anterior view of superficial skeletal muscles.

Figure 8.16
Posterior view of superficial skeletal muscles.

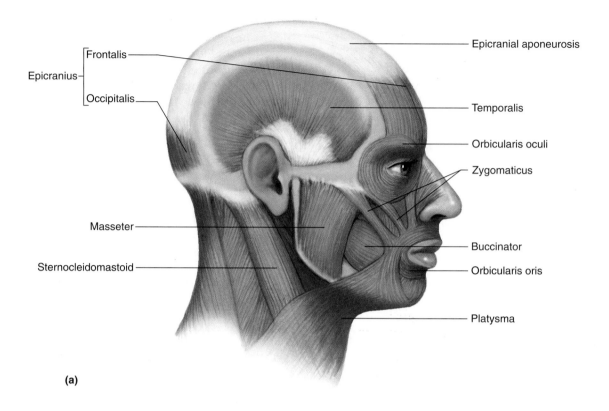

(a)

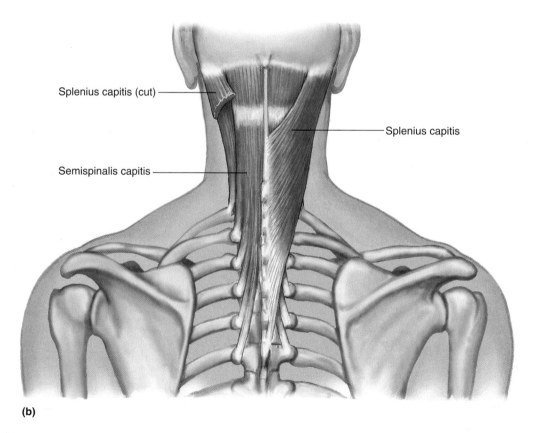

(b)

Figure 8.17

Muscles of the face and neck. (*a*) Muscles of facial expression and mastication. (*b*) Posterior view of muscles that move the head.

anger, fear, disgust, and pain. As a group, the muscles of facial expression join the bones of the skull to connective tissue in various regions of the overlying skin. They include:

epicranius (ep″ĭ-kra′ne-us) Composed of two parts, the *frontalis* (frun-ta′lis) and the *occipitalis* (ok-sip″ĭ-ta′lis).
orbicularis oculi (or-bik′u-la-rus ok′u-li)
orbicularis oris (or-bik′u-la-rus o′ris)
buccinator (buk′sĭ-na″tor)
zygomaticus (zi″go-mat′ik-us)
platysma (plah-tiz′mah)

Table 8.4 lists the origins, insertions, and actions of the muscles of facial expression. (The muscles that move the eyes are listed in chapter 10, pp. 277–278.)

The human body has more than 600 distinct skeletal muscles. The face alone includes 60 muscles, more than 40 of which are used to frown, and 20 to smile. Thinner than a thread and barely visible, the stapedius in the middle ear is the body's smallest muscle. In contrast is the gluteus maximus, the largest muscle, located in the buttock. The sartorius, which pulls on the thigh, is the longest muscle in the body.

Muscles of Mastication

Muscles attached to the mandible produce chewing movements. Two pairs of these muscles elevate the mandible, a motion used in biting. These muscles are the *masseter* (mas-se′ter) and the *temporalis* (tem-po-ra′lis) (fig. 8.17*a*). Table 8.5 lists the origins, insertions, and actions of the muscles of mastication. The Topic of Interest on page 196 describes a recently identified muscle thought to be associated with chewing.

Grinding the teeth, a common response to stress, may strain the temporomandibular joint—the articulation between the mandibular condyle of the mandible and the mandibular fossa of the temporal bone. This condition, called temporomandibular joint syndrome (TMJ syndrome), may produce headache, earache, and pain in the jaw, neck, or shoulder.

Muscles That Move the Head

Head movements result from the actions of paired muscles in the neck and upper back. These muscles flex, extend, and rotate the head. They include (fig. 8.17):

sternocleidomastoid (ster″no-kli″do-mas′toid)
splenius capitis (sple′ne-us kap′ĭ-tis)
semispinalis capitis (sem″e-spi-na′lis kap′ĭ-tis)

Table 8.6 lists the origins, insertions, and actions of muscles that move the head.

Muscles That Move the Pectoral Girdle

The muscles that move the pectoral girdle are closely associated with those that move the arm. A number of these chest and shoulder muscles connect the scapula to nearby bones and move the scapula upward, downward, forward, and backward. They include (figs. 8.18 and 8.19):

Table 8.4	Muscles of Facial Expression		
Muscle	Origin	Insertion	Action
Epicranius	Occipital bone	Skin and muscles around eye	Raises eyebrow
Orbicularis oculi	Maxillary and frontal bones	Skin around eye	Closes eye
Orbicularis oris	Muscles near the mouth	Skin of lips	Closes and protrudes lips
Buccinator	Outer surfaces of maxilla and mandible	Orbicularis oris	Compresses cheeks inward
Zygomaticus	Zygomatic bone	Orbicularis oris	Raises corner of mouth
Platysma	Fascia in upper chest	Lower border of mandible	Draws angle of mouth downward

Table 8.5	Muscles of Mastication		
Muscle	Origin	Insertion	Action
Masseter	Lower border of zygomatic arch	Lateral surface of mandible	Elevates mandible
Temporalis	Temporal bone	Coronoid process and lateral surface of mandible	Elevates mandible

Topic of Interest

A New Muscle Discovered?

An unusual view of a cadaver provided a new perspective that may have revealed a previously undiscovered muscle. Two dentists were examining a cadaver's skull whose eyes had been dissected out when they discovered what they believe is a new muscle in the head. The muscle, named the *sphenomandibularis*, extends about an inch and a half from behind the eyes to the inside of the jawbone and may help produce the movements of chewing. The muscle has a unique combination of the five characteristics of muscles: origin, insertion, innervation, blood vessel supply, and specific function.

In traditional dissection from the side, the new muscle's origin and insertion are not visible, so it may have appeared to be part of the larger and overlying temporalis muscle. Although the sphenomandibularis inserts on the inner side of the jawbone, as does the temporalis, it originates differently, on the sphenoid bone.

Following their discovery of the sphenomandibularis in the cadaver head, the dentists quickly identified it in twenty-five other cadavers, and by using computer-aided dissection of the Visible Human data sets, which are collections of scans and sectioned images of the human body. Other researchers soon found it in live patients undergoing magnetic resonance imaging scans. The discovery of a new muscle illustrates that anatomy is an ever-changing science.

trapezius (trah-pe′ze-us)
rhomboid major (rom-boid′)
levator scapulae (le-va′tor scap′u-lē)
serratus anterior (ser-ra′tus an-te′re-or)
pectoralis minor (pek″to-ra′lis)

Table 8.7 lists the origins, insertions, and actions of the muscles that move the pectoral girdle.

Muscles That Move the Arm

The arm is one of the more freely movable parts of the body. Muscles that connect the humerus to various regions of the pectoral girdle, ribs, and vertebral column make these movements possible (figs. 8.18, 8.19, 8.20, and 8.21). These muscles can be grouped according to their

Table 8.6	Muscles That Move the Head		
Muscle	Origin	Insertion	Action
Sternocleidomastoid	Anterior surface of sternum and upper surface of clavicle	Mastoid process of temporal bone	Pulls head to one side, pulls head toward chest, or raises sternum
Splenius capitis	Spinous processes of lower cervical and upper thoracic vertebrae	Mastoid process of temporal bone	Rotates head, bends head to one side, or brings head into an upright position
Semispinalis capitis	Processes of lower cervical and upper thoracic vertebrae	Occipital bone	Extends head, bends head to one side, or rotates head

Table 8.7	Muscles That Move the Pectoral Girdle		
Muscle	Origin	Insertion	Action
Trapezius	Occipital bone and spines of cervical and thoracic vertebrae	Clavicle; spine and acromion process of scapula	Rotates scapula and raises arm; raises scapula; pulls scapula medially or pulls scapula and shoulder downward
Rhomboid major	Spines of upper thoracic vertebrae	Medial border of scapula	Raises and adducts scapula
Levator scapulae	Transverse processes of cervical vertebrae	Medial margin of scapula	Elevates scapula
Serratus anterior	Outer surfaces of upper ribs	Ventral surface of scapula	Pulls scapula anteriorly and downward
Pectoralis minor	Sternal ends of upper ribs	Coracoid process of scapula	Pulls scapula anteriorly and downward or raises ribs

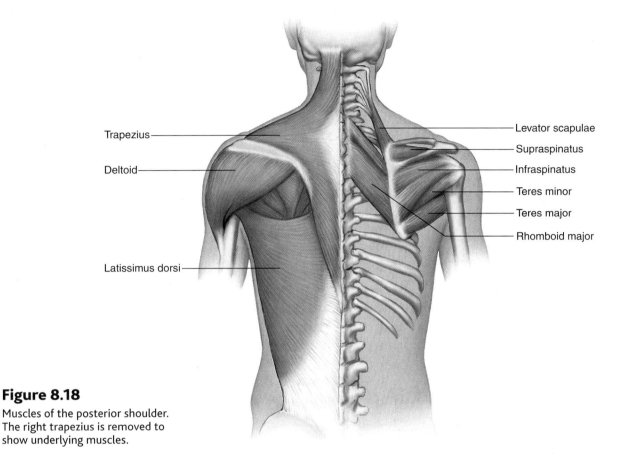

Figure 8.18

Muscles of the posterior shoulder. The right trapezius is removed to show underlying muscles.

Trapezius

Deltoid

Latissimus dorsi

Levator scapulae

Supraspinatus

Infraspinatus

Teres minor

Teres major

Rhomboid major

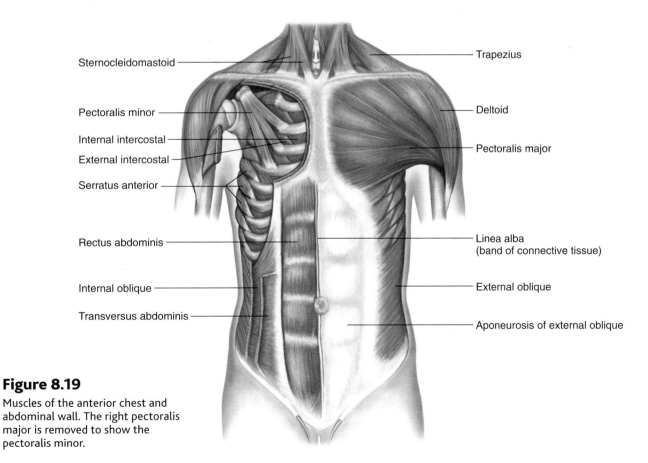

Figure 8.19

Muscles of the anterior chest and abdominal wall. The right pectoralis major is removed to show the pectoralis minor.

Sternocleidomastoid

Pectoralis minor

Internal intercostal

External intercostal

Serratus anterior

Rectus abdominis

Internal oblique

Transversus abdominis

Trapezius

Deltoid

Pectoralis major

Linea alba
(band of connective tissue)

External oblique

Aponeurosis of external oblique

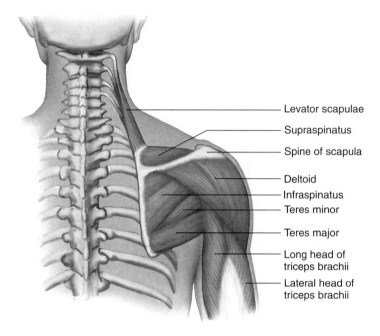

Figure 8.20
Muscles of the posterior surface of the scapula and arm.

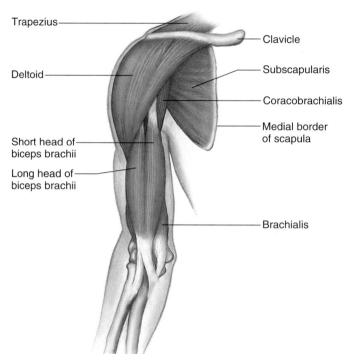

Figure 8.21
Muscles of the anterior shoulder and arm, with the rib cage removed.

primary actions—flexion, extension, abduction, and rotation—as follows:

Flexors
coracobrachialis (kor″ah-ko-bra′ke-al-is)
pectoralis major (pek″to-ra′lis)

Extensors
teres major (te′rēz)
latissimus dorsi (lah-tis′ĭ-mus dor′si)

Abductors
supraspinatus (su″prah-spi′na-tus)
deltoid (del′toid)

Rotators
subscapularis (sub-scap′u-lar-is)
infraspinatus (in″frah-spi′na-tus)
teres minor (te′rēz)

Table 8.8 lists the origins, insertions, and actions of muscles that move the arm.

Table 8.8	Muscles That Move the Arm		
Muscle	**Origin**	**Insertion**	**Action**
Coracobrachialis	Coracoid process of scapula	Shaft of humerus	Flexes and adducts arm
Pectoralis major	Clavicle, sternum, and costal cartilages of upper ribs	Intertubercular groove of humerus	Pulls arm anteriorly and across chest, rotates humerus, or adducts arm
Teres major	Lateral border of scapula	Intertubercular groove of humerus	Extends humerus or adducts and rotates arm medially
Latissimus dorsi	Spines of sacral, lumbar, and lower thoracic vertebrae, iliac crest, and lower ribs	Intertubercular groove of humerus	Extends and adducts arm and rotates humerus inwardly, or pulls shoulder downward and posteriorly
Supraspinatus	Posterior surface of scapula	Greater tubercle of humerus	Abducts arm
Deltoid	Acromion process, spine of scapula, and clavicle	Deltoid tuberosity of humerus	Abducts arm, extends or flexes humerus
Subscapularis	Anterior surface of scapula	Lesser tubercle of humerus	Rotates arm medially
Infraspinatus	Posterior surface of scapula	Greater tubercle of humerus	Rotates arm laterally
Teres minor	Lateral border of scapula	Greater tubercle of humerus	Rotates arm laterally

Muscles That Move the Forearm

Muscles that connect the radius or ulna to the humerus or pectoral girdle produce most of the forearm movements. A group of muscles located along the anterior surface of the humerus flexes the elbow, and a single posterior muscle extends this joint. Other muscles move the radioulnar joint and rotate the forearm.

Muscles that move the forearm include (figs. 8.20, 8.21, and 8.22):

Flexors
biceps brachii (bi′seps bra′ke-i)
brachialis (bra′ke-al-is)
brachioradialis (bra″ke-o-ra″de-a′lis)

Extensor
triceps brachii (tri′seps bra′ke-i)

Rotators
supinator (su′pĭ-na-tor)
pronator teres (pro-na′tor te′rēz)
pronator quadratus (pro-na′tor kwod-ra′tus)

Table 8.9 lists the origins, insertions, and actions of muscles that move the forearm.

Muscles That Move the Hand

Many muscles move the hand. They originate from the distal end of the humerus and from the radius and ulna. The two major groups of these muscles are flexors on the anterior side of the forearm and extensors on the posterior side. These muscles include (figs. 8.22 and 8.23):

Flexors
flexor carpi radialis (flex′sor kar-pi′ ra″de-a′lis)
flexor carpi ulnaris (flex′sor kar-pi′ ul-na′ris)
palmaris longus (pal-ma′ris long′gus)
flexor digitorum profundus (flex′sor dij″ĭ-to′rum pro-fun′dus)

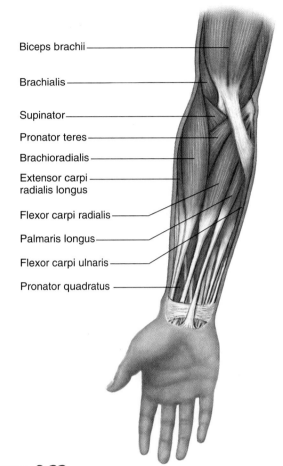

Figure 8.22
Muscles of the anterior forearm.

Extensors
extensor carpi radialis longus (eks-ten′sor kar-pi′ ra″de-a′lis long′gus)
extensor carpi radialis brevis (eks-ten′sor kar-pi′ ra″de-a′lis brev′ĭs)
extensor carpi ulnaris (eks-ten′sor kar-pi′ ul-na′ris)
extensor digitorum (eks-ten′sor dij″ĭ-to′rum)

Table 8.9	Muscles That Move the Forearm		
Muscle	**Origin**	**Insertion**	**Action**
Biceps brachii	Coracoid process and tubercle above glenoid cavity of scapula	Radial tuberosity of radius	Flexes forearm at elbow and rotates hand laterally
Brachialis	Anterior shaft of humerus	Coronoid process of ulna	Flexes forearm at elbow
Brachioradialis	Distal lateral end of humerus	Lateral surface of radius above styloid process	Flexes forearm at elbow
Triceps brachii	Tubercle below glenoid cavity and lateral and medial surfaces of humerus	Olecranon process of ulna	Extends forearm at elbow
Supinator	Lateral epicondyle of humerus and crest of ulna	Lateral surface of radius	Rotates forearm laterally
Pronator teres	Medial epicondyle of humerus and coronoid process of ulna	Lateral surface of radius	Rotates forearm medially
Pronator quadratus	Anterior distal end of ulna	Anterior distal end of radius	Rotates forearm medially

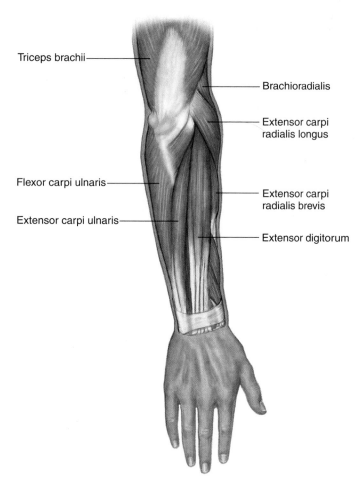

Figure 8.23

Muscles of the posterior forearm.

Labels on figure:
- Triceps brachii
- Brachioradialis
- Extensor carpi radialis longus
- Flexor carpi ulnaris
- Extensor carpi radialis brevis
- Extensor carpi ulnaris
- Extensor digitorum

Table 8.10 lists the origins, insertions, and actions of muscles that move the hand.

Muscles of the Abdominal Wall

Bone supports the walls of the chest and pelvic regions, but not those of the abdomen. Instead, the anterior and lateral walls of the abdomen are composed of layers of broad, flattened muscles. These muscles connect the rib cage and vertebral column to the pelvic girdle. A band of tough connective tissue called the **linea alba** extends from the xiphoid process of the sternum to the symphysis pubis (see fig. 8.19). It is an attachment for some of the abdominal wall muscles.

Contraction of these muscles decreases the size of the abdominal cavity and increases the pressure inside. These actions help press air out of the lungs during forceful exhalation and aid in the movements of defecation, urination, vomiting, and childbirth.

The abdominal wall muscles include (see fig. 8.19):

external oblique (eks-ter′nal o-blēk′)
internal oblique (in-ter′nal o-blēk′)
transversus abdominis (trans-ver′sus ab-dom′ĭ-nis)
rectus abdominis (rek′tus ab-dom′ĭ-nis)

Table 8.11 lists the origins, insertions, and actions of muscles of the abdominal wall.

Muscles of the Pelvic Outlet

Two muscular sheets—a deeper **pelvic diaphragm** and a more superficial **urogenital diaphragm**—span the outlet of the pelvis. The pelvic diaphragm forms the floor of the pelvic cavity, and the urogenital diaphragm fills the space within the pubic arch (see fig. 7.29, p. 159). The muscles of the male and female pelvic outlets include (fig. 8.24):

Table 8.10	Muscles That Move the Hand		
Muscle	**Origin**	**Insertion**	**Action**
Flexor carpi radialis	Medial epicondyle of humerus	Base of second and third metacarpals	Flexes and abducts wrist
Flexor carpi ulnaris	Medial epicondyle of humerus and olecranon process	Carpal and metacarpal bones	Flexes and adducts wrist
Palmaris longus	Medial epicondyle of humerus	Fascia of palm	Flexes wrist
Flexor digitorum profundus	Anterior surface of ulna	Bases of distal phalanges in fingers 2–5	Flexes distal joints of fingers
Extensor carpi radialis longus	Distal end of humerus	Base of second metacarpal	Extends wrist and abducts hand
Extensor carpi radialis brevis	Lateral epicondyle of humerus	Base of second and third metacarpals	Extends wrist and abducts hand
Extensor carpi ulnaris	Lateral epicondyle of humerus	Base of fifth metacarpal	Extends and adducts wrist
Extensor digitorum	Lateral epicondyle of humerus	Posterior surface of phalanges in fingers 2–5	Extends fingers

Table 8.11	Muscles of the Abdominal Wall		
Muscle	**Origin**	**Insertion**	**Action**
External oblique	Outer surfaces of lower ribs	Outer lip of iliac crest and linea alba	Tenses abdominal wall and compresses abdominal contents
Internal oblique	Crest of ilium and inguinal ligament	Cartilages of lower ribs, linea alba, and crest of pubis	Tenses abdominal wall and compresses abdominal contents
Transversus abdominis	Costal cartilages of lower ribs, processes of lumbar vertebrae, lip of iliac crest, and inguinal ligament	Linea alba and crest of pubis	Tenses abdominal wall and compresses abdominal contents
Rectus abdominis	Crest of pubis and symphysis pubis	Xiphoid process of sternum and costal cartilages	Tenses abdominal wall and compresses abdominal contents; also flexes vertebral column

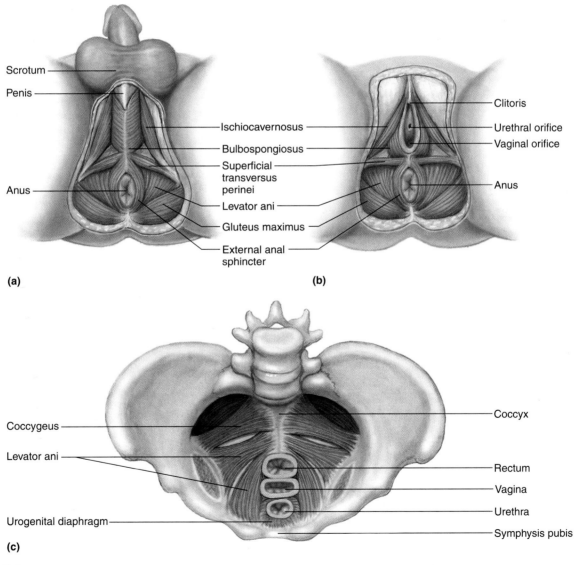

(a) **(b)** **(c)**

Figure 8.24
External view of muscles of (*a*) the male pelvic outlet and (*b*) the female pelvic outlet. (*c*) Internal view of the female pelvic and urogenital diaphragms.

Pelvic diaphragm
levator ani (le-va′tor ah-ni′)

Urogenital diaphragm
superficial transversus perinei (su″per-fish′al
    trans-ver′sus per″ĭ-ne′i)
bulbospongiosus (bul″bo-spon″je-o′sus)
ischiocavernosus (is″ke-o-kav″er-no′sus)

Table 8.12 lists the origins, insertions, and actions of pelvic outlet muscles.

Muscles That Move the Thigh

Muscles that move the thigh are attached to the femur and to some part of the pelvic girdle. These muscles occur in anterior and posterior groups. Muscles of the anterior group primarily flex the thigh; those of the posterior group extend, abduct, or rotate the thigh. The muscles in these groups include (figs. 8.25, 8.26, and 8.27):

Anterior group
psoas major (so′as)
iliacus (il′e-ak-us)

Posterior group
gluteus maximus (gloo′te-us mak′si-mus)
gluteus medius (gloo′te-us me′de-us)
gluteus minimus (gloo′te-us min′ĭ-mus)
tensor fasciae latae (ten′sor fash′e-e lah-tē)

Still another group of muscles attached to the femur and pelvic girdle adduct the thigh. They include (figs. 8.25 and 8.27):

adductor longus (ah-duk′tor long′gus)
adductor magnus (ah-duk′tor mag′nus)
gracilis (gras′il-is)

Table 8.13 lists the origins, insertions, and actions of muscles that move the thigh.

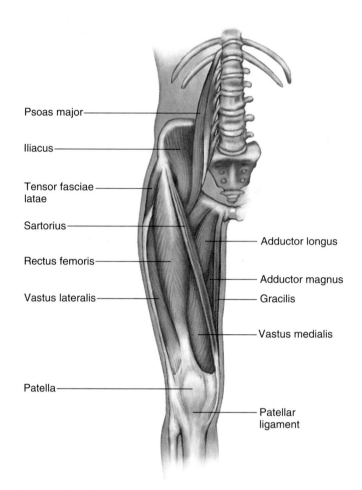

Figure 8.25
Muscles of the anterior right thigh. (Note that the vastus intermedius is a deep muscle not visible in this view.)

Muscles That Move the Leg

Muscles that move the leg connect the tibia or fibula to the femur or to the pelvic girdle. They can be separated into two major groups—those that flex the knee and those that extend the knee. Muscles of these groups

Table 8.12	Muscles of the Pelvic Outlet		
Muscle	**Origin**	**Insertion**	**Action**
Levator ani	Pubic bone and ischial spine	Coccyx	Supports pelvic viscera and provides sphincter-like action in anal canal and vagina
Superficial transversus perinei	Ischial tuberosity	Central tendon	Supports pelvic viscera
Bulbospongiosus	Central tendon	Males: Urogenital diaphragm and fascia of the penis	Males: Assists emptying of urethra
		Females: Pubic arch and root of clitoris	Females: Constricts vagina
Ischiocavernosus	Ischial tuberosity	Pubic arch	Assists function of bulbospongiosus

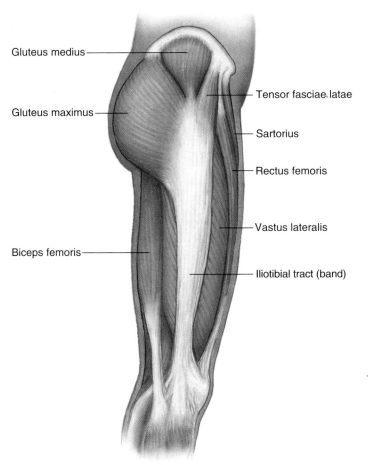

Figure 8.26

Muscles of the lateral right thigh.

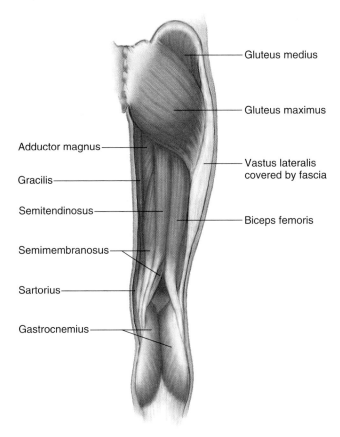

Figure 8.27

Muscles of the posterior right thigh.

include the hamstring group and the quadriceps femoris group (figs. 8.25, 8.26, and 8.27):

Flexors
biceps femoris (bi′seps fem′or-is)
semitendinosus (sem″e-ten′dĭ-no-sus)

semimembranosus (sem″e-mem′brah-no-sus)
sartorius (sar-to′re-us)

Extensor
quadriceps femoris group (kwod′rĭ-seps fem′or-is)
        Composed of four parts—the rectus femoris, vastus lateralis, vastus medialis, and vastus intermedius.

Table 8.13	Muscles That Move the Thigh		
Muscle	**Origin**	**Insertion**	**Action**
Psoas major	Lumbar intervertebral discs, bodies and transverse processes of lumbar vertebrae	Lesser trochanter of femur	Flexes thigh
Iliacus	Iliac fossa of ilium	Lesser trochanter of femur	Flexes thigh
Gluteus maximus	Sacrum, coccyx, and posterior surface of ilium	Posterior surface of femur and fascia of thigh	Extends thigh
Gluteus medius	Lateral surface of ilium	Greater trochanter of femur	Abducts and rotates thigh medially
Gluteus minimus	Lateral surface of ilium	Greater trochanter of femur	Abducts and rotates thigh medially
Tensor fasciae latae	Anterior iliac crest	Fascia of thigh	Abducts, flexes, and rotates thigh medially
Adductor longus	Pubic bone near symphysis pubis	Posterior surface of femur	Adducts, flexes, and rotates thigh laterally
Adductor magnus	Ischial tuberosity	Posterior surface of femur	Adducts, extends, and rotates thigh laterally
Gracilis	Lower edge of symphysis pubis	Medial surface of tibia	Adducts thigh, flexes and rotates lower limb medially

Table 8.14 lists the origins, insertions, and actions of muscles that move the leg.

Muscles That Move the Foot

A number of muscles that move the foot are located in the leg. They attach the femur, tibia, and fibula to bones of the foot, move the foot upward (dorsiflexion) or downward (plantar flexion), and turn the sole of the foot medial (inversion) or lateral (eversion). These muscles include (figs. 8.28, 8.29, and 8.30):

Dorsal flexors
tibialis anterior (tib″e-a′lis an-te′re-or)
fibularis (peroneus) tertius (fib″u-la′ris ter′shus)
extensor digitorum longus (eks-ten′sor dij″ĭ-to′rum
    long′gus)

Plantar flexors
gastrocnemius (gas″trok-ne′me-us)
soleus (so′le-us)
flexor digitorum longus (flek′sor dij″ĭ-to′rum long′gus)

Invertor
tibialis posterior (tib″e-a′lis pos-tēr′e-or)

Evertor
fibularis (peroneus) longus (fib″u-la′ris long′gus)

Table 8.15 lists the origins, insertions, and actions of muscles that move the foot.

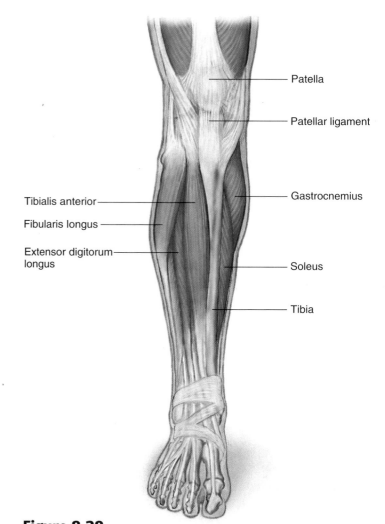

Patella

Patellar ligament

Gastrocnemius

Tibialis anterior

Fibularis longus

Extensor digitorum longus

Soleus

Tibia

Figure 8.28
Muscles of the anterior right leg.

Table 8.14	Muscles That Move the Leg		
Muscle	**Origin**	**Insertion**	**Action**
Sartorius	Anterior superior iliac spine	Medial surface of tibia	Flexes leg and thigh, abducts thigh, rotates thigh laterally, and rotates leg medially
Hamstring group			
Biceps femoris	Ischial tuberosity and posterior surface of femur	Head of fibula and lateral condyle of tibia	Flexes leg, extends thigh
Semitendinosus	Ischial tuberosity	Medial surface of tibia	Flexes leg, extends thigh
Semimembranosus	Ischial tuberosity	Medial condyle of tibia	Flexes leg, extends thigh
Quadriceps femoris group			
Rectus femoris	Spine of ilium and margin of acetabulum	Patella by the tendon, which continues as patellar ligament to tibial tuberosity	Extends leg at knee
Vastus lateralis	Greater trochanter and posterior surface of femur	Patella by the tendon, which continues as patellar ligament to tibial tuberosity	Extends leg at knee
Vastus medialis	Medial surface of femur	Patella by the tendon, which continues as patellar ligament to tibial tuberosity	Extends leg at knee
Vastus intermedius	Anterior and lateral surfaces of femur	Patella by the tendon, which continues as patellar ligament to tibial tuberosity	Extends leg at knee

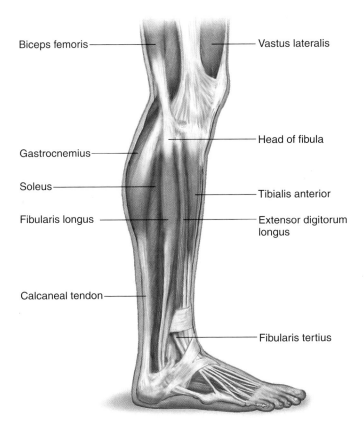

Figure 8.29

Muscles of the lateral right leg. (Note that the tibialis posterior is a deep muscle not visible in this view.)

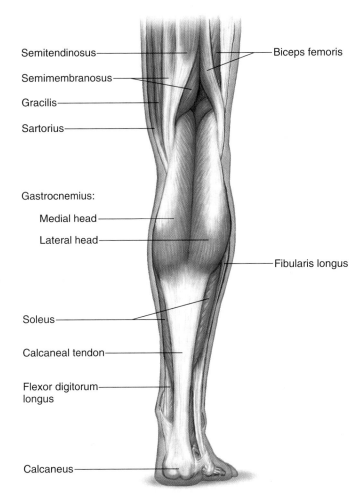

Figure 8.30

Muscles of the posterior right leg.

Table 8.15	Muscles That Move the Foot		
Muscle	**Origin**	**Insertion**	**Action**
Tibialis anterior	Lateral condyle and lateral surface of tibia	Tarsal bone (cuneiform) and first metatarsal	Dorsiflexion and inversion of foot
Fibularis tertius	Anterior surface of fibula	Dorsal surface of fifth metatarsal	Dorsiflexion and eversion of foot
Extensor digitorum longus	Lateral condyle of tibia and anterior surface of fibula	Dorsal surfaces of second and third phalanges of the four lateral toes	Dorsiflexion and eversion of foot and extension of toes
Gastrocnemius	Lateral and medial condyles of femur	Posterior surface of calcaneus	Plantar flexion of foot and flexion of leg at knee
Soleus	Head and shaft of fibula and posterior surface of tibia	Posterior surface of calcaneus	Plantar flexion of foot
Flexor digitorum longus	Posterior surface of tibia	Distal phalanges of the four lateral toes	Plantar flexion and inversion of foot, and flexion of the four lateral toes
Tibialis posterior	Lateral condyle and posterior surface of tibia, and posterior surface of fibula	Tarsal and metatarsal bones	Plantar flexion and inversion of foot
Fibularis longus	Lateral condyle of tibia and head and shaft of fibula	Tarsal and metatarsal bones	Plantar flexion and eversion of foot; also supports arch

Check Your Recall

29. What information is imparted in a muscle's name?
30. Which muscles provide facial expressions? Ability to chew? Head movements?
31. Which muscles move the pectoral girdle? Abdominal wall? Pelvic outlet? The arm, forearm, and hand? The thigh, leg, and foot?

Clinical Terms Related to the Muscular System

contracture (kon-trak′cher) Condition of great resistance to the stretch of a muscle.

convulsion (kun-vul′shun) Series of involuntary contractions of various voluntary muscles.

electromyography (e-lek″tro-mi-og′rah-fe) Technique for recording electrical changes in muscle tissues.

fibrillation (fi″brĭ-la′shun) Spontaneous contractions of individual muscle fibers, producing rapid and uncoordinated activity within a muscle.

fibrosis (fi-bro′sis) Degenerative disease in which fibrous connective tissue replaces skeletal muscle tissue.

fibrositis (fi″bro-si′tis) Inflammation of fibrous connective tissues, especially in the muscle fascia. This disease is also called *muscular rheumatism.*

muscular dystrophies (mus′ku-lar dis′tro-fez) Group of inherited disorders in which deficiency of dystrophin or associated glycoproteins collapses muscle cells, leading to progressive loss of function.

myalgia (mi-al′je-ah) Pain from any muscular disease or disorder.

myasthenia gravis (mi″as-the′ne-ah gra′vis) Chronic disease in which muscles are weak and easily fatigued because of malfunctioning neuromuscular junctions.

myokymia (mi″o-ki′me-ah) Persistent quivering of a muscle.

myology (mi-ol′o-je) Study of muscles.

myoma (mi-o′mah) Tumor composed of muscle tissue.

myopathy (mi-op′ah-the) Any muscular disease.

myositis (mi″o-si′tis) Inflammation of skeletal muscle tissue.

myotomy (mi-ot′o-me) Cutting of muscle tissue.

myotonia (mi″o-to′ne-ah) Prolonged muscular spasm.

paralysis (pah-ral′ĭ-sis) Loss of ability to move a body part.

paresis (pah-re′sis) Partial or slight paralysis of muscles.

shin splints (shin splints) Soreness on the front of the leg due to straining the anterior leg muscles, often as a result of walking up and down hills.

torticollis (tor″tĭ-kol′is) Condition in which the neck muscles, such as the sternocleidomastoids, contract involuntarily. It is more commonly called *wryneck.*

Clinical Connection

During summer and fall in the 1940s and early 1950s, thousands of children in the United States developed a viral infection called *acute paralytic poliomyelitis,* or polio. Usually, the virus remained in the throat or the small intestine lining, or traveled to the tonsils and lymph nodes, but when it entered the spinal cord and concentrated in cells that control muscle contraction, paralysis could develop in just days. When fever first struck a child, there was no way to predict the consequences.

Polio survivors vividly recall their treatment. Because of the infectious nature of polio, patients were quarantined. Many had their limbs splinted or their entire bodies immobilized in casts; others wore braces or had surgery to restore muscle function. An early type of respirator called an iron lung enabled patients to breathe when their respiratory muscles could not work. The survivors learned to live with permanent disabilities by training other muscles to take over the functions of damaged ones. But a few decades later, people who'd had polio as children began to experience muscle weakness, great fatigue, muscle and joint pain, difficulty sleeping and breathing, and headache—the symptoms of *postpolio syndrome.* The precise cause of this new collection of symptoms isn't known.

Despite decades of vaccination against polio in many nations, the disease still exists, in places where vaccine was not available or used, and possibly from live vaccine strains that have mutated into pathogenic strains. The goal of the World Health Organization was to eradicate polio by the year 2000, but instead the number of cases began to rise. In 2003, more than 800 cases of polio occurred, more than half of them in northern Nigeria where people declined vaccination because they believed rumors that the United States had intentionally altered the vaccine to cause female infertility, in an attempt to control Nigerian population growth. The disease also reappeared in ten countries near Nigeria. However, widespread vaccination resumed in late 2004, following the urging of many governments. Still, the disease remains endemic in some nations and outbreaks have occurred in others. In 2006, 1,441 were reported, from Nigeria, India, Bangladesh, Pakistan, Afghanistan, Somalia, Ethiopia, Congo, and Angola.

Muscular System

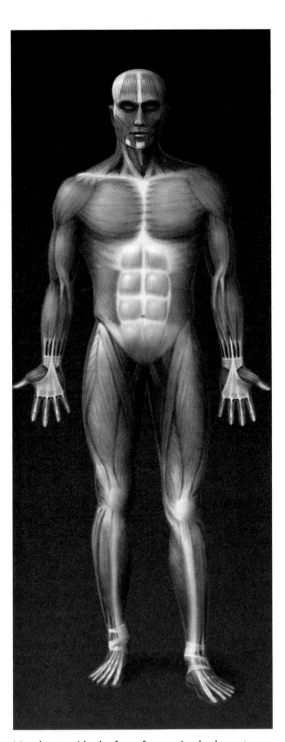

Muscles provide the force for moving body parts.

Integumentary System

The skin increases heat loss during skeletal muscle activity.

Lymphatic System

Muscle action pumps lymph through lymphatic vessels.

Skeletal System

Bones provide attachments that allow skeletal muscles to cause movement.

Digestive System

Skeletal muscles are important in swallowing. The digestive system absorbs nutrients needed for muscle contraction.

Nervous System

Neurons control muscle contractions.

Respiratory System

Breathing depends on skeletal muscles. The lungs provide oxygen for body cells and excrete carbon dioxide.

Endocrine System

Hormones help increase blood flow to exercising skeletal muscles.

Urinary System

Skeletal muscles help control expulsion of urine from the urinary bladder.

Cardiovascular System

The heart pumps as a result of cardiac muscle contraction. Blood flow delivers oxygen and nutrients and removes wastes.

Reproductive System

Skeletal muscles are important in sexual activity.

SUMMARY OUTLINE

8.1 Introduction (p. 177)

The three types of muscle tissue are skeletal, smooth, and cardiac.

8.2 Structure of a Skeletal Muscle (p. 177)

Individual muscles are the organs of the muscular system. They include skeletal muscle tissue, nervous tissue, blood, and connective tissues.

1. Connective tissue coverings
   a. Fascia covers skeletal muscles.
   b. Other connective tissues attach muscles to bones or to other muscles.
   c. A network of connective tissue extends throughout the muscular system.
2. Skeletal muscle fibers
   a. Each skeletal muscle fiber is a single muscle cell, which is the unit of contraction.
   b. The cytoplasm contains mitochondria, sarcoplasmic reticulum, and myofibrils of actin and myosin.
   c. The organization of actin and myosin filaments produces striations.
   d. Transverse tubules extend inward from the cell membrane and associate with the sarcoplasmic reticulum.
3. Neuromuscular junction
   a. Motor neurons stimulate muscle fibers to contract.
   b. In response to a nerve impulse, the end of a motor neuron axon secretes a neurotransmitter, which stimulates the muscle fiber to contract.
4. Motor units
   a. One motor neuron and the muscle fibers associated with it constitute a motor unit.
   b. All the muscle fibers of a motor unit contract together.

8.3 Skeletal Muscle Contraction (p. 181)

Muscle fiber contraction results from a sliding movement of actin and myosin filaments.

1. Role of myosin and actin
   a. Cross-bridges of myosin filaments form linkages with actin filaments.
   b. The reaction between actin and myosin filaments generates the force of contraction.
2. Stimulus for contraction
   a. Acetylcholine released from the distal end of a motor neuron axon stimulates a skeletal muscle fiber.
   b. Acetylcholine causes the muscle fiber to conduct an impulse over the surface of the fiber that reaches deep within the fiber through the transverse tubules.
   c. A muscle impulse signals the sarcoplasmic reticulum to release calcium ions.
   d. Linkages form between actin and myosin, and the myosin cross-bridges pull on actin filaments, shortening the fiber.
   e. The muscle fiber relaxes when cross-bridges release from actin (ATP is needed, but is not broken down) and when calcium ions are actively transported (requiring ATP breakdown) back into the sarcoplasmic reticulum.
   f. Acetylcholinesterase breaks down acetylcholine.
3. Energy sources for contraction
   a. ATP supplies the energy for muscle fiber contraction.
   b. Creatine phosphate stores energy that can be used to synthesize ATP.
   c. ATP is needed for muscle relaxation.
4. Oxygen supply and cellular respiration
   a. Aerobic respiration requires oxygen.
   b. Red blood cells carry oxygen to body cells.
   c. Myoglobin in muscle cells temporarily stores oxygen.
5. Oxygen debt
   a. During rest or moderate exercise, muscles receive enough oxygen to respire aerobically.
   b. During strenuous exercise, oxygen deficiency may cause lactic acid to accumulate.
   c. Oxygen debt is the amount of oxygen required to convert accumulated lactic acid to glucose and to restore supplies of ATP and creatine phosphate.
6. Muscle fatigue
   a. A fatigued muscle loses its ability to contract.
   b. Muscle fatigue is usually due to accumulation of lactic acid.
7. Heat production
   a. More than half of the energy released in cellular respiration is lost as heat.
   b. Muscle action is an important source of body heat.

8.4 Muscular Responses (p. 187)

1. Threshold stimulus is the minimal stimulus required to elicit a muscular contraction.
2. Recording a muscle contraction
   a. A twitch is a single, short contraction reflecting stimulation of a muscle fiber.
   b. A myogram is a recording of an electrically stimulated isolated muscle.
   c. The latent period, the time between stimulus and responding muscle contraction, is followed by a period of contraction and a period of relaxation.
3. Summation
   a. A rapid series of stimuli may produce summation of twitches.
   b. Forceful, sustained contraction without relaxation is a tetanic contraction.
4. Recruitment of motor units
   a. A whole muscle is composed of many motor units controlled by different motor neurons which respond to different thresholds of stimulation.
   b. At a low intensity of stimulation, small numbers of motor units contract.
   c. At increasing intensities of stimulation, other motor units are recruited until the muscle contracts with maximal tension.
5. Sustained contractions
   a. Summation and recruitment together can produce a sustained contraction of increasing strength.
   b. Even when a muscle is at rest, its fibers usually remain partially contracted.

8.5 Smooth Muscle (p. 189)

The contractile mechanism of smooth muscle is similar to that of skeletal muscle.

1. Smooth muscle fibers
   a. Smooth muscle cells contain filaments of actin and myosin.
   b. Types include multiunit smooth muscle and visceral smooth muscle.
   c. Visceral smooth muscle displays rhythmicity and is self-exciting.
2. Smooth muscle contraction
   a. Two neurotransmitters—acetylcholine and norepinephrine—and hormones affect smooth muscle function.
   b. Smooth muscle can maintain a contraction longer with a given amount of energy than can skeletal muscle.
   c. Smooth muscles can change length without changing tension.

8.6 Cardiac Muscle (p. 190)

1. Like skeletal muscle cells, cardiac muscle cells have many actin and myosin filaments.
2. Cardiac muscle twitches last longer than skeletal muscle twitches.
3. Intercalated discs connect cardiac muscle cells.
4. A network of fibers contracts as a unit.
5. Cardiac muscle is self-exciting and rhythmic.

8.7 Skeletal Muscle Actions (p. 190)

The type of movement a skeletal muscle produces depends on the way the muscle attaches on either side of a joint.

1. Origin and insertion
   a. The immovable end of a skeletal muscle is its origin, and the movable end is its insertion.
   b. Some muscles have more than one origin.
2. Interaction of skeletal muscles
   a. Skeletal muscles function in groups.
   b. A prime mover is responsible for most of a movement. Synergists aid prime movers. Antagonists can resist the action of a prime mover.
   c. Smooth movements depend on antagonists giving way to the actions of prime movers.

8.8 Major Skeletal Muscles (p. 193)

1. Muscles of facial expression
   a. These muscles lie beneath the skin of the face and scalp and are used to communicate feelings through facial expression.
   b. They include the epicranius, orbicularis oculi, orbicularis oris, buccinator, zygomaticus, and platysma.
2. Muscles of mastication
   a. These muscles attach to the mandible and are used in chewing.
   b. They include the masseter and temporalis.
3. Muscles that move the head
   a. Muscles in the neck and upper back move the head.
   b. They include the sternocleidomastoid, splenius capitis, and semispinalis capitis.
4. Muscles that move the pectoral girdle
   a. Most of these muscles connect the scapula to nearby bones and closely associate with muscles that move the arm.
   b. They include the trapezius, rhomboid major, levator scapulae, serratus anterior, and pectoralis minor.
5. Muscles that move the arm
   a. These muscles connect the humerus to various regions of the pectoral girdle, ribs, and vertebral column.
   b. They include the coracobrachialis, pectoralis major, teres major, latissimus dorsi, supraspinatus, deltoid, subscapularis, infraspinatus, and teres minor.
6. Muscles that move the forearm
   a. These muscles connect the radius and ulna to the humerus or pectoral girdle.
   b. They include the biceps brachii, brachialis, brachioradialis, triceps brachii, supinator, pronator teres, and pronator quadratus.
7. Muscles that move the hand
   a. These muscles arise from the distal end of the humerus and from the radius and ulna.
   b. They include the flexor carpi radialis, flexor carpi ulnaris, palmaris longus, flexor digitorum profundus, extensor carpi radialis longus, extensor carpi radialis brevis, extensor carpi ulnaris, and extensor digitorum.
8. Muscles of the abdominal wall
   a. These muscles connect the rib cage and vertebral column to the pelvic girdle.
   b. They include the external oblique, internal oblique, transversus abdominis, and rectus abdominis.
9. Muscles of the pelvic outlet
   a. These muscles form the floor of the pelvic cavity and fill the space within the pubic arch.
   b. They include the levator ani, superficial transversus perinei, bulbospongiosus, and ischiocavernosus.
10. Muscles that move the thigh
    a. These muscles attach to the femur and to some part of the pelvic girdle.
    b. They include the psoas major, iliacus, gluteus maximus, gluteus medius, gluteus minimus, tensor fasciae latae, adductor longus, adductor magnus, and gracilis.
11. Muscles that move the leg
    a. These muscles connect the tibia or fibula to the femur or pelvic girdle.
    b. They include the biceps femoris, semitendinosus, semimembranosus, sartorius, and the quadriceps femoris group.
12. Muscles that move the foot
    a. These muscles attach the femur, tibia, and fibula to bones of the foot.
    b. They include the tibialis anterior, fibularis tertius, extensor digitorum longus, gastrocnemius, soleus, flexor digitorum longus, tibialis posterior, and fibularis longus.

CHAPTER ASSESSMENTS

8.1 Introduction

1. The three types of muscle tissue are _____, _____, and _____. (p. 177)

8.2 Structure of a Skeletal Muscle

2. Describe the difference between a tendon and an aponeurosis. (p. 177)
3. Describe how connective tissue associates with skeletal muscle. (p. 177)
4. List the major parts of a skeletal muscle fiber, and describe the function of each part. (p. 177)
5. Describe a neuromuscular junction. (p. 180)
6. A neurotransmitter _____. (p. 180)
   a. binds actin filaments, causing them to slide
   b. travels across a synapse from a neuron to a muscle cell
   c. ferries ATP across a synapse
   d. travels across a synapse from a muscle cell to a neuron.
   e. is a contractile protein that is part of a skeletal muscle fiber.
7. Define *motor unit*. (p. 180)

8.3 Skeletal Muscle Contraction

8. List the major events of muscle fiber contraction and relaxation. (p. 181)
9. Describe how ATP and creatine phosphate interact. (p. 183)
10. Describe how muscles obtain oxygen. (p. 184)
11. Describe how an oxygen debt may develop. (p. 184)
12. Explain how muscles may become fatigued. (p. 185)
13. Explain how skeletal muscle function affects the maintenance of body temperature. (p. 186)

8.4 Muscular Responses

14. Define *threshold stimulus*. (p. 187)
15. Sketch a myogram of a single muscular twitch, and identify the latent period, period of contraction, and period of relaxation. (p. 187)
16. Explain motor unit *recruitment*. (p. 188)
17. Explain how skeletal muscle stimulation produces a sustained contraction. (p. 188)

18. Distinguish between tetanic contraction and muscle tone. (p. 188)

8.5 Smooth Muscle

19. Distinguish between multiunit and visceral smooth muscle fibers. (p. 189)

20. Compare smooth and skeletal muscle contractions. (p. 189)

8.6 Cardiac Muscle

21. Make a table comparing contraction mechanisms of cardiac and skeletal muscle fibers. (p. 190)

8.7 Skeletal Muscle Actions

22. Distinguish between a muscle's origin and its insertion. (p. 192)

23. Define *prime mover, synergist,* and *antagonist.* (p. 192)

8.8 Major Skeletal Muscles

24. Match the muscles on the left to the descriptions and functions on the right. (pp. 195–205)

(1) Buccinator	A.	Inserted on coronoid process of mandible
(2) Epicranius	B.	Draws corner of mouth upward
(3) Orbicularis oris	C.	Can raise and adduct scapula
(4) Platysma	D.	Can pull head into an upright position
(5) Rhomboid major	E.	Raises eyebrow
(6) Splenius capitis	F.	Compresses cheeks
(7) Temporalis	G.	Extends over neck from chest to face
(8) Zygomaticus	H.	Closes lips
(9) Biceps brachii	I.	Extends forearm at elbow
(10) Brachialis	J.	Pulls shoulder back and downward
(11) Deltoid	K.	Abducts arm
(12) Latissimus dorsi	L.	Inserted on radial tuberosity
(13) Pectoralis major	M.	Pulls arm forward and across chest
(14) Pronator teres	N.	Rotates forearm medially
(15) Teres minor	O.	Inserted on coronoid process of ulna
(16) Triceps brachii	P.	Rotates arm laterally
(17) Biceps femoris	Q.	Inverts foot
(18) External oblique	R.	Member of quadriceps femoris group
(19) Gastrocnemius	S.	Plantar flexor of foot
(20) Gluteus maximus	T.	Compresses contents of abdominal cavity
(21) Gluteus medius	U.	Extends thigh
(22) Gracilis	V.	Hamstring muscle
(23) Rectus femoris	W.	Adducts thigh
(24) Tibialis anterior	X.	Abducts thigh

25. Which muscles can you identify in the bodies of these models? (pp. 195–205)

INTEGRATIVE ASSESSMENTS/ CRITICAL THINKING

→ OUTCOME 8.2

1. Discuss how connective tissue is part of the muscular system.

→ OUTCOME 8.3

2. As lactic acid and other substances accumulate in an active muscle they stimulate pain receptors, and the muscle may feel sore. How might the application of heat or substances that dilate blood vessels relieve such soreness?

→ OUTCOMES 8.3, 8.4

3. A woman takes her daughter to a sports medicine specialist and requests that she determine the percentage of fast- and slow-twitch fibers in the girl's leg muscles. The parent wants to know if the healthy girl should try out for soccer or cross-country running. Do you think this is a valid reason to test muscle tissue? Why or why not?

4. Following an injury to a nerve, the muscle it supplies with motor nerve fibers may become paralyzed. How would you explain to a patient the importance of moving the disabled muscles passively or contracting them using electrical stimulation?

→ OUTCOMES 8.5, 8.8

5. What steps might be taken to minimize atrophy of the skeletal muscles in patients confined to bed for prolonged times?

WEB CONNECTIONS

Visit the text website at **aris.mhhe.com** for additional quizzes, interactive learning exercises, and more.

AP|R MUSCULAR SYSTEM

Anatomy & Physiology | REVEALED includes cadaver photos that allow you to peel away layers of the human body to reveal structures beneath the surface. This program also includes animations, radiologic imaging, audio pronunciation, and practice quizzing. To learn more visit **www.aprevealed.com**.

9

Nervous System

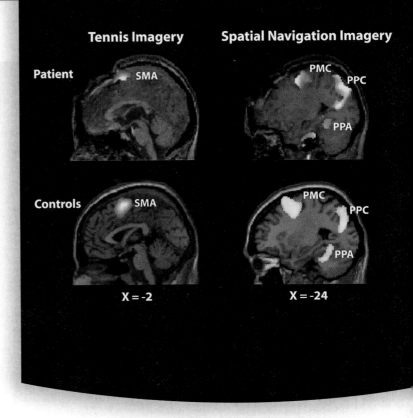

Tennis Imagery **Spatial Navigation Imagery**

Patient   SMA   PMC   PPC   PPA

Controls   SMA   PMC   PPC   PPA

X = -2   X = -24

A woman who had suffered brain injury in a traffic accident and was in a persistent vegetative state was asked to imagine herself playing tennis and walking through the rooms of her home, while undergoing neuroimaging with functional MRI. The patterns in which her brain lit up matched those of 12 healthy individuals as they completed the same tasks.

"ISLANDS OF AWARENESS" IN THE VEGETATIVE BRAIN. The 23-year-old had been in a persistent vegetative state for five months after sustaining traumatic brain injury in a car accident. She was awake, but apparently not aware, and unable to communicate in any way. To an observer, she had no sense of her own existence and did not react to sight or sound. But British researchers decided to take a different type of look at the young woman—from her point of view.

The investigators used functional MRI (fMRI), a form of neuro-imaging that measures regional blood flow. If a patient in a persistent vegetative state is given a particular stimulus, and the appropriate part of the brain lights up, then perhaps she is responding—just not in a way that can be directly observed. This is exactly what happened with the young accident victim. In a preliminary experiment, fMRI tracked her response to speech. First she heard a sentence that made sense, and then a sentence that had the same cadence as the first but was all nonsense words. Her brain lit up, in the speech-processing centers, only when the sentence had meaning. When she heard a sentence that included a homonym—a word that could have either of two meanings—an additional brain region lit up, presumably because she had to choose the correct meaning. Then the researchers asked her to imagine herself in two settings: playing tennis and walking through all of the rooms of her house. Healthy individuals asked the same questions served as controls. Both the young woman's brain and the control brains lit up in exactly the same areas.

The researchers had identified what they called "islands of awareness" in the brain of this supposedly completely unaware young woman. Although she did not have the most severe degree of brain injury, and her brain may not be able to coordinate those islands of awareness, the study suggests that neuroimaging may be a valuable tool in assessing consciousness in people who cannot communicate their self-awareness.

Learning Outcomes   *After studying this chapter, you should be able to do the following:*

9.1 Introduction

1. Distinguish between the two types of cells that comprise nervous tissue. (p. 212)

2. Name the two major groups of nervous system organs. (p. 212)

9.2 General Functions of the Nervous System

3. Explain the general functions of the nervous system. (p. 213)

9.3 Neuroglial Cells

4. State the functions of neuroglial cells in the central nervous system. (p. 214)

5. Distinguish among the types of neuroglial cells in the central nervous system. (p. 214)

6. Describe the Schwann cells of the peripheral nervous system. (p. 214)

9.4 Neurons

7. Describe the general structure of a neuron. (p. 214)

8. Explain how differences in structure and function are used to classify neurons. (p. 218)

9.5 The Synapse

9. Explain how information passes from one neuron to another. (p. 220)

9.6 Cell Membrane Potential

10. Explain how a membrane becomes polarized. (p. 221)

11. Describe the events that lead to the conduction of a nerve impulse. (p. 224)

9.7 Nerve Impulses

12. Compare nerve impulse conduction in myelinated and unmyelinated neurons in terms of the all-or-none response. (p. 225)

9.8 Synaptic Transmission

13. Identify the changes in membrane potential associated with excitatory and inhibitory neurotransmitters. (p. 225)

Aids to Understanding Words   *(Appendix A on page 567 has a complete list of Aids to Understanding Words.)*

ax- [axis] *ax*on: Cylindrical nerve fiber that carries impulses away from a neuron cell body.

dendr- [tree] *dendr*ite: Branched nerve cell process that serves as a receptor surface of a neuron.

funi- [small cord or fiber] *funi*culus: Major nerve tract or bundle of myelinated nerve cell axons within the spinal cord.

gangli- [a swelling] *gangli*on: Mass of neuron cell bodies.

-lemm [rind or peel] neuri*lemm*a: Sheath that surrounds the myelin of a nerve cell axon.

mening- [membrane] *mening*es: Membranous coverings of the brain and spinal cord.

moto- [moving] *moto*r neuron: Neuron that stimulates a muscle to contract or a gland to secrete.

peri- [around] *peri*pheral nervous system: Portion of the nervous system that consists of nerves branching from the brain and spinal cord.

plex- [interweaving] choroid *plex*us: Mass of specialized capillaries associated with spaces in the brain.

sens- [feeling] *sens*ory neuron: Neuron that conducts impulses into the brain or spinal cord.

syn- [together] *syn*apse: Junction between two neurons.

ventr- [belly or stomach] *ventr*icle: Fluid-filled space within the brain.

9.1 INTRODUCTION

Feeling, thinking, remembering, moving, and being aware of the world require activity from the nervous system. This vast collection of cells also helps coordinate all other body functions to maintain homeostasis and to enable the body to respond to changing conditions. Information from within and outside the body is brought to the brain and spinal cord, which then stimulates responses from muscles and glands.

Recall from chapter 5 (p. 111) that nervous tissue consists of masses of nerve cells, or **neurons.** These cells are the structural and functional units of the nervous system and are specialized to react to physical and chemical changes in their surroundings (fig. 9.1). Neurons transmit information in the form of electrochemical changes, called **nerve impulses,** to other neurons and to cells outside the nervous system.

Neurons typically have a rounded area called the **cell body,** and two types of extensions: dendrites and axons. **Dendrites,** which may be numerous, receive electrochemical messages. **Axons** are extensions that send information in the form of nerve impulses. Usually a neuron has only one axon. Figure 9.1 depicts these major parts of a neuron.

Nerves are bundles of axons. Nervous tissue also includes **neuroglial cells** that provide physical support, insulation, and nutrients for neurons. During development before birth, neuroglial cells release and relay signals that guide the differentiation of neurons from progenitor cells (see chapter 3, p. 70).

The organs of the nervous system can be divided into two groups. One group, consisting of the brain and spinal cord, forms the **central nervous system (CNS).** The other, composed of the nerves (peripheral nerves) that connect the central nervous system to other body parts, is called the **peripheral nervous system (PNS)** (fig. 9.2). Together, these systems provide three general functions: sensory, integrative, and motor.

Check Your Recall

1. What are the two major types of cells that form nervous tissue?

2. What are the two major subdivisions of the nervous system?

9.2 GENERAL FUNCTIONS OF THE NERVOUS SYSTEM

The *sensory function* of the nervous system derives from **sensory receptors** (sen′so-re re-sep′torz) at the ends of peripheral neurons (see chapter 10, p. 261). These receptors gather information by detecting changes inside and outside the body. Sensory receptors monitor external environmental factors, such as light and sound intensities, and conditions of the body's internal environment, such as temperature and oxygen level.

Sensory receptors convert environmental information into nerve impulses, which are then transmitted over peripheral nerves to the central nervous system. There, the signals are integrated; that is, they are brought together, creating sensations, adding to memory, or helping produce thoughts that translate sensations into

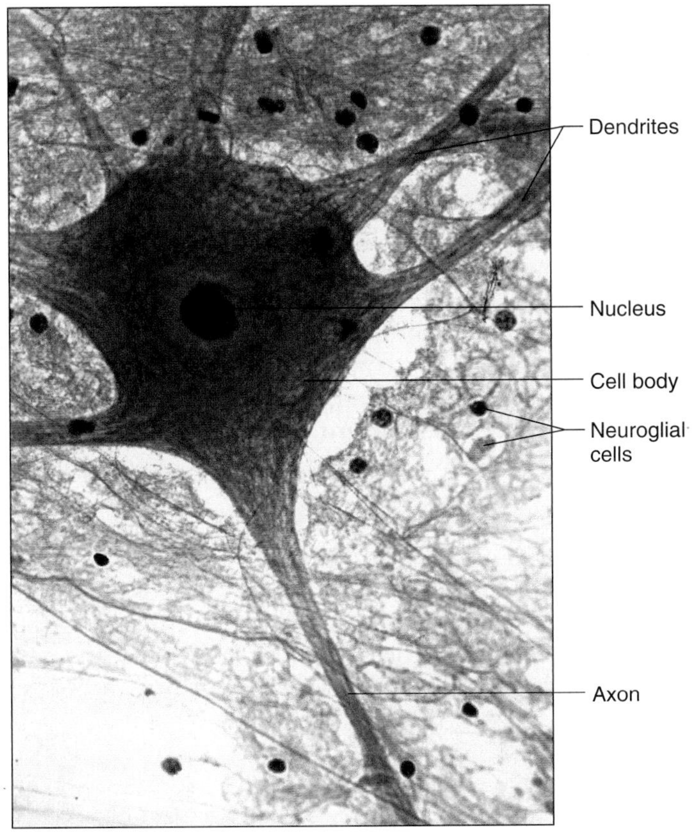

Figure 9.1

Neurons are the structural and functional units of the nervous system (600×). The dark spots in the area surrounding the neuron are neuroglial cells. Note the dendrites and the single axon of the neuron.

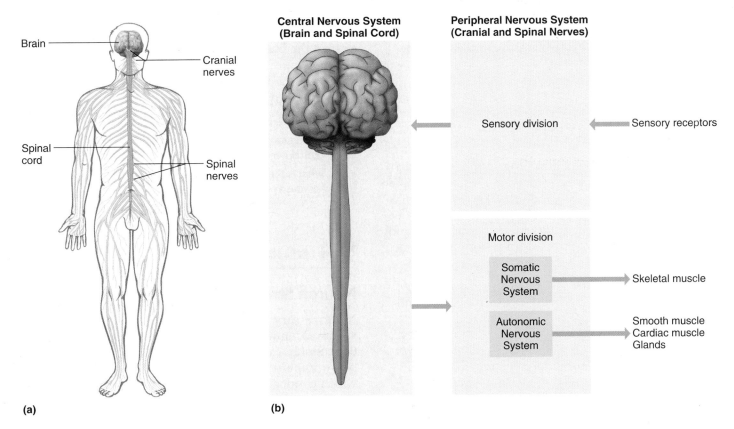

Figure 9.2

Nervous system. (*a*) The nervous system includes the central nervous system (brain and spinal cord) and the peripheral nervous system (cranial nerves and spinal nerves). (*b*) The nervous system receives information from sensory receptors and initiates responses through effector organs (muscles and glands).

perceptions. As a result of this *integrative function,* we make conscious or subconscious decisions, and then we use *motor functions* to act on them.

The motor functions of the nervous system employ peripheral neurons, which carry impulses from the central nervous system to responsive structures called **effectors** (e-fek'torz). Effectors, which are outside the nervous system, include muscles that contract and glands that secrete when stimulated by nerve impulses.

The motor functions of the peripheral nervous system can be divided into two categories. Those that are consciously controlled comprise the **somatic nervous system,** which controls skeletal muscle. In contrast, the **autonomic nervous system** controls effectors that are involuntary, such as the heart, smooth muscle in blood vessels, and various glands.

The nervous system can detect changes outside and within the body, make decisions based on the information received, and stimulate muscles or glands to respond. Typically, these responses counteract the effects of the changes detected, and in this way, the nervous system helps maintain homeostasis.

Check Your Recall

3. How do sensory receptors collect information?
4. How does the central nervous system integrate incoming information?
5. What are the two types of motor functions of the nervous system?

9.3 NEUROGLIAL CELLS

Neurons cannot exist without neuroglial cells (neuroglia), which fill spaces, provide structural frameworks, produce the components of the electrical insulator **myelin** (mi'ĕ-lin), and carry on phagocytosis. In the central nervous system, neuroglial cells greatly outnumber neurons, and can divide, whereas neurons do not normally divide. Neuroglia are of the following types (fig. 9.3):

1. **Microglial cells** are scattered throughout the central nervous system. They support neurons and phagocytize bacterial cells and cellular debris.
2. **Oligodendrocytes** align along nerve fibers. They provide insulating layers of myelin, called a *myelin sheath,* around axons within the brain and spinal cord.
3. **Astrocytes,** commonly found between neurons and blood vessels, provide structural support, join parts by their abundant cellular processes, and help regulate the concentrations of nutrients and ions within the tissue. Astrocytes also form scar tissue that fills spaces following injury to the CNS.

4. **Ependymal cells** form an epithelia-like membrane that covers specialized brain parts (choroid plexuses) and forms the inner linings that enclose spaces within the brain (ventricles) and spinal cord (central canal).

The peripheral nervous system includes neuroglial cells called **Schwann cells** that form a myelin sheath around axons.

Excess neuroglial cells can harm health. Fast-growing gliomas are brain tumors consisting of rapidly-dividing neuroglia (neurons do not divide). Immediately after a spinal cord injury, destruction of neuroglia strips axons of myelin. Subsequent overgrowth of neuroglia forms scars, which impede recovery of function.

In most of the body, capillaries (the smallest blood vessels) are "leaky," allowing small molecules to enter or leave the bloodstream. The cells that form capillaries in the brain, in contrast, are much more tightly connected, thanks partly to astrocytes. This specialized architecture creates a "blood–brain barrier" that shields delicate brain tissue from chemical fluctuations, blocking entry to many substances. The barrier can allow for selective drug delivery, such as preventing some antihistamines from entering the brain so they do not cause drowsiness. But this presents a trade-off—many drugs needed to treat the brain cannot get there.

Check Your Recall

6. List the functions of the cells that support neurons.
7. Distinguish among the types of neuroglial cells in the central nervous system.
8. What is the function of Schwann cells in the peripheral nervous system?

9.4 NEURONS

Neuron Structure

Neurons vary considerably in size and shape, but they all have common features. These include a cell body; the tubular, cytoplasm-filled dendrites, which conduct nerve impulses to the neuron cell body; and an axon, which conducts impulses away.

The neuron cell body consists of granular cytoplasm, a cell membrane, and organelles such as mitochondria, lysosomes, a Golgi apparatus, and a network of fine threads called **neurofibrils** (nu"ro-fi'brilz), which extends into the axon. Scattered throughout the cytoplasm are many membranous sacs called **chromatophilic substance**

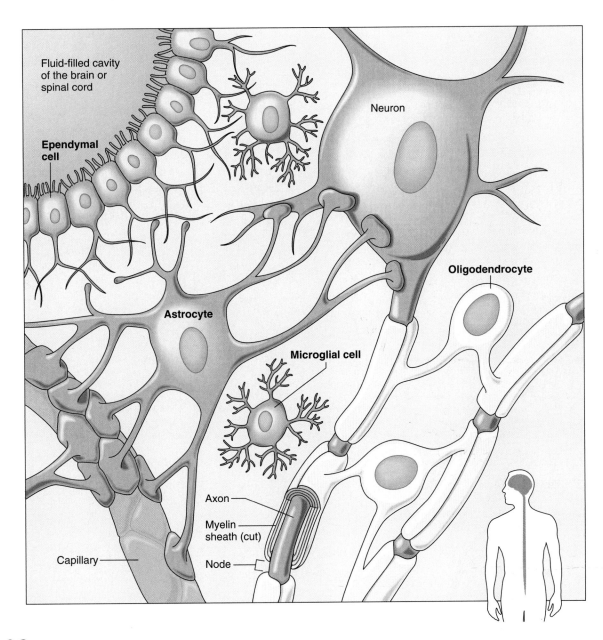

Figure 9.3

Types of neuroglial cells in the central nervous system include the microglial cell, oligodendrocyte, astrocyte, and ependymal cell. (Ependymal cells have cilia into early childhood. In adults, cilia remain only on ependymal cells in the ventricles of the brain.)

(Nissl bodies). These are similar to rough endoplasmic reticulum in other cells (fig. 9.4). Ribosomes attached to chromatophilic substance function in protein synthesis, as they do elsewhere. Near the center of the cell body is a large, spherical nucleus with a conspicuous nucleolus.

Dendrites are usually short and highly branched. These processes, together with the membrane of the cell body, are the neuron's main receptive surfaces with which axons from other neurons communicate.

In most neurons the axon arises from a slight elevation of the cell body called the *axonal hillock*. The axon conducts nerve impulses away from the cell body. Many mitochondria, microtubules, and neurofibrils are in the axon cytoplasm. An axon originates as a single structure but may give off side branches (collaterals). Its end may branch into many fine extensions that contact the receptive surfaces of other cells.

Larger axons of peripheral neurons are enclosed in **myelin sheaths** (mi'ĕ-lin shēthz) produced by Schwann cells (figs. 9.4 and 9.5). These cells wind tightly around axons, somewhat like a bandage wrapped around a finger, coating them with many layers of cell membrane that have little or no cytoplasm between them. The portions of the Schwann cells that contain most of the cytoplasm and the nuclei remain outside the myelin sheath and comprise a **neurilemma** (nu"rĭ-lem'ah),

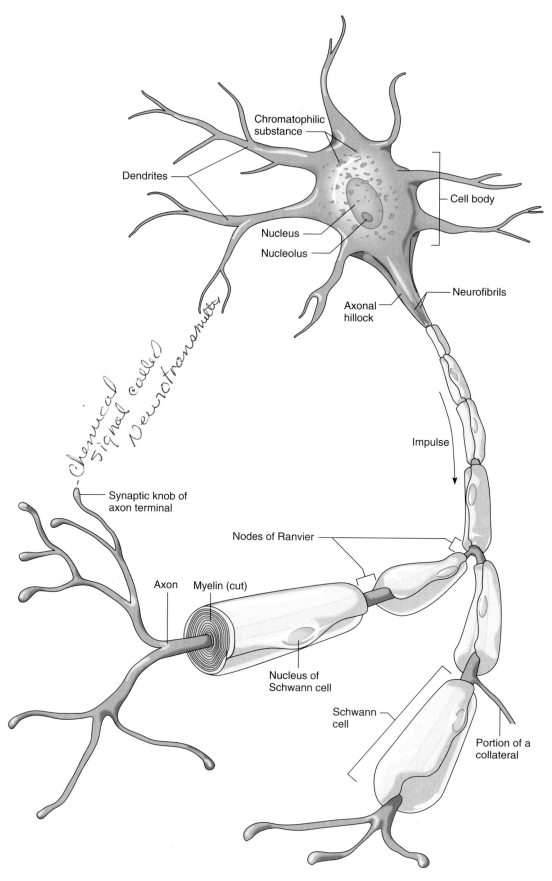

Chromatophilic
substance

Dendrites

Cell body

Nucleus

Nucleolus

Neurofibrils

Axonal
hillock

Impulse

Chemical
Signal called
Neurotransmitter

Synaptic knob of
axon terminal

Nodes of Ranvier

Axon   Myelin (cut)

Nucleus of
Schwann cell

Schwann
cell

Portion of a
collateral

Figure 9.4

A common neuron.

or neurilemmal sheath, which surrounds the myelin sheath. Narrow gaps between Schwann cells are called **nodes of Ranvier** (nō-dz uv ron'vee-ay) (fig. 9.5).

Axons with myelin sheaths are called *myelinated,* and those that lack sheaths are *unmyelinated.* Myelin is also found in the CNS, where groups of myelinated axons appear white, and masses of such axons form the *white matter.* Unmyelinated axons and neuron cell bodies form *gray matter* within the CNS.

Myelin begins to form on axons during the fourteenth week of prenatal development. Yet many of the axons in newborns are not completely myelinated. As a result, an infant's nervous system cannot function as effectively as that of an older child or adult. Infants' responses to stimuli are coarse and undifferentiated, and may involve the whole body. All myelinated axons begin to develop sheaths by the time a child starts to walk, and myelination continues into adolescence. Deficiencies of essential nutrients during the developmental years may limit myelin formation, which may impair nervous system function later in life.

When peripheral nerves are damaged, their axons can regenerate. The neurilemma plays an important role in this process. In contrast, CNS axons are myelinated by oligodendrocytes, which do not provide a neurilemma. Damaged CNS neurons usually do not regenerate.

The brain harbors small collections of neural stem cells that can divide to give rise to new neurons or neuroglial cells, depending upon their chemical surroundings. Neural stem cells are found in the hippocampus and near the brain's ventricles.

To picture the relative sizes of a typical neuron's parts, imagine that the cell body is the size of a tennis ball. The axon would then be a mile long and half an inch thick. The dendrites would fill a large bedroom.

Classification of Neurons

Neurons differ in the structure, size, and shape of their cell bodies. They also vary in the length and size of their axons and dendrites and in the number of connections they make with other neurons.

On the basis of structural differences, neurons are classified into three major groups (fig. 9.6). Each type of neuron is specialized to send a nerve impulse in one direction.

CNS - do not provide Neurilemma so damaged ones do not regenerate. PNS can regenerate due to having neurilemma.

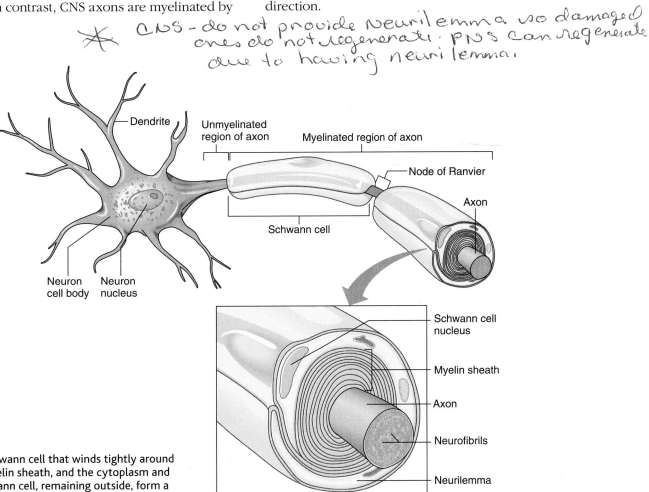

Figure 9.5

The portion of a Schwann cell that winds tightly around an axon forms a myelin sheath, and the cytoplasm and nucleus of the Schwann cell, remaining outside, form a neurilemma, or neurilemmal sheath.

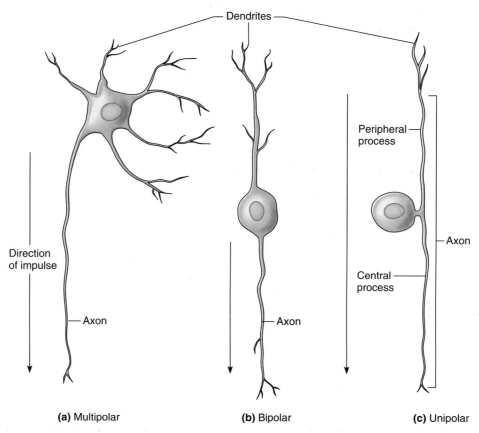

Figure 9.6

Structural types of neurons include (*a*) the multipolar neuron, (*b*) the bipolar neuron, and (*c*) the unipolar neuron.

1. **Multipolar neurons** have many processes arising from their cell bodies. Only one process of each neuron is an axon; the rest are dendrites. Most neurons whose cell bodies lie within the brain or spinal cord are multipolar.

2. **Bipolar neurons** have only two processes, one arising from each end of the cell body. These processes are structurally similar, but one is an axon and the other a dendrite. Neurons in specialized parts of the eyes, nose, and ears are bipolar.

3. **Unipolar neurons** have a single process extending from the cell body. A short distance from the cell body, this process divides into two branches, which really function as a single axon. One branch (the peripheral process) is associated with dendrites near a peripheral body part. The other branch (the central process) enters the brain or spinal cord. The cell bodies of some unipolar neurons aggregate in specialized masses of nervous tissue called **ganglia** (gang′gle-ah) (singular, *ganglion*), which are located outside the brain and spinal cord.

Neurons also vary in function. They may carry impulses into the brain or spinal cord, conduct impulses from neuron to neuron within the brain or spinal cord, or transmit impulses out of the brain or spinal cord.

On the basis of functional differences, neurons are grouped as follows (fig. 9.7):

1. **Sensory neurons** (afferent neurons) carry nerve impulses from peripheral body parts into the brain or spinal cord. Sensory neurons either have specialized *receptor ends* at the tips of their dendrites, or they have dendrites that are closely associated with *receptor cells* in the skin or in sensory organs.

   Changes that occur inside or outside the body stimulate receptor ends or receptor cells, triggering sensory nerve impulses. The impulses travel along the sensory neuron axons, which lead to the brain or spinal cord, where other neurons process the impulses. Most sensory neurons are unipolar; some are bipolar.

2. **Interneurons** (also called *association* or *internuncial neurons*) lie entirely within the brain or spinal cord. They are multipolar and link other neurons. Interneurons transmit impulses from one part of the brain or spinal cord to another. That is, they may direct incoming sensory impulses to appropriate parts for processing and interpreting. Other incoming impulses are transferred to motor neurons. The cell bodies of some interneurons aggregate in specialized masses of nervous tissue called **nuclei** (singular,

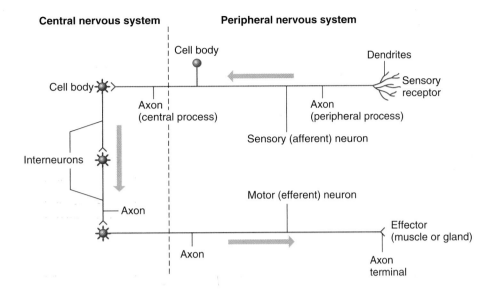

Central nervous system   **Peripheral nervous system**

Figure 9.7

Neurons are classified by function as well as structure. Sensory (afferent) neurons carry information into the central nervous system (CNS), interneurons are completely within the CNS, and motor (efferent) neurons carry instructions to the peripheral nervous system (PNS).

nucleus). Nuclei are similar to ganglia, but are located within the central nervous system.

3. **Motor neurons** (efferent neurons) are multipolar and carry nerve impulses out of the brain or spinal cord to effectors. Motor impulses stimulate muscles to contract and glands to release secretions.

Neurons deprived of oxygen change shape as their nuclei shrink, and they eventually disintegrate. Oxygen deficiency can result from lack of blood flow (ischemia) through nerve tissue, an abnormally low blood oxygen concentration (hypoxemia), or toxins that prevent neurons from using oxygen by blocking aerobic respiration.

Check Your Recall

9. Distinguish between a dendrite and an axon.
10. Describe the components of a neuron.
11. Describe how a myelin sheath forms.
12. Explain why axons of peripheral nerves can regenerate, but axons of central nervous system nerves cannot.
13. Name three groups of neurons based on structure and three groups based on function.

9.5 THE SYNAPSE

Nerve impulses travel along complex **nerve pathways.** The junction between any two communicating neurons is called a **synapse** (sin'aps). The neurons at a synapse are not in direct physical contact, but are separated by a gap called a *synaptic cleft*. Communication along a nerve pathway must cross these gaps (fig. 9.8).

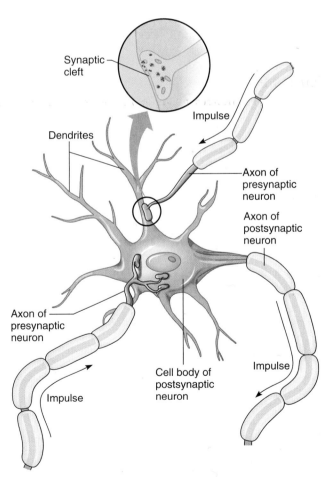

Figure 9.8

Synapses separate neurons. For an impulse to continue from one neuron to another, it must cross the synaptic cleft at a synapse. A synapse is usually between an axon and a dendrite or between an axon and a cell body.

When you receive a text message, the person writing the message is the sender and you are the receiver. Similarly, the neuron carrying the impulse into the synapse is the sender, or *presynaptic neuron.* The neuron that receives this input at the synapse is the receiver, or *postsynaptic neuron.* The process of crossing the synaptic cleft with this message is called *synaptic transmission.* The Topic of Interest box on page 221 discusses some factors that affect synaptic transmission.

Synaptic transmission is a one-way process carried out by biochemicals called **neurotransmitters.** The distal ends of axons have one or more extensions called *synaptic knobs,* absent in dendrites, which contain many membranous sacs, called *synaptic vesicles.* When a nerve impulse reaches a synaptic knob, some of the synaptic vesicles release neurotransmitter (figs. 9.9 and 9.10). The neurotransmitter diffuses across the synaptic cleft and reacts with specific receptors on the postsynaptic neuron membrane.

The action of neurotransmitter on a postsynaptic cell is either excitatory (turning a process on) or inhibitory (turning a process off). The net effect on the postsynaptic cell depends on the combined effect of the excitatory and inhibitory inputs from as few as 1 and as many as 10,000 presynaptic neurons.

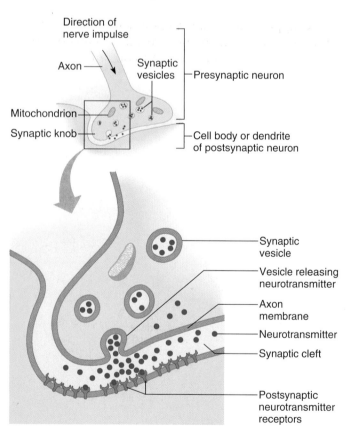

Figure 9.9

Action across a synapse. When a nerve impulse reaches the synaptic knob at the end of an axon, synaptic vesicles release a neurotransmitter that diffuses acoross the synaptic cleft and binds to specific receptors on the postsynaptic membrane.

9.6 CELL MEMBRANE POTENTIAL

The surface of a cell membrane (including a nonstimulated or *resting* neuron) is usually electrically charged, or *polarized,* with respect to the inside. This polarization arises from an unequal distribution of positive and negative ions between sides of the membrane, and it is particularly important in the conduction of muscle and nerve impulses. A characteristic change in neuron membrane polarization and return to the resting state, called an *action potential,* forms a nerve impulse that is propagated along an axon.

Distribution of Ions

Because of the active transport of sodium and potassium ions, cells throughout the body have a greater concentration of sodium ions (Na^+) outside and a greater concentration of potassium ions (K^+) inside (see chapter 3, p. 64). The cytoplasm of these cells has many large, negatively charged particles, including phosphate ions (PO_4^{-3}), sulfate ions (SO_4^{-2}), and proteins, that cannot diffuse across the cell membranes.

Chapter 3 (p. 53) introduced cell membranes as selectively permeable phospholipid bilayers. The distribution of ions inside and outside cells is determined in part by channels in the cell membranes (see chapter 3, pp. 54–55). Some channels are always open, and others can be opened or closed. Furthermore, channels can be selective; that is, a channel may allow one kind of ion to pass through and exclude other kinds (fig. 9.11).

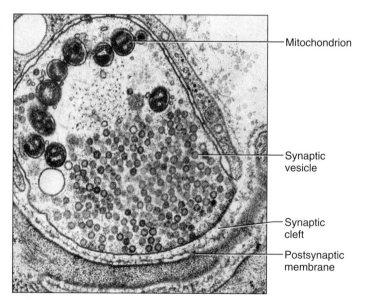

Figure 9.10

This transmission electron micrograph of a synaptic knob shows abundant synaptic vesicles, which are filled with neurotransmitter molecules.

Factors Affecting Synaptic Transmission

Nerve impulses reaching synaptic knobs too rapidly can exhaust neurotransmitter supplies, and impulse conduction ceases until more neurotransmitters are synthesized. This happens during an epileptic seizure. Abnormal and too rapid impulses originate from certain brain cells and reach skeletal muscle fibers, stimulating violent contractions. In time, the synaptic knobs run out of neurotransmitters, and the seizure subsides.

A drug called Dilantin (diphenylhydantoin) treats seizure disorders by increasing the effectiveness of the sodium active transport mechanism. More sodium ions transported from inside the neurons stabilize membrane thresholds against too rapid stimulation.

Many other drugs affect synaptic transmission. For example, caffeine in coffee, tea, and cola drinks stimulates nervous system activity by lowering the thresholds at synapses so that neurons are more easily excited. Antidepressants called "selective serotonin reuptake inhibitors" keep the neurotransmitter serotonin in synapses longer, compensating for a still little-understood deficit that presumably causes depression.

Potassium ions pass through cell membranes much more easily than sodium ions. This makes potassium ions a major contributor to membrane polarization. Calcium ions are less able to cross the resting cell membrane than either sodium ions or potassium ions, and have a special role in nerve function, described later.

Resting Potential

Sodium and potassium ions follow the laws of diffusion discussed in chapter 3 (p. 61) and show a net movement from high concentration to low concentration as permeabilities permit. Because a resting cell membrane is more permeable to potassium ions than to sodium ions, potassium ions diffuse out of the cell more rapidly than sodium ions can diffuse in (fig. 9.12*a*). Every millisecond, more positive charges leave the cell by diffusion than enter it. As a result, the outside of the cell membrane gains a slight surplus of positive charges, and the inside is left with a slight surplus of impermeant negative charges (fig. 9.12*b*).

The difference in electrical charge between two regions is called a *potential difference.* In a resting nerve cell, the potential difference between the region inside the membrane and the region outside the membrane is called a **resting potential.** As long as a nerve cell membrane is undisturbed, the membrane remains in this polarized state. At the same time, the cell continues to expend energy to drive the Na^+/K^+ "pumps" that actively transport sodium and potassium ions in opposite directions. The pump maintains the concentration gradients responsible for diffusion of these ions in the first place (fig. 9.12*c*).

Potential Changes

Nerve cells are excitable; that is, they can respond to changes in their surroundings. Some nerve cells, for example, are specialized to detect changes in temper-

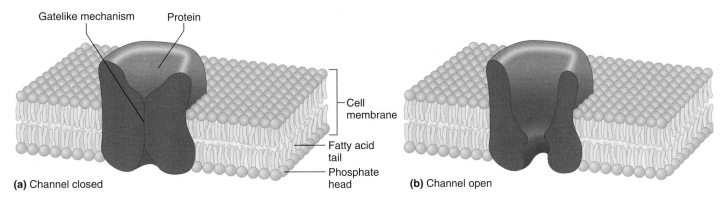

(a) Channel closed

(b) Channel open

Cell membrane
Fatty acid tail
Phosphate head

Figure 9.11

Cell membrane polarization is necessary for nerve transmission, and depends upon the movements of ions through channels. A gatelike mechanism can (*a*) close or (*b*) open some of the channels in cell membranes through which ions pass.

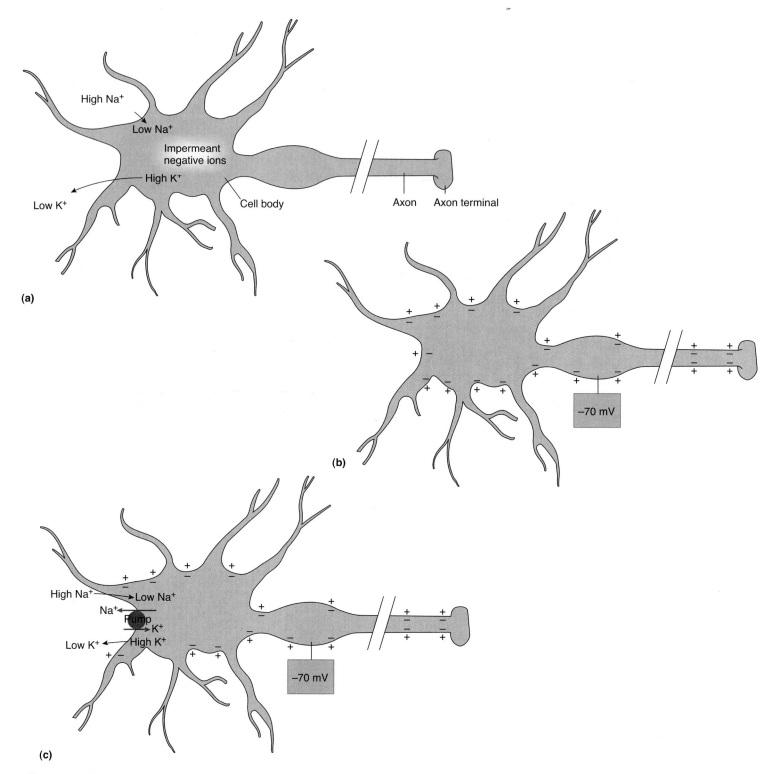

Figure 9.12

The resting potential. (*a*) Conditions that lead to the resting potential. (*b*) In the resting neuron, the inside of the membrane is negative relative to the outside. (*c*) The Na⁺/K⁺ pump maintains the concentration gradients for Na⁺ and K⁺ ions.

ature, light, or pressure from outside the body. Many neurons respond to neurotransmitters from other neurons. Such changes (or stimuli) usually affect the resting potential in a particular region of a nerve cell membrane. If the membrane's resting potential decreases

(as the inside of the membrane becomes less negative when compared to the outside), the membrane is said to be *depolarized* (fig. 9.13*a*).

Local potential changes are graded. This means that the magnitude of change in the resting potential is directly

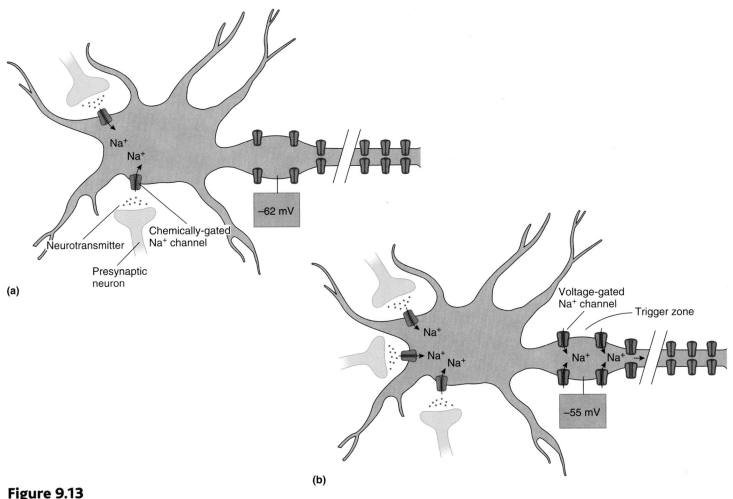

(a)

(b)

Figure 9.13

Action potentials. (*a*) A subthreshold depolarization will not result in an action potential. (*b*) Stimulation from multiple presynaptic neurons may cause the postsynaptic neuron to reach threshold, opening voltage-gated channels at the trigger zone.

proportional to the intensity of the stimulus. That is, if the membrane is being depolarized, the greater the stimulus, the greater the depolarization. If neurons are depolarized sufficiently, the membrane potential reaches a level called the **threshold potential,** which is approximately –55 millivolts. If threshold is reached, an **action potential** results, which is the basis for the nerve impulse.

Action Potential

At the threshold potential, permeability suddenly changes at the trigger zone of the neuron being stimulated. Channels highly selective for sodium ions open and allow sodium to diffuse freely inward (figs. 9.13*b* and 9.14*b*). This movement is aided by the negative electrical condition on the inside of the membrane, which attracts the positively charged sodium ions.

As sodium ions diffuse inward, the membrane loses its negative electrical charge and becomes depolarized. At almost the same time, however, membrane channels open that allow potassium ions to pass through, and as these positive ions diffuse outward, the inside of the membrane becomes negatively charged once more (fig. 9.14*c*). The membrane potential may briefly become overly negative (*hyperpolarization*), but the membrane quickly returns to the resting potential (*repolarization*), and it remains in this state until stimulated again.

This rapid sequence of depolarization and repolarization, which takes about one-thousandth of a second, is the action potential. Because only a small fraction of the sodium and potassium ions move through the membrane during an action potential, many action potentials can occur, and resting potentials be reestablished, before the original concentrations of these ions change significantly. Also, active transport within the membrane maintains the original concentrations of sodium and potassium ions on either side.

Check Your Recall

14. Describe the events that occur at a synapse.

15. Summarize how a nerve fiber becomes polarized.

16. List the major events of an action potential.

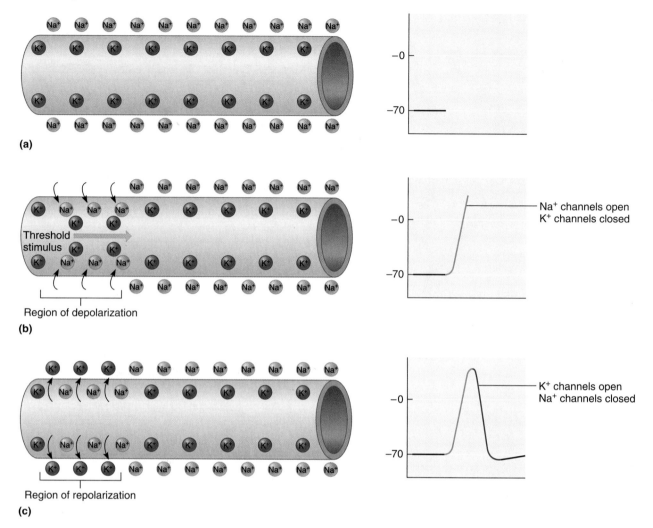

Figure 9.14

Action potential. (*a*) At rest, the membrane potential is negative. (*b*) When the membrane reaches threshold, sodium channels open, some sodium (Na⁺) diffuses in, and the membrane is depolarized. (*c*) Soon afterward, potassium channels open, potassium (K⁺) diffuses out, and the membrane is repolarized. (For simplicity, negative ions are not shown.)

9.7 NERVE IMPULSES

When an action potential occurs in one region of a nerve cell membrane, it causes a bioelectric current to flow to adjacent portions of the membrane. This *local current* stimulates the adjacent membrane to its threshold level and triggers another action potential. This, in turn, stimulates the next adjacent region. A wave of action potentials moves down the axon to the end. This propagation of action potentials along a nerve axon constitutes the nerve impulse (fig. 9.15). Table 9.1 summarizes the events leading to the conduction of a nerve impulse.

Certain local anesthetic drugs, such as those used in dentistry, decrease membrane permeability to sodium ions. Such a drug in the fluids surrounding an axon interrupts impulses from passing through the affected region and reaching the brain, preventing sensations of touch and pain.

Table 9.1	Events Leading to the Conduction of a Nerve Impulse

1. Neuron membrane maintains resting potential.

2. Threshold stimulus is received.

3. Sodium channels in the trigger zone of the neuron open.

4. Sodium ions diffuse inward, depolarizing the membrane.

5. Potassium channels in the membrane open.

6. Potassium ions diffuse outward, repolarizing the membrane.

7. The resulting action potential causes a local bioelectric current that stimulates adjacent portions of the membrane.

8. A wave of action potentials travels the length of the axon as a nerve impulse.

Region of
action potential

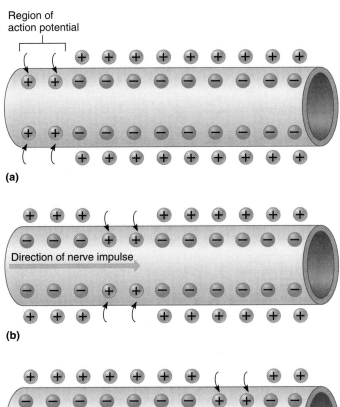

(a)

Direction of nerve impulse

(b)

(c)

Figure 9.15

A nerve impulse. (*a*) An action potential in one region stimulates the adjacent region, and (*b*) and (*c*) a wave of action potentials (a nerve impulse) moves along the axon.

Impulse Conduction

An unmyelinated axon conducts an impulse over its entire surface. A myelinated axon functions differently because myelin insulates and prevents almost all ion flow through the membrane it encloses. The myelin sheath would prevent the conduction of a nerve impulse altogether if the sheath was continuous. However, nodes of Ranvier between Schwann cells interrupt the sheath (see fig. 9.4). Action potentials occur at these nodes, where the exposed axon membrane has sodium and potassium channels. A nerve impulse traveling along a myelinated axon appears to jump from node to node. This type of impulse conduction, termed saltatory, is many times faster than conduction on an unmyelinated axon.

The speed of nerve impulse conduction is proportional to the diameter of the axon—the greater the diameter, the faster the impulse. For example, an impulse on a relatively thick, myelinated axon, such as that of a motor neuron associated with a skeletal mus-

cle, might travel 120 meters per second. An impulse on a thin, unmyelinated axon, such as that of a sensory neuron associated with the skin, might move only 0.5 meter per second.

All-or-None Response

Nerve impulse conduction is an *all-or-none response.* That is, if a neuron responds at all, it responds completely. Thus, a nerve impulse is conducted whenever a stimulus of threshold intensity or above is applied to an axon, and all impulses carried on that axon are of the same strength. A greater intensity of stimulation does not produce a stronger impulse, but rather, more impulses per second.

For a very short time following a nerve impulse, a threshold stimulus will not trigger another impulse on an axon. This brief period, called the *refractory period,* limits the frequency of impulses in a neuron. It also ensures that the impulse proceeds in only one direction—down the axon. Although a frequency of 700 impulses per second is possible, 100 impulses per second is more common.

> ### Check Your Recall
> 17. What is the relationship between action potentials and nerve impulses?
> 18. Explain how impulse conduction differs in myelinated and unmyelinated nerve fibers.
> 19. Define *all-or-none response* as it relates to nerve impulse conduction.

9.8 SYNAPTIC TRANSMISSION

Neurotransmitters have various effects when they diffuse across the synaptic cleft and react with specific receptor molecules in the postsynaptic neuron membrane.

Excitatory and Inhibitory Actions

Neurotransmitters that increase postsynaptic membrane permeability to sodium ions will bring the postsynaptic membrane closer to threshold and may trigger nerve impulses. Such neurotransmitters are **excitatory.** Neurotransmitters that make it less likely that threshold will be reached are called **inhibitory,** because they decrease the chance that a nerve impulse will occur.

The synaptic knobs of a thousand or more neurons may communicate with the dendrites and cell body of a single postsynaptic neuron. Neurotransmitters released by some of these knobs have an excitatory action, but those from other knobs have an inhibitory action. The effect on the postsynaptic neuron depends on which presynaptic knobs are activated from moment to moment. In other words, if more excitatory than inhibitory neurotransmitters are released, the postsynaptic neuron's threshold may be reached, and a nerve impulse will be triggered. Conversely, if most of the neurotransmitters released are inhibitory, threshold may not be reached.

Neurotransmitters

About fifty types of neurotransmitters have been identified in the nervous system. Some neurons release only one type, while others produce two or three kinds. The neurotransmitters include *acetylcholine,* which stimulates skeletal muscle contractions (see chapter 8, p. 181); a group of compounds called *monoamines* (such as epinephrine, norepinephrine, dopamine, and serotonin), which form from modified amino acids; several *amino acids* (such as glycine, glutamic acid, aspartic acid, and gamma-aminobutyric acid—GABA); and a large group of *neuropeptides,* which are short chains of amino acids. Acetylcholine and norepinephrine are excitatory. Dopamine, GABA, and glycine are inhibitory. Neurotransmitters are usually synthesized in the cytoplasm of the synaptic knobs and stored in the synaptic vesicles. Some neurotransmitters and their actions are listed in table 9.2.

When an action potential reaches the membrane of a synaptic knob, it increases the membrane's permeability to calcium ions by opening calcium ion channels in the membrane. Consequently, calcium ions diffuse inward, and in response, some synaptic vesicles fuse with the membrane and release their contents into the synaptic cleft. A vesicle that has released its neurotransmitter eventually breaks away from the membrane and reenters the cytoplasm, where it can pick up more neurotransmitter.

After being released, some neurotransmitters are decomposed by enzymes. For example, the enzyme *acetylcholinesterase* decomposes acetylcholine and is present in the synapse and on the postsynaptic membrane of neuromuscular junctions, which control skeletal muscle contraction. Other neurotransmitters are transported back into the synaptic knob that released them (reuptake) or into nearby neurons or neuroglial cells. Decomposition or removal of neurotransmitters prevents continuous stimulation of postsynaptic neurons. Table 9.3 summarizes the events leading to the release of a neurotransmitter.

Check Your Recall

20. Distinguish between the actions of excitatory and inhibitory neurotransmitters.
21. What types of chemicals function as neurotransmitters?
22. What are possible fates of neurotransmitters?

Table 9.2	Some Neurotransmitters and Representative Actions	
Neurotransmitter	**Location**	**Major Actions**
Acetylcholine	CNS	Controls skeletal muscle actions
	PNS	Stimulates skeletal muscle contraction at neuromuscular junctions. May excite or inhibit at autonomic nervous system synapses
Monoamines		
Norepinephrine	CNS	Creates a sense of feeling good; low levels may lead to depression
	PNS	May excite or inhibit autonomic nervous system actions, depending on receptors
Dopamine	CNS	Creates a sense of feeling good; deficiency in some brain areas is associated with Parkinson disease
	PNS	Limited actions in autonomic nervous system; may excite or inhibit, depending on receptors
Serotonin	CNS	Primarily inhibitory; leads to sleepiness; action is blocked by LSD, enhanced by selective serotonin reuptake inhibitor drugs
Histamine	CNS	Release in hypothalamus promotes alertness
Amino acids		
GABA	CNS	Generally inhibitory
Glutamic acid	CNS	Generally excitatory
Neuropeptides		
Substance P	PNS	Excitatory; pain perception
Endorphins, enkephalins	CNS	Generally inhibitory; reduce pain by inhibiting substance P release
Gases		
Nitric oxide	PNS	Vasodilation
	CNS	May play a role in memory

Table 9.3	Events Leading to the Release of a Neurotransmitter

1. Action potential passes along an axon and over the surface of its synaptic knob.

2. Synaptic knob membrane becomes more permeable to calcium ions, and they diffuse inward.

3. In the presence of calcium ions, synaptic vesicles fuse to synaptic knob membrane.

4. Synaptic vesicles release their neurotransmitter into synaptic cleft.

5. Synaptic vesicles reenter cytoplasm of axon and pick up more neurotransmitter.

9.9 IMPULSE PROCESSING

The way the nervous system processes and responds to nerve impulses reflects, in part, the organization of neurons and their axons within the brain and spinal cord.

Neuronal Pools

Neurons within the CNS are organized into **neuronal pools.** These are groups of neurons that make hundreds of synaptic connections with each other and work together to perform a common function. Each pool receives input from neurons (which may be part of other pools), and each pool generates output. Neuronal pools may have excitatory or inhibitory effects on other pools or on peripheral effectors.

Facilitation

As a result of incoming impulses and neurotransmitter release, a particular neuron of a neuronal pool may receive excitatory and inhibitory input. If the net effect of the input is excitatory, threshold may be reached, and an outgoing impulse triggered. If the net effect is excitatory but subthreshold, an impulse is not triggered, but the neuron is more excitable to incoming stimulation than before, a state called **facilitation** (fah-sil″ĭ-ta′shun).

Convergence

Any single neuron in a neuronal pool may receive impulses from two or more incoming axons. Axons originating from different parts of the nervous system and leading to the same neuron exhibit **convergence** (kon-ver′jens) (fig. 9.16*a*).

Convergence makes it possible for impulses arriving from different sources to have an additive effect on a neuron. For example, if a neuron is facilitated by receiving subthreshold stimulation from one input neuron,

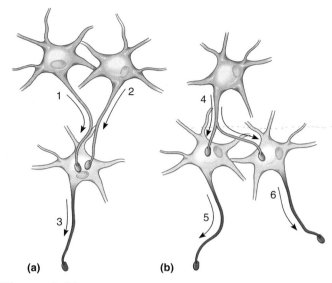

Figure 9.16

Impulse processing in neuronal pools. (*a*) Axons of neurons 1 and 2 converge to the cell body of neuron 3. (*b*) The axon of neuron 4 diverges to the cell bodies of neurons 5 and 6.

it may reach threshold if it receives additional stimulation from a second input neuron. As a result, a nerve impulse may travel to a particular effector and evoke a response.

Incoming impulses often bring information from several sensory receptors that detect changes. Convergence allows the nervous system to collect a variety of kinds of information, process it, and respond to it in a special way.

Divergence

Impulses leaving a neuron of a neuronal pool often exhibit **divergence** (di-ver′jens) by passing into several other output neurons (fig. 9.16*b*). For example, an impulse from one neuron may stimulate two others; each of these, in turn, may stimulate several others, and so forth. Divergence can amplify an impulse—that is, spread it to more neurons within the pool. As a result of divergence, an impulse originating from a single neuron in the CNS may be amplified so that impulses reach enough motor units within a skeletal muscle to cause forceful contraction (see chapter 8, p. 188). Similarly, an impulse originating from a sensory receptor may diverge and reach several different regions of the CNS, where the resulting impulses are processed and acted upon.

Check Your Recall

23. Define *neuronal pool.*

24. Distinguish between convergence and divergence.

9.10 TYPES OF NERVES

Recall from section 9.1 that nerves are bundles of axons. An axon is often referred to as a nerve fiber. Because of this, we will refer to the neuron processes that bring sensory information into the CNS as **sensory fibers,** or **afferent fibers.** In contrast, **motor fibers** or **efferent fibers** carry impulses from the CNS to effectors (muscles or glands). A nerve is a cordlike bundle (or group of bundles) of nerve fibers within layers of connective tissue (fig. 9.17).

> The terminology used to describe muscle and nerve fibers is somewhat inconsistent. "Muscle fiber" refers to a muscle cell, whereas "nerve fiber" refers to an axon, which is part of a cell. However, names for the associated connective tissues are similar. Both muscle and nerve fibers are bundled into fascicles. Recall from figure 8.1 that epimysium and perimysium connective tissue separates muscle tissue into compartments. Similarly, a nerve is defined by an outer *epineurium,* with *perineurium* surrounding a nerve fascicle within the nerve, and *endoneurium* surrounding an individual nerve fiber.

Like neurons, nerves that conduct impulses to the brain or spinal cord are called **sensory nerves,** and those that carry impulses to muscles or glands are termed **motor nerves.** Most nerves include both sensory and motor fibers and are called **mixed nerves.**

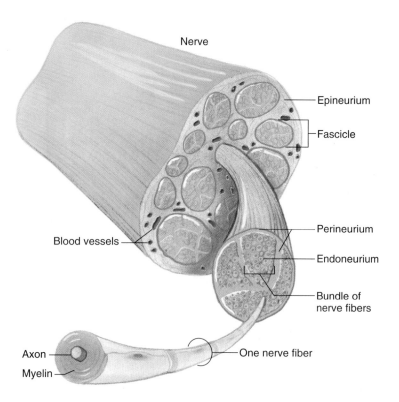

Figure 9.17
Connective tissue binds a bundle of nerve fibers, forming a fascicle. Many fascicles form a nerve.

Labels: Nerve; Epineurium; Fascicle; Perineurium; Endoneurium; Bundle of nerve fibers; Blood vessels; One nerve fiber; Axon; Myelin

Check Your Recall

25. What is a nerve?

26. How does a mixed nerve differ from a sensory nerve? From a motor nerve?

9.11 NERVE PATHWAYS

Recall from section 9.5 that the routes nerve impulses follow as they travel through the nervous system are called *nerve pathways.* The simplest of these pathways includes only a few neurons and is called a **reflex** (re′fleks) **arc.** It constitutes the structural and functional basis for involuntary actions called **reflexes.**

Reflex Arcs

A reflex arc begins with a receptor at the end of a sensory (or afferent) neuron. This neuron usually leads to several interneurons within the CNS, which serve as a processing center, or *reflex center.* These interneurons can connect with interneurons in other parts of the nervous system. They also communicate with motor (or efferent) neurons, whose axons pass outward from the CNS to effectors, usually muscles or glands (fig. 9.18).

Reflex Behavior

Reflexes are automatic responses to changes (stimuli) within or outside the body. They help maintain homeostasis by controlling many involuntary processes, such as heart rate, breathing rate, blood pressure, and digestion. Reflexes also carry out the automatic actions of swallowing, sneezing, coughing, and vomiting.

The *patellar reflex* (knee-jerk reflex) is an example of a simple reflex involving a pathway of only two neurons—a sensory neuron communicating directly with a motor neuron. Striking the patellar ligament just below

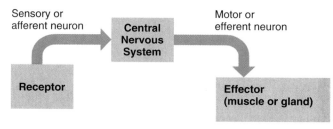

Labels: Sensory or afferent neuron; Central Nervous System; Motor or efferent neuron; Receptor; Effector (muscle or gland)

Figure 9.18
A reflex arc is the simplest nerve pathway. It involves a sensory neuron that sends a message to the CNS, and a motor neuron that sends the message from the CNS to a muscle or gland.

the patella initiates this reflex. The quadriceps femoris muscle group, which is attached to the patella by a tendon, is pulled slightly, stimulating stretch receptors in these muscles. These receptors, in turn, trigger impulses that pass along the axon of a sensory neuron into the spinal cord. Within the spinal cord, the sensory axon makes a synapse with a motor neuron. An impulse is then triggered along the axon of the motor neuron and travels back to the quadriceps femoris group. The muscle group contracts in response, and the reflex is completed as the leg extends (fig. 9.19).

The patellar reflex helps maintain upright posture. If the knee begins to bend from the force of gravity when a person is standing still, the quadriceps femoris group is stretched, the reflex is triggered, and the leg straightens again.

Another type of reflex, called a *withdrawal reflex,* occurs when a person unexpectedly touches a body part to something painful, such as stepping on a tack. This activates skin receptors and sends sensory impulses to the spinal cord. There, the impulses pass to the interneurons of a reflex center and are directed to motor neurons. The motor neurons transmit signals to flexor muscles in the injured part, and the muscles contract in response. At the same time, the antagonistic extensor muscles are inhibited, and the foot is rapidly and unconsciously withdrawn from the painful stimulus. Concurrent with the withdrawal reflex, other interneurons carry sensory impulses to the brain and the person becomes aware of the experience and may feel pain (fig. 9.20). A withdrawal reflex is protective because it may limit tissue damage caused by touching something harmful. Table 9.4 summarizes the parts of a reflex arc.

> Because normal reflexes depend on normal neuron functions, reflexes provide information about the condition of the nervous system. For instance, an anesthesiologist may try to initiate a reflex in a patient being anesthetized to determine how well the anesthetic drug is affecting nerve functions. A neurologist may test reflexes when nervous system injury has occurred to determine the location and extent of damage.

Check Your Recall

27. What is a nerve pathway?
28. List the parts of a reflex arc.
29. Define *reflex*.
30. List the actions that occur during a withdrawal reflex.

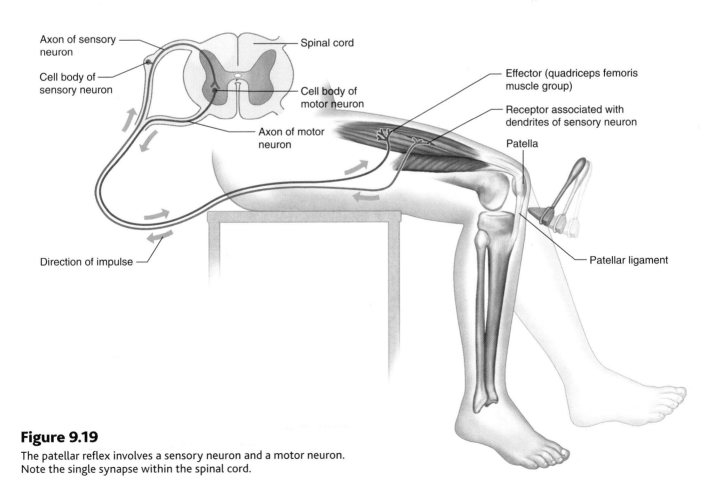

Figure 9.19
The patellar reflex involves a sensory neuron and a motor neuron. Note the single synapse within the spinal cord.

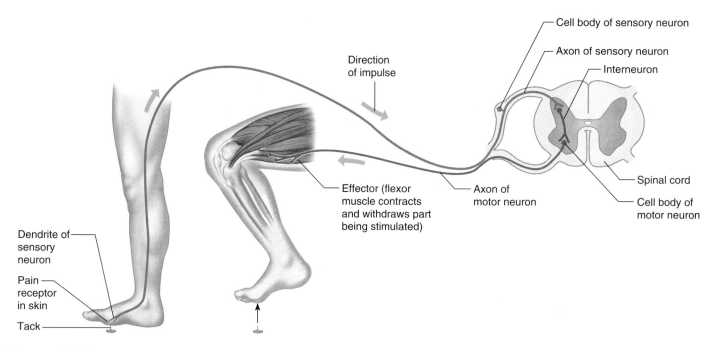

Figure 9.20

A withdrawal reflex involves a sensory neuron, an interneuron, and a motor neuron.

Table 9.4	Parts of a Reflex Arc	
Part	**Description**	**Function**
Receptor	Receptor end of a dendrite or a specialized receptor cell in a sensory organ	Senses specific type of internal or external change
Sensory neuron	Dendrite, cell body, and axon of a sensory neuron	Transmits nerve impulse from receptor into brain or spinal cord
Interneuron	Dendrite, cell body, and axon of a neuron within the brain or spinal cord	Conducts nerve impulse from sensory neuron to motor neuron
Motor neuron	Dendrite, cell body, and axon of a motor neuron	Transmits nerve impulse from brain or spinal cord out to effector
Effector	Muscle or gland	Responds to stimulation by motor neuron and produces reflex or behavioral action

9.12 MENINGES

Bones, membranes, and fluid surround the organs of the CNS. The brain lies within the cranial cavity of the skull, and the spinal cord occupies the vertebral canal within the vertebral column. Layered membranes called **meninges** (mĕ-nin′jēz) (singular, *meninx*) lie between these bony coverings and the soft tissues of the CNS, protecting the brain and spinal cord (fig. 9.21*a*).

The meninges have three layers—dura mater, arachnoid mater, and pia mater (fig. 9.21*b*). The **dura mater** (du′rah ma′ter) is the outermost layer of the meninges. It is composed primarily of tough, white, fibrous connective tissue and contains many blood vessels and nerves. It attaches to the inside of the cranial cavity and forms the internal periosteum of the surrounding skull bones. In some regions, the dura mater extends inward between lobes of the brain and forms partitions that support and protect these parts.

The dura mater continues into the vertebral canal as a strong, tubular sheath that surrounds the spinal cord. It terminates as a blind sac below the end of the cord. The membrane around the spinal cord is not attached directly to the vertebrae but is separated by an *epidural space,* which lies between the dural sheath and the bony walls (fig. 9.22). This space contains loose connective and adipose tissues, which pad the spinal cord.

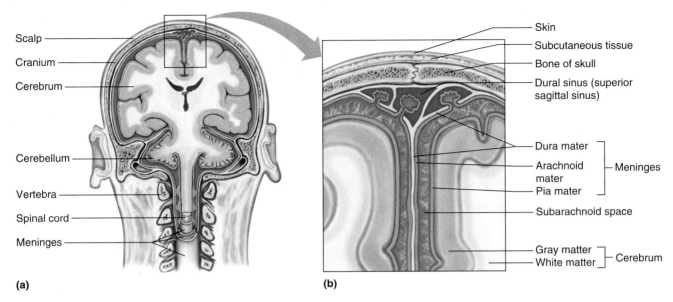

Figure 9.21

Meninges. (*a*) Membranes called meninges enclose the brain and spinal cord. (*b*) The meninges include three layers: dura mater, arachnoid mater, and pia mater.

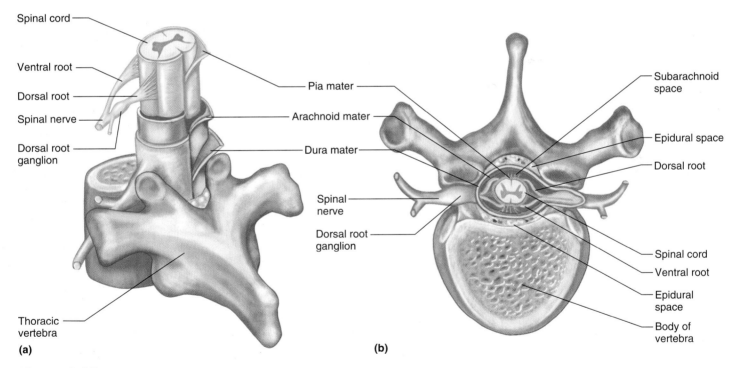

Figure 9.22

Meninges of the spinal cord. (*a*) The dura mater ensheaths the spinal cord. (*b*) Tissues forming a protective pad around the cord fill the epidural space between the dural sheath and the bone of the vertebra.

The **arachnoid mater** is a thin, weblike membrane without blood vessels that lies between the dura and pia maters. Between the arachnoid and pia maters is a *subarachnoid space* that contains the clear, watery **cerebrospinal fluid (CSF).** The **pia mater** (pi′ah ma′ter) is very thin and contains many nerves and blood vessels that nourish underlying cells of the brain and spinal cord. This layer hugs the surfaces of these organs and follows their irregular contours, passing over high areas and dipping into depressions.

A blow to the head may break some blood vessels associated with the brain, and escaping blood may collect beneath the dura mater. Such a *subdural hematoma* increases pressure between the rigid bones of the skull and the soft tissues of the brain. Unless the accumulating blood is evacuated, compression of the brain may lead to functional losses or even death.

Check Your Recall

31. Describe the meninges.
32. State the location of cerebrospinal fluid.

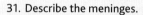

9.13 SPINAL CORD

The **spinal cord** is a slender nerve column that passes downward from the brain into the vertebral canal. Although continuous with the brain, the spinal cord begins where nervous tissue leaves the cranial cavity at the level of the foramen magnum. The spinal cord tapers to a point and terminates near the intervertebral disc that separates the first and second lumbar vertebrae (fig. 9.23).

Structure of the Spinal Cord

The spinal cord consists of thirty-one segments, each of which gives rise to a pair of **spinal nerves.** These nerves branch to various body parts and connect them with the CNS (see fig. 9.35).

In the neck region, a thickening in the spinal cord, called the *cervical enlargement,* supplies nerves to the upper limbs. A similar thickening in the lower back, the *lumbar enlargement,* gives off nerves to the lower limbs (fig. 9.23).

Two grooves, a deep *anterior median fissure* and a shallow *posterior median sulcus,* extend the length of the spinal cord, dividing it into right and left halves (fig. 9.24). A cross section of the cord reveals a core of gray matter within white matter. The pattern of gray matter roughly resembles a butterfly with its wings spread. The upper and lower wings of gray matter are called the *posterior horns* and *anterior horns,* respectively. Between them on either side in the thoracic and upper lumbar segments is a protrusion of gray matter called the *lateral horn.*

Neurons with large cell bodies located in the anterior horns give rise to motor fibers that pass out through spinal nerves to skeletal muscles. However, the majority of neurons in the gray matter of the spinal cord are interneurons.

Gray matter divides the white matter of the spinal cord into three regions on each side—the *anterior, lat-*

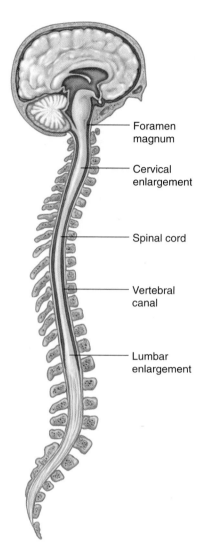

Foramen magnum

Cervical enlargement

Spinal cord

Vertebral canal

Lumbar enlargement

Figure 9.23

The spinal cord begins at the level of the foramen magnum and ends near the intervertebral disc between the first and second lumbar vertebrae.

eral, and *posterior funiculi* (fig. 9.24*a*). Each funiculus consists of longitudinal bundles of myelinated nerve fibers that comprise major nerve pathways called **nerve tracts.**

A horizontal bar of gray matter in the middle of the spinal cord, the *gray commissure,* connects the wings of the gray matter on the right and left sides. This bar surrounds the **central canal,** which contains cerebrospinal fluid.

Functions of the Spinal Cord

The spinal cord has two major functions—conducting nerve impulses and serving as a center for spinal reflexes. The nerve tracts of the spinal cord consist of axons that provide a two-way communication system between the brain and the body parts outside the nervous system. The tracts that carry sensory information to

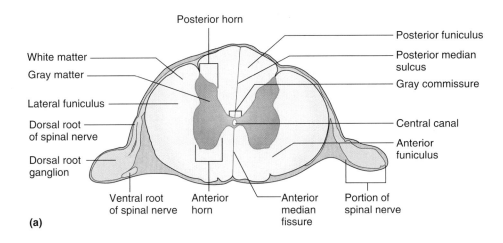

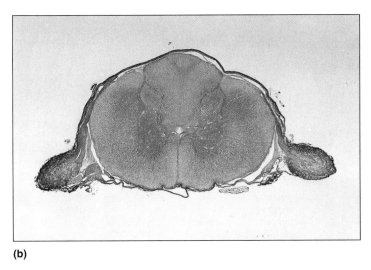

(b)

Figure 9.24

The spinal cord. (*a*) A cross section of the spinal cord. (*b*) Identify the parts of the spinal cord in this micrograph (7.5×).

the brain are called **ascending tracts** (fig. 9.25); those that conduct motor impulses from the brain to muscles and glands are called **descending tracts** (fig. 9.26).

Typically, all the axons within a given tract originate from neuron cell bodies located in the same part of the nervous system and terminate together in some other part. The names that identify nerve tracts often reflect these common origins and terminations. For example, a *spinothalamic tract* begins in the spinal cord and carries sensory impulses associated with the sensations of pain, touch, and temperature to the thalamus of the brain. A *corticospinal tract* originates in the cortex of the brain and carries motor impulses downward through the spinal cord and spinal nerves. These impulses control skeletal muscle movements.

Corticospinal tracts are also called *pyramidal tracts* after the pyramid-shaped areas in the medulla oblongata of the brain through which they pass. Other descending tracts, called *extrapyramidal tracts,* control motor activities associated with maintaining balance and posture.

In addition to providing a pathway for nerve tracts, the spinal cord functions in many reflexes, including the patellar and withdrawal reflexes described previously. These are called **spinal reflexes** because their reflex arcs pass through the spinal cord.

Some axons extend from the base of the spinal cord to the toes. If you stub your toe, a sensory message reaches the spinal cord in less than one-hundredth of a second.

Check Your Recall

33. Describe the structure of the spinal cord.

34. Describe the general functions of the spinal cord.

35. Distinguish between an ascending and a descending tract.

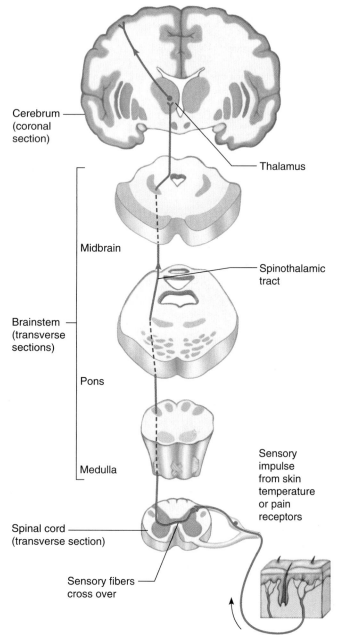

Figure 9.25

Ascending tracts. Sensory impulses originating in skin receptors cross over in the spinal cord and ascend to the brain. Other sensory tracts cross over in the medulla oblongata.

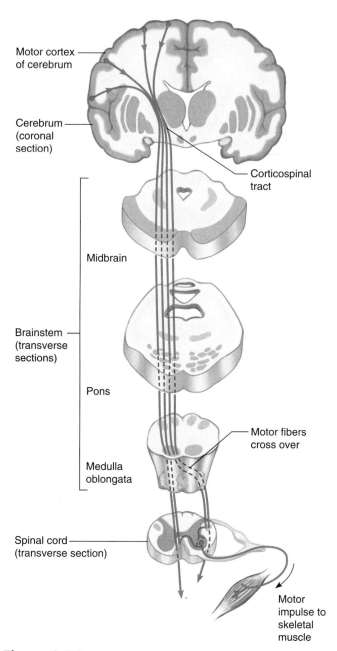

Figure 9.26

Descending tracts. Motor fibers of the corticospinal tract begin in the cerebral cortex, cross over in the medulla oblongata, and descend in the spinal cord. There, they synapse with neurons whose fibers lead to the spinal nerves that supply skeletal muscles.

9.14 BRAIN

The **brain** is composed of about 100 billion (10^{11}) multipolar neurons which communicate with one another and with neurons in other parts of the nervous system. As figure 9.27 shows, the brain can be divided into four major portions—the cerebrum, the diencephalon, the brainstem, and the cerebellum. The *cerebrum,* the largest part, includes nerve centers associated with sensory and motor functions and provides higher mental functions, including memory and reasoning. The *diencephalon* also processes sensory information. Nerve pathways in the *brainstem* connect various parts of the nervous system and regulate certain visceral activities. The *cerebellum* includes centers that coordinate voluntary muscular movements.

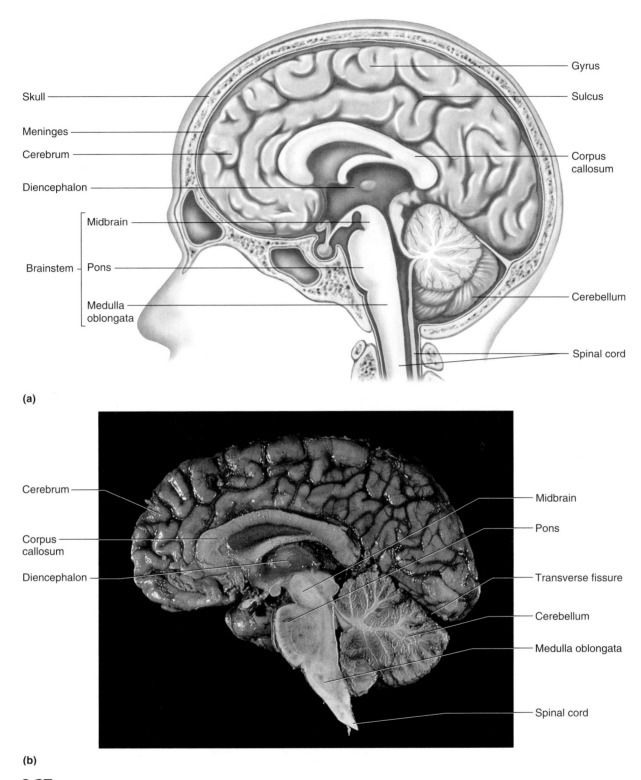

(a)

(b)

Figure 9.27

The major portions of the brain are the cerebrum, the diencephalon, the brainstem, and the cerebellum.

Structure of the Cerebrum

The **cerebrum** (ser'ĕ-brum) consists of two large masses called the left and right **cerebral hemispheres** (ser″ĕ-bral hem'ĭ-sfērz), which are essentially mirror images of each other. A deep bridge of nerve fibers called the **corpus callosum** (kor'pus kah-lo'sum) connects the cerebral hemispheres. A layer of dura mater (falx cerebri) separates them.

The surface of the cerebrum has many ridges (convolutions) or **gyri** (ji'ri), singular *gyrus*, separated by grooves. A shallow groove is called a **sulcus** (sul′kus), and a deep groove is called a **fissure.** Although the structural organization of these elevations and depressions is complex, they form distinct patterns in all normal brains. For example, a *longitudinal fissure* separates the right and left cerebral hemispheres, a *transverse fissure* separates the cerebrum from the cerebellum, and several sulci divide each hemisphere into lobes.

The lobes of the cerebral hemispheres are named after the skull bones they underlie (fig. 9.28). They include:

1. **Frontal lobe** The frontal lobe forms the anterior portion of each cerebral hemisphere. It is bordered posteriorly by a *central sulcus,* which extends from the longitudinal fissure at a right angle, and inferiorly by a *lateral sulcus,* which extends from the undersurface of the brain along its sides.
2. **Parietal lobe** The parietal lobe is posterior to the frontal lobe and separated from it by the central sulcus.
3. **Temporal lobe** The temporal lobe lies below the frontal and parietal lobes and is separated from them by the lateral sulcus.
4. **Occipital lobe** The occipital lobe forms the posterior portion of each cerebral hemisphere and is separated from the cerebellum by a shelflike extension of dura mater (tentorium cerebelli). The boundary between the occipital lobe and the parietal and temporal lobes is not distinct.

5. **Insula** (in′su-lah) The insula is located deep within the lateral sulcus and is covered by parts of the frontal, parietal, and temporal lobes. A *circular sulcus* separates the insula from the other lobes.

A thin layer of gray matter called the **cerebral cortex** (ser″ĕ-bral kor′teks) is the outermost portion of the cerebrum. This layer covers the gyri and dips into the sulci and fissures. It contains nearly 75% of all the neuron cell bodies in the nervous system.

Just beneath the cerebral cortex is a mass of white matter that makes up the bulk of the cerebrum. This mass contains bundles of myelinated axons that connect neuron cell bodies of the cortex with other parts of the nervous system. Some of these fibers pass from one cerebral hemisphere to the other by way of the corpus callosum, and others carry sensory or motor impulses from portions of the cortex to nerve centers in the brain or spinal cord.

Functions of the Cerebrum

The cerebrum provides higher brain functions. It has centers for interpreting sensory impulses arriving from sense organs and centers for initiating voluntary muscular movements. The cerebrum stores the information that comprises memory and utilizes it to reason. Intelligence and personality also stem from cerebral activity.

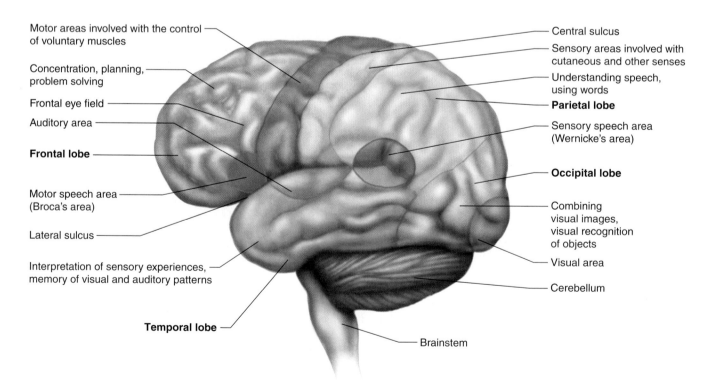

Motor areas involved with the control of voluntary muscles

Concentration, planning, problem solving

Frontal eye field

Auditory area

Frontal lobe

Motor speech area (Broca's area)

Lateral sulcus

Interpretation of sensory experiences, memory of visual and auditory patterns

Temporal lobe

Central sulcus

Sensory areas involved with cutaneous and other senses

Understanding speech, using words

Parietal lobe

Sensory speech area (Wernicke's area)

Occipital lobe

Combining visual images, visual recognition of objects

Visual area

Cerebellum

Brainstem

Figure 9.28

The cerebral cortex. Major portions of the cortex, called lobes, are named for the skull bones they lie beneath. The figure also depicts motor, sensory, and association areas of the left cerebral cortex.

Functional Regions of the Cerebral Cortex

Specific regions of the cerebral cortex perform specific functions. Although functions overlap among regions, the cortex can be divided into motor, sensory, and association areas.

The primary **motor areas** of the cerebral cortex lie in the frontal lobes, just in front of the central sulcus (fig. 9.28). The nervous tissue in these regions contains many large *pyramidal cells,* named for their pyramid-shaped cell bodies. These cells are also termed *upper motor neurons,* because of their location.

Impulses from the pyramidal cells travel downward through the brainstem and into the spinal cord on the corticospinal tracts (see fig. 9.26). Here they form synapses with *lower motor neurons* whose axons leave the spinal cord and reach skeletal muscle fibers. Most of the axons in these tracts cross over from one side of the brain to the other within the brainstem. As a result, the motor area of the right cerebral hemisphere generally controls skeletal muscles on the left side of the body, and vice versa.

In addition to the primary motor areas, certain other regions of the frontal lobe affect motor functions. For example, anterior and lateral to the primary motor cortex is a region called the *frontal eye field.* The motor cortex in this area controls voluntary movements of the eyes and eyelids. Another region just in front of the primary motor area controls the muscular movements of the hands and fingers that make skills such as writing possible.

Sensory areas located in several lobes of the cerebrum interpret impulses that arrive from sensory receptors, producing feelings or sensations. For example, sensations from all parts of the skin (cutaneous senses) arise in the anterior portions of the parietal lobes along the central sulcus (fig. 9.28). The posterior parts of the occipital lobes affect vision (visual area), and the temporal lobes contain the centers for hearing (auditory area). The sensory areas for taste are located near the bases of the central sulci along the lateral sulci, and the sense of smell arises from centers deep within the cerebrum.

Like motor fibers, sensory fibers cross over either in the spinal cord or in the brainstem (see fig. 9.25). Thus, the centers in the right cerebral hemisphere interpret impulses originating from the left side of the body, and vice versa.

Association areas are neither primarily sensory nor primarily motor. They connect with one another and with other brain structures. These areas analyze and interpret sensory experiences and oversee memory, reasoning, verbalizing, judgment, and emotion. Association areas occupy the anterior portions of the frontal lobes and are widespread in the lateral portions of the parietal, temporal, and occipital lobes (fig. 9.28).

Association areas of the frontal lobes control a number of higher intellectual processes. These include concentrating, planning, complex problem solving, and judging the possible consequences of behavior. Association areas of the parietal lobes help in understanding speech and choosing words to express thoughts and feelings. Two connected areas that accomplish this are the *sensory speech area* or Wernicke's (ver'nĭ-kēz) area and the *motor speech area* or Broca's (bro'kahz) area. In most people these are found in the left hemisphere. The sensory speech area is located in the parietal lobe near the temporal lobe, just posterior to the lateral sulcus. This area receives input from both the visual cortex and auditory cortex, and is necessary for understanding written and spoken language. The motor speech area is in the frontal lobe, just anterior to the primary motor cortex and superior to the lateral sulcus. This area generates the movements of muscles necessary for speech (fig. 9.28).

These regions also provide memory of visual scenes, music, and other complex sensory patterns. Association areas of the occipital lobes that are adjacent to the visual centers are important in analyzing visual patterns and combining visual images with other sensory experiences, as when one recognizes another person or an object.

Check Your Recall

36. List the major divisions of the brain.

37. Describe the cerebral cortex.

38. Describe the major functions of the cerebrum.

39. Locate the major functional regions of the cerebral cortex.

The effects of injuries to the cerebral cortex depend on the location and extent of the damage. The abilities that become impaired can indicate the site of damage. For example, injury to the motor areas of one frontal lobe causes partial or complete paralysis on the opposite side of the body.

A person with damage to the association areas of the frontal lobe may have difficulty concentrating on complex mental tasks and may appear disorganized and easily distracted. A person who suffers damage to association areas of the temporal lobes may have trouble recognizing printed words or arranging words into meaningful thoughts.

Hemisphere Dominance

Both cerebral hemispheres participate in basic functions, such as receiving and analyzing sensory impulses, controlling skeletal muscles, and storing memory. However, in most persons, one side of the cerebrum is the **dominant hemisphere,** controlling the ability to use and understand language.

In most people the left hemisphere is dominant for the language-related activities of speech, writing, and reading, and for complex intellectual functions requiring verbal, analytical, and computational skills. In other persons, the right hemisphere is dominant for language-related abilities, or the hemispheres are equally dominant. Broca's area in the dominant hemisphere controls the muscles that function in speaking.

In addition to carrying on basic functions, the non-dominant hemisphere specializes in nonverbal functions, such as motor tasks that require orientation of the body in space, understanding and interpreting musical patterns, and nonverbal visual experiences. The nondominant hemisphere also controls emotional and intuitive thinking.

Nerve fibers of the corpus callosum, which connect the cerebral hemispheres, allow the dominant hemisphere to control the motor cortex of the nondominant hemisphere (see fig. 9.27). These fibers also transfer sensory information reaching the nondominant hemisphere to the dominant one, where the information can be used in decision making.

Deep within each cerebral hemisphere are several masses of gray matter called **basal nuclei** (basal ganglia) (fig. 9.29). They are the *caudate nucleus,* the *putamen,* and the *globus pallidus.* Their neuron cell bodies serve as relay stations for motor impulses originating in the cerebral cortex and passing into the brainstem and spinal cord. The basal nuclei modify the pattern of these motor impulses and thereby help control various skeletal muscle activities. Neurons of the basal nuclei

respond to the inhibitory neurotransmitter dopamine released from nearby cells.

The signs of Parkinson disease and Huntington disease result from altered activity of basal nuclei neurons. In Parkinson disease, nearby neurons release less dopamine, and the basal nuclei become overactive, inhibiting movement. In Huntington disease, basal nuclei neurons gradually deteriorate, resulting in unrestrained movement.

Ventricles and Cerebrospinal Fluid

Within the cerebral hemispheres and brainstem is a series of interconnected cavities called **ventricles** (fig. 9.30). These spaces are continuous with the central canal of the spinal cord, and like it, they contain cerebrospinal fluid.

The largest ventricles are the *lateral ventricles* (first and second ventricles), which extend into the cerebral hemispheres and occupy portions of the frontal, temporal, and occipital lobes. A narrow space that constitutes the *third ventricle* is in the midline of the brain, beneath the corpus callosum. This ventricle communicates with the lateral ventricles through openings (interventricular foramina) in its anterior end. The *fourth ventricle* is in the brainstem just anterior to the cerebellum. A narrow canal, the *cerebral aqueduct,* connects it to the third ventricle and passes lengthwise through the brainstem. The fourth ventricle is continuous with

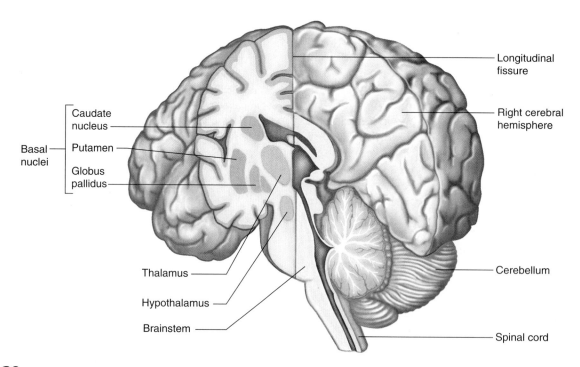

Figure 9.29

A coronal section of the left cerebral hemisphere reveals some of the basal nuclei.

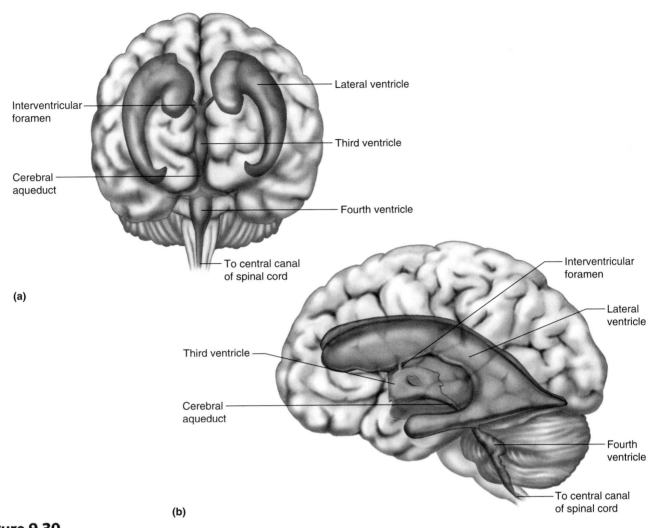

Figure 9.30

Ventricles in the brain. (*a*) Anterior view of the ventricles within the cerebral hemispheres and brainstem. (*b*) Lateral view.

the central canal of the spinal cord and has openings in its roof that lead into the subarachnoid space of the meninges.

Tiny, reddish, cauliflower-like masses of specialized capillaries from the pia mater, called **choroid plexuses** (plek'sus-ez), secrete cerebrospinal fluid (fig. 9.31). These structures project into the ventricles. Most of the cerebrospinal fluid is formed in the lateral ventricles. From there, it circulates slowly into the third and fourth ventricles and into the central canal of the spinal cord. Cerebrospinal fluid also enters the subarachnoid space of the meninges through the wall of the fourth ventricle near the cerebellum and completes its circuit by being reabsorbed into the blood.

Cerebrospinal fluid completely surrounds the brain and spinal cord because it occupies the subarachnoid space of the meninges. In effect, these organs float in the fluid, which supports and protects them by absorbing forces that might otherwise jar and damage them. Cerebrospinal fluid also maintains a stable ionic concentration in the CNS and provides a pathway to the blood for wastes.

Because cerebrospinal fluid is secreted and reabsorbed continuously, the fluid pressure in the ventricles normally remains relatively constant. An infection, a tumor, or a blood clot can interfere with fluid circulation, increasing pressure within the ventricles and thus in the cranial cavity (intracranial pressure). This can injure the brain by forcing it against the rigid skull.

A *lumbar puncture* (spinal tap) is used to measure the pressure of cerebrospinal fluid. In this procedure, a fine, hollow needle is inserted into the subarachnoid space between the third and fourth or between the fourth and fifth lumbar vertebrae. An instrument called a *manometer* then measures the pressure.

Check Your Recall

40. What is hemisphere dominance?

41. What are the major functions of the dominant hemisphere? The nondominant one?

42. Where are the ventricles of the brain?

43. Describe the circulation of cerebrospinal fluid.

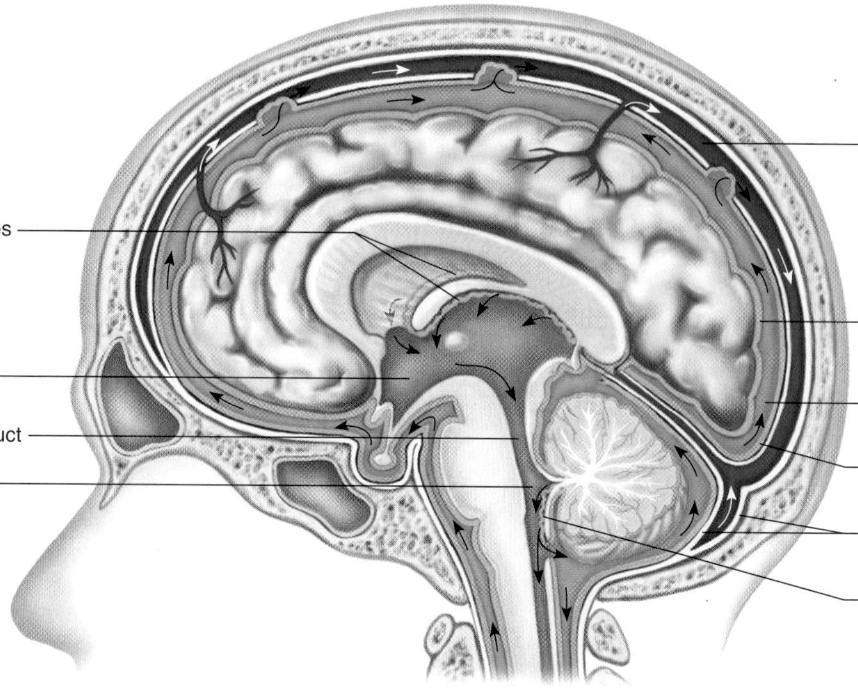

Figure 9.31

The choroid plexuses in the walls of the ventricles secrete cerebrospinal fluid. The fluid circulates through the ventricles and central canal, enters the subarachnoid space, and is reabsorbed into the blood.

Diencephalon

The **diencephalon** (di″en-sef′ah-lon) is located between the cerebral hemispheres and above the midbrain. It surrounds the third ventricle and is composed largely of gray matter. Within the diencephalon, a dense mass called the **thalamus** bulges into the third ventricle from each side (see figs. 9.29 and 9.32*b*). Another region of the diencephalon that includes many nuclei (masses of gray matter) is the **hypothalamus.** It lies below the thalamus and forms the lower walls and floor of the third ventricle.

Other parts of the diencephalon include: (1) the **optic tracts** and the **optic chiasma** that is formed by optic nerve fibers crossing over each other; (2) the **infundibulum,** a conical process behind the optic chiasma to which the pituitary gland attaches; (3) the **posterior pituitary gland,** which hangs from the floor of the hypothalamus; (4) the **mammillary bodies,** which appear as two rounded structures behind the infundibulum; and (5) the **pineal gland** (pin′e-al gland), a cone-shaped structure attached to the upper portion of the diencephalon (see chapter 11, p. 307).

The thalamus is a central relay station for sensory impulses ascending from other parts of the nervous system to the cerebral cortex. It receives all sensory impulses (except those associated with the sense of smell) and channels them to the appropriate regions of the cortex for interpretation. In addition, all regions of the cerebral cortex can communicate with the thalamus by means of descending fibers. The cerebral cortex pinpoints the ori-

gin of sensory stimulation, and the thalamus produces a general awareness of certain sensations, such as pain, touch, and temperature.

Nerve fibers connect the hypothalamus to the cerebral cortex, thalamus, and other parts of the brainstem. The hypothalamus maintains homeostasis by regulating a variety of visceral activities and by linking the nervous and endocrine systems. The hypothalamus regulates:

1. Heart rate and arterial blood pressure
2. Body temperature
3. Water and electrolyte balance
4. Control of hunger and body weight
5. Control of movements and glandular secretions of the stomach and intestines
6. Production of neurosecretory substances that stimulate the pituitary gland to secrete hormones
7. Sleep and wakefulness

Structures in the general region of the diencephalon also control emotional responses. For example, portions of the cerebral cortex in the medial parts of the frontal and temporal lobes interconnect with a number of deep masses of gray matter, including the hypothalamus, thalamus, and basal nuclei. Together, these structures comprise a complex called the **limbic system.**

The limbic system controls emotional experience and expression. It can modify the way a person acts by producing such feelings as fear, anger, pleasure, and sorrow. The limbic system recognizes upsets in a person's physical or psychological condition that might threaten

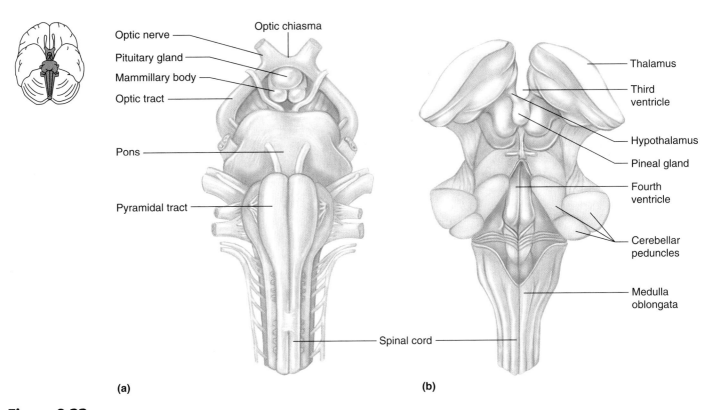

Optic nerve
Pituitary gland
Mammillary body
Optic tract
Optic chiasma
Pons
Pyramidal tract

Thalamus
Third ventricle
Hypothalamus
Pineal gland
Fourth ventricle
Cerebellar peduncles
Medulla oblongata

Spinal cord

(a)   (b)

Figure 9.32

The brainstem. (*a*) Ventral view of the brainstem. (*b*) Dorsal view of the brainstem with the cerebellum removed, exposing the fourth ventricle.

life. By causing pleasant or unpleasant feelings about experiences, the limbic system guides a person into behavior that is likely to increase the chance of survival.

Brainstem

The **brainstem** is a bundle of nervous tissue that connects the cerebrum to the spinal cord. It consists of numerous tracts of nerve fibers and several nuclei. The parts of the brainstem include the midbrain, pons, and medulla oblongata (figs. 9.27 and 9.32).

Midbrain

The **midbrain** is a short section of the brainstem between the diencephalon and the pons (see fig. 9.27). It contains bundles of myelinated axons that join lower parts of the brainstem and spinal cord with higher parts of the brain. Two prominent bundles of axons on the underside of the midbrain are the corticospinal tracts and are the main motor pathways between the cerebrum and lower parts of the nervous system.

The midbrain includes several masses of gray matter that serve as reflex centers. For example, the midbrain contains the centers for certain visual reflexes, such as those responsible for moving the eyes to view something as the head turns. It also contains the auditory reflex centers that enable a person to move the head to hear sounds more distinctly.

Pons

The **pons** (ponz) is a rounded bulge on the underside of the brainstem, where it separates the midbrain from the medulla oblongata (see fig. 9.27). The dorsal portion of the pons consists largely of longitudinal nerve fibers, which relay impulses to and from the medulla oblongata and the cerebrum. The ventral portion of the pons has large bundles of transverse nerve fibers, which transmit impulses from the cerebrum to centers within the cerebellum.

Several nuclei of the pons relay sensory impulses from peripheral nerves to higher brain centers. Other nuclei function with centers of the medulla oblongata to maintain the basic rhythm of breathing (see chapter 16, pp. 457–458).

Medulla Oblongata

The **medulla oblongata** (mĕ-dul′ah ob″long-gah′tah) extends from the pons to the foramen magnum of the skull (see fig. 9.27). Its dorsal surface flattens to form the floor of the fourth ventricle, and its ventral surface is marked by the corticospinal tracts, most of whose fibers cross over at this level (see fig. 9.26).

All of the ascending and descending nerve fibers connecting the brain and spinal cord must pass through the medulla oblongata because of its location. As in the spinal cord, the white matter of the medulla oblongata surrounds a central mass of gray matter. Here, however, nerve fibers separate the gray matter into nuclei, some of which relay ascending impulses to the other side of the brainstem and then on to higher brain centers. Other nuclei within the medulla oblongata control vital visceral activities. These centers include:

1. **Cardiac center** Impulses originating in the cardiac center are transmitted to the heart on peripheral nerves, altering heart rate.
2. **Vasomotor center** Certain cells of the vasomotor center initiate impulses that travel to smooth muscles in the walls of certain blood vessels and stimulate them to contract. This constricts the blood vessels (vasoconstriction), elevating blood pressure. Other cells of the vasomotor center produce the opposite effect—dilating blood vessels (vasodilation) and consequently dropping blood pressure.
3. **Respiratory center** The respiratory center adjusts the rate and depth of breathing and acts with the pons to maintain the basic rhythm of breathing.

Still other nuclei within the medulla oblongata are centers for the reflexes associated with coughing, sneezing, swallowing, and vomiting.

Reticular Formation

Scattered throughout the medulla oblongata, pons, and midbrain is a complex network of nerve fibers associated with tiny islands of gray matter. This network, the **reticular formation** (rĕ-tik′u-lar for-ma′shun) (reticular activating system), extends from the upper portion of the spinal cord into the diencephalon. Its nerve fibers join centers of the hypothalamus, basal nuclei, cerebellum, and cerebrum with fibers in all the major ascending and descending tracts.

When sensory impulses reach the reticular formation, it responds by activating the cerebral cortex into a state of wakefulness. Without this arousal, the cortex remains unaware of stimulation and cannot interpret sensory information or carry on thought processes. Thus, decreased activity in the reticular formation results in sleep. If the reticular formation is injured so that it cannot function, the person remains unconscious and cannot be aroused, even with strong stimulation. This is called a comatose state. Barbiturate drugs, which dampen CNS activity, affect the reticular formation (see the Topic of Interest on page 245).

Check Your Recall

44. What are the major functions of the thalamus? The hypothalamus?
45. How may the limbic system influence behavior?
46. List the structures of the brainstem.
47. What vital reflex centers are located in the brainstem?
48. What is the function of the reticular formation?

Cerebellum

The **cerebellum** (ser″ĕ-bel′um) is a large mass of tissue located below the occipital lobes of the cerebrum and posterior to the pons and medulla oblongata (see fig. 9.27). It consists of two lateral hemispheres partially separated by a layer of dura mater (falx cerebelli) and connected in the midline by a structure called the *vermis*. Like the cerebrum, the cerebellum is composed primarily of white matter, with a thin layer of gray matter, the **cerebellar cortex,** on its surface.

The cerebellum communicates with other parts of the CNS by means of three pairs of nerve tracts called *cerebellar peduncles* (figs. 9.32 and 9.33). One pair (the inferior peduncles) brings sensory information concerning the position of the limbs, joints, and other body parts to the cerebellum. Another pair (the middle peduncles) transmits signals from the cerebral cortex to the cerebellum concerning the desired positions of these parts. After integrating and analyzing this information, the cerebellum sends correcting impulses via a third pair (the superior peduncles) to the midbrain. These corrections are incorporated into motor impulses that travel downward through the pons, medulla oblongata, and spinal cord in the appropriate patterns to move the body in the desired way.

The cerebellum is a reflex center for integrating sensory information concerning the position of body parts and for coordinating complex skeletal muscle movements. It also helps maintain posture. Damage to the cerebellum is likely to result in tremors, inaccurate movements of voluntary muscles, loss of muscle tone, a reeling walk, and loss of equilibrium.

Check Your Recall

49. Where is the cerebellum located?
50. What are the major functions of the cerebellum?

9.15 PERIPHERAL NERVOUS SYSTEM

The peripheral nervous system (PNS) consists of nerves that branch out from the CNS and connect it to other body parts. The PNS includes the cranial nerves, which arise from the brain, and the spinal nerves, which arise from the spinal cord.

The PNS can also be subdivided into the somatic and autonomic nervous systems. Generally, the **somatic** (so-mat′ik) **nervous system** consists of the cranial and spinal nerve fibers that connect the CNS to the skin and skeletal muscles; it oversees conscious activities. The **autonomic** (aw″to-nom′ik) **nervous system** includes fibers that connect the CNS to viscera, such as the heart, stomach, intestines, and glands; it controls unconscious

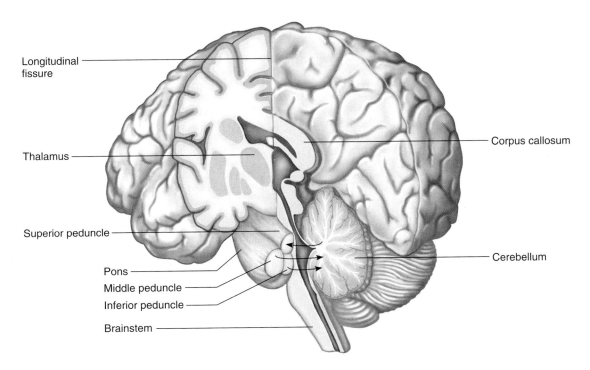

Figure 9.33
The cerebellum, which is located below the occipital lobes of the cerebrum, communicates with other parts of the nervous system by means of the cerebellar peduncles.

activities. Table 9.5 outlines the subdivisions of the nervous system (see fig. 9.2).

Cranial Nerves

Twelve pairs of **cranial nerves** arise from the underside of the brain (fig. 9.34). Except for the first pair, which begins in the cerebrum, these nerves originate from the brainstem. They pass from their sites of origin through foramina of the skull and lead to parts of the head, neck, and trunk.

Most of the cranial nerves are mixed nerves, but some of those associated with special senses, such as smell and vision, contain only sensory fibers. Other cranial nerves that affect muscles and glands are composed primarily of motor fibers.

Table 9.5	Subdivisions of the Nervous System

1. Central nervous system (CNS)
   a. Brain
   b. Spinal cord
2. Peripheral nervous system (PNS)
   a. Cranial nerves arising from the brain and brainstem
      (1) Somatic fibers connecting to skin and skeletal muscles
      (2) Autonomic fibers connecting to viscera
   b. Spinal nerves arising from the spinal cord
      (1) Somatic fibers connecting to skin and skeletal muscles
      (2) Autonomic fibers connecting to viscera

Sensory fibers present in the cranial nerves have neuron cell bodies that are outside the brain, usually in groups called *ganglia*. On the other hand, motor neuron cell bodies are typically located within the gray matter of the brain.

Numbers or names designate the cranial nerves. The numbers indicate the order in which the nerves arise from the front to the back of the brain, and the names describe their primary functions or the general distribution of their fibers (fig. 9.34).

The first pair of cranial nerves, the **olfactory nerves (I),** are associated with the sense of smell and contain axons only of sensory neurons. These bipolar neurons, located in the lining of the upper nasal cavity, serve as *olfactory receptor cells.* Axons from these receptors pass upward through the cribriform plates of the ethmoid bone, carrying impulses to the olfactory neurons in the *olfactory bulbs,* which are extensions of the cerebral cortex located just beneath the frontal lobes (see fig. 10.4, p. 266). Sensory impulses travel from the olfactory bulbs along *olfactory tracts* to cerebral centers where they are interpreted. The result of this interpretation is the sensation of smell.

The second pair of cranial nerves, the **optic nerves (II),** lead from the eyes to the brain and are associated with vision. The sensory nerve cell bodies of these nerve fibers are in ganglion cell layers within the eyes, and their axons pass through the *optic foramina* of the orbits and continue into the visual nerve pathways of the brain (see chapter 10, p. 284). Sensory impulses

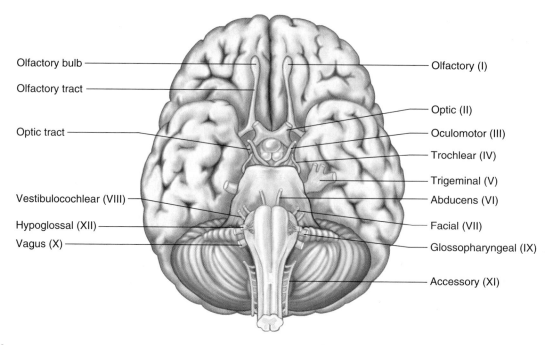

Olfactory bulb

Olfactory tract

Optic tract

Vestibulocochlear (VIII)

Hypoglossal (XII)

Vagus (X)

Olfactory (I)

Optic (II)

Oculomotor (III)

Trochlear (IV)

Trigeminal (V)

Abducens (VI)

Facial (VII)

Glossopharyngeal (IX)

Accessory (XI)

Figure 9.34

The cranial nerves, except for the first pair, arise from the brainstem. They are identified by numbers indicating their order, by their function, or by the general distribution of their fibers.

transmitted on the optic nerves are interpreted in the visual cortices of the occipital lobes.

The third pair of cranial nerves, the **oculomotor nerves (III),** arise from the midbrain and pass into the orbits of the eyes. One component of each nerve connects to the voluntary muscles that raise the eyelid and to four of the six muscles that move the eye. A second component of each oculomotor nerve is part of the autonomic nervous system and supplies involuntary muscles within the eyes. These muscles adjust the amount of light entering the eyes and focus the lenses.

The fourth pair of cranial nerves, the **trochlear nerves (IV),** arise from the midbrain and are the smallest cranial nerves. Each nerve carries motor impulses to a fifth voluntary muscle that moves the eye and is not innervated by the oculomotor nerve.

The fifth pair of cranial nerves, the **trigeminal nerves (V),** are the largest cranial nerves and arise from the pons. They are mixed nerves, with the sensory portions more extensive than the motor portions. Each sensory component includes three large branches, called the ophthalmic, maxillary, and mandibular divisions.

The *ophthalmic division* of the trigeminal nerves consists of sensory fibers that bring impulses to the brain from the surface of the eyes, the tear glands, and the skin of the anterior scalp, forehead, and upper eyelids. The fibers of the *maxillary division* carry sensory impulses from the upper teeth, upper gum, and upper lip, as well as from the mucous lining of the palate and the skin of the face. The *mandibular division* includes both motor and sensory fibers. The sensory branches transmit

impulses from the scalp behind the ears, the skin of the jaw, the lower teeth, the lower gum, and the lower lip. The motor branches supply the muscles of mastication and certain muscles in the floor of the mouth.

The sixth pair of cranial nerves, the **abducens nerves (VI),** are quite small and originate from the pons near the medulla oblongata. Each nerve enters the orbit of the eye and supplies motor impulses to the remaining muscle that moves the eye.

The seventh pair of cranial nerves, the **facial nerves (VII),** arise from the lower part of the pons and emerge on the sides of the face. Their sensory branches are associated with taste receptors on the anterior two-thirds of the tongue, and some of their motor fibers transmit impulses to the muscles of facial expression. Still other motor fibers of these nerves function in the autonomic nervous system and stimulate secretions from tear glands and salivary glands.

The eighth pair of cranial nerves, the **vestibulocochlear nerves (VIII),** are sensory nerves that arise from the medulla oblongata. Each of these nerves has two distinct parts—a vestibular branch and a cochlear branch.

The neuron cell bodies of the *vestibular branch* fibers are located in ganglia associated with parts of the inner ear. These parts contain the receptors involved with reflexes that help maintain equilibrium. The neuron cell bodies of the *cochlear branch* fibers are located in the parts of the inner ear that house the hearing receptors. Impulses from these branches pass through the pons and medulla oblongata on their way to the temporal lobes, where they are interpreted.

Topic of Interest

Drug Abuse

Drug abuse is the chronic self-administration of a drug in doses high enough to cause *addiction*—a physical or psychological dependence in which the user is preoccupied with locating and taking the drug. Stopping drug use causes intense, unpleasant withdrawal symptoms. Prolonged and repeated abuse of a drug may also result in *drug tolerance,* in which the physiological response to a particular dose of the drug becomes less intense over time. Drug tolerance results as the drug increases synthesis of certain liver enzymes, which metabolize the drug more rapidly, so that the addict needs the next dose sooner. Drug tolerance also arises from physiological changes that lessen the drug's effect on its target cells.

The most commonly abused drugs are CNS depressants ("downers"), CNS stimulants ("uppers"), hallucinogens, and anabolic steroids (see Topic of Interest in chapter 8, p. 186).

CNS depressants include barbiturates, benzodiazepines, opiates, and cannabinoids. *Barbiturates* act uniformly throughout the brain, but the reticular formation is particularly sensitive to their effects. CNS depression occurs due to inhibited secretion of certain excitatory and inhibitory neurotransmitters. Effects range from mild calming of the nervous system (sedation) to sleep, loss of sensory sensations (anesthesia), respiratory distress, cardiovascular collapse, and death.

The *benzodiazepines,* such as diazepam, depress activity in the limbic system and the reticular formation. Low doses relieve anxiety, and higher doses cause sedation, sleep, or anesthesia. These drugs increase either the activity or release of the inhibitory neurotransmitter GABA. When benzodiazepines are metabolized, they may form other biochemicals that have depressing effects.

The *opiates* include heroin (which has no legal use in the United States), codeine, morphine, meperidine, and methadone. These drugs stimulate certain receptors (opioid receptors) in the CNS, and when taken in prescribed dosages, they sedate and relieve pain (analgesia). Opiates cause both physical and psychological dependence. Effects of overdose include a feeling of well-being (euphoria), respiratory distress, convulsions, coma, and possible death. On the other hand, these drugs are very important in treating chronic, severe pain. For example, cancer patients find pain relief with oxycodone, which is taken twice daily in a timed-release pill. Many people abuse this drug, and by breaking the pills, release high doses rapidly, which can be deadly.

The *cannabinoids* include marijuana and hashish, both derived from the hemp plant. Hashish is several times more potent than marijuana. These drugs depress higher brain centers and release lower brain centers from the normal inhibitory influence of the higher centers. This induces an anxiety-free state, characterized by euphoria and a distorted perception of time and space. *Hallucinations* (sensory perceptions that have no external stimuli), respiratory distress, and vasomotor depression may occur with higher doses.

CNS stimulants include amphetamines and cocaine (including "crack"). These drugs have great abuse potential and may quickly produce psychological dependence. Cocaine, especially when smoked or inhaled, produces euphoria but may also change personality, cause seizures, and constrict certain blood vessels, leading to sudden death from stroke or cardiac arrhythmia. Cocaine's very rapid effect, and perhaps its addictiveness, reflect its rapid entry and metabolism in the brain. Cocaine arrives at the basal nuclei in 4 to 6 minutes and is cleared mostly within 30 minutes. The drug inhibits transporter molecules that remove dopamine from synapses after it is released. "Ecstasy" is a type of amphetamine.

Hallucinogens alter perceptions. They cause *illusions,* which are distortions of vision, hearing, taste, touch, and smell; *synesthesia,* such as "hearing" colors or "feeling" sounds; and hallucinations. The most commonly abused and most potent hallucinogen is lysergic acid diethylamide (LSD). LSD may act as an excitatory neurotransmitter. A person under the influence of LSD may greatly overestimate physical capabilities, such as believing he or she can fly off the top of a high building. Phencyclidine (PCP) is another commonly abused hallucinogen. Its use can lead to prolonged psychosis that may provoke assault, murder, and suicide.

The ninth pair of cranial nerves, the **glossopharyngeal nerves (IX),** are associated with the tongue and pharynx. These mixed nerves arise from the medulla oblongata, with predominantly sensory fibers. These sensory fibers carry impulses from the linings of the pharynx, tonsils, and posterior third of the tongue to the brain. Fibers in the motor component innervate muscles of the pharynx that function in swallowing.

The tenth pair of cranial nerves, the **vagus nerves (X),** originate in the medulla oblongata and extend downward through the neck into the chest and abdomen. These nerves are mixed, containing both somatic and autonomic branches, with autonomic fibers predominant. Certain somatic motor fibers carry impulses to muscles of the larynx that are associated with speech and swallowing. Autonomic motor fibers of the vagus nerves supply the heart and many smooth muscles and glands in the thorax and abdomen.

The eleventh pair of cranial nerves, the **accessory nerves (XI),** originate in the medulla oblongata and the spinal cord; thus, they have both cranial and spinal branches. Each *cranial branch* joins a vagus nerve and carries impulses to muscles of the soft palate, pharynx, and larynx. The *spinal branch* descends into the neck and supplies motor fibers to the trapezius and sterno-cleidomastoid muscles.

The twelfth pair of cranial nerves, the **hypoglossal nerves (XII),** arise from the medulla oblongata and pass into the tongue. They include motor fibers that carry impulses to muscles that move the tongue in speaking, chewing, and swallowing. Table 9.6 summarizes the functions of the cranial nerves.

Check Your Recall

51. Define *peripheral nervous system.*
52. Distinguish between somatic and autonomic nerve fibers.
53. Name the cranial nerves, and list the major functions of each.

The consequences of a cranial nerve injury depend on the injury's location and extent. Damage to one member of a nerve pair limits loss of function to the affected side, but injury to both nerves affects both sides. If a nerve is severed completely, functional loss is total; if the cut is incomplete, loss may be partial.

Table 9.6	**Functions of Cranial Nerves**	
Nerve	**Type**	**Function**
I Olfactory	Sensory	Sensory fibers transmit impulses associated with the sense of smell.
II Optic	Sensory	Sensory fibers transmit impulses associated with the sense of vision.
III Oculomotor	Primarily motor	Motor fibers transmit impulses to muscles that raise eyelids, move eyes, adjust the amount of light entering the eyes, and focus lenses.
		Some sensory fibers transmit impulses associated with the condition of muscles.
IV Trochlear	Primarily motor	Motor fibers transmit impulses to muscles that move the eyes.
		Some sensory fibers transmit impulses associated with the condition of muscles.
V Trigeminal	Mixed	
Ophthalmic division		Sensory fibers transmit impulses from the surface of the eyes, tear glands, scalp, forehead, and upper eyelids.
Maxillary division		Sensory fibers transmit impulses from the upper teeth, upper gum, upper lip, lining of the palate, and skin of the face.
Mandibular division		Sensory fibers transmit impulses from the skin of the jaw, lower teeth, lower gum, and lower lip.
		Motor fibers transmit impulses to muscles of mastication and to muscles in the floor of the mouth.
VI Abducens	Primarily motor	Motor fibers transmit impulses to muscles that move the eyes.
		Some sensory fibers transmit impulses associated with the condition of muscles.
VII Facial	Mixed	Sensory fibers transmit impulses associated with taste receptors of the anterior tongue. Motor fibers transmit impulses to muscles of facial expression, tear glands, and salivary glands.
VIII Vestibulocochlear	Sensory	
Vestibular branch		Sensory fibers transmit impulses associated with the sense of equilibrium.
Cochlear branch		Sensory fibers transmit impulses associated with the sense of hearing.
IX Glossopharyngeal	Mixed	Sensory fibers transmit impulses from the pharynx, tonsils, posterior tongue, and carotid arteries.
		Motor fibers transmit impulses to muscles of the pharynx used in swallowing and to salivary glands.
X Vagus	Mixed	Somatic motor fibers transmit impulses to muscles associated with speech and swallowing; autonomic motor fibers transmit impulses to the heart, smooth muscles, and glands in the thorax and abdomen.
		Sensory fibers transmit impulses from the pharynx, larynx, esophagus, and viscera of the thorax and abdomen.
XI Accessory	Primarily motor	
Cranial branch		Motor fibers transmit impulses to muscles of the soft palate, pharynx, and larynx.
Spinal branch		Motor fibers transmit impulses to muscles of the neck and back.
XII Hypoglossal	Primarily motor	Motor fibers transmit impulses to muscles that move the tongue.

Spinal Nerves

Thirty-one pairs of **spinal nerves** originate from the spinal cord (fig. 9.35). They are mixed nerves that provide two-way communication between the spinal cord and parts of the upper and lower limbs, neck, and trunk.

Spinal nerves are not named individually, but are grouped according to the level from which they arise. Each nerve is numbered in sequence. On each vertebra the vertebral notches, the major parts of the interver-

tebral foramina, are associated with the inferior portion of their respective vertebrae. For this reason, each spinal nerve, as it passes through the intervertebral foramen, is associated with the vertebra above it. The cervical spinal nerves are an exception, because spinal nerve C1 passes superior to the vertebra C1. Thus, although there are seven cervical vertebrae, there are eight pairs of *cervical nerves* (numbered C1 to C8), twelve pairs of *thoracic nerves* (numbered T1 to T12), five pairs of *lumbar nerves* (numbered L1 to L5), five

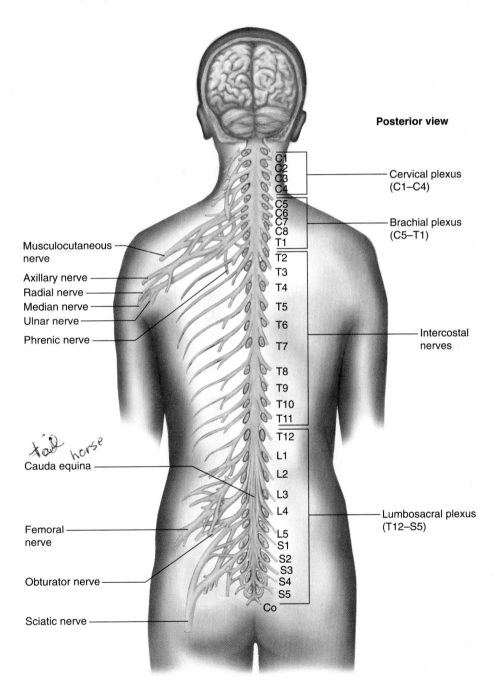

Posterior view

Figure 9.35

The anterior branches of the spinal nerves in the thoracic region give rise to intercostal nerves. Those in other regions combine to form complex networks called plexuses. (Note that there are eight pairs of cervical nerves, one pair originating above the first cervical vertebra and the eighth pair originating below the seventh cervical vertebra.)

pairs of *sacral nerves* (numbered S1 to S5), and one pair of *coccygeal nerves* (Co).

The adult spinal cord ends at the level between the first and second lumbar vertebrae. The lumbar, sacral, and coccygeal nerves descend beyond the end of the cord, forming a structure called the *cauda equina* (horse's tail).

Each spinal nerve emerges from the cord by two short branches, or *roots,* which lie within the vertebral column. The **dorsal root** (posterior or sensory root) can be identified by an enlargement called the *dorsal root ganglion* (see fig. 9.22*a*). This ganglion contains the cell bodies of the sensory neurons whose axons (peripheral process) conduct impulses inward from the peripheral body parts. The axons of these neurons extend through the dorsal root and into the spinal cord (central process), where they form synapses with dendrites of other neurons (see fig. 9.6). The **ventral root** (anterior or motor root) of each spinal nerve consists of axons from the motor neurons whose cell bodies are within the gray matter of the cord.

A ventral root and a dorsal root unite to form a spinal nerve, which extends outward from the vertebral canal through an *intervertebral foramen* (see fig. 7.17, p. 149). Just beyond its foramen, each spinal nerve divides into several parts.

Except in the thoracic region, the main portions of the spinal nerves combine to form complex networks called **plexuses** instead of continuing directly to peripheral body parts (fig. 9.35). In a plexus, spinal nerve fibers are sorted and recombined so that fibers that innervate a particular peripheral body part reach it in the same nerve, even though the fibers originate from different spinal nerves.

Cervical Plexuses

The **cervical plexuses** lie deep in the neck on either side and form from the branches of the first four cervical nerves. Fibers from these plexuses supply the muscles and skin of the neck. In addition, fibers from the third, fourth, and fifth cervical nerves pass into the right and left **phrenic nerves,** which conduct motor impulses to the muscle fibers of the diaphragm.

Brachial Plexuses

Branches of the lower four cervical nerves and the first thoracic nerve give rise to the **brachial plexuses.** These networks of nerve fibers are located deep within the shoulders between the neck and axillae (armpits). The major branches emerging from the brachial plexuses supply the muscles and skin of the arm, forearm, and hand, and include the **musculocutaneous, ulnar, median, radial,** and **axillary nerves.**

Lumbosacral Plexuses

The **lumbosacral plexuses** are formed on either side by the last thoracic nerve and the lumbar, sacral, and coccygeal nerves. These networks of nerve fibers extend from the lumbar region of the back into the pelvic cavity, giving rise to a number of motor and sensory fibers associated with the muscles and skin of the lower abdominal wall, external genitalia, buttocks, thighs, legs, and feet. The major branches of these plexuses include the **obturator, femoral,** and **sciatic nerves.**

The anterior branches of the thoracic spinal nerves do not enter a plexus. Instead, they enter spaces between the ribs and become **intercostal nerves.** These nerves supply motor impulses to the intercostal muscles and the upper abdominal wall muscles. They also receive sensory impulses from the skin of the thorax and abdomen.

Check Your Recall

54. How are spinal nerves grouped?

55. Describe how a spinal nerve joins the spinal cord.

56. Name and locate the major nerve plexuses.

Spinal nerves may be injured in a variety of ways, including stabs, gunshot wounds, birth injuries, dislocations and fractures of the vertebrae, and pressure from tumors in surrounding tissues. The nerves of the cervical plexuses, for example, are sometimes compressed by a sudden bending of the neck called *whiplash,* which may occur during rear-end automobile collisions. Whiplash may cause continuing headaches and pain in the neck and skin, which are supplied by the cervical nerves.

9.16 AUTONOMIC NERVOUS SYSTEM

The **autonomic nervous system** is the portion of the PNS that functions independently (autonomously) and continuously without conscious effort. This system controls visceral functions by regulating the actions of smooth muscles, cardiac muscles, and glands. It regulates heart rate, blood pressure, breathing rate, body temperature, and other visceral activities that maintain homeostasis. Portions of the autonomic nervous system respond to emotional stress and prepare the body to meet the demands of strenuous physical activity.

General Characteristics

Reflexes in which sensory signals originate from receptors within the viscera and the skin regulate autonomic activities. Nerve fibers transmit these signals to nerve centers within the brain or spinal cord. In response, motor impulses travel out from these centers on peripheral nerve fibers within cranial and spinal nerves.

Typically, peripheral nerve fibers lead to ganglia outside the CNS. The impulses they carry are integrated within these ganglia and relayed to viscera (muscles and glands) that respond by contracting, releasing secretions, or being inhibited. The integrative function of the ganglia provides the autonomic system with a degree of independence from the brain and spinal cord.

The autonomic nervous system includes two divisions—the **sympathetic** (sim″pah-thet′ik) and **parasympathetic** (par″ah-sim″pah-thet′ik) **divisions.** Some viscera have nerve fibers from each division. In such cases, impulses on one set of fibers may activate an organ, while impulses on the other set inhibit it. Thus, the divisions may act antagonistically, alternately activating or inhibiting the actions of some viscera.

The functions of the autonomic divisions are mixed; that is, each activates some organs and inhibits others. However, the divisions have important functional differences. The sympathetic division prepares the body for energy-expending, stressful, or emergency situations, as part of the *fight or flight* response. Conversely, the parasympathetic division is most active under ordinary, restful conditions. It also counterbalances the effects of the sympathetic division and restores the body to a resting state following a stressful experience. For example, during an emergency, the sympathetic division increases

heart and breathing rates; following the emergency, the parasympathetic division decreases these activities.

Autonomic Nerve Fibers

The neurons of the autonomic nervous system are motor neurons. However, unlike the motor pathways of the somatic nervous system, which usually include a single neuron between the brain or spinal cord and a skeletal muscle, those of the autonomic system include two neurons (fig. 9.36). The cell body of one neuron is located in the brain or spinal cord. Its axon, the **preganglionic fiber** (pre″gang-gle-on′ik fi′ber), leaves the CNS and synapses with one or more neurons whose cell bodies are housed within an autonomic ganglion. The axon of such a second neuron is called a **postganglionic fiber** (pōst″gang-gle-on′ik fi′ber), and it extends to a visceral effector.

Sympathetic Division

In the sympathetic division, the preganglionic fibers originate from neurons in the gray matter of the spinal cord (fig. 9.37). Their axons leave the cord through

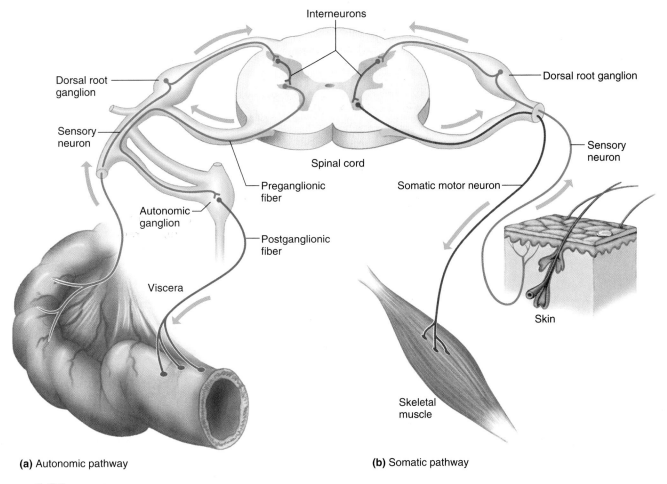

(a) Autonomic pathway   **(b)** Somatic pathway

Figure 9.36

Motor pathways. (*a*) Autonomic pathways include two neurons between the CNS and an effector. (*b*) Somatic pathways usually have a single neuron between the CNS and an effector. Note that in both cases the motor fibers pass through the ventral root of the spinal cord.

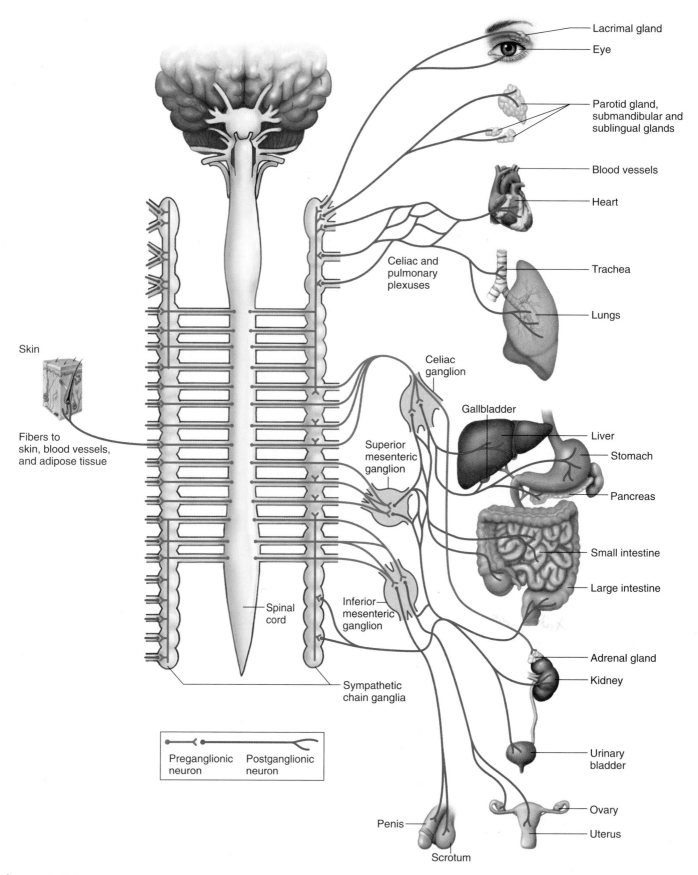

Lacrimal gland

Eye

Parotid gland, submandibular and sublingual glands

Blood vessels

Heart

Celiac and pulmonary plexuses

Trachea

Lungs

Skin

Celiac ganglion

Gallbladder

Fibers to skin, blood vessels, and adipose tissue

Superior mesenteric ganglion

Liver

Stomach

Pancreas

Small intestine

Large intestine

Spinal cord

Inferior mesenteric ganglion

Adrenal gland

Kidney

Sympathetic chain ganglia

Preganglionic neuron

Postganglionic neuron

Urinary bladder

Ovary

Penis

Uterus

Scrotum

Figure 9.37

The preganglionic fibers of the sympathetic division of the autonomic nervous system arise from the thoracic and lumbar regions of the spinal cord (T1–L2). Note that the adrenal medulla is innervated directly by a preganglionic fiber.

the ventral roots of spinal nerves in the first thoracic through the second lumbar segments. After traveling a short distance, these fibers leave the spinal nerves, and each enters a member of a chain of sympathetic ganglia (*paravertebral ganglia*). One of these sympathetic chains extends longitudinally along each side of the vertebral column.

Within paravertebral ganglia, preganglionic fibers form synapses with second neurons. The axons of these neurons, the postganglionic fibers, typically return to spinal nerves and extend to visceral effectors.

Parasympathetic Division

The preganglionic fibers of the parasympathetic division arise from the brainstem and sacral region of the spinal cord (fig. 9.38). From there, they lead outward in cranial or sacral nerves to ganglia located near or within various viscera. The relatively short postganglionic fibers continue from the ganglia to specific muscles or glands within these viscera.

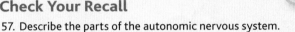

> #### Check Your Recall
> 57. Describe the parts of the autonomic nervous system.
> 58. Distinguish between the divisions of the autonomic nervous system.
> 59. Describe a sympathetic nerve pathway and a parasympathetic nerve pathway.

Autonomic Neurotransmitters

The preganglionic fibers of the sympathetic and parasympathetic divisions all secrete *acetylcholine* and are therefore called **cholinergic fibers** (ko″lin-er′jik

fi′berz). The parasympathetic postganglionic fibers are also cholinergic. One exception, parasympathetic neurons that secrete nitric oxide, is described in chapter 19 (p. 514). However, most sympathetic postganglionic neurons secrete *norepinephrine* (noradrenalin) and are called **adrenergic fibers** (ad″ren-ur′jik fi′berz) (fig. 9.39). The different postganglionic neurotransmitters cause the different effects that the sympathetic and parasympathetic divisions have on their effector organs.

Most organs receive innervation from both sympathetic and parasympathetic divisions, usually with opposing actions. For example, parasympathetic activity increases activity of the digestive system, whereas sympathetic activity decreases it. Similarly, sympathetic stimulation increases heart rate, but parasympathetic action slows heart rate.

Some viscera are controlled primarily by one division or the other. That is, the divisions are not always actively antagonistic. For example, the sympathetic division regulates the diameter of most blood vessels, which lack parasympathetic innervation. Smooth muscles in the walls of these vessels are continuously stimulated and thus are in a state of partial contraction (sympathetic tone). Decreasing sympathetic stimulation increases (dilates) the diameter of the vessels, which relaxes their muscular walls. Conversely, increasing sympathetic stimulation constricts the vessels. Table 9.7 summarizes the effects of stimulation by adrenergic and cholinergic fibers on some visceral effectors.

Control of Autonomic Activity

The brain and spinal cord largely control the autonomic nervous system, despite the system's independence resulting from the integrative function of its ganglia. For example, control centers in the medulla oblongata

Table 9.7	Effects of Neurotransmitter Substances on Visceral Effectors or Actions	
Visceral Effector or Action	**Response to Adrenergic Stimulation (Sympathetic)**	**Response to Cholinergic Stimulation (Parasympathetic)**
Pupil of the eye	Dilation	Constriction
Heart rate	Increases	Decreases
Bronchioles of lungs	Dilation	Constriction
Muscles of intestinal wall	Slows peristaltic action	Speeds peristaltic action
Intestinal glands	Secretion decreases	Secretion increases
Blood distribution	More blood to skeletal muscles; less blood to digestive organs	More blood to digestive organs; less blood to skeletal muscles
Blood glucose concentration	Increases	Decreases
Salivary glands	Secretion decreases	Secretion increases
Tear glands	No action	Secretion
Muscles of gallbladder	Relaxation	Contraction
Muscles of urinary bladder	Relaxation	Contraction

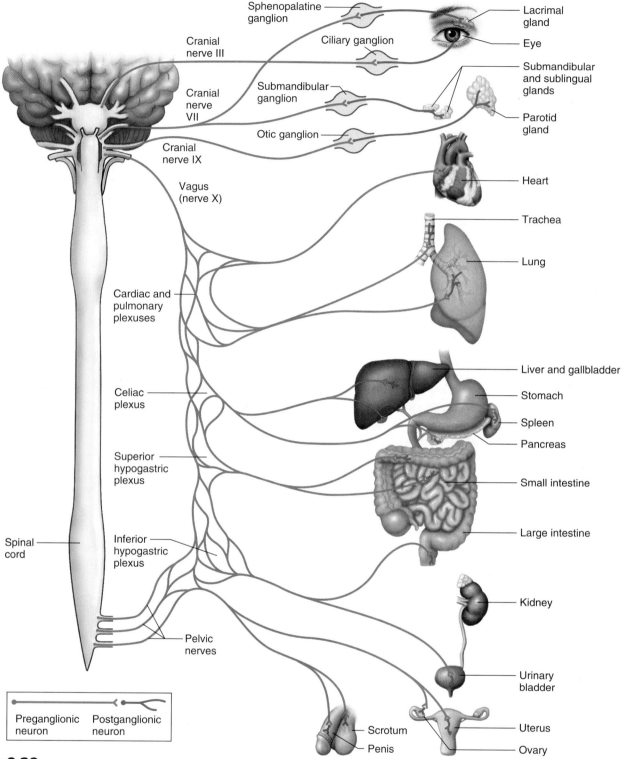

Figure 9.38

The preganglionic fibers of the parasympathetic division of the autonomic nervous system arise from the brainstem and sacral region of the spinal cord.

for cardiac, vasomotor, and respiratory activities receive sensory impulses from viscera on vagus nerve fibers and use autonomic nerve pathways to stimulate motor responses in muscles and glands. Similarly, the hypothalamus helps regulate body temperature, hunger, thirst, and water and electrolyte balance by influencing autonomic pathways.

More complex centers in the brain, including the limbic system and the cerebral cortex, control the autonomic nervous system during emotional stress. These

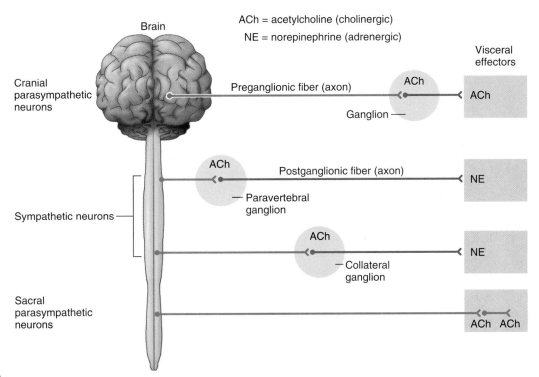

Figure 9.39

Most sympathetic fibers are adrenergic and secrete norepinephrine at the ends of the postganglionic fiber; parasympathetic fibers are cholinergic and secrete acetylcholine at the ends of the postganglionic fibers. Two arrangements of parasympathetic postganglionic fibers are seen in both the cranial and sacral portions. Similarly, sympathetic paravertebral and collateral ganglia are seen in both the thoracic and lumbar portions of the nervous system. (Note: This representation does not show dendrites.)

structures utilize autonomic pathways to regulate emotional expression and behavior.

Check Your Recall

60. Which neurotransmitters operate in the autonomic nervous system?
61. How do the divisions of the autonomic nervous system regulate visceral activities?
62. How are autonomic activities controlled?

Clinical Terms Related to the Nervous System

analgesia (an″al-je′ze-ah) Loss or reduction in the ability to sense pain, but without loss of consciousness.

analgesic (an″al-je′sik) Pain-relieving drug.

anesthesia (an″es-the′ze-ah) Loss of feeling.

aphasia (ah-fa′ze-ah) Disturbance or loss of the ability to use words or to understand them, usually due to damage to cerebral association areas.

apraxia (ah-prak′se-ah) Impairment in the ability to use objects.

ataxia (ah-tak′se-ah) Partial or complete inability to coordinate voluntary movements.

cerebral palsy (ser′ĕ-bral pawl′ze) Partial paralysis and lack of muscular coordination caused by damage to the cerebrum.

coma (ko′mah) Unconscious condition in which a person does not respond to stimulation.

cordotomy (kor-dot′o-me) Surgical procedure that severs a nerve tract within the spinal cord to relieve intractable pain.

craniotomy (kra″ne-ot′o-me) Surgical procedure that opens part of the skull.

electroencephalogram (EEG) (e-lek″tro-en-sef′ah-lo-gram″) Recording of the brain's electrical activity.

encephalitis (en″sef-ah-li′tis) Inflammation of the brain and meninges, producing drowsiness and apathy.

epilepsy (ep′ĭ-lep″se) Disorder of the central nervous system that temporarily disturbs brain impulses, producing convulsive seizures and loss of consciousness.

hemiplegia (hem″ĭ-ple′je-ah) Paralysis of one side of the body and the limbs on that side.

Huntington disease (hunt′ing-tun diz-ēz′) Inherited disorder of the brain causing involuntary, dancelike movements and personality changes.

laminectomy (lam″ĭ-nek′to-me) Surgical removal of the posterior arch of a vertebra, usually to relieve the symptoms of a ruptured intervertebral disc pressing on a spinal nerve.

monoplegia (mon″o-ple′je-ah) Paralysis of a single limb.

multiple sclerosis (mul′tĭ-pl skle-ro′sis) Loss of myelin and the appearance of scarlike patches throughout the brain or spinal cord or both.

neuralgia (nu-ral′je-ah) Sharp, recurring pain associated with a nerve; usually caused by inflammation or injury.

neuritis (nu-ri′tis) Inflammation of a nerve.

paraplegia (par″ah-ple′je-ah) Paralysis of both lower limbs.

quadriplegia (kwod″rĭ-ple′je-ah) Paralysis of all four limbs.

vagotomy (va-got′o-me) Surgical severing of a vagus nerve—for example, to reduce acid secretion in a patient with ulcers nonresponsive to other treatment.

Nervous System

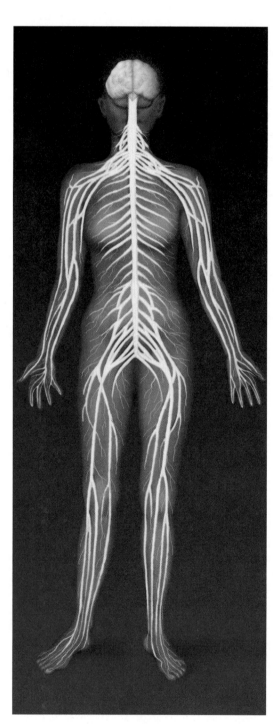

Neurons carry impulses that allow body systems to communicate.

Integumentary System

Sensory receptors provide the nervous system with information about the outside world.

Skeletal System

Bones protect the brain and spinal cord and help maintain plasma calcium, which is important to neuron function.

Muscular System

Nerve impulses control movement and carry information about the position of body parts.

Endocrine System

The hypothalamus controls secretion of many hormones.

Cardiovascular System

Nerve impulses help control blood flow and blood pressure.

Lymphatic System

Stress may impair the immune response.

Digestive System

The nervous system can influence digestive function.

Respiratory System

The nervous system alters respiratory activity to control oxygen levels and blood pH.

Urinary System

Nerve impulses affect urine production and elimination.

Reproductive System

The nervous system plays a role in egg and sperm formation, sexual pleasure, childbirth, and nursing.

Clinical Connection

In the 1970s, researchers discovered that the body produces its own compounds that bind to the same receptors on brain neurons as do opiate drugs such as morphine. The opiates presumably exert their effects by interfering with this system. In 1992, researchers discovered a compound in the body that binds the same receptors on brain neurons as does the active ingredient in marijuana. The existence of this compound, called anandamide, explains why marijuana exerts psychoactive and other effects. Anandamide normally is released from postsynaptic neurons stimulated by calcium ion influx, and then binds to the presynaptic cell, temporarily shutting down neurotransmission. When a person takes in the plant version of anandamide through smoke, these neural connections are overwhelmed, and somehow this interference produces marijuana's effects on mood and thinking.

Like the opiates, which come from the poppy plant, marijuana has medicinal applications, and is available by prescription in several states to treat pain and to stimulate appetite. In 1985, the Food and Drug Administration approved Marinol, a drug based on the most active chemical in marijuana, to treat nausea and vomiting resulting from cancer chemotherapy. In 1992, the agency approved Marinol to treat AIDS-related anorexia. A drug is being developed from a different chemical found in marijuana smoke called ajulemic acid. It relieves pain, but without psychoactive effects.

SUMMARY OUTLINE

9.1 Introduction (p. 212)
1. Nervous tissue includes neurons, which are the structural and functional units of the nervous system, and neuroglial cells.
2. Organs of the nervous system are divided into the central and peripheral nervous systems.

9.2 General Functions of the Nervous System (p. 213)
1. Sensory functions receive stimulation from receptors that detect internal and external changes.
2. Integrative functions collect sensory information and make decisions that motor functions carry out.
3. Motor functions stimulate effectors to respond.

9.3 Neuroglial Cells (p. 214)
1. Neuroglial cells in the central nervous system include microglial cells, oligodendrocytes, astrocytes, and ependymal cells.
2. In the peripheral nervous system, Schwann cells form myelin sheaths.

9.4 Neurons (p. 214)
1. A neuron includes a cell body, dendrites, and an axon.
2. Dendrites and the cell body provide receptive surfaces.
3. A single axon arises from the cell body and may be enclosed in a myelin sheath and a neurilemma.
4. Classification of neurons
   a. Neurons are classified structurally as multipolar, bipolar, or unipolar.
   b. Neurons are classified functionally as sensory neurons, interneurons, or motor neurons.

9.5 The Synapse (p. 219)
A synapse is a junction between two neurons.
1. A presynaptic neuron carries an impulse into a synapse; a postsynaptic neuron responds.
2. Axons have synaptic knobs at their distal ends, which secrete neurotransmitters.
3. A neurotransmitter is released when a nerve impulse reaches the end of an axon.
4. A neurotransmitter reaching the postsynaptic neuron membrane is either excitatory or inhibitory.

9.6 Cell Membrane Potential (p. 220)
A cell membrane is usually polarized as a result of unequal ion distribution.
1. Distribution of ions
   a. Pores and channels in cell membranes that allow passage of some ions but not others set up differences in the concentrations of specific ions inside and outside a neuron.
   b. Potassium ions pass more easily through cell membranes than do sodium ions.
2. Resting potential
   a. A high concentration of sodium ions is outside a cell membrane, and a high concentration of potassium ions is inside.
   b. Many negatively charged ions are inside a cell.
   c. In a resting cell, more positive ions leave than enter, so the outside of the cell membrane develops a positive charge, while the inside develops a negative charge.
3. Potential changes
   a. Stimulation of a cell membrane affects the membrane's resting potential.
   b. When its resting potential becomes less negative, a membrane becomes depolarized.
   c. Potential changes are subject to summation.
   d. Achieving threshold potential triggers an action potential.
4. Action potential
   a. At threshold, sodium channels open, and sodium ions diffuse inward, depolarizing the membrane.
   b. About the same time, potassium channels open, and potassium ions diffuse outward, repolarizing the membrane.
   c. This rapid change in potential is an action potential.
   d. Many action potentials can occur before active transport reestablishes the resting potential.

9.7 Nerve Impulses (p. 224)

A wave of action potentials is a nerve impulse.

1. Impulse conduction
   a. Unmyelinated axons conduct impulses over their entire surfaces.
   b. Myelinated axons conduct impulses more rapidly.
   c. Axons with larger diameters conduct impulses faster than those with smaller diameters.
2. All-or-none response
   a. A nerve impulse is conducted in an all-or-none manner whenever a stimulus of threshold intensity is applied to an axon.
   b. All the impulses conducted on an axon are of the same strength.

9.8 Synaptic Transmission (p. 225)

1. Excitatory and inhibitory actions
   a. Neurotransmitters that trigger nerve impulses are excitatory. Those that inhibit impulses are inhibitory.
   b. The net effect of synaptic knobs communicating with a neuron depends on which knobs are activated from moment to moment.
2. Neurotransmitters
   a. The nervous system produces many different neurotransmitters.
   b. Neurotransmitters include acetylcholine, monoamines, amino acids, and peptides.
   c. A synaptic knob releases neurotransmitters when an action potential increases membrane permeability to calcium ions.
   d. After being released, neurotransmitters are decomposed or removed from synaptic clefts.

9.9 Impulse Processing (p. 227)

How the nervous system processes and responds to nerve impulses reflects the organization of neurons in the brain and spinal cord.

1. Neuronal pools
   a. Neurons form pools within the central nervous system.
   b. Each pool receives impulses, processes them, and conducts impulses away.
2. Facilitation
   a. Each neuron in a pool may receive excitatory and inhibitory stimuli.
   b. A neuron is facilitated when it receives subthreshold stimuli and becomes more excitable.
3. Convergence
   a. Impulses from two or more incoming axons may converge on a single neuron.
   b. Convergence enables impulses from different sources to have an additive effect on a neuron.
4. Divergence
   a. Impulses leaving a pool may diverge by passing into several output neurons.
   b. Divergence amplifies impulses.

9.10 Types of Nerves (p. 228)

1. Nerves are cordlike bundles (fascicles) of nerve fibers (axons).
2. Nerves are sensory, motor, or mixed, depending on which type of fibers they contain.

9.11 Nerve Pathways (p. 228)

A nerve pathway is the route an impulse follows through the nervous system.

1. A reflex arc usually includes a sensory neuron, a reflex center composed of interneurons, and a motor neuron.

2. Reflex behavior
   a. Reflexes are automatic, subconscious responses to changes.
   b. They help maintain homeostasis.
   c. Two neurons carry out the patellar reflex.
   d. Withdrawal reflexes are protective.

9.12 Meninges (p. 230)

1. Bone and meninges surround the brain and spinal cord.
2. The meninges are the dura mater, arachnoid mater, and pia mater.
3. Cerebrospinal fluid fills the space between the arachnoid and pia maters.

9.13 Spinal Cord (p. 232)

The spinal cord is a nerve column that extends from the brain into the vertebral canal.

1. Structure of the spinal cord
   a. Each of the spinal cord's thirty-one segments gives rise to a pair of spinal nerves.
   b. The spinal cord has a cervical enlargement and a lumbar enlargement.
   c. A central core of gray matter lies within white matter.
   d. White matter consists of bundles of myelinated axons.
2. Functions of the spinal cord
   a. The spinal cord provides a two-way communication system between the brain and other body parts and serves as a center for spinal reflexes.
   b. Ascending tracts carry sensory impulses to the brain. Descending tracts carry motor impulses to muscles and glands.

9.14 Brain (p. 234)

The brain is subdivided into the cerebrum, diencephalon, brainstem, and cerebellum.

1. Structure of the cerebrum
   a. The cerebrum consists of two cerebral hemispheres connected by the corpus callosum.
   b. The cerebral cortex is a thin layer of gray matter near the surface.
   c. White matter consists of myelinated axons that connect neurons within the nervous system and communicate with other body parts.
2. Functions of the cerebrum
   a. The cerebrum provides higher brain functions.
   b. The cerebral cortex consists of sensory, motor, and association areas.
   c. One cerebral hemisphere usually dominates for certain intellectual functions.
3. Ventricles and cerebrospinal fluid
   a. Ventricles are interconnected cavities within the cerebral hemispheres and brainstem.
   b. Cerebrospinal fluid fills the ventricles.
   c. The choroid plexuses in the walls of the ventricles secrete cerebrospinal fluid.
4. Diencephalon
   a. The diencephalon contains the thalamus, which is a central relay station for incoming sensory impulses, and the hypothalamus, which maintains homeostasis.
   b. The limbic system produces emotions and modifies behavior.
5. Brainstem
   a. The brainstem consists of the midbrain, pons, and medulla oblongata.
   b. The midbrain contains reflex centers associated with eye and head movements.

c. The pons transmits impulses between the cerebrum and other parts of the nervous system and contains centers that help regulate the rate and depth of breathing.

d. The medulla oblongata transmits all ascending and descending impulses and contains several vital and nonvital reflex centers.

e. The reticular formation filters incoming sensory impulses, arousing the cerebral cortex into wakefulness when significant impulses arrive.

6. Cerebellum
   a. The cerebellum consists of two hemispheres.
   b. It functions primarily as a reflex center for integrating sensory information required in the coordination of skeletal muscle movements and the maintenance of equilibrium.

9.15 Peripheral Nervous System (p. 242)

The peripheral nervous system consists of cranial and spinal nerves that branch from the brain and spinal cord to all body parts. It is subdivided into the somatic and autonomic systems.

1. Cranial nerves
   a. Twelve pairs of cranial nerves connect the brain to parts in the head, neck, and trunk.
   b. Most cranial nerves are mixed, but some are purely sensory, and others are primarily motor.
   c. The names of the cranial nerves indicate their primary functions or the general distributions of their fibers.
   d. Some cranial nerves are somatic, and others are autonomic.
2. Spinal nerves
   a. Thirty-one pairs of spinal nerves originate from the spinal cord.
   b. These mixed nerves provide a two-way communication system between the spinal cord and parts of the upper and lower limbs, neck, and trunk.
   c. Spinal nerves are grouped according to the levels from which they arise, and they are numbered in sequence.
   d. Each spinal nerve emerges by a dorsal and a ventral root.
   e. Each spinal nerve divides into several branches just beyond its foramen.
   f. Most spinal nerves combine to form plexuses in which nerve fibers are sorted and recombined so that those fibers associated with a particular part reach it together.

9.16 Autonomic Nervous System (p. 248)

The autonomic nervous system functions without conscious effort. It regulates the visceral activities that maintain homeostasis.

1. General characteristics
   a. Autonomic functions are reflexes controlled from nerve centers in the brain and spinal cord.
   b. The autonomic nervous system consists of two divisions—the sympathetic and the parasympathetic.
   c. The sympathetic division responds to stressful and emergency conditions.
   d. The parasympathetic division is most active under ordinary conditions.
2. Autonomic nerve fibers
   a. Autonomic nerve fibers are motor fibers.
   b. Sympathetic fibers leave the spinal cord and synapse in paravertebral ganglia.
   c. Parasympathetic fibers begin in the brainstem and sacral region of the spinal cord and synapse in ganglia near viscera.
3. Autonomic neurotransmitters
   a. Sympathetic and parasympathetic preganglionic fibers secrete acetylcholine.
   b. Parasympathetic postganglionic fibers secrete acetylcholine. Sympathetic postganglionic fibers secrete norepinephrine.
   c. The different effects of the autonomic divisions are due to the different neurotransmitters the postganglionic fibers release.
   d. The two divisions usually have opposite actions.
4. Control of autonomic activity
   a. The autonomic nervous system is somewhat independent.
   b. Control centers in the medulla oblongata and hypothalamus utilize autonomic nerve pathways.
   c. The limbic system and cerebral cortex control the autonomic system during emotional stress.

CHAPTER ASSESSMENTS

9.1 Introduction

1. The general function of neurons is to _____, whereas the general functions of neuroglia are to _____. (p. 212)

2. Match the neuron part on the left to its description on the right. (p. 212)

   (1) dendrite   A. A cell process that sends information
   (2) axon   B. One of usually several cell processes that receive information
   (3) cell body
             C. The rounded part of a neuron

3. Explain the relationship between the CNS and the PNS. (p. 212)

9.2 General Functions of the Nervous System

4. List the general functions of the nervous system. (p. 213)

9.3 Neuroglial Cells

5. Match the types of neuroglial cells on the left to their functions on the right. (p. 214)

   (1) ependymal cells   A. Form a myelin sheath around peripheral nerves
   (2) oligodendrocytes
   (3) astrocytes   B. Phagocytize cellular debris and bacteria
   (4) Schwann cells   C. Line inner parts of ventricles and spinal cord
   (5) microglial cells   D. Form scar tissue and regulate ion and nutrient concentrations in the CNS
             E. Form a myelin sheath around neurons in the CNS

9.4 Neurons

6. Describe three structures found in neurons that are also in other cell types, and describe two structures that are unique to neurons. (p. 214)

7. The part of a Schwann cell that contributes to the myelin sheath is the _____, and the part that contributes to the neurilemma is the _____. (p. 215)

8. Distinguish between myelinated and unmyelinated axons. (p. 217)

9. Distinguish among multipolar, bipolar, and unipolar neurons. (p. 218)

10. Distinguish among sensory neurons, interneurons, and motor neurons. (p. 218)

11. Distinguish between ganglia and nuclei. (p. 218)

9.5 The Synapse

12. Define *synapse*. (p. 219)

13. Explain how information passes from one neuron to another. (p. 220)

9.6 Cell Membrane Potential

14. Explain how a membrane becomes polarized. (p. 220)

15. Describe how ions associated with nerve cell membranes are distributed. (p. 220)

16. Define *resting potential*. (p. 221)

17. Explain the relationship between threshold potential and an action potential. (p. 223)

18. List the events that occur during an action potential. (p. 223)

9.7 Nerve Impulses

19. Choose the correct sequence of events along an axon: (p. 224)
    a. Resting potentials are propagated along a stimulated axon, causing an overall action potential.
    b. A threshold stimulus opens K^+ channels and the ions diffuse in, depolarizing the cell membrane. Then Na^+ channels open, Na^+ exits, and the cell membrane repolarizes, generating an action potential that stimulates adjacent cell membrane, forming the nerve impulse.
    c. A threshold stimulus opens Na^+ channels and the ions diffuse in, depolarizing the cell membrane. Then K^+ channels open, K^+ exits, and the cell membrane repolarizes, generating an action potential that stimulates adjacent cell membrane, forming the nerve impulse.
    d. A threshold stimulus opens Na^+ channels and the ions diffuse in, depolarizing the cell membrane. Then K^+ channels open, K^+ exits, and the cell membrane repolarizes, generating a nerve impulse that stimulates adjacent cell membrane, forming the action potential.
    e. Action potentials occur at different points along an axon, then join to generate a nerve impulse.

20. Explain why a myelin sheath covering an entire axon would inhibit conduction of a nerve impulse. (p. 225)

21. "All-or-none" response in nerve impulse conduction means that _____. (p. 225)

9.8 Synaptic Transmission

22. Distinguish between excitatory and inhibitory actions of neurotransmitters. (p. 225)

23. Neurotransmitters are synthesized in _____ and are stored in _____. (p. 226)

24. Match the neurotransmitter on the left to its description on the right. (p. 226)

    (1) monoamine   A. Short chains of amino acids
    (2) acetylcholine   B. A modified amino acid
    (3) neuropeptide   C. An amino acid
    (4) GABA   D. Stimulates skeletal muscle contraction

25. Explain what happens to neurotransmitters after they are released. (p. 226)

9.9 Impulse Processing

26. Describe the components of a neuronal pool. (p. 227)

27. "Facilitation in a neuronal pool" refers to _____. (p. 227)

28. Distinguish between convergence and divergence in a neuronal pool. (p. 227)

9.10 Types of Nerves

29. Describe how sensory, motor, and mixed nerves differ. (p. 228)

9.11 Nerve Pathways

30. Distinguish between a reflex arc and a reflex. (p. 228)

31. Describe the components of a reflex arc and their functions. (p. 228)

32. List three body functions that reflexes control. (p. 228)

9.12 Meninges

33. Match the layer of the meninges on the left to its description on the right. (p. 230)

    (1) dura mater
    (2) arachnoid mater
    (3) pia mater

    A. The thin, innermost layer, containing blood vessels and nerves
    B. The tough, outermost layer, consisting mostly of connective tissue
    C. The lacy membrane, lacking blood vessels, sandwiched between the other two layers

9.13 Spinal Cord

34. Describe the structure of the spinal cord. (p. 232)

35. Distinguish between the ascending and descending tracts of the spinal cord. (p. 233)

9.14 Brain

36. Name the four major parts of the brain and describe their general functions. (p. 234)

37. The area of the brain that connects parts of the nervous system to particular visceral activities is the: (p. 234)
    a. cerebrum
    b. cerebellum
    c. brainstem
    d. diencephalon
    e. corpus callosum

38. The structure that connects the cerebral hemispheres is the _____. (p. 235)

39. Distinguish between a sulcus and a fissure. (p. 236)

40. Relate the lobes of the cerebral hemispheres to the skull bones. (p. 236)

41. Locate the motor, sensory, and association areas of the cerebral cortex, and describe the general functions of each. (p. 237)

42. Define *hemisphere dominance*. (p. 237)

43. The function of the basal nuclei is to _____. (p. 238)

44. Locate the ventricles in the brain. (p. 238)

45. Explain how cerebrospinal fluid is produced and how it functions. (p. 239)

46. The part of the diencephalon that regulates hunger, weight, water and electrolyte balance, sleep and wakefulness, temperature, arterial blood pressure, heart rate, production of substances that stimulate the pituitary gland, and movement and secretion in areas of the digestive tract is the: (p. 240)
    a. thalamus
    b. pineal gland
    c. infundibulum
    d. hypothalamus
    e. mammillary bodies

47. Define *limbic system*, and explain its functions. (p. 240)

48. The parts of the brainstem are the _____, _____, and _____. (p. 241)

49. List the functions of the three parts of the brainstem. (p. 241)

50. Vomiting is controlled by: (p. 242)
    a. the reticular formation
    b. a nucleus within the medulla oblongata
    c. the midbrain
    d. the pons
    e. the thalamus

51. Describe what happens to the body when the reticular formation receives sensory impulses, and what happens when it does not receive stimulation. (p. 242)

52. Describe the functions of the cerebellum. (p. 242)

9.15 Peripheral Nervous System

53. Distinguish between the somatic nervous system and the autonomic nervous system. (p. 242)

54. Distinguish between cranial nerves and spinal nerves. (pp. 243, 247)

55. Match the cranial nerves (on the left) to the body parts or functions that they affect (on the right). More than one nerve pair may correspond to the same structure or function. (pp. 243–246)

(1) olfactory nerves (I)	A. Vision
(2) optic nerves (II)	B. Hearing and equilibrium
(3) oculomotor nerves (III)	C. Muscles of the larynx, pharynx, soft palate, sternocleidomastoid and trapezius muscles
(4) trochlear nerves (IV)	
(5) trigeminal nerves (V)	
(6) abducens nerves (VI)	D. Heart, various smooth muscles and glands in the thorax and abdomen
(7) facial nerves (VII)	
(8) vestibulocochlear nerves (VIII)	
(9) glossopharyngeal nerves (IX)	E. Taste, facial expressions, secretion of tears and saliva
(10) vagus nerves (X)	
(11) accessory nerves (XI)	F. Sense of smell
(12) hypoglossal nerves (XII)	G. Tongue movements and swallowing
	H. Face and scalp
	I. Eye movements

56. Explain how the spinal nerves are classified and numbered. (p. 247)

57. Describe the structure of a spinal nerve. (p. 248)

58. Define *plexus,* and locate the major plexuses of the spinal nerves. (p. 248)

9.16 Autonomic Nervous System

59. Describe the general functions of the autonomic nervous system. (p. 248)

60. Distinguish between the sympathetic and parasympathetic divisions of the autonomic nervous system. (p. 249)

61. Distinguish between preganglionic and postganglionic nerve fibers. (p. 249)

62. The effects of the sympathetic and parasympathetic autonomic divisions differ because _____. (p. 251)

63. List two ways in which the CNS controls autonomic activities. (p. 251)

INTEGRATIVE ASSESSMENTS/ CRITICAL THINKING

OUTCOMES 9.1, 9.13, 9.14

1. Discuss the rationale for considering the nervous system in two parts, central and peripheral.

OUTCOMES 9.3, 9.4

2. State two reasons why rapidly growing brain cancers are composed of neuroglial cells rather than neurons.

OUTCOMES 9.3, 9.4, 9.7, 9.13, 9.14

3. In multiple sclerosis, nerve fibers in the CNS lose their myelin. Explain why this loss affects skeletal muscle function.

OUTCOMES 9.4, 9.5, 9.11, 9.13, 9.14

4. List four skills encountered in everyday life that depend on nervous system function, and list the part of the nervous system responsible for each.

OUTCOMES 9.6, 9.8

5. Compare and contrast the roles of potassium and calcium ions in nerve transmission.

OUTCOMES 9.11, 9.13

6. The biceps-jerk reflex is carried out by motor neurons that exit the spinal cord in the fifth spinal nerve (C5). The triceps-jerk reflex uses motor neurons in the seventh spinal nerve (C7). Describe how these reflexes might be tested to help pinpoint damage in a patient with a neck injury.

OUTCOMES 9.11, 9.14

7. Describe the roles of the cerebrum and cerebellum in athletics.

OUTCOMES 9.13, 9.14

8. Describe expected functional losses in a patient who has suffered injury to the right occipital lobe of the cerebral cortex compared to injury in the right temporal lobe of the cerebral cortex.

OUTCOME 9.14

9. Select three parts of the brain and explain why they are essential for survival.

WEB CONNECTIONS

Visit the text website at **aris.mhhe.com** for additional quizzes, interactive learning exercises, and more.

AP R NERVOUS SYSTEM

Anatomy & Physiology | REVEALED includes cadaver photos that allow you to peel away layers of the human body to reveal structures beneath the surface. This program also includes animations, radiologic imaging, audio pronunciation, and practice quizzing. To learn more visit **www.aprevealed.com**.

10

The Senses

THE SOUND OF MUSIC. The band Nirvana and singer Tori Amos have each recorded the song "Smells Like Teen Spirit." In the original, Nirvana version, Kurt Cobain's voice is loud and brash, as is the instrumentation; in contrast, Tori Amos's song is slow and subdued. Yet it is easy to tell that these are the same songs. What isn't easy is figuring out how the brain does this.

Most of the neurons in the auditory cortex sense a certain range of frequencies of incoming sound waves. However, certain neurons are "pitch-sensitive," which means that they can recognize the same note, whether it comes from an oboe or an elephant. This property of sound, called pitch, is a vibration frequency from objects that vibrate periodically. The vibration is complex—plucking a string on an instrument vibrates the entire string, but also vibrates parts of it, creating a complex sound. Pitch-sensitive neurons recognize the "fundamental" vibration, which is the lowest one coming from the entire vibrating object, corresponding to plucking the entire string.

In experiments to identify and localize pitch-sensitive neurons, researchers placed electrodes over the auditory cortices of marmoset monkeys, who hear the same range of sounds as humans. When the monkeys listened to sounds that shared the fundamental vibration, even though different sources made the sounds, the same neurons fired action potentials. Moreover, the pitch-sensitive neurons in the monkey brains were in the same part of the auditory cortex that is damaged in humans who lose the ability to distinguish pitches after suffering a

Experiments in which monkeys listened to music suggest how the human brain processes pitch.

stroke. However, we don't yet know how the brain learns and matches the temporal combination of notes that make up a melody—which is how we perceive that Kurt Cobain and Tori Amos sang the same song. Presumably memory is part of the picture, which may explain why we can remember lyrics to a song many years after last hearing it but cannot remember what we learned in a class just a day ago.

In an evolutionary sense, pitch sensitivity enabled our ancestors in the treetops to recognize and distinguish the rhythmic sounds from other animals. In our own species, that ability has given us the gift of music.

Learning Outcomes   *After studying this chapter, you should be able to do the following:*

Aids to Understanding Words

(Appendix A on page 567 has a complete list of Aids to Understanding Words.)

choroid [skinlike] *choroid* coat: Middle, vascular layer of the eye.

cochlea [snail] *cochlea:* Coiled tube in the inner ear.

iris [rainbow] *iris:* Colored, muscular part of the eye.

labyrinth [maze] *labyrinth:* Complex system of connecting chambers and tubes of the inner ear.

lacri- [tears] *lacri*mal gland: Tear gland.

macula [spot] *macula* lutea: Yellowish spot on the retina.

olfact- [to smell] *olfact*ory: Pertaining to the sense of smell.

scler- [hard] *scler*a: Tough, outer protective layer of the eye.

tympan- [drum] *tympan*ic membrane: Eardrum.

vitre- [glass] *vitre*ous humor: Clear, jellylike substance within the eye.

10.1 INTRODUCTION

How dull life would be without sight and sound, smell and taste, touch and balance. Our senses are necessary not only for us to enjoy life, but to survive. *Sensory receptors* detect environmental changes and trigger nerve impulses that travel on sensory pathways into the central nervous system for processing and interpretation. The body reacts with a particular feeling or sensation.

Sensory receptors vary greatly but fall into two major categories. Receptors associated with the *somatic senses* of touch, pressure, temperature, and pain form one group. These receptors are widely distributed throughout the skin and deeper tissues, and are structurally simple. Receptors of the second type are parts of complex, specialized sensory organs that provide the *special senses* of smell, taste, hearing, equilibrium, and vision.

10.2 RECEPTORS, SENSATIONS, AND PERCEPTION

Sensory receptors are diverse but share certain features. Each type of receptor is particularly sensitive to a distinct kind of environmental change and is much less sensitive to other forms of stimulation. The raw form in which these receptors send information to the brain is called sensation. The way our brains interpret this information is called perception.

Types of Receptors

Sensory receptors are categorized into five types according to their sensitivities: **Chemoreceptors** (ke″mo-re-sep′torz) are stimulated by changes in the chemical concentration of substances; **pain receptors** (pān re-sep′torz) by tissue damage; **thermoreceptors** (ther′mo-re-sep′torz) by changes in temperature; **mechanoreceptors** (mek″ah-no-re-sep′torz) by changes in pressure or movement; and **photoreceptors** (fo″to-re-sep′torz) by light energy.

Sensations and Perception

A **sensation** occurs when the brain becomes aware of sensory impulses. A *perception* occurs when the brain interprets those sensory impulses. Because all the nerve impulses that travel away from sensory receptors into the central nervous system are alike, the resulting sensation depends on which region of the brain receives the impulse. For example, impulses reaching one region are always interpreted as sounds, and those reaching another are always sensed as touch. (Some receptors, such as those that measure oxygen levels in the blood, do not trigger sensations.)

At the same time that a sensation forms, the cerebral cortex causes the feeling to seem to come from the stimulated receptors. This process is called **projection** (pro-jek′shun) because the brain projects the sensation back to its apparent source. Projection allows a person to pinpoint the region of stimulation; thus, the eyes seem to see, and the ears seem to hear.

Sensory Adaptation

The brain must prioritize the sensory input it receives, or it would be overwhelmed by unimportant information. For example, until this sentence prompts you to think about it, you are probably unaware of the pressure of your clothing against your skin, or the background noise in the room. This ability to ignore unimportant stimuli is called **sensory adaptation** (sen′so-re ad″ap-ta′shun), and it may result from receptors becoming unresponsive (*peripheral adaptation*) or inhibition along the central nervous system pathways leading to the sensory regions of the cerebral cortex (*central adaptation*).

Check Your Recall

1. List five general types of sensory receptors.
2. Explain how a sensation occurs.
3. What is sensory adaptation?

10.3 GENERAL SENSES

General senses are widespread, and are associated with receptors in the skin, muscles, joints, and viscera. They include the senses of touch and pressure, temperature, and pain.

Touch and Pressure Senses

The senses of touch and pressure derive from three kinds of receptors (fig. 10.1). These receptors sense mechanical forces that deform or displace tissues. Touch and pressure receptors include:

1. **Free nerve endings** These receptors are common in epithelial tissues, where their free ends extend between epithelial cells. They are responsible for the sensation of itching.
2. **Tactile (Meissner's) corpuscles** These are small, oval masses of flattened connective tissue cells within connective tissue sheaths. Two or more sensory nerve fibers branch into each corpuscle and end within it as tiny knobs.

Tactile corpuscles are abundant in the hairless portions of the skin, such as the lips, fingertips, palms, soles, nipples, and external genital organs. They respond to the motion of objects that barely contact the skin, interpreting impulses from them as the sensation of light touch.

3. **Lamellated (Pacinian) corpuscles** These sensory bodies are relatively large structures composed of connective tissue fibers and cells. They are common in the deeper dermal and subcutaneous tissues and in muscle tendons and joint ligaments. Lamellated corpuscles respond to heavy pressure and are associated with the sensation of deep pressure.

Temperature Senses

Temperature sensation depends on two types of free nerve endings in the skin. Those that respond to warmer temperatures are called *warm receptors,* and those that respond to colder temperatures are called *cold receptors.*

Warm receptors are most sensitive to temperatures above 25°C (77°F) and become unresponsive at tem-

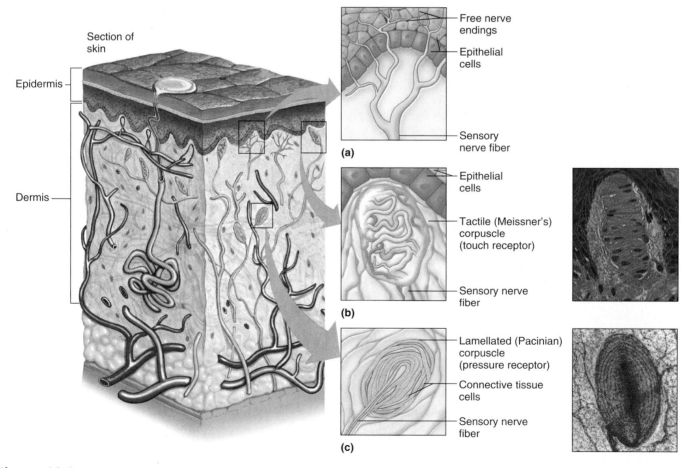

Figure 10.1

Touch and pressure receptors include (*a*) free ends of sensory nerve fibers, (*b*) tactile corpuscles (with 225× micrograph), and (*c*) lamellated corpuscles (with 50× micrograph).

peratures above 45°C (113°F). Temperatures near and above 45°C stimulate pain receptors, producing a burning sensation.

Cold receptors are most sensitive to temperatures between 10°C (50°F) and 20°C (68°F). Temperatures below 10°C stimulate pain receptors, producing a freezing sensation.

Both warm and cold receptors adapt rapidly. Within about a minute of continuous stimulation, the sensation of warmth or cold begins to fade.

Sense of Pain

Certain receptors that consist of free nerve endings sense pain. These receptors are widely distributed throughout the skin and internal tissues, except in the nervous tissue of the brain, which lacks pain receptors.

Pain receptors protect the body because tissue damage stimulates them. Pain sensation is usually perceived as unpleasant, and it signals a person to act to remove the stimulation. Pain receptors adapt poorly, if at all. Once a pain receptor is activated, even by a single stimulus, it may send impulses into the central nervous system for some time. Thus, pain may persist.

The way in which tissue damage stimulates pain receptors is poorly understood. Injuries likely promote the release of certain chemicals that build up and stimulate pain receptors. Deficiency of oxygen-rich blood (ischemia) in a tissue or stimulation of certain mechanoreceptors also triggers pain sensations. For example, the pain elicited during a muscle cramp stems from sustained contraction that squeezes capillaries and interrupts blood flow. Stimulation of mechanical-sensitive pain receptors also contributes to the sensation.

> Injuries to bones, tendons, or ligaments stimulate pain receptors that may also contract nearby skeletal muscles. The contracting muscles may become ischemic, which may trigger still other pain receptors within the muscle tissue, further increasing muscular contraction.

Visceral Pain

As a rule, pain receptors are the only receptors in viscera whose stimulation produces sensations. Pain receptors in these organs respond differently to stimulation than those associated with surface tissues. For example, localized damage to intestinal tissue during surgical procedures may not elicit pain sensations, even in a conscious person. However, when visceral tissues are subjected to more widespread stimulation, as when intestinal tissues are stretched or smooth muscles in intestinal walls undergo spasms, a strong pain sensation may follow. Once again, the resulting pain seems to stem from stimulation of mechanoreceptors and from decreased blood flow accompanied by lower tissue oxy-

gen concentration and accumulation of pain-stimulating chemicals via chemoreceptors. The Topic of Interest on page 265 discusses blood flow and headaches.

Visceral pain may feel as if it is coming from some part of the body other than the part being stimulated, a phenomenon called **referred pain.** For example, pain originating in the heart may be referred to the left shoulder or left upper limb (fig. 10.2). Referred pain may arise from common nerve pathways that carry sensory impulses from skin areas as well as viscera. For example, pain impulses from the heart travel over the same nerve pathways as those from the skin of the left shoulder and left upper limb (fig. 10.3). Consequently, during a heart attack, the cerebral cortex may incorrectly interpret the source of the pain impulses as the left shoulder or upper limb, rather than the heart.

> ## Check Your Recall
> 4. Describe the three types of touch and pressure receptors.
> 5. Describe the receptors that sense temperature.
> 6. What types of stimuli excite pain receptors?
> 7. What is referred pain?

Pain Nerve Fibers

Nerve fibers that conduct impulses away from pain receptors are of two main types: acute pain fibers and chronic pain fibers. *Acute pain fibers* are relatively thin, myelinated nerve fibers. They conduct nerve impulses rapidly and are associated with the sensation of sharp pain, which typically originates from a restricted area of the skin and seldom continues after the pain-producing stimulus stops. *Chronic pain fibers* are thin, unmyelinated nerve fibers. They conduct impulses more slowly and produce a dull, aching sensation that may be diffuse and difficult to pinpoint. Such pain may continue for some time after the original stimulus ceases. Acute pain is usually sensed as coming only from the skin; chronic pain is felt in deeper tissues as well.

An event that stimulates pain receptors usually triggers impulses on both acute and chronic pain fibers. This causes a dual sensation—a sharp, pricking pain, followed shortly by a dull, aching one. The aching pain is usually more intense and may worsen with time. Chronic pain can cause prolonged suffering.

Pain impulses that originate from the head reach the brain on sensory fibers of cranial nerves. All other pain impulses travel on the sensory fibers of spinal nerves, and they pass into the spinal cord by way of the dorsal roots of these spinal nerves. Within the spinal cord, neurons process pain impulses in the gray matter of the dorsal horn, and the impulses are transmitted to the brain. Here, most pain fibers terminate in the reticular formation

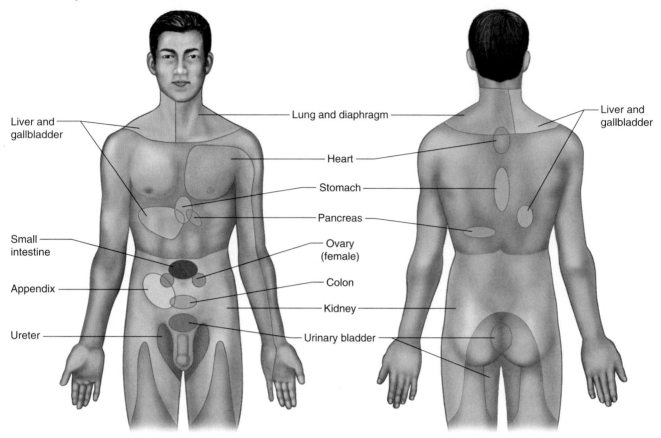

Figure 10.2

Referred pain feels as if it is coming from a different body part than the one being stimulated. Visceral pain may be felt at the surface regions indicated in the illustration.

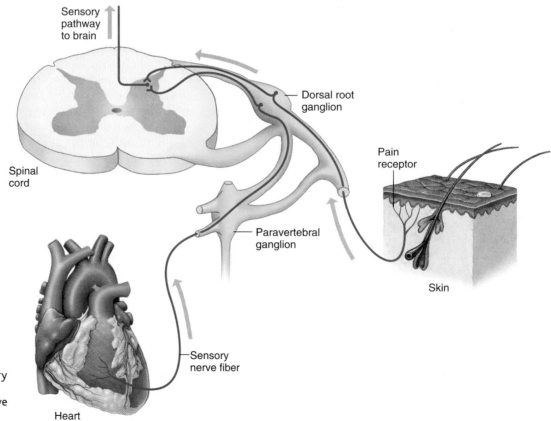

Figure 10.3

Pain originating in the heart may feel as if it is coming from the skin because sensory impulses from the heart and the skin follow common nerve pathways to the brain.

Topic of Interest

Headache

Headaches are a common type of pain. Although the nervous tissue of the brain lacks pain receptors, nearly all the other tissues of the head, including the meninges and blood vessels, are richly innervated.

Many headaches are associated with stressful life situations that cause fatigue, emotional tension, anxiety, or frustration. These conditions can trigger various physiological changes, such as prolonged contraction of the skeletal muscles in the forehead, sides of the head, or back of the neck. Such contractions may stimulate pain receptors and produce a *tension headache*. More severe *vascular headaches* accompany constriction or dilation of the cranial blood vessels. For example, the throbbing headache of a "hangover" from drinking too much alcohol may be due to blood pulsating through dilated cranial vessels.

Migraine is another form of vascular headache. In this disorder, certain cranial blood vessels constrict, producing a localized cerebral blood deficiency. This causes a variety of symptoms, such as seeing patterns of bright light that obstruct vision or feeling numbness in the limbs or face. Typically, vasoconstriction subsequently leads to vasodilation of the affected vessels, causing a severe headache, which usually affects one side of the head and may last for several hours. Effective drug treatment is now available for migraines.

There are other causes of headaches. These include sensitivity to food additives, high blood pressure, increased intracranial pressure due to a tumor or to blood escaping from a ruptured vessel, decreased cerebrospinal fluid pressure following a lumbar puncture, and sensitivity to or withdrawal from certain drugs.

(see chapter 9, p. 242). From there, other neurons conduct impulses to the thalamus, hypothalamus, and cerebral cortex.

Regulation of Pain Impulses

Awareness of pain arises when pain impulses reach the thalamus—that is, even before they reach the cerebral cortex. The cerebral cortex, however, determines pain intensity, locates the pain source, and mediates emotional and motor responses to the pain.

Areas of gray matter in the midbrain, pons, and medulla oblongata regulate movement of pain impulses from the spinal cord. Impulses from special neurons in these brain areas descend in the lateral funiculus (see chapter 9, p. 232) to various levels of the spinal cord. These impulses stimulate the ends of certain nerve fibers to release biochemicals that can block pain signals by inhibiting presynaptic nerve fibers in the posterior horn of the spinal cord.

The inhibiting substances released in the posterior horn include neuropeptides called *enkephalins* and the monoamine *serotonin* (see chapter 9, p. 226). Enkephalins can suppress acute and chronic pain impulses and thus can relieve severe pain, much as morphine and other opiate drugs do. In fact, enkephalins bind to the same receptor sites on neuron membranes as does morphine. Serotonin stimulates other neurons to release enkephalins.

Endorphins are another group of neuropeptides with pain-suppressing, morphinelike actions. Endorphins are found in the pituitary gland and the hypothalamus. Enkephalins and endorphins are released in response to extreme pain and provide natural pain control.

Neuropathic pain is an overreaction to a stimulus that would ordinarily cause pain, or a pain response to a normally innocuous stimulus. Reflex sympathetic dystrophy is a form of neuropathic pain that causes an intense burning sensation in a hand or foot, even if the extremity is paralyzed or has been amputated. During the Civil War, it was called "causalgia." Union Army Surgeon S. Weir Mitchell described causalgia as "the most terrible of all tortures."

Check Your Recall

8. Describe two types of pain fibers.
9. How do acute pain and chronic pain differ?
10. What parts of the brain interpret pain impulses?
11. How do neuropeptides help control pain?

10.4 SPECIAL SENSES

Special senses are those whose sensory receptors are within large, complex sensory organs in the head. These senses and their respective organs include the following:

- Smell ⟶ Olfactory organs
- Taste ⟶ Taste buds
- Hearing ⟶ Ears
- Equilibrium ⟶ Ears
- Sight ⟶ Eyes

10.5 SENSE OF SMELL

The sense of smell is associated with complex sensory structures in the upper region of the nasal cavity.

Olfactory Receptors

Smell (olfactory) receptors and taste receptors are chemoreceptors, which means that chemicals dissolved in liquids stimulate them. Smell and taste function closely together and aid in food selection because we usually smell food at the same time we taste it.

Olfactory Organs

The **olfactory organs,** which contain the olfactory receptors, are yellowish-brown masses of epithelium about the size of postage stamps that cover the upper parts of the nasal cavity, the superior nasal conchae, and a portion of the nasal septum. **Olfactory receptor cells** are bipolar neurons surrounded by columnar epithelial cells (fig. 10.4). Hairlike cilia cover tiny knobs at the distal ends of these neurons' dendrites. In any particular such neuron, the cilia harbor many copies of one type of olfactory receptor protein. Chemicals called odorant molecules stimulate different sets of olfactory receptor proteins, and therefore different sets of olfactory receptor cells, to send a signal of a detected odor to the brain. Odorant molecules enter the nasal cavity as gases, but they must dissolve at least partially in the watery fluids that surround the cilia before receptors can detect them.

Olfactory Nerve Pathways

Stimulated olfactory receptor cells send nerve impulses along their axons. These fibers (which form the first cranial nerves) synapse with neurons located in enlargements called **olfactory bulbs.** These structures lie on either side of the crista galli of the ethmoid bone (see fig. 7.14, p. 145). Within the olfactory bulbs, the impulses are analyzed, and as a result, additional impulses travel along the **olfactory tracts** to the limbic system (see chapter 9, p. 240). The major interpreting areas for these impulses in the olfactory cortex lie within the temporal lobes and at the bases of the frontal lobes, anterior to the hypothalamus.

Humans smell the world using about 12 million olfactory receptor cells. Bloodhounds have 4 billion such cells—and hence a much better sense of smell. Of the 1,000 or so genes that encode human olfactory receptor proteins, about 425 have mutated into inactivity. In monkeys, apes, dogs, and mice, a much higher proportion of their olfactory receptor genes remain active—and these animals rely more on the sense of smell to identify food than humans do. Evolution has apparently diminished the human sense of smell compared to that of other mammals.

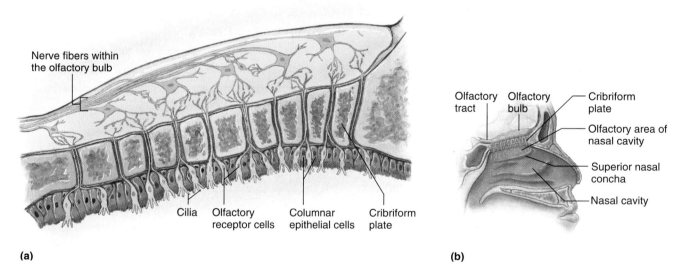

(a)

(b)

Figure 10.4

Olfactory receptors convey the sense of smell. (a) Columnar epithelial cells support olfactory receptor cells, which have cilia at their distal ends. The actual olfactory receptors, which are proteins, are on the cilia. Binding of odorants to these receptors in distinctive patterns conveys the information that the brain interprets as an odor. (b) The olfactory area is associated with the superior nasal concha.

Olfactory Stimulation

When odorant molecules bind to olfactory receptor proteins in olfactory receptor cell membranes, a biochemical pathway is activated that culminates in an influx of sodium ions, which may trigger an action potential if the depolarization reaches threshold. The action potentials from this and other olfactory receptor cells travel to the olfactory bulbs in the brain, where the sensation of smell arises.

The several hundred types of olfactory receptor cells can code for many thousands of odors when they signal the brain in groups. That is, an odorant molecule stimulates a distinct set of receptor types. Experiments have shown that although an olfactory receptor cell has only one type of olfactory receptor, that receptor can bind several different types of odorant molecules. In addition, any one odorant molecule can bind several different receptors. The brain interprets this binding information as a combinatorial olfactory code. In a simplified example, if there are ten odor receptors, parsley might stimulate receptors 3, 4, and 8, while chocolate might stimulate receptors 1, 5, and 10. Researchers have recently identified olfactory and taste receptors by searching human genome sequence information for proteins that reside in cell membranes and are found only in certain receptor cells.

Because the olfactory organs are located high in the nasal cavity above the usual pathway of inhaled air, a person may have to sniff and force air up to the receptor areas to smell a faint odor. Astronauts on the first space flights could not smell their food because they had to squeeze the food from tubes directly into their mouths—odorant molecules were not inhaled as they usually are.

The sense of smell adapts rapidly, but adaptation to one scent will not diminish sensitivity to new odors. For example, a person visiting a fish market might at first be acutely aware of the fishy smell, but then that odor fades. If a second person enters the fish market wearing a strong perfume, the person already there will detect the scent.

> Partial or complete loss of smell is called *anosmia*. It may result from inflammation of the nasal cavity lining due to a respiratory infection, tobacco smoking, or using certain drugs, such as cocaine.

Check Your Recall

12. Where are olfactory receptors located?
13. Trace the pathway of an olfactory impulse from a receptor to the cerebrum.

10.6 SENSE OF TASTE

Taste buds are the special organs of taste (fig. 10.5). The 10,000 or so taste buds are located primarily on the surface of the tongue and are associated with tiny elevations called *papillae*. About 1,000 taste buds are scattered in the roof of the mouth and walls of the throat.

Taste Receptors

Each taste bud includes 50 to 150 modified epithelial cells which function as receptor cells, the **taste cells** (gustatory cells). Each taste cell is replaced every three days. The taste bud also includes epithelial supporting cells. The entire structure is spherical, with an opening, the **taste pore,** on its free surface. Tiny projections called **taste hairs** protrude from the outer ends of the taste cells and extend from the taste pore. These taste hairs are the sensitive parts of the receptor cells.

Interwoven among the taste cells and wrapped around them is a network of nerve fibers. Stimulation of a receptor cell triggers an impulse on a nearby nerve fiber, and the impulse then travels into the brain.

> Cats and dogs may be satisfied with less varied diets than humans because cats have only about 473 taste buds and dogs about 1,700.

Before a particular chemical can be tasted, it must dissolve in the watery fluid surrounding the taste buds. The salivary glands provide this fluid. Food molecules bind to specific receptor proteins embedded in taste hairs on the taste cells. The pattern of receptor types that bind food molecules and generate sensory impulses on nearby nerve fibers is interpreted as a particular taste sensation. Therefore, the chemical senses of smell and taste arise from molecules from the environment that bind receptors on neurons specialized as sensory receptors.

The taste cells in all taste buds appear alike microscopically, but are of at least five types. Each type is most sensitive to a particular kind of chemical stimulus, producing at least five primary taste (gustatory) sensations.

Taste Sensations

The five primary taste sensations are:

1. *Sweet,* such as table sugar
2. *Sour,* such as a lemon
3. *Salty,* such as table salt
4. *Bitter,* such as caffeine or quinine
5. *Umami* (a Japanese term meaning "delicious"), a response to certain amino acids and their chemical relatives, such as monosodium glutamate.

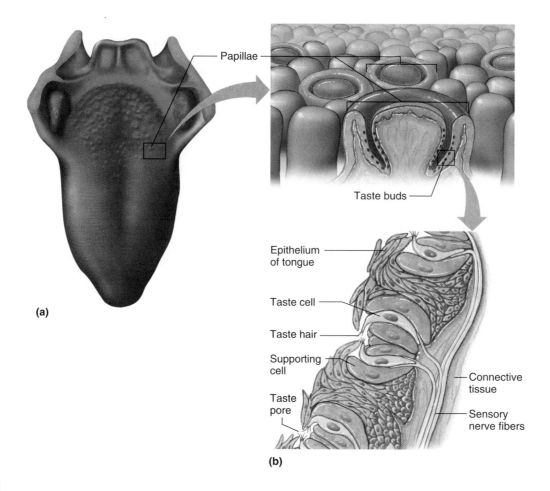

(a)

(b)

Figure 10.5

Taste receptors. (*a*) Taste buds on the surface of the tongue are associated with nipplelike elevations called papillae. (*b*) A taste bud contains taste cells and has an opening, the taste pore, at its free surface.

Some investigators recognize other taste sensations—*alkaline* and *metallic*.

A flavor results from one or a combination of the primary sensations. Experiencing flavors involves tasting, which reflects the concentrations of stimulating chemicals, as well as smelling and feeling the texture and temperature of foods. Furthermore, the chemicals in some foods—chili peppers and ginger, for instance—may stimulate pain receptors, which cause a burning sensation. In fact, the chemical in chili peppers that tastes "hot"—capsaicin—actually stimulates heat receptors.

Each taste cell is thought to respond to one taste sensation only, with distinct receptors, according to experiments done to date. Taste cells with each of the five taste sensations are found in all areas of the tongue, but they are distributed such that each sensation seems to arise most strongly from a particular region. The tip of the tongue is most sensitive to sweet stimuli, the margins of the tongue most sensitive to sourness, the back of the tongue more likely to detect bitter substances, and responsiveness to salt quite widely distributed.

Taste sensation, like the sense of smell, undergoes adaptation rapidly. Moving bits of food over the surface of the tongue to stimulate different receptors at different moments keeps us from losing taste due to sensory adaptation.

Taste Nerve Pathways

Sensory impulses from taste receptor cells in the tongue travel on fibers of the facial, glossopharyngeal, and vagus nerves into the medulla oblongata. From there, the impulses ascend to the thalamus and are directed to the gustatory cortex, which is located in the parietal lobe of the cerebrum, along a deep portion of the lateral sulcus (see fig. 9.28, p. 236).

Check Your Recall

14. Why is saliva necessary for the sense of taste?

15. Name the five primary taste sensations.

16. Trace a sensory impulse from a taste receptor to the cerebral cortex.

10.7   SENSE OF HEARING

The organ of hearing, the ear, has outer, middle, and inner parts. The ear also functions in the sense of equilibrium.

Outer (External) Ear

The outer ear consists of three parts. The first is an outer, funnel-like structure called the **auricle** (aw'ri-kl) (pinna). The second is an S-shaped tube called the **external acoustic meatus** (me-a'tus), or external auditory canal, that leads inward through the temporal bone for about 2.5 centimeters (fig. 10.6).

The transmission of vibrations through matter produces sound. Vibrating strings or reeds produce the sounds of some musical instruments, and vibrating vocal folds in the larynx produce the voice. The auricle of the ear helps collect sound waves traveling through the air and directs them into the external acoustic meatus. The meatus terminates with the **eardrum** (tympanic membrane).

The eardrum is a semitransparent membrane covered by a thin layer of skin on its outer surface and by mucous membrane on the inside. It has an oval margin and is cone-shaped, with the apex of the cone directed inward. The attachment of one of the auditory ossicles (the malleus) maintains the eardrum's cone shape.

Sound waves that enter the external acoustic meatus change the pressure on the eardrum, which moves back and forth in response and thus reproduces the vibrations of the sound wave source.

Middle Ear

The middle ear, or *tympanic cavity,* is an air-filled space in the temporal bone. It contains three small bones called **auditory ossicles** (aw'di-to"re os'i-klz): the *malleus,* the *incus,* and the *stapes* (fig. 10.7). Tiny ligaments attach them to the wall of the tympanic cavity, and they are covered by mucous membrane. These bones bridge the eardrum and the inner ear, transmitting vibrations between these parts. Specifically, the malleus attaches to the eardrum, and when the eardrum vibrates, the malleus vibrates in unison. The malleus causes the incus to vibrate, and the incus passes the movement on to the stapes. Ligaments hold the stapes to an opening in the wall of the tympanic cavity called the **oval window,** which leads into the inner ear. Vibration of the stapes at the oval window moves a fluid within the inner ear, which stimulates the hearing receptors.

In addition to transmitting vibrations, the auditory ossicles help increase (amplify) the force of vibrations as they pass from the eardrum to the oval window. Because the ossicles transmit vibrations from the relatively large surface of the eardrum to a much smaller area at the oval window, the vibrational force concentrates as it

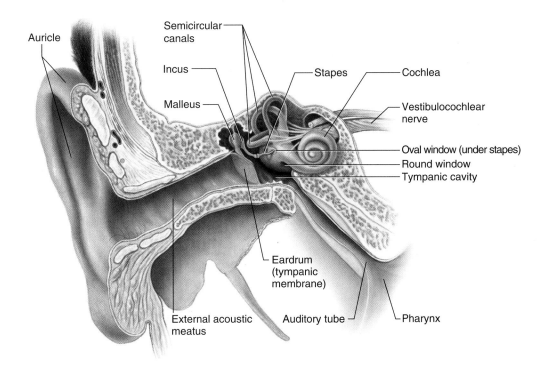

Figure 10.6

Major parts of the ear. The outer ear includes the auricle, external acoustic meatus, and eardrum. The middle ear includes the auditory ossicles (malleus, incus, and stapes) and the oval window. The inner ear includes the semicircular canals and the cochlea.

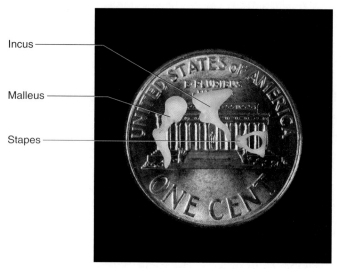

Incus

Malleus

Stapes

Figure 10.7

The auditory ossicles—the malleus, incus, and stapes—are bones that bridge the eardrum and the inner ear (2.5×) (see fig. 10.6). Comparison to a penny emphasizes their tiny size.

moves from the outer to the inner ear. As a result, the pressure (per square millimeter) that the stapes applies on the oval window is many times greater than the pressure that sound waves exert on the eardrum.

Auditory Tube

An **auditory tube** (aw'di-to"re tūb) (eustachian tube) connects each middle ear to the throat. This tube conducts air between the tympanic cavity and the outside of the body by way of the throat (nasopharynx) and mouth. The auditory tube helps maintain equal air pressure on both sides of the eardrum, which is necessary for normal hearing.

The function of the auditory tube is noticeable during rapid changes in altitude. As a person moves from a high altitude to a lower one, air pressure on the outside of the eardrum increases. This may push the eardrum inward, impairing hearing. When the air pressure difference is great enough, air movement through the auditory tube equalizes the pressure on both sides of the eardrum, and the membrane moves back into its regular position. This produces a popping sound, which restores normal hearing.

Because auditory tube mucous membranes connect directly with middle ear linings, mucous membrane infections of the throat may spread through these tubes and cause middle ear infection. Pinching a nostril when blowing the nose may force material from the throat up the auditory tube and into the middle ear.

Inner (Internal) Ear

The inner ear is a complex system of communicating chambers and tubes called a **labyrinth** (lab'ĭ-rinth). Each ear has two parts to the labyrinth—the *osseous labyrinth* and the *membranous labyrinth* (fig. 10.8). The osseous labyrinth is a bony canal in the temporal bone. The membranous labyrinth is a tube that lies within the osseous labyrinth and has a similar shape. Between the osseous and membranous labyrinths is a fluid called *perilymph,* which is secreted by cells in the wall of the bony canal. The membranous labyrinth contains another fluid, called *endolymph.*

The parts of the labyrinths include three **semicircular canals,** which provide a sense of equilibrium (discussed in section 10.8 on page 273), and a **cochlea** (kok'le-ah), which functions in hearing. The cochlea has a bony core and a thin, bony shelf that winds around the core like the threads of a screw. The shelf divides the osseous labyrinth of the cochlea into upper and lower compartments. The upper compartment, called the *scala vestibuli,* leads from the oval window to the apex of the spiral. The lower compartment, the *scala tympani,* extends from the apex of the cochlea to a membrane-covered opening in the wall of the inner ear called the **round window** (fig. 10.8).

The portion of the membranous labyrinth within the cochlea is called the *cochlear duct.* It lies between the two bony compartments and ends as a closed sac at the apex of the cochlea. The cochlear duct is separated from the scala vestibuli by a *vestibular membrane* (Reissner's membrane) and from the scala tympani by a *basilar membrane* (fig. 10.9).

The basilar membrane has many thousands stiff, elastic fibers, which lengthen from the base of the cochlea to its apex. Sound vibrations entering the perilymph at the oval window travel along the scala vestibuli and pass through the vestibular membrane and into the endolymph of the cochlear duct, where they move the basilar membrane.

After passing through the basilar membrane, the vibrations enter the perilymph of the scala tympani. Movements of the membrane covering the round window dissipate the vibrations into the tympanic cavity.

The **spiral organ** (organ of Corti) has hearing receptors. It is located on the upper surface of the basilar membrane and stretches from the apex to the base of the cochlea (fig. 10.9). The receptor cells, called *hair cells,* are organized in rows and have many hairlike processes that project into the endolymph of the cochlear duct. Above these hair cells is a *tectorial membrane* attached to the bony shelf of the cochlea, passing over the receptor cells and contacting the tips of their hairs.

As sound vibrations pass through the inner ear, the hairs shear back and forth against the tectorial membrane, and the resulting mechanical deformation of the

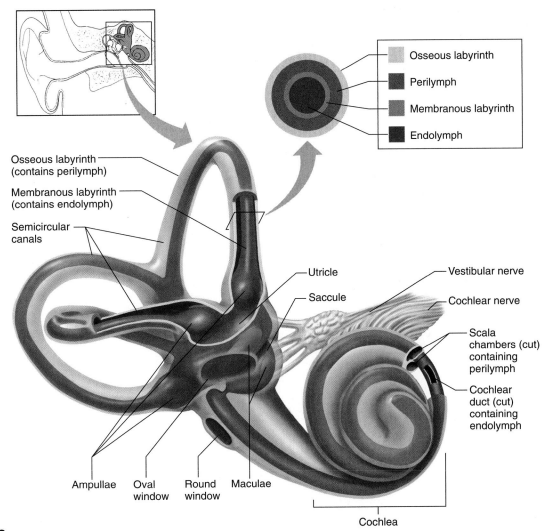

Figure 10.8

A closer look at the inner ear. Perilymph separates the osseous labyrinth of the inner ear from the membranous labyrinth, which contains endolymph.

hairs stimulates the hair cells (figs. 10.9 and 10.10). Some of the cells, however, have slightly different sensitivities to deformation. Thus, two frequencies of sound waves excite different sets of hair cells.

Hair cells are epithelial, but function somewhat like neurons. For example, when a hair cell is at rest, its membrane is polarized. When it is stimulated, selective ion channels open, depolarizing the membrane and making it more permeable to calcium ions. The hair cell has no axon or dendrites, but it has neurotransmitter-containing vesicles near its base. As calcium ions diffuse into the cell, some of these vesicles fuse with the cell membrane and release a neurotransmitter. The neurotransmitter stimulates the ends of nearby sensory nerve fibers, and in response, they transmit impulses along the cochlear branch of the vestibulocochlear nerve to the auditory cortex of the temporal lobe of the brain.

The ear of a young person with normal hearing can detect sound waves with frequencies ranging from 20 to 20,000 or more vibrations per second. The range of greatest sensitivity is 2,000–3,000 vibrations per second. Table 10.1 summarizes the steps of hearing.

Units called *decibels* (dB) measure sound intensity on a logarithmic scale. The decibel scale begins at 0 dB, which is the intensity of the sound that is least perceptible by a normal human ear. A sound of 10 dB is 10 times as intense as the least perceptible sound; a sound of 20 dB is 100 times as intense; and a sound of 30 dB is 1,000 times as intense. A whisper has an intensity of about 40 dB, normal conversation measures 60–70 dB, and heavy traffic or a ringing telephone produces about 80 dB. A sound of 120 dB, common at a rock concert, produces discomfort, and a sound of 140 dB, such as that emitted by a jet plane at takeoff, causes pain.

Frequent or prolonged exposure to sounds with intensities above 85 dB can damage hearing receptors and cause permanent hearing loss. Many rock musicians suffer hearing loss due to years of exposure to loud sounds. More common sources of sounds that, if prolonged, damage hearing are boom boxes, car alarms, and leaf blowers.

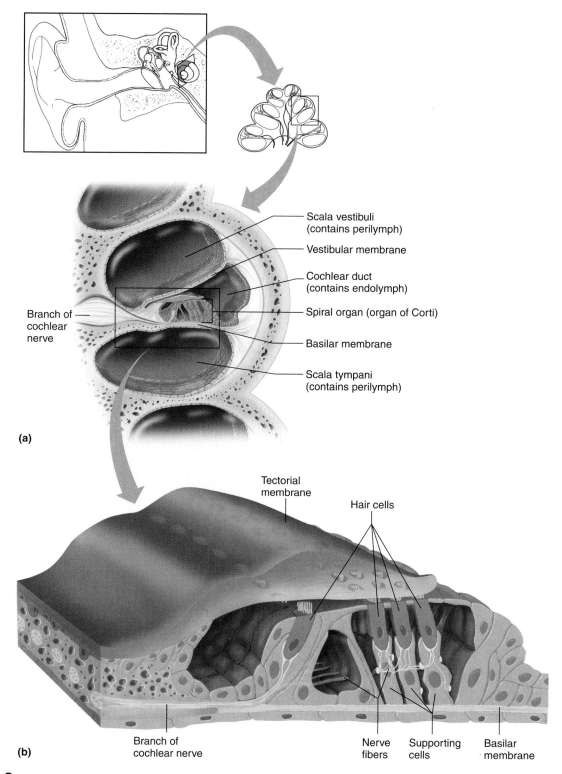

(a)

Scala vestibuli
(contains perilymph)

Vestibular membrane

Cochlear duct
(contains endolymph)

Spiral organ (organ of Corti)

Basilar membrane

Scala tympani
(contains perilymph)

Branch of cochlear nerve

Tectorial membrane

Hair cells

(b)

Branch of cochlear nerve

Nerve fibers

Supporting cells

Basilar membrane

Figure 10.9

The cochlea. (*a*) Cross section of the cochlea. (*b*) The spiral organ and the tectorial membrane.

Auditory Nerve Pathways

The nerve fibers associated with hearing enter the auditory nerve pathways, which pass into the auditory cortices of the temporal lobes of the cerebrum, where they are interpreted. On the way, some of these fibers cross over, so that impulses arising from each ear are interpreted on both sides of the brain. Consequently, damage to a temporal lobe on one side of the brain does not necessarily cause complete hearing loss in the ear on that side.

Figure 10.10

Scanning electron micrograph of hair cells in the spiral organ (3,800×).

Table 10.1	Steps in the Generation of Sensory Impulses from the Ear

1. Sound waves enter external acoustic meatus.
2. Waves of changing pressures cause eardrum to reproduce vibrations coming from sound wave source.
3. Auditory ossicles amplify and transmit vibrations to end of stapes.
4. Movement of stapes at oval window transmits vibrations to perilymph in scala vestibuli.
5. Vibrations pass through the vestibular membrane and enter endolymph of cochlear duct.
6. Different frequencies of vibration in endolymph stimulate different sets of receptor cells.
7. As a receptor cell depolarizes, its membrane becomes more permeable to calcium ions.
8. Inward diffusion of calcium ions causes vesicles at the base of the receptor cell to release neurotransmitter.
9. Neurotransmitter stimulates ends of nearby sensory neurons.
10. Sensory impulses are triggered on fibers of the cochlear branch of vestibulocochlear nerve.
11. Auditory cortex of temporal lobe interprets sensory impulses.

A variety of factors can cause partial or complete hearing loss, including interference with the transmission of vibrations to the inner ear (*conductive deafness*) or damage to the cochlea, auditory nerve, or auditory nerve pathways (*sensorineural deafness*). Conductive deafness may be due to plugging of the external acoustic meatus or to changes in the eardrum or auditory ossicles. For example, the eardrum may harden as a result of disease and become less responsive to sound waves, or disease or injury may tear or perforate the eardrum. Sensorineural deafness can be caused by loud sounds, tumors in the central nervous system, brain damage as a result of vascular accidents, or use of certain drugs.

10.8 SENSE OF EQUILIBRIUM

The sense of equilibrium is really two senses—static equilibrium and dynamic equilibrium—that come from different sensory organs. The organs of **static equilibrium** (stat′ik e″kwĭ-lib′re-um) sense the position of the head, maintaining stability and posture when the head and body are still. When the head and body suddenly move or rotate, the organs of **dynamic equilibrium** (di-nam′ik e″kwĭ-lib′re-um) detect such motion and aid in maintaining balance.

Static Equilibrium

The organs of static equilibrium are located within the **vestibule,** a bony chamber between the semicircular canals and the cochlea. The membranous labyrinth inside the vestibule consists of two expanded chambers—a **utricle** (u′trĭ-kl) and a **saccule** (sak′ūl) (see fig. 10.8).

Each of these chambers has a tiny structure called a **macula** (mak′u-lah). Maculae have many hair cells, which serve as sensory receptors. When the head is upright, the hairs of the hair cells project upward into a mass of gelatinous material, which has grains of calcium carbonate (otoliths) embedded in it. These particles add weight to the gelatinous structure.

The head bending forward, backward, or to one side stimulates hair cells. Such movements tilt the gelatinous masses of the maculae, and as they sag in response to gravity, the hairs projecting into them bend. This action stimulates the hair cells, and they signal the nerve fibers associated with them in a manner similar to that of hearing receptors. The nerve impulses travel into the central nervous system on the vestibular branch of the vestibulocochlear nerve, informing the brain of the head's new position. The brain responds by sending motor impulses to skeletal muscles, which contract or relax to maintain balance (fig. 10.11).

Dynamic Equilibrium

The organs of dynamic equilibrium are the three semicircular canals located in the labyrinth. They detect motion of the head and aid in balancing the head and body during sudden movement. These canals lie at right

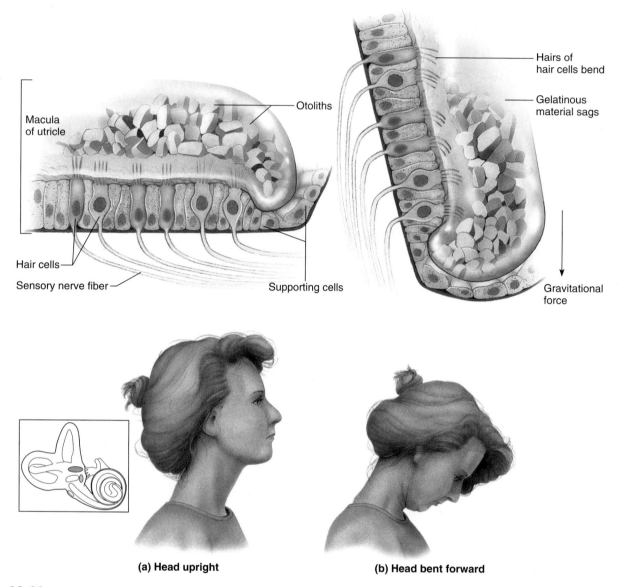

Macula of utricle

Otoliths

Hairs of hair cells bend

Gelatinous material sags

Hair cells

Sensory nerve fiber

Supporting cells

Gravitational force

(a) Head upright

(b) Head bent forward

Figure 10.11

The maculae respond to changes in head position. (*a*) Macula of the utricle with the head in an upright position. (*b*) Macula of the utricle with the head bent forward.

angles to each other, and each corresponds to a different anatomical plane (see fig. 10.8).

Suspended in the perilymph of the osseous portion of each semicircular canal is a membranous canal that ends in a swelling called an **ampulla** (am-pul'ah), which houses the sensory organs of the semicircular canals. Each of these organs, called a **crista ampullaris** (kris'tah am-pul'ar-is), contains a number of sensory hair cells and supporting cells (fig. 10.12). Like the hairs of the maculae, the hair cells extend upward into a dome-shaped, gelatinous mass called the *cupula*.

Rapid turns of the head or body stimulate the hair cells of the crista ampullaris (fig. 10.13). At such times, the semicircular canals move with the head or body, but the fluid inside the membranous canals remains stationary. This bends the cupula in one or more of the canals in a direction opposite that of the head or body movement, and the hairs embedded in it also bend. The stimulated hair cells signal their associated nerve fibers, sending impulses to the brain. The brain interprets these impulses as a movement in a particular direction.

Parts of the cerebellum are particularly important in interpreting impulses from the semicircular canals. Analysis of such information allows the brain to predict the consequences of rapid body movements, and by modifying signals to appropriate skeletal muscles, the cerebellum can maintain balance.

Other sensory structures aid in maintaining equilibrium. For example, certain mechanoreceptors (proprioceptors), particularly those associated with the joints of

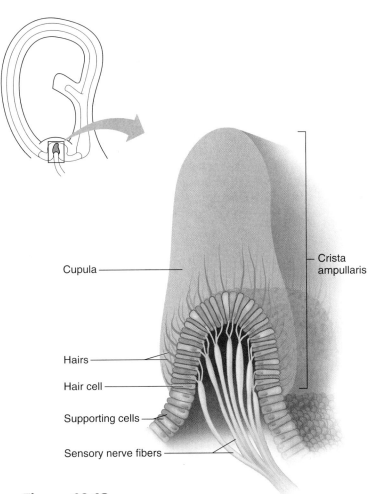

(a) Head in still position

(b) Head rotating

Figure 10.12

A crista ampullaris is located within the ampulla of each semicircular canal.

the neck, inform the brain about the position of body parts. In addition, the eyes detect changes in posture that result from body movements. Such visual information is so important that even if the organs of equilibrium are damaged, a person may be able to maintain normal balance by keeping the eyes open and moving slowly.

The nausea, vomiting, dizziness, and headache of *motion sickness* arise from senses that don't make sense. The eyes of a person reading in a moving car, for example, signal the brain that the person is stationary, because the print doesn't move. However, receptors in the skin detect bouncing, swaying, starting and stopping, as the inner ear detects movement. The contradiction triggers the symptoms. Similarly, in a passenger of an airplane flying through heavy turbulence, skin receptors register the chaos outside, but the eyes focus on the immobile seats and surroundings.

　　To prevent or lessen the misery of motion sickness, focus on the horizon or an object in the distance ahead. Medications are available by pill (diphenhydramine and meclizine) and, for longer excursions, in a skin patch (scopalamine). Ginger root may ease nausea too—try ginger ale.

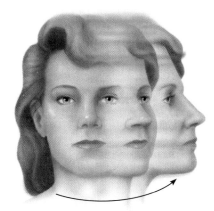

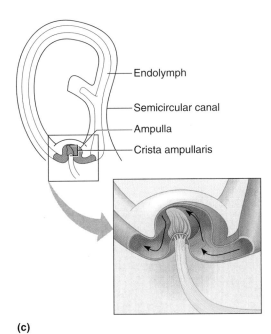

(c)

Figure 10.13

Equilibrium. (*a*) When the head is stationary, the cupula of the crista ampullaris remains upright. (*b*) and (*c*) When the head is moving rapidly, the cupula bends opposite the motion of the head, stimulating sensory receptors.

10.9 SENSE OF SIGHT

The eye, the organ containing visual receptors, provides vision, with the assistance of *accessory organs*. These accessory organs include the eyelids and lacrimal apparatus, which protect the eye, and a set of extrinsic muscles, which move the eye.

Visual Accessory Organs

The eye, lacrimal gland, and associated extrinsic muscles are housed within the orbital cavity, or orbit, of the skull. Each orbit is lined with the periosteum of various bones, and also contains fat, blood vessels, nerves, and connective tissues.

Each **eyelid** has four layers—skin, muscle, connective tissue, and conjunctiva. The skin of the eyelid, which is the thinnest skin of the body, covers the lid's outer surface and fuses with its inner lining near the margin of the lid. The eyelids are moved by the *orbicularis oculi* muscle (see fig. 8.17*a*, p. 194), which acts as a sphincter and closes the lids when it contracts, and by the *levator palpebrae superioris* muscle, which raises the upper lid and thus helps open the eye (fig. 10.14). The **conjunctiva** is a mucous membrane that lines the inner surfaces of the eyelids and folds back to cover the anterior surface of the eyeball, except for its central portion (cornea).

The *lacrimal apparatus* consists of the **lacrimal gland,** which secretes tears, and a series of ducts that carry tears into the nasal cavity (fig. 10.15). The gland is located in the orbit and secretes tears continuously. The tears exit through tiny tubules and flow downward and medially across the eye.

Two small ducts (the superior and inferior canaliculi) collect tears, which flow into the *lacrimal sac,* located in a deep groove of the lacrimal bone, and then into the

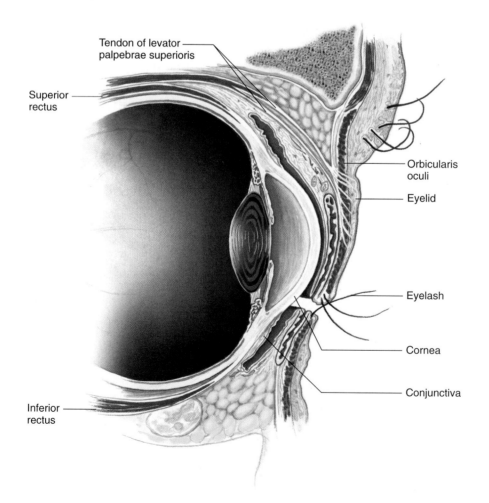

Superior rectus

Tendon of levator palpebrae superioris

Orbicularis oculi

Eyelid

Eyelash

Cornea

Conjunctiva

Inferior rectus

Figure 10.14

Sagittal section of the closed eyelids and anterior portion of the eye.

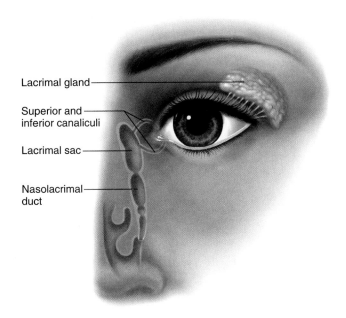

Figure 10.15

The lacrimal apparatus consists of a tear-secreting gland and a series of ducts.

nasolacrimal duct, which empties into the nasal cavity. Secretion by the lacrimal gland moistens and lubricates the surface of the eye and the lining of the lids. Tears also have an enzyme (*lysozyme*) that is an antibacterial agent, reducing the risk of eye infections.

The **extrinsic muscles** arise from the bones of the orbit and insert by broad tendons on the eye's tough outer surface. Six extrinsic muscles move the eye in various directions. Any given eye movement may utilize more than one extrinsic muscle, but each muscle is associated with one primary action. Figure 10.16 illustrates the locations of these extrinsic muscles, and table 10.2 lists their functions, as well as the functions of the eyelid muscles.

One eye deviating from the line of vision may result in double vision (diplopia). If this condition persists, the brain may suppress the image from the deviated eye. As a result, the turning eye may become blind (suppression amblyopia). Treating eye deviation early in life with exercises, eyeglasses, and surgery can prevent such monocular (one eye) blindness.

Check Your Recall

23. Explain how the eyelid moves.
24. Describe the conjunctiva.
25. What is the function of the lacrimal apparatus?

Structure of the Eye

The eye is a hollow, spherical structure about 2.5 centimeters in diameter. Its wall has three distinct layers—an outer (fibrous) layer, a middle (vascular) layer, and an

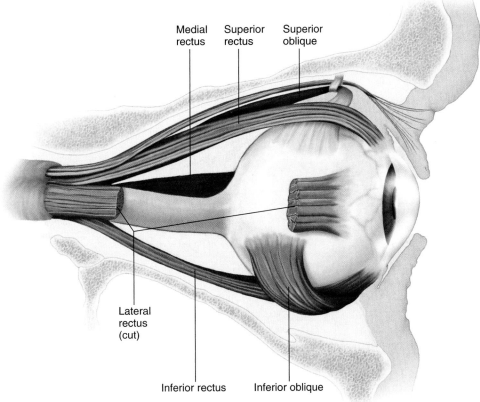

Figure 10.16

Extrinsic muscles of the right eye (lateral view).

Table 10.2	Muscles Associated with the Eyelids and Eyes		
Name	**Innervation**	**Function**	
Muscles of the Eyelids			
Orbicularis oculi	Facial nerve (VII)	Closes eye	
Levator palpebrae superioris	Oculomotor nerve (III)	Opens eye	
Extrinsic Muscles of the Eyes			
Superior rectus	Oculomotor nerve (III)	Rotates eye upward and toward midline	
Inferior rectus	Oculomotor nerve (III)	Rotates eye downward and toward midline	
Medial rectus	Oculomotor nerve (III)	Rotates eye toward midline	
Lateral rectus	Abducens nerve (VI)	Rotates eye away from midline	
Superior oblique	Trochlear nerve (IV)	Rotates eye downward and away from midline	
Inferior oblique	Oculomotor nerve (III)	Rotates eye upward and away from midline	

inner (nervous) layer. The spaces within the eye are filled with fluids that support its wall and internal parts that help maintain its shape. Figure 10.17 shows the major parts of the eye.

Outer Layer

The anterior sixth of the outer layer bulges forward as the transparent **cornea** (kor′ne-ah), which is the window of the eye and helps focus entering light rays. The cornea is composed largely of connective tissue with a thin surface layer of epithelium. It is transparent because it contains few cells and no blood vessels, and the cells and collagenous fibers form unusually regular patterns.

Along its circumference, the cornea is continuous with the **sclera** (skle′rah), the white portion of the eye.

The sclera makes up the posterior five-sixths of the outer layer and is opaque due to many large, disorganized, collagenous and elastic fibers. The sclera protects the eye and is an attachment for the extrinsic muscles. In the back of the eye, the **optic nerve** and certain blood vessels pierce the sclera.

Worldwide, the most common cause of blindness is loss of transparency of the cornea. A corneal transplant (penetrating keratoplasty) can treat this condition by replacing the central two-thirds of the defective cornea with a similar-sized portion of cornea from a donor eye. Because corneal tissues lack blood vessels, transplanted tissue is usually not rejected. The success rate of the procedure is very high.

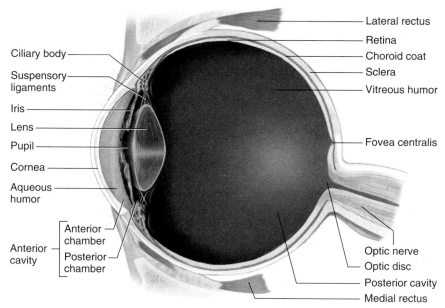

Figure 10.17
Transverse section of the right eye (superior view).

Middle Layer

The middle layer includes the choroid coat, ciliary body, and iris (fig. 10.17). The **choroid coat** (ko′roid kōt), in the posterior five-sixths of the globe of the eye, is loosely joined to the sclera and is honeycombed with blood vessels, which nourish surrounding tissues. The choroid coat also has many pigment-producing melanocytes. The melanin that these cells produce absorbs excess light and thus helps keep the inside of the eye dark.

The **ciliary body** (sil′e-er″e bod′e), which is the thickest part of the middle layer, extends forward from the choroid coat and forms an internal ring around the front of the eye. Within the ciliary body are many radiating folds called *ciliary processes* and groups of muscle fibers that constitute the *ciliary muscles.*

Many strong but delicate fibers, called *suspensory ligaments,* extend inward from the ciliary processes and hold the transparent **lens** in position (fig. 10.18). The distal ends of these fibers attach along the margin of a thin capsule that surrounds the lens. The body of the lens lies directly behind the iris and pupil and is composed of differentiated epithelial cells called *lens fibers.* The cytoplasm of these cells is the transparent substance of the lens.

The ciliary muscles and suspensory ligaments, along with the structure of the lens itself, enable the lens to adjust shape to facilitate focusing, a phenomenon called **accommodation** (ah-kom″o-da′shun). The lens is enclosed by a clear capsule composed largely of elastic fibers. This elastic nature keeps the lens under constant tension, and enables it to assume a globular shape. The suspensory ligaments attached to the margin of the capsule are also under tension. When they pull outward, flattening the capsule and the lens inside, the lens focuses on distant objects (fig. 10.19*a*). However, if the tension on the suspensory ligaments relaxes, the elastic lens capsule rebounds, and the lens surface becomes more convex—focused for viewing closer objects (fig. 10.19*b*).

The ciliary muscles control the actions of the suspensory ligaments in accommodation. For example, one set of these muscle fibers extends back from fixed points in the sclera to the choroid coat. When the fibers contract, the choroid coat is pulled forward, and the ciliary body shortens. This relaxes the suspensory ligaments, and the lens thickens in response (see fig. 10.19*b*). When the ciliary muscles relax, tension on the suspensory ligaments increases, and the lens becomes thinner and less convex again (see fig. 10.19*a*).

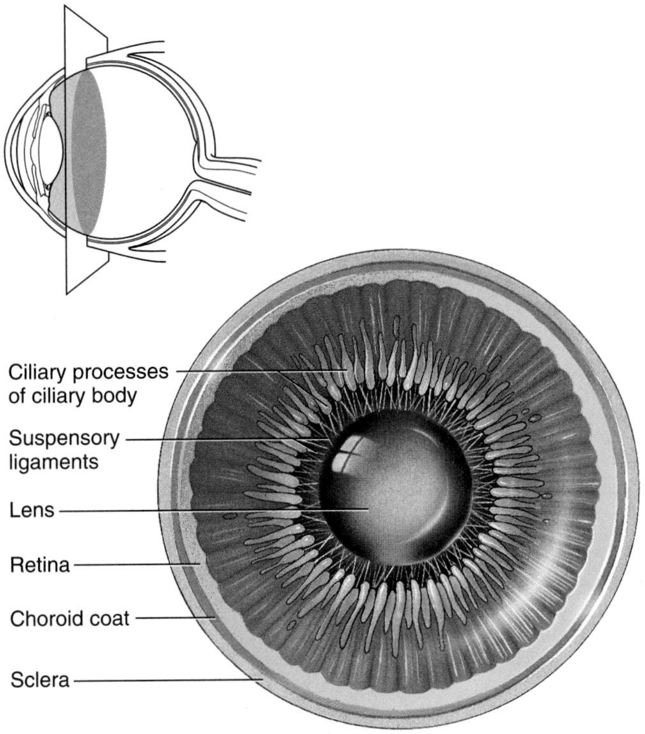

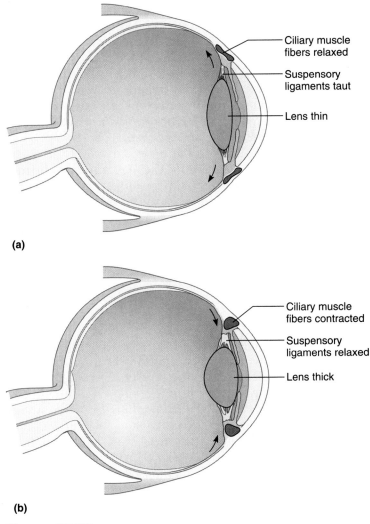

(a)

Ciliary muscle fibers relaxed

Suspensory ligaments taut

Lens thin

Ciliary muscle fibers contracted

Suspensory ligaments relaxed

Lens thick

(b)

Figure 10.19

Accommodation. (*a*) The lens thins as the ciliary muscle fibers relax. (*b*) The lens thickens as the ciliary muscle fibers contract.

Ciliary processes of ciliary body

Suspensory ligaments

Lens

Retina

Choroid coat

Sclera

Figure 10.18

Lens and ciliary body viewed from behind.

A common eye disorder, particularly in older people, is *cataract*. The lens or its capsule slowly becomes cloudy and opaque. Without treatment, cataracts eventually cause blindness. In the past, cataracts were treated surgically, with a two-week recovery period. Today, cataracts are treated on an outpatient basis with a laser.

Check Your Recall

26. Describe the outer and middle layers of the eye.
27. What factors contribute to the transparency of the cornea?
28. How does the shape of the lens change during accommodation?
29. Why would reading something for a long time cause eye fatigue, while looking at something distant is restful?

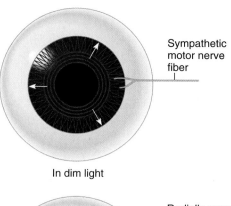

In dim light

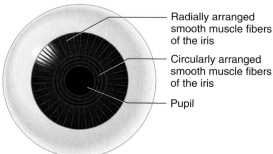

In normal light

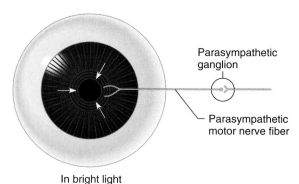

In bright light

The **iris** (i′ris) is a thin diaphragm composed mostly of connective tissue and smooth muscle fibers. From the outside, the iris is the colored portion of the eye. The iris extends forward from the periphery of the ciliary body and lies between the cornea and lens (see fig. 10.17). The iris divides the space (anterior cavity) separating these parts into an *anterior chamber* (between the cornea and the iris) and a *posterior chamber* containing the lens (between the iris and the vitreous body).

The epithelium on the inner surface of the ciliary body secretes a watery fluid called **aqueous humor** (a′kwe-us hu′mor) into the posterior chamber. The fluid circulates from this chamber through the **pupil** (pu′pil), a circular opening in the center of the iris, and into the anterior chamber. Aqueous humor fills the space between the cornea and lens, helps nourish these parts, and aids in maintaining the shape of the front of the eye. It subsequently leaves the anterior chamber through veins and a special drainage canal, the scleral venous sinus (canal of Schlemm) located in its wall at the junction of the cornea and the sclera.

The smooth muscle fibers of the iris are organized into two groups, a *circular set* and a *radial set*. These muscles control the size of the pupil, through which light passes as it enters the eye. The circular set of muscle fibers acts as a sphincter. When it contracts, the pupil gets smaller, and less light enters. Bright light stimulates the circular muscles to contract, which decreases the intensity of light entering the eye. Conversely, when the radial muscle fibers contract, the pupil's diameter increases, and more light enters (fig. 10.20). Dim light stimulates the radial muscles to contract, which dilates the pupil, allowing more light to enter the eye.

Figure 10.20

Dim light stimulates the radial muscles of the iris to contract, and the pupil dilates. Bright light stimulates the circular muscles of the iris to contract, and the pupil constricts.

An eye disorder called *glaucoma* develops when the rate of aqueous humor formation exceeds the rate of its removal. As fluid accumulates in the anterior chamber of the eye, fluid pressure rises and is transmitted to all parts of the eye. In time, the building pressure squeezes shut blood vessels that supply the receptor cells of the retina. Cells that are robbed of nutrients and oxygen in this way may die, and permanent blindness can result.

When diagnosed early, glaucoma can usually be treated successfully with drugs, laser therapy, or surgery, all of which promote the outflow of aqueous humor. Since glaucoma in its early stages typically produces no symptoms, discovery of the condition usually depends on measuring intraocular pressure, using an instrument called a *tonometer*.

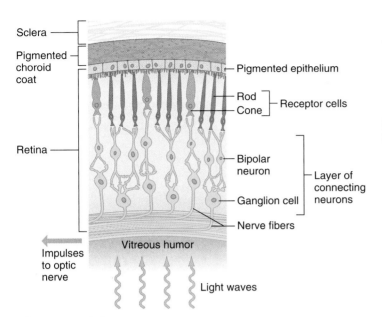

Figure 10.21

The retina consists of several cell layers. Light waves penetrate a layer of connecting neurons to impinge on the rods and cones, which are the visual receptors. The pigmented epithelium absorbs stray light rays.

Inner Layer

The inner layer consists of the **retina** (ret′ĭ-nah), which contains the visual receptor cells (photoreceptors). This nearly transparent sheet of tissue is continuous with the optic nerve in the back of the eye and extends forward as the inner lining of the eyeball. It ends just behind the margin of the ciliary body.

The retina is thin and delicate, but its structure is quite complex. It has a number of distinct layers, as figures 10.21 and 10.22 illustrate.

In the central region of the retina is a yellowish spot called the *macula lutea*. A depression in its center, called the **fovea centralis** (fo′ve-ah sen-tral′is), is in the region of the retina that produces the sharpest vision (see figs. 10.17 and 10.23).

The fovea centralis of the human eye has 150,000 cones per square millimeter. In contrast, a bird of prey's eye has about a million cones per square millimeter.

Just medial to the fovea centralis is an area called the **optic disc** (op′tik disk) (fig. 10.23). Here, nerve fibers from the retina leave the eye and join the optic nerve. A central artery and vein also pass through the optic disc. These vessels are continuous with the capillary networks of the retina, and along with vessels in the underlying choroid coat, they supply blood to the cells of the inner layer. Because the optic disc region lacks receptor cells, it is commonly known as the *blind spot* of the eye.

The space bounded by the lens, ciliary body, and retina is the largest compartment of the eye and is called the *posterior cavity* (see fig. 10.17). It is filled with a transparent, jellylike fluid called **vitreous humor** (vit′re-us hu′mor), which along with collagenous fibers comprises the *vitreous body*. The vitreous body supports the internal parts of the eye and helps maintain its shape.

As a person ages, tiny, dense clumps of gel or deposits of crystal-like substances form in the vitreous humor. When these clumps cast shadows on the retina, the person sees small, moving specks in the field of vision. Such specks, known as *floaters*, are most apparent when looking at a plain background, such as the sky or a wall.

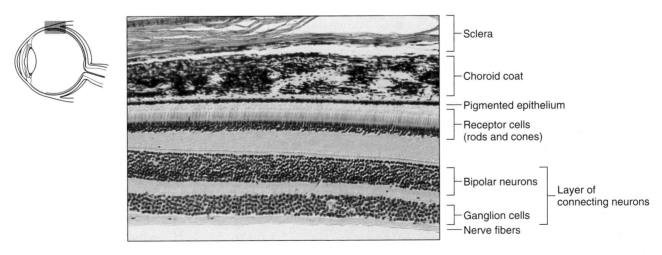

Figure 10.22

Retinal structure. Note the layers of cells and nerve fibers in this light micrograph of the retina (75×).

Figure 10.23

The retina. (*a*) Major features of the retina. (*b*) Nerve fibers leave the retina of the eye in the area of the optic disc (arrow) to form the optic nerve in this magnified view of the retina (53×).

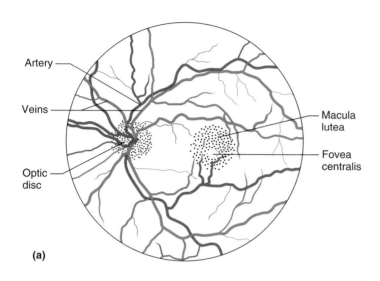

Artery

Veins

Macula lutea

Fovea centralis

Optic disc

(a)

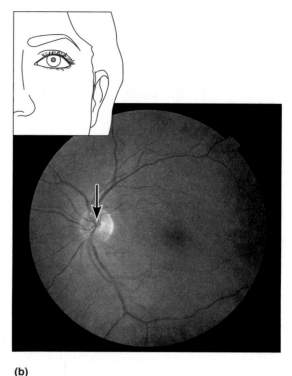

(b)

Check Your Recall

30. Explain the source of aqueous humor, and trace its path through the eye.
31. How does the pupil respond to changes in light intensity?
32. Describe the structure of the retina.

Light Refraction

When a person sees an object, either the object is giving off light, or light waves from another source are reflected from it. These light waves enter the eye, and an image of the object is focused on the retina. Focusing bends the light waves, a phenomenon called **refraction** (re-frak'shun).

Refraction occurs when light waves pass at an oblique angle from a medium of one optical density into a medium of a different optical density. This occurs at the curved surface between the air and the cornea and at the curved surface of the lens itself. A lens with a *convex* surface (as in the eye) causes light waves to converge (fig. 10.24).

The convex surface of the cornea refracts light waves from outside objects. The convex surface of the lens and, to a lesser extent, the surfaces of the fluids within the chambers of the eye then refract the light again.

If eye shape is normal, light waves focus sharply on the retina, much as a motion picture image is focused on a screen for viewing. Unlike the motion picture image, however, the image that forms on the retina is upside down and reversed from left to right. The visual cortex interprets the image in its proper position.

Check Your Recall

33. What is refraction?
34. What parts of the eye provide refracting surfaces?

Visual Receptors

Visual receptor cells are modified neurons of two distinct kinds, as figure 10.21 illustrates. One group, called *rods* (rodz), have long, thin projections at their ends, and provide black and white vision. The other group, *cones* (kōnz), have short, blunt projections, and provide color vision. Rods are of one type, cones of three.

Rods and cones are in a deep portion of the retina, closely associated with a layer of pigmented epithelium

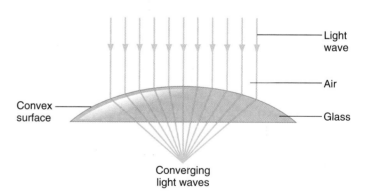

Light wave

Air

Convex surface

Glass

Converging light waves

Figure 10.24

A lens with a convex surface causes light waves to converge. The lens of the eye functions the same way.

(see fig. 10.22). The epithelial pigment absorbs light waves not absorbed by the receptor cells, and together with the pigment of the choroid coat, keeps light from reflecting off surfaces inside the eye. Projections from receptors, which are loaded with light-sensitive visual pigments, extend into this pigmented layer.

Visual receptors are stimulated only when light reaches them. A light image focused on an area of the retina stimulates some receptors, and impulses travel from them to the brain. However, the impulse leaving each activated receptor provides only a fragment of the information required for the brain to interpret a complete scene.

Rods and cones provide different aspects of vision. Rods are hundreds of times more sensitive to light than cones and therefore can provide vision in dim light, without color. Cones detect color.

> A human eye has 125 million rods and 7 million cones.
> A cat has three types of cone cells, but sees mostly pastels.
> A dog has two types of cone cells, and its visual world is much like that of a person with colorblindness.

Rods and cones also differ in the sharpness of the perceived images, or visual acuity. Cones provide sharp images, and rods provide more general outlines of objects. Rods give less precise images because nerve fibers from many rods converge, their impulses transmitted to the brain on the same nerve fiber (fig. 10.25*a*). Thus, if a point of light stimulates a rod, the brain cannot tell which one of many receptors has been stimulated. Convergence of impulses is less common among cones. When a cone is stimulated, the brain can pinpoint the stimulation more accurately (fig. 10.25*b*).

The fovea centralis, the area of sharpest vision, lacks rods but contains densely packed cones with few or no converging fibers (see fig. 10.17). Also in the fovea centralis, the overlying layers of the retina and the retinal blood vessels are displaced to the sides, more fully exposing receptors to incoming light. Consequently, to view something in detail, a person moves the eyes so that the important part of an image falls on the fovea centralis.

Visual Pigments

Both rods and cones contain light-sensitive pigments that decompose when they absorb light energy. The light-sensitive biochemical in rods is called **rhodopsin** (ro-dop′sin), or *visual purple*. In the presence of light, rhodopsin molecules are broken down into a colorless protein called *opsin* and a yellowish substance called *retinal* (retinene) that is synthesized from vitamin A.

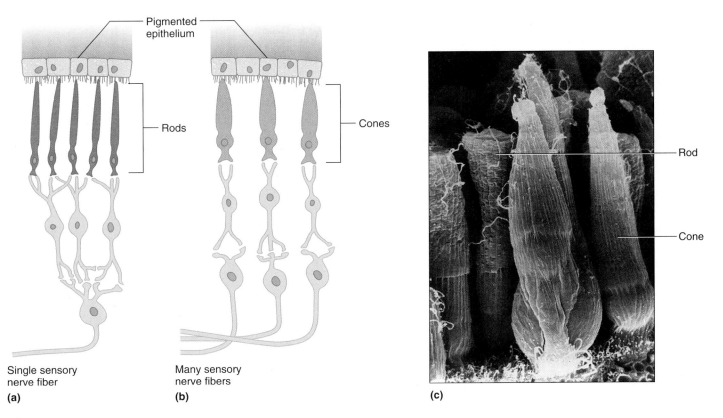

Figure 10.25

Rods and cones are visual receptors. (*a*) A single sensory nerve fiber transmits impulses from several rods to the brain. (*b*) Separate sensory nerve fibers transmit impulses from cones to the brain. (*c*) Scanning electron micrograph of rods and cones (1,350×).

Decomposition of rhodopsin molecules activates an enzyme that initiates a series of reactions altering the permeability of the rod cell membrane. As a result, a complex pattern of nerve impulses originates in the retina. The impulses travel away from the retina along the optic nerve into the brain, where they are interpreted as vision.

In bright light, nearly all of the rhodopsin in the rods of the retina decomposes, greatly reducing rod sensitivity. In dim light, however, regeneration of rhodopsin from opsin and retinal is faster than rhodopsin breakdown. ATP provides the energy required for this regeneration (see chapter 4, p. 80).

> Poor vision in dim light, called night blindness, results from vitamin A deficiency. Lack of the vitamin reduces the supply of retinal, rhodopsin production falls, and rod sensitivity is low. Supplementing the diet with vitamin A is used to treat night blindness.

As in rods, the light-sensitive pigments in cones are composed of retinal and protein. In cones, however, three different opsin proteins, different from that found in rods, combine with retinal to form the three cone pigments. The three types of cones each contain one of these three visual pigments.

The wavelength of light determines the color that the brain perceives from it. For example, the shortest wavelengths of visible light are perceived as violet, and the longest are perceived as red. One type of cone pigment (erythrolabe) is most sensitive to red light waves, another (chlorolabe) to green light waves, and a third (cyanolabe) to blue light waves. The color a person perceives depends on which set of cones or combination of sets the light in a given image stimulates. If all three sets of cones are stimulated, the person senses the light as white, and if none are stimulated, the person senses black. Different forms of colorblindness result from lack of different types of cone pigments.

Visual Nerve Pathways

Visual nerve pathways bring nerve impulses from the retina to the visual cortex, where they are perceived as vision. The pathways begin as the axons of the retinal neurons leave the eyes to form the *optic nerves* (fig. 10.26). Just anterior to the pituitary gland, these nerves give rise to the X-shaped *optic chiasma* (op'tik ki-az'mah), and within the chiasma, some of the fibers cross over. More specifically, the fibers from the nasal (medial) half of each retina cross over, but those from the temporal (lateral) sides do not. Thus, fibers from the nasal half of the left eye and the temporal half of the right eye form the *right optic tract*, and fibers from the nasal half of the right eye and the temporal half of the left eye form the *left optic tract*.

Just before the nerve fibers reach the thalamus, a few of them enter nuclei that function in various visual

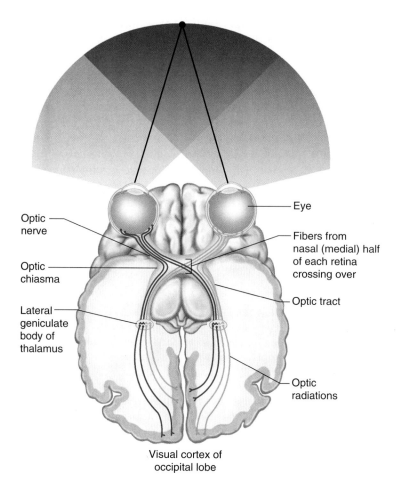

Figure 10.26
A visual pathway includes the optic nerve, optic chiasma, optic tract, and optic radiations.

reflexes. Most of the fibers, however, enter the thalamus and synapse in its posterior portion (lateral geniculate body). From this region, the visual impulses enter nerve pathways called *optic radiations,* which lead to the visual cortex of the occipital lobes.

> ### Check Your Recall
> 35. Distinguish between the rods and cones of the retina.
> 36. Explain the roles of visual pigments.
> 37. Trace a nerve impulse from the retina to the visual cortex.

Clinical Terms Related to the Senses

amblyopia (am"ble-o'pe-ah) Dim vision not due to a refractive disorder or lesion.
anopia (an-o'pe-ah) Absence of an eye.
audiometry (aw"de-om'ĕ-tre) Measurement of auditory acuity for various frequencies of sound waves.
blepharitis (blef"ah-ri'tis) Inflammation of the eyelid margins.
causalgia (kaw-zal'je-ah) Persistent, burning pain usually associated with injury to a limb. Also called neuropathic pain.

conjunctivitis (kon-junk″tĭ-vi′tis) Inflammation of the conjunctiva.

diplopia (dĭ-plo′pe-ah) Double vision.

emmetropia (em″ĕ-tro′pe-ah) Normal condition of the eyes; eyes with no refractive defects.

enucleation (e-nu″kle-a′shun) Removal of the eyeball.

exophthalmos (ek″sof-thal′mos) Abnormal protrusion of the eyes.

hemianopsia (hem″e-an-op′se-ah) Defective vision affecting half of the visual field.

hyperalgesia (hi″per-al-je′ze-ah) Heightened sensitivity to pain.

iridectomy (ir″ĭ-dek′to-me) Surgical removal of part of the iris.

iritis (i-ri′tis) Inflammation of the iris.

keratitis (ker″ah-ti′tis) Inflammation of the cornea.

labyrinthectomy (lab″ĭ-rin-thek′to-me) Surgical removal of the labyrinth.

labyrinthitis (lab″ĭ-rin-thi′tis) Inflammation of the labyrinth.

Ménière's disease (men″e-ārz′ dĭ-zēz′) Inner ear disorder that causes ringing in the ears, increased sensitivity to sounds, dizziness, and hearing loss.

neuralgia (nu-ral′je-ah) Pain resulting from inflammation of a nerve or a group of nerves.

neuritis (nu-ri′tis) Inflammation of a nerve.

nystagmus (nis-tag′mus) Involuntary oscillation of the eyes.

otitis media (o-ti′tis me′de-ah) Inflammation of the middle ear.

otosclerosis (o″to-skle-ro′sis) Formation of spongy bone in the inner ear, which often causes deafness by fixing the stapes to the oval window.

pterygium (tĕ-rij′e-um) Abnormally thickened patch of conjunctiva that extends over part of the cornea.

retinitis pigmentosa (ret″ĭ-ni′tis pig″men-to′sa) Inherited, progressive retinal sclerosis characterized by pigment deposits in the retina and by retinal atrophy.

retinoblastoma (ret″ĭ-no-blas-to′mah) Inherited, highly malignant tumor arising from immature retinal cells.

tinnitus (tĭ-ni′tus) Ringing or buzzing in the ears.

trachoma (trah-ko′mah) Bacterial disease of the eye that causes conjunctivitis, which may lead to blindness.

tympanoplasty (tim″pah-no-plas′te) Surgical reconstruction of the middle ear bones and the establishment of continuity from the eardrum to the oval window.

uveitis (u″ve-i′tis) Inflammation of the uvea, the region of the eye that includes the iris, ciliary body, and choroid coat.

vertigo (ver′tĭ-go) Sensation of dizziness.

Clinical Connection

"The song was full of glittering orange diamonds."
"The paint smelled blue."
"The sunset was salty."
"The pickle tasted like a rectangle."

To 1 in 500,000 people with a condition called synesthesia, sensation and perception mix, so that the brain perceives a stimulus to one sense as coming from another. Most commonly, letters, numbers, or periods of time evoke specific colors. These associations are involuntary, are very specific, and persist over a lifetime. For example, a person might report that 3 is always mustard yellow, or Thursday a very dark brown.

Synesthesia seems to be inherited and is more common in women and among creative individuals, such as artists and writers. One of the authors of this book (R.L.) has it—to her, days and months are specific colors. People have reported the condition to psychologists and physicians for at least 200 years.

PET (positron emission tomography) scanning reveals a physical basis to synesthesia. Brain scans of six nonsynesthetes were compared with those of six synesthetes who associate words with colors. The researchers monitored blood flow in the cerebral cortex while a list of words was read aloud to both groups. While blood flow was increased in word-processing areas for both groups, areas important in vision and color processing were also lit up in those with synesthesia.

Another form of mixed senses is when one sense compensates for another that is impaired. For example, an adolescent who has been blind since age three uses sound to see. In a variation of a sense called echolocation that dolphins and bats use, this young man makes clicking sounds with his tongue, and can then sense the echoes of the sounds hitting objects. His brain converts the echo patterns into visualizations of distance to objects. He can even distinguish among glass, wood, metal, and flesh.

SUMMARY OUTLINE

10.1 Introduction (p. 261)
Sensory receptors sense changes in their surroundings.

10.2 Receptors, Sensations, and Perception (p. 261)

1. Types of receptors
   a. Each type of receptor is most sensitive to a distinct type of stimulus.
   b. The major types of receptors are chemoreceptors, pain receptors, thermoreceptors, mechanoreceptors, and photoreceptors.
2. Sensations
   a. Sensations are feelings resulting from sensory stimulation.
   b. A particular part of the cerebral cortex interprets every impulse reaching it in a specific way.
   c. The cerebral cortex projects a sensation back to the region of stimulation.
3. Sensory adaptation may involve receptors becoming unresponsive or inhibition along the CNS pathways leading to the sensory regions of the cerebral cortex.

10.3 General Senses (p. 262)
General senses are associated with receptors in the skin, muscles, joints, and viscera.

1. Touch and pressure senses
   a. Free ends of sensory nerve fibers are receptors for the sensation of itching.
   b. Tactile corpuscles are receptors for the sensation of light touch.
   c. Lamellated corpuscles are receptors for the sensation of heavy pressure.

2. Temperature senses
Temperature receptors include two sets of free nerve endings that are warm and cold receptors.
3. Sense of pain
   a. Pain receptors are free nerve endings that tissue damage stimulates.
   b. Visceral pain
      (1) Pain receptors are the only receptors in viscera that provide sensations.
      (2) Pain sensations produced from visceral receptors may feel as if they are coming from some other body part, called referred pain.
      (3) Visceral pain may be referred because sensory impulses from the skin and viscera travel on common nerve pathways.
   c. Pain nerve fibers
      (1) The two main types of pain fibers are acute pain fibers and chronic pain fibers.
      (2) Acute pain fibers conduct nerve impulses rapidly. Chronic pain fibers conduct impulses more slowly.
      (3) Pain impulses are processed in the gray matter of the spinal cord and ascend to the brain.
      (4) Within the brain, pain impulses pass through the reticular formation before being conducted to the cerebral cortex.
   d. Regulation of pain impulses
      (1) Awareness of pain occurs when pain impulses reach the thalamus.
      (2) The cerebral cortex determines pain intensity and locates its source.
      (3) Impulses descending from the brain stimulate neurons to release pain-relieving neuropeptides, such as enkephalins.

10.4 Special Senses (p. 265)

Special senses have receptors within large, complex sensory organs of the head.

10.5 Sense of Smell (p. 266)

1. Olfactory receptors
   a. Olfactory receptors are chemoreceptors that are stimulated by chemicals dissolved in liquid.
   b. Olfactory receptors function with taste receptors and aid in food selection.
2. Olfactory organs
   a. Olfactory organs consist of receptors and supporting cells in the nasal cavity.
   b. Olfactory receptor cells are bipolar neurons with cilia.
3. Olfactory nerve pathways
   Nerve impulses travel from the olfactory receptor cells through the olfactory nerves, olfactory bulbs, and olfactory tracts to interpreting centers in the temporal and frontal lobes of the cerebrum.
4. Olfactory stimulation
   a. Olfactory impulses may result when odorant molecules bind cell surface olfactory receptors on cilia of receptor cells. The binding pattern encodes a specific odor, which is interpreted in the brain.
   b. The sense of smell adapts rapidly.

10.6 Sense of Taste (p. 267)

1. Taste receptors
   a. Taste buds consist of taste (receptor) cells and supporting cells.
   b. Taste cells have taste hairs.
   c. Taste hair surfaces have receptors to which chemicals bind, stimulating nerve impulses.
2. Taste sensations
   a. The five primary taste sensations are sweet, sour, salty, bitter, and umami.
   b. Various taste sensations result from the stimulation of at least two sets of taste receptors.
   c. A single taste receptor cell detects only one of the five tastes, but receptors corresponding to different tastes are scattered on the tongue.
3. Taste nerve pathways
   a. Sensory impulses from taste receptors travel on fibers of the facial, glossopharyngeal, and vagus nerves.
   b. These impulses are carried to the medulla oblongata and then ascend to the thalamus, from which they travel to the gustatory cortex in the parietal lobes.

10.7 Sense of Hearing (p. 269)

1. Outer ear
   The outer ear collects sound waves of vibrating objects.
2. Middle ear
   Auditory ossicles of the middle ear conduct sound waves from the eardrum to the oval window of the inner ear.
3. Auditory tube
   Auditory tubes connect the middle ears to the throat and help maintain equal air pressure on both sides of the eardrums.
4. Inner ear
   a. The inner ear is a complex system of connected tubes and chambers—the osseous and membranous labyrinths.
   b. The spiral organ has hearing receptors that are stimulated by vibrations in the fluids of the inner ear.
   c. Different frequencies of vibrations stimulate different sets of receptor cells.
5. Auditory nerve pathways
   a. Auditory nerves carry impulses to the auditory cortices of the temporal lobes.
   b. Some auditory nerve fibers cross over, so that impulses arising from each ear are interpreted on both sides of the brain.

10.8 Sense of Equilibrium (p. 273)

1. Static equilibrium
   Static equilibrium maintains the stability of the head and body when they are motionless.
2. Dynamic equilibrium
   a. Dynamic equilibrium balances the head and body when they are moved or rotated suddenly.
   b. Other structures that help maintain equilibrium include the eyes and mechanoreceptors associated with certain joints.

10.9 Sense of Sight (p. 276)

1. Visual accessory organs
   Visual accessory organs include the eyelids, lacrimal apparatus, and extrinsic muscles of the eyes.
2. Structure of the eye
   a. The wall of the eye has an outer (fibrous), a middle (vascular), and an inner (nervous) layer.
      (1) The outer layer is protective, and its transparent anterior portion (cornea) refracts light entering the eye.
      (2) The middle layer is vascular and contains pigments that keep the inside of the eye dark.
      (3) The inner layer contains the visual receptor cells.

b. The lens is a transparent, elastic structure. Ciliary muscles control its shape.

c. The lens must thicken to focus on close objects.

d. The iris is a muscular diaphragm that controls the amount of light entering the eye.

e. Spaces within the eye are filled with fluids that help maintain its shape.

3. Light refraction

The cornea and lens refract light waves to focus an image on the retina.

4. Visual receptors

a. Visual receptors are rods and cones.

b. Rods are responsible for colorless vision in dim light, and cones provide color vision.

5. Visual pigments

a. A light-sensitive pigment in rods decomposes in the presence of light and triggers a complex series of reactions that initiate nerve impulses.

b. Color vision comes from three sets of cones containing different light-sensitive pigments.

6. Visual nerve pathways

a. Nerve fibers from the retina form the optic nerves.

b. Some fibers cross over in the optic chiasma.

c. Most of the fibers enter the thalamus and synapse with others that continue to the visual cortex in the occipital lobes.

CHAPTER ASSESSMENTS

10.1 Introduction

1. Distinguish between somatic senses and special senses. (p. 261)

10.2 Receptors, Sensations, and Perception

2. Match each sensory receptor to the type of stimulus to which it is likely to respond. (p. 261)

   (1) chemoreceptor   A. Approaching headlights
   (2) pain receptor   B. A change in blood pressure
   (3) thermoreceptor   C. The smell of roses
   (4) mechanoreceptor   D. An infected tooth
   (5) photoreceptor   E. A cool breeze

3. Explain the difference between a sensation and a perception. (p. 261)

4. Explain the projection of a sensation. (p. 261)

5. You fill up the tub to take a hot bath, but the water is too hot to the touch. You try a second and third time, and within a few seconds it feels fine. Which of the following is the most likely explanation? (p. 261)

   a. The water has cooled down unusually quickly.
   b. Your ability to sense heat has adapted.
   c. Your nervous system is suddenly not functioning properly.
   d. Your ability to sense cold has adapted.

10.3 General Senses

6. Describe the functions of free nerve endings, tactile corpuscles, and lamellated corpuscles. (p. 262)

7. Explain why pain may be referred, and provide an example. (p. 263)

10.4 Special Senses

8. Identify the location of the receptors for smell, taste, hearing, equilibrium, and sight. (p. 265)

10.5 Sense of Smell

9. Which two of the following are part of the olfactory organs? (p. 266)

   a. Olfactory receptors
   b. Columnar epithelial cells in the nasal mucosa
   c. The nose
   d. The brain

10. Trace a nerve impulse from an olfactory receptor to the interpreting center of the cerebrum. (p. 266)

10.6 Sense of Taste

11. Salivary glands are important in taste because (p. 267)

    a. they provide the fluid in which food molecules dissolve.
    b. the taste receptors are located in salivary glands.
    c. salivary glands are part of the brain.
    d. they lubricate the teeth.

12. Name the five primary taste sensations. (p. 267)

13. Trace the pathway of a nerve impulse from a taste receptor to the interpreting center of the cerebrum. (p. 268)

10.7 Sense of Hearing

14. Match the ear area with the associated structure. (p. 269)

    (1) outer ear   A. Cochlea
    (2) middle ear   B. Eardrum
    (3) inner ear   C. Auditory ossicles

15. Trace the path of sound waves from the external acoustic meatus to the hearing receptors. (p. 269)

16. Describe the functions of the auditory ossicles. (p. 269)

17. The function of the auditory tube is to: (p. 270)

    a. equalize air pressure on both sides of the eardrum.
    b. transmit sound vibrations to the eardrum.
    c. contain the hearing receptors.

18. Distinguish between the osseous and membranous labyrinths. (p. 270)

19. Describe the cochlea and its function. (p. 270)

20. Trace a nerve impulse from the spiral organ to the interpreting centers of the cerebrum. (pp. 270–272)

21. Which of the following best describes hearing receptor "hair cells"? (p. 271)

    a. They are neurons.
    b. They lack ion channels.
    c. They are epithelial, but function like neurons.
    d. They are built of the protein keratin.

22. Explain how a hearing receptor stimulates a sensory neuron. (p. 271)

10.8 Sense of Equilibrium

23. Contrast static equilibrium and dynamic equilibrium. (p. 273)

24. Describe the organs of static and dynamic equilibrium and their functions. (p. 273)

10.9 Sense of Sight

25. Match the visual accessory organ with its function: (p. 276)

    (1) eyelid   A. Move the eye
    (2) conjunctiva   B. Covers the eye
    (3) lacrimal gland   C. Lines the eyelids
    (4) extrinsic muscles D. Produces tears

26. Name the three layers of the eye wall and describe the functions of each layer. (p. 278)

27. Explain why looking at a close object causes fatigue, in terms of how accommodation is accomplished. (p. 279)

28. Explain the mechanisms of pupil constriction and pupil dilation. (p. 280)

29. All of the following are compartments within the eye. In which one is vitreous humor found? (p. 281)
    a. Anterior chamber
    b. Posterior chamber
    c. Anterior cavity
    d. Posterior cavity

30. Distinguish between the fovea centralis and the optic disc. (p. 281)

31. Explain how light is focused on the retina. (p. 282)

32. Distinguish between rods and cones. (p. 282)

33. Explain why cone vision is generally more acute than rod vision. (p. 283)

34. Describe the function of rhodopsin (p. 283)

35. Explain why rod vision may be more important under dim light conditions. (p. 284)

36. Describe the relationship between light wavelength and color vision. (p. 284)

37. Trace a nerve impulse from the retina to the visual cortex. (p. 284)

INTEGRATIVE ASSESSMENTS/ CRITICAL THINKING

➡ OUTCOMES 10.1, 10.2

1. We have relatively few sensory systems. How, then, do we experience such a huge and diverse number of sensory perceptions?

➡ OUTCOMES 10.2, 10.3

2. Why are some serious injuries, like a bullet entering the abdomen, relatively painless, but others, such as a burn, considerably more painful?

➡ OUTCOMES 10.2, 10.5

3. Loss of the sense of smell often precedes the major symptoms of Alzheimer disease and Parkinson disease. What additional information is needed to use this association to prevent or treat these diseases?

4. Describe how the taste of a medicine might be modified from sour to sweet, so that children would be more willing to take it.

➡ OUTCOMES 10.2, 10.7, 10.8

5. People who are deaf due to cochlear damage do not suffer from motion sickness. Why not?

➡ OUTCOMES 10.2, 10.8

6. Labyrinthitis is an inflammation of the inner ear. What symptoms would you expect in a patient with this disorder?

➡ OUTCOMES 10.2, 10.9

7. PET (positron emission tomography) scans of the brains of people who have been blind since birth reveal high neural activity in the visual centers of the cerebral cortex when these people read Braille. However, when sighted individuals run their fingers over the raised letters of Braille, the visual centers do not show increased activity. Explain these experimental results.

WEB CONNECTIONS

Visit the text website at **aris.mhhe.com** for additional quizzes, interactive learning exercises, and more.

AP R   NERVOUS SYSTEM

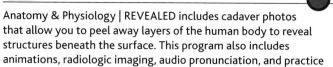

Anatomy & Physiology | REVEALED includes cadaver photos that allow you to peel away layers of the human body to reveal structures beneath the surface. This program also includes animations, radiologic imaging, audio pronunciation, and practice quizzing. To learn more visit **www.aprevealed.com**.

11
Endocrine System

SMELLY TEE SHIRTS. The endocrine system produces hormones, which are biochemicals that spread messages within an individual. Less well understood are pheromones, which are chemical signals sent between individuals of a species. In insects and rodents, pheromones often stimulate mating behavior. Experiments suggest that this may be the case with humans, too.

Mice and rats choose mates that are dissimilar to themselves in a group of genes that provide immunity. Their sense of smell helps them discern appropriate mates. Biologists hypothesize that choosing mates based on scent may protect offspring in two ways—it prevents close relatives from mating, and it may team immune systems with different types of strengths.

Researchers have traced mouse social and mating behavior to receptors in the olfactory epithelium, in the nasal cavity. The receptors—called trace amine–associated receptors—are attuned to molecules in mouse urine that direct social behavior. The genes that encode the receptors are also found in humans.

To test whether heterosexual humans use the sense of smell to respond to pheromones in mate selection as rodents do, researchers in Switzerland recruited forty-nine young women and forty-four young men. Each participant donated DNA, which was typed for genes that affect mating in rodents. The women used nasal spray for two weeks to clear their nasal passages. The men wore the same tee shirt on two consecutive days, using no deodorant or soap and avoiding contact with anything smelly that could linger. Each woman was then given three tee shirts from men genetically similar to her and three tee shirts from men genetically dissimilar to her, not knowing which shirts came from which men.

The endocrine system produces hormones, which act within an individual. Humans may also produce pheromones, which affect other individuals and may play a role in mate selection, as they do in rodents and insects.

The women rated the shirts on intensity, pleasantness, and sexiness. Like the mice and rats, women preferred the sweaty tees from the men least like them.

Another, more specific experiment supported these findings. Women were given vials of fluid to sniff that either contained or did not contain a component of male sweat called androstadienone. Although they didn't know which samples they were sniffing, the women consistently reported mood elevation and sexual arousal when they smelled the sweat. In addition, their saliva had increased amounts of cortisol, a hormone that raises blood pressure and blood sugar level, when they smelled androstadienone, suggesting that it might be a human pheromone.

Despite the mounting scientific evidence for human pheromones, the many websites that sell them should be approached with caution. A definitive human pheromone has not yet been described.

Learning Outcomes   *After studying this chapter, you should be able to do the following:*

Aids to Understanding Words   *(Appendix A on page 567 has a complete list of Aids to Understanding Words.)*

-crin [to secrete] endo*crin*e: Pertaining to internal secretions.

diuret- [to pass urine] *diuret*ic: Substance that promotes urine production.

endo- [within] *endo*crine gland: Gland that releases its secretion internally into a body fluid.

exo- [outside] *exo*crine gland: Gland that releases its secretion to the outside through a duct.

hyper- [above] *hyper*thyroidism: Condition resulting from an above-normal secretion of thyroid hormone.

hypo- [below] *hypo*thyroidism: Condition resulting from a below-normal secretion of thyroid hormone.

para- [beside] *para*thyroid glands: Set of glands on the surface of the thyroid gland.

toc- [birth] oxy*toc*in: Hormone that stimulates the uterine muscles to contract during childbirth.

-tropic [influencing] adrenocortico*tropic* hormone: Hormone that influences secretions from the adrenal cortex.

11.1 INTRODUCTION

Regulating the functions of the human body to maintain homeostasis is an enormous job. Two organ systems function coordinately to enable body parts to communicate with each other and to adjust constantly to changing incoming signals. The nervous system is one biological communication system; it utilizes nerve impulses carried on nerve fibers. The other is the endocrine system.

The **endocrine system** includes cells, tissues, and organs, collectively called endocrine glands, that secrete substances called **hormones** (hor′mōnz) into the internal environment. The hormones diffuse from the interstitial fluid into the bloodstream, and eventually act on cells called **target cells** some distance away.

Some glands secrete substances into the interstitial fluid, but because these secretions are rapidly broken down, they do not reach the bloodstream and are not hormones by the traditional definition. However, they do function similarly as messenger molecules and are sometimes referred to as "local hormones." These include **paracrine** secretions, which affect only neighboring cells, and **autocrine** secretions, which affect only the secreting cell itself.

A different group of glands, called exocrine glands, secrete outside the body through tubes or ducts that lead to the surface. Sweat, secreted by sweat glands and reaching the surface of the skin, is one example of an exocrine secretion (see chapter 5, p. 101).

Check Your Recall

1. What are the components of the endocrine system?
2. How do paracrine and autocrine secretions function differently than traditionally defined hormones?
3. Distinguish between endocrine and exocrine glands.

11.2 GENERAL CHARACTERISTICS OF THE ENDOCRINE SYSTEM

Both the endocrine system and the nervous system oversee cell-to-cell communication using chemical signals that bind to receptor molecules. Table 11.1 summarizes some similarities and differences between the nervous and endocrine systems. In contrast to the nervous system, in which neurons release neurotransmitter molecules into synapses, the glandular cells of the endocrine system

Table 11.1	**A Comparison Between the Nervous System and the Endocrine System**	
	Nervous System	**Endocrine System**
Cells	Neurons	Glandular epithelium
Chemical signal	Neurotransmitter	Hormone
Specificity of response	Receptors on postsynaptic cell	Receptors on target cell
Speed of onset	Seconds	Seconds to hours
Duration of action	Very brief unless neuronal activity continues	May be brief or may last for days even if secretion ceases

release hormones into the bloodstream, which carries these messenger molecules everywhere (fig. 11.1). However, the endocrine system is no less precise, because only target cells can respond to a hormone. A hormone's target cells have specific receptors that other cells lack. These receptors are proteins or glycoproteins with binding sites for a specific hormone.

Endocrine glands and their hormones help regulate metabolic processes. They control the rates of certain chemical reactions, aid in the transport of substances across cell membranes, and help regulate water and electrolyte balances. They also play vital roles in reproduction, development, and growth.

Specialized small groups of cells produce some hormones. However, the major endocrine glands are the pituitary gland, thyroid gland, parathyroid glands, adrenal glands, pancreas, pineal gland, thymus gland, and reproductive glands (testes and ovaries) (fig. 11.2).

Check Your Recall

4. Explain how the nervous and endocrine systems are alike and how they are different.

5. What determines whether a cell is a target cell for a particular hormone?

6. State some functions of hormones.

11.3 HORMONE ACTION

Most hormones are either steroids (or steroidlike substances) synthesized from cholesterol, or they are amines, peptides, proteins, or glycoproteins synthesized from amino acids (table 11.2). Hormones can stimulate changes in target cells even in extremely low concentrations.

Table 11.2	Types of Hormones	
Type of Compound	**Formed From**	**Examples**
Steroids	Cholesterol	Estrogen, testosterone, aldosterone, cortisol
Amines	Amino acids	Norepinephrine, epinephrine
Peptides	Amino acids	Antidiuretic hormone, oxytocin, thyrotropin-releasing hormone
Proteins	Amino acids	Parathyroid hormone, growth hormone, prolactin
Glycoproteins	Protein and carbohydrate	Follicle-stimulating hormone, luteinizing hormone, thyroid-stimulating hormone

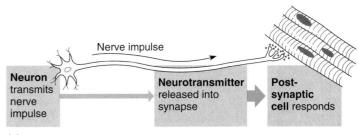

(a)

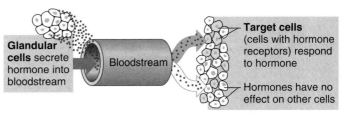

(b)

Figure 11.1

Chemical communication takes place in both the nervous system and the endocrine system. In both cases, cells respond to chemicals released from other cells. (*a*) Neurons release neurotransmitters into a synapse, affecting postsynaptic cells. (*b*) Glands release hormones into the bloodstream. Blood carries hormone molecules throughout the body, but only target cells respond.

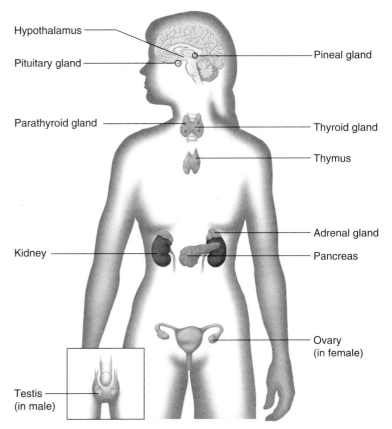

Figure 11.2

Locations of the major endocrine glands. The pituitary, thyroid, parathyroid, adrenal glands, and the pancreas are the main topics of this chapter. The functions of the other glands are described in more detail in subsequent chapters.

Steroid Hormones

Steroid molecules consist of complex rings of carbon and hydrogen atoms, and some oxygen atoms. Steroids differ according to the kinds and numbers of atoms attached to these rings and the ways they are joined.

Steroid hormones are insoluble in water. They are carried in the bloodstream weakly bound to plasma proteins in a way that allows them to be released in sufficient quantity in the vicinity of their target cells. However, unlike amine, peptide, and protein hormones, steroid hormones are soluble in lipids.

Because lipids make up the bulk of cell membranes, and steroid molecules are lipid-soluble, these hormones can diffuse into cells relatively easily and may enter any cell in the body. When a steroid hormone molecule enters a target cell, the following events occur (fig. 11.3):

1. The lipid-soluble steroid hormone diffuses through the cell membrane.
2. The steroid hormone binds a specific protein molecule—the receptor for that hormone.

3. The resulting hormone-receptor complex binds within the nucleus to particular regions of the target cell's DNA and activates transcription of specific genes into messenger RNA (mRNA) molecules.
4. The mRNA molecules leave the nucleus and enter the cytoplasm.
5. The mRNA molecules associate with ribosomes to direct the synthesis of specific proteins.

The newly synthesized proteins, which may be enzymes, transport proteins, or even hormone receptors, carry out the specific effects associated with the particular steroid hormone.

Nonsteroid Hormones

Nonsteroid hormones, such as amines, peptides, and proteins, usually bind receptors in target cell membranes. Each of these receptor molecules is a protein with a *binding site* and an *activity site*. A hormone molecule delivers its message to its target cell by uniting with the binding site of its receptor. This combination

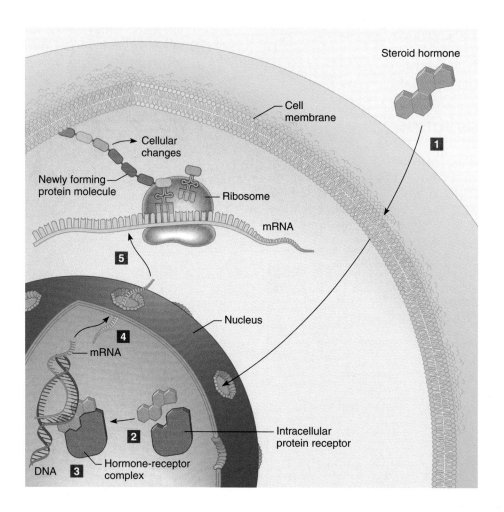

Figure 11.3

Steroid hormones. (*1*) A steroid hormone crosses a cell membrane and (*2*) combines with a protein receptor, usually in the nucleus. (*3*) The hormone-receptor complex activates synthesis of specific messenger RNA (mRNA) molecules. (*4*) The mRNA molecules leave the nucleus and enter the cytoplasm (*5*), where they guide synthesis of the encoded proteins.

stimulates the receptor's activity site to interact with other membrane proteins. Receptor binding may alter the function of enzymes or membrane transport mechanisms, changing the concentrations of other cellular components. The hormone that triggers this cascade of biochemical activity is called a *first messenger*. The biochemicals in the cell that induce changes in response to the hormone's binding are called *second messengers*. The entire process of chemical communication, from outside cells to inside, is called signal transduction.

The second messenger associated with one group of hormones is *cyclic adenosine monophosphate*, or **cyclic AMP (cAMP)** (sī′klik ay em pee). This mechanism works as follows (fig. 11.4):

1. A hormone binds its receptor.
2. The resulting hormone-receptor complex activates a protein called a *G protein*.
3. The G protein activates an enzyme called *adenylate cyclase*, which is a membrane protein.
4. Activated adenylate cyclase catalyzes the circularization of ATP in the cytoplasm into cAMP.

5. cAMP activates another set of enzymes, called protein kinases, which transfer phosphate groups from ATP to their substrate molecules, specific proteins within the cell. This action, called phosphorylation, alters the shapes of these substrate molecules, thereby activating them.

The activated proteins then alter various cellular processes, bringing about the characteristic effect of the hormone.

The type of membrane receptors present and the kinds of protein substrate molecules in a cell determine the cell's response to a hormone. Such responses to second messenger activation include altering membrane permeabilities, activating enzymes, promoting synthesis of certain proteins, stimulating or inhibiting specific metabolic pathways, moving the cell, and initiating secretion of hormones or other substances.

Another enzyme (phosphodiesterase) quickly inactivates cAMP, so that its action is short-lived. For this reason, a continuing response of a target cell requires a continuing signal from hormone molecules binding the target cell's membrane receptors.

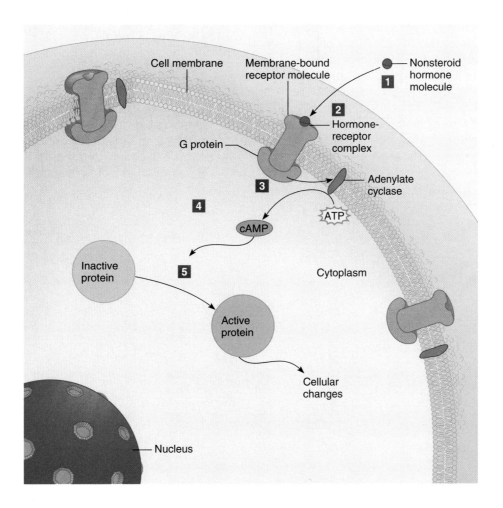

Figure 11.4

Nonsteroid hormone action. (*1*) Body fluids carry nonsteroid hormone molecules to the target cell, where (*2*) they bind receptor molecules on the cell membrane. (*3*) This activates molecules of adenylate cyclase, which (*4*) catalyze conversion of ATP into cyclic adenosine monophosphate (cAMP). (*5*) The cAMP promotes a series of reactions leading to the cellular changes associated with the hormone's action.

A number of other second messengers work in much the same way. These include diacylglycerol (DAG) and inositol triphosphate (IP$_3$).

> Abnormal or missing G proteins cause a variety of disorders, including colorblindness, precocious puberty, retinitis pigmentosa, and several thyroid problems.

Prostaglandins

A group of biochemicals called **prostaglandins** (pros"tah-glan'dinz) also regulates cells. Prostaglandins are lipids synthesized from a fatty acid (arachidonic acid) in cell membranes. A great variety of cells produce prostaglandins, including those of the liver, kidneys, heart, lungs, thymus, pancreas, brain, and reproductive organs. Prostaglandins usually act more locally than hormones, often affecting only the organ where they are produced.

Prostaglandins are potent and are present in very small quantities. They are not stored in cells, but rather synthesized just before release. They are rapidly inactivated.

Prostaglandins produce diverse and even opposite effects. Some prostaglandins, for example, relax smooth muscles in the airways of the lungs and in blood vessels, while others contract smooth muscles in the walls of the uterus and intestines. Prostaglandins stimulate hormone secretion from the adrenal cortex and inhibit secretion of hydrochloric acid from the stomach wall. They also influence the movements of sodium ions and water molecules in the kidneys, help regulate blood pressure, and have powerful effects on male and female reproductive physiology.

Check Your Recall ✓

7. How does a steroid hormone promote cellular changes? How does a nonsteroid hormone do the same?
8. What is a second messenger?
9. What are prostaglandins?
10. What kinds of effects do prostaglandins produce?

11.4 CONTROL OF HORMONAL SECRETIONS

Hormones are continually excreted in the urine and broken down by various enzymes, primarily in the liver. Therefore, increasing or decreasing the blood levels of a hormone requires increasing or decreasing secretion. Not surprisingly, hormone secretion is precisely regulated.

Generally, hormone secretion is controlled in three ways, all of which employ *negative feedback* (see chapter 1, p. 6). In each case, an endocrine gland or the system controlling it detects the concentration of the hormone the gland secretes, a process the hormone controls, or an action the hormone has on the internal environment (fig. 11.5). The three mechanisms of hormone control are described below.

1. The hypothalamus regulates the anterior pituitary gland's release of hormones that stimulate other endocrine glands to release hormones. Its location near the thalamus and the third ventricle allows the hypothalamus to constantly receive information

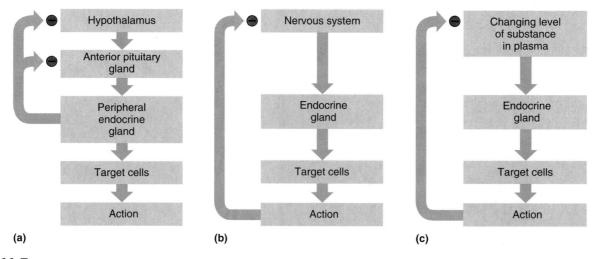

(a)   (b)   (c)

Figure 11.5

Control of the endocrine system occurs in three ways: (*a*) The hypothalamus and anterior pituitary stimulate other endocrine glands; (*b*) the nervous system stimulates a gland directly; or (*c*) changes in the internal environment stimulate glands directly. (● indicates negative feedback inhibition.)

about the internal environment from neural connections and cerebrospinal fluid (fig. 11.5*a*).

2. The nervous system stimulates some glands directly. The adrenal medulla, for example, secretes its hormones in response to sympathetic nerve impulses (fig. 11.5*b*).

3. Another group of glands responds directly to changes in the composition of the internal environment. For example, when the blood glucose level rises, the pancreas secretes insulin, and when the blood glucose level falls, it secretes glucagon, as discussed later in the chapter (fig. 11.5*c*).

In each of these cases, as hormone levels rise in the blood and the hormone exerts its effects, negative feedback inhibits the system, and hormone secretion decreases. Then, as hormone levels in the blood decrease and the hormone's effects are no longer taking place, inhibition of the system is lifted, and secretion of that hormone increases again. As a result of negative feedback, hormone levels in the bloodstream remain relatively stable, tending to fluctuate slightly above and below an average value (fig. 11.6).

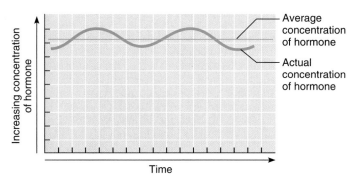

Figure 11.6

As a result of negative feedback, hormone concentrations remain relatively stable, although they may fluctuate slightly above and below average concentrations.

Check Your Recall

11. Explain three examples of control of hormonal secretion.

12. Describe a negative feedback system that controls hormone secretion.

11.5 PITUITARY GLAND

The **pituitary gland** (hypophysis) is located at the base of the brain, where a pituitary stalk (infundibulum) attaches it to the hypothalamus. The gland is about 1 centimeter in diameter and consists of an **anterior pituitary** (pĭ-tu′ĭ-tār″e), or anterior lobe, and a **posterior pituitary,** or posterior lobe (fig. 11.7).

> In the fetus, a narrow region develops between the anterior and posterior lobes of the pituitary gland. Called the *intermediate lobe* (pars intermedia), it produces melanocyte-stimulating hormone (MSH), which regulates the synthesis of melanin—the pigment in skin and in portions of the eyes and brain. This intermediate lobe seems to disappear during fetal development, but its secretory cells persist and become part of the two remaining lobes.

The brain controls most of the pituitary gland's activities. For example, the posterior pituitary releases hormones when nerve impulses from the hypothalamus signal the axon ends of neurosecretory cells in the pos-

terior pituitary (fig. 11.8). On the other hand, **releasing hormones** from the hypothalamus control secretion from the anterior pituitary (fig. 11.8). These releasing hormones travel in a capillary network associated with the hypothalamus. The capillaries merge to form the **hypophyseal portal veins,** which pass downward along the pituitary stalk and give rise to a capillary network in the anterior pituitary. Thus, the hypothalamus releases substances that the blood carries directly to the anterior pituitary.

Upon reaching the anterior pituitary, each of the hypothalamic releasing hormones acts on a specific population of cells. Some of the resulting actions are inhibitory, but most stimulate the anterior pituitary to release hormones that stimulate secretions from peripheral endocrine glands. In many of these cases, important negative feedback regulates hormone levels in the bloodstream.

Check Your Recall

13. Where is the pituitary gland located?

14. Explain how the hypothalamus controls the actions of the posterior and anterior lobes of the pituitary gland.

Anterior Pituitary Hormones

The anterior pituitary is enclosed in a capsule of dense, collagenous connective tissue and consists largely of epithelial tissue organized in blocks around many thin-walled blood vessels. So far, researchers have identified five types of secretory cells within the epithelium. Four of these each secrete a different hormone—growth hormone (GH), prolactin (PRL), thyroid-stimulating hormone (TSH), and adrenocorticotropic hormone (ACTH). The fifth type of cell secretes follicle-stimulating hormone (FSH) and luteinizing hormone (LH). (In males, luteinizing hormone is known as interstitial cell stimulating hormone, or ICSH.)

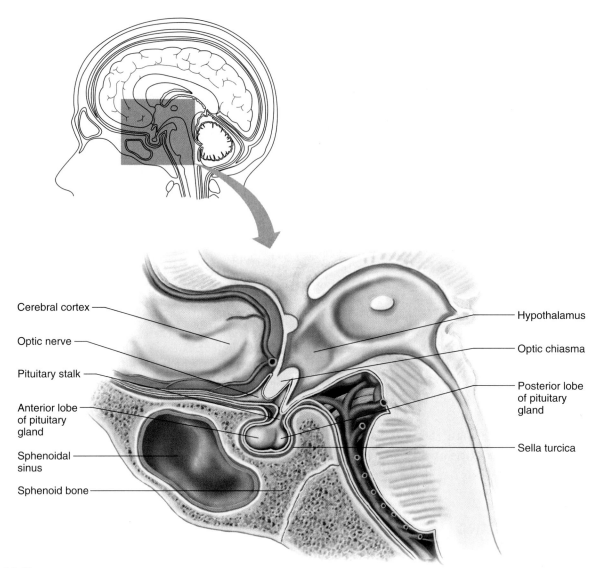

Figure 11.7

The pituitary gland is attached to the hypothalamus and lies in the sella turcica of the sphenoid bone.

Growth hormone (GH) stimulates cells to increase in size and divide more frequently. It also enhances the movement of amino acids across cell membranes and speeds the rate at which cells utilize carbohydrates and fats. The hormone's effect on amino acids is important in stimulating growth.

Two hormones from the hypothalamus control GH secretion: GH-releasing hormone and GH release-inhibiting hormone. Nutritional state also influences control of GH. For example, more GH is released during periods of protein deficiency and abnormally low blood glucose concentration. Conversely, when blood protein and glucose concentrations increase, GH secretion decreases.

Prolactin (pro-lak′tin) **(PRL)** stimulates and sustains a woman's milk production following the birth of an infant (see chapter 20, p. 555). No normal physiological role in human males has been firmly established,

although PRL may help to maintain normal sperm production. In contrast, abnormally elevated levels can disrupt sexual function in both sexes.

Thyroid-stimulating hormone (TSH) controls thyroid gland secretions, described in section 11.6 on page 299. The hypothalamus partially regulates TSH secretion by producing *thyrotropin-releasing hormone (TRH)* (fig. 11.9). Circulating thyroid hormones inhibit release of TRH and TSH. As the blood concentration of thyroid hormones increases, secretion of TRH and TSH decreases.

Check Your Recall

15. How does growth hormone affect protein synthesis?

16. What is the function of prolactin?

17. How is secretion of thyroid-stimulating hormone regulated?

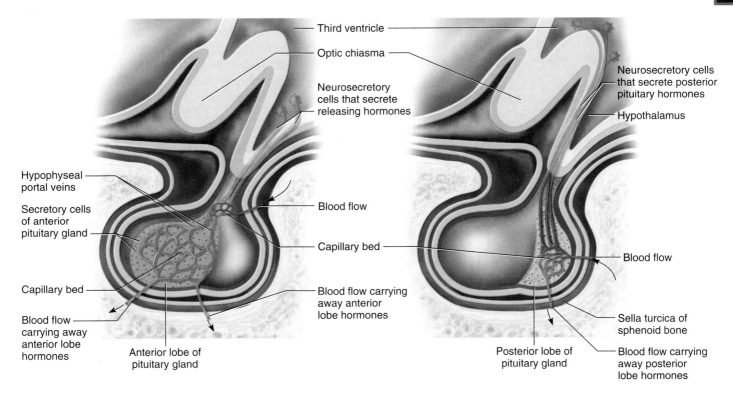

Figure 11.8

Secretion of pituitary hormones. Releasing hormones from neurosecretory cells in the hypothalamus stimulate secretory cells of the anterior lobe of the pituitary gland to secrete hormones. Other neurosecretory cells in the hypothalamus release their hormones directly into capillaries of the posterior lobe of the pituitary gland.

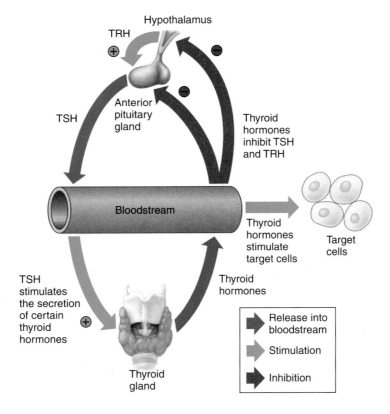

Figure 11.9

Thyrotropin-releasing hormone (TRH) from the hypothalamus stimulates the anterior pituitary gland to release thyroid-stimulating hormone (TSH), which stimulates the thyroid gland to release hormones. These thyroid hormones reduce the secretion of TSH and TRH by negative feedback. (⊕ = stimulation; ⊖ = inhibition)

Insufficient secretion of growth hormone (GH) during childhood limits growth, causing hypopituitary dwarfism. Body parts are normally proportioned, and mental development is normal. However, abnormally low GH secretion usually accompanies deficient secretion of other anterior pituitary hormones, leading to additional hormone deficiency symptoms. Hormone therapy can help.

Oversecretion of GH during childhood causes gigantism, in which height may exceed 8 feet. This rare condition is usually a result of a pituitary gland tumor, which may also cause oversecretion of other pituitary hormones. As a result, a person with gigantism often has several metabolic disturbances.

Acromegaly is the overproduction of growth hormone in adulthood. The many symptoms attesting to the wide effects of this hormone include enlarged heart, bones, thyroid gland, facial features, hands, feet, and head. Early symptoms include headache, joint pain, fatigue, and depression.

Adrenocorticotropic hormone (ad-re″no-kor″te-ko-trōp′ik hor′mōn) **(ACTH)** controls the manufacture and secretion of certain hormones from the outer layer, or *cortex,* of the adrenal gland. These hormones are discussed in section 11.8 on page 304. ACTH secretion is regulated in part by *corticotropin-releasing hormone (CRH),* which the hypothalamus releases in response to decreased concentrations of adrenal cortical hormones. Also, stress may increase ACTH secretion by stimulating the release of CRH.

Follicle-stimulating hormone (fol′ĭ-kl stim′u-la″ting hor′mōn) **(FSH)** and **luteinizing hormone** (lu′te-in-iz″ing hor′mōn) **(LH)** are *gonadotropins,* which means they exert their actions on the gonads, or reproductive organs. Gonads are the **testes** (tes′tēz) in the male and the **ovaries** (o′vah-rēz) in the female. Chapter 19, pages 516 and 520–521, discusses the functions of these gonadotropins and the ways they interact.

Check Your Recall

18. What is the function of adrenocorticotropic hormone?
19. What is a gonadotropin?

Posterior Pituitary Hormones

The posterior pituitary consists mostly of nerve fibers and neuroglial cells, unlike the anterior pituitary, which is composed primarily of glandular epithelial cells. The neuroglial cells support the nerve fibers, which originate in the hypothalamus.

Specialized neurons in the hypothalamus produce the two hormones associated with the posterior pituitary—**antidiuretic hormone** (an″tĭ-di″u-ret′ik hor′mōn) **(ADH)** and **oxytocin** (ok″sĭ-to′sin) **(OT)** (see fig. 11.8). These hormones travel down axons through the pituitary stalk to the posterior lobe, and vesicles (secretory granules) near the ends of the axons store them. Nerve

impulses from the hypothalamus release the hormones into the blood. Thus, though synthesized in the hypothalamus, posterior pituitary hormones are so named because that is where they enter the bloodstream.

A *diuretic* is a chemical that increases urine production, whereas an *antidiuretic* decreases urine formation. ADH produces an antidiuretic effect by reducing the volume of water the kidneys excrete. In this way, ADH regulates the water concentration of body fluids.

The hypothalamus regulates ADH secretion. Certain neurons in this part of the brain, called *osmoreceptors,* sense changes in the osmotic pressure of body fluids. Dehydration due to lack of water intake increasingly concentrates blood solutes. Osmoreceptors, sensing the resulting increase in osmotic pressure, signal the posterior pituitary to release ADH, which travels in the blood to the kidneys. As a result, the kidneys produce less urine, conserving water. On the other hand, drinking too much water dilutes body fluids, inhibiting ADH release. The kidneys excrete more dilute urine until the water concentration of body fluids returns to normal.

If an injury or tumor damages any parts of the ADH-regulating mechanism, too little ADH may be synthesized or released, producing *diabetes insipidus.* An affected individual may produce as much as 25–30 liters of very dilute urine per day, and solute concentrations in body fluids rise.

OT contracts smooth muscles in the uterine wall and stimulates uterine contractions in the later stages of childbirth. Stretching of uterine and vaginal tissues late in pregnancy triggers OT release during childbirth. In the breast, OT contracts certain cells associated with the milk-producing glands and their ducts. In lactating breasts, this action forces liquid from the milk glands into the milk ducts and ejects the milk from the breasts for breastfeeding. In addition, OT is an antidiuretic, but it is much weaker than ADH. Table 11.3 reviews the hormones of the pituitary gland.

If the uterus is not contracting sufficiently to expel a fully developed fetus, commercial preparations of oxytocin are sometimes used to stimulate uterine contractions, inducing labor. Such preparations are also often administered to the mother following childbirth to ensure that uterine muscles contract enough to squeeze broken blood vessels closed, minimizing the risk of hemorrhage.

Check Your Recall

20. What is the function of antidiuretic hormone?
21. How is secretion of antidiuretic hormone controlled?
22. What effects does oxytocin produce in females?

Table 11.3	Hormones of the Pituitary Gland	
Hormone	**Action**	**Source of Control**
Anterior Lobe		
Growth hormone (GH)	Stimulates an increase in the size and division rate of body cells; enhances movement of amino acids across membranes	Growth hormone-releasing hormone and growth hormone release-inhibiting hormone from hypothalamus
Prolactin (PRL)	Sustains milk production after birth	Secretion restrained by prolactin release-inhibiting hormone and stimulated by prolactin-releasing factor from hypothalamus
Thyroid-stimulating hormone (TSH)	Controls secretion of hormones from thyroid gland	Thyrotropin-releasing hormone (TRH) from hypothalamus
Adrenocorticotropic hormone (ACTH)	Controls secretion of certain hormones from adrenal cortex	Corticotropin-releasing hormone (CRH) from hypothalamus
Follicle-stimulating hormone (FSH)	In females, responsible for the development of egg-containing follicles in ovaries and stimulates follicular cells to secrete estrogen; in males, stimulates production of sperm cells	Gonadotropin-releasing hormone from hypothalamus
Luteinizing hormone (LH)	Promotes secretion of sex hormones; plays a role in releasing an egg cell in females	Gonadotropin-releasing hormone from hypothalamus
Posterior Lobe		
Antidiuretic hormone (ADH)	Causes kidneys to conserve water; in high concentration, increases blood pressure	Hypothalamus in response to changes in water concentration in body fluids
Oxytocin (OT)	Contracts muscles in the uterine wall; contracts muscles associated with milk-secreting glands	Hypothalamus in response to stretching of uterine and vaginal walls and stimulation of breasts

11.6 THYROID GLAND

The **thyroid gland** (thi'roid gland) is a very vascular structure that consists of two large lobes connected by a broad *isthmus* (is'mus). It is just below the larynx on either side and in front of the trachea (fig. 11.10 and reference plate 4, p. 26).

Structure of the Gland

A capsule of connective tissue covers the thyroid gland, which is made up of many secretory parts called *follicles*. The cavities within these follicles are lined with a single layer of cuboidal epithelial cells and filled with a clear, viscous substance called *colloid*. The follicular cells produce and secrete hormones that may either be stored in the colloid or released into the blood in nearby capillaries.

Thyroid Hormones

The follicular cells of the thyroid gland synthesize two hormones—**thyroxine** (thi-rok'sin) (tetraiodothyronine), also known as T_4 because it contains four atoms of iodine, and **triiodothyronine** (tri"i-o"do-thi'ro-nēn), known as T_3 because it includes three atoms of iodine. Thyroxine and triiodothyronine have similar actions,

although triiodothyronine is five times more potent. These hormones help regulate the metabolism of carbohydrates, lipids, and proteins. They increase the rate at which cells release energy from carbohydrates, increase the rate of protein synthesis, and stimulate breakdown and mobilization of lipids. They are the major factors determining how many calories the body must consume at rest in order to maintain life, which is known as the *basal metabolic rate (BMR)*. Thyroid hormones are required for normal growth and development, and are essential to nervous system maturation.

Up to 80% of the iodine in the body is in the thyroid gland. Here, the concentration of iodine is 25 times that in the bloodstream.

Follicular cells require iodine salts (iodides) to produce thyroxine and triiodothyronine. Foods normally provide iodides, and after the iodides have been absorbed from the intestine, blood transports them to the thyroid gland. An efficient active transport mechanism moves the iodides into the follicular cells, where they are used to synthesize the hormones. The hypothalamus and pituitary gland control release of thyroid hormones. Once in the blood, thyroxine and triiodothyronine combine with

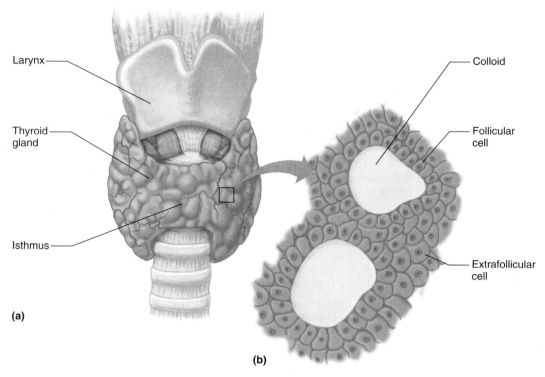

Figure 11.10

Thyroid gland. (*a*) The thyroid gland consists of two lobes connected anteriorly by an isthmus. (*b*) Follicular cells secrete thyroid hormones. Extrafollicular cells secrete calcitonin.

proteins in the blood (plasma proteins) and are transported to body cells.

A third hormone, **calcitonin** (kal′sĭ-to′nin), is often not considered a thyroid hormone because the gland's extrafollicular (other than follicle) cells produce it. Along with parathyroid hormone (PTH) from the parathyroid glands, calcitonin regulates the concentrations of blood calcium and phosphate ions (see fig. 7.8, p. 139).

Blood concentration of calcium ions regulates calcitonin release. As this concentration increases, so does calcitonin secretion. Calcitonin inhibits the bone-resorbing activity of osteoclasts (see chapter 7, p. 139) and increases the kidneys' excretion of calcium and phosphate ions—

actions that lower the blood calcium and phosphate ion concentrations. Table 11.4 reviews the actions and controls of the thyroid hormones.

Check Your Recall

23. Where is the thyroid gland located?
24. Which hormones of the thyroid gland affect carbohydrate metabolism and protein synthesis?
25. How does the thyroid gland influence the concentrations of blood calcium and phosphate ions?

Table 11.4	Hormones of the Thyroid Gland	
Hormone	**Action**	**Source of Control**
Thyroxine (T_4)	Increases rate of energy release from carbohydrates; increases rate of protein synthesis; accelerates growth; stimulates activity in nervous system	Thyroid-stimulating hormone from the anterior pituitary gland
Triiodothyronine (T_3)	Same as above, but five times more potent than thyroxine	Thyroid-stimulating hormone from the anterior pituitary gland
Calcitonin	Lowers blood calcium and phosphate ion concentrations by inhibiting release of calcium and phosphate ions from bones and by increasing excretion of these ions by kidneys	Blood calcium concentration

Many thyroid disorders produce overactivity (*hyperthyroidism*) or underactivity (*hypothyroidism*) of the glandular cells. One form of hypothyroidism appears in infants whose thyroid glands do not function normally. An affected child may appear normal at birth because the mother provided an adequate supply of thyroid hormones for the child *in utero*. But when the infant's own thyroid gland does not produce sufficient quantities of these hormones, a condition called *cretinism* develops. Symptoms include stunted growth, abnormal bone formation, retarded mental development, low body temperature, and sluggishness. Without treatment within a month or so following birth, the child may suffer permanent mental retardation.

Hyperthyroidism produces an elevated metabolic rate, restlessness, and overeating. The eyes protrude (exopthalmia) because of swelling in the tissues behind them, and the thyroid gland enlarges, producing a bulge in the neck called a *goiter.*

11.7 PARATHYROID GLANDS

The **parathyroid glands** (par"ah-thi'roid glandz) are on the posterior surface of the thyroid gland, as figure 11.11 shows. Usually, there are four parathyroid glands—a superior and an inferior gland associated with each of the thyroid's lateral lobes.

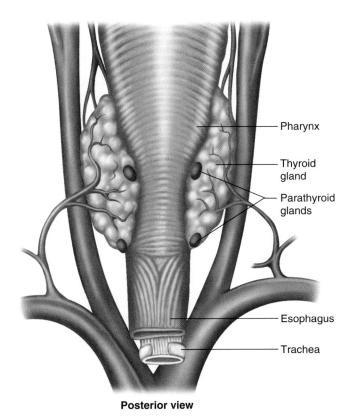

Posterior view

Figure 11.11

The parathyroid glands are embedded in the posterior surface of the thyroid gland.

Structure of the Glands

A thin capsule of connective tissue covers each small, yellowish-brown parathyroid gland. The body of the gland consists of many tightly packed secretory cells closely associated with capillary networks.

Parathyroid Hormone

The parathyroid glands secrete **parathyroid hormone (PTH),** which increases blood calcium concentration and decreases blood phosphate ion concentration. PTH affects the bones, kidneys, and intestine.

The extracellular matrix of bone tissue is rich in mineral salts, including calcium phosphate (see chapter 7, p. 139). PTH inhibits the activity of osteoblasts and stimulates osteoclasts to resorb bone and release calcium and phosphate ions into the blood. At the same time, PTH causes the kidneys to conserve blood calcium and to excrete more phosphate ions in the urine. It also stimulates calcium absorption from food in the intestine, further increasing blood calcium concentration.

Negative feedback between the parathyroid glands and the blood calcium concentration regulates PTH secretion. As blood calcium concentration drops, more PTH is secreted; as blood calcium concentration rises, less PTH is released (fig. 11.12).

To summarize, calcitonin and PTH activities maintain stable blood calcium concentration. Calcitonin decreases an above-normal blood calcium concentration, while PTH increases a below-normal blood calcium concentration (see also fig. 7.8, p. 139).

Check Your Recall

26. Where are the parathyroid glands?
27. How does parathyroid hormone help regulate concentrations of blood calcium and phosphate ions?

A tumor in a parathyroid gland may cause *hyperparathyroidism*, which increases PTH secretion. This stimulates osteoclast activity, and as bone tissue is resorbed, the bones soften, deform, and more easily fracture spontaneously. In addition, excess calcium and phosphate released into body fluids may be deposited in abnormal places, causing new problems, such as kidney stones.

Injury to the parathyroids or their surgical removal can cause *hypoparathyroidism,* in which decreased PTH secretion reduces osteoclast activity. Although the bones remain strong, the blood calcium concentration decreases. The nervous system may become abnormally excitable, triggering spontaneous impulses. As a result, muscles may undergo tetanic contractions, possibly leading to respiratory failure and death.

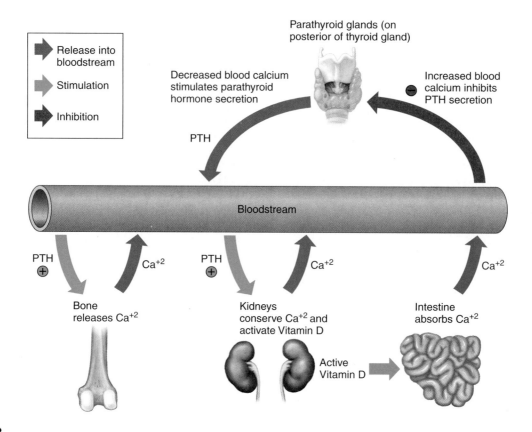

Figure 11.12

Parathyroid hormone (PTH) stimulates bone to release calcium (Ca^{+2}) and the kidneys to conserve calcium. It indirectly stimulates the intestine to absorb calcium. The resulting increase in blood calcium concentration inhibits secretion of PTH by negative feedback. (⊕ = stimulation; ⊖ = inhibition)

11.8 ADRENAL GLANDS

The **adrenal glands** (ah-dre′nal glandz) are closely associated with the kidneys (fig. 11.13 and reference plate 6, p. 28). A gland sits atop each kidney like a cap and is embedded in the mass of adipose tissue that encloses the kidney.

Structure of the Glands

Each adrenal gland is very vascular and consists of two parts: The central portion is the **adrenal medulla** (ah-dre′nal me-dul′ah), and the outer part is the **adrenal cortex** (ah-dre′nal kor′teks). These regions are not sharply divided, but they are functionally distinct glands that secrete different hormones.

The adrenal medulla consists of irregularly shaped cells organized in groups around blood vessels. These cells are intimately connected with the sympathetic division of the autonomic nervous system. Adrenal medullary cells are actually modified postganglionic neurons. Preganglionic autonomic nerve fibers lead to them from the central nervous system (see chapter 9, p. 249).

The adrenal cortex, which makes up the bulk of the adrenal gland, is composed of closely packed masses of epithelial cells, organized in layers. These layers form an outer (glomerulosa), middle (fasciculata), and inner (reticularis) zone of the cortex (fig. 11.13b). As in the adrenal medulla, the cells of the adrenal cortex are well supplied with blood vessels.

Check Your Recall

28. Where are the adrenal glands?
29. Describe the two portions of an adrenal gland.

Hormones of the Adrenal Medulla

The cells of the adrenal medulla secrete two closely related hormones—**epinephrine** (ep″ĭ-nef′rin) (adrenalin) and **norepinephrine** (nor″ep-ĭ-nef′rin) (noradrenalin). These hormones have similar molecular structures and physiological functions. In fact, epinephrine, which makes up 80% of the adrenal medullary secretion, is synthesized from norepinephrine.

The effects of the adrenal medullary hormones resemble those of sympathetic neurons stimulating their effectors. Hormonal effects, however, last up to ten times longer because hormones are broken down more slowly

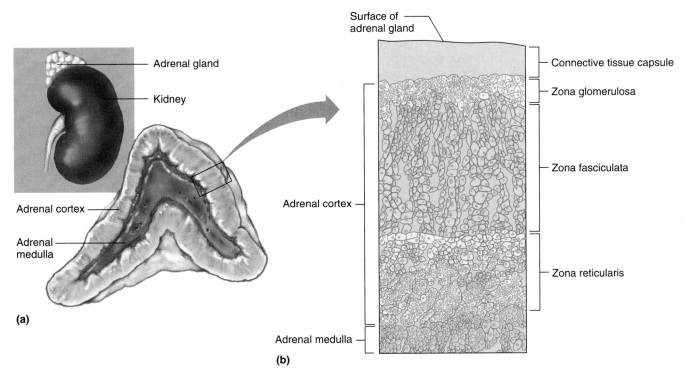

Figure 11.13

Adrenal glands. (*a*) An adrenal gland consists of an outer cortex and an inner medulla. (*b*) The cortex consists of three layers, or zones, of cells.

than neurotransmitters. Epinephrine and norepinephrine increase heart rate, the force of cardiac muscle contraction, breathing rate, and blood glucose level. They also elevate blood pressure and decrease digestive activity.

Impulses arriving on sympathetic nerve fibers stimulate the adrenal medulla to release its hormones at the same time that sympathetic impulses are stimulating other effectors. These sympathetic impulses originate in the hypothalamus in response to stress. Thus, adrenal medullary secretions function with the sympathetic division of the autonomic nervous system in preparing the body for energy-expending action, sometimes called "fight-or-flight responses." Table 11.5 compares some of the effects of the adrenal medullary hormones.

Tumors in the adrenal medulla can increase hormonal secretion. Release of norepinephrine usually predominates, prolonging sympathetic responses—high blood pressure, increased heart rate, elevated blood sugar, and so forth. Surgical removal of the tumor corrects the condition.

Check Your Recall

30. Name the hormones the adrenal medulla secretes.
31. What effects do hormones from the adrenal medulla produce?
32. What stimulates release of hormones from the adrenal medulla?

Table 11.5	Comparative Effects of Epinephrine and Norepinephrine	
Part or Function Affected	**Epinephrine**	**Norepinephrine**
Heart	Increases rate and force of contraction	Increases rate and force of contraction
Blood vessels	Dilates vessels in skeletal muscle, decreasing resistance to blood flow	Increases blood flow to skeletal muscles, resulting from constriction of blood vessels in skin and viscera
Systemic blood pressure	Increases somewhat due to increased cardiac output	Increases greatly due to vasoconstriction
Airways	Dilates	Dilates slightly
Reticular formation of brain	Activates	Produces little effect
Liver	Promotes breakdown of glycogen to glucose, increasing blood sugar concentration	Produces little effect on blood sugar concentration
Metabolic rate	Increases	Increases

Hormones of the Adrenal Cortex

The cells of the adrenal cortex produce more than thirty different steroids, including several hormones. Unlike the adrenal medullary hormones, without which a person can still survive, some adrenal cortical hormones are vital. Without them, a person usually dies within a week unless extensive electrolyte therapy is provided. The most important adrenal cortical hormones are aldosterone, cortisol, and certain sex hormones.

Aldosterone

Cells in the outer zone of the adrenal cortex synthesize **aldosterone** (al-dos'ter-ōn"). This hormone is a *mineralocorticoid* (min"er-al-o-kor'tĭ-koid) because it helps regulate the concentration of mineral electrolytes. More specifically, aldosterone causes the kidney to conserve sodium ions and excrete potassium ions. By conserving sodium ions, aldosterone stimulates water retention indirectly by osmosis, helping to maintain blood volume and blood pressure.

A decrease in the blood concentration of sodium ions or an increase in the blood concentration of potassium ions stimulates the cells that secrete aldosterone. The kidneys also indirectly stimulate aldosterone secretion if blood pressure falls (see chapter 17, p. 482).

Cortisol

Cortisol (kor'tĭ-sol) (hydrocortisone) is a *glucocorticoid* (gloo"ko-kor'tĭ-koid), which means it affects glucose metabolism. It is produced in the middle zone of the adrenal cortex and, like aldosterone, is a steroid. Cortisol also influences protein and fat metabolism.

The more important actions of cortisol include:

1. Inhibition of protein synthesis in tissues, increasing the blood concentration of amino acids.
2. Promotion of fatty acid release from adipose tissue, increasing the utilization of fatty acids as an energy source and decreasing the use of glucose.
3. Stimulation of liver cells to synthesize glucose from noncarbohydrates, such as circulating amino acids and glycerol, increasing the blood glucose concentration.

These actions of cortisol help keep blood glucose concentration within the normal range between meals, because a few hours without food can exhaust the supply of liver glycogen, a major source of glucose.

Negative feedback controls cortisol release. This is much like control of thyroid hormones, involving the hypothalamus, anterior pituitary gland, and adrenal cortex. The hypothalamus secretes corticotropin-releasing hormone (CRH) into the hypophyseal portal veins, which carry CRH to the anterior pituitary, stimulating it to secrete ACTH. In turn, ACTH stimulates the adrenal cortex to release cortisol. Cortisol inhibits the release of CRH and ACTH, and as concentrations of these fall, cortisol production drops (fig. 11.14).

The set point of the feedback mechanism controlling cortisol secretion changes from time to time, altering hormone output to meet the demands of changing conditions. For example, under stress—injury, disease, extreme temperature, or emotional upset—nerve impulses send the brain information concerning the stressful condition. In response, brain centers signal the hypothalamus to release more CRH, elevating cortisol concentration until the stress subsides (fig. 11.14).

Adrenal Sex Hormones

Cells in the inner zone of the adrenal cortex produce sex hormones. These hormones are male types (adrenal androgens), but some are converted to female hormones (estrogens) in the skin, liver, and adipose tissue. Adrenal sex hormones may supplement the supply of sex hormones from the gonads and stimulate early development of reproductive organs. Table 11.6 summarizes the characteristics of the adrenal cortical hormones.

Hyposecretion of adrenal cortical hormones leads to *Addison disease,* a condition characterized by decreased blood sodium, increased blood potassium, low blood glucose concentration (hypoglycemia), dehydration, low blood pressure, and increased skin pigmentation. Without treatment with mineralocorticoids and glucocorticoids, Addison disease can be lethal in days because of severe disturbances in electrolyte balance.

Hypersecretion of adrenal cortical hormones, which may be associated with an adrenal tumor or with the anterior pituitary oversecreting ACTH, causes *Cushing syndrome.* This condition alters carbohydrate and protein metabolism and electrolyte balance. For example, when mineralocorticoids and glucocorticoids are overproduced, blood glucose concentration remains high, depleting tissue protein. Also, too much sodium is retained, increasing tissue fluids, and the skin becomes puffy. At the same time, increase in adrenal sex hormone production may cause masculinizing effects in a female, such as beard growth and deepening of the voice.

Check Your Recall

33. Name the most important hormones of the adrenal cortex.
34. What is the function of aldosterone?
35. What actions does cortisol produce?
36. How are the blood concentrations of aldosterone and cortisol regulated?

11.9 PANCREAS

The **pancreas** (pan'kre-as) consists of two major types of secretory tissues. This organization reflects the pancreas's dual function as an exocrine gland that secretes

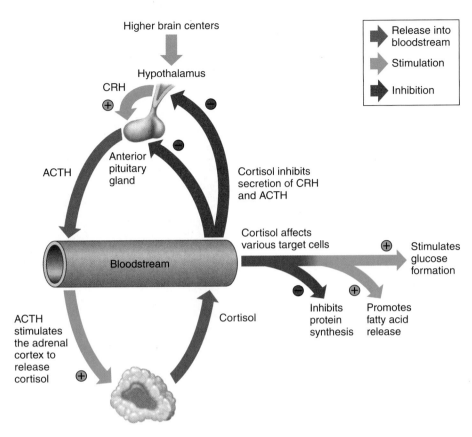

Higher brain centers

Hypothalamus

CRH

Anterior pituitary gland

ACTH

Cortisol inhibits secretion of CRH and ACTH

Bloodstream

Cortisol affects various target cells

Stimulates glucose formation

Inhibits protein synthesis

Promotes fatty acid release

ACTH stimulates the adrenal cortex to release cortisol

Cortisol

Release into bloodstream

Stimulation

Inhibition

Figure 11.14

Negative feedback regulates cortisol secretion, similar to the regulation of thyroid hormone secretion (see fig. 11.9). (⊕ = stimulation; ⊖ = inhibition)

Table 11.6	Hormones of the Adrenal Cortex	
Hormone	**Action**	**Factor Regulating Secretion**
Aldosterone	Helps regulate concentration of extracellular electrolytes by conserving sodium ions and excreting potassium ions	Electrolyte concentrations in body fluids
Cortisol	Decreases protein synthesis, increases fatty acid release, and stimulates glucose synthesis from noncarbohydrates	Corticotropin-releasing hormone from hypothalamus and adrenocorticotropic hormone from anterior pituitary
Adrenal androgens	Supplement sex hormones from the gonads; may be converted to estrogens in females	

digestive juice and an endocrine gland that releases hormones (fig. 11.15 and reference plate 6, p. 28).

Structure of the Gland

The pancreas is an elongated, somewhat flattened organ posterior to the stomach and behind the parietal peritoneum. A duct joins the pancreas to the duodenum (the first section of the small intestine) and transports pancreatic digestive juice to the intestine. The pancreas's dual nature begins in the embryo. First,

ducts form whose walls harbor progenitor cells (see fig. 3.23, p. 71). Some of the progenitor cells divide to yield daughter cells that specialize as exocrine cells, and others divide to yield cells that differentiate into endocrine cells. The two functions are elaborated as the gland develops further.

The endocrine portion of the pancreas consists of groups of cells that are closely associated with blood vessels. These groups form "islands" of cells called *pancreatic islets* (islets of Langerhans) (figs. 11.15 and 11.16). The pancreatic islets include two distinct types

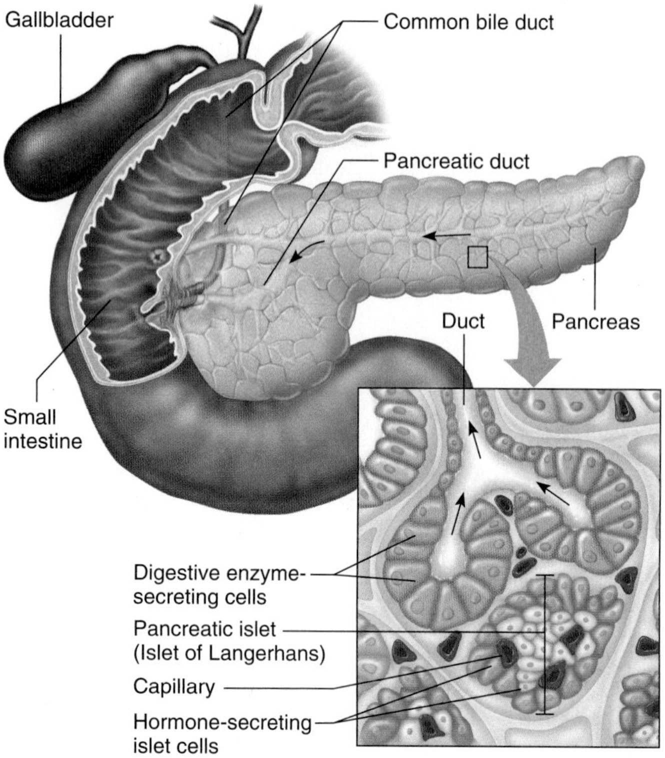

Gallbladder

Common bile duct

Pancreatic duct

Duct Pancreas

Small intestine

Digestive enzyme-secreting cells

Pancreatic islet (Islet of Langerhans)

Capillary

Hormone-secreting islet cells

Figure 11.15

The hormone-secreting cells of the pancreas are grouped in clusters, or islets, that are in close proximity to blood vessels. Other pancreatic cells secrete digestive enzymes into ducts.

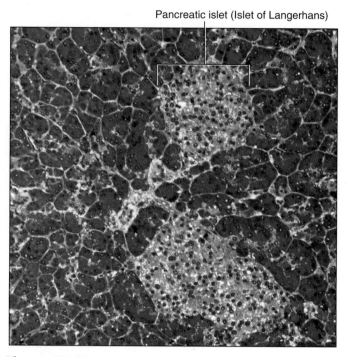

Pancreatic islet (Islet of Langerhans)

Figure 11.16

Light micrograph of a pancreatic islet (200×).

of cells—alpha cells, which secrete the hormone glucagon, and beta cells, which secrete the hormone insulin. The Topic of Interest on pages 308–309 discusses diabetes mellitus, a disorder that affects the beta cells. Chapter 15 (p. 414) discusses the digestive functions of the pancreas.

Hormones of the Pancreatic Islets

Glucagon (gloo′kah-gon) stimulates the liver to break down glycogen and convert certain noncarbohydrates, such as amino acids, into glucose, raising blood sugar concentration. Glucagon much more effectively elevates blood glucose than does epinephrine.

A negative feedback system regulates glucagon secretion. Low blood glucose concentration stimulates alpha cells to release glucagon. When blood glucose concentration rises, glucagon secretion falls. This control prevents hypoglycemia when glucose concentration is relatively low, such as between meals, or when glucose is used rapidly, such as during periods of exercise.

The main effect of **insulin** (in′su-lin) is exactly opposite that of glucagon. Insulin stimulates the liver to form glycogen from glucose and inhibits conversion of noncarbohydrates into glucose. Insulin also has the special effect of promoting facilitated diffusion (see

chapter 3, p. 62) of glucose across cell membranes that have insulin receptors, such as those of cardiac muscle, adipose tissue, and resting skeletal muscle. (Glucose uptake by exercising skeletal muscle does not require insulin.) These actions of insulin decrease blood glucose concentration. In addition, insulin secretion promotes transport of amino acids into cells, increases the rate of protein synthesis, and stimulates adipose cells to synthesize and store fat.

A negative feedback system sensitive to blood glucose concentration regulates insulin secretion. When blood glucose concentration is high, as after a meal, beta cells release insulin. Insulin helps prevent too high a blood glucose concentration by promoting glycogen formation in the liver and entrance of glucose into adipose and muscle cells. When glucose concentration falls, as occurs between meals or during the night, insulin secretion decreases.

As insulin output decreases, less and less glucose enters adipose and muscle cells. Cells that lack insulin receptors, such as nerve cells, can then use the glucose that remains in the blood. At the same time that insulin is decreasing, glucagon secretion is increasing. Therefore, insulin and glucagon function coordinately to maintain a relatively stable blood glucose concentration, despite great variation in the amount of carbohydrates a person eats (fig. 11.17).

Nerve cells, including those of the brain, obtain glucose by a facilitated diffusion mechanism that does not require insulin but rather depends only on the blood glucose concentration. For this reason, nerve cells are

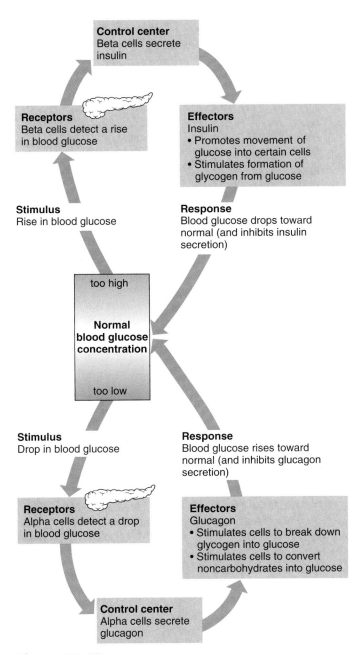

Figure 11.17

Insulin and glucagon function together to help maintain a relatively stable blood glucose concentration. Negative feedback responding to blood glucose concentration controls the levels of both hormones.

particularly sensitive to changes in blood glucose concentration, and conditions that cause such changes—oversecretion of insulin leading to decreased blood glucose, for example—are likely to alter brain functions.

Cancer cells that develop from nonendocrine tissues sometimes inappropriately synthesize and secrete great amounts of peptide hormones or peptide hormonelike chemicals. For example, in endocrine paraneoplastic syndrome, a person with a non-endocrine cancer overproduces ADH, ACTH, a PTH-like substance, or an insulin-like substance.

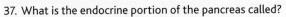

Check Your Recall

37. What is the endocrine portion of the pancreas called?
38. What is the function of glucagon?
39. What is the function of insulin?
40. How are glucagon and insulin secretion controlled?
41. Why are nerve cells particularly sensitive to changes in blood glucose concentration?

11.10 OTHER ENDOCRINE GLANDS

Other glands that produce hormones and thus are parts of the endocrine system include the pineal gland, the thymus, reproductive organs, and certain glands of the digestive tract, heart, and kidneys.

The **pineal gland** (pin'e-al gland) is a small structure located deep between the cerebral hemispheres, where it attaches to the upper portion of the thalamus near the roof of the third ventricle (see fig. 11.2). The pineal gland secretes the hormone **melatonin** (mel"ah-to'nin) in response to light conditions outside the body. Nerve impulses originating in the retinas of the eyes send this information to the pineal gland. In the dark, nerve impulses from the eyes decrease, and melatonin secretion increases.

Melatonin acts on certain brain regions that function as a "biological clock," and may thereby help to regulate **circadian rhythms** (ser"kah-de'an rithmz), which are patterns of repeated activity associated with the environmental cycles of day and night. The changing levels of melatonin throughout the 24-hour day may enable the body to distinguish day from night. Circadian rhythms include the sleep-wake rhythm and seasonal cycles of fertility in many mammals. The Topic of Interest on page 310 discusses biological rhythms.

The fact that melatonin secretion responds to day length explains why traveling across several time zones produces the temporary insomnia of jet lag. Melatonin supplements are advertised as preventing jet lag, based on anecdotal reports and small studies. However, the first large study, conducted on 257 doctors traveling from Norway to New York, testing three nightly doses of melatonin supplement versus placebo, showed no effect at all in preventing or alleviating jet lag.

Although the mechanism of melatonin action is poorly understood, the hormone inhibits the secretion of gonadotropins from the anterior pituitary and may help regulate the female reproductive cycle. It may also control the onset of puberty.

The **thymus** (thi'mus), which lies in the mediastinum posterior to the sternum and between the lungs, is

Topic of Interest

Diabetes Mellitus

Diabetes mellitus is a metabolic derangement that arises from lack of insulin or inability of cells to recognize it. Before insulin was isolated and its mechanism understood in 1921, a diagnosis of type 1 diabetes mellitus meant that a child had, at best, three years to live. By 1922, children began to be treated with insulin. They rapidly recovered (figure 11A). As people began living longer by receiving insulin, the disease's longer-term effects on other organs—such as the eye, heart, kidney, and peripheral nerves—became noticeable.

Insulin deficiency disturbs carbohydrate, protein, and fat metabolism. Because insulin helps glucose cross some cell membranes, movement of glucose into adipose and resting skeletal muscle cells becomes impaired in diabetes. At the same time, formation of glycogen, which is a long chain of glucose molecules, declines. As a result, blood sugar concentration rises (hyperglycemia). When it reaches a certain level, the kidneys begin to excrete the excess. Glucose in the urine (glycosuria) raises the urine's osmotic pressure, and too much water and electrolytes are excreted. Excess urine output causes dehydration and extreme thirst (polydipsia).

Diabetes mellitus also hampers protein and fat synthesis. Glucose-starved cells increasingly use proteins for energy, and as a result, tissues waste away as weight drops, hunger increases, exhaustion becomes overwhelming, children stop growing, and wounds do not heal. Changes in fat metabolism cause fatty acids and ketone bodies to accumulate in

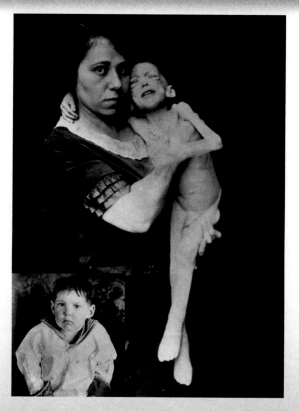

Figure 11A

Before and after insulin treatment: The boy in his mother's arms is three years old but weighs only 15 pounds because of type 1 diabetes mellitus. The inset shows the same child after just two months of receiving insulin—his weight had doubled.

relatively large in young children but shrinks with age (see fig. 11.2). This gland secretes a group of hormones called **thymosins** (thi′mo-sinz) that affect the production and differentiation of certain white blood cells (lymphocytes). In this way, the thymus plays an important role in immunity, discussed in chapter 14 (p. 381).

The reproductive organs that secrete important hormones include the ovaries, which produce estrogens and progesterone; the **placenta** (plah-sen′tah), which produces estrogens, progesterone, and gonadotropin; and the testes, which produce testosterone. These glands and their secretions are discussed in chapter 19 (pp. 515–516 and 524) and chapter 20 (p. 545).

The digestive glands that secrete hormones are associated with the linings of the stomach and small intestine. Chapter 15 (pp. 413 and 415) describes these structures and their secretions.

Other organs outside of the endocrine system produce hormones. The heart, for example, secretes *atrial natriuretic peptide*, a hormone that stimulates urinary sodium excretion (see chapter 17, p. 478). The kidneys secrete a red blood cell growth hormone called *erythropoietin* (see chapter 12, p. 320).

Check Your Recall

42. Where is the pineal gland located?
43. What is the function of the pineal gland?
44. Where is the thymus located?
45. Which reproductive organs secrete hormones?
46. Which other organs secrete hormones?

the blood, which lowers pH (acidosis). Dehydration and acidosis may harm brain cells, causing disorientation, coma, and, eventually, death.

Type 1 Diabetes Mellitus

The two common forms of diabetes mellitus are type 1 (insulin-dependent or juvenile diabetes) and type 2 (non-insulin-dependent or maturity-onset diabetes). Type 1 usually appears before age twenty and is an autoimmune disease: the immune system destroys the beta cells of the pancreas (see chapter 14, p. 393).

Treatment for type 1 diabetes is still to give insulin, but delivery has improved so that treatment better mimics normal pancreatic function. Before 1978, for example, people with diabetes used insulin from pigs. Then genetically modified bacteria began to supply the human version of the hormone, to which allergy is far less likely.

People with type 1 diabetes typically inject insulin several times a day, or receive the hormone from an implanted insulin pump. Replacing a diabetic pancreas with a healthy transplanted organ is too difficult to be practical—the surgery is complex, the supply of organs very limited, and immune rejection difficult to prevent or control. Instead, much research has focused on islet transplantation.

Since 2000, several hundred people have received islet transplants in a procedure called the Edmonton protocol, which introduces islets into a vein in the liver. By a year after transplant, 50–68% of patients do not need to receive additional insulin, but by five years after the procedure, fewer

than 10% of total patients are free of daily insulin supplementation. The procedure itself is risky—12% of patients hemorrhage, and 4% develop blood clots in the liver vein. These risks plus the apparent short-term nature of effectiveness have prompted physicians to carefully evaluate which patients are most likely to benefit from the few years of insulin independence that the procedure may offer.

Type 2 Diabetes Mellitus

About 85–90% of people with diabetes mellitus have type 2, in which the beta cells produce insulin, but body cells lose the ability to recognize it. The condition usually develops gradually after age forty and has milder symptoms than type 1 diabetes. Most affected individuals are overweight when symptoms begin. Treatment includes controlling the diet, exercising, and maintaining a desirable body weight. Several oral drugs can help control glucose levels, which can delay the onset of diabetes-related complications.

People with either type of diabetes must monitor their blood glucose level at least daily, and do what they can to regulate it, to forestall complications, which include coronary artery disease, peripheral nerve damage, and retinal damage. Evidence suggests that these complications may begin even before blood glucose level indicates disease. The American Diabetes Association now recognizes "prediabetes" as blood glucose levels above the normal range but not yet indicative of type 2 diabetes. About 20 million people in the U.S. between the ages of 40 and 74 fall into this category.

11.11 STRESS AND HEALTH

Survival depends on the maintenance of homeostasis. Therefore, factors that change the body's internal environment can threaten life. When the body senses danger, nerve impulses to the hypothalamus trigger physiological responses that preserve homeostasis. These responses include increased activity in the sympathetic division of the autonomic nervous system and increased secretion of adrenal and other hormones. A factor that can stimulate such a response is called a *stressor*, and the condition it produces in the body is called **stress.**

Types of Stress

Stressors include physical factors, such as exposure to extreme heat or cold, decreased oxygen concentration,

infections, injuries, prolonged heavy exercise, and loud sounds. Stressors also include psychological factors, such as thoughts about real or imagined dangers, personal losses, and unpleasant social interactions. Feelings of anger, fear, grief, anxiety, depression, and guilt can also produce psychological stress. Sometimes, even pleasant stimuli, such as friendly social contact, feelings of joy and happiness, or sexual arousal, may be stressful.

Responses to Stress

Physiological responses to stress consist of reactions called the *stress response* or *general adaptation syndrome*, which is under hypothalamic control. These reactions occur in two stages: the immediate "alarm" stage and the longer term "resistance" stage. Initially, the hypothalamus activates mechanisms that prepare the body for "fight or

Topic of Interest

Biological Rhythms

Biological rhythms are changes that systematically recur in organisms. In complex animals, they include the daily ebb and flow of biochemical levels in blood, reproductive cycles, and migration schedules. The period of any rhythm is the duration of one complete cycle. The frequency of a rhythm is the number of cycles per time unit. The study of biological rhythms is called *chronobiology.*

Three common types of rhythms in humans are ultradian, infradian, and circadian rhythms. *Ultradian rhythms* have periods shorter than 24 hours and include the cardiac cycle and the breathing cycle. Periods of *infradian rhythms,* such as the reproductive cycle, are longer than 24 hours. Periods of *circadian rhythms,* such as the sleep-wake cycle, are approximately 24 hours.

Both external (exogenous) and internal (endogenous) factors regulate human biological rhythms. Exogenous factors are environmental components, such as daily temperature changes and the light-dark cycle. Endogenous factors include "clock" genes. Many members of an extended family in Utah, for example, have "advanced sleep phase syndrome" due to a mutation in gene called "period." The effect is striking—they promptly fall asleep at 7:30 each night and awaken suddenly at 4:30 A.M. The same gene alters sleep patterns in fruit flies and golden hamsters, which are commonly used in chronobiology research. The mutation in all three species disrupts a signal that synchronizes the sleep-wake cycle to daily sunrise and sunset. The cells that respond to environmental light and dark signals are located in a part of the brain called the suprachiasmatic nuclei. Understanding how clock genes function may lead to new treatments for jet lag, insomnia, and a form of advanced sleep phase syndrome that is common among older individuals.

The sleep-wake cycle is the most obvious circadian rhythm in humans. It is largely controlled by the pattern of daylight and night, but under laboratory conditions of constant light or dark, the human body eventually follows an approximately 25-hour cycle. Other circadian rhythms in humans affect body temperature, cardiovascular functioning, and hormone secretion. Body temperature is mostly endogenously regulated, but light exposure and physical activity help keep this rhythm on a 24- rather than 25-hour cycle. Body temperature is usually lowest between 4 and 6 A.M., and then increases and peaks between 5 and 11 P.M. It drops during the late evening hours and into the night. Cardiovascular functioning is least efficient between 6 and 9 A.M. Platelet cohesion, blood pressure, and pulse rate are typically highest 2 hours after awakening, which may explain why heart attacks and strokes are more likely to occur between 8 and 10 A.M. than at other times.

Hormones may have ultradian and infradian as well as circadian rhythms. Plasma cortisol, for example, surges and peaks at about 6 A.M., and then gradually declines to its minimum level in late evening before increasing again in the early morning. Growth hormone secretion peaks during the night. Antidiuretic hormone secretion is greater at night, when it decreases urine formation.

---

flight." These responses include raising blood concentrations of glucose, glycerol, and fatty acids; increasing heart rate, blood pressure, and breathing rate; dilating air passages; shunting blood from the skin and digestive organs to the skeletal muscles; and increasing epinephrine secretion from the adrenal medulla (fig. 11.18).

In the resistance response, the hypothalamus releases CRH, which stimulates the anterior pituitary to secrete ACTH, which increases cortisol secretion. Cortisol increases blood amino acid concentration, fatty acid release, and glucose formation from noncarbohydrates. Thus, while the alarm responses prepare the body for physical action to alleviate the stress, cortisol supplies cells with biochemicals required during stress (fig. 11.18).

Other hormones whose secretions increase with stress include glucagon, GH, and ADH. Glucagon and GH mobilize energy sources, such as glucose, glycerol, fatty acids, and amino acids. ADH stimulates the kidneys to retain water, which increases blood volume—particularly important if a person is bleeding or sweating heavily.

Increased cortisol secretion may be accompanied by a decrease in the number of certain white blood cells (lymphocytes), which lowers resistance to infectious diseases and some cancers. Also, excess cortisol production may raise the risk of developing high blood pressure, atherosclerosis, and gastrointestinal ulcers.

The Clinical Connection on pages 311 and 313 discusses a specific type of long-term stress response, called post-traumatic stress disorder.

Check Your Recall

47. What is stress?

48. Distinguish between physical stress and psychological stress.

49. Describe the stress response.

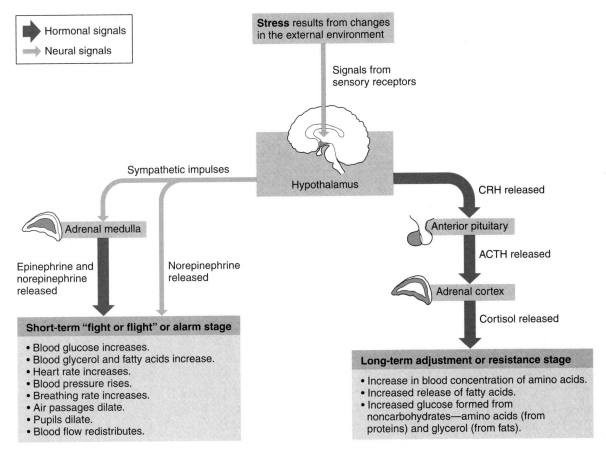

Figure 11.18

Stress response. During stress, the hypothalamus helps prepare the body for "fight or flight" by triggering sympathetic impulses to various organs. It also stimulates epinephrine release, intensifying the sympathetic responses. The hypothalamus also secretes corticotropin-releasing hormone, which sets into motion more lasting responses to stress.

Clinical Terms Related to the Endocrine System

adrenalectomy (ah-dre″nah-lek′to-me) Surgical removal of the adrenal glands.

adrenogenital syndrome (ah-dre″no-jen′ĭ-tal sin′drōm) A group of symptoms associated with changes in sexual characteristics as a result of increased secretion of adrenal androgens.

diabetes insipidus (di″ah-be′tēz in-sip′ĭdus) Large output of dilute urine caused by the posterior pituitary's decreased secretion of antidiuretic hormone.

diabetes mellitus (di″ah-be′tēz mel′i-tus) Condition due to insulin deficiency or the inability to respond to insulin that disturbs carbohydrate, protein, and lipid metabolism.

exophthalmia (ek″sof-thal′mē-ah) Abnormal protrusion of the eyes.

goiter (goi′ter) Bulge in the neck resulting from an enlarged thyroid gland.

hirsutism (her′sūt-izm) Excess hair growth, especially in women.

hypercalcemia (hi″per-kal-se′me-ah) Excess blood calcium.

hyperglycemia (hi″per-gli-se′me-ah) Excess blood glucose.

hypocalcemia (hi″po-kal-se′me-ah) Low blood calcium.

hypoglycemia (hi″po-gli-se′me-ah) Low blood glucose.

hypophysectomy (hi-pof″ĭ-sek′to-me) Surgical removal of the pituitary gland.

parathyroidectomy (par″ah-thi″roi-dek′to-me) Surgical removal of the parathyroid glands.

pheochromocytoma (fe-o-kro″mo-si-to′mah) Type of tumor in the adrenal medulla.

thymectomy (thi-mek′to-me) Surgical removal of the thymus.

thyroidectomy (thi″roi-dek′to-me) Surgical removal of the thyroid gland.

thyroiditis (thi″roi-di′tis) Inflammation of the thyroid gland.

virilism (vir′ĭ-lizm) Masculinization of a female.

Clinical Connection

Post-traumatic Stress Disorder

Once called "battle fatigue," *post-traumatic stress disorder* (PTSD) was named after more than a million veterans of the Vietnam war reported flashbacks, nightmares, angry outbursts, depression, and sudden memories that would paralyze them with fear. Their experiences go well beyond normal memory—a veteran in the throes

Endocrine System

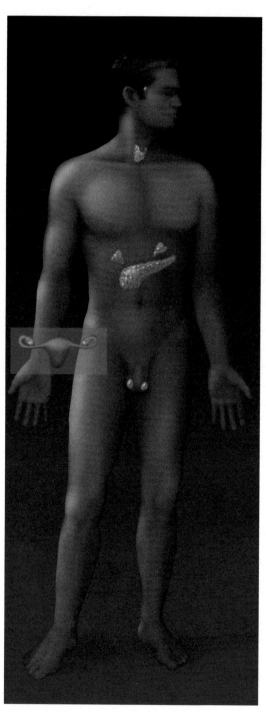

Integumentary System

Melanocytes produce skin pigment in response to hormonal stimulation.

Lymphatic System

Hormones stimulate lymphocyte production.

Skeletal System

Hormones act on bones to control calcium balance.

Digestive System

Hormones help control digestive system activity.

Muscular System

Hormones help increase blood flow to exercising muscles.

Respiratory System

Decreased oxygen causes hormonal stimulation of red blood cell production; red blood cells transport oxygen and carbon dioxide.

Nervous System

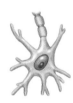

Neurons control the secretions of the anterior and posterior pituitary glands and the adrenal medulla.

Urinary System

Hormones act on the kidneys to help control water and electrolyte balance.

Cardiovascular System

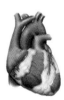

Hormones are carried in the bloodstream; some have direct actions on the heart and blood vessels.

Reproductive System

Sex hormones play a major role in development of secondary sex characteristics, egg, and sperm.

Glands secrete hormones that have a variety of effects on cells, tissues, organs, and organ systems.

of PTSD sees, hears, and smells the long-ago trauma-inducing scene, recalling such details as weather conditions and who stood where. Researchers hypothesize that the elevated levels of stress hormones (epinephrine and norepinephrine) that accompany trauma strengthen consolidation of memories, leading to the incredibly vivid recollections.

PTSD is an anxiety disorder that is triggered by experiencing a frightening event and recalling it, or feeling threatened. For example, in the weeks following the destruction of the World Trade Center on September 11, 2001, 130,000 residents of Manhattan were officially diagnosed with the disorder—the number of affected individuals was actually probably much greater. But people who were not at the scene that day also developed symptoms of PTSD due to the anxiety that swept the nation in the aftermath.

All forms of violence, accidents, abuse, and natural disasters can cause PTSD. About a third of war veterans develop symptoms. PTSD affects millions, and can strike at any age, usually within a few months of the triggering event, but sometimes years later. PTSD may be difficult to diagnose because affected individuals often seek health care for associated physical symptoms, such as chest pain, stomachache, or headache. Unless the health care professional specifically asks about traumatic experiences, the connection of bodily complaints to the traumatized mind may not be obvious.

PTSD may be caused by overactivity of the amygdala, a small structure in the brain where feelings of fear arise. What may have evolved to be a protective response may cause PTSD when extreme or prolonged. Imaging studies reveal that the hippocampus, the seat of memory, functions differently in people with PTSD, perhaps explaining the vivid flashbacks.

A combination of treatment approaches helps some people with PTSD. Drugs may treat associated depression or help alleviate insomnia or anxiety. Studies on schoolchildren who experience natural disasters such as hurricanes indicate that talking about the scary situation soon after it occurs can lessen the risk of developing PTSD. This is why counselors are typically called to schools dealing with a crisis. Conversely, exposure therapy, in which a person talks about a traumatic event in detail under safe conditions, confronting the fear, may actually worsen the condition in some cases.

SUMMARY OUTLINE

11.1 Introduction (p. 290)

The endocrine and nervous systems maintain homeostasis.

1. The endocrine system is a network of glands that secrete hormones, which travel in the bloodstream and affect the functioning of target cells.
2. Paracrine secretions act locally, and autocrine secretions act on the cells that produce them.
3. Exocrine glands secrete through tubes or ducts.

11.2 General Characteristics of the Endocrine System (p. 290)

1. The nervous and endocrine systems both exert precise effects.
2. Hormones are secreted from glands to help regulate metabolic processes of target cells.

11.3 Hormone Action (p. 291)

Endocrine glands secrete hormones that affect target cells with specific receptors. Hormones are very potent.

1. Chemically, hormones are steroids, amines, peptides, proteins, or glycoproteins.
2. Steroid hormones
   a. Steroid hormones enter a target cell and bind receptors, forming complexes in the nucleus.
   b. These complexes activate specific genes, so that specific proteins are synthesized.
3. Nonsteroid hormones
   a. Nonsteroid hormones bind receptors in the target cell membrane.
   b. The hormone-receptor complex signals a G protein to stimulate a membrane protein, such as adenylate cyclase, to induce formation of second messenger molecules.
   c. A second messenger, such as cyclic adenosine monophosphate (cAMP), diacylglycerol (DAG), or inositol triphosphate (IP_3), activates protein kinases.
   d. Protein kinases activate protein substrate molecules, which in turn change a cellular process.
4. Prostaglandins
   a. Prostaglandins act on the cells of the organs that produce them.
   b. Prostaglandins are present in small quantities and have powerful hormonelike effects.

11.4 Control of Hormonal Secretions (p. 294)

The concentration of each hormone in body fluids is regulated.

1. Some endocrine glands secrete hormones in response to releasing hormones the hypothalamus secretes.
2. Other glands secrete their hormones in response to nerve impulses.
3. Some glands respond to levels of a substance in the bloodstream.
4. These control mechanisms employ negative feedback.
   a. In a negative feedback system, a gland senses the concentration of a substance it regulates.
   b. When the concentration of the regulated substance reaches a certain point, it inhibits the gland.
   c. As the gland secretes less hormone, the controlled substance also decreases.
   d. Negative feedback systems maintain relatively stable hormone concentrations.

11.5 Pituitary Gland (p. 295)

The pituitary gland has an anterior lobe and a posterior lobe. The hypothalamus controls most pituitary secretions.

1. Anterior pituitary hormones
   a. The anterior pituitary secretes growth hormone (GH), prolactin (PRL), thyroid-stimulating hormone (TSH), adrenocorticotropic hormone (ACTH), follicle-stimulating hormone (FSH), and luteinizing hormone (LH).

b. Growth hormone
  (1) GH stimulates cells to enlarge and divide more frequently.
  (2) GH-releasing hormone and GH release-inhibiting hormone from the hypothalamus control GH secretion.
c. PRL stimulates and sustains a woman's milk production.
d. Thyroid-stimulating hormone
  (1) TSH controls secretion of hormones from the thyroid gland.
  (2) The hypothalamus secretes thyrotropin-releasing hormone (TRH), which regulates TSH secretion.
e. Adrenocorticotropic hormone
  (1) ACTH controls secretion of hormones from the adrenal cortex.
  (2) The hypothalamus secretes corticotropin-releasing hormone (CRH), which regulates ACTH secretion.
f. FSH and LH are gonadotropins.
2. Posterior pituitary hormones
a. The posterior lobe of the pituitary gland consists largely of neuroglial cells and nerve fibers.
b. The hypothalamus produces the hormones of the posterior pituitary.
c. Antidiuretic hormone (ADH)
  (1) ADH reduces the volume of water the kidneys excrete.
  (2) The hypothalamus regulates ADH secretion.
d. Oxytocin (OT)
  (1) OT contracts muscles in the uterine wall.
  (2) OT also contracts cells associated with producing and ejecting milk.

11.6 Thyroid Gland (p. 299)

The thyroid gland in the neck consists of two lobes.
1. Structure of the gland
a. The thyroid gland consists of many follicles.
b. The follicles are fluid-filled and store hormones.
2. Thyroid hormones
a. Thyroxine and triiodothyronine increase the metabolic rate of cells, enhance protein synthesis, and stimulate lipid utilization.
b. Calcitonin decreases blood calcium level and increases blood phosphate ion concentration.

11.7 Parathyroid Glands (p. 301)

The parathyroid glands are on the posterior surface of the thyroid gland.
1. Each parathyroid gland consists of secretory cells that are well supplied with capillaries.
2. Parathyroid hormone (PTH)
a. PTH increases blood calcium level and decreases blood phosphate ion concentration.
b. A negative feedback mechanism operates between the parathyroid glands and the blood.

11.8 Adrenal Glands (p. 302)

The adrenal glands are located atop the kidneys.
1. Structure of the glands
a. Each gland consists of an adrenal medulla and an adrenal cortex.
b. These parts are functionally distinct, and secrete different hormones.

2. Hormones of the adrenal medulla
a. The adrenal medulla secretes epinephrine and norepinephrine, which have similar effects.
b. Sympathetic impulses stimulate secretion of these hormones.
3. Hormones of the adrenal cortex
a. The adrenal cortex produces several steroid hormones.
b. Aldosterone is a mineralocorticoid that causes the kidneys to conserve sodium ions and water and to excrete potassium ions.
c. Cortisol is a glucocorticoid that affects carbohydrate, protein, and fat metabolism.
d. Adrenal sex hormones
  (1) These hormones are of the male type but may be converted to female hormones.
  (2) They may supplement the sex hormones the gonads produce.

11.9 Pancreas (p. 304)

The pancreas secretes digestive juices as well as hormones.
1. Structure of the gland
a. The pancreas is attached to the small intestine.
b. The pancreatic islets secrete glucagon and insulin.
2. Hormones of the pancreatic islets
a. Glucagon stimulates the liver to produce glucose from glycogen and noncarbohydrates.
b. Insulin moves glucose across some cell membranes, stimulates glucose and fat storage, and promotes protein synthesis.
c. Nerve cells do not require insulin to obtain glucose.

11.10 Other Endocrine Glands (p. 307)

1. Pineal gland
a. The pineal gland attaches to the thalamus.
b. It secretes melatonin in response to varying light conditions.
c. Melatonin may help regulate the female reproductive cycle by inhibiting gonadotropin secretion from the anterior pituitary.
2. Thymus
a. The thymus lies behind the sternum and between the lungs.
b. It secretes thymosins, which affect the production of certain lymphocytes that function in immunity.
3. Reproductive organs
a. The ovaries secrete estrogens and progesterone.
b. The placenta secretes estrogens, progesterone, and gonadotropin.
c. The testes secrete testosterone.
4. Digestive glands
Certain glands of the stomach and small intestine secrete hormones.
5. Other hormone-producing organs
Other organs, such as the heart and the kidneys, also produce hormones.

11.11 Stress and Health (p. 309)

Stress occurs when the body responds to stressors that threaten the maintenance of homeostasis. Stress responses include increased activity of the sympathetic nervous system and increased secretion of adrenal hormones.
1. Types of stress
a. Physical stress results from environmental factors that are harmful or potentially harmful to tissues.

b. Psychological stress results from thoughts about real or imagined dangers.

2. Responses to stress
   a. Responses to stress maintain homeostasis.
   b. The hypothalamus controls the stress response.

CHAPTER ASSESSMENTS

11.1 Introduction

1. Contrast endocrine glands and exocrine glands. (p. 290)

2. Define *hormone* and *target cell*. (p. 290)

11.2 General Characteristics of the Endocrine System

3. Compare and contrast the nervous and endocrine systems. (p. 290)

4. Explain the specificity of a hormone for its target cell. (p. 291)

5. Functions of hormones include which of the following? (p. 291)
   a. Control rates of certain chemical reactions
   b. Transport substances across cell membranes
   c. Help regulate water and electrolyte balances
   d. Play a role in reproduction
   e. All of the above

11.3 Hormone Action

6. List the steps of steroid hormone action. (p. 292)

7. List the steps in the action of most nonsteroid hormones. (p. 292)

8. Explain how prostaglandins are similar to hormones and how they are different. (p. 294)

11.4 Control of Hormonal Secretions

9. Draw diagrams of the three mechanisms by which hormone secretion is controlled, including negative feedback. (p. 294)

11.5 Pituitary Gland

10. Describe the location and structure of the pituitary gland. (p. 295)

11. Explain the two ways in which the brain controls pituitary gland activity. (p. 295)

12. Releasing hormones come from which one of the following? (p. 295)
    a. Thyroid gland   c. Posterior pituitary gland
    b. Anterior pituitary gland   d. Hypothalamus

13. List the hormones secreted by the anterior pituitary. (p. 295)

14. Match the following hormones with their actions. More than one hormone can correspond to the same function. (pp. 296–298)

    (1) growth hormone   A. Milk production
    (2) thyroid-stimulating hormone   B. Cell division
    (3) prolactin   C. Metabolic rate
    (4) adrenocorticotropic hormone   D. Exerts action on gonads
    (5) follicle-stimulating hormone   E. Controls secretion of adrenal
    (6) luteinizing hormone   cortex hormones

15. Describe the control of growth hormone secretion. (p. 296)

16. Prolactin does which of the following? (p. 296)
    a. Stimulates breast milk secretion
    b. Stimulates breast milk production
    c. Inhibits breast milk secretion
    d. Inhibits breast milk production

17. Diagram the control of thyroid hormone secretion. (p. 296)

18. Describe the anatomical differences between the anterior and posterior lobes of the pituitary gland. (p. 298)

19. Describe the functions of the posterior pituitary hormones. (p. 298)

20. Under which of the following conditions would you expect an increase in antidiuretic hormone secretion? (p. 298)
    a. An individual ingests excess water.
    b. The posterior pituitary is removed from an individual because of a tumor.
    c. An individual is rescued after three days in the desert without food or water.

11.6 Thyroid Gland

21. Describe the location and structure of the thyroid gland. (p. 299)

22. Match the hormones from the thyroid gland with their descriptions. (p. 299)

    (1) thyroxine   A. Most potent at controlling metabolism
    (2) triiodothyronine   B. Regulates blood calcium
    (3) calcitonin   C. Has four iodine atoms

23. List the source of control for each thyroid hormone. (p. 300)

11.7 Parathyroid Gland

24. Describe the location and structure of the parathyroid glands. (p. 301)

25. Explain the general function of parathyroid hormone. (p. 301)

26. Draw a diagram that shows how the secretion of parathyroid hormone is regulated. (p. 301)

11.8 Adrenal Glands

27. Distinguish between the adrenal medulla and the adrenal cortex. (p. 302)

28. Match the adrenal hormones with their source and actions. (pp. 302–304)

    (1) cortisol   A. Cortex; sodium retention
    (2) aldosterone   B. Cortex; fatty acid release
    (3) epinephrine   C. Medulla; fight-or-flight response

29. Draw a diagram illustrating the regulation of cortisol secretion. (p. 304)

11.9 Pancreas

30. Describe the location and structure of the pancreas. (p. 305)

31. List the hormones secreted by the pancreatic islets, the type of cell that secretes each, and the actions of these hormones. (p. 306)

32. Draw a diagram that shows how the secretion of pancreatic hormones is regulated. (p. 306)

11.10 Other Endocrine Glands

33. Describe the location and general function of the pineal gland. (p. 307)

34. Describe the location and general function of the thymus. (p. 307)

35. Name five additional hormone-secreting organs. (p. 308)

11.11 Stress and Health

36. Define *stress*. (p. 309)

37. List the similarities and differences between the short-term alarm stage of stress and the long-term resistance stage. (p. 309)

INTEGRATIVE ASSESSMENTS/ CRITICAL THINKING

OUTCOMES 11.3, 11.4, 11.5, 11.6

1. When reactor 4 at the Chernobyl Nuclear Power Station in Ukraine exploded at 1:23 P.M. on April 26, 1986, a great plume of radioactive isotopes erupted into the air and spread for thousands of miles. Most of the isotopes emitted immediately following the blast were of the element iodine. Which of the glands of the endocrine system would be most seriously—and immediately—affected by the blast, and how do you think this would become evident in the nearby population?

OUTCOMES 11.4, 11.5

2. Growth hormone is administered to people who have pituitary dwarfism. Parents wanting their normal children to be taller have requested the treatment for them. Do you think this is a wise request? Why or why not?

3. What hormone supplements would an adult whose anterior pituitary has been removed require?

OUTCOMES 11.4, 11.5, 11.6, 11.11

4. How might the lifestyle of a patient with hyperthyroidism be modified to minimize the drain on body energy resources?

OUTCOMES 11.4, 11.8

5. The adrenal cortex of a patient who has lost a large volume of blood will increase secretion of aldosterone. What effect will this increased secretion have on the patient's blood concentrations of sodium and potassium ions?

OUTCOMES 11.4, 11.9

6. Why might oversecretion of insulin actually reduce glucose uptake by nerve cells?

WEB CONNECTIONS

Visit the text website at **aris.mhhe.com** for additional quizzes, interactive learning exercises, and more.

AP R ENDOCRINE SYSTEM

Anatomy & Physiology | REVEALED includes cadaver photos that allow you to peel away layers of the human body to reveal structures beneath the surface. This program also includes animations, radiologic imaging, audio pronunciation, and practice quizzing. To learn more visit **www.aprevealed.com**.

12

Blood

Protection from infection. Universal precautions can be challenging to carry out in disaster situations, such as the outbreak of Marburg hemorrhagic fever in Angola in the spring of 2005. This health-care worker is wearing personal protective equipment to shield her from body fluids containing the deadly virus.

UNIVERSAL PRECAUTIONS. Blood can be more than cells, nutrients, proteins, and water—a single drop from an infected individual can harbor billions of viruses. In the wake of the AIDS epidemic, in 1988 the U.S. Centers for Disease Control and Prevention (CDC) devised "universal precautions," which are specific measures that health-care workers should take to prevent transmission of blood-borne infectious agents. The CDC singled out HIV and the hepatitis B virus. The guidelines grew out of earlier suggestions for handling patients suspected to have been exposed to viruses. The term *universal* refers to the assumption that *any* patient may have been exposed to a pathogen that can be transmitted in a body fluid.

Attention to safety in the health-care setting can prevent transmission of infectious diseases. The World Health Organization estimates that 4–7% of new infections worldwide are transmitted via unsafe injections. Specific recommendations include:

- Use of personal protection equipment, such as gloves, goggles, and masks.
- Engineering controls, such as fume hoods and sharps containers.
- Work-practice controls, such as enforcing hand washing before and after performing procedures.

Universal precautions were designed for, and work well in, preventing transmission of viral illnesses in settings that are already relatively safe. Unfortunately they did not help several pediatric nurses who tended to their neighbors infected with the Marburg virus during an outbreak in the isolated town of Uige in Angola, South Africa. The nurses and hundreds of others died.

Marburg virus causes a hemorrhagic fever. Headache, fever, vomiting, and diarrhea begin three to nine days after exposure to the virus.

Then the person bleeds from all body openings, internally and under the skin. The resulting drop in blood pressure kills 30% to 90% of infected individuals within a week, and anyone contacting their blood is in danger of infection.

The Angola Marburg outbreak was the worst in recorded history. It had a 90% mortality rate and affected mostly children. Epidemiologists suspect that contaminated medical equipment caused the rapid and deadly spread. Nontrained clinic workers re-used needles, and some people used needles and intravenous equipment in their homes. The disease also spread fast in crowded hospitals where family members provided much of the nursing care. Universal precautions might not have contained the virus, which spreads in vomit, sweat, and saliva as well as blood. However, universal precautions are critical for containing outbreaks under less dire circumstances.

Learning Outcomes   *After studying this chapter, you should be able to do the following:*

12.1 Introduction

1. Describe the general characteristics of blood, and discuss its major functions. (p. 318)

2. Distinguish among the formed elements and liquid portion of blood. (p. 318)

12.2 Blood Cells

3. Explain the significance of red blood cell counts. (p. 319)

4. Summarize the control of red blood cell production. (p. 319)

5. Distinguish among the five types of white blood cells, and give the function(s) of each type. (p. 323)

12.3 Blood Plasma

6. Describe the functions of each of the major components of blood plasma. (p. 325)

12.4 Hemostasis

7. Define hemostasis, and explain the mechanisms that help achieve it. (p. 328)

8. Review the major steps in blood coagulation. (p. 330)

12.5 Blood Groups and Transfusions

9. Explain blood typing and how it is used to avoid adverse reactions to blood transfusions. (p. 332)

10. Describe how blood reactions may occur between fetal and maternal tissues. (p. 335)

Aids to Understanding Words   (*Appendix A on page 567 has a complete list of Aids to Understanding Words.*)

agglutin- [to glue together] *agglutin*ation: Clumping of red blood cells.

bil- [bile] *bil*irubin: Pigment excreted in the bile.

embol- [stopper] *embol*ism: Obstruction of a blood vessel.

erythr- [red] *erythr*ocyte: Red blood cell.

hema- [blood] *hema*tocrit: Percentage of red blood cells in a given volume of blood.

hemo- [blood] *hemo*globin: Red pigment responsible for the color of blood.

leuko- [white] *leuko*cyte: White blood cell.

-osis [abnormal condition] leuko*cytosis*: Condition in which white blood cells are overproduced.

-poie [make, produce] erythro*poie*tin: Hormone that stimulates the production of red blood cells.

-stasis [halt] hemo*stasis*: Arrest of bleeding from damaged blood vessels.

thromb- [clot] *thromb*ocyte: Blood platelet involved in the formation of a blood clot.

12.1 INTRODUCTION

Blood signifies life, and for good reason—it has many vital functions. This complex mixture of cells, cell fragments, and dissolved biochemicals transports nutrients, oxygen, wastes, and hormones; helps maintain the stability of the interstitial fluid; and distributes heat. The blood, heart, and blood vessels form the cardiovascular system and link the body's internal and external environments.

Blood is a type of connective tissue whose cells are suspended in a liquid extracellular matrix. Blood is vital in transporting substances between body cells and the external environment, thereby promoting homeostasis.

Whole blood is slightly heavier and three to four times more viscous than water. Its cells, which form mostly in red bone marrow, include red blood cells that transport gases and white blood cells that fight disease. Blood also contains cellular fragments called blood platelets that help control blood loss. Together, the cells and platelets are termed "formed elements" of the blood, in contrast to the liquid portion called **plasma** (plaz′mah) (fig. 12.1).

A blood sample is usually about 45% red blood cells by volume. This percentage is called the **hematocrit (HCT).** The white blood cells and platelets account for less than 1%. The remaining blood sample, about 55%, is the plasma, a clear, straw-colored liquid. Plasma is a complex mixture of water, amino acids, proteins, carbohydrates, lipids, vitamins, hormones, electrolytes, and cellular wastes (fig. 12.1).

Blood volume varies with body size, changes in fluid and electrolyte concentrations, and the amount of adipose tissue. An average-size adult has a blood volume of about 5 liters (5.3 quarts).

Men have more blood than women. Men have 5–6 liters (1.500 gallons), compared to 4–5 liters (0.875 gallons) for women.

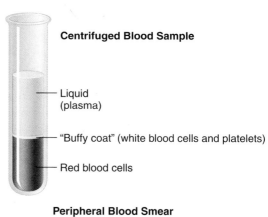

Centrifuged Blood Sample

Liquid (plasma)

"Buffy coat" (white blood cells and platelets)

Red blood cells

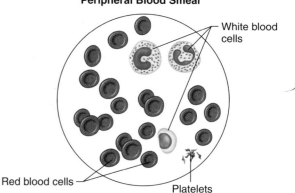

Peripheral Blood Smear

White blood cells

Red blood cells

Platelets

Figure 12.1

Blood consists of a liquid portion called plasma and a solid portion (the formed elements) that includes red blood cells, white blood cells, and platelets. (Note: When blood components are separated by centrifugation, the white blood cells and platelets form a thin layer, called the "buffy coat," between the plasma and the red blood cells, which accounts for about 1% of the total blood volume.) Blood cells and platelets can be seen under a light microscope when a blood sample is smeared onto a glass slide.

Check Your Recall

1. What are the major components of blood?
2. What factors affect blood volume?

12.2 BLOOD CELLS

Red Blood Cells

Red blood cells, or **erythrocytes** (ĕ-rith′ro-sītz), are biconcave discs. This shape is an adaptation for transporting gases; it increases the surface area through which gases can diffuse (fig. 12.2). The red blood cell's shape also places the cell membrane closer to oxygen-carrying **hemoglobin** within the cell.

Each red blood cell is about one-third hemoglobin by volume. This protein imparts the color of blood. When hemoglobin binds oxygen, the resulting *oxyhemoglobin* is bright red, and when oxygen is released, the resulting *deoxyhemoglobin* is darker. Blood rich in deoxyhemoglobin may appear bluish when it is viewed through blood vessel walls.

A person experiencing prolonged oxygen deficiency (hypoxia) may become *cyanotic.* The skin and mucous membranes appear bluish due to an abnormally high blood concentration of deoxyhemoglobin. Exposure to low temperature may also result in cyanosis. Such exposure constricts superficial blood vessels, which slows blood flow, allowing removal of more oxygen than usual from the blood flowing through the vessels.

Red blood cells have nuclei during their early stages of development, but extrude them as the cells mature, providing more space for hemoglobin. Because they lack nuclei, mature red blood cells cannot synthesize proteins or divide. Because they also lack mitochondria, red blood cells produce ATP through glycolysis only and use none of the oxygen they carry.

Red Blood Cell Counts

The number of red blood cells in a microliter (µL or mcL or 1 mm³) of blood is called the *red blood cell count* (*RBCC* or *RCC*). The typical range for adult males is 4,600,000–6,200,000 cells per microliter, and that for adult females is 4,200,000–5,400,000 cells per microliter.

Since increasing the number of circulating red blood cells increases the blood's *oxygen-carrying capacity,* changes in this number may affect health. For this reason, red blood cell counts are routinely consulted to help diagnose and evaluate the courses of various diseases.

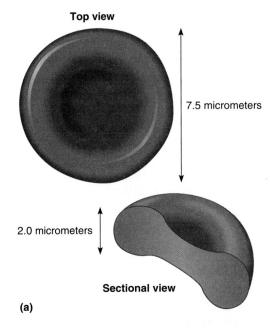

Top view

7.5 micrometers

2.0 micrometers

Sectional view

(a)

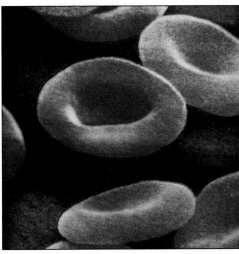

(b)

Figure 12.2

Red blood cells. (a) The biconcave shape of a red blood cell makes it efficient at transporting gases. (b) Falsely colored scanning electron micrograph of human red blood cells (5,000×).

Check Your Recall

3. Describe a red blood cell.
4. What is the function of hemoglobin?
5. How does a red blood cell change as it matures?
6. What is the typical red blood cell count for an adult male? For an adult female?

Red Blood Cell Production and Its Control

Recall from chapter 7 (p. 138) that red blood cell formation (erythropoiesis) initially occurs in the yolk sac, liver, and spleen. After birth, these cells are produced almost exclusively by tissue lining the spaces in bones, filled with red bone marrow.

The average life span of a red blood cell is 120 days. Many of these cells are removed from the circulation each day, and yet the number of cells in the circulating blood remains relatively stable. This observation suggests a homeostatic control of the rate of red blood cell production.

> The combined surface area of all the red blood cells in the human body is roughly 2,000 times as great as the body's exterior surface.

The hormone **erythropoietin** (ĕ-rith″ro-poi′ĕ-tin) controls the rate of red blood cell formation through *negative feedback*. The kidneys, and to a lesser extent the liver, release erythropoietin in response to prolonged oxygen deficiency (fig. 12.3). At high altitudes, for example, where the amount of oxygen in the air is reduced, oxygen delivery to the tissues initially decreases. This drop in oxygen triggers the release of erythropoietin, which travels via the blood to the red bone marrow and stimulates red blood cell production.

After a few days, many newly formed red blood cells appear in the circulating blood. The increased rate of production continues until the number of erythrocytes in the circulation is sufficient to supply tissues with oxygen. When the availability of oxygen returns to normal, erythropoietin release decreases, and the rate of red blood cell production returns to normal as well. Figure 12.4 illustrates the stages in the formation of red blood cells from hemocytoblasts.

Dietary Factors Affecting Red Blood Cell Production

B-complex vitamins—*vitamin B_{12}* and *folic acid*—significantly influence red blood cell production. These vitamins are necessary for DNA synthesis, so all cells with nuclei require them to grow and divide. Blood-cell-forming (hematopoietic) tissue is especially vulnerable to deficiency of either of these vitamins because many of its cells are dividing.

Hemoglobin synthesis and normal red blood cell production require iron. The small intestine absorbs iron slowly from food. The body reuses much of the iron released by the decomposition of hemoglobin from damaged red blood cells. Therefore, the diet need only supply small amounts of iron.

Too few red blood cells or too little hemoglobin causes *anemia.* This reduces the oxygen-carrying capacity of the blood, and the affected person may appear pale and lack energy. A pregnant woman may become anemic if she doesn't eat iron-rich foods because her blood volume increases due to fluid retention to accommodate the requirements of the fetus. This increased blood volume decreases the hematocrit.

In contrast to anemia, in an inherited disorder called hemochromatosis, the small intestine absorbs iron at ten times the normal rate. Iron builds up in organs, to toxic levels. Treatment is simple: periodic blood removal.

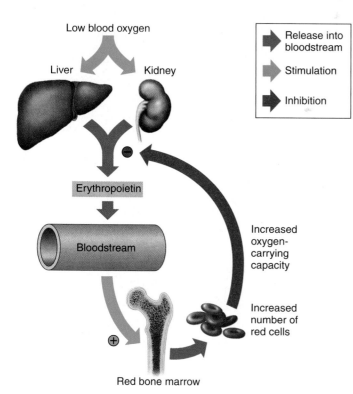

Figure 12.3

Low blood oxygen causes the kidneys and liver to release erythropoietin. Erythropoietin stimulates target cells in red bone marrow to increase the production of red blood cells that carry oxygen to tissues.

In *sickle cell disease,* a single DNA base change causes an incorrect amino acid to be incorporated into globin, causing hemoglobin to crystallize in a low-oxygen environment. The abnormal hemoglobin molecules bend the red blood cells into a sickle shape, which blocks circulation in small blood vessels, causing excruciating joint pain and damaging many organs. As the spleen works harder to recycle the abnormally short-lived red blood cells, infection becomes likely.

Most children with sickle cell disease are diagnosed at birth and receive antibiotics daily for years to prevent infection. Hospitalization for blood transfusions may be necessary if the person experiences painful sickling "crises" of blocked circulation.

A bone marrow or umbilical cord stem cell transplant can completely cure sickle cell disease but has a 15% risk of fatality. A new treatment is an old drug, used to treat cancer, called hydroxyurea. It activates production of a slightly different form of hemoglobin that is normally present only in the fetus. The fetal hemoglobin slows sickling, which gives red blood cells more time to reach the lungs—where fresh oxygen restores the cells' normal shapes.

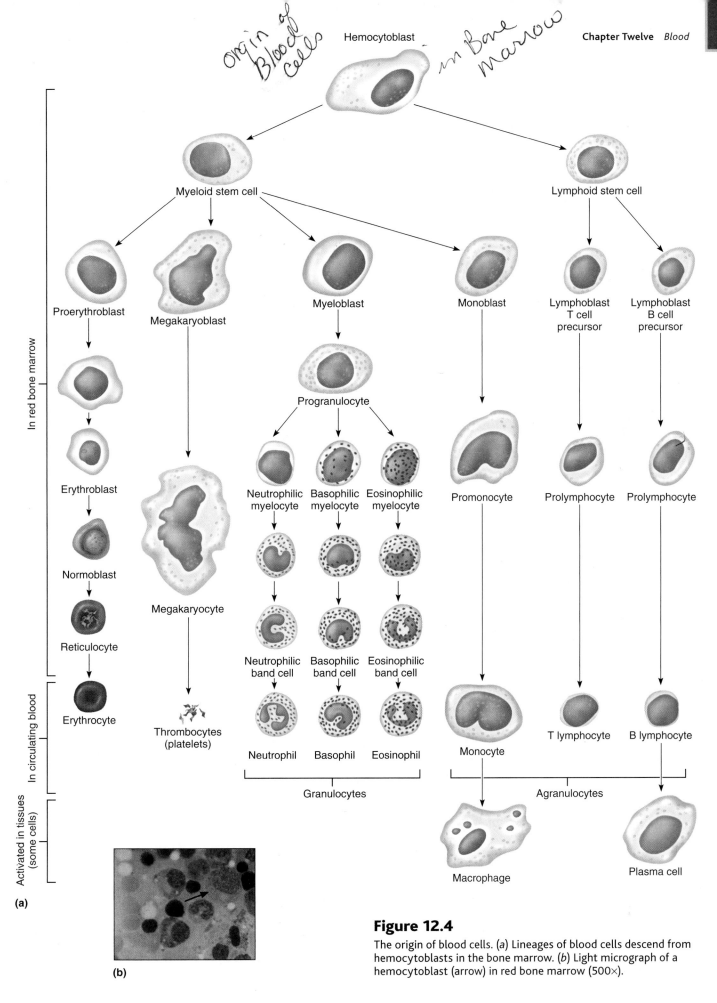

Origin of Blood Cells

in Bone Marrow

Hemocytoblast

Myeloid stem cell

Lymphoid stem cell

In red bone marrow

Proerythroblast

Megakaryoblast

Myeloblast

Monoblast

Lymphoblast T cell precursor

Lymphoblast B cell precursor

Erythroblast

Progranulocyte

Normoblast

Neutrophilic myelocyte

Basophilic myelocyte

Eosinophilic myelocyte

Promonocyte

Prolymphocyte

Prolymphocyte

Reticulocyte

Megakaryocyte

Neutrophilic band cell

Basophilic band cell

Eosinophilic band cell

In circulating blood

Erythrocyte

Thrombocytes (platelets)

Neutrophil

Basophil

Eosinophil

Monocyte

T lymphocyte

B lymphocyte

Granulocytes

Agranulocytes

Activated in tissues (some cells)

Macrophage

Plasma cell

(a)

(b)

Figure 12.4

The origin of blood cells. (*a*) Lineages of blood cells descend from hemocytoblasts in the bone marrow. (*b*) Light micrograph of a hemocytoblast (arrow) in red bone marrow (500×).

Destruction of Red Blood Cells

Red blood cells are quite elastic and flexible, and they readily bend as they pass through small blood vessels. As these cells age, however, they become more fragile and may be damaged simply by passing through capillaries, particularly those in active muscles that must withstand contractile forces. **Macrophages** phagocytize and destroy damaged red blood cells, primarily in the liver and spleen. Recall from chapter 5, page 104, that macrophages are large, phagocytic, wandering cells.

Hemoglobin molecules liberated from red blood cells are broken down into subunits of *heme,* an iron-containing portion, and *globin,* a protein. The heme further decomposes into iron and a greenish pigment called **biliverdin.** The blood may transport the iron, combined with a protein, to the hematopoietic tissue in red bone marrow to be reused in synthesizing new hemoglobin. About 80% of the iron is stored in the liver in the form of an iron-protein complex. Biliverdin eventually is converted to an orange pigment called **bilirubin.** Biliverdin and bilirubin are excreted in the bile as bile pigments (see chapter 15, p. 419). Figure 12.5 summarizes the life cycle of a red blood cell.

Newborns can develop *physiologic jaundice* a few days after birth. In this condition and other forms of jaundice (icterus), accumulation of bilirubin turns the skin and eyes yellowish.

Physiologic jaundice may be the result of immature liver cells that ineffectively excrete bilirubin into the bile. Treatment includes exposure to fluorescent light, which breaks down bilirubin in the tissues, and feedings that promote bowel movements. In hospital nurseries, babies being treated for physiologic jaundice lie under "bili lights," clad only in diapers and protective goggles. The healing effect of fluorescent light was discovered in the 1950s, when an astute nurse noted that jaundiced babies improved after sun exposure, except in the areas their diapers covered.

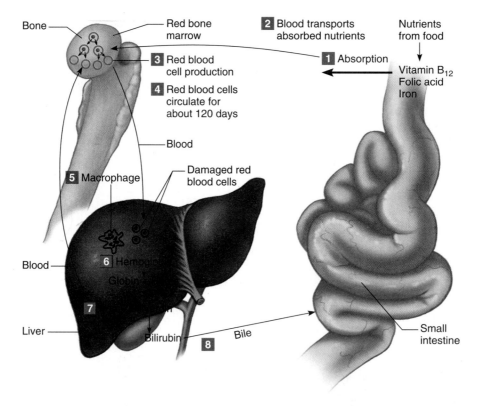

Figure 12.5

Life cycle of a red blood cell. (*1*) The small intestine absorbs essential nutrients. (*2*) Blood transports nutrients to red bone marrow. (*3*) In the marrow, red blood cells arise from the division of less specialized progenitor cells. (*4*) Mature red blood cells are released into the bloodstream, where they circulate for about 120 days. (*5*) Macrophages destroy damaged red blood cells in the spleen and liver. (*6*) Hemoglobin liberated from red blood cells is broken down into heme and globin. (*7*) Iron from heme returns to red bone marrow and is reused. (*8*) Biliverdin and bilirubin are excreted in bile.

Check Your Recall

11. What happens to damaged red blood cells?
12. What are the products of hemoglobin breakdown?

White Blood Cells

White blood cells, or **leukocytes** (lu'ko-sītz), protect against disease. Leukocytes develop from hemocyto-blasts in the red bone marrow (see fig. 12.4) in response to hormones, much as red blood cells form from precursors upon stimulation from erythropoietin. These hormones fall into two groups—**interleukins** and **colony-stimulating factors (CSFs).** Interleukins are numbered, while most colony-stimulating factors are named for the cell population they stimulate.

Blood transports white blood cells to sites of infection. White blood cells may then leave the bloodstream, as described on page 324.

Normally, five types of white blood cells are in circulating blood. They differ in size, the nature of their cytoplasm, the shape of the nucleus, and their staining characteristics, and are named for these distinctions. For example, leukocytes with granular cytoplasm are called **granulocytes,** whereas those without cytoplasmic granules are called **agranulocytes** (see fig. 12.4).

A typical granulocyte is about twice the size of a red blood cell. Members of this group include neutrophils, eosinophils, and basophils. Granulocytes develop in red bone marrow as do red blood cells, but have short life spans, averaging about 12 hours.

Neutrophils (nu'tro-filz) have fine cytoplasmic granules that appear light purple in neutral stain. The nucleus of an older neutrophil is lobed and consists of two to five sections (segments, so these cells are sometimes called *segs*) connected by thin strands of chromatin. Younger neutrophils are also called *bands* because their nuclei are C-shaped (fig. 12.6). Neutrophils account for 54–62% of the leukocytes in a typical blood sample from an adult.

Eosinophils (e"o-sin'o-filz) contain coarse, uniformly sized cytoplasmic granules that appear deep red in acid stain (fig. 12.7). The nucleus usually has only two lobes (termed bilobed). Eosinophils make up 1–3% of the total number of circulating leukocytes.

Basophils (ba'so-filz) are similar to eosinophils in size and in the shape of their nuclei, but they have fewer, more irregularly shaped cytoplasmic granules that become deep blue in basic stain (fig. 12.8). Basophils usually account for less than 1% of the circulating leukocytes.

The leukocytes of the agranulocyte group include monocytes and lymphocytes. Monocytes generally arise from red bone marrow. Lymphocytes differentiate in the organs of the lymphatic system, as well as in the red bone marrow (see chapter 14, p. 385).

Monocytes (mon'o-sītz), the largest blood cells, are two to three times greater in diameter than red blood cells (fig. 12.9). Their nuclei vary in shape and are round, kidney-shaped, oval, or lobed. They usually make up 3–9% of the leukocytes in a blood sample and live for several weeks or even months.

Lymphocytes (lim'fo-sītz) are usually only slightly larger than red blood cells. A typical lymphocyte has a large, round nucleus surrounded by a thin rim of cytoplasm (fig. 12.10). These cells account for 25–33% of circulating leukocytes. Lymphocytes may live for years.

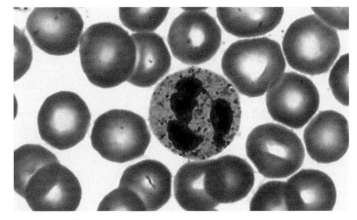

Figure 12.6

A neutrophil has a lobed nucleus with two to five components (2,000×). This blood cell type has abundant lysosomes, which contain enzymes that break down parts of phagocytized bacteria.

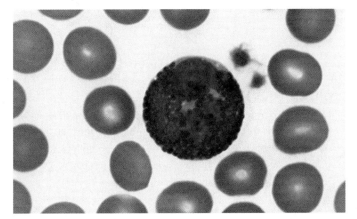

Figure 12.7

An eosinophil has red-staining cytoplasmic granules (2,000×). This type of white blood cell kills certain parasites and helps to control inflammation and allergic reactions.

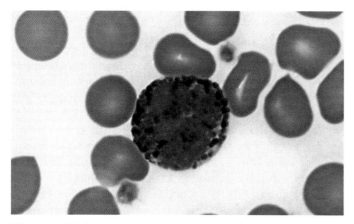

Figure 12.8

A basophil has cytoplasmic granules that stain deep blue (2,000×). This type of white blood cell produces heparin, which prevents inappropriate blood clotting, and histamines, which increase circulation to injured tissues. Basophils also take part in certain allergic reactions.

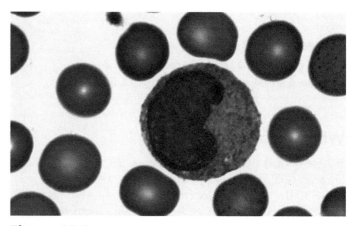

Figure 12.9

A monocyte is the largest of the blood cells (2,000×). It may leave the bloodstream and become a macrophage, which is a wandering, phagocytic cell that destroys damaged red blood cells.

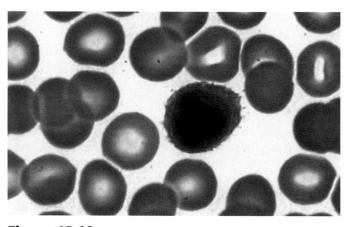

Figure 12.10

A lymphocyte, the smallest of the white blood cells, has a large, round nucleus (2,000×). Lymphocytes carry out the immune response and are discussed further in chapter 14.

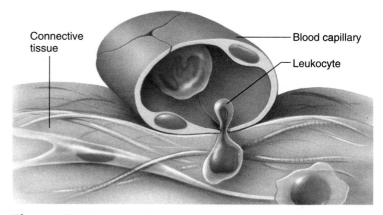

Connective tissue · Blood capillary · Leukocyte

Figure 12.11

In a type of movement called diapedesis, leukocytes squeeze between the endothelial cells of a capillary wall and enter the tissue space outside the blood vessel.

Check Your Recall

13. Which hormones are necessary for differentiation of white blood cells from hemocytoblasts in the red bone marrow?

14. Distinguish between granulocytes and agranulocytes.

15. List the five types of white blood cells, and explain how they differ from one another.

Functions of White Blood Cells

White blood cells protect against infection in various ways. Some leukocytes phagocytize bacterial cells in the body, and others produce proteins (*antibodies*) that destroy or disable foreign particles.

Leukocytes can squeeze between the cells that form blood vessel walls. This movement, called *diapedesis*, allows the white blood cells to leave the circulation (fig. 12.11). Once outside the blood, they move through interstitial spaces using a form of self-propulsion called *amoeboid motion.*

The most mobile and active phagocytic leukocytes are neutrophils and monocytes. Neutrophils cannot ingest particles much larger than bacterial cells, but monocytes can engulf large objects. Both of these phagocytes contain many *lysosomes,* which are organelles filled with digestive enzymes that break down organic molecules in captured bacteria. Neutrophils and monocytes may become so engorged with digestive products and bacterial toxins that they die.

Eosinophils are only weakly phagocytic, but they are attracted to and can kill certain parasites. Eosinophils also help control inflammation and allergic reactions by removing biochemicals associated with these reactions.

Some of the cytoplasmic granules of basophils contain a blood-clot-inhibiting substance called *heparin,* and other granules contain *histamine.* Basophils release heparin, which helps prevent intravascular blood clot formation, and they release histamines, which increase blood flow to injured tissues. Basophils also play major roles in certain allergic reactions.

Lymphocytes are important in *immunity.* Some, for example, produce antibodies that attack specific foreign substances that enter the body. Chapter 14 (pp. 385–391) discusses immunity.

Check Your Recall

16. How do white blood cells fight infection?
17. How do white blood cells reach microorganisms that are outside blood vessels?
18. Which white blood cells are the most active phagocytes?

White Blood Cell Counts

The number of white blood cells in a microliter of human blood, called the *white blood cell count* (*WBCC* or *WCC*), normally is 4,500–10,000 cells. Because this number may change in response to abnormal conditions, white blood cell counts are of clinical interest. For example, a rise in the number of circulating white blood cells may indicate infection. A total number of white blood cells exceeding 10,000 per mm^3 of blood constitutes **leukocytosis,** indicating acute infection, such as appendicitis. White blood cell count is greatly elevated in leukemia, as the Topic of Interest on pages 328–329 describes.

A total white blood cell count below 5,000 per mm^3 of blood is called **leukopenia.** Such a deficiency may accompany typhoid fever, influenza, measles, mumps, chickenpox, AIDS, or poliomyelitis.

A *differential white blood cell count* (*DIFF*) lists percentages of the types of leukocytes in a blood sample. This test is useful because the relative proportions of white blood cells may change in particular diseases. For instance, the number of neutrophils usually increases during bacterial infections, and the number of eosinophils may increase during certain parasitic infections and allergic reactions. In AIDS, the number of a certain type of lymphocyte drops sharply.

Check Your Recall

19. What is the normal human white blood cell count?
20. Distinguish between leukocytosis and leukopenia.
21. What is a differential white blood cell count?

Blood Platelets

Platelets (plăt′letz), or **thrombocytes** (throm′bo-sĭtz), are not complete cells. They arise from very large cells in red bone marrow, called **megakaryocytes** (meg″ah-kar′e-o-sĭtz), that fragment like a shattered plate, releasing small sections of cytoplasm—the platelets—into the circulation. The larger fragments of the megakaryocytes shrink and become platelets as they pass through blood vessels in the lungs. Megakaryocytes, and therefore platelets, develop from hemocytoblasts (see fig. 12.4) in response to the hormone **thrombopoietin** (throm″bo-poi′ĕ-tin).

Each platelet lacks a nucleus and is less than half the size of a red blood cell. It is capable of amoeboid movement and may live for about ten days. In normal blood, the *platelet count* varies from 130,000 to 360,000 per microliter. Platelets help close breaks in damaged blood vessels and initiate formation of blood clots, as section 12.4 explains on page 328. Table 12.1 summarizes the characteristics of blood cells and platelets.

Check Your Recall

22. What is the normal blood platelet count?
23. What is the function of blood platelets?

12.3 BLOOD PLASMA

Plasma is the clear, straw-colored, liquid portion of the blood in which the cells and platelets are suspended. It is approximately 92% water and contains a complex mixture of organic and inorganic biochemicals. The functions of plasma constituents include transporting nutrients, gases, and vitamins; helping regulate fluid and electrolyte balance; and maintaining a favorable pH.

Plasma Proteins

Plasma proteins (plaz′mah pro′tēnz) are the most abundant of the dissolved substances (solutes) in plasma. These proteins remain in the blood and interstitial fluids, and ordinarily are not used as energy sources. The three main types of plasma proteins—albumins, globulins, and fibrinogen—differ in composition and function.

Albumins (al-bu′minz) are the smallest of the plasma proteins, yet account for about 60% of these proteins by weight. They are synthesized in the liver, and because they are so plentiful, albumins are an important determinant of the *osmotic pressure* of the plasma. Recall from chapter 3 (p. 63) that the presence of solute that cannot cross a selectively permeable membrane creates an osmotic pressure and that water always

Table 12.1	Cellular Components of Blood		
Component	Description	Number Present	Function
Red blood cell (erythrocyte)	Biconcave disc without a nucleus; about one-third hemoglobin	4,200,000–6,200,000 per microliter	Transports oxygen and carbon dioxide
White blood cell (leukocyte)		5,000–10,000 per microliter	Destroys pathogenic microorganisms and parasites and removes worn cells
Granulocytes	About twice the size of red blood cells; cytoplasmic granules are present		
1. Neutrophil	Nucleus with two to five lobes; cytoplasmic granules stain light purple in neutral stain	54–62% of white blood cells	Phagocytizes small particles
2. Eosinophil	Bilobed nucleus, cytoplasmic granules stain red in acid stain	1–3% of white blood cells	Kills parasites and helps control inflammation and allergic reactions
3. Basophil	Bilobed nucleus, cytoplasmic granules stain blue in basic stain	Less than 1% of white blood cells	Releases heparin and histamine
Agranulocytes	Cytoplasmic granules are absent		
1. Monocyte	Two to three times larger than a red blood cell; nuclear shape varies from spherical to lobed	3–9% of white blood cells	Phagocytizes large particles
2. Lymphocyte	Only slightly larger than a red blood cell; its nucleus nearly fills cell	25–33% of white blood cells	Provides immunity
Platelet (thrombocyte)	Cytoplasmic fragment	130,000–360,000 per microliter	Helps control blood loss from broken vessels

diffuses toward a greater osmotic pressure. Because plasma proteins are too large to pass through the capillary walls, they create an osmotic pressure that tends to hold water in the capillaries, despite the fact that blood pressure tends to force water out of capillaries by filtration (see chapter 3, pp. 63–64). The term *colloid osmotic pressure* is used to describe this osmotic effect caused by the plasma proteins.

By maintaining the colloid osmotic pressure of plasma, albumins and other plasma proteins help regulate water movement between the blood and the tissues. In doing so, they help control blood volume, which, in turn, directly affects blood pressure (see chapter 13, p. 359).

> If the concentration of plasma proteins falls, tissues swell—a condition called *edema*. This may result from starvation or a protein-deficient diet, either of which requires the body to use protein for energy, or from an impaired liver that cannot synthesize plasma proteins. As the concentration of plasma proteins drops, so does colloid osmotic pressure, sending fluids into interstitial spaces.

Globulins (glob′u-linz), which make up about 36% of the plasma proteins, can be further subdivided into *alpha, beta,* and *gamma globulins.* The liver synthesizes alpha and beta globulins, which have a variety of functions, including transport of lipids and fat-soluble vita-

mins. Lymphatic tissues produce the gamma globulins, which are a type of antibody (see chapter 14, p. 390).

Fibrinogen (fi-brin′o-jen), which constitutes about 4% of the plasma proteins, functions in blood coagulation, as discussed in section 12.4 on page 330. Synthesized in the liver, fibrinogen is the largest of the plasma proteins. Table 12.2 summarizes the characteristics of the plasma proteins.

Table 12.2	Plasma Proteins		
Protein	Percentage of Total	Origin	Function
Albumin	60%	Liver	Helps maintain colloid osmotic pressure
Globulin	36%		
Alpha globulins		Liver	Transport lipids and fat-soluble vitamins
Beta globulins		Liver	Transport lipids and fat-soluble vitamins
Gamma globulins		Lymphatic tissues	Constitute a type of antibody
Fibrinogen	4%	Liver	Plays a key role in blood coagulation

Gases and Nutrients

The most important *blood gases* are oxygen and carbon dioxide. Plasma also contains a considerable amount of dissolved nitrogen, which ordinarily has no physiological function. Chapter 16 (pp. 461–463) discusses the blood gases and their transport.

The *plasma nutrients* include amino acids, simple sugars, nucleotides, and lipids absorbed from the digestive tract. For example, plasma transports glucose from the small intestine to the liver, where glucose can be stored as glycogen or converted to fat. If blood glucose concentration drops below the normal range, glycogen may be broken down into glucose, as described in chapter 11 (p. 306). Plasma also carries recently absorbed amino acids to the liver, where they can be used to manufacture proteins, or deaminated and used as an energy source (see chapter 15, p. 432).

Plasma lipids include fats (triglycerides), phospholipids, and cholesterol. Because lipids are not water-soluble and plasma is almost 92% water, these lipids are carried in the plasma by joining with proteins, forming lipoprotein complexes.

Nonprotein Nitrogenous Substances

Molecules that have nitrogen atoms but are not proteins comprise a group called **nonprotein nitrogenous substances.** In plasma, this group includes amino acids, urea, and uric acid. Amino acids come from protein digestion and amino acid absorption. Urea and uric acid are products of protein and nucleic acid catabolism, respectively, and are excreted in the urine.

Plasma Electrolytes

Blood plasma contains a variety of *electrolytes* that are absorbed from the intestine or released as by-products of cellular metabolism. They include sodium, potassium, calcium, magnesium, chloride, bicarbonate, phosphate, and sulfate ions. Sodium and chloride ions are the most abundant. Bicarbonate ions are important in maintaining the osmotic pressure and pH of plasma, and like other plasma constituents, they are regulated so that their blood concentrations remain relatively stable. Chapter 18 (p. 492) discusses these electrolytes in connection with water and electrolyte balance. Figure 12.12 summarizes the composition of blood.

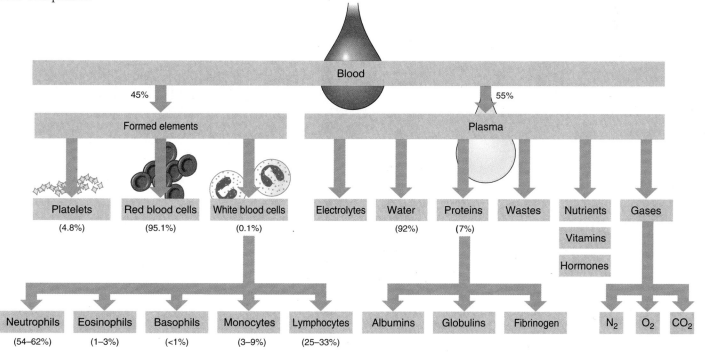

Figure 12.12

Blood composition. Blood is a complex mixture of formed elements in a liquid extracellular matrix, plasma.

Leukemia

When twenty-three-year-old magazine editor Erin Zammett Ruddy had a routine physical examination in late 2001, she expected reassurance that her healthy lifestyle had indeed been keeping her healthy (figure 12A). After all, she felt great. What she got, a few days later, was a shock. Instead of having 4,500 to 10,000 white blood cells per microliter of blood, she had more than ten times that number—and many of the cells were cancerous. Erin had chronic myeloid leukemia (CML). Her red bone marrow was flooding her circulation with too many granulocytes, most of them poorly differentiated.

Another type of leukemia is lymphoid, in which the cancer cells are lymphocytes, produced in lymph nodes. Both myeloid and lymphoid leukemia can cause fatigue, headaches, nosebleeds and other bleeding, frequent respiratory infections, fever, bone pain, bruising, and other signs of slow blood clotting. The symptoms arise from the disrupted proportions of the blood's formed elements and their malfunction (figure 12B).

Immature white blood cells increase the risk of infection. Leukemic cells crowd out red blood cells and their precursors in the red marrow, causing anemia and resulting fatigue. Platelet deficiency (thrombocytopenia) slows clotting time, causing bruises and bleeding. Finally, spread of the cancer cells outside the marrow painfully weakens the surrounding bone. Eventually, without treatment, cancer cells spread outside the cardiovascular system, causing other tissues that would normally not produce white blood cells to do so.

Leukemia is also classified as acute or chronic. An acute condition appears suddenly, symptoms progress rapidly, and without treatment, death occurs in a few months. Chronic forms begin more slowly and may remain unde-

Figure 12A

"My third bone marrow biopsy—you never get used to the pain," said Erin Zammett Ruddy. Gleevec has treated her leukemia. Bone marrow biopsies are required at regular intervals, even after successful treatment, to be certain that the disease has not returned.

tected for months or even years or, in rare cases, decades. Without treatment, life expectancy after symptoms develop is about three years.

Erin was unlucky to get cancer, but lucky to have gotten it just as a new drug was flying through clinical trials. She became one of the first patients to take Gleevec, which is now standard treatment for CML and several other cancers. Unlike traditional cancer treatments that destroy any cell that divides rapidly, Gleevec targets only the cancer cells by nestling into ATP-binding sites on a type of enzyme called a tyrosine kinase, which blocks the

12.4 HEMOSTASIS

Hemostasis (he"mo-sta′sis) is the stoppage of bleeding, which is vitally important when blood vessels are damaged. Following an injury to the blood vessels, several actions may help limit or prevent blood loss, including blood vessel spasm, platelet plug formation, and blood coagulation.

Blood Vessel Spasm

Cutting or breaking a smaller blood vessel stimulates the smooth muscles in its walls to contract, a phenomenon called a **vasospasm,** and blood loss lessens almost immediately. A vasospasm may completely close the ends of a severed vessel.

Vasospasm may last only a few minutes, but the effect of the direct stimulation usually continues for about 30 minutes. By then, a *platelet plug* has formed, and blood

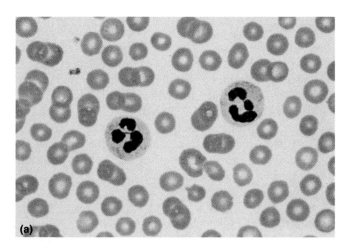

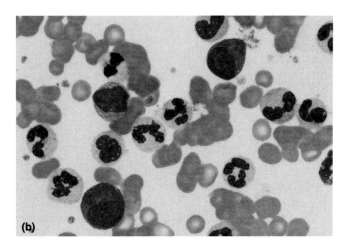

Figure 12B

Leukemia and blood cells. (*a*) Normal blood cells (700×). (*b*) Blood cells from a person with granulocytic leukemia, a type of myeloid leukemia (700×). Note the increased number of leukocytes.

message to divide. After passing safety tests, the drug worked so dramatically in vanquishing the cancer that it was approved in just ten weeks. People with leukemia have other options too—since the 1970s standard chemotherapies have helped many patients, and bone marrow and stem cell transplants can cure the condition.

Another way that leukemia treatment is improving is refining diagnosis, based on identifying the proteins that leukemia cells produce. This can predict which drugs are most likely to be effective, and which will cause intolerable side effects or not work in particular individuals. For example, some people with acute lymphoblastic leukemia (ALL), diagnosed on the basis of the appearance of the cancer cells in a blood smear, do not respond to standard chemo-

therapy. However, DNA microarray (also called DNA chip) technology revealed that the cells of patients who do not improve produce different proteins than the cancer cells of patients who do respond to the drugs used to treat ALL—the nonresponders have a different and newly recognized form of leukemia, called mixed-lineage leukemia. These people respond to different drugs.

As for Erin, she went off Gleevec to become a mother and is doing well. If her leukemia should return, and if Gleevec becomes ineffective, she will have newer drugs to try that bind more strongly to cancer cells or that fit slightly altered resistant cancer cells. Follow her progress on her blog: www.glamour.com/lifestyle/blogs/editor.

is coagulating. Also, platelets release **serotonin,** which contracts smooth muscles in the blood vessel walls. This vasoconstriction further helps reduce blood loss.

Platelet Plug Formation

Platelets adhere to any rough surface and to the collagen in connective tissue. When a blood vessel breaks, platelets adhere to the collagen underlying the endothelial lining of blood vessels. Platelets also adhere to each other, forming a platelet plug in the vascular break. A plug may control blood loss from a small break, but a larger break may require a blood clot to halt bleeding. Figure 12.13 shows the steps in platelet plug formation.

Check Your Recall

32. What is hemostasis?
33. How does a blood vessel spasm help control bleeding?
34. Describe the formation of a platelet plug.

Blood Coagulation

Coagulation (ko-ag″u-la′shun), the most effective hemostatic mechanism, is the formation of a *blood clot*. Blood coagulation is complex and utilizes many biochemicals called *clotting factors*. Some of these factors

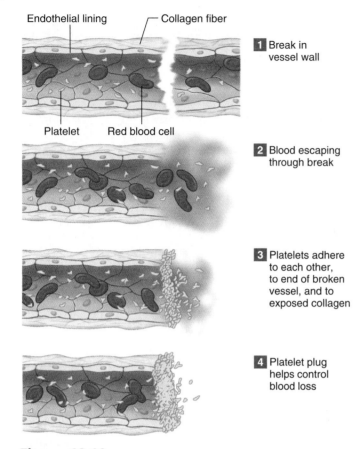

Endothelial lining — Collagen fiber

1 Break in vessel wall

Platelet   Red blood cell

2 Blood escaping through break

3 Platelets adhere to each other, to end of broken vessel, and to exposed collagen

4 Platelet plug helps control blood loss

Figure 12.13

Steps in platelet plug formation.

promote coagulation, and others inhibit it. Whether or not blood coagulates depends on the balance between these two groups of factors. Normally, anticoagulants prevail, and the blood does not clot. However, as a result of injury (trauma), biochemicals that favor coagulation may increase in concentration, and the blood may coagulate.

The major event in blood clot formation is the conversion of the soluble plasma protein fibrinogen into insoluble threads of the protein **fibrin.** Formation of fibrin takes several steps. First, damaged tissues release *tissue thromboplastin,* initiating a series of reactions that results in the production of *prothrombin activator.* This series of changes requires calcium ions as well as certain proteins and phospholipids. As its name suggests, prothrombin activator acts on prothrombin (see figure 12.16).

Prothrombin is an alpha globulin that the liver continually produces and is thus a normal constituent of plasma. Prothrombin activator converts prothrombin into **thrombin,** which in turn catalyzes a reaction that cuts fibrinogen into pieces of fibrin, which join, forming long threads. Certain other proteins also enhance fibrin formation.

Once fibrin threads form, they stick to the exposed surfaces of damaged blood vessels, creating a meshwork

that entraps blood cells and platelets (fig. 12.14). The resulting mass is a blood clot, which may block a vascular break and prevent further blood loss. The clear, yellow liquid that remains after the clot forms is called **serum.** Serum is plasma minus clotting factors.

The amount of prothrombin activator that appears in the blood is directly proportional to the degree of tissue damage. Once a blood clot begins to form, it promotes more clotting because thrombin also acts directly on blood clotting factors other than fibrinogen, causing prothrombin to form more thrombin. This is an example of positive feedback, in which the original action stimulates more of the action. Such a positive feedback mechanism produces unstable conditions and can operate for only a short time without disrupting the stable internal environment (see chapter 1, p. 7).

Laboratory tests commonly used to evaluate the blood coagulation mechanisms include *prothrombin time (PT)* and *partial thromboplastin time (PTT).* Both tests measure the time for fibrin threads to form in a sample of plasma.

Normally, blood flow prevents formation of a massive clot by rapidly carrying excess thrombin away. As a result, its concentration is too low in any one place to promote further clotting. Consequently, blood coagulation usually occurs only in blood that is standing still (or moving slowly). Clotting ceases where a clot contacts circulating blood.

Fibroblasts (see chapter 5, p. 102) invade blood clots that form in ruptured vessels, producing fibrous connective tissue throughout, which helps strengthen and seal vascular breaks. Many clots, including those that form in tissues as a result of blood leakage (hematomas), disappear in time. This dissolution requires activation of

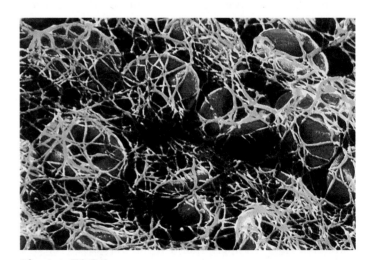

Figure 12.14

A scanning electron micrograph of fibrin threads forming a blood clot (2,800×).

a plasma protein, *plasminogen,* to *plasmin,* a protein-splitting enzyme that can digest fibrin threads and other proteins associated with clots. Plasmin formation may dissolve a whole clot; however, clots that fill large blood vessels are seldom removed naturally.

A blood clot abnormally forming in a vessel is a **thrombus** (throm′bus). A clot that dislodges, or a fragment of a clot that breaks loose and is carried away by the blood flow, is called an **embolus** (em′bo-lus). Generally, emboli continue to move until they reach narrow places in vessels, where they may lodge and block blood flow.

A blood clot forming in a vessel that supplies a vital organ, such as the heart (coronary thrombosis) or the brain (cerebral thrombosis), kills tissues the vessel serves (*infarction*) and may be fatal. A blood clot that travels and then blocks a vessel that supplies a vital organ, such as the lungs (pulmonary embolism), affects the portion of the organ the blocked blood vessel supplies. The Genetics Connection on page 333 discusses several blood clotting disorders.

> Drugs based on "clot-busting" biochemicals can be lifesavers. *Tissue plasminogen activator* (tPA) may restore blocked coronary or cerebral circulation if given within four hours of a heart attack or stroke. A drug derived from bacteria called *streptokinase* may also be successful, for a fraction of the cost. Another plasminogen activator used as a drug is *urokinase,* an enzyme produced in certain kidney cells. Heparin and coumadin are drugs that interfere with clot formation.

Abnormal clot formations are often associated with conditions that change the endothelial linings of vessels. For example, in *atherosclerosis,* accumulations of fatty deposits alter arterial linings, sometimes initiating inappropriate clotting (fig. 12.15). Figure 12.16 summarizes the three primary hemostatic mechanisms: blood vessel spasm, platelet plug formation, and blood coagulation.

Check Your Recall

35. Review the major steps in blood clot formation.
36. What prevents the formation of massive clots throughout the cardiovascular system?
37. Distinguish between a thrombus and an embolus.

12.5 BLOOD GROUPS AND TRANSFUSIONS

Early attempts to transfer blood from one person to another produced varied results. Sometimes, the recipient improved. Other times, the recipient suffered a blood transfusion reaction in which the red blood cells clumped, obstructing vessels and producing great pain and organ damage.

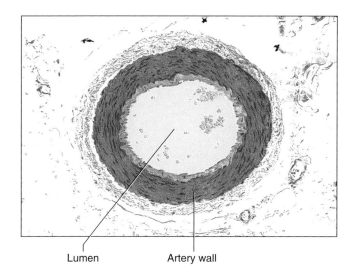

(a)

Lumen   Artery wall

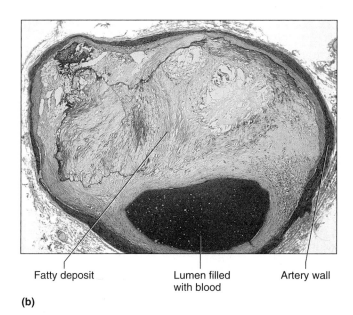

Fatty deposit   Lumen filled   Artery wall
                  with blood

(b)

Figure 12.15

Artery cross sections. (*a*) Light micrograph of a normal artery (90×). (*b*) Atherosclerosis has greatly changed the inner wall of this artery. Not only is blood flow impeded, but the uneven inner surface can snag platelets, triggering coagulation.

Eventually, scientists determined that blood is of differing types and that only certain combinations of blood types are compatible. These discoveries led to the development of procedures for typing blood. Today, safe transfusions of whole blood depend on properly matching the blood types of donors and recipients.

Antigens and Antibodies

Agglutination is the clumping of red blood cells following a transfusion reaction. Red blood cell surface molecules called **antigens** (an′tĭ-jenz), also called

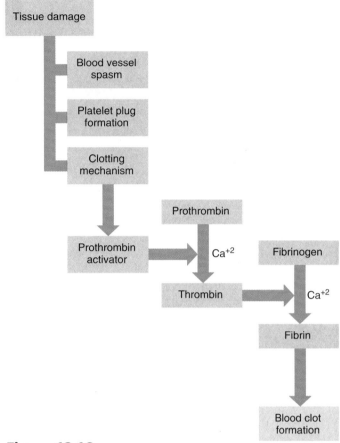

Figure 12.16

The three mechanisms of hemostasis: blood vessel spasm, platelet plug formation, and blood coagulation.

agglutinogens, react with protein **antibodies** (an′tĭ-bod″ēz), also called *agglutinins,* carried in plasma.

Only a few of the more than 260 antigens on red blood cell membranes can produce serious transfusion reactions. These include the antigens of the ABO group and those of the Rh group. Avoiding the mixture of certain kinds of antigens and antibodies prevents adverse transfusion reactions.

A mismatched blood transfusion quickly produces telltale signs of agglutination—anxiety, breathing difficulty, facial flushing, headache, and severe pain in the neck, chest, and lumbar area. Red blood cells burst, releasing free hemoglobin. Macrophages phagocytize the hemoglobin, converting it to bilirubin, which may sufficiently accumulate to cause the yellow skin of jaundice. Free hemoglobin in the kidneys may ultimately cause them to fail.

ABO Blood Group

The *ABO blood group* is based on the presence (or absence) of two major protein antigens on red blood cell membranes—antigen A and antigen B. A person's erythrocytes have on their surfaces one of four antigen combinations: only A, only B, both A and B, or neither A nor B. The resulting ABO blood type, because it reflects a protein combination, is inherited.

A person with only antigen A has *type A blood.* A person with only antigen B has *type B blood.* An individual with both antigen A and B has *type AB blood.* A person with neither antigen A nor B has *type O blood.* Thus, all people have one of four possible ABO blood types—A, B, AB, or O.

In the United States, the most common ABO blood types are O (47%) and A (41%). Rarer are type B (9%) and type AB (3%). These percentages vary in subpopulations and over time, reflecting changes in the genetic structure of populations.

Certain antibodies that affect the ABO blood group are synthesized in the plasma about two to eight months following birth. Specifically, whenever antigen A is absent in red blood cells, an antibody called *anti-A* is produced, and whenever antigen B is absent, an antibody called *anti-B* is produced. Therefore, persons with type A blood have anti-B antibody in their plasma; those with type B blood have anti-A antibody; those with type AB blood have neither antibody; and those with type O blood have both anti-A and anti-B antibodies (fig. 12.17 and table 12.3). The antibodies anti-A and anti-B are large and do not cross the placenta. Thus, a pregnant woman and her fetus may be of different ABO blood types, and agglutination in the fetus will not occur.

An antibody of one type will react with an antigen of the same type and clump red blood cells (fig. 12.18); therefore, such combinations must be avoided. The major concern in blood transfusion procedures is that the cells in the donated blood not clump due to antibodies in the recipient's plasma. For this reason, a person with type A (anti-B) blood must not receive blood of type B or AB, either of which would clump in the presence of anti-B in the recipient's type A blood. Likewise, a person with type B (anti-A) blood must not receive type A or AB blood, and a person with type O

Table 12.3	Antigens and Antibodies of the ABO Blood Group	
Blood Type	**Antigen**	**Antibody**
A	A	Anti-B
B	B	Anti-A
AB	A and B	Neither anti-A nor anti-B
O	Neither A nor B	Both anti-A and anti-B

Genetics Connection

Coagulation Disorders

Hemophilia

Abnormalities of different clotting factors cause different forms of the bleeding disorder hemophilia, but hemophilia A is the most common. Factor VIII is deficient or absent. Symptoms of the hemophilias include severe hemorrhage following minor injuries, frequent nosebleeds, large intramuscular hematomas, and blood in the urine. The pattern of inheritance of hemophilia A is such that most affected individuals are male.

Hemophilia has left its mark on history. One of the earliest descriptions is in the Talmud, a second-century B.C. Jewish document, which reads, "If she circumcised her first child and he died, and a second one also died, she must not circumcise her third child." England's Queen Victoria (1819–1901) passed the hemophilia gene to several of her children, eventually spreading the condition to the royal families of Russia, Germany, and Spain. Hemophilia achieved notoriety when factor VIII pooled from blood donations was discovered to transmit HIV in 1985. Ninety percent of people with severe hemophilia who used such pooled factor VIII in the few years prior to that time developed AIDS.

von Willebrand Disease

The tendency to bleed and bruise easily may be a sign of *von Willebrand disease,* an inherited clotting disorder that is usually less severe, but much more common, than hemophilia. Affected persons lack a plasma protein, von Willebrand factor, that is secreted by the endothelial cells lining blood vessels. Von Willebrand factor enables platelets to adhere to damaged blood vessel walls, a key step preceding actual clotting. Sometimes, the condition can cause spontaneous bleeding from the mucous membranes of the gastrointestinal and urinary tracts. A person might not become aware of symptoms until excessive bleeding follows an injury. Von Willebrand disease is equally likely to affect males and females.

Immune Thrombocytopenic Purpura (ITP)

In ITP the immune system attacks platelets, dropping the count from 130,000 to 360,000 per microliter of blood to fewer than 10,000 to 30,000 per microliter. This autoimmune disorder affects three times as many females as males.

Symptoms of ITP include small purple marks on the skin where tiny blood vessels have broken, bleeding gums, nosebleeds, and bruising. Some people with ITP may never experience symptoms and may discover the condition only when a routine blood count reveals the platelet deficiency. Rarely, ITP causes a fatal brain bleed. Muscle aches, depression, and fatigue are also associated with the disorder. A form of ITP called gestational thrombocytopenia affects about 1 in 20 women late in pregnancy but clears up afterward.

Most people with ITP can live a normal life but must be careful to avoid activities that might cause injury, such as contact sports. Precautions must be taken during dental procedures and surgery. Various drugs are used to control ITP.

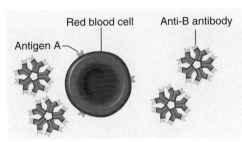

Type A blood

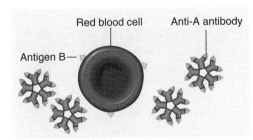

Type B blood

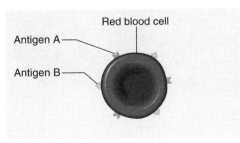

Type AB blood

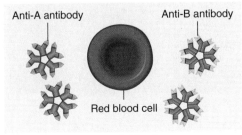

Type O blood

Figure 12.17

Different combinations of antigens and antibodies distinguish blood types. (Cells and antibodies not drawn to scale.)

(anti-A and anti-B) blood must not receive type A, B, or AB blood.

Because type AB blood lacks both anti-A and anti-B antibodies, an AB person can receive a transfusion of blood of any type. For this reason, type AB persons are sometimes called *universal recipients*. However, type A (anti-B) blood, type B (anti-A) blood, and type O (anti-A and anti-B) blood still contain antibodies (either anti-A and/or anti-B) that could agglutinate type AB cells if transfused rapidly. Consequently, even for AB individuals, using donor blood of the same type as the recipient is best (table 12.4).

Because type O blood lacks antigens A and B, this type could theoretically be transfused into persons with blood of any other type. Therefore, persons with type O blood are sometimes called *universal donors*. Type O blood, however, does contain both anti-A and anti-B antibodies. If type O blood is given to a person with blood type A, B, or AB, it should be transfused slowly so that the recipient's larger blood volume will dilute it, minimizing the chance of an adverse reaction.

Table 12.4	Preferred and Permissible Blood Types for Transfusions	
Blood Type of Recipient	Preferred Blood Type of Donor	Permissible Blood Type of Donor (in Extreme Emergency)
A	A	O
B	B	O
AB	AB	A, B, O
O	O	No alternate types

Blood in the umbilical cord at birth is rich in stem cells that can be used to treat a variety of disorders, including leukemias, sickle cell disease and other hemoglobin abnormalities, and certain inborn errors of metabolism. More than 7,000 people have received cord blood transplants since 1988. The United States and the United Kingdom recently established umbilical cord blood banks that will serve the donors but also be available for free to anyone who needs the blood and its valuable stem cells.

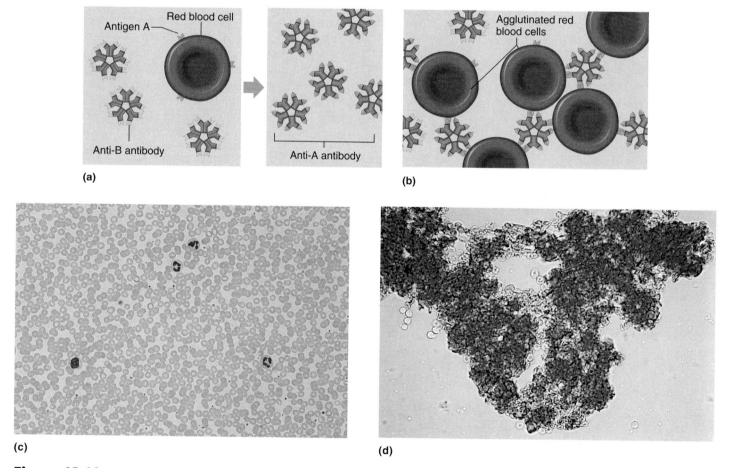

(a)

(b)

(c)

(d)

Figure 12.18

Agglutination. (*a*) If red blood cells with antigen A are added to blood containing anti-A antibody, (*b*) the antibodies react with the antigens, causing clumping (agglutination). (*c*) Nonagglutinated blood (210×). (*d*) Agglutinated blood (220×). (Cells and antibodies in *a* and *b* not drawn to scale.)

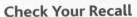
Rh Blood Group

The *Rh blood group* was named after the rhesus monkey in which it was first studied. In humans, this group includes several Rh antigens (factors). The most prevalent of these is *antigen D.* If the Rh antigen is present on the red blood cell membranes, the blood is said to be *Rh-positive.* Conversely, if the red blood cells lack Rh antigen, the blood is called *Rh-negative.*

Only 15% of the U.S. population is Rh-negative. Therefore, AB⁻ blood is the rarest type, and O⁺ the most common.

As in the case of antigens A and B, the presence (or absence) of Rh antigen is an inherited trait. But unlike anti-A and anti-B, antibodies that react with Rh antigen (*anti-Rh antibodies*) do not appear spontaneously. Instead, they form only in Rh-negative persons in response to special stimulation.

If an Rh-negative person receives a transfusion of Rh-positive blood, the Rh antigen stimulates the recipient to begin producing anti-Rh antibodies. Generally, this initial transfusion has no serious consequences, but if the Rh-negative person—who is now sensitized to

Rh-positive blood—receives another transfusion of Rh-positive blood some months later, the donated red cells are likely to agglutinate.

A related condition may occur when an Rh-negative woman is pregnant with an Rh-positive fetus for the first time. Such a pregnancy may be uneventful; however, at birth (or if a miscarriage occurs), the placental membranes that separated the maternal blood from the fetal blood during the pregnancy tear, and some of the infant's Rh-positive blood cells may enter the maternal circulation. These Rh-positive cells may then stimulate the maternal tissues to begin producing anti-Rh antibodies.

If a woman who has already developed anti-Rh antibodies becomes pregnant with a second Rh-positive fetus, these antibodies, called hemolysins, cross the placental membrane and destroy the fetal red blood cells (fig. 12.19). The fetus then develops a condition called *hemolytic disease of the fetus and newborn* (erythroblastosis fetalis).

Hemolytic disease of the fetus and newborn is extremely rare today because physicians carefully track Rh status. An Rh-negative woman who might carry an Rh-positive fetus is given an injection of a drug called RhoGAM. This injection is actually composed of anti-Rh antibodies, which bind to and shield any Rh-positive fetal cells that might contact the woman's cells and sensitize her immune system. RhoGAM must be given within 72 hours of possible contact with Rh-positive cells—including giving birth, terminating a pregnancy, miscarrying, or undergoing amniocentesis (a prenatal test in which a needle is inserted into the uterus).

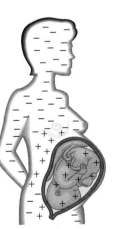

Figure 12.19

Rh incompatibility. If a man who is Rh-positive and a woman who is Rh-negative conceive a child who is Rh-positive, the woman's body may manufacture antibodies that attack future Rh-positive offspring.

Rh-negative woman with Rh-positive fetus

Cells from Rh-positive fetus enter woman's bloodstream

Woman becomes sensitized— antibodies (◇) form to fight Rh-positive blood cells

In the next Rh-positive pregnancy, maternal antibodies attack fetal red blood cells

Clinical Terms Related to the Blood

anisocytosis (an-i"so-si-to'sis) Abnormal variation in the size of erythrocytes.

antihemophilic plasma (an"ti-he"mo-fil'ik plaz'mah) Normal blood plasma that has been processed to preserve an antihemophilic factor.

citrated whole blood (sit'rāt-ed hōl blud) Normal blood to which a solution of acid citrate has been added to prevent coagulation.

dried plasma (drīd plaz'mah) Normal blood plasma that has been vacuum-dried to prevent the growth of microorganisms.

hemorrhagic telangiectasia (hem"o-raj'ik tel-an"je-ek-ta'ze-ah) Inherited tendency to bleed from localized lesions of the capillaries.

heparinized whole blood (hep'er-ĭ-nīzd" hōl blud) Normal blood to which a solution of heparin has been added to prevent coagulation.

macrocytosis (mak"ro-si-to'sis) Abnormally large erythrocytes.

microcytosis (mi"kro-si-to'sis) Abnormally small erythrocytes.

neutrophilia (nu"tro-fil'e-ah) Increase in the number of circulating neutrophils.

packed red cells Concentrated suspension of red blood cells from which the plasma has been removed.

pancytopenia (pan"si-to-pe'ne-ah) Abnormal depression of all the cellular components of blood.

poikilocytosis (poi"kĭ-lo-si-to'sis) Irregularly shaped erythrocytes.

purpura (per'pu-rah) Spontaneous bleeding into the tissues and through the mucous membranes.

septicemia (sep"ti-se'me-ah) Growth of disease-causing microorganisms or presence of their toxins in the blood.

spherocytosis (sfēr"o-si-to'sis) Hemolytic anemia caused by defective proteins supporting the cell membranes of red blood cells. The cells are abnormally spherical.

thalassemia (thal"ah-se'me-ah) Group of hereditary hemolytic anemias resulting from very thin, fragile erythrocytes.

Clinical Connection

Thrombotic thrombocytopenic purpura (TTP) was first described in 1924, but the underlying cause was not discovered until 2001. In this rare disorder, platelets adhere to abnormally large clumps of the plasma protein von Willebrand factor. When the clumps lodge in narrow blood vessels in major organs, symptoms begin, usually in young adulthood. Neurological symptoms include headache, confusion, changes in speech, and altered consciousness. The kidneys may fail. The deficiency of platelets causes bleeding beneath the skin, which leads to characteristic red bruises. Anemia results from the shattering of red blood cells, which overtaxes the spleen. Treatment, which is 80% to 90% effective, includes removing the spleen and cleansing the plasma using a technique called plasmaphoresis.

A hereditary form of TTP results from absence or malfunction of an enzyme that normally cuts von Willebrand factor protein aggregates down to size. It is likely that noninherited forms of the illness also affect the size of these plasma protein aggregates. The missing enzyme is very similar, in structure and apparently in function, to a component of snake venom that causes bleeding in a bite victim.

SUMMARY OUTLINE

12.1 Introduction (p. 318)

Blood is a type of connective tissue consisting of red blood cells, white blood cells, and platelets suspended in a liquid, plasma, extracellular matrix. It transports substances between body cells and the external environment, and helps maintain a stable internal environment.

1. Blood can be separated into formed elements and liquid portions.
   a. The formed elements portion is mostly red blood cells.
   b. The liquid plasma includes water, gases, nutrients, hormones, electrolytes, and cellular wastes.
2. Blood volume varies with body size, fluid and electrolyte balance, and adipose tissue content.

12.2 Blood Cells (p. 319)

1. Red blood cells
   a. Red blood cells are biconcave discs with shapes that increase surface area.
   b. Red blood cells contain hemoglobin, which combines with oxygen.
2. Red blood cell counts
   a. The red blood cell count equals the number of cells per microliter of blood.
   b. The average count ranges from approximately 4 to 6 million cells per microliter of blood.
   c. Red blood cell count determines the oxygen-carrying capacity of the blood. It is used to diagnose and evaluate the courses of certain diseases.
3. Red blood cell production and its control
   a. Red bone marrow produces red blood cells.
   b. In health, the number of red blood cells remains relatively stable.
   c. Erythropoietin controls the rate of red blood cell formation by negative feedback.
4. Dietary factors affecting red blood cell production
   a. Availability of vitamin B_{12} and folic acid influences red blood cell production.
   b. Hemoglobin synthesis requires iron.
5. Destruction of red blood cells
   a. Macrophages in the liver and spleen phagocytize damaged red blood cells.
   b. Hemoglobin molecules decompose, and some of the iron they contain is recycled.
   c. Hemoglobin releases biliverdin and bilirubin pigments.
6. White blood cells
   a. White blood cells develop from hemocytoblasts in red bone marrow, in response to interleukins and colony-stimulating factors.

b. Granulocytes include neutrophils, eosinophils, and basophils.
c. Agranulocytes include monocytes and lymphocytes.
7. Functions of white blood cells
   a. Neutrophils and monocytes phagocytize foreign particles.
   b. Eosinophils kill parasites and help control inflammation and allergic reactions.
   c. Basophils release heparin, which inhibits blood clotting, and histamine to increase blood flow to injured tissues.
   d. Lymphocytes produce antibodies that attack specific foreign substances.
8. White blood cell counts
   a. Normal total white blood cell counts vary from 4,500 to 10,000 cells per microliter of blood.
   b. The number of white blood cells may change in response to abnormal conditions, such as infections, emotional disturbances, or excessive loss of body fluids.
   c. A differential white blood cell count indicates the percentages of various types of leukocytes.
9. Blood platelets
   a. Blood platelets, which develop in the red bone marrow in response to thrombopoietin, are fragments of giant cells.
   b. The normal platelet count varies from 130,000 to 360,000 platelets per microliter of blood.
   c. Platelets help close breaks in blood vessels.

12.3 Blood Plasma (p. 325)

Plasma transports gases and nutrients, helps regulate fluid and electrolyte balance, and helps maintain stable pH.
1. Plasma proteins
   a. Plasma proteins remain in blood and interstitial fluids, and are not normally used as energy sources.
   b. Three major types exist.
      (1) Albumins help maintain the colloid osmotic pressure.
      (2) Globulins transport lipids and fat-soluble vitamins and include antibodies that provide immunity.
      (3) Fibrinogen functions in blood clotting.
2. Gases and nutrients
   a. Gases in plasma include oxygen, carbon dioxide, and nitrogen.
   b. Plasma nutrients include simple sugars, amino acids, and lipids.
      (1) The liver stores glucose as glycogen and releases glucose whenever blood glucose concentration falls.
      (2) Amino acids are used to synthesize proteins and are deaminated for use as energy sources.
      (3) Lipoproteins function in the transport of lipids.
3. Nonprotein nitrogenous substances
   a. Nonprotein nitrogenous substances are composed of molecules that contain nitrogen atoms but are not proteins.
   b. They include amino acids, urea, and uric acid.
4. Plasma electrolytes
   a. Plasma electrolytes include ions of sodium, potassium, calcium, magnesium, chloride, bicarbonate, phosphate, and sulfate.
   b. Bicarbonate ions are important in maintaining the osmotic pressure and pH of plasma.

12.4 Hemostasis (p. 328)

Hemostasis is the stoppage of bleeding.
1. Blood vessel spasm
   a. Smooth muscles in blood vessel walls reflexly contract following injury.
   b. Platelets release serotonin, which stimulates vasoconstriction and helps maintain vessel spasm.
2. Platelet plug formation
   a. Platelets adhere to rough surfaces and exposed collagen.
   b. Platelets adhere to each other at injury sites and form platelet plugs in broken vessels.
3. Blood coagulation
   a. Blood clotting is the most effective means of hemostasis.
   b. Clot formation depends on the balance between factors that promote clotting and those that inhibit clotting.
   c. The basic event of coagulation is the conversion of soluble fibrinogen into insoluble fibrin.
   d. Biochemicals that promote clotting include prothrombin activator, prothrombin, and calcium ions.
   e. A thrombus is an abnormal blood clot in a vessel. An embolus is a clot or fragment of a clot that moves in a vessel.

12.5 Blood Groups and Transfusions (p. 331)

Blood can be typed on the basis of cell surface antigens.
1. Antigens and antibodies
   a. Agglutination is the clumping of red blood cells following a transfusion reaction.
   b. Red blood cell membranes may contain specific antigens, and blood plasma may contain antibodies against certain of these antigens.
2. ABO blood group
   a. Blood is grouped according to the presence or absence of antigens A and B.
   b. Mixing red blood cells that contain an antigen with plasma that contains the corresponding antibody results in an adverse transfusion reaction.
3. Rh blood group
   a. Rh antigen is present on the red blood cell membranes of Rh-positive blood. Rh antigen is absent in Rh-negative blood.
   b. Mixing Rh-positive red blood cells with plasma that contains anti-Rh antibodies agglutinates the positive cells.
   c. Anti-Rh antibodies in maternal blood may cross the placental tissues and react with the red blood cells of an Rh-positive fetus.

CHAPTER ASSESSMENTS

12.1 Introduction

1. Major functions of blood include: (p. 318)
   a. nutrient, hormone, oxygen, and waste transport.
   b. helping maintain the stability of interstitial fluid.
   c. heat distribution.
   d. all of the above.
2. Formed elements in blood are _____, _____, and _____. (p. 318)
3. The liquid portion of blood is _____. (p. 318)

12.2 Blood Cells

4. Describe a red blood cell. (p. 319)
5. Contrast oxyhemoglobin and deoxyhemoglobin. (p. 319)
6. Connect the significance of red blood cell counts with the function of red blood cells. (p. 319)
7. Describe the life cycle of a red blood cell, beginning with its production and ending with its destruction. (p. 319)
8. List dietary factors affecting red blood cell production. (p. 320)
9. Name five types of leukocytes, identifying which are granulocytes and which are agranulocytes, and list the major function(s) of each type. (p. 323)

10. _____ are fragments of megakaryocytes that function in _____. (p. 325)

12.3 Blood Plasma

11. The most abundant component of plasma is: (p. 325)
    a. vitamins.
    b. oxygen.
    c. proteins.
    d. water.
    e. electrolytes.

12. List three types of plasma proteins, and indicate the major functions of each type. (p. 325)

13. List the gases and nutrients found in plasma. (p. 327)

14. Define *nonprotein nitrogenous substances,* and name those commonly present in plasma. (p. 327)

15. The most abundant plasma electrolytes are _____ and _____. (p. 327)

12.4 Hemostasis

16. _____ is the stoppage of bleeding. (p. 328)

17. Explain how blood vessel spasm is stimulated following an injury. (p. 328)

18. Platelets adhering to form a plug may control blood loss from a _____ break, but a larger break may require a _____ to halt bleeding. (p. 329)

19. Describe the major steps leading to the formation of a blood clot. (p. 330)

20. Contrast thrombus and embolus. (p. 331)

12.5 Blood Groups and Transfusions

21. An individual with B antigens and anti-A antibodies is ABO blood type _____. (p. 332)

22. Explain why the individual described in question 21 should not receive a transfusion with type AB blood. (p. 332)

23. Distinguish between Rh-positive and Rh-negative blood. (p. 335)

24. Define *hemolytic disease of the fetus and newborn* and explain how this condition may develop. (p. 335)

INTEGRATIVE ASSESSMENTS/ CRITICAL THINKING

OUTCOME 12.2

1. Erythropoietin is available as a drug. Why would athletes abuse it?

2. Researchers are developing several types of chemicals to be used as temporary red blood cell substitutes. What characteristics should a red blood cell substitute have?

3. If a patient with inoperable cancer is treated using a drug that reduces the rate of cell division, how might the patient's white blood cell count change? How might the patient's environment be modified to compensate for the effects of these changes?

OUTCOMES 12.2, 12.5

4. Why can a person receive platelets donated by anyone, but must receive a particular type of whole blood?

OUTCOME 12.5

5. How might a technique to remove A and B antigens from red blood cells be used to increase the supply of donated blood?

WEB CONNECTIONS

Visit the text website at **aris.mhhe.com** for additional quizzes, interactive learning exercises, and more.

AP R CARDIOVASCULAR SYSTEM

Anatomy & Physiology | REVEALED includes cadaver photos that allow you to peel away layers of the human body to reveal structures beneath the surface. This program also includes animations, radiologic imaging, audio pronunciation, and practice quizzing. To learn more visit **www.aprevealed.com**.

13

Cardiovascular System

CARDIOVASCULAR DEFIBRILLATORS. A man rushing to catch a flight at a busy airport stops suddenly, looks about in confusion, and collapses. People gather around him, as a woman runs to a device mounted on a nearby wall. It is an automated external defibrillator (AED), and looks like a laptop computer. The woman learned how to use it in a cardiopulmonary resuscitation class. She brings it over to the man, opens it, and places electrode pads over the man's chest, as indicated in a drawing on the inner cover of the defibrillator. Then the device speaks, "Analyzing heart rhythm," it declares as a computer assesses the heart rhythm. After a short pause, the device says, "charging, stand clear," and then "push button." The woman does so, and the device delivers a shock to the man's chest. It assesses the heart rhythm again, and instructs the woman to deliver a second shock. Soon, the man recovers, just as emergency technicians arrive.

The automated external defibrillators found in airports, malls, schools, health clubs, and other public places can save the life of a person suffering sudden cardiac arrest. One study conducted at Chicago's O'Hare and Midway airports found that over a ten-month period, automated external defibrillators saved 64% of the people they were used on. Without defibrillation, only 5–7% of people survive sudden cardiac arrest. Each minute, the odds of survival shrink by 10%, and after six minutes, brain damage is irreversible. Sudden cardiac arrest can result from an accelerated heartbeat (ventricular tachycardia) or a chaotic and irregular heartbeat (ventricular fibrillation). The electrical malfunction that usually causes these conditions may result from an artery blocked with plaque or from buildup of scar tissue from a previous myocardial infarction (heart attack).

An implantable cardioverter defibrillator delivers a shock to a heart whose ventricles are contracting wildly, restoring a normal heartbeat.

For people who know they have an inherited disorder that causes sudden cardiac arrest (by having suffered an event and then had genetic tests), a device called an implantable cardioverter defibrillator (ICD) can be placed under the skin of the chest in a one-hour procedure. Like the AED, the ICD monitors heart rhythm, and when the telltale deviations of ventricular tachycardia or ventricular fibrillation begin, it delivers a shock.

ICDs have been so successful in preventing subsequent cardiac arrests that in some countries they are offered to people at high risk for the condition. The two major risk factors are having had a previous myocardial infarction and having a low ejection fraction, which is the volume of blood pumped with each heartbeat. Normal ejection fraction is 50–60%; low is below 30–40%. Scarring lowers the ejection fraction. An echocardiogram, which is an ultrasound scan of the heart, can reveal the ejection fraction.

Learning Outcomes   *After studying this chapter, you should be able to do the following:*

11. Describe the mechanisms that aid in
    returning venous blood to the heart.
    (p. 361)

13.6 Paths of Circulation

12. Compare the pulmonary and systemic
    circuits of the cardiovascular system.
    (p. 362)

**13.7–13.8 Arterial System–Venous
System**

13. Identify and locate the major arteries and
    veins. (p. 362–370)

Aids to Understanding Words

(Appendix A on page 567 has a complete list of Aids to Understanding Words.)

brady- [slow] *brady*cardia: Abnormally slow
heartbeat.

diastol- [dilation] *diastol*ic pressure: Blood
pressure when the ventricle of the heart is
relaxed.

-gram [something written] electrocardio*gram*:
Recording of the electrical changes in the
myocardium during a cardiac cycle.

papill- [nipple] *papill*ary muscle: Small mound of
muscle projecting into a ventricle of the heart.

syn- [together] *syn*cytium: Mass of merging cells
that act together.

systol- [contraction] *systol*ic pressure: Blood
pressure resulting from a single ventricular
contraction.

tachy- [rapid] *tachy*cardia: Abnormally fast
heartbeat.

13.1 INTRODUCTION

The heart pumps 7,000 liters of blood through the body
each day, contracting some 2.5 billion times in an average lifetime. This muscular pump forces blood through
arteries, which connect to smaller-diameter vessels
called arterioles. Arterioles branch into the tiniest tubes,
the capillaries, which are sites of nutrient, electrolyte,
gas, and waste exchange. Capillaries converge into
venules, which in turn converge into veins that return
blood to the heart, completing the closed system of

blood circulation. These structures—the pump and its
vessels—form the cardiovascular system.

The **pulmonary** (pul′mo-ner″e) **circuit** sends oxygen-
depleted (deoxygenated) blood to the lungs to pick up
oxygen and unload carbon dioxide. The **systemic** (sis-
tem′ik) **circuit** sends oxygen-rich (oxygenated) blood
and nutrients to all body cells and removes wastes. Without circulation, tissues would lack a supply of oxygen and
nutrients, and wastes would accumulate. Such deprived
cells soon begin irreversible change, which quickly leads
to their death. Figure 13.1 shows the general pattern of
blood transport in the cardiovascular system.

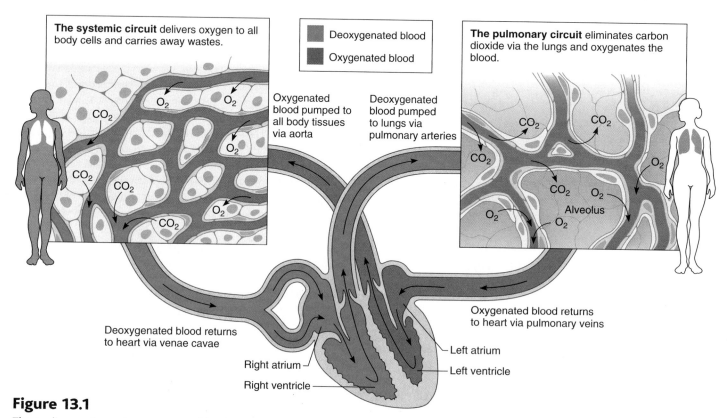

Figure 13.1

The cardiovascular system transports blood between the body cells and organs such as the lungs, intestines, and kidneys that communicate
with the external environment. Vessels in the pulmonary circuit carry blood from the heart to the lungs and back to the heart, replenishing
oxygen and releasing the metabolic waste CO_2. Vessels of the systemic circuit supply all of the other cells.

13.2 STRUCTURE OF THE HEART

The heart is a hollow, cone-shaped, muscular pump. It lies within the thoracic cavity and rests on the diaphragm (fig. 13.2).

⸱he Heart

y size. An average adult's
⸱rs long and 9 centimeters

e mediastinum, bordered
⸱riorly by the vertebral col-
sternum. The *base* of the
⸱ral large blood vessels, l<u>ies</u>
⸱ heart's distal end extends
terminating as a bluntly
he fifth intercostal space.

Coverings of the Heart

The **pericardium** (per″ĭ-kar′de-um) encloses the heart and the proximal ends of the large blood vessels to which it attaches. The pericardium consists of an outer bag, the fibrous pericardium. The fibrous pericardium is dense connective tissue. It is attached to the central portion of the diaphragm, the posterior of the sternum, the vertebral column, and the large blood vessels emerging from the heart.

The fibrous pericardium surrounds a more delicate, double-layered sac. The innermost layer of this sac, the *visceral pericardium* (epicardium), covers the heart. At the base of the heart, the visceral pericardium turns back on itself to become the *parietal pericardium,* which forms the inner lining of the fibrous pericardium (figs. 13.2 and 13.4; see reference plate 3, p. 25). Between the parietal and visceral layers of the pericardium is a space, the *pericardial cavity,* that

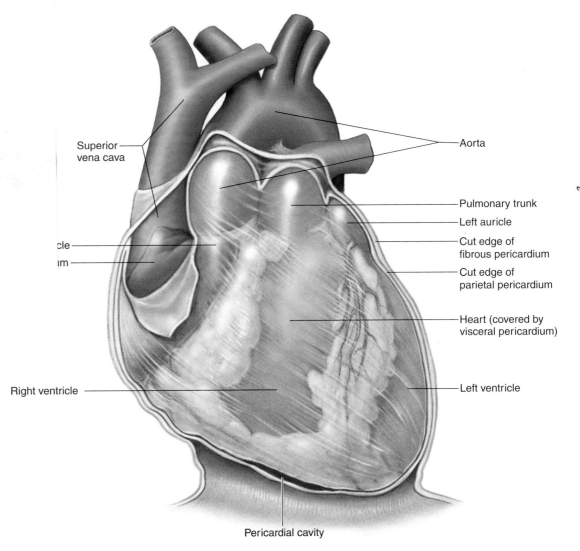

Superior vena cava

Aorta

Pulmonary trunk

Left auricle

Cut edge of fibrous pericardium

Cut edge of parietal pericardium

cle

ım

Heart (covered by visceral pericardium)

Right ventricle

Left ventricle

Pericardial cavity

Figure 13.2

The heart is within the mediastinum and is enclosed by a layered pericardium.

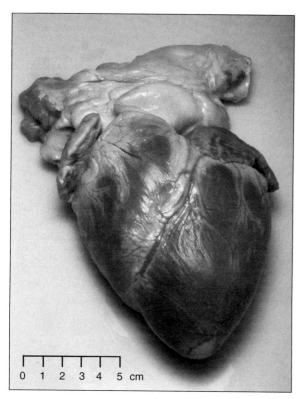

Figure 13.3

Anterior view of a human heart. (Because this photo is not life-size, a proportionately reduced ruler is included to help the student grasp the true size of the organ.)

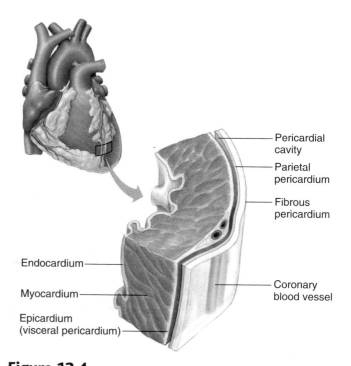

Figure 13.4

The heart wall has three layers: an endocardium, a myocardium, and an epicardium.

contains a small volume of serous fluid (fig. 13.4). This fluid reduces friction between the pericardial membranes as the heart moves within them.

In *pericarditis,* inflammation of the pericardium due to viral or bacterial infection produces adhesions that attach the layers of the pericardium to each other. This condition is very painful and interferes with heart movements.

Check Your Recall

1. Where is the heart located?
2. Distinguish between the visceral pericardium and the parietal pericardium.

Wall of the Heart

The wall of the heart is composed of three distinct layers—an outer epicardium, a middle myocardium, and an inner endocardium (fig. 13.4). The **epicardium** (ep″ĭ-kar′de-um), which corresponds to the visceral pericardium, protects the heart by reducing friction. It is a serous membrane that consists of connective tissue beneath epithelium. Its deeper portion often contains

adipose tissue, particularly along the paths of coronary arteries and cardiac veins that carry blood through the myocardium.

The thick middle layer, or **myocardium** (mi″o-kar′de-um), consists mostly of cardiac muscle tissue that pumps blood out of the heart chambers. The muscle fibers are organized in planes, separated by connective tissue richly supplied with blood capillaries, lymph capillaries, and nerve fibers.

The inner layer, or **endocardium** (en″do-kar′de-um), consists of epithelium and connective tissue that contains many elastic and collagenous fibers. The endocardium also contains blood vessels and some specialized cardiac muscle fibers, called *Purkinje fibers,* described in section 13.3 on page 349. The endocardium is continuous with the inner linings of blood vessels attached to the heart.

Heart Chambers and Valves

Internally, the heart is divided into four hollow chambers—two on the left and two on the right (fig. 13.5). The upper chambers, called **atria** (a′tre-ah; singular, *atrium*), have thin walls and receive blood returning to the heart. Small, earlike projections called *auricles* extend anteriorly from the atria. The lower chambers, the **ventricles** (ven′trĭ-klz), receive blood from the atria and contract to force blood out of the heart into arteries.

A solid, wall-like **septum** separates the atrium and ventricle on the right side from their counterparts on the

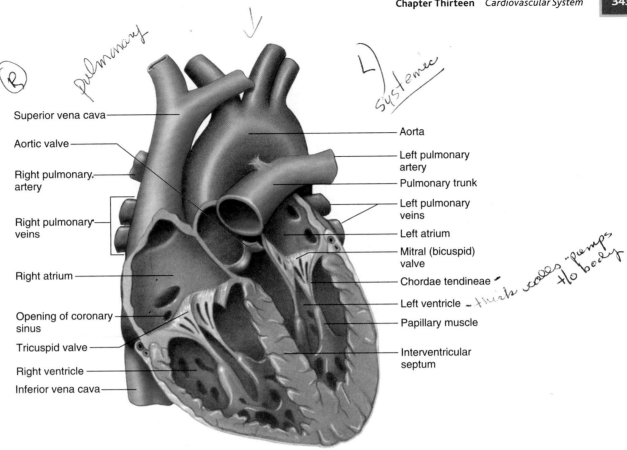

Pulmonary

R

L
Systemic

Superior vena cava

Aortic valve

Right pulmonary artery

Right pulmonary veins

Right atrium

Opening of coronary sinus

Tricuspid valve

Right ventricle

Inferior vena cava

Aorta

Left pulmonary artery

Pulmonary trunk

Left pulmonary veins

Left atrium

Mitral (bicuspid) valve

Chordae tendineae

Left ventricle — thick walls "pumps" to body

Papillary muscle

Interventricular septum

Figure 13.5

Coronal section of the heart showing the connection between the left ventricle and the aorta, as well as the four hollow chambers.

left. As a result, blood from one side of the heart never mixes with blood from the other side (except in the fetus, see chapter 20, p. 553). An *atrioventricular valve* (A-V valve), the tricuspid on the right and the mitral on the left, ensures one-way blood flow between the atria and the ventricles.

The right atrium receives blood from two large veins—the *superior vena cava* and the *inferior vena cava*. A smaller vein, the *coronary sinus,* also drains blood into the right atrium from the myocardium of the heart itself.

The large **tricuspid valve,** which has three tapered projections called *cusps* as its name implies, lies between the right atrium and the right ventricle (fig. 13.5). The valve permits blood to move from the right atrium into the right ventricle and prevents backflow.

Strong, fibrous strings called **chordae tendineae** (kor′de ten′dĭ-ne) attach to the cusps of the tricuspid valve on the ventricular side. These strings originate from small mounds of cardiac muscle tissue, the **papillary muscles,** that project inward from the walls of the ventricle. The papillary muscles contract when the ventricle contracts. As the tricuspid valve closes, these muscles pull on the chordae tendineae and prevent the cusps from swinging back into the atrium.

The right ventricle has a thinner muscular wall than the left ventricle (fig. 13.5). This right chamber pumps blood a short distance to the lungs against a relatively low resistance to blood flow. The left ventricle, on the other hand, must force blood to all the other parts of the body against a much greater resistance to flow.

When the muscular wall of the right ventricle contracts, the blood inside its chamber is put under increasing pressure, and the tricuspid valve closes passively. As a result, the only exit for the blood is through the *pulmonary trunk,* which divides to form the left and right *pulmonary arteries* that lead to the lungs. At the base of this trunk is a **pulmonary valve** with three cusps. This valve allows blood to leave the right ventricle and prevents backflow into the ventricular chamber (fig. 13.6).

The left atrium receives blood from the lungs through four *pulmonary veins*—two from the right lung and two from the left lung. Blood passes from the left atrium into the left ventricle through the **mitral valve** (shaped like a miter, a type of headpiece), or bicuspid valve, which prevents blood from flowing back into the left atrium from the ventricle (see fig. 13.5). As with the tricuspid valve, the papillary muscles and the chordae tendineae prevent the cusps of the mitral valve from

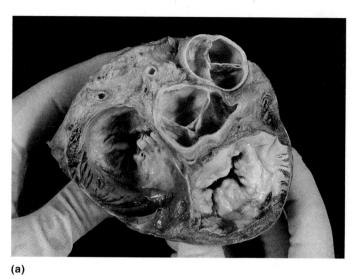

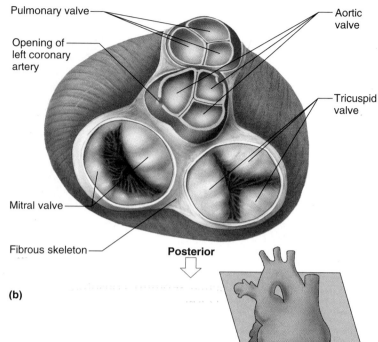

Figure 13.6

Heart valves. (*a*) Photograph of a transverse section through the heart, showing the four valves (superior view). (*b*) The skeleton of the heart consists of fibrous rings to which the heart valves are attached (superior view).

swinging back into the left atrium during ventricular contraction.

When the left ventricle contracts, the mitral valve closes passively, and the only exit is through a large artery, the **aorta** (a-or′tah). At the base of the aorta is the **aortic valve** (a-or′tik valv), which has three cusps. The aortic valve opens and allows blood to leave the left ventricle as it contracts. When the ventricular muscles relax, this valve closes and prevents blood from backing up into the ventricle (fig. 13.6).

The mitral and tricuspid valves are called atrioventricular valves because they are between atria and ventricles. The pulmonary and aortic valves are called "semilunar" because of the half-moon shapes of their cusps. Table 13.1 summarizes the locations and functions of the heart valves.

Mitral valve prolapse (MVP) affects up to 6% of the U.S. population. In this condition, one (or both) of the cusps of the mitral valve stretches and bulges into the left atrium during ventricular contraction. The valve usually continues to function adequately, but sometimes blood regurgitates into the left atrium. Through a stethoscope, a regurgitating MVP sounds like a click at the end of ventricular contraction, and then a murmur as blood goes back through the valve into the left atrium. Symptoms of MVP include chest pain, palpitations, fatigue, and anxiety.

The mitral valve can be damaged by certain species of *Streptococcus* bacteria. Endocarditis, an inflammation of the endocardium due to such an infection, looks like a plantlike growth on the valve. People with MVP are particularly susceptible to endocarditis. They take antibiotics before undergoing dental work to prevent *Streptococcus* infection in the mouth from migrating through the blood to the heart and causing infection.

Table 13.1	Heart Valves	
Valve	**Location**	**Function**
Tricuspid valve	Opening between right atrium and right ventricle	Prevents blood from moving from right ventricle into right atrium during ventricular contraction
Pulmonary valve	Entrance to pulmonary trunk	Prevents blood from moving from pulmonary trunk into right ventricle during ventricular relaxation
Mitral (bicuspid) valve	Opening between left atrium and left ventricle	Prevents blood from moving from left ventricle into left atrium during ventricular contraction
Aortic valve	Entrance to aorta	Prevents blood from moving from aorta into left ventricle during ventricular relaxation

Skeleton of the Heart

Rings of dense connective tissue surround the pulmonary trunk and aorta at their proximal ends. These rings provide firm attachments for the heart valves and for muscle fibers; they also prevent the outlets of the atria and ventricles from dilating during contraction. The fibrous rings, together with other masses of dense connective tissue in the portion of the septum between the ventricles (interventricular septum), constitute the *skeleton of the heart* (fig. 13.6b).

Path of Blood Through the Heart

Blood that is low in oxygen and high in carbon dioxide enters the right atrium through the venae cavae and coronary sinus. As the right atrial wall contracts, the blood passes through the tricuspid valve and enters the chamber of the right ventricle (fig. 13.7). When the right ventricular wall contracts, the tricuspid valve closes, and blood moves through the pulmonary valve and into the pulmonary trunk and its branches (pulmonary arteries).

From the pulmonary arteries, blood enters the capillaries associated with the alveoli (microscopic air sacs) of the lungs. Gas exchanges occur between blood in the capillaries and air in the alveoli. The freshly oxygenated blood, low in carbon dioxide, returns to the heart through the pulmonary veins that lead to the left atrium.

The left atrial wall contracts, and blood moves through the mitral valve and into the chamber of the left ventricle. When the left ventricular wall contracts,

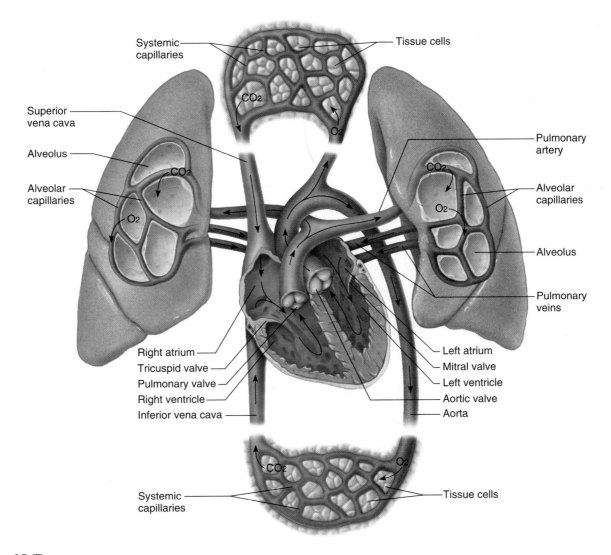

Figure 13.7

The right ventricle forces blood to the lungs, whereas the left ventricle forces blood to all other body parts. (Structures are not drawn to scale.)

the mitral valve closes, and blood moves through the aortic valve and into the aorta and its branches.

Blood Supply to the Heart

The first two branches of the aorta, called the right and left **coronary arteries,** supply blood to the tissues of the heart. Their openings lie just beyond the aortic valve (fig. 13.8).

A thrombus or embolus that blocks or narrows a coronary artery branch deprives myocardial cells of oxygen, producing ischemia and a painful condition called *angina pectoris.* The pain usually occurs during physical activity, when oxygen requirements exceed oxygen supply. Pain lessens with rest. Emotional disturbance may also trigger angina pectoris.

Angina pectoris may cause a sensation of heavy pressure, tightening, or squeezing in the chest. The pain is usually felt behind the sternum or in the anterior portion of the upper thoracic cavity, but may radiate to the neck, jaw, throat, left shoulder, left upper limb, back, or upper abdomen. Other symptoms include profuse perspiration (diaphoresis), difficulty breathing (dyspnea), nausea, or vomiting.

A blood clot completely obstructing a coronary artery or one of its branches (coronary thrombosis) kills part of the heart. This is a *myocardial infarction (MI),* more commonly known as a heart attack.

The heart must beat continually to supply blood to body tissues. To do this, myocardial cells require a constant supply of freshly oxygenated blood. Branches of the coronary arteries feed the many capillaries of the myocardium (fig. 13.9). The smaller branches of these arteries usually have connections (anastomoses) between vessels that provide alternate pathways for blood, called collateral circulation. These detours in circulation may supply oxygen and nutrients to the myocardium when a coronary artery is blocked.

Branches of the **cardiac veins,** whose paths roughly parallel those of the coronary arteries, drain blood that has passed through myocardial capillaries. As figure 13.9*b* shows, these veins join an enlarged vein on the heart's posterior surface—the **coronary sinus**—which empties into the right atrium (see fig. 13.5).

In *heart transplantation,* the recipient's failing heart is removed, except for the posterior walls of the right and left atria and their connections to the venae cavae and pulmonary veins. The donor heart is prepared similarly and is attached to the atrial cuffs remaining in the recipient's thoracic cavity. Finally, the recipient's aorta and pulmonary arteries are connected to those of the donor heart.

Check Your Recall

6. Review the path of blood through the heart.

7. Which vessels supply blood to the myocardium?

8. How does blood return from the cardiac tissues to the right atrium?

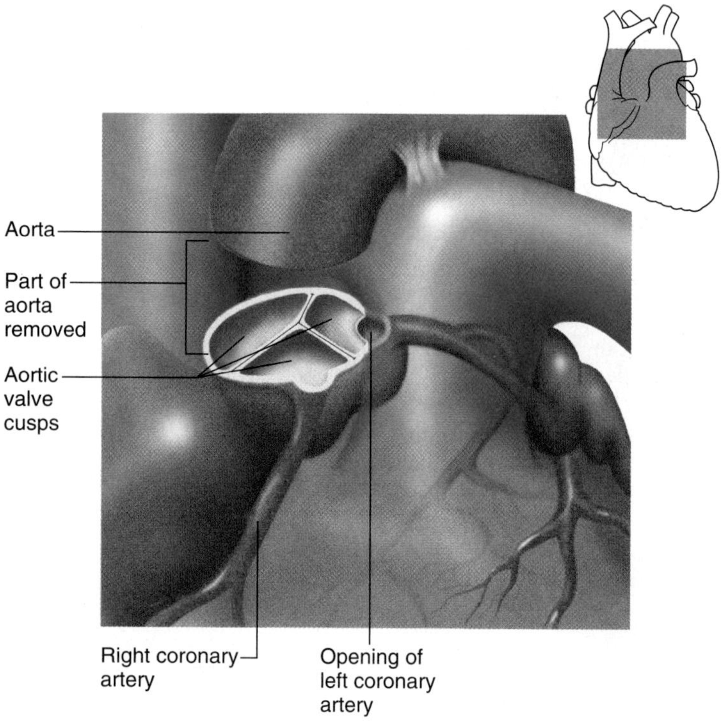

13.3 HEART ACTIONS

The heart chambers function in a coordinated fashion. Their actions are regulated so that atria contract, called atrial **systole** (sis'to-le), while ventricles relax, called ventricular **diastole** (di-as'to-le); then ventricles contract (ventricular systole) while atria relax (atrial diastole). Then the atria and ventricles both relax for a brief interval. This series of events constitutes a complete heartbeat, or **cardiac cycle** (kar'de-ak si'kl).

Cardiac Cycle

During a cardiac cycle, pressure within the heart chambers rises and falls and it is these changes that open and close the valves, much like a door being blown open or closed by the wind. When pressure in the ventricles is low, early in diastole, the pressure difference between the atria and ventricles causes the A-V valves to open and the ventricles to fill. About 70% of the returning blood enters the ventricles prior to contraction. When the atria contract, the remaining 30% of returning blood

Aorta

Part of aorta removed

Aortic valve cusps

Right coronary artery

Opening of left coronary artery

Figure 13.8
The openings of the coronary arteries lie just beyond the aortic valve.

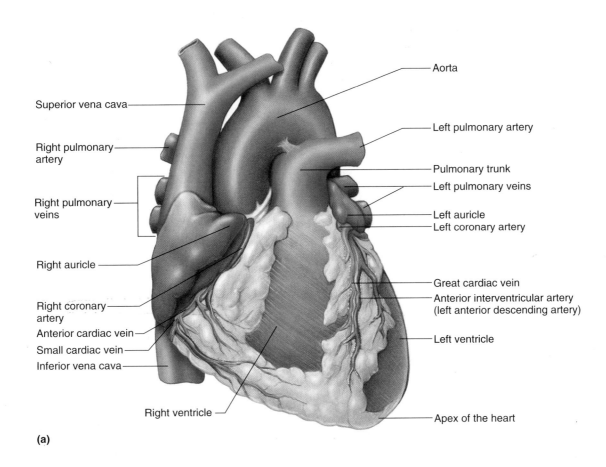

Superior vena cava

Right pulmonary artery

Right pulmonary veins

Right auricle

Right coronary artery

Anterior cardiac vein

Small cardiac vein

Inferior vena cava

Right ventricle

Aorta

Left pulmonary artery

Pulmonary trunk

Left pulmonary veins

Left auricle

Left coronary artery

Great cardiac vein

Anterior interventricular artery (left anterior descending artery)

Left ventricle

Apex of the heart

(a)

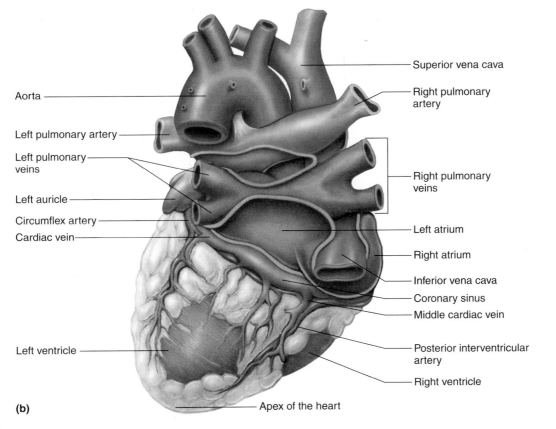

Aorta

Left pulmonary artery

Left pulmonary veins

Left auricle

Circumflex artery

Cardiac vein

Left ventricle

Superior vena cava

Right pulmonary artery

Right pulmonary veins

Left atrium

Right atrium

Inferior vena cava

Coronary sinus

Middle cardiac vein

Posterior interventricular artery

Right ventricle

Apex of the heart

(b)

Figure 13.9

Blood vessels associated with the surface of the heart. (*a*) Anterior view. (*b*) Posterior view.

is pushed into the ventricles (fig. 13.10*a*). Then, as the ventricles contract, ventricular pressure rises sharply, and as soon as the ventricular pressure exceeds the atrial pressure, the A-V valves close. At the same time, the papillary muscles contract, and by pulling on the chordae tendineae, they prevent the cusps of the A-V valves from bulging too far into the atria.

During ventricular contraction, the A-V valves remain closed. The atria are now relaxed, and pressure in the atria is quite low, even lower than venous pressure. As a result, blood flows into the atria from the large, attached veins. That is, as the ventricles are contracting, the atria are filling, already preparing for the next cardiac cycle (fig. 13.10*b*).

When ventricular pressure exceeds the pressure in the pulmonary trunk (right side) and aorta (left side), the pulmonary and aortic valves open; blood is ejected from each valve's respective ventricle into these arteries. As blood flows out of the ventricles, ventricular pressure begins to drop, and it drops even further as the ventricles begin to relax. When ventricular pressure is lower than blood pressure in the aorta and pulmonary trunk, the pressure difference is reversed, and the semilunar valves close. The ventricles continue to relax, and as soon as ventricular pressure is less than atrial pressure, the A-V valves open, and the ventricles begin to fill once more. Atria and ventricles are both relaxed for a brief interval. The graph in Appendix C (p. 569) summarizes some of the changes that occur during a cardiac cycle.

Heart Sounds

A heartbeat heard through a stethoscope sounds like *lubb-dupp*. These sounds are due to vibrations in the heart tissues associated with the closing of the valves.

The first part of a heart sound (*lubb*) occurs during ventricular contraction, when the A-V valves are closing. The second part (*dupp*) occurs during ventricular relaxation, when the pulmonary and aortic valves are closing.

Heart sounds provide information concerning the condition of the heart valves. For example, inflammation of the endocardium (endocarditis) may erode the edges of the valvular cusps. As a result, the cusps may not close completely, and some blood may leak back through the valve, producing an abnormal sound called a *murmur*. The seriousness of a murmur depends on the degree of valvular damage. Many heart murmurs are harmless. For people who have serious problems, open heart surgery may repair or replace damaged valves.

Cardiac Muscle Fibers

Cardiac muscle fibers function much like those of skeletal muscles, but the fibers connect in branching networks. Stimulation to any part of the network sends impulses throughout the heart, which contracts as a unit.

A mass of merging cells that function as a unit is called a **functional syncytium** (funk'shun-al sin-sish'e-um). Two such structures are in the heart—in the atrial walls and in the ventricular walls. Portions of the heart's fibrous skeleton separate these masses of cardiac muscle fibers from each other, except for a small area in the right atrial floor. In this region, the *atrial syncytium* and the *ventricular syncytium* are connected by fibers of the cardiac conduction system.

Check Your Recall

9. Describe the pressure changes in the atria and ventricles during a cardiac cycle.
10. What causes heart sounds?
11. What is a functional syncytium?

Cardiac Conduction System

Throughout the heart are clumps and strands of specialized cardiac muscle tissue whose fibers contain only a few myofibrils. Instead of contracting, these areas ini-

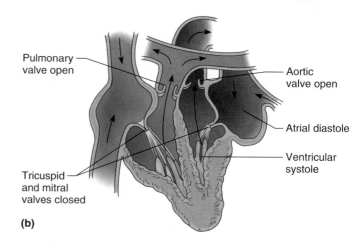

(a)   (b)

Figure 13.10

A cardiac cycle. The atria (*a*) empty during atrial systole and (*b*) fill with blood during atrial diastole.

tiate and distribute impulses throughout the myocardium. They comprise the **cardiac conduction system** (kar′de-ak kon-duk′shun sis′tem), which coordinates the events of the cardiac cycle (fig. 13.11).

A key portion of this conduction system is the **S-A node** or **sinoatrial node,** a small, elongated mass of specialized cardiac muscle tissue just beneath the epicardium. It is located in the right atrium near the opening of the superior vena cava, and its fibers are continuous with those of the atrial syncytium.

The cells of the S-A node can reach threshold on their own, and their membranes contact one another. Without stimulation from nerve fibers or any other outside agents, the nodal cells initiate impulses that spread into the surrounding myocardium and stimulate cardiac muscle fibers to contract.

S-A node activity is rhythmic. The S-A node initiates one impulse after another, seventy to eighty times a minute in an adult. Because it generates the heart's rhythmic contractions, it is often called the **pacemaker.**

Figure 13.12 traces the path of a cardiac impulse. As a cardiac impulse travels from the S-A node into the atrial syncytium, the right and left atria begin to contract almost simultaneously. The cardiac impulse does not pass directly into the ventricular syncytium, which is separated from the atrial syncytium by the fibrous skeleton of the heart. Instead, the impulse passes along fibers (junctional fibers) of the conduction system that lead to a mass of specialized cardiac muscle tissue called the **A-V node** or **atrioventricular node.** This node is

located in the inferior portion of the septum that separates the atria (interatrial septum) and just beneath the endocardium. It provides the only normal conduction pathway between the atrial and ventricular syncytia.

The junctional fibers that conduct the cardiac impulse into the A-V node have very small diameters, and because small fibers conduct impulses slowly, they delay impulse transmission. The impulse is delayed further as it moves through the A-V node, allowing more time for the atria to contract completely so that they empty all their blood into the ventricles prior to ventricular contraction.

Once the cardiac impulse reaches the distal side of the A-V node, it passes into a group of large fibers that make up the **A-V bundle** (bundle of His). The A-V bundle enters the upper part of the interventricular septum and divides into right and left bundle branches that lie just beneath the endocardium. About halfway down the septum, the branches give rise to enlarged **Purkinje fibers** (pur-kin′je fi′berz).

Purkinje fibers spread from the interventricular septum into the papillary muscles, which project inward from ventricular walls and then continue downward to the apex of the heart. There they curve around the tips of the ventricles and pass upward over the lateral walls of these chambers. Along the way, the Purkinje fibers give off many small branches, which become continuous with cardiac muscle fibers.

The muscle fibers in ventricular walls form irregular whorls. When impulses on the Purkinje fibers stimulate

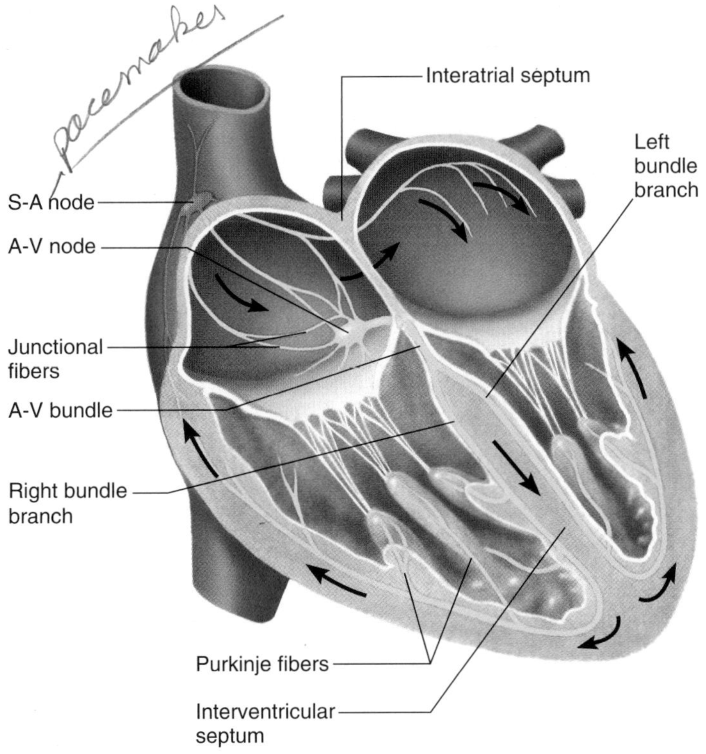

pacemaker

Interatrial septum

Left bundle branch

S-A node

A-V node

Junctional fibers

A-V bundle

Right bundle branch

Purkinje fibers

Interventricular septum

Figure 13.11
The cardiac conduction system coordinates the cardiac cycle.

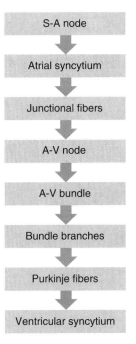

S-A node

Atrial syncytium

Junctional fibers

A-V node

A-V bundle

Bundle branches

Purkinje fibers

Ventricular syncytium

Figure 13.12
Components of the cardiac conduction system.

these muscle fibers, the ventricular walls contract with a twisting motion (fig. 13.13). This action squeezes blood out of the ventricular chambers and forces it into the aorta and pulmonary trunk.

Check Your Recall

12. What kinds of tissues make up the cardiac conduction system?

13. How is a cardiac impulse initiated?

14. How is a cardiac impulse transmitted from the right atrium to the other heart chambers?

Electrocardiogram

An **electrocardiogram** (e-lek″tro-kar′de-o-gram″), or **ECG,** is a recording of the electrical changes that occur in the myocardium during a cardiac cycle. (This pattern occurs as action potentials stimulate cardiac muscle fibers to contract, but it is not the same as individual action potentials.) Because body fluids can conduct electrical currents, such changes can be detected on the surface of the body.

To record an ECG, electrodes are placed on the skin and connected by wires to an instrument that responds to very weak electrical changes by moving a pen or stylus on a moving strip of paper. Up-and-down movements of the pen correspond to electrical changes in the myocardium. Because the paper moves past the pen at a known rate, the distance between pen deflections indicates the time between phases of the cardiac cycle.

As figure 13.14*a* illustrates, a normal ECG pattern includes several deflections, or *waves,* during each cardiac cycle. Between cycles, the muscle fibers remain polarized, with no detectable electrical changes, and the pen does not move but simply marks along the baseline. When the S-A node triggers a cardiac impulse, atrial fibers depolarize, producing an electrical change. The pen moves, and at the end of the electrical change, returns to the base position. This first pen movement produces a *P wave,* corresponding to depolarization of the atrial fibers that will lead to contraction of the atria (fig. 13.14*b*).

When the cardiac impulse reaches ventricular fibers, they rapidly depolarize. Because ventricular walls are thicker than those of the atria, the electrical change is greater, and the pen deflects more. When the electrical change ends, the pen returns to the baseline, leaving a mark called the *QRS complex.* This mark consists of a *Q wave,* an *R wave,* and an *S wave,* and corresponds to depolarization of ventricular fibers just prior to the contraction of the ventricular walls.

The electrical changes occurring as the ventricular muscle fibers repolarize slowly produce a *T wave* as the pen deflects again, ending the ECG pattern. The record of atrial repolarization seems to be missing from the pattern because atrial fibers repolarize at the same time that ventricular fibers depolarize. Thus, the QRS complex obscures the recording of atrial repolarization.

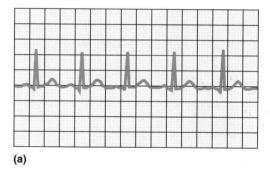

(a)

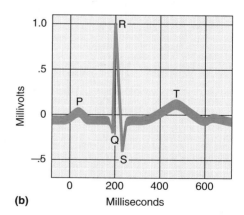

(b)

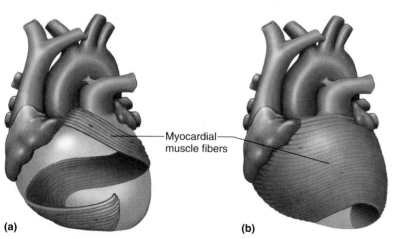

(a)   (b)

Figure 13.13

The muscle fibers within the ventricular walls form patterns of whorls. The fibers of groups (*a*) and (*b*) surround both ventricles in these anterior views of the heart.

—Myocardial muscle fibers

Figure 13.14

An electrocardiogram records electrical changes in the myocardium during a cardiac cycle. (*a*) A normal ECG. (*b*) In an ECG pattern, the P wave results from a depolarization of the atria, the QRS complex results from a depolarization of the ventricles, and the T wave results from a repolarization of the ventricles.

Physicians use ECG patterns to assess the heart's ability to conduct impulses. For example, the time period between the beginning of a P wave and the beginning of a QRS complex, called the *P-Q interval* (or if the initial portion of the QRS complex is upright, the P-R interval), indicates the time for the cardiac impulse to travel from the S-A node through the A-V node. Ischemia or other problems affecting the fibers of the A-V conduction pathways can increase this P-Q interval. Similarly, injury to the A-V bundle can extend the QRS complex, because it may take longer for an impulse to spread throughout the ventricular walls (fig. 13.15).

Check Your Recall

15. What is an electrocardiogram?

16. Which cardiac events do the P wave, QRS complex, and T wave represent?

Regulation of the Cardiac Cycle

The volume of blood pumped changes to accommodate cellular requirements. For example, during strenuous exercise, skeletal muscles require more blood, and the heart rate increases in response. Since the S-A node normally controls heart rate, changes in this rate are often a response to factors that affect the S-A node, such as the motor impulses carried on the parasympathetic and sympathetic nerve fibers (see chapter 9, p. 249).

The parasympathetic fibers that innervate the heart arise from neurons in the medulla oblongata (fig. 13.16). Most of these fibers branch to the S-A and A-V nodes. When the nerve impulses reach nerve fiber endings, they secrete acetylcholine, which decreases S-A and A-V nodal activity. As a result, the heart rate decreases.

Parasympathetic fibers carry impulses continually to the S-A and A-V nodes, "braking" heart action. Conse-quently, parasympathetic activity can change heart rate in either direction. An increase in the impulses slows the heart rate, and a decrease in the impulses releases the parasympathetic "brake" and increases the heart rate.

Sympathetic fibers reach the heart and join the S-A and A-V nodes as well as other areas of the atrial and ventricular myocardium. The endings of these fibers secrete norepinephrine in response to nerve impulses, which increases the rate and force of myocardial contractions.

Reflexes called *baroreceptor reflexes* involving the *cardiac control center* of the medulla oblongata maintain balance between the inhibitory effects of parasympathetic fibers and the excitatory effects of sympathetic fibers. This center receives sensory impulses from throughout the cardiovascular system and relays motor impulses to the heart in response. For example, receptors sensitive to stretch are located in certain regions of the aorta (aortic arch) and in the carotid arteries (carotid sinuses) (fig. 13.16). These receptors, called *baroreceptors* (pressoreceptors), can detect changes in blood pressure. Rising pressure stretches the receptors, and they signal the cardioinhibitor center in the medulla oblongata. In response, the medulla oblongata sends parasympathetic impulses to the heart, decreasing heart rate. This action helps lower blood pressure toward normal.

Impulses from the cerebrum or hypothalamus also influence the cardiac control center. Such impulses may decrease heart rate, as occurs when a person faints following an emotional upset, or they may increase heart rate during a period of anxiety.

Two other factors that influence heart rate are temperature change and certain ions. Rising body temperature increases heart action, which is why heart rate usually increases during fever. On the other hand, abnormally low body temperature decreases heart action.

Of the ions that influence heart action, the most important are potassium (K^+) and calcium (Ca^{+2}) ions. In hyperkalemia, excess extracellular potassium ions decrease the rate and force of contractions. In hypokalemia, deficient extracellular potassium ions may cause a potentially life-threatening abnormal heart rhythm (arrhythmia).

In hypercalcemia, excess extracellular calcium ions increase heart actions, posing the danger that the heart will contract for an abnormally long time. Conversely, in hypocalcemia, low extracellular calcium concentration depresses heart action.

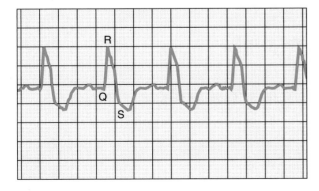

Figure 13.15
A prolonged QRS complex may result from damage to the A-V bundle fibers.

Check Your Recall

17. How do parasympathetic and sympathetic impulses help control heart rate?

18. How do changes in body temperature affect heart rate?

19. Describe the effects on the heart of abnormal concentrations of potassium and calcium ions.

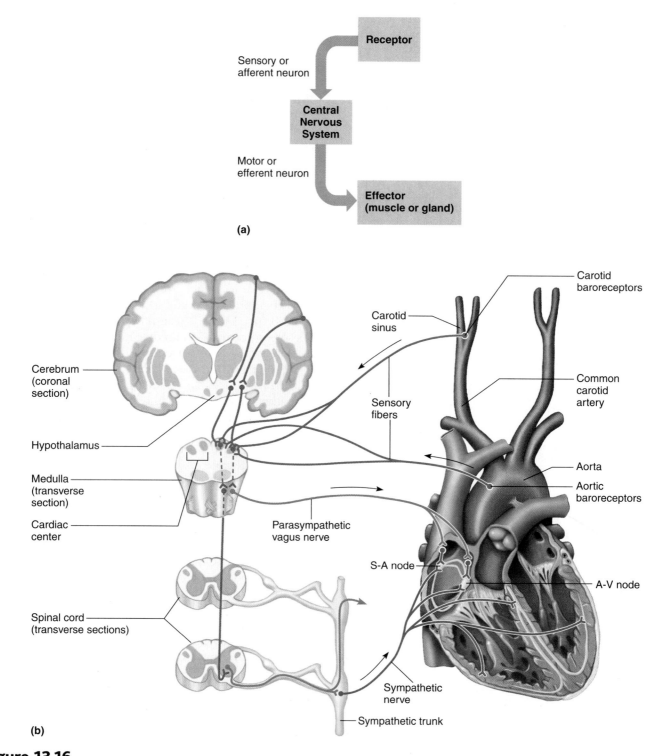

Figure 13.16

Baroreceptor reflex. (*a*) Schematic of a general reflex arc. (*b*) Autonomic nerve impulses alter the activities of the S-A and A-V nodes.

13.4 BLOOD VESSELS

The blood vessels form a closed circuit of tubes that carries blood from the heart to the body cells and back again. These vessels include arteries, arterioles, capillaries, venules, and veins.

There are about 62,000 miles of blood vessels in the human body—enough to stretch 2 1/2 times around the world.

Arteries and Arterioles

Arteries are strong, elastic vessels that are adapted for carrying blood away from the heart under high pressure. These vessels subdivide into progressively thinner tubes and eventually give rise to finer, branched **arterioles** (ar-te′re-olz).

The wall of an artery consists of three distinct layers (fig. 13.17*a*). The innermost layer (*tunica interna*) is composed of a layer of simple squamous epithelium, called *endothelium,* that rests on a connective tissue membrane that is rich in elastic and collagenous fibers. Endothelium helps prevent blood clotting by providing a smooth surface that allows blood cells and platelets to flow through without being damaged and by secreting biochemicals that inhibit platelet aggregation. Endothelium also may help regulate local blood flow by secreting substances that dilate or constrict blood vessels. For example, endothelium releases the gas nitric oxide, which relaxes the smooth muscle of the vessel. The Topic of Interest on page 354 describes atherosclerosis, in which fatty deposits accumulate on the inner walls of arteries.

The middle layer (*tunica media*) makes up the bulk of the arterial wall. It includes smooth muscle fibers, which encircle the tube, and a thick layer of elastic connective tissue.

The outer layer (*tunica externa*) is relatively thin and chiefly consists of connective tissue with irregular elastic and collagenous fibers. This layer attaches the artery to the surrounding tissues.

If the wall of an artery weakens so that blood pressure dilates a region of it, a pulsating sac called an *aneurysm* may form. Aneurysms tend to grow. If the resulting sac develops by a longitudinal splitting of the middle layer of the arterial wall, it is called a *dissecting aneurysm.* An aneurysm may cause symptoms by pressing on nearby organs, or it may rupture and produce a great loss of blood.

Aneurysms may also result from trauma, high blood pressure, infections, inherited disorders such as Marfan syndrome, or congenital defects in blood vessels. Common sites of aneurysms include the thoracic and abdominal aorta and an arterial circle at the base of the brain (circle of Willis).

The sympathetic branches of the autonomic nervous system innervate smooth muscle in artery and arteriole walls. Impulses on these *vasomotor fibers* stimulate the smooth muscles to contract, reducing the diameter of the vessel. This is called **vasoconstriction** (vas″o-kon-strik′shun). If vasomotor impulses are inhibited, the muscle fibers relax, and the diameter of the

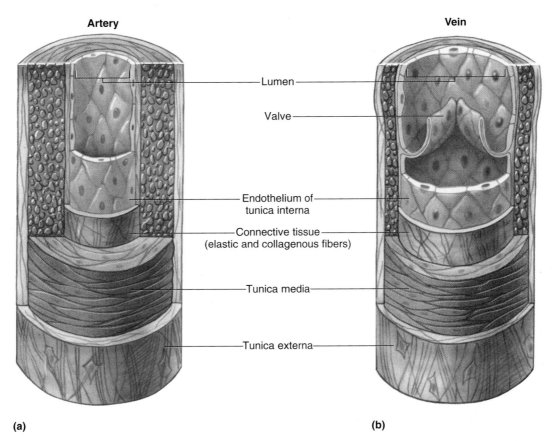

Artery   **Vein**

Lumen

Valve

Endothelium of
tunica interna

Connective tissue
(elastic and collagenous fibers)

Tunica media

Tunica externa

(a)   (b)

Figure 13.17

Blood vessels. (*a*) The wall of an artery. (*b*) The wall of a vein.

Topic of Interest

Atherosclerosis

In the arterial disease *atherosclerosis* (ath"er-o-sklĕ-ro'sis), deposits of fatty materials, particularly cholesterol, form within and on the inner lining of the arterial walls. Such deposits, called *plaque*, protrude into the lumens of vessels and interfere with blood flow (fig. 13A). Furthermore, plaque often forms a surface texture that can initiate formation of a blood clot, increasing the risk of developing thrombi or emboli that cause blood deficiency (ischemia) or tissue death (necrosis) downstream from the obstruction. "Atherosclerosis" means "paste-hardness" in Greek.

The walls of plaque-laden arteries may degenerate, losing their elasticity and becoming hardened, or *sclerotic*. In this stage of the disease, called *arteriosclerosis*, a sclerotic vessel may rupture under the force of blood pressure.

Risk factors for developing atherosclerosis include a fatty diet, elevated blood pressure, tobacco smoking, obesity, and lack of physical exercise. Genetic factors may also increase susceptibility to atherosclerosis.

For millions of people who cannot control cholesterol with diet and exercise, drugs called statins can inhibit a liver enzyme that the body uses to synthesize cholesterol. Statins dramatically reduce LDL cholesterol, moderately increase HDL cholesterol, and regulate triglyceride levels, all of which lower the risk of atherosclerosis.

Several invasive treatments attempt to clear clogged arteries. In *percutaneous transluminal angioplasty,* a thin, plastic catheter with a tiny deflated balloon at its tip is passed through a tiny incision in the skin and into the lumen of the affected blood vessel. Once in position at the blockage, the balloon is inflated for several minutes, and presses the plaque against the arterial wall, widening the arterial lumen and restoring blood flow. Sometimes the catheter also introduces a stent, which is a coiled steel tube that opens as the balloon inflates, helping to keep the lumen clear. However, blockage can recur if the underlying cause of cholesterol buildup is not addressed.

Laser energy is also used to destroy atherosclerotic plaque and to channel through arterial obstructions to increase blood flow. In *laser angioplasty,* the light energy of a laser is transmitted through a bundle of optical fibers passed through a small incision in the skin and into the lumen of an obstructed artery.

Another invasive procedure for treating arterial obstruction is *bypass graft surgery.* A surgeon uses a portion of a vein from the patient's lower limb or elsewhere to connect a healthy artery to the affected artery at a point beyond the obstruction. Blood from the healthy artery then bypasses the narrowed region of the affected artery, supplying the tissues downstream. The vein is connected backward, so that its valves do not impede blood flow.

A new treatment for atherosclerosis is *fibroblast growth factor,* a body chemical given as a drug. It stimulates new blood vessels to grow in the heart, a process called angiogenesis.

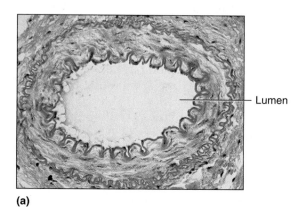

(a)

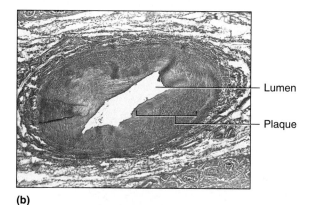

(b)

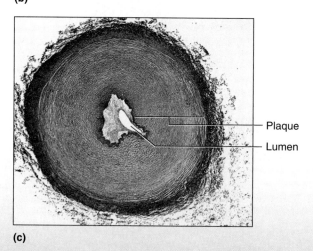

(c)

Figure 13A

Development of atherosclerosis. (*a*) Normal arteriole. (*b*) and (*c*) Accumulation of plaque on the inner wall of the arteriole. Plaque includes cholesterol and other lipids, calcium, and cellular debris.

vessel increases. This is called **vasodilation** (vas″o-di-la′shun). Changes in the diameters of arteries and arterioles greatly influence blood flow and blood pressure.

The walls of the larger arterioles have three layers, similar to those of arteries. These walls thin as arterioles approach capillaries. The wall of a very small arteriole consists only of an endothelial lining and some smooth muscle fibers, surrounded by a small amount of connective tissue (fig. 13.18).

Check Your Recall

20. Describe the wall of an artery.
21. What is the function of smooth muscle in the arterial wall?
22. How is the structure of an arteriole different from that of an artery?

Capillaries

Capillaries (kap′ĭ-lar″ēz), the smallest-diameter blood vessels, connect the smallest arterioles and the smallest venules. Capillaries are extensions of the inner linings of arterioles in that their walls are composed of endothelium (fig. 13.18). These thin walls form the semipermeable layer through which substances in the blood are exchanged for substances in the tissue fluid surrounding body cells.

The openings in capillary walls are thin slits where endothelial cells overlap (fig. 13.19). The sizes of these openings and, consequently, the permeability of the capillary wall vary from tissue to tissue. For example, the openings are smaller in capillaries of smooth, skeletal, and cardiac muscle than they are in capillaries associated with endocrine glands, the kidneys, and the lining of the small intestine.

Capillary density reflects tissues' rates of metabolism. Muscle and nerve tissues, which use abundant oxygen and nutrients, are richly supplied with capillaries. Tissues with slow metabolic rates, such as cartilage, the epidermis, and the cornea, lack capillaries.

The patterns of capillaries also differ in various body parts. For example, some capillaries pass directly from arterioles to venules, but others lead to highly branched networks (fig. 13.20).

Smooth muscles that encircle capillary entrances regulate blood distribution in capillary pathways. These

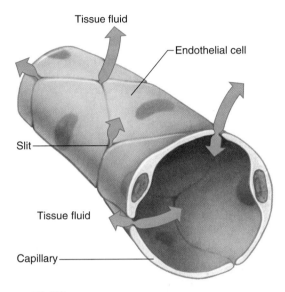

Figure 13.19

In capillaries, substances are exchanged between the blood and tissue fluid through openings (slits) separating endothelial cells.

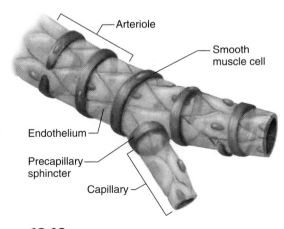

Figure 13.18

The smallest arterioles have only a few smooth muscle fibers in their walls. Capillaries lack these fibers.

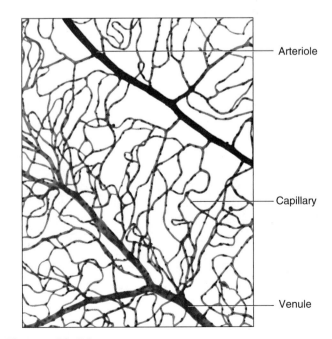

Figure 13.20

Light micrograph of a capillary network (100×).

muscles form *precapillary sphincters,* which may close a capillary by contracting or open it by relaxing (see . g. 13.18). A precapillary sphincter responds to the demands of the cells the capillary supplies. When these cells have low concentrations of oxygen and nutrients, the sphincter relaxes; when cellular requirements have been met, the sphincter may contract again. In this way, blood flow can follow different pathways through a tissue to meet the changing cellular requirements.

Routing of blood flow to different parts of the body is due to vasoconstriction and vasodilation of arterioles and precapillary sphincters. During exercise, for example, blood enters the capillary networks of the skeletal muscles, where the cells have increased oxygen and nutrient requirements. At the same time, blood can bypass some of the capillary networks in the digestive tract tissues, where demand for blood is less immediate.

Check Your Recall

23. Describe a capillary wall.
24. What is the function of a capillary?
25. What controls blood flow into capillaries?

Exchanges in Capillaries

Gases, nutrients, and metabolic by-products are exchanged between the blood in capillaries and the tissue fluid surrounding body cells. The substances exchanged move through capillary walls by diffusion, filtration, and osmosis (see chapter 3, pp. 61–64).

Because blood entering systemic capillaries carries high concentrations of oxygen and nutrients, these substances diffuse through the capillary walls and enter the tissue fluid. Conversely, the concentrations of carbon dioxide and other wastes are generally greater in the tissues, and such wastes diffuse into the capillary blood.

Plasma proteins generally remain in the blood because they are too large to diffuse through the membrane pores or slitlike openings between the endothelial cells of most capillaries. Also, these bulky proteins are not soluble in the lipid portions of capillary cell membranes.

Whereas diffusion depends on concentration gradients, filtration forces molecules through a membrane with hydrostatic pressure. In capillaries, the blood pressure generated when ventricle walls contract provides the force for filtration.

Blood pressure also moves blood through the arteries and arterioles. This pressure decreases as the distance from the heart increases, because of friction (peripheral resistance) between the blood and the vessel walls. For this reason, blood pressure is greater in the arteries than in the arterioles, and greater in the arterioles than in the capillaries. Blood pressure is similarly greater at the arteriolar end of a capillary than at the venular end. Therefore, the filtration effect occurs primarily at the arteriolar ends of capillaries.

The presence of an impermeant solute on one side of a cell membrane creates an osmotic pressure. Because plasma proteins are trapped within the capillaries, they create an osmotic pressure that draws water into the capillaries. The term *colloid osmotic pressure* is often used to describe this osmotic effect due solely to the plasma proteins.

The effect of capillary blood pressure, which favors filtration, opposes the actions of the plasma colloid osmotic pressure, which favors reabsorption. At the arteriolar end of capillaries, the blood pressure is higher than the colloid osmotic pressure, so filtration predominates. At the venular end, the colloid osmotic pressure is essentially unchanged, but the blood pressure has decreased due to resistance through the capillary, so reabsorption predominates (fig. 13.21).

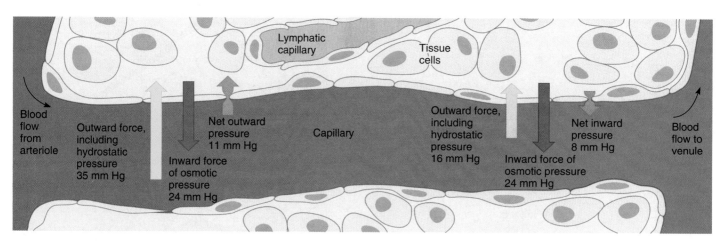

Figure 13.21

Water and other substances leave capillaries because of a net outward pressure at the capillaries' arteriolar ends. Water enters at the capillaries' venular ends because of a net inward pressure. Substances move in and out along the length of the capillaries according to their respective concentration gradients.

Normally, more fluid leaves the capillaries than returns to them. Closed-ended vessels called lymphatic capillaries collect the excess fluid and return it through lymphatic vessels to the venous circulation. Chapter 14 (pp. 378–379) discusses this mechanism.

Sometimes unusual events may increase blood flow to capillaries, and excess fluid enters the spaces between tissue cells. This may occur in response to certain chemicals, such as histamine, that vasodilate the arterioles near capillaries and increase capillary permeability. Enough fluid may leak out of the capillaries to overwhelm lymphatic drainage. Affected tissues become swollen (edematous) and painful.

Check Your Recall

26. What forces affect the exchange of substances between blood and tissue fluid?

27. Why is the fluid movement out of a capillary greater at its arteriolar end than at its venular end?

Venules and Veins

Venules (ven'ūlz) are the microscopic vessels that continue from the capillaries and merge to form **veins** (vānz). The veins, which carry blood back to the atria, follow pathways that roughly parallel those of the arteries.

The walls of veins are similar to those of arteries in that they are composed of three distinct layers (see fig. 13.17*b*). However, the middle layer of the venous wall is poorly developed. Consequently, veins have thinner walls that have less smooth muscle and less elastic connective tissue than those of comparable arteries, but their lumens have a greater diameter (fig. 13.22).

Many veins, particularly those in the upper and lower limbs, have flaplike *valves,* which project inward from their linings. Valves are usually composed of two leaflets that close if blood begins to back up in a vein (fig. 13.23). These valves aid in returning blood to the heart because they open if blood flow is toward the heart, but close if it is in the opposite direction.

Veins also function as blood reservoirs. For example, in hemorrhage accompanied by a drop in arterial blood pressure, sympathetic nerve impulses reflexly stimulate the muscular walls of the veins. The resulting venous constrictions help maintain blood pressure by returning more blood to the heart. This mechanism ensures a nearly normal blood flow even when as much as 25% of blood volume is lost. Table 13.2 summarizes the characteristics of blood vessels.

Check Your Recall

28. How does the structure of a vein differ from that of an artery?

29. How does venous circulation help maintain blood pressure when hemorrhaging causes blood loss?

13.5 BLOOD PRESSURE

Blood pressure is the force blood exerts against the inner walls of blood vessels. Although this force occurs throughout the vascular system, the term *blood pressure* most commonly refers to pressure in arteries supplied by branches of the aorta.

Contraction of the human heart creates enough pressure to squirt blood 30 feet.

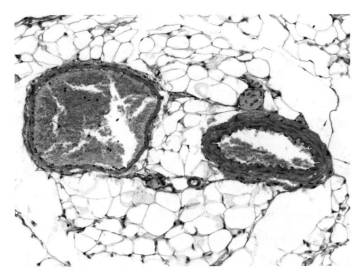

Figure 13.22
Note the structural differences in these cross sections of a vein (left) and an artery (right) (60×).

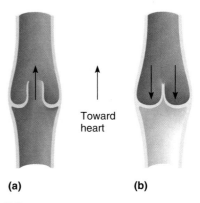

Figure 13.23
Venous valves (*a*) allow blood to move toward the heart, but (*b*) prevent blood from moving backward away from the heart.

Table 13.2	Characteristics of Blood Vessels	
Vessel	Type of Wall	Function
Artery	Thick, strong wall with three layers—an endothelial lining, a middle layer of smooth muscle and elastic connective tissue, and an outer layer of connective tissue	Carries blood under relatively high pressure from heart to arterioles
Arteriole	Thinner wall than an artery but with three layers; smaller arterioles have an endothelial lining, some smooth muscle tissue, and a small amount of connective tissue	Connects an artery to a capillary; helps control blood flow into a capillary by vasoconstricting or vasodilating
Capillary	Single layer of squamous epithelium	Provides a membrane through which nutrients, gases, and wastes are exchanged between the blood and tissue fluid; connects an arteriole to a venule
Venule	Thinner wall than in an arteriole, less smooth muscle and elastic connective tissue	Connects a capillary to a vein
Vein	Thinner wall than an artery but with similar layers; the middle layer is more poorly developed; some have flaplike valves	Carries blood under relatively low pressure from a venule to the heart; valves prevent backflow of blood; serves as a blood reservoir

Arterial Blood Pressure

Arterial blood pressure rises and falls in a pattern corresponding to the phases of the cardiac cycle. That is, contracting ventricles (ventricular systole) squeeze blood out and into the pulmonary trunk and aorta, which sharply increases the pressures in these arteries. The maximum pressure during ventricular contraction is called the **systolic pressure** (sis-tol′ik presh′ur). When the ventricles relax (ventricular diastole), the arterial pressure drops, and the lowest pressure that remains in the arteries before the next ventricular contraction is termed the **diastolic pressure** (di-a-stol′ik presh′ur). A sphygmonanometer is used to measure arterial blood pressure. The results of this blood pressure measurement are reported as a fraction, normally about 120/80. In this notation, the upper number indicates the arterial systolic pressure in mm Hg (SP), and the lower number indicates the arterial diastolic pressure in mm Hg (DP). Figure 13.24 shows how these pressures decrease as distance from the left ventricle increases.

The surge of blood entering the arterial system during a ventricular contraction distends the elastic arterial walls, but the pressure drops almost immediately as the contraction ends, and the arterial walls recoil. This alternate expanding and recoiling of the arterial wall can be felt as a *pulse* in an artery that runs close to the surface. The radial artery is commonly used to take a person's pulse. Other sites where an arterial pulse is easily detected include the carotid, brachial, and femoral arteries.

Check Your Recall

30. What is blood pressure?
31. Distinguish between systolic and diastolic blood pressure.
32. What causes a pulse in an artery?

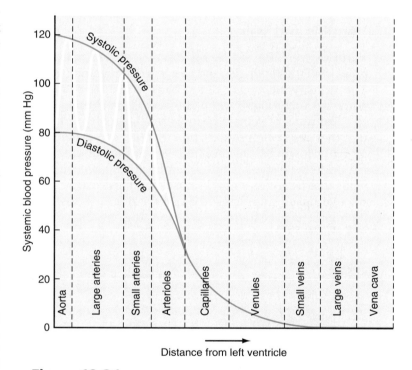

Figure 13.24

Blood pressure decreases as the distance from the left ventricle increases. Systolic pressure occurs during maximal ventricular contraction. Diastolic pressure occurs when the ventricles relax.

Factors That Influence Arterial Blood Pressure

Arterial blood pressure depends on a variety of factors. These include heart action which includes heart rate and stroke volume, blood volume, peripheral resistance, and blood viscosity (fig. 13.25).

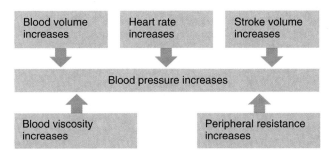

Figure 13.25
Some of the factors that influence arterial blood pressure.

Heart Action

In addition to producing blood pressure by forcing blood into the arteries, heart action determines how much blood enters the arterial system with each ventricular contraction. The volume of blood discharged from the ventricle with each contraction is called the **stroke volume** and equals about 70 milliliters in an average-weight male at rest. The volume discharged from the ventricle per minute is called the **cardiac output,** calculated by multiplying the stroke volume by the heart rate in beats per minute (cardiac output = stroke volume × heart rate). Thus, if the stroke volume is 70 milliliters and the heart rate is 72 beats per minute, the cardiac output is 5,040 milliliters per minute.

Blood pressure varies with cardiac output. If either stroke volume or heart rate increases, so does cardiac output, and as a result, blood pressure initially rises. Conversely, if stroke volume or heart rate decreases, cardiac output decreases, and blood pressure also initially decreases.

Blood Volume

Blood volume equals the sum of the formed elements and plasma volumes in the vascular system. Although the blood volume varies somewhat with age, body size, and sex, it is usually about 5 liters for adults, or 8% of body weight in kilograms.

Blood pressure is normally directly proportional to blood volume within the cardiovascular system. Thus, any changes in blood volume can initially alter blood pressure. For example, if a hemorrhage reduces blood volume, blood pressure initially drops. If a transfusion restores normal blood volume, normal blood pressure may be reestablished. Blood volume can also fall if the fluid balance is upset, as happens in dehydration. Fluid replacement can reestablish normal blood volume and pressure.

> Black licorice contains a compound called glycyrrhizic acid that can raise blood pressure if a person eats more than 3 ounces of black licorice a day for several months. Too much of the acid raises blood pressure by interfering with sodium and water balance.

Peripheral Resistance

Friction between the blood and the walls of blood vessels produces a force called **peripheral resistance** (pĕ-rif′er-al re-zis′tans), which hinders blood flow. Blood pressure must overcome this force if the blood is to continue flowing. Therefore, factors that alter the peripheral resistance change blood pressure. For example, contracting smooth muscles in arteriolar walls increase the peripheral resistance by constricting these vessels. Blood backs up into the arteries supplying the arterioles, and the arterial pressure rises. Dilation of arterioles has the opposite effect—peripheral resistance lessens, and arterial blood pressure drops in response.

Blood Viscosity

Viscosity (vis-kos′ĭ-te) is the ease with which a fluid's molecules flow past one another. The greater the viscosity, the greater the resistance to flowing.

Blood cells and plasma proteins increase blood viscosity. The greater the blood's resistance to flowing, the greater is the force needed to move it through the vascular system. Thus, it is not surprising that blood pressure rises as blood viscosity increases and drops as viscosity decreases.

> **Check Your Recall**
> 33. How are cardiac output and blood pressure related?
> 34. How does blood volume affect blood pressure?
> 35. What is the relationship between peripheral resistance and blood pressure? Between blood viscosity and blood pressure?

Control of Blood Pressure

Blood pressure (BP) is determined by cardiac output (CO) and peripheral resistance (PR) according to this relationship:

$$BP = CO \times PR$$

Maintenance of normal arterial pressure therefore requires regulation of these two factors. For example, cardiac output, depending on the volume of blood discharged from the ventricle (stroke volume), is affected by the blood volume entering the ventricle. Entering blood mechanically stretches myocardial fibers in the ventricular wall. Within limits, the longer these fibers, the greater is the force with which they contract.

The relationship between fiber length (due to stretching of the cardiac muscle cell just before contraction) and force of contraction is called the *Frank-Starling law of the heart.* This becomes important, for example, during exercise when much more blood returns to the heart from the veins. The more blood that enters the heart from the veins, the greater the ventricular distension, the stronger the contraction, the greater the stroke volume, and the greater the cardiac output. Conversely, the less blood that returns from the veins, the less the ventricle distends, the weaker the ventricular contraction, the lesser the stroke volume and cardiac output. This mechanism ensures that the volume of blood discharged from the heart is equal to the volume entering its chambers.

Baroreceptors in the walls of the aorta and carotid arteries sense changes in blood pressure. If arterial pressure increases, nerve impulses travel from the baroreceptors to the cardiac center of the medulla oblongata. This center relays parasympathetic impulses to the S-A node in the heart, and the heart rate decreases in response. As a result of this *cardioinhibitor reflex,* cardiac output falls, and blood pressure decreases toward the normal level (fig. 13.26). Conversely, decreasing arterial blood pressure initiates the *cardioaccelerator reflex,* which sends sympathetic impulses to the S-A node. As a result, the heart beats faster, increasing cardiac output and arterial pressure. Other factors that increase heart rate and blood pressure include exercise, a rise in body temperature, and emotional responses, such as fear and anger. The Topic of Interest on page 361 discusses the effects of exercise on cardiovascular functioning.

Peripheral resistance also controls blood pressure. Changes in arteriole diameters regulate peripheral resistance. Because blood vessels with smaller diameters offer a greater resistance to blood flow, factors that cause arteriole vasoconstriction increase peripheral resistance increasing blood pressure, and factors causing vasodilation decrease resistance decreasing blood pressure.

The *vasomotor center* of the medulla oblongata continually sends sympathetic impulses to smooth muscles in the arteriole walls, keeping them in a state of tonic contraction, which helps maintain the peripheral resistance associated with normal blood pressure. Because the vasomotor center responds to changes in blood pressure, it can increase peripheral resistance by increasing its outflow of sympathetic impulses, or it can decrease such resistance by decreasing its sympathetic outflow. In the latter case, the vessels vasodilate as sympathetic stimulation decreases.

Whenever arterial blood pressure suddenly increases, baroreceptors in the aorta and carotid arteries signal the vasomotor center, and the sympathetic outflow to the arterioles falls. The resulting vasodilation decreases peripheral resistance, and blood pressure decreases toward the normal level.

Certain chemicals, including carbon dioxide, oxygen, and hydrogen ions, also influence peripheral resistance by affecting precapillary sphincters and smooth muscle in arteriole walls. For example, increasing blood carbon dioxide, decreasing blood oxygen, and lowering blood pH relaxes smooth muscle in the systemic circulation. This increases local blood flow to tissues with high metabolic rates, such as exercising skeletal muscles. In addition, epinephrine and norepinephrine vasoconstrict many systemic vessels, increasing peripheral resistance.

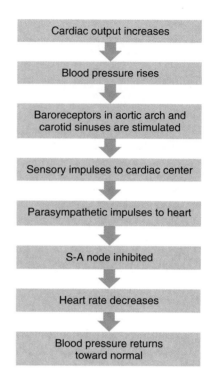

Figure 13.26

If blood pressure rises, baroreceptors initiate the cardioinhibitor reflex, which lowers the blood pressure.

Hypertension, or high blood pressure, is persistently elevated arterial pressure. It is one of the more common diseases of the cardiovascular system in industrialized nations.

High blood pressure with unknown cause is called *essential* (also primary or idiopathic) *hypertension.* Elevated pressure can be secondary, caused by another problem, such as kidney disease, high sodium intake, obesity, psychological stress, and arteriosclerosis. In arteriosclerosis, decreased elasticity of arterial walls and narrowed arterial lumens increase blood pressure.

The consequences of prolonged, uncontrolled hypertension can be very serious. As the left ventricle works harder to pump sufficient blood, the myocardium thickens, enlarging the heart. If coronary blood vessels cannot support this overgrowth, parts of the heart muscle die and are replaced with fibrous tissue. Eventually, the enlarged and weakened heart dies.

Hypertension also contributes to the development of atherosclerosis. Plaque accumulation in arteries may cause a *coronary thrombosis* or *coronary embolism.* Similar changes in brain arteries increase the chances of a *cerebral vascular accident (CVA),* or *stroke,* due to a cerebral thrombosis, embolism, or hemorrhage.

Treatment of hypertension varies. Exercising regularly, controlling body weight, reducing stress, and limiting sodium in the diet may be sufficient to control blood pressure. If not, drug treatment includes diuretics and/or inhibitors of sympathetic nerve activity.

Exercise and the Cardiovascular System

The cardiovascular system adapts to exercise. The conditioned athlete experiences increases in heart pumping efficiency, blood volume, blood hemoglobin concentration, and the number of mitochondria in muscle fibers. All of these adaptations improve oxygen delivery to, and utilization by, muscle tissue.

An athlete's heart typically changes in response to these increased demands, and may enlarge 40% or more. Myocardial mass increases, ventricular cavities expand, and the ventricle walls thicken. Stroke volume increases, and heart rate decreases, as does blood pressure. The lowest heart rate recorded in an athlete was 25 beats per minute! To a physician unfamiliar with a conditioned cardiovascular system, a trained athlete may appear abnormal.

The cardiovascular system responds beautifully to a slow, steady buildup in exercise frequency and intensity. It does not react well to sudden demands—as when a person who never exercises suddenly shovels snow or runs 3 miles. Although sedentary people have a two- to sixfold increased risk of cardiac arrest while exercising than when not, people in shape have little or no excess risk while exercising.

For exercise to benefit the cardiovascular system, the heart rate must be elevated to 70–85% of its "theoretical maximum" for at least half an hour three times a week. You can calculate your theoretical maximum by subtracting your age from 220. If you are eighteen years old, your theoretical maximum is 202 beats per minute. Then, 70–85% of this value is 141–172 beats per minute. Some good activities for raising the heart rate are tennis, skating, skiing, handball, vigorous dancing, hockey, basketball, biking, and fast walking.

It is wise to consult a physician before starting an exercise program. People over age thirty are advised to have a stress test, which is an electrocardiogram taken while exercising. (The standard electrocardiogram is taken at rest.) An arrhythmia that appears only during exercise may indicate heart disease that has not yet produced symptoms.

The American Heart Association suggests that after a physical exam, a sedentary person wishing to start an exercise program begin with 30 minutes of activity (perhaps broken into two 15-minute sessions at first) at least five times per week.

Check Your Recall

36. What factors affect cardiac output?
37. What is the function of baroreceptors in the walls of the aorta and carotid arteries?
38. How does the vasomotor center control peripheral resistance?

Venous Blood Flow

Blood pressure decreases as blood moves through the arterial system and into the capillary networks, so that little pressure remains at the venular ends of capillaries (see fig. 13.24). Instead, blood flow through the venous system is only partly the direct result of heart action and depends on other factors, such as skeletal muscle contraction, breathing movements, and vasoconstriction of veins (*venoconstriction*).

Contracting skeletal muscles press on nearby vessels, squeezing the blood inside. As skeletal muscles press on veins with valves, some blood moves from one valve section to another (see fig. 13.23). This massaging action of contracting skeletal muscles helps push blood through the venous system toward the heart.

Respiratory movements also move venous blood. During inspiration, the pressure within the thoracic cavity is reduced as the diaphragm contracts and the rib cage moves upward and outward. At the same time, the pressure within the abdominal cavity is increased as the diaphragm presses downward on the abdominal viscera. Consequently, blood is squeezed out of abdominal veins and forced into thoracic veins. During exercise, these respiratory movements act with skeletal muscle contractions to increase the return of venous blood to the heart.

Venoconstriction also returns venous blood to the heart. When venous pressure is low, sympathetic reflexes stimulate smooth muscles in the walls of veins to contract. The veins also provide a blood reservoir that can adapt its capacity to changes in blood volume. If some blood is lost and blood pressure falls, venoconstriction can force blood out of this reservoir. In both of these examples, venoconstriction helps maintain blood pressure by forcing more blood toward the heart.

Check Your Recall

39. What is the function of venous valves?
40. How do skeletal muscles and respiratory movements affect venous blood flow?
41. What factors stimulate venoconstriction?

13.6 PATHS OF CIRCULATION

Recall from fig. 13.1 that the blood vessels can be divided into two major pathways. The *pulmonary circuit* (pul'mo-ner"e ser'kit) or pulmonary circulation consists of vessels that carry blood from the heart to the lungs and back to the heart. The *systemic circuit* (sis-tem'ik ser'kit) or systemic circulation carries blood from the heart to all other parts of the body and back again. The systemic circuit includes the coronary circulation.

The following sections describe the circulatory pathways of an adult. Chapter 20 (pp. 552–554) describes the somewhat different fetal pathways.

Pulmonary Circuit

Blood enters the pulmonary circuit as it leaves the right ventricle through the pulmonary trunk. The pulmonary trunk extends upward and posteriorly from the heart. About 5 centimeters above its origin, the pulmonary trunk divides into the right and left pulmonary arteries (see fig. 13.5), which penetrate the right and left lungs, respectively. After repeated divisions, the pulmonary arteries give rise to arterioles that continue into the capillary networks associated with the walls of the alveoli, where gas is exchanged between the blood and the air (see chapter 16, pp. 460–461).

From the pulmonary capillaries, blood enters the venules, which merge to form small veins, and these veins in turn converge to form still larger veins. Four pulmonary veins, two from each lung, return blood to the left atrium. This completes the vascular loop of the pulmonary circuit.

Systemic Circuit

Freshly oxygenated blood moves from the left atrium into the left ventricle. Contraction of the left ventricle forces this blood into the systemic circuit, which includes the aorta and its branches that lead to all the body tissues, as well as the companion system of veins that returns blood to the right atrium.

Check Your Recall

42. Distinguish between the pulmonary and systemic circuits of the cardiovascular system.

43. Trace the path of blood through the pulmonary circuit from the right ventricle.

13.7 ARTERIAL SYSTEM

The **aorta** is the largest-diameter artery in the body. It extends upward from the left ventricle, arches over the heart to the left, and descends just anterior and to the left of the vertebral column. Figure 13.27 shows the aorta and its main branches.

Principal Branches of the Aorta

The first portion of the aorta is called the *ascending aorta*. Located at its base are the three cusps of the aortic valve, and opposite each cusp is a swelling in the aortic wall called an **aortic sinus.** The right and left coronary arteries arise from two of these sinuses (see fig. 13.8).

Three major arteries originate from the *aortic arch* (arch of the aorta): the **brachiocephalic** (brăk"e-o-sĕ-fal'ik) **artery,** the left **common carotid** (kah-rot'id) **artery,** and the left **subclavian** (sub-kla've-an) **artery.**

The upper part of the *descending aorta* is left of the midline. It gradually extends medially and finally lies directly in front of the vertebral column at the level of the twelfth thoracic vertebra. The portion of the descending aorta above the diaphragm is the **thoracic aorta.** It branches into the thoracic wall and thoracic viscera.

Below the diaphragm, the descending aorta becomes the **abdominal aorta,** and it branches into the abdominal wall and various abdominal organs. Branches to abdominal organs include: the **celiac** (se'le-ak) **artery,** which gives rise to the *gastric, splenic,* and *hepatic arteries;* the **superior** (supplies small intestine and superior portion of large intestine) and **inferior** (supplies inferior portion of large intestine) **mesenteric** (mes"en-ter' ik) **arteries;** and the **suprarenal** (soo"prah-re'nal) **arteries, renal** (re'nal) **arteries,** and **gonadal** (go'nad-al) **arteries,** which supply blood to the adrenal glands, kidneys, and ovaries or testes, respectively. The abdominal aorta ends near the brim of the pelvis, where it divides into right and left **common iliac** (il'e-ak) **arteries.** These vessels supply blood to lower regions of the abdominal wall, the pelvic organs, and the lower extremities. Table 13.3 summarizes the major branches of the aorta.

Arteries to the Neck, Head, and Brain

Branches of the subclavian and common carotid arteries supply blood to structures within the neck, head, and brain (fig. 13.28). The main divisions of the subclavian artery to these regions include the vertebral and thyrocervical arteries. The common carotid artery communicates with these regions by means of the internal and external carotid arteries.

The **vertebral arteries** pass upward through the foramina of the transverse processes of the cervical vertebrae and enter the skull through the foramen magnum. These vessels supply blood to the vertebrae and to their associated ligaments and muscles.

In the cranial cavity, the vertebral arteries unite to form a single *basilar artery.* This vessel passes along

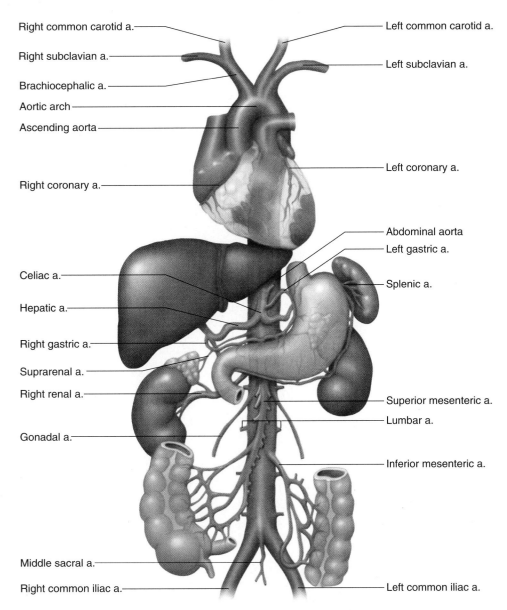

Right common carotid a.
Right subclavian a.
Brachiocephalic a.
Aortic arch
Ascending aorta
Right coronary a.
Celiac a.
Hepatic a.
Right gastric a.
Suprarenal a.
Right renal a.
Gonadal a.
Middle sacral a.
Right common iliac a.

Left common carotid a.
Left subclavian a.
Left coronary a.
Abdominal aorta
Left gastric a.
Splenic a.
Superior mesenteric a.
Lumbar a.
Inferior mesenteric a.
Left common iliac a.

Figure 13.27
Major branches of the aorta. (*a.* stands for *artery.*)

the ventral brainstem and gives rise to branches leading to the pons, midbrain, and cerebellum. The basilar artery ends by dividing into two *posterior cerebral arteries* that supply portions of the occipital and temporal lobes of the cerebrum. The posterior cerebral arteries also help form the **cerebral arterial circle** (circle of Willis) at the base of the brain, which connects the vertebral artery and internal carotid artery systems (fig. 13.29). The union of these systems provides alternate pathways for blood to circumvent blockages and reach brain tissues. It also equalizes blood pressure in the brain's blood supply.

The **thyrocervical** (thi″ro-ser′vĭ-kal) **arteries** are short vessels. At the thyrocervical axis, these vessels give off branches to the thyroid gland, parathyroid glands,

larynx, trachea, esophagus, and pharynx, as well as to muscles in the neck, shoulder, and back.

The left and right *common carotid arteries* diverge into the internal and external carotid arteries. The **external carotid artery** courses upward on the side of the head, giving off branches to structures in the neck, face, jaw, scalp, and base of the skull. The **internal carotid artery** follows a deep course upward along the pharynx to the base of the skull. Entering the cranial cavity, it provides the major blood supply to the brain. Near the base of the internal carotid arteries are enlargements called **carotid sinuses** that, like aortic sinuses, contain baroreceptors controlling blood pressure. Table 13.4 summarizes the major branches of the external and internal carotid arteries.

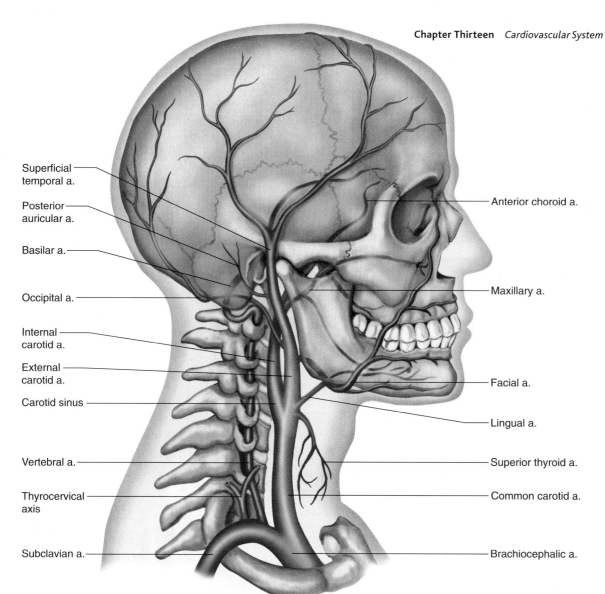

Figure 13.28

The major arteries of
the head and neck.
Note that the clavicle
has been removed.
(*a.* stands for *artery.*)

Labels (clockwise from upper left):
Superficial temporal a.
Posterior auricular a.
Basilar a.
Occipital a.
Internal carotid a.
External carotid a.
Carotid sinus
Vertebral a.
Thyrocervical axis
Subclavian a.
Anterior choroid a.
Maxillary a.
Facial a.
Lingual a.
Superior thyroid a.
Common carotid a.
Brachiocephalic a.

Table 13.3	Major Branches of the Aorta	
Portion of Aorta	**Branch**	**General Regions or Organs Supplied**
Ascending aorta	Right and left coronary arteries	Heart
Arch of the aorta	Brachiocephalic artery	Right upper limb, right side of head
	Left common carotid artery	Left side of head
	Left subclavian artery	Left upper limb
Descending aorta		
Thoracic aorta	Bronchial artery	Bronchi
	Pericardial artery	Pericardium
	Esophageal artery	Esophagus
	Mediastinal artery	Mediastinum
	Posterior intercostal artery	Thoracic wall
Abdominal aorta	Celiac artery	Organs of upper digestive tract
	Phrenic artery	Diaphragm
	Superior mesenteric artery	Portions of small and large intestines
	Suprarenal artery	Adrenal gland
	Renal artery	Kidney
	Gonadal artery	Ovary or testis
	Inferior mesenteric artery	Lower portions of large intestine
	Lumbar artery	Posterior abdominal wall
	Middle sacral artery	Sacrum and coccyx
	Common iliac artery	Lower abdominal wall, pelvic organs, and lower limb

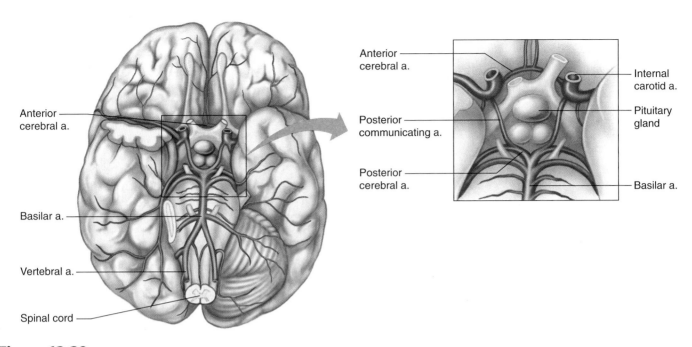

Figure 13.29

The cerebral arterial circle (circle of Willis) is formed by the anterior and posterior cerebral arteries, which join the internal carotid arteries. (*a.* stands for *artery.*)

Table 13.4	Major Branches of the External and Internal Carotid Arteries	
Artery	**Branch**	**General Region or Organs Supplied**
External carotid artery	Superior thyroid artery	Larynx and thyroid gland
	Lingual artery	Tongue and salivary glands
	Facial artery	Pharynx, palate, chin, lips, and nose
	Occipital artery	Posterior scalp, meninges, and neck muscles
	Posterior auricular artery	Ear and lateral scalp
	Maxillary artery	Teeth, jaw, cheek, and eyelids
	Superficial temporal artery	Parotid salivary gland and surface of the face and scalp
Internal carotid artery	Ophthalmic artery	Eye and eye muscles
	Anterior choroid artery	Choroid plexus and brain
	Anterior cerebral artery	Frontal and parietal lobes of the brain

Arteries to the Shoulder and Upper Limb

The subclavian artery, after giving off branches to the neck, continues into the arm (fig. 13.30). It passes between the clavicle and the first rib, and becomes the axillary artery. The **axillary artery** supplies branches to structures in the axilla and chest wall and becomes the **brachial artery,** which follows the humerus to the elbow. It gives rise to a *deep brachial artery* that curves posteriorly around the humerus and supplies the triceps brachii. Within the elbow, the brachial artery divides into an ulnar artery and a radial artery.

The **ulnar artery** leads downward on the ulnar side of the forearm to the wrist. Some of its branches supply the elbow joint, and some supply blood to muscles in the forearm.

The **radial artery** travels along the radial side of the forearm to the wrist, supplying the lateral muscles of the forearm. As the radial artery nears the wrist, it approaches the surface and provides a convenient vessel for taking the pulse (radial pulse).

At the wrist, the branches of the ulnar and radial arteries join to form a network of vessels. Arteries arising from this network supply blood to the hand.

Arteries to the Thoracic and Abdominal Walls

Blood reaches the thoracic wall through several vessels. The **internal thoracic artery,** a branch of the subclavian artery, gives off two *anterior intercostal*

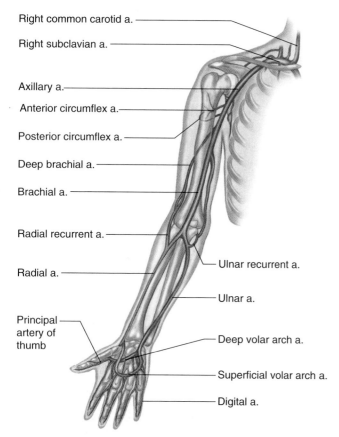

Right common carotid a.

Right subclavian a.

Axillary a.

Anterior circumflex a.

Posterior circumflex a.

Deep brachial a.

Brachial a.

Radial recurrent a.

Radial a.

Principal artery of thumb

Ulnar recurrent a.

Ulnar a.

Deep volar arch a.

Superficial volar arch a.

Digital a.

Figure 13.30

The major arteries to the shoulder and upper limb. (*a.* stands for *artery.*)

(in″ter-kos′tal) *arteries* that supply the intercostal muscles and mammary glands. The *posterior intercostal arteries* arise from the thoracic aorta and enter the intercostal spaces. They supply the intercostal muscles, the vertebrae, the spinal cord, and the deep muscles of the back.

Branches of the *internal thoracic* and *external iliac arteries* provide blood to the anterior abdominal wall. Paired vessels originating from the abdominal aorta, including the *phrenic* and *lumbar arteries,* supply blood to structures in the posterior and lateral abdominal wall.

Arteries to the Pelvis and Lower Limb

The abdominal aorta divides to form the **common iliac** (il′e-ak) **arteries** at the level of the pelvic brim, and these vessels provide blood to the pelvic organs, gluteal region, and lower limbs (fig. 13.31). Each common iliac artery divides into an internal and an external branch. The **internal iliac artery** gives off many branches to pelvic muscles and visceral structures, as well as to the gluteal muscles and the external reproductive organs. The **external iliac artery** provides the main blood supply to the lower limbs. It passes downward along the brim of the pelvis and branches to sup-

ply the muscles and skin in the lower abdominal wall. Midway between the symphysis pubis and the anterior superior iliac spine of the ilium, the external iliac artery becomes the femoral artery.

The **femoral** (fem′or-al) **artery,** which approaches the anterior surface of the upper thigh, branches to muscles and superficial tissues of the thigh. These branches also supply the skin of the groin and the lower abdominal wall.

As the femoral artery reaches the proximal border of the space behind the knee, it becomes the **popliteal** (pop-lit′e-al) **artery.** Branches of this artery supply blood to the knee joint and to certain muscles in the thigh and calf. The popliteal artery diverges into the anterior and posterior tibial arteries.

The **anterior tibial artery** passes downward between the tibia and fibula, giving off branches to the skin and muscles in the anterior and lateral regions of the leg. This vessel continues into the foot as the *dorsalis pedis artery* (dorsal pedis artery), which supplies blood to the foot. The **posterior tibial artery,** the larger of the two popliteal branches, descends beneath the calf muscles, branching to the skin, muscles, and other tissues of the leg along the way.

Check Your Recall

44. Name the portions of the aorta.

45. Name the vessels that arise from the aortic arch.

46. Name the branches of the thoracic and abdominal aorta.

47. Which vessels supply blood to the head? To the upper limb? To the abdominal wall? To the lower limb?

13.8 VENOUS SYSTEM

Venous circulation returns blood to the heart after blood and body cells exchange gases, nutrients, and wastes.

Characteristics of Venous Pathways

Venous vessels begin as capillaries merge into venules, venules merge into small veins, and small veins meet to form larger ones. Unlike the arterial pathways, however, the vessels of the venous system are difficult to follow. This is because they connect in irregular networks, so many unnamed tributaries may join to form a large vein.

The pathways of larger veins are clearer. These veins typically parallel the courses of named arteries, and often bear the same names as their arterial counterparts. For example, the renal vein parallels the renal artery, and the common iliac vein accompanies the common iliac artery.

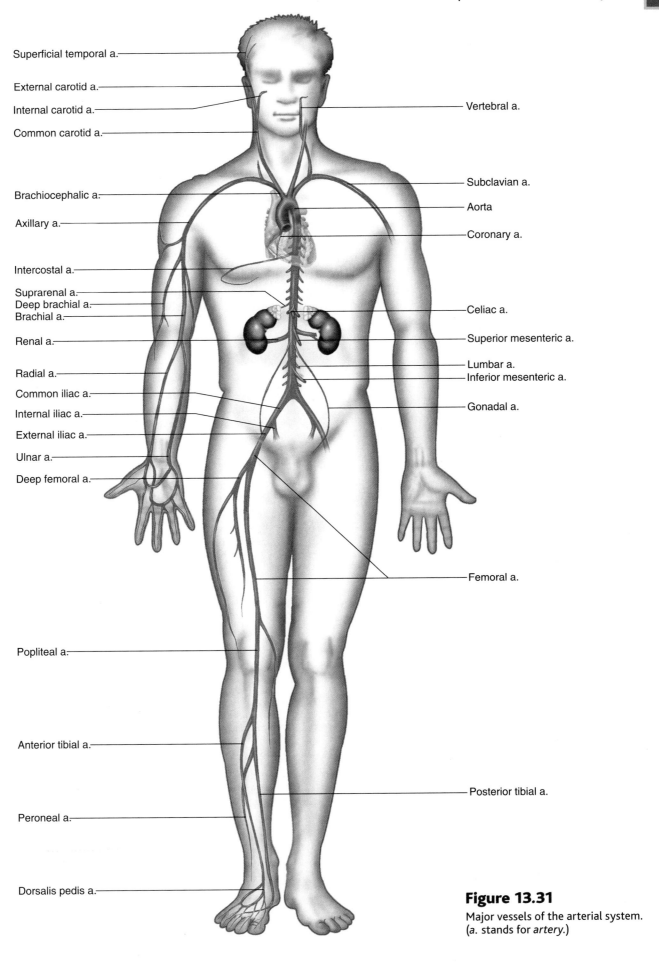

Superficial temporal a.

External carotid a.

Internal carotid a.

Common carotid a.

Vertebral a.

Brachiocephalic a.

Axillary a.

Subclavian a.

Aorta

Coronary a.

Intercostal a.

Suprarenal a.
Deep brachial a.
Brachial a.

Celiac a.

Renal a.

Superior mesenteric a.

Radial a.

Lumbar a.
Inferior mesenteric a.

Common iliac a.

Internal iliac a.

External iliac a.

Gonadal a.

Ulnar a.

Deep femoral a.

Femoral a.

Popliteal a.

Anterior tibial a.

Posterior tibial a.

Peroneal a.

Dorsalis pedis a.

Figure 13.31

Major vessels of the arterial system.
(*a.* stands for *artery.*)

The veins that carry blood from the lungs and myocardium back to the heart have already been described. The veins from all the other parts of the body converge into two major pathways, the **superior** and **inferior venae cavae,** which lead to the right atrium.

Veins from the Brain, Head, and Neck

The **external jugular** (jug′u-lar) **veins** drain blood from the face, scalp, and superficial regions of the neck. These vessels descend on either side of the neck and empty into the *right* and *left subclavian veins* (fig. 13.32).

The **internal jugular veins,** which are somewhat larger than the external jugular veins, arise from numerous veins and venous sinuses of the brain and from deep veins in parts of the face and neck. They descend through the neck and join the subclavian veins. These unions of the internal jugular and subclavian veins form large **brachiocephalic veins** on each side. The vessels then merge and give rise to the superior vena cava, which enters the right atrium.

Veins from the Upper Limb and Shoulder

A set of deep veins and a set of superficial ones drain the upper limb. The deep veins generally parallel the arteries in each region and have similar names, such as the radial veins, ulnar veins, brachial veins, and axillary vein. The superficial veins connect in complex networks just beneath the skin. They also communicate with the deep vessels of the upper limb, providing many alternate pathways through which blood can leave the tissues (fig. 13.33). The main vessels of the superficial network are the basilic and cephalic veins.

The **basilic** (bah-sil′ik) **vein** ascends from the forearm to the middle of the arm, where it penetrates deeply and joins the *brachial vein*. The basilic and brachial veins merge, forming the *axillary vein*.

The **cephalic** (sĕ-fal′ik) **vein** courses upward from the hand to the shoulder. In the shoulder, it pierces the tissues and empties into the axillary vein. Beyond the axilla, the axillary vein becomes the subclavian vein.

In the bend of the elbow, a *median cubital vein* ascends from the cephalic vein on the lateral side of the forearm to the basilic vein on the medial side. This large vein is usually visible. It is often used as a site for *venipuncture*, when it is necessary to remove a blood sample or to add fluids to blood.

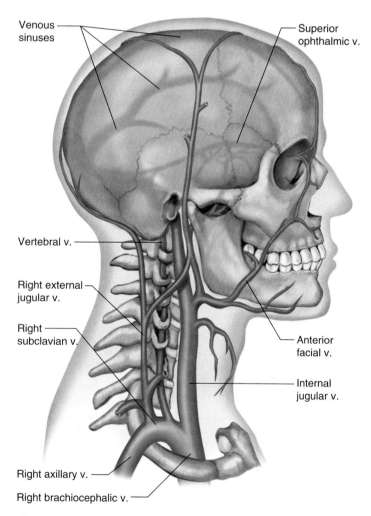

Figure 13.32
The major veins of the brain, head, and neck. Note that the clavicle has been removed. (*v.* stands for *vein.*)

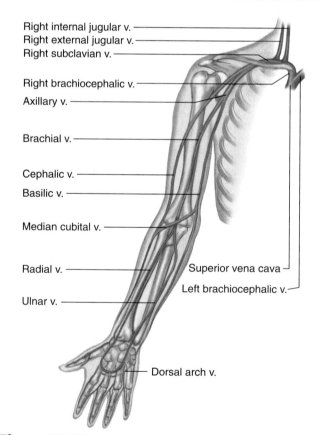

Figure 13.33
The major veins of the upper limb and shoulder. (*v.* stands for *vein.*) Though drawn as one vessel, many of the peripheral veins are in pairs.

Veins from the Abdominal and Thoracic Walls

Tributaries of the brachiocephalic and azygos veins drain the abdominal and thoracic walls. For example, the *brachiocephalic vein* receives blood from the *internal thoracic vein,* which generally drains the tissues the internal thoracic artery supplies. Some *intercostal veins* also empty into the brachiocephalic vein.

The **azygos** (az'ĭ-gos) **vein** originates in the dorsal abdominal wall and ascends through the mediastinum on the right side of the vertebral column to join the superior vena cava. It drains most of the muscular tissue in the abdominal and thoracic walls.

Tributaries of the azygos vein include the *posterior intercostal veins* on the right side, which drain the intercostal spaces, and the *superior* and *inferior hemiazygos veins,* which receive blood from the posterior intercostal veins on the left. The right and left *ascending lumbar veins,* with tributaries that include vessels from the lumbar and sacral regions, also connect to the azygos system.

Veins from the Abdominal Viscera

Most veins carry blood directly to the atria of the heart. Veins that drain the abdominal viscera are exceptions (fig. 13.34). They originate in the capillary networks of the stomach, intestines, pancreas, and spleen and carry blood from these organs through a **hepatic portal** (por'tal) **vein** to the liver. This unique venous pathway is called the **hepatic portal system.**

Tributaries of the hepatic portal vein include:

1. Right and left *gastric veins* from the stomach.
2. *Superior mesenteric vein* from the small intestine, ascending colon, and transverse colon.
3. *Splenic vein* from a convergence of several veins draining the spleen, the pancreas, and a portion of the stomach. Its largest tributary, the *inferior mesenteric vein,* brings blood upward from the descending colon, sigmoid colon, and rectum.

About 80% of the blood flowing to the liver in the hepatic portal system comes from capillaries in the

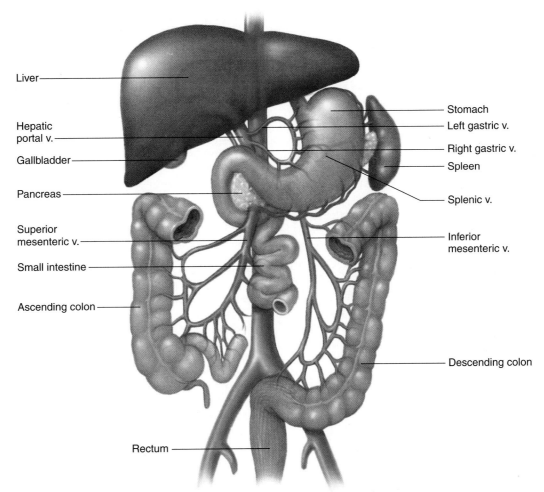

Figure 13.34

Veins that drain the abdominal viscera. (*v.* stands for *vein.*)

stomach and intestines, and is oxygen-poor but nutrient-rich. As discussed in chapter 15 (pp. 416 and 418), the liver handles these nutrients in a variety of ways. It regulates blood glucose concentration by polymerizing excess glucose into glycogen for storage or by breaking down glycogen into glucose when blood glucose concentration drops below normal. The liver helps regulate blood concentrations of recently absorbed amino acids and lipids by modifying them into forms cells can use, by oxidizing them, or by changing them into storage forms. The liver also stores certain vitamins and detoxifies harmful substances. Blood in the hepatic portal vein nearly always contains bacteria that have entered through intestinal capillaries. Large *Kupffer cells* lining small vessels in the liver called hepatic sinusoids phagocytize these microorganisms, removing them from portal blood before it leaves the liver.

After passing through the hepatic sinusoids of the liver, blood in the hepatic portal system travels through a series of merging vessels into **hepatic veins.** These veins empty into the inferior vena cava, returning the blood to the general circulation.

Veins from the Lower Limb and Pelvis

Veins that drain blood from the lower limb are divided into deep and superficial groups, as is the case in the upper limb (fig. 13.35). The deep veins of the leg, such as the *anterior* and *posterior tibial veins,* are named for the arteries they accompany. At the level of the knee, these vessels form a single trunk, the **popliteal vein.** This vein continues upward through the thigh as the **femoral vein,** which in turn becomes the **external iliac vein.**

The superficial veins of the foot, leg, and thigh connect to form a complex network beneath the skin. These vessels drain into two major trunks—the small and great saphenous veins. The **small saphenous** (sah-fe′nus) **vein** ascends along the back of the calf, enters the popliteal fossa, and joins the popliteal vein. The **great saphenous vein,** which is the longest vein in the body, ascends in front of the medial malleolus and extends upward along the medial side of the leg and thigh. In the thigh, it penetrates deeply and joins the femoral vein. Near its termination, the great saphenous vein receives tributaries from a number of vessels that drain the upper thigh, groin, and lower abdominal wall.

In addition to communicating freely with each other, the saphenous veins communicate extensively with the deep veins of the leg and thigh. Blood can thus return to the heart from the lower extremities by several routes.

In the pelvic region, vessels leading to the **internal iliac vein** carry blood away from the organs of the reproductive, urinary, and digestive systems. The internal iliac veins unite with the right and left external iliac veins to form the **common iliac veins.** These vessels, in turn, merge to produce the inferior vena cava.

> *Varicose veins* have abnormal dilations. They result from increased blood pressure in the saphenous veins due to gravity, as occurs when a person stands for a prolonged period.

Check Your Recall

48. Name the veins that return blood to the right atrium.
49. Which major veins drain blood from the head? From the upper limbs? From the abdominal viscera? From the lower limbs?

Clinical Terms Related to the Cardiovascular System

anastomosis (ah-nas″to-mo′sis) Connection between two blood vessels, sometimes produced surgically.

angiospasm (an′je-o-spazm″) Muscular spasm in the wall of a blood vessel.

arteriography (ar″te-re-og′rah-fe) Injection of radiopaque solution into the vascular system for X-ray examination of arteries.

asystole (a-sis′to-le) Failure of the myocardium to contract.

cardiac tamponade (kar′de-ak tam″po-nād′) Compression of the heart by fluid accumulating within the pericardial cavity.

congestive heart failure (kon-jes′tiv hart fāl′yer) Inability of the left ventricle to pump adequate blood to cells.

cor pulmonale (kor pul-mo-na′le) Pulmonary hypertension and hypertrophy of the right ventricle.

embolectomy (em″bo-lek′to-me) Removal of an embolus through an incision in a blood vessel.

endarterectomy (en″dar-ter-ek′to-me) Removal of the inner wall of an artery to reduce an arterial occlusion.

palpitation (pal″pĭ-ta′shun) Awareness of a heartbeat that is unusually rapid, strong, or irregular.

pericardiectomy (per″ĭ-kar″de-ek′to-me) Excision of the pericardium.

phlebitis (flĕ-bi′tis) Inflammation of a vein, usually in the lower limbs.

phlebotomy (flĕ-bot′o-me) Incision or puncture of a vein to withdraw blood.

sinus rhythm (si′nus rithm) The normal cardiac rhythm regulated by the S-A node.

thrombophlebitis (throm″bo-flĕ-bi′tis) Formation of a blood clot in a vein in response to inflammation of the venous wall.

valvotomy (val-vot′o-me) Incision of a valve.

venography (ve-nog′rah-fe) Injection of radiopaque solution into the vascular system for X-ray examination of veins.

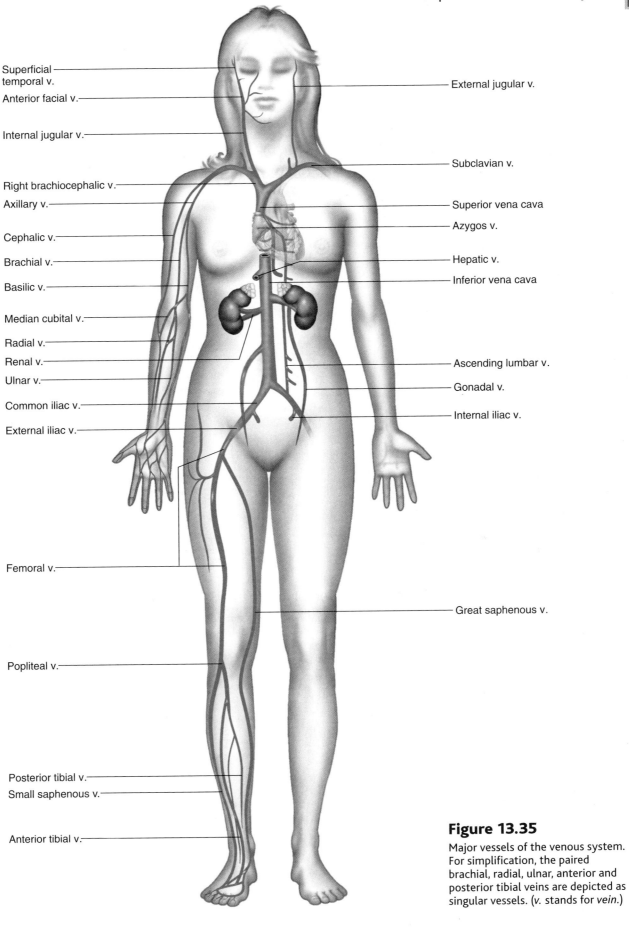

Superficial temporal v.

Anterior facial v.

Internal jugular v.

Right brachiocephalic v.

Axillary v.

Cephalic v.

Brachial v.

Basilic v.

Median cubital v.

Radial v.

Renal v.

Ulnar v.

Common iliac v.

External iliac v.

Femoral v.

Popliteal v.

Posterior tibial v.

Small saphenous v.

Anterior tibial v.

External jugular v.

Subclavian v.

Superior vena cava

Azygos v.

Hepatic v.

Inferior vena cava

Ascending lumbar v.

Gonadal v.

Internal iliac v.

Great saphenous v.

Figure 13.35

Major vessels of the venous system. For simplification, the paired brachial, radial, ulnar, anterior and posterior tibial veins are depicted as singular vessels. (*v.* stands for *vein.*)

Cardiovascular System

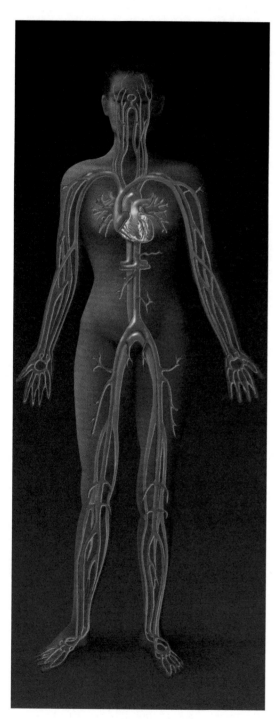

Integumentary System

Changes in skin blood flow are important in temperature control.

Lymphatic System

The lymphatic system returns tissue fluids to the bloodstream.

Skeletal System

Bones help control plasma levels of calcium ions, which influence heart action.

Digestive System

The digestive system breaks down nutrients into forms readily absorbed by the bloodstream.

Muscular System

Blood flow increases to exercising skeletal muscle, delivering oxygen and nutrients and removing wastes. Muscle actions help the blood circulate.

Respiratory System

The respiratory system oxygenates the blood and removes carbon dioxide. Respiratory movements help the blood circulate.

Nervous System

The brain depends on blood flow for survival. The nervous system helps control blood flow and blood pressure.

Urinary System

The kidneys clear the blood of wastes and substances present in the body. The kidneys help control blood pressure and blood volume.

Endocrine System

Hormones are carried in the bloodstream. Some hormones directly affect the heart and blood vessels.

Reproductive System

Blood pressure is important in normal function of the sex organs.

The heart pumps blood through as many as 60,000 miles of blood vessels delivering nutrients to, and removing wastes from, all body cells.

Clinical Connection

On July 2, 2001, fifty-year-old Robert Tools became the first person to receive an implantable replacement heart, developed at the University of Louisville. At the time, Tools was weeks from death due to congestive heart failure, but after the seven-hour surgery, he enjoyed five more months of life, thanks to the device that replaced the functioning of his ventricles. The 2-pound, titanium and plastic device consists of an internal motor-driven hydraulic pump, a battery and electronic package, and an external battery pack. The electronics component manages the rate and force of the pump's actions, tailoring them to the patient's condition. The internal battery allows the patient to be free of connections for up to an hour, and the two external batteries allow free movement for up to two hours. The batteries are recharged at night, while the patient sleeps and the system is plugged into an electrical outlet.

The implantable replacement heart was granted approval by the U.S. Food and Drug Administration in 2006 as a "humanitarian use device." This means that it is used to treat conditions that affect fewer than 4,000 people a year. Recipients must also not be candidates for other procedures, such as heart transplantation, and must have less than a month to live. The approval was granted after testing on only fourteen individuals because the device was considered safe and probably beneficial, extending life by a few months and improving the quality of life. Maximal survival was seventeen months. A second generation of implantable replacement hearts is currently being tested. It is smaller and is designed to provide up to five years of life.

SUMMARY OUTLINE

13.1 Introduction (p. 340)

The cardiovascular system, consisting of the heart and blood vessels, provides oxygen and nutrients to tissues and removes wastes.

13.2 Structure of the Heart (p. 341)

1. Size and location of the heart
   a. The heart is about 14 centimeters long and 9 centimeters wide.
   b. It is located within the mediastinum and rests on the diaphragm.
2. Coverings of the heart
   a. A layered pericardium encloses the heart.
   b. The pericardial cavity is a space between the parietal and visceral layers of the pericardium.
3. Wall of the heart
   The wall of the heart has three layers—an epicardium, a myocardium, and an endocardium.
4. Heart chambers and valves
   a. The heart is divided into two atria and two ventricles.
   b. Right chambers and valves
      (1) The right atrium receives blood from the venae cavae and coronary sinus.
      (2) The tricuspid valve separates the right atrium from the right ventricle.
      (3) A pulmonary valve guards the base of the pulmonary trunk.
   c. Left chambers and valves
      (1) The left atrium receives blood from the pulmonary veins.
      (2) The mitral valve separates the left atrium from the left ventricle.
      (3) An aortic valve guards the base of the aorta.
5. Skeleton of the heart
   The skeleton of the heart consists of fibrous rings that enclose the bases of the pulmonary artery and aorta and masses of dense connective tissues in the septum between the ventricles.
6. Path of blood through the heart
   a. Blood low in oxygen and high in carbon dioxide enters the right side of the heart and is pumped into the pulmonary circulation.
   b. After blood is oxygenated in the lungs and some carbon dioxide is removed, it returns to the left side of the heart.
7. Blood supply to the heart
   a. The coronary arteries supply blood to the myocardium.
   b. Blood returns to the right atrium through the cardiac veins and coronary sinus.

13.3 Heart Actions (p. 346)

1. Cardiac cycle
   a. The atria contract (atrial systole) while the ventricles relax (ventricular diastole). The ventricles contract (ventricular systole) while the atria relax (atrial diastole).
   b. Pressure within the chambers rises and falls in repeated cycles.
2. Heart sounds
   Heart sounds are due to the vibrations produced when the valves close.
3. Cardiac muscle fibers
   a. Cardiac muscle fibers connect to form a functional syncytium.
   b. If any part of the syncytium is stimulated, the whole structure contracts as a unit.
4. Cardiac conduction system
   a. This system initiates and conducts impulses throughout the myocardium.
   b. Impulses from the S-A node pass slowly to the A-V node. Impulses travel rapidly along the A-V bundle and Purkinje fibers.
5. Electrocardiogram (ECG)
   a. An ECG records electrical changes in the myocardium during a cardiac cycle.
   b. The pattern contains several waves.
      (1) The P wave represents atrial depolarization.
      (2) The QRS complex represents ventricular depolarization.
      (3) The T wave represents ventricular repolarization.
6. Regulation of the cardiac cycle
   a. Physical exercise, body temperature, and the concentration of various ions affect heartbeat.
   b. Branches of sympathetic and parasympathetic nerve fibers innervate the S-A and A-V nodes.

c. The cardiac control center in the medulla oblongata regulates autonomic impulses to the heart.

13.4 Blood Vessels (p. 352)

Blood vessels form a closed circuit of tubes that carry blood from the heart to body cells and back again.

1. Arteries and arterioles
   a. Arteries are adapted to carry blood under high pressure away from the heart.
   b. The walls of arteries and arterioles consist of layers of endothelium, smooth muscle, and connective tissue.
   c. Autonomic fibers that can stimulate vasoconstriction or vasodilation innervate smooth muscle in vessel walls.
2. Capillaries
   a. Capillaries connect arterioles and venules.
   b. The capillary wall is a single layer of cells that forms a semipermeable membrane.
   c. Openings in capillary walls, where endothelial cells overlap, vary in size from tissue to tissue.
   d. Precapillary sphincters regulate capillary blood flow.
3. Exchanges in capillaries
   a. Capillary blood and tissue fluid exchange gases, nutrients, and metabolic by-products.
   b. Diffusion provides the most important means of transport.
   c. Filtration, which is due to the hydrostatic pressure of blood, causes a net outward movement of fluid at the arteriolar end of a capillary.
   d. Osmosis due to colloid osmotic pressure causes a net inward movement of fluid at the venular end of a capillary.
4. Venules and veins
   a. Venules continue from capillaries and merge to form veins.
   b. Veins carry blood to the heart.
   c. Venous walls are similar to arterial walls, but are thinner and contain less smooth muscle and elastic tissue.

13.5 Blood Pressure (p. 357)

Blood pressure is the force blood exerts against the insides of blood vessels.

1. Arterial blood pressure
   a. Arterial blood pressure rises and falls with the phases of the cardiac cycle.
   b. Systolic pressure is produced when the ventricle contracts. Diastolic pressure is the pressure in the arteries when the ventricle relaxes.
2. Factors that influence arterial blood pressure
   Arterial blood pressure increases as cardiac output, blood volume, peripheral resistance, or blood viscosity increases.
3. Control of blood pressure
   a. Blood pressure is controlled in part by the mechanisms that regulate cardiac output and peripheral resistance.
   b. The more blood that enters the heart, the stronger the ventricular contraction, the greater the stroke volume, and the greater the cardiac output.
   c. The baroreceptor reflexes involving the cardiac control center of the medulla oblongata regulate heart rate.
4. Venous blood flow
   a. Venous blood flow depends on skeletal muscle contraction, breathing movements, and venoconstriction.
   b. Many veins contain flaplike valves that prevent blood from backing up.

13.6 Paths of Circulation (p. 362)

1. Pulmonary circuit
   The pulmonary circuit consists of vessels that carry blood from the right ventricle to the lungs and back to the left atrium.
2. Systemic circuit
   a. The systemic circuit consists of vessels that lead from the left ventricle to the body cells (including those of the heart itself) and back to the heart.
   b. It includes the aorta and its branches.

13.7 Arterial System (p. 362)

1. Principal branches of the aorta
   a. The aorta is the largest artery with respect to diameter.
   b. Its major branches include the coronary, brachiocephalic, left common carotid, and left subclavian arteries.
   c. The branches of the descending aorta include the thoracic and abdominal groups.
   d. The abdominal aorta diverges into the right and left common iliac arteries.
2. Arteries to the neck, head, and brain
   These include branches of the subclavian and common carotid arteries.
3. Arteries to the shoulder and upper limb
   a. The subclavian artery passes into the upper limb, and in various regions is called the axillary and brachial artery.
   b. Branches of the brachial artery include the ulnar and radial arteries.
4. Arteries to the thoracic and abdominal walls
   a. Branches of the subclavian artery and thoracic aorta supply the thoracic wall.
   b. Branches of the abdominal aorta and other arteries supply the abdominal wall.
5. Arteries to the pelvis and lower limb
   The common iliac arteries supply the pelvic organs, gluteal region, and lower limbs.

13.8 Venous System (p. 366)

1. Characteristics of venous pathways
   a. Veins return blood to the heart.
   b. Larger veins usually parallel the paths of major arteries.
2. Veins from the brain, head, and neck
   a. Jugular veins drain these regions.
   b. Jugular veins unite with subclavian veins to form the brachiocephalic veins.
3. Veins from the upper limb and shoulder
   a. Sets of superficial and deep veins drain these regions.
   b. Deep veins parallel arteries with similar names.
4. Veins from the abdominal and thoracic walls
   Tributaries of the brachiocephalic and azygos veins drain these walls.
5. Veins from the abdominal viscera
   a. Blood from the abdominal viscera enters the hepatic portal system and is carried to the liver.
   b. From the liver, hepatic veins carry blood to the inferior vena cava.
6. Veins from the lower limb and pelvis
   a. Sets of deep and superficial veins drain these regions.
   b. The deep veins include the tibial veins, and the superficial veins include the saphenous veins.

CHAPTER ASSESSMENTS

13.1 Introduction

1. The cardiovascular system includes the: (p. 340)
   - a. heart
   - b. arteries
   - c. veins
   - d. capillaries
   - e. all of the above

13.2 Structure of the Heart

2. Describe the pericardium. (p. 341)

3. Compare the layers of the cardiac wall. (p. 342)

4. Draw a heart and label the chambers and valves. (p. 342)

5. Blood flows through the vena cavae and coronary sinus into the right atrium through the _____, to the right ventricle through the pulmonary valve to the pulmonary trunk into the right and left _____ to the lungs, then through the pulmonary veins into the _____ through the mitral valve to the _____ and through the _____ to the aorta. (p. 345)

6. List the vessels through which blood flows from the aorta to the myocardium and back to the right atrium (p. 346)

13.3 Heart Actions

7. Describe a cardiac cycle, including the pressure changes in the atria and ventricles during the cardiac cycle. (p. 346)

8. Distinguish between the roles of the S-A node and the A-V node. (p. 349)

9. Explain how the cardiac conduction system coordinates the events of the cardiac cycle. (p. 349)

10. Describe and explain the normal ECG pattern. (p. 350)

11. Discuss how the nervous system regulates the cardiac cycle. (p. 351)

13.4 Blood Vessels

12. Distinguish between an artery and an arteriole. (p. 353)

13. Explain control of vasodilation and vasoconstriction. (p. 353)

14. Describe the structure and function of a capillary. (p. 355)

15. Relate how diffusion functions in the exchange of substances between the blood and tissues. (p. 356)

16. Explain why water and dissolved substances leave the arteriolar end of a capillary and enter the venular end. (p. 356)

17. Distinguish between a vein and a venule. (p. 357)

13.5 Blood Pressure

18. Arterial blood pressure reaches its maximum when the ventricles contract. This point is called _____. (p. 358)

19. Name several factors that influence blood pressure, and explain how each produces its effect. (p. 359)

20. Describe the control of blood pressure. (p. 359)

21. Which of the following is *not* one of the major factors that promote the flow of venous blood? (p. 361)
    - a. skeletal muscle contraction
    - b. breathing
    - c. arterial blood pressure
    - d. venoconstriction

13.6 Paths of Circulation

22. Distinguish between the pulmonary and systemic circuits of the cardiovascular system. (p. 362)

13.7–13.8 Arterial System–Venous System

23. Describe the aorta, and name its principal branches. (p. 362)

24. Discuss the relationship between the major venous pathways and the major arterial pathways to the head, upper limbs, abdominal viscera, and lower limbs. (p. 366)

INTEGRATIVE ASSESSMENTS/ CRITICAL THINKING

OUTCOMES 13.2, 13.3

1. What structures and properties should an artificial heart have?

OUTCOME 13.4

2. If you were asked to invent a blood vessel (artery, capillary, or vein) substitute, what materials might you use to build it? Include synthetic as well as natural materials.

OUTCOMES 13.4, 13.5, 13.8

3. Cirrhosis of the liver, a disease commonly associated with alcoholism, obstructs blood flow through hepatic blood vessels. As a result, blood backs up and capillary pressure greatly increases in organs drained by the hepatic portal system. What effects might this increasing capillary pressure produce, and which organs would it affect?

OUTCOME 13.7

4. If a cardiologist inserts a catheter into a patient's right femoral artery, which arteries will the tube have to pass through in order to reach the entrance to the left coronary artery?

OUTCOME 13.8

5. If a patient develops a blood clot in the femoral vein of the left lower limb and a portion of the clot breaks loose, where is the blood flow likely to carry the embolus? What symptoms are likely?

WEB CONNECTIONS

Visit the text website at **aris.mhhe.com** for additional quizzes, interactive learning exercises, and more.

AP|R CARDIOVASCULAR SYSTEM

Anatomy & Physiology | REVEALED includes cadaver photos that allow you to peel away layers of the human body to reveal structures beneath the surface. This program also includes animations, radiologic imaging, audio pronunciations, and practice quizzing. To learn more visit **www.aprevealed.com**.

14

Lymphatic System and Immunity

PEANUT ALLERGY. The young woman was admitted to the emergency department for sudden onset of difficulty breathing. Although she had a history of asthma, this was something different, for she was also flushed and had vomited. An astute medical student taking a quick history from the woman's roommates discovered that she had just eaten cookies from a vending machine in their dorm. Suspecting that the cookies may have contained peanuts, the student alerted the attending physician, who treated the woman for suspected peanut allergy—giving oxygen, an antihistamine, a steroid drug, and epinephrine. She recovered.

Peanut allergy is common and on the rise, but only in certain westernized countries. In the United States, 6% to 8% of children under the age of four and 2% of the population over ten years of age are allergic to peanuts. About 30,000 people react each year, and about 200 die.

Peculiarities of peanuts and our fondness for them may explain why allergy is apparently increasing. Three glycoproteins in peanuts are allergens, causing the misdirected immune response that constitutes an allergy. These glycoproteins are highly concentrated in the peanut, and when eaten, they disturb the intestinal lining in such a way that they enter the circulation rapidly, without being digested. As a result, many allergens gain quick and easy access to immune system cells located beneath the intestinal lining.

Compounding the rapidity with which peanut allergens flood the bloodstream is the fact that people in the United States eat many peanuts. Virtually everyone has eaten a peanut by two years of age, usually in peanut butter. This is sufficient exposure to set the stage for later allergy in genetically predisposed individuals. The fact that the

Peculiarities of peanuts, combined with our fondness for them, sets the stage for allergy, a misplaced immune reaction.

average age of first allergic reaction to peanuts is fourteen months suggests that the initial exposure—necessary to "prime" the immune system for future response—happens early either through breast milk or in the uterus. This hypothesis is consistent with the popularity of peanuts among pregnant and nursing women. Countries where peanuts are rarely eaten, such as Denmark and Norway, have very low incidence of peanut allergy.

The method of peanut preparation in the United States—dry roasting—may make the three glycoproteins that evoke the allergic response more active. In China, where peanuts are equally popular but are boiled or fried, allergy is rare. However, children of Chinese immigrants in the United States have the same incidence of peanut allergy as other children in the U.S., supporting the idea that method of preparation contributes to allergenicity.

Learning Outcomes   *After studying this chapter, you should be able to do the following:*

14.1 Introduction

1. Describe the general functions of the lymphatic system. (p. 377)

14.2 Lymphatic Pathways

2. Identify the locations of the major lymphatic pathways. (p. 378)

14.3 Tissue Fluid and Lymph

3. Describe how tissue fluid and lymph form, and explain the function of lymph. (p. 378)

14.4 Lymph Movement

4. Explain how lymphatic circulation is maintained. (p. 379)

14.5 Lymph Nodes

5. Describe a lymph node and its major functions. (p. 380)

14.6 Thymus and Spleen

6. Discuss the locations and functions of the thymus and spleen. (p. 381)

14.7 Body Defenses Against Infection

7. Distinguish between innate (nonspecific) and adaptive (specific) defenses. (p. 383)

14.8 Innate (Nonspecific) Defenses

8. List seven innate body defense mechanisms, and describe the action of each mechanism. (p. 384)

14.9 Adaptive (Specific) Defenses, or Immunity

9. Explain how two major types of lymphocytes are formed and activated, and how they function in immune mechanisms. (p. 385)

10. Discuss the origins and actions of the five different types of immunoglobulins. (p. 390)

11. Distinguish between primary and secondary immune responses. (p. 391)

12. Distinguish between active and passive immunity. (p. 392)

13. Explain how allergic reactions, tissue rejection reactions, and autoimmunity arise from immune mechanisms. (p. 392)

Aids to Understanding Words   *(Appendix A on page 567 has a complete list of Aids to Understanding Words.)*

-gen [be produced] aller*gen:* Substance that stimulates an allergic response.

humor- [fluid] *humor*al immunity: Immunity resulting from antibodies in body fluids.

immun- [free] *immun*ity: Resistance to (freedom from) a specific disease.

inflamm- [set on fire] *inflamm*ation: Localized redness, heat, swelling, and pain in tissues.

nod- [knot] *nod*ule: Small mass of lymphocytes surrounded by connective tissue.

patho- [disease] *patho*gen: Disease-causing agent.

14.1 INTRODUCTION

The **lymphatic** (lim-fat′ik) **system,** like the cardiovascular system, includes a network of vessels that transports fluids. The lymphatic system is composed of a vast collection of cells and biochemicals that travel in lymphatic vessels, and the organs and glands that produce them. A major function of lymphatic vessels is to transport excess fluid away from interstitial spaces in most tissues and return it to the bloodstream (fig. 14.1). Without the lymphatic system, this fluid would accumulate in tissue spaces. Special lymphatic capillaries, called *lacteals* (lak′te-alz), are located in the lining of the small intestine, where they absorb digested fats and transport them to the venous circulation.

The lymphatic system has a second major function—it enables us to live in a world with different types of organisms, some of which take up residence in or on the human body and may cause infectious diseases. Cells and biochemicals of the lymphatic system launch both generalized and targeted attacks against "foreign" particles, enabling the body to destroy infectious microorganisms and viruses. This immunity against disease also protects against toxins and cancer cells. When the immune response is abnormal, persistent infection, cancer, autoimmune disorders, and allergies may result.

14.2 LYMPHATIC PATHWAYS

The **lymphatic pathways** begin as lymphatic capillaries. These tiny tubes merge to form larger lymphatic vessels, which in turn lead to larger vessels that unite with the veins in the thorax.

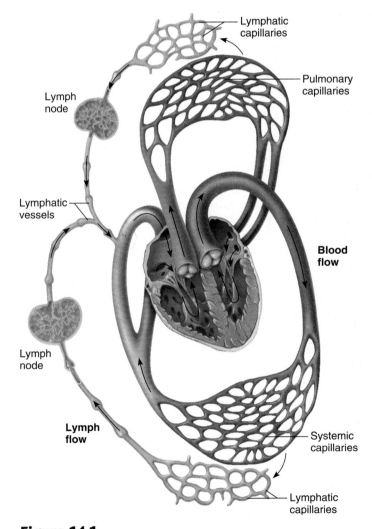

Figure 14.1

Schematic representation of lymphatic vessels transporting fluid from interstitial spaces to the bloodstream.

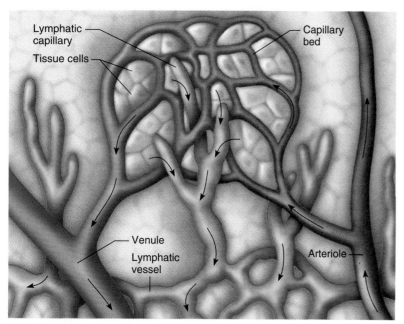

Figure 14.2
Lymphatic capillaries are microscopic, closed-ended tubes that originate in the interstitial spaces of most tissues.

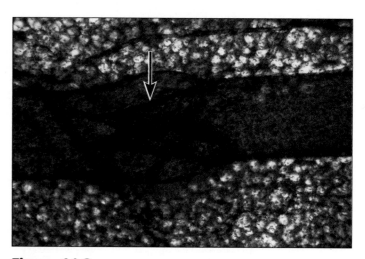

Figure 14.3
Light micrograph of the flaplike valve (arrow) in a lymphatic vessel (25×).

Lymphatic Capillaries

Lymphatic capillaries are microscopic, closed-ended tubes (fig. 14.2). They extend into interstitial spaces, forming complex networks that parallel those of blood capillaries. The walls of lymphatic capillaries, like those of blood capillaries, are formed from a single layer of squamous epithelial cells. These thin walls allow tissue fluid to enter lymphatic capillaries. The fluid inside lymphatic capillaries is called **lymph** (limf).

Lymphatic Vessels

The walls of **lymphatic vessels** are similar to those of veins, but thinner. Also like veins, lymphatic vessels have flaplike valves that help prevent backflow of lymph (fig. 14.3).

The larger lymphatic vessels lead to specialized organs called **lymph nodes** (limf nōdz). After leaving the nodes, the vessels merge to form still larger lymphatic trunks.

Lymphatic Trunks and Collecting Ducts

Lymphatic trunks, which drain lymph from the lymphatic vessels, are named for the regions they serve. They join one of two **collecting ducts**—the thoracic duct or the right lymphatic duct (fig. 14.4*a*).

The **thoracic duct** is the larger and longer collecting duct. It receives lymph from the lower limbs and abdominal regions, left upper limb, and left side of the thorax, head, and neck, and empties into the left subclavian vein near the junction of the left jugular vein. The **right lymphatic duct** receives lymph from the right side of the head and neck, right upper limb, and right thorax, and empties into the right subclavian vein near the junction of the right jugular vein.

After leaving the two collecting ducts, lymph enters the venous system and becomes part of the plasma just before blood returns to the right atrium. Figure 14.5 summarizes the typical lymphatic pathway.

The skin has many lymphatic capillaries. Consequently, if the skin is broken or if something is injected into it (such as venom from a stinging insect), foreign substances rapidly enter the lymphatic system.

Check Your Recall

1. What are the general functions of the lymphatic system?
2. Distinguish between the thoracic duct and the right lymphatic duct.

14.3 TISSUE FLUID AND LYMPH

Lymph is essentially tissue fluid that has entered a lymphatic capillary. Thus, lymph formation depends upon tissue fluid formation.

Tissue Fluid Formation

Recall from chapter 13 (pp. 356–357) that tissue fluid originates from blood plasma and is composed of water and dissolved substances that leave blood capillaries.

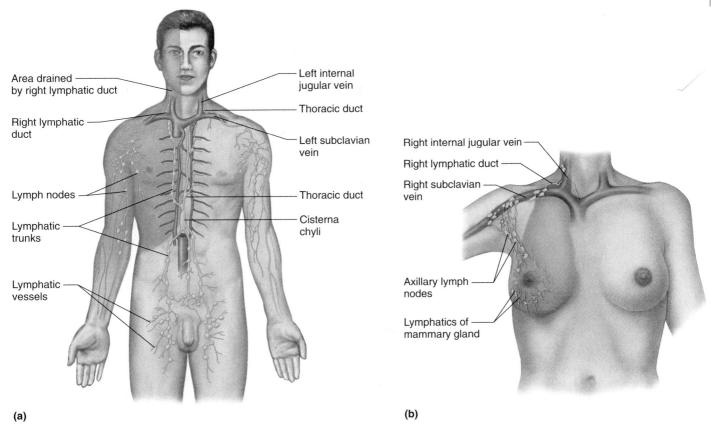

Figure 14.4

Lymphatic pathways. (*a*) The right lymphatic duct drains lymph from the upper right side of the body, whereas the thoracic duct drains lymph from the rest of the body. (*b*) Lymph drainage of the right breast illustrates a localized function of the lymphatic system. Surgery to treat breast cancer can disrupt this drainage, causing painful swelling (edema) in the arm.

Capillary blood pressure filters water and small molecules from the plasma. The resulting fluid is very similar in composition to the blood plasma (including nutrients, gases, and hormones), with the important exception of the plasma proteins, which are generally too large to pass through the capillary walls. The osmotic effect of these (called the *plasma colloid osmotic pressure*) helps draw fluid back into the capillaries by osmosis.

Lymph Formation and Function

Filtration from the plasma normally exceeds reabsorption, leading to the net formation of tissue fluid. This increases the tissue fluid hydrostatic pressure somewhat, favoring movement of tissue fluid into lymphatic capillaries, forming lymph (see fig. 14.2). Lymph returns to the bloodstream most of the small proteins that the blood capillaries filtered. At the same time, lymph transports foreign particles, such as bacteria or viruses, to lymph nodes.

Check Your Recall

3. What is the relationship between tissue fluid and lymph?
4. How do plasma proteins in tissue fluid affect lymph formation?
5. What are the major functions of lymph?

14.4 LYMPH MOVEMENT

The hydrostatic pressure of tissue fluid drives the entry of lymph into lymphatic capillaries. However, muscular activity largely influences the movement of lymph through the lymphatic vessels. Lymph, like venous blood, is under low hydrostatic pressure and may not flow readily through lymphatic vessels without help from contraction of skeletal muscles, contraction of the smooth muscle in the walls of the larger lymphatic trunks, and pressure changes associated with breathing.

Contracting skeletal muscles compress lymphatic vessels and move the lymph inside lymphatic vessels. Valves in these vessels prevent backflow, so lymph can only move toward a collecting duct. Additionally, the smooth muscle in the walls of larger lymphatic trunks can contract and compress the lymph inside, forcing the fluid onward.

Breathing aids lymph circulation by creating a relatively low pressure in the thoracic cavity during inhalation. At the same time, the contracting diaphragm increases the pressure in the abdominal cavity. Together, these actions squeeze lymph out of the abdominal vessels and force it into the thoracic vessels. Once again, valves within lymphatic vessels prevent lymph backflow.

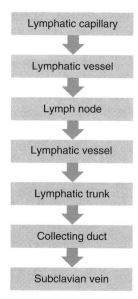

Figure 14.5

The lymphatic pathway.

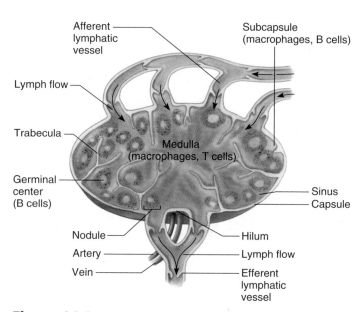

Figure 14.6

A section of a lymph node.

The continuous movement of fluid from interstitial spaces into blood and lymphatic capillaries stabilizes the volume of fluid in these spaces. Conditions that interfere with lymph movement cause tissue fluid to accumulate within the interstitial spaces, producing *edema,* or swelling. This may happen when surgery removes lymphatic tissue, obstructing certain lymphatic vessels. For example, a surgeon removing a cancerous breast tumor may also remove nearby axillary lymph nodes to prevent associated lymphatic vessels from transporting cancer cells to other sites. Removing the lymphatic tissue can obstruct drainage from the upper limb, causing edema (see fig. 14.4*b*).

Check Your Recall

6. What factors promote lymph flow?
7. What is the consequence of lymphatic obstruction?

14.5 LYMPH NODES

Lymph nodes (lymph glands) are located along the lymphatic pathways. They contain large numbers of *lymphocytes* and *macrophages* that fight invading microorganisms.

Structure of a Lymph Node

Lymph nodes vary in size and shape, but are usually less than 2.5 centimeters long and somewhat bean-shaped (figs. 14.6 and 14.7). Blood vessels and nerves join a

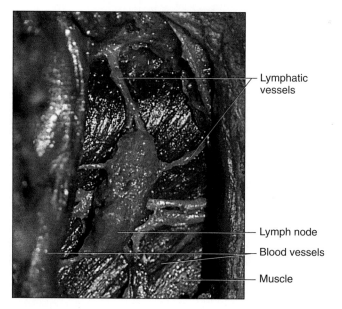

Figure 14.7

Lymph enters and leaves a lymph node through lymphatic vessels.

lymph node through the indented region of the node, called the **hilum.** The lymphatic vessels leading to a node (afferent vessels) enter separately at various points on its convex surface, but the lymphatic vessels leaving the node (efferent vessels) exit from the hilum.

A *capsule* of connective tissue encloses each lymph node and subdivides it into compartments. Masses of B cells and macrophages in the cortex, called **lymph nodules** (lymph follicles), are the functional units of the lymph node. The spaces within a node, called **lymph sinuses,** provide a complex network of chambers and channels through which lymph circulates. Macrophages are most highly concentrated in the lymph sinuses.

Lymph nodules occur singly or in groups associated with the mucous membranes of the respiratory and digestive tracts. The *tonsils,* described in chapter 15 (pp. 405–406), are partially encapsulated lymph nodules. Aggregations of nodules called *Peyer's patches* are scattered throughout the mucosal lining of the distal portion of the small intestine.

Locations of Lymph Nodes

Lymph nodes are generally in groups or chains along the paths of the larger lymphatic vessels throughout the body, but are absent in the central nervous system. Figure 14.8 shows the locations of the major lymph nodes.

Functions of Lymph Nodes

Lymph nodes have two primary functions: (1) filtering potentially harmful particles from lymph before returning it to the bloodstream, and (2) monitoring

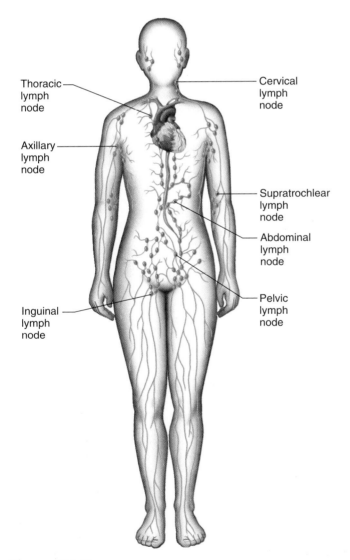

Figure 14.8
Locations of major lymph nodes.

Thoracic lymph node

Axillary lymph node

Inguinal lymph node

Cervical lymph node

Supratrochlear lymph node

Abdominal lymph node

Pelvic lymph node

body fluids (immune surveillance) provided by lymphocytes and macrophages. Along with red bone marrow, the lymph nodes are centers for lymphocyte production. Lymphocytes attack invading viruses, bacteria, and other parasitic cells that lymphatic vessels bring to the nodes. Macrophages in the nodes engulf and destroy foreign substances, damaged cells, and cellular debris.

> Superficial lymphatic vessels inflamed by bacterial infection appear as red streaks beneath the skin, a condition called *lymphangitis.* Inflammation of the lymph nodes, called *lymphadenitis,* often follows. Affected nodes enlarge and may be quite painful.

Check Your Recall

8. Distinguish between a lymph node and a lymph nodule.
9. What are the major functions of the lymph nodes?

14.6 THYMUS AND SPLEEN

Two other lymphatic organs whose functions are similar to those of the lymph nodes are the thymus and the spleen.

Thymus

The **thymus** (thī′mus) is a soft, bilobed structure enclosed in a connective tissue capsule and located anterior to the aorta and posterior to the upper part of the sternum (fig. 14.9*a*). The thymus is relatively large during infancy and early childhood, but shrinks after puberty and may be quite small in an adult. In elderly persons, adipose and connective tissues replace lymphatic tissue in the thymus.

Connective tissues extend inward from the surface of the thymus, subdividing the gland into *lobules* (fig. 14.9*b*). The lobules contain many lymphocytes. Most of these cells (thymocytes) are inactive; however, some mature into *T lymphocytes,* which leave the thymus and provide immunity. Epithelial cells in the thymus secrete hormones called *thymosins,* which stimulate maturation of T lymphocytes after they leave the thymus and migrate to other lymphatic tissues.

> By age seventy years, the thymus is one-tenth the size it was at the age of ten, and the immune system is only 25% as powerful.

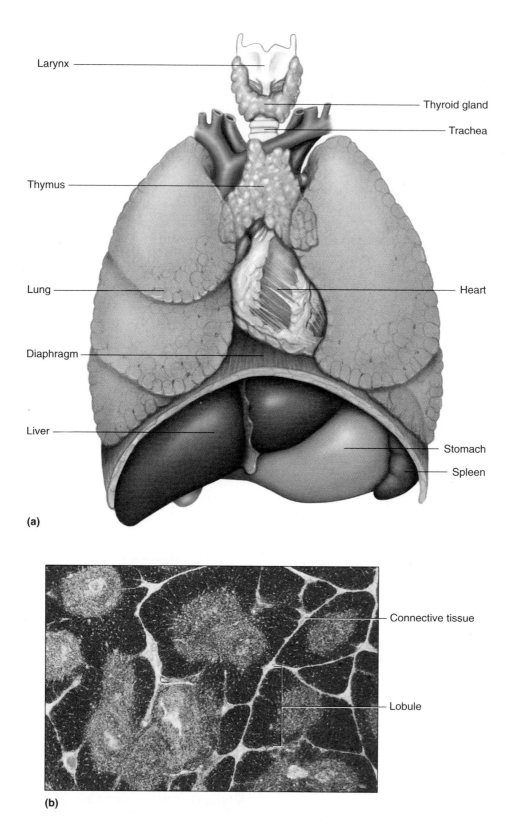

(a)

Larynx

Thyroid gland

Trachea

Thymus

Lung

Diaphragm

Heart

Liver

Stomach

Spleen

Connective tissue

Lobule

(b)

Figure 14.9

Thymus and spleen. (*a*) The thymus is bilobed and located between the lungs and superior to the heart. The spleen is located inferior to the diaphragm and posterior and lateral to the stomach. (*b*) A cross section of the thymus (20×). Note how the gland is subdivided into lobules.

Spleen

The **spleen** (splēn), the largest lymphatic organ, is in the upper left portion of the abdominal cavity, just inferior to the diaphragm and posterior and lateral to the stomach (fig. 14.9*a*). The spleen resembles a large lymph node and is subdivided into lobules. However, unlike the sinuses of a lymph node, the spaces (venous sinuses) of the spleen are filled with blood instead of lymph.

The tissues within splenic lobules are of two types (fig. 14.10). The *white pulp* is distributed throughout the spleen in tiny islands. This tissue is composed of splenic nodules, which are similar to those in lymph nodes and contain many lymphocytes. The *red pulp,* which fills the remaining spaces of the lobules, surrounds the venous sinuses. This pulp contains numerous red blood cells, which impart its color, plus many lymphocytes and macrophages.

Blood capillaries within the red pulp are quite permeable. Red blood cells can squeeze through the pores in these capillary walls and enter the venous sinuses. The older, more fragile red blood cells may rupture as they make this passage, and the resulting cellular debris is removed by phagocytic macrophages within the splenic sinuses. These macrophages also engulf and destroy foreign particles, such as bacteria, that may be carried in the blood as it flows through the splenic sinuses. Thus, the spleen filters blood much as the lymph nodes filter lymph.

Check Your Recall

10. Why are the thymus and the spleen considered organs of the lymphatic system?

11. What are the major functions of the thymus and the spleen?

14.7 BODY DEFENSES AGAINST INFECTION

The presence and multiplication of a disease-causing agent, or **pathogen** (path′o-jen), may cause an **infection.** Pathogens include viruses, bacteria, fungi, and protozoans.

The human body can prevent the entry of pathogens or destroy them if they enter. Some mechanisms are quite general and protect against many types of pathogens, providing **innate (nonspecific) defense.** These mechanisms include species resistance, mechanical barriers, chemical barriers (enzyme action, interferon, and complement), natural killer cells, inflammation, phagocytosis, and fever. Other defense mechanisms are very precise, targeting specific pathogens and providing **adaptive (specific) defense,** or **immunity** (ĭ-mu′nĭ-te). Specialized lymphocytes that recognize foreign molecules (nonself antigens) in the body act against them. Innate and adaptive defense mechanisms work together to protect the body against infection. While the innate defenses respond quite rapidly, slower-to-respond adaptive defenses begin as well.

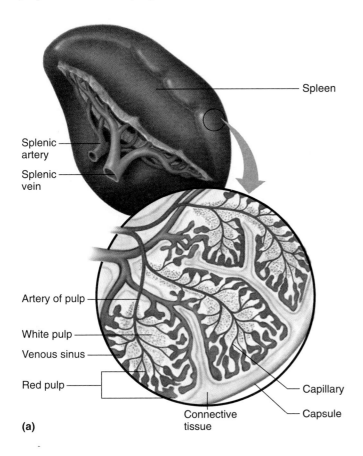

(a)

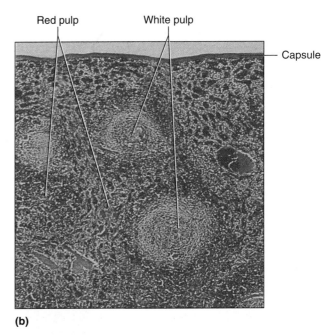

(b)

Figure 14.10

Spleen. (*a*) The spleen resembles a large lymph node. (*b*) Light micrograph of the spleen (15×).

14.8 INNATE (NONSPECIFIC) DEFENSES

Species Resistance

Species resistance refers to the fact that a given kind of organism, or *species* (such as the human species, *Homo sapiens*), develops diseases that are unique to it. A species may be resistant to diseases that affect other species because its tissues somehow fail to provide the temperature or chemical environment that a particular pathogen requires. For example, the infectious agents that cause measles, mumps, gonorrhea, and syphilis infect humans, but not other animal species.

Mechanical Barriers

The skin as well as the mucous membranes lining the passageways of the respiratory, digestive, urinary, and reproductive systems create **mechanical barriers** against some infectious agents. Along with the hair that traps infectious agents associated with the skin and mucous membranes is the fluid (sweat and mucus) that rinses away microorganisms. These barriers provide a *first line of defense*. As long as the skin and mucous membranes remain intact, they can keep out many pathogens. The rest of the innate defenses discussed in this section are part of the *second line of defense*.

Chemical Barriers

Enzymes in body fluids provide a **chemical barrier** to pathogens. Gastric juice, for example, contains the protein-splitting enzyme pepsin and has a low pH due to the presence of hydrochloric acid (HCl) (see chapter 15, p. 412). The combined effect of pepsin and HCl is lethal to many pathogens that enter the stomach. Similarly, tears contain the enzyme lysozyme, which has an antibacterial action against certain pathogens that may get onto eye surfaces. Finally, the accumulation of salt from perspiration kills some types of bacteria on the skin.

Lymphocytes and fibroblasts produce hormonelike peptides called **interferons** in response to viruses or tumor cells. Once released from the virus-infected cell, interferon binds to receptors on uninfected cells, stimulating them to synthesize proteins that block replication of a variety of viruses. Thus, the effect of interferon is nonspecific. Interferons also stimulate phagocytosis and enhance the activity of other cells that help the body resist infections and the growth of tumors.

Complement (kom′ple-ment) is a group of proteins, in plasma and other body fluids, that interact in a series of reactions or cascade. Activation of complement stimulates inflammation, attracts phagocytes, and enhances phagocytosis.

Natural Killer (NK) Cells

Natural killer (NK) **cells** are a small population of lymphocytes, distinctly different from the lymphocytes that provide adaptive (specific) defense mechanisms (discussed later in this chapter). NK cells defend the body against various viruses and cancer cells by secreting cytolytic ("cell-cutting") substances called **perforins** that lyse the cell membrane, destroying the infected cell. NK cells also secrete chemicals that enhance inflammation.

Inflammation

Inflammation (in″flah-ma′shun) is a tissue response to injury or infection, producing localized redness, swelling, heat, and pain. The redness is a result of blood vessel dilation that increases blood flow and volume within the affected tissues. This effect, coupled with an increase in the permeability of nearby capillaries, swells tissues (edema). The heat comes from blood from deeper body parts, which is generally warmer than that near the surface. Pain results from stimulation of nearby pain receptors.

Infected cells release chemicals that attract white blood cells to inflammation sites, where they phagocytize pathogens. Local heat speeds up phagocytic activity. In bacterial infections, the resulting mass of white blood cells, bacterial cells, and damaged tissue may form a thick fluid called **pus.**

Body fluids also collect in inflamed tissues. These fluids contain fibrinogen and other blood-clotting factors. Clotting forms a network of fibrin threads in the affected region. Later, fibroblasts may arrive and secrete matrix components until the area is enclosed in a connective tissue sac. Partitioning off the infected area inhibits the spread of pathogens and toxins to adjacent tissues.

Phagocytosis

Recall from chapter 12 (p. 324) that blood's most active phagocytic cells are *neutrophils* and *monocytes*. These cells can leave the bloodstream by squeezing between the cells of blood vessel walls (diapedesis). Chemicals released from injured tissues attract these cells (chemotaxis). Neutrophils engulf and digest smaller particles; monocytes phagocytize larger ones.

Monocytes give rise to *macrophages* (histiocytes), which become fixed in various tissues and attach to the inner walls of blood vessels and lymphatic vessels. These relatively nonmotile phagocytic cells, which can divide and produce new macrophages, are found in the lymph nodes, spleen, liver, and lungs. Monocytes, macrophages, and neutrophils constitute the **mononuclear phagocytic system** (reticuloendothelial system).

Phagocytosis removes foreign particles from the lymph as it moves from the interstitial spaces to the bloodstream. Phagocytes in the blood vessels and in the tissues of the spleen, liver, or bone marrow remove particles that reach the blood.

Fever

Elevated body temperature due to **fever** offers powerful protection. Higher body temperature causes the liver and spleen to sequester iron, which reduces the level of iron in the blood. Because bacteria and fungi require more iron as temperature rises, their growth and reproduction in a fever-ridden body slow and may cease. Also, phagocytic cells attack more vigorously when the temperature rises. For these reasons, low-grade fever of short duration may be a desired response, not a symptom to be treated aggressively with medications.

> ### Check Your Recall
> 12. What is an infection?
> 13. Explain seven innate (nonspecific) defense mechanisms.

14.9 ADAPTIVE (SPECIFIC) DEFENSES, OR IMMUNITY

The *third line of defense,* immunity, is resistance to specific pathogens or to their toxins or metabolic byproducts. Lymphocytes and macrophages that recognize and remember specific foreign molecules carry out adaptive immune responses.

Antigens

Antigens may be proteins, polysaccharides, glycoproteins, or glycolipids, usually located on a cell's surface. Before birth, cells inventory the proteins and other large molecules in the body, learning to recognize them as "self." The lymphatic system responds to nonself, or foreign, antigens, but not normally to self antigens. Receptors on lymphocyte surfaces enable these cells to recognize foreign antigens.

The antigens that are most effective in eliciting an immune response are large and complex, with few repeating parts. Sometimes, a smaller molecule that cannot by itself stimulate an immune response combines with a larger one, which makes it able to do so. Such a small molecule is called a **hapten** (hap'ten). Stimulated lymphocytes react either to the hapten or to the larger molecule of the combination. Haptens are found in certain drugs such as penicillin, in household and industrial chemicals, in dust particles, and in products of animal skins (dander).

Lymphocyte Origins

During fetal development (before birth), red bone marrow releases relatively unspecialized precursors to lymphocytes into the circulation (figs. 14.11 and 14.12). About half of these cells reach the thymus, where they specialize into **T lymphocytes,** or **T cells.** Later, some of these T cells constitute 70–80% of the circulating lymphocytes in blood. Other T cells reside in lymphatic organs and are particularly abundant in the lymph nodes, thoracic duct, and white pulp of the spleen.

Other lymphocytes remain in the red bone marrow until they differentiate into **B lymphocytes,** or **B cells.** The blood distributes B cells, which constitute 20–30% of circulating lymphocytes. B cells settle in lymphatic organs along with T cells and are abundant in the lymph nodes, spleen, bone marrow, and intestinal lining (see fig. 14.11).

> ### Check Your Recall
> 14. What is immunity?
> 15. What is the difference between an antigen and a hapten?
> 16. How do T cells and B cells originate?

Lymphocyte Functions

T cells and B cells respond in different ways to antigens they recognize. T cells attach to foreign, antigen-bearing cells, such as bacterial cells, and interact directly—that is, by cell-to-cell contact. This is called the **cellular immune response,** or cell-mediated immunity.

T cells (and some macrophages) also synthesize and secrete polypeptides called *cytokines* that enhance certain cellular responses to antigens. For example, *interleukin-1* and *interleukin-2* stimulate the synthesis of several cytokines from other T cells. In addition, interleukin-1 helps activate T cells, whereas interleukin-2 causes T cells to proliferate and activates a specific type of T cell (cytotoxic T cells). Other cytokines, called *colony-stimulating factors (CSFs),* stimulate leukocyte production in red bone marrow, cause B cells to grow and mature, and activate macrophages. T cells may also secrete toxins that kill their antigen-bearing target cells, as well as growth-inhibiting factors that prevent target-cell growth, and interferon that inhibits the proliferation of viruses and tumor cells.

B cells attack foreign antigens in a different way. When stimulated, they divide to give rise to cells that differentiate into **plasma cells,** which produce and secrete large globular proteins called **antibodies,** also known as **immunoglobulins** (im″u-no-glob′u-linz). Body fluids carry antibodies, which then react in various ways to destroy specific antigens or antigen-bearing particles. This antibody-mediated immune response is called the **humoral immune response** ("humoral" refers to fluid).

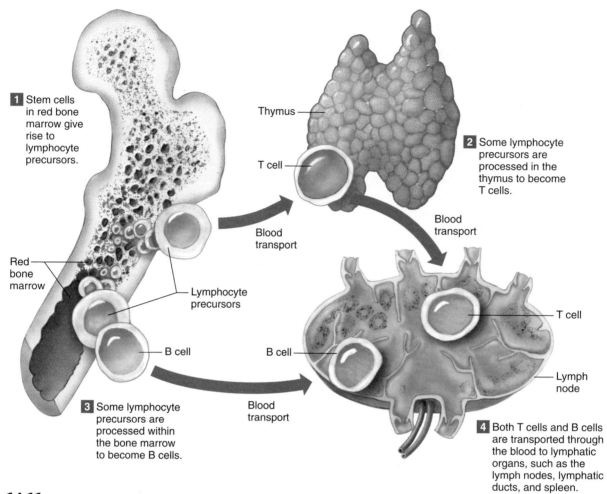

1 Stem cells in red bone marrow give rise to lymphocyte precursors.

Thymus

T cell

2 Some lymphocyte precursors are processed in the thymus to become T cells.

Blood transport

Blood transport

Red bone marrow

Lymphocyte precursors

T cell

B cell

B cell

Lymph node

3 Some lymphocyte precursors are processed within the bone marrow to become B cells.

Blood transport

4 Both T cells and B cells are transported through the blood to lymphatic organs, such as the lymph nodes, lymphatic ducts, and spleen.

Figure 14.11

Bone marrow releases relatively unspecialized lymphocyte precursors, which after processing specialize as T cells (T lymphocytes) or B cells (B lymphocytes). Note that in the fetus, the medullary cavity contains red marrow.

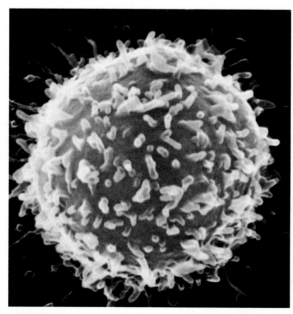

Figure 14.12

Scanning electron micrograph of a human circulating lymphocyte (7,000×).

Each person has millions of varieties of T and B cells. Because the members of each variety originate from a single early cell, they are all alike, forming a **clone** (klōn) of cells (identical cells originating from division of a single cell). The members of each variety have a particular type of antigen receptor on their cell membranes that can respond only to a specific antigen. Table 14.1 compares the characteristics of T cells and B cells.

T Cells and the Cellular Immune Response

A lymphocyte must be activated before it can respond to an antigen. T cell activation requires the presence of processed fragments of antigen attached to the surface of another kind of cell, called an **antigen-presenting cell** (accessory cell). Macrophages, B cells, and several other cell types can be antigen-presenting cells.

T cell activation begins when a macrophage phagocytizes a bacterium, digesting it in its lysosomes. Some bacterial antigens exit the lysosomes and move to the macrophage's surface. Here, they are displayed on the cell membrane near certain protein molecules that are

Table 14.1	A Comparison of T Cells and B Cells	
	T Cells	**B Cells**
Origin of Undifferentiated Cell	Red bone marrow	Red bone marrow
Site of Differentiation	Thymus	Red bone marrow
Primary Locations	Lymphatic tissues, 70–80% of the circulating lymphocytes	Lymphatic tissues, 20–30% of the circulating lymphocytes
Primary Functions	Provides cellular immune response in which T cells interact directly with the antigens or antigen-bearing agents to destroy them	Provides humoral immune response in which B cells interact indirectly, producing antibodies that destroy the antigens or antigen-bearing agents

part of a group of proteins called the *major histocompatibility complex (MHC).*

MHC antigens help T cells recognize that an antigen is foreign, not self. A specialized type of T cell, called a *helper T cell,* contacts the displayed foreign antigen (fig. 14.13). If the displayed antigen fits and combines with the helper T cell's antigen receptors, the helper cell becomes activated. Once activated, the helper T cell

stimulates a B cell to produce antibodies that are specific for the displayed antigen.

A second type of T cell is a *cytotoxic T cell,* which recognizes and combines with nonself antigens that cancerous cells or virally infected cells display on their surfaces near certain MHC proteins. Cytokines from helper T cells activate the cytotoxic T cell. Next, the cytotoxic T cell proliferates, enlarging its clone of cells. Cytotoxic T cells then

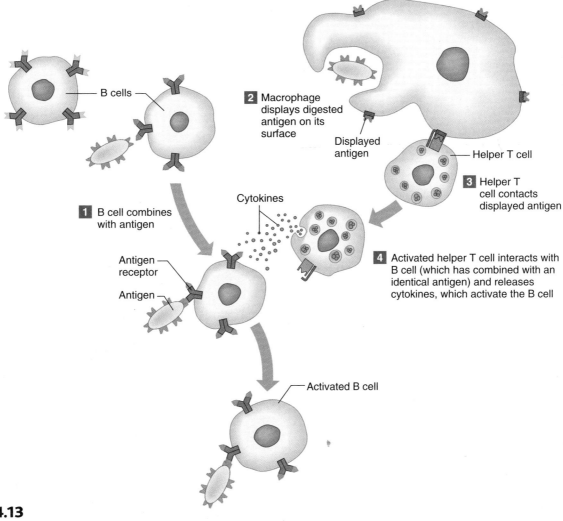

Figure 14.13

T cell and B cell activation.

bind to the surfaces of antigen-bearing cells, where they release a protein that cuts porelike openings, destroying these cells. In this way, cytotoxic T cells continually monitor body cells, recognizing and eliminating tumor cells and cells infected with viruses. Cytotoxic T cells provide much of the immune system's defense against HIV infection, discussed in the Topic of Interest on page 389. Unfortunately, the virus eventually kills these cells.

Some T cells do not respond to a foreign antigen on first exposure, but remain as *memory cells*. These memory cells immediately divide to yield more cytotoxic T cells and helper T cells upon subsequent exposure to the same antigen, often even before symptoms arise.

Check Your Recall

17. What are the functions of T cells and B cells?
18. How do T cells become activated?
19. What is the function of cytokines?
20. How do cytotoxic T cells destroy cells bearing foreign antigens?

B Cells and the Humoral Immune Response

Sometimes a B cell may become activated when it encounters an antigen whose molecular shape fits the shape of the B cell's antigen receptors. In response to the receptor-antigen combination, the B cell divides repeatedly, expanding its clone. However, most of the time B cell activation requires T cell "help."

When an activated helper T cell encounters a B cell that has already bound an identical foreign antigen, the helper cell releases certain cytokines that stimulate the B cell to proliferate, enlarging its clone of antibody-producing cells (fig. 14.14). The cytokines also attract macrophages and leukocytes into inflamed tissues and help keep them there.

Some members of the activated B cell's clone differentiate further into *memory cells*. Like memory T cells, these memory B cells respond rapidly to subsequent exposure to a specific antigen.

Other members of the activated B cell's clone differentiate further into antibody-secreting *plasma cells*. These antibodies are similar in structure to the antigen receptor

Figure 14.14

An activated B cell proliferates after stimulation by cytokines released from helper T cells. The B cell's clone enlarges. Some cells of the clone give rise to antibody-secreting plasma cells and others to dormant memory cells.

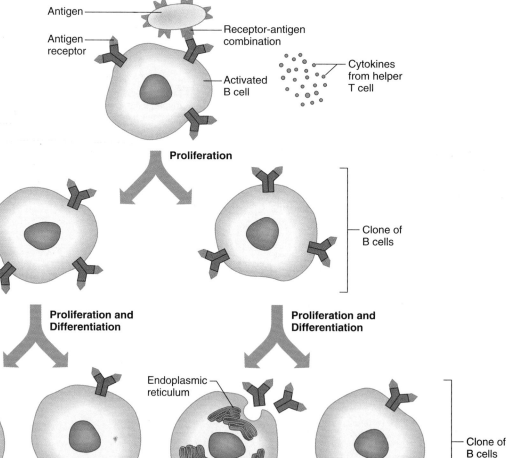

Topic of Interest

Immunity Breakdown: AIDS

In the early 1980s, health care workers from large cities in the United States began seeing otherwise healthy young men who had rare infections and cancers. The conditions were opportunistic, taking advantage of a weakened immune system. Soon, a portrait of an apparently new lethal disease emerged—acquired immune deficiency syndrome or AIDS.

The human immunodeficiency virus (HIV), which causes AIDS, can be present for a decade or longer before symptoms begin, such as recurrent fever, weakness, and weight loss. Then, usually after another healthy period, infections begin. However, a small percentage of people who are HIV positive have remained healthy, and people with rare mutations that remove the T cell receptors to which HIV binds cannot become infected at all.

Transmission of HIV requires contact with a body fluid containing abundant virus, such as blood or semen. Levels of the virus in sweat, tears, and saliva are so low that transmission is highly unlikely. Whether or not a person becomes infected depends on the amount of infected fluid contacted, the site of exposure in the body, the individual's health, and genetic background. Table 14A lists some of the ways that HIV infection can and cannot be spread.

HIV infection gradually shuts down the immune system. First, HIV enters macrophages, impairing this first line of defense. In these cells and later in helper T cells, the virus adheres with a surface protein, called gp120, to two receptors on the host cell surface, called CD4 and CCR5. Once the virus enters the cell, a viral enzyme, reverse transcriptase, catalyzes construction of a DNA strand complementary to the viral RNA sequence (the virus has RNA as its genetic material). The initial viral DNA strand replicates to form a DNA double helix, which enters the cell's nucleus and inserts into a chromosome. The viral DNA sequences are then transcribed and translated. The cell fills with viral pieces, which are assembled into complete new viral particles that burst from the cell.

Once infected helper T cells start to die at a high rate, bacterial infections begin, because B cells aren't activated to produce antibodies. Much later in infection, HIV variants arise that bind to receptors on cytotoxic T cells, killing them too. Loss of these cells renders the body very vulnerable to other infections and to cancers.

HIV replicates quickly, mutates often, and can hide, twisting and altering its surface features in ways that evade recognition and attack by antibodies or cytotoxic T cells. For several years, the bone marrow produces 2 billion new T and B cells a day, countering the million to billion new HIV particles that infected cells release daily.

So genetically diverse is the population of HIV in a human host that within days of initial infection, viral variants can arise that resist the drugs used to treat HIV infection and AIDS. HIV's changeable nature has important clinical implications. Combining drugs that act in different ways minimizes the number of viruses and delays symptom onset and progression. The first drugs developed block viral replication. A second class of drugs, called protease inhibitors, prevent HIV from processing its proteins to a functional size. A third class of drugs, called entry inhibitors, either block the binding and/or fusing of HIV to T cell surfaces or keep the virus out even if it can bind to the cell. However, because viral genetic material mutates at a rapid rate, viral variants usually emerge that resist these drugs. The drugs kill sensitive viruses, leaving only resistant ones, and they take over. More than 200 drugs are also available to treat AIDS-associated opportunistic infections and cancers.

Developing a vaccine against HIV has been extremely difficult because components of the immune system recognize only small parts of the virus, and the virus often alters these very parts. More than two dozen candidate vaccines are currently being tested. Most of the vaccines target molecules that are on the viral envelope and/or a protein that the virus uses to package itself into particles as it leaves cells.

HIV's highly changeable nature suggests that it might not be possible to develop a widely effective vaccine, or any one drug that works for a lifetime. However, a goal already realized for many people is to keep viral levels low enough so that health, and possibly even life span, can approach normal.

Table 14A	HIV Transmission
How HIV Is Transmitted	
Sexual contact, particularly anal intercourse, but also vaginal intercourse and oral sex	
Contaminated needles (intravenous drug use, injection of anabolic steroids, accidental needle stick in medical setting)	
During birth from infected mother	
Breast milk from infected mother	
Receiving infected blood or other tissue (precautions usually prevent this)	
How HIV Is Not Transmitted	
Casual contact (social kissing, hugging, handshakes)	
Objects (toilet seats, deodorant sticks, doorknobs)	
Mosquitoes	
Sneezing and coughing	
Sharing food	
Swimming in the same water	
Donating blood	

molecules on the original B cell's surface, accounting for the specificity of the antibody-antigen reaction. Thus, specific antibodies can combine with their corresponding foreign antigens and react against them. Table 14.2 summarizes the steps leading to antibody production as a result of B cell and T cell activities.

A plasma cell, during its brief life span, secretes up to 2,000 identical antibodies per second.

An individual's B cells can produce an estimated 10 million to 1 billion different varieties of antibodies, each reacting against a specific antigen. The enormity and diversity of the antibody response defends against many pathogens.

Table 14.2	Steps in Antibody Production

B Cell Activities

1. Antigen-bearing agents enter tissues.

2. B cell encounters an antigen that fits its antigen receptors.

3. Either alone or more often in conjunction with helper T cells, the B cell is activated. The B cell proliferates, enlarging its clone.

4. Some of the newly formed B cells differentiate further to become plasma cells.

5. Plasma cells synthesize and secrete antibodies whose molecular structure is similar to the activated B cell's antigen receptors.

6. Antibodies combine with antigen-bearing agents, helping to destroy them.

T Cell Activities

1. Antigen-bearing agents enter tissues.

2. An accessory cell, such as a macrophage, phagocytizes the antigen-bearing agent, and the macrophage's lysosomes digest the agent.

3. Antigens from the digested antigen-bearing agents are displayed on the membrane of the accessory cell.

4. Helper T cell becomes activated when it encounters a displayed antigen that fits its antigen receptors.

5. Activated helper T cell releases cytokines when it encounters a B cell that has previously combined with an identical antigen-bearing agent.

6. Cytokines stimulate the B cell to proliferate.

7. Some of the newly formed B cells give rise to cells that differentiate into antibody-secreting plasma cells.

8. Antibodies combine with antigen-bearing agents, helping to destroy them.

Types of Antibodies

Antibodies (immunoglobulins) are soluble, globular proteins that constitute the *gamma globulin* fraction of plasma proteins (see chapter 12, p. 326). Of the five major types of immunoglobulins, the most abundant are immunoglobulin G, immunoglobulin A, and immunoglobulin M.

Immunoglobulin G (IgG) is in plasma and tissue fluids and is particularly effective against bacteria, viruses, and toxins. It also activates *complement*.

Immunoglobulin A (IgA) is commonly found in exocrine gland secretions. It is in breast milk, tears, nasal fluid, gastric juice, intestinal juice, bile, and urine.

Immunoglobulin M (IgM) is a type of antibody present in plasma in response to contact with certain antigens in foods or bacteria. The antibodies anti-A and anti-B, described in chapter 12 (p. 332), are examples of IgM. IgM also activates complement.

Immunoglobulin D (IgD) is found on the surfaces of most B cells, especially those of infants. IgD is important in activating B cells.

Immunoglobulin E (IgE) is in exocrine secretions with IgA. It is associated with allergic reactions, which are described later in this section (see pp. 392–393).

A newborn does not yet have its own antibodies, but does retain IgG that passed through the placenta from the mother. These maternal antibodies protect the infant against some illnesses to which the mother is immune. At about the same time that the maternal antibody supply falls, the infant begins to manufacture its own antibodies. The newborn also receives IgA from colostrum, a substance secreted from the mother's breasts for the first few days after birth. Antibodies in colostrum protect against certain digestive and respiratory infections.

Check Your Recall

21. How are B cells activated?

22. How does the antibody response protect against diverse infections?

23. Which immunoglobulins are most abundant, and how do they differ from each other?

Antibody Actions

In general, antibodies directly attack antigens, activate complement, or stimulate localized (inflammation) changes that help prevent the spread of pathogens or cells bearing foreign antigens.

In a direct attack, antibodies combine with antigens, causing them to clump (agglutinate) or to form insoluble substances (precipitate). Such actions make it easier for phagocytic cells to recognize and engulf the antigen-bearing agents and eliminate them. In other instances, antibodies cover the toxic portions of antigen molecules and neutralize their effects. However, under normal conditions, direct antibody attack is not as important as complement activation in protecting against infection.

When certain IgG or IgM antibodies combine with antigens, they expose reactive sites on antibody molecules. This triggers a series of reactions, leading to activation of the complement proteins, which in turn produce a variety of effects. These include: coating the antigen-antibody complexes (opsonization), making the complexes more susceptible to phagocytosis; attracting macrophages and neutrophils into the region (chemotaxis); rupturing membranes of foreign cells (lysis); clumping antigen-bearing cells; and altering the molecular structure of viruses, making them harmless (fig. 14.15). Other proteins promote inflammation, which helps prevent the spread of infectious agents.

Check Your Recall

24. In what general ways do antibodies function?
25. How is complement activated?
26. What is the function of complement?

Immune Responses

Activation of B cells or T cells after first encountering the antigens for which they are specialized to react constitutes a **primary immune response.** During such a response, plasma cells release antibodies (IgM, followed by IgG) into the lymph. The antibodies are transported to the blood and then throughout the body, where they help destroy antigen-bearing agents. Production and release of antibodies continues for several weeks.

Following a primary immune response, some of the B cells produced during proliferation of the clone remain dormant as memory cells (see fig. 14.14). If the identical antigen is encountered again, clones of these memory cells enlarge, and can respond rapidly with IgG to the antigen to which they were previously sensitized. These memory B cells, along with the memory T cells, produce a **secondary immune response.**

As a result of a primary immune response, detectable concentrations of antibodies usually appear in the blood plasma five to ten days after exposure to antigens. If the identical antigen is encountered later, a secondary immune response may produce additional antibodies within a day or two (fig. 14.16). Although newly formed antibodies may persist in the body for only a few months or years, memory cells live much longer. A secondary immune response may be very long-lasting.

Check Your Recall

27. How do primary and secondary immune responses differ?

Superantigens are foreign antigens that elicit unusually vigorous lymphocyte responses. The bacterium *Staphylococcus aureus* produces two such superantigens. One type causes food poisoning until digestive enzymes destroy it. The second type causes toxic shock syndrome, a potentially fatal condition producing high fever, diarrhea, vomiting, confusion, and plummeting blood pressure.

Practical Classification of Immunity

Adaptive, also known as acquired, immunity can arise in response to natural events or be induced artificially by injecting or orally administering a suspension of killed

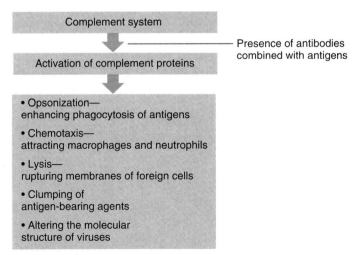

Figure 14.15

Actions of the complement system.

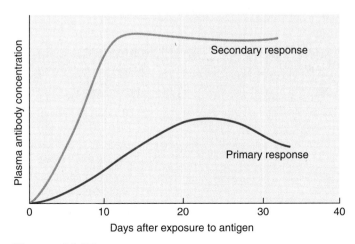

Figure 14.16

A primary immune response produces less vigorous antibody production than does a secondary immune response.

or weakened pathogens or molecules unique to them. Both naturally and artificially acquired immunities can be either active or passive. Active immunity results when the person produces an immune response (including memory cells) to the antigen; it is long-lasting. Passive immunity occurs when a person receives antibodies produced by another individual. Since the person does not produce an immune response, passive immunity is short-term, and the individual will be susceptible upon exposure to the antigen at some later date.

Naturally acquired active immunity occurs when a person exposed to a pathogen develops a disease. Resistance to that pathogen is the result of a primary immune response.

A **vaccine** (vak′sēn) is a preparation that produces another type of active immunity. A vaccine might consist of bacteria or viruses that have been killed or weakened so that they cannot cause a serious infection, or only molecules unique to the pathogens. A vaccine might also be a toxoid, which is a toxin from an infectious organism that has been chemically altered to destroy its dangerous effects. Whatever its composition, a vaccine includes the antigens that stimulate a primary immune response, but does not produce the severe symptoms of disease. A vaccine causes a person to develop *artificially acquired active immunity*. Developing a vaccine that is effective against HIV, the virus that causes AIDS, has been very difficult, as the Topic of Interest on page 389 discusses. The Genetics Connection on pages 394–395 discusses inherited immune deficiencies.

Some people were afraid of the first vaccinations, which were derived from cows. They feared that their vaccinated children might behave like cows. Today vaccines still provoke fear. In Nigeria, for example, polio has returned because many parents feared the vaccine had been adulterated to cause AIDS or female infertility.

Sometimes a person who has been exposed to infection needs protection against a disease-causing microorganism but lacks the time to develop active immunity. An injection of antiserum (ready-made antibodies) may help. These antibodies may be obtained from gamma globulin separated from the plasma of persons who have already developed immunity against the particular disease. A gamma globulin injection provides *artificially acquired passive immunity*.

During pregnancy, certain antibodies (IgG) pass from the maternal blood into the fetal bloodstream. As a result, the fetus acquires limited immunity against pathogens that the pregnant woman has developed active immunities against. The fetus thus has *naturally acquired passive immunity*, which may last for six months to a year after birth. Table 14.3 summarizes the types of acquired immunity.

Check Your Recall

28. Distinguish between active and passive immunity.
29. What is a vaccine?

Allergic Reactions

An allergic response is an immune response to a nonharmful substance, such as chocolate. Allergic reactions are similar to immune responses because they sensitize lymphocytes, and antibodies they produce may combine with antigens. However, unlike normal immune responses, allergic reactions can damage tissues. Antigens that trigger allergic responses are called **allergens** (al′er-jenz).

Vaccines stimulate active immunity against a variety of diseases, including typhoid fever, cholera, whooping cough, diphtheria, tetanus, polio, chickenpox, measles (rubeola), German measles (rubella), mumps, influenza, hepatitis A, hepatitis B, and bacterial pneumonia. Vaccine distribution is not equitable worldwide. Many thousands of people in underdeveloped countries die of infectious diseases for which vaccines are available in other nations.

Table 14.3	Practical Classification of Immunity	
Type	**Mechanism**	**Result**
Naturally acquired active immunity	Exposure to live pathogens	Stimulation of an immune response with symptoms of a disease
Artificially acquired active immunity	Exposure to a vaccine containing weakened or dead pathogens or their components	Stimulation of an immune response without the severe symptoms of a disease
Artificially acquired passive immunity	Injection of gamma globulin containing antibodies	Short-term immunity without stimulating an immune response
Naturally acquired passive immunity	Antibodies passed to fetus from pregnant woman with active immunity or to newborn through breast milk from a woman with active immunity	Short-term immunity for newborn without stimulating an immune response

Allergies can be classified by how quickly they follow exposure to the allergen. A *delayed-reaction allergy* results from repeated exposure of the skin to certain chemicals—commonly, household or industrial chemicals or ingredients of some cosmetics. After repeated contact, the presence of the foreign substance activates T cells, many of which collect in the skin. The T cells and the macrophages they attract release chemical factors, which in turn cause eruptions and inflammation of the skin (dermatitis). This reaction is called *delayed* because it usually takes about 48 hours to occur. A delayed-reaction allergy may affect anyone.

An *immediate-reaction allergy* occurs within minutes after contact with an allergen, and affects people who have an inherited tendency to overproduce IgE antibodies in response to certain antigens. IgE normally comprises a tiny fraction of plasma proteins.

An immediate-reaction allergy activates B cells, which become sensitized when the allergen is first encountered. Subsequent exposures to the allergen trigger allergic reactions. In the initial exposure, IgE attaches to the membranes of widely distributed mast cells and basophils. When a subsequent allergen-antibody reaction occurs, these cells release allergy mediators such as *histamine, prostaglandin D$_2$,* and *leukotrienes.* These substances cause a variety of physiological effects, including dilation of blood vessels, increased vascular permeability that swells tissues, contraction of bronchial and intestinal smooth muscles, and increased mucus production. The result is a severe inflammation reaction that is responsible for the symptoms of the allergy, such as hives, hay fever, asthma, eczema, or gastric disturbances.

Anaphylactic shock is a severe form of immediate-reaction allergy in which mast cells release allergy mediators throughout the body. The person may at first feel an inexplicable apprehension, and then suddenly the entire body itches and breaks out in red hives. Vomiting and diarrhea may follow. The face, tongue, and larynx begin to swell, and breathing becomes difficult. Unless the person receives an injection of epinephrine (adrenalin) and sometimes a tracheotomy (an incision into the windpipe to restore breathing), he or she will lose consciousness and may die within 5 minutes. Anaphylactic shock most often results from an allergy to penicillin or insect stings. Fortunately, thanks to prompt medical attention and avoidance of allergens by people who know they have allergies, fewer than 100 people a year actually die from anaphylactic shock. The peanut allergy described in the chapter-opening vignette on page 376 causes many of the symptoms of anaphylactic shock, but usually not the sensation of the throat closing.

Transplantation and Tissue Rejection

Transplantation of tissues or an organ, such as the skin, kidney, heart, or liver, from one person to another can replace a nonfunctional, damaged, or lost body part. However, the recipient's immune system may recognize the donor's cell surfaces as foreign and attempt to destroy the transplanted tissue, causing a **tissue rejection reaction.**

Tissue rejection resembles the cellular immune response against a nonself antigen. The greater the antigenic difference between the cell surface molecules (MHC antigens, discussed on pages 386–387) of the recipient tissues and the donor tissues, the more rapid and severe the rejection reaction. Matching donor and recipient tissues can minimize the rejection reaction.

Immunosuppressive drugs are used to reduce rejection of transplanted tissues. These drugs interfere with the recipient's immune response by suppressing formation of antibodies or production of T cells, thereby dampening the humoral and cellular immune responses. Unfortunately, the use of immunosuppressive drugs can leave a recipient unprotected against infections. It is not uncommon for a patient to survive a transplant, but die of infection because of a weakened immune system.

Donated organs need to be transplanted quickly. How long can donated organs last outside the body?

- A heart lasts three to five hours.
- A liver lasts ten hours.
- A kidney lasts twenty-four to forty-eight hours.

Autoimmunity

Sometimes the immune system fails to distinguish self from nonself, producing **autoantibodies** and cytotoxic T cells that attack and damage the body's tissues and organs. This attack against self is called **autoimmunity.** The specific nature of an autoimmune disorder reflects the cell types that are the target of the immune attack. Type 1 (insulin dependent) diabetes mellitus, rheumatoid arthritis, and systemic lupus erythematosus are autoimmune disorders. About 5% of the population has an autoimmune disorder.

Why might the immune system attack body tissues? Perhaps a virus, while replicating within a human cell, "borrows" proteins from the host cell's surface and incorporates them onto its own surface. When the immune system "learns" the surface of the virus in order to destroy it, it also learns to attack the human cells that normally bear the particular protein. Another explanation of autoimmunity is that somehow T cells never learn to distinguish self from nonself. A third possible route of autoimmunity is when a nonself antigen coincidentally resembles a self antigen. This is what happens when an infection by *Streptococcus* bacteria triggers inflammation of heart valves, as mentioned in chapter 13 (p. 344).

Conquering Inherited Immune Deficiency— Children Who Made Medical History

The importance of T cells in establishing and maintaining immunity is obvious in AIDS, which is acquired, and in severe combined immune deficiency (SCID), which is inherited. The 20 types of SCID disrupt receptors on T cells or hamper cytokine production. SCID is called "combined" because both T and B cells are affected. Medical technology has evolved to treat SCID. Here is a look at some young pioneers.

David

David Vetter was born in Texas in 1971 without a thymus gland. He therefore could not make mature T cells or activate B cells. He spent his short life in a vinyl bubble that protected him from infection, awaiting a treatment that never came (fig. 14A). Before AIDS, living without immunity was very unusual.

As David reached adolescence, he desperately wanted to leave the bubble. After receiving a bone marrow transplant, he did so. But the transplant hadn't worked, and within days, David began vomiting and developed diarrhea, both signs of infection. He soon died.

Laura

For her first few years, Laura Cay Boren didn't know what it felt like to be well (fig. 14B). Ever since her birth in July 1982, she fought infection after infection. Colds landed her in the hospital with pneumonia, and routine vaccines caused severe skin abscesses. Laura had inherited a form of SCID in which the body lacks an enzyme, adenosine deaminase (ADA). Lack of ADA blocks a biochemical pathway that breaks down a metabolic toxin, which instead builds up and destroys T cells. The T cells in turn can no longer activate B cells. Immunity fails.

Laura underwent two bone marrow transplants, which temporarily restored her immune defenses, and blood transfusions helped. But by the end of 1985, Laura was near death. Then she was chosen to receive experimental injections of ADA altered to remain in the bloodstream long enough to help T cells survive. Within hours of the first treatment, Laura's ADA level increased twenty-fold. After three months, her blood was free of toxins. After six months, her immune function neared normal, and stayed that way, with weekly ADA shots. By the following year, Laura began school. She is healthy today.

Ashi

In the late 1980s, the DeSilvas did not think that their little girl, Ashanthi ("Ashi"), would survive. She suffered near-continual coughs and colds, so fatigued that she could walk only a few steps without becoming winded, her father Raj recalls. "We took her to so many doctors that I stopped

Figure 14A
David Vetter was born without a thymus. Because his T cells could not mature, he was virtually defenseless against infection.

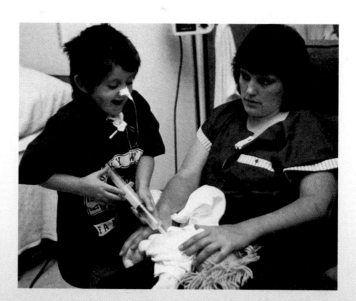

Figure 14B
Laura Cay Boren spent much of her life in hospitals until she received the enzyme her body lacks, adenosine deaminase (ADA). Here, she pretends to inject her doll as her mother looks on.

counting. One doctor after another would say it was asthma, an allergy, or bronchitis."

Raj's brother, an immunologist, suggested the blood tests that would eventually reveal that Ashi had inherited SCID due to ADA deficiency. By then, the condition was so well understood that it was first in line for gene therapy—and Ashi became the first recipient.

On September 14, 1990, at 12:52 P.M., four-year-old Ashi sat up in bed at the National Institutes of Health in Bethesda, Maryland, and began receiving her own white blood cells intravenously. Earlier, doctors had removed the cells and patched them with normal ADA genes. Within weeks, Ashi began to make her own, functional T cells. Although she required further treatments as the bolstered cells died off, today she is well, anticipating a career in the music industry after college. Over the years, she has told audiences at medical conferences about the gene therapy that restored her health. Figure 14C shows her at a meeting when she was 17, where she introduced the head of the research, Dr. Blaese: "Our duty on Earth is to help others. I thank you from the bottom of my heart for all you have enabled me to do."

Andrew

Crystal and Leonard Gobea had already lost a baby to ADA deficiency when they learned in 1993 that the fetus Crystal was carrying had also inherited the condition. They were asked to allow Andrew to participate in a new type of gene therapy, with two other newborns, that would replace defective T cells with T cells taken from their umbilical cord blood and given the missing ADA gene (fig. 14D).

The three children were also given ADA injections to maintain health, as doctors monitored immune function. Since the defective cells were replaced early, the hope was that the new ones would take over the immune system. Over the next few years, the percentage of T cells bearing the functional ADA gene rose in each child, and two enjoyed good health; the third stayed well with ADA injections.

Along with gene therapy successes have been setbacks. In 2003, three boys in France developed leukemia after the healing gene for SCID had been inserted into a cancer-causing gene. The boys recovered, but the problem focused attention on the unpredictability of gene therapy.

Figure 14C
Ashi DeSilva often publicly thanks Dr. Michael Blaese, who pioneered the gene therapy that restored her immune system.

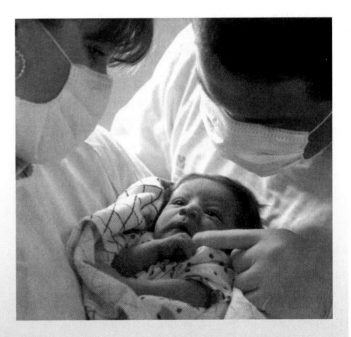

Figure 14D
Newborn Andrew Gobea received the ADA gene in stem cells taken from his umbilical cord. Gradually, corrected T cells accumulated, restoring immunity.

Clinical Terms Related to the Lymphatic System and Immunity

allograft (al'o-graft) Transplantation of tissue from an individual of one species to another individual of the same species.

asplenia (ah-sple'ne-ah) Absence of a spleen.

autograft (aw'to-graft) Transplantation of tissue from one part of the body to another part of the same body.

immunocompetence (im''u-no-kom'pe-tens) Ability to produce an immune response to antigens.

immunodeficiency (im''u-no-de-fish'en-se) Inability to produce an immune response.

lymphadenectomy (lim-fad''ĕ-nek'to-me) Surgical removal of lymph nodes.

lymphadenopathy (lim-fad''ĕ-nop'ah-the) Enlargement of lymph nodes.

lymphadenotomy (lim-fad''ĕ-not'o-me) Incision of a lymph node.

lymphocytopenia (lim''fo-si''to-pe'ne-ah) Too few lymphocytes in the blood.

lymphocytosis (lim''fo-si''to'sis) Too many lymphocytes in the blood.

lymphoma (lim-fo'mah) Tumor composed of lymphatic tissue.

lymphosarcoma (lim''fo-sar-ko'mah) Cancer within the lymphatic tissue.

splenectomy (sple-nek'to-me) Surgical removal of the spleen.

splenitis (sple-ni'tis) Inflammation of the spleen.

splenomegaly (sple''no-meg'ah-le) Enlargement of the spleen.

splenotomy (sple-not'o-me) Incision of the spleen.

thymectomy (thi-mek'to-me) Surgical removal of the thymus.

thymitis (thi-mi'tis) Inflammation of the thymus.

xenograft (zeen'-o-graft) Transplantation of tissue from one species into a recipient of another species.

Clinical Connection

Some disorders thought to be autoimmune may have a stranger cause—fetal cells persisting in a woman's circulation, for decades. In response to an as yet unknown trigger, the fetal cells, perhaps "hiding" in a tissue such as skin, emerge and stimulate antibody production. The resulting antibodies and symptoms appear to be an autoimmune disorder. The presence of more than one genetically distinct cell population in an individual is called microchimerism ("small mosaic"). Microchimerism that reflects the retention of cells from a fetus may explain the higher prevalence of autoimmune disorders among women. It is seen in a disorder called scleroderma, which means "hard skin."

Scleroderma, which typically begins between ages forty-five and fifty-five, is described as "the body turning to stone." Symptoms include fatigue, swollen joints, stiff fingers, and a masklike face. The hardening may also affect blood vessels, the lungs, and the esophagus. Clues that scleroderma is a delayed response to persisting fetal cells include the following observations:

- It is much more common among women.
- Symptoms resemble those of graft-versus-host disease (GVHD), in which transplanted tissue produces chemicals that destroy the recipient's tissues. Antigens on cells in scleroderma lesions match those that cause GVHD.
- Mothers who have scleroderma and their sons have cell surfaces that are more similar than those of unaffected mothers and their sons. Perhaps the similarity of cell surfaces enabled the fetal cells to escape destruction by the woman's immune system. (Female fetal cells probably have the same effect, but these cells cannot be distinguished from maternal cells by the presence of the Y chromosome.)

Perhaps other disorders considered autoimmune actually reflect an immune system response to lingering fetal cells.

SUMMARY OUTLINE

14.1 Introduction (p. 377)

The lymphatic system is closely associated with the cardiovascular system. It transports excess tissue fluid to the bloodstream, absorbs fats, and helps defend the body against disease-causing agents.

14.2 Lymphatic Pathways (p. 377)

1. Lymphatic capillaries
   a. Lymphatic capillaries are microscopic, closed-ended tubes that extend into interstitial spaces.
   b. They receive lymph through their thin walls.

2. Lymphatic vessels
   a. Lymphatic vessels have walls similar to those of veins, only thinner, and possess valves that prevent backflow of lymph.
   b. Larger lymphatic vessels lead to lymph nodes and then merge into lymphatic trunks.

3. Lymphatic trunks and collecting ducts
   a. Lymphatic trunks lead to two collecting ducts—the thoracic duct and the right lymphatic duct.
   b. Collecting ducts join the subclavian veins.

Lymphatic System

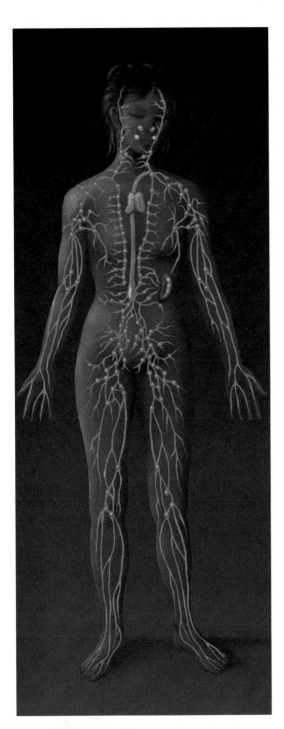

Integumentary System

The skin is a first line of defense against infection.

Cardiovascular System

The lymphatic system returns tissue fluid to the bloodstream. Lymph originates as tissue fluid, formed by the action of blood pressure.

Skeletal System

Cells of the immune system originate in the bone marrow.

Digestive System

Lymph plays a major role in the absorption of fats.

Muscular System

Muscle action helps pump lymph through the lymphatic vessels.

Respiratory System

Cells of the immune system patrol the respiratory system to defend against infection.

Nervous System

Stress may impair the immune response.

Urinary System

The kidneys control the volume of extracellular fluid, including lymph.

Endocrine System

Hormones stimulate lymphocyte production.

Reproductive System

Special mechanisms inhibit the female immune system in its attack of sperm as foreign invaders.

The lymphatic system is an important link between tissue fluid and the plasma; it also plays a major role in the response to infection.

14.3 Tissue Fluid and Lymph (p. 378)

1. Tissue fluid formation
   a. Tissue fluid originates from blood plasma.
   b. It generally lacks large proteins, but some smaller proteins are filtered out of blood capillaries into interstitial spaces.
   c. As the protein concentration of tissue fluid increases, colloid osmotic pressure increases.
2. Lymph formation and function
   a. Increasing pressure within interstitial spaces forces some tissue fluid into lymphatic capillaries, and this fluid becomes lymph.
   b. Lymph returns protein molecules to the bloodstream and transports foreign particles to lymph nodes.

14.4 Lymph Movement (p. 379)

1. Lymph is under low pressure and may not flow readily without external aid.
2. Lymph is moved by the contraction of skeletal muscles, contraction of smooth muscle in the walls of large lymphatic trunks, and low pressure in the thorax created by breathing movements.

14.5 Lymph Nodes (p. 380)

1. Structure of a lymph node
   a. Lymph nodes are subdivided into nodules.
   b. Nodules contain masses of lymphocytes and macrophages.
2. Locations of lymph nodes
   Lymph nodes aggregate in groups or chains along the paths of larger lymphatic vessels.
3. Functions of lymph nodes
   a. Lymph nodes filter potentially harmful foreign particles from lymph.
   b. Lymph nodes are centers for the production of lymphocytes, and they also contain phagocytic cells.

14.6 Thymus and Spleen (p. 381)

1. Thymus
   a. The thymus is composed of lymphatic tissue subdivided into lobules.
   b. It slowly shrinks after puberty.
   c. Some lymphocytes leave the thymus and provide immunity.
2. Spleen
   a. The spleen resembles a large lymph node subdivided into lobules.
   b. Spaces within splenic lobules are filled with blood.
   c. The spleen contains many macrophages, which filter foreign particles and damaged red blood cells from blood.

14.7 Body Defenses Against Infection (p. 383)

The body has innate and adaptive defenses against infection.

14.8 Innate (Nonspecific) Defenses (p. 384)

1. Species resistance
   Each species is resistant to certain diseases that may affect other species.
2. Mechanical barriers
   Mechanical barriers include the skin and mucous membranes, which block the entrance of some pathogens.
3. Chemical barriers
   a. Enzymes in gastric juice and tears kill some pathogens.
   b. Interferons stimulate uninfected cells to synthesize antiviral proteins that stimulate phagocytosis, block proliferation of viruses, and enhance activity of cells that help resist infections and stifle tumor growth.
   c. Activation of complement proteins in plasma stimulates inflammation, attracts phagocytes, and enhances phagocytosis.
4. Natural killer cells
   Natural killer cells secrete perforins, which destroy cancer cells and cells infected with viruses.
5. Inflammation
   a. Inflammation is a tissue response to injury or infection, and includes localized redness, swelling, heat, and pain.
   b. Chemicals released by damaged tissues attract white blood cells to the site.
   c. Connective tissue may form a sac around injured tissue and thus block the spread of pathogens.
6. Phagocytosis
   a. The most active phagocytes in blood are neutrophils and monocytes. Monocytes give rise to macrophages, which remain fixed in tissues.
   b. Phagocytic cells are associated with the linings of blood vessels in the bone marrow, liver, spleen, lungs, and lymph nodes.
   c. Phagocytes remove foreign particles from tissues and body fluids.
7. Fever
   Higher body temperature and the resulting decrease in blood iron level and increase in phagocytic activity hamper infection.

14.9 Adaptive (Specific) Defenses, or Immunity (p. 385)

1. Antigens
   a. Before birth, body cells inventory "self" proteins and other large molecules.
   b. After inventory, lymphocytes develop receptors that allow them to differentiate between nonself (foreign) and self antigens.
   c. Nonself antigens combine with T cell and B cell surface receptors and stimulate these cells to cause an immune reaction.
   d. Haptens are small molecules that can combine with larger ones, becoming antigenic.
2. Lymphocyte origins
   a. Lymphocytes originate in red bone marrow and are released into the blood.
   b. Some reach the thymus, where they mature into T cells.
   c. Others, the B cells, mature in the red bone marrow.
   d. Both T cells and B cells reside in lymphatic tissues and organs.
3. Lymphocyte functions
   a. Some T cells interact with antigen-bearing agents directly, providing the cellular immune response.
   b. T cells secrete cytokines, such as interleukins, that enhance cellular responses to antigens.
   c. T cells may also secrete substances that are toxic to their target cells.
   d. B cells interact with antigen-bearing agents indirectly, providing the humoral immune response.
   e. Varieties of T cells and B cells number in the millions.
   f. The members of each variety respond only to a specific antigen.
   g. As a group, the members of each variety form a clone.
4. T cells and the cellular immune response
   a. T cells are activated when an antigen-presenting cell displays a foreign antigen.
   b. When a macrophage acts as an accessory cell, it phagocytizes an antigen-bearing agent, digests the agent, and displays the

antigens on its cell membrane in association with certain MHC proteins.
  c. A helper T cell becomes activated when it encounters displayed antigens for which it is specialized to react.
  d. An activated helper T cell contacts a B cell that carries the foreign antigen the T cell previously encountered on an antigen-presenting cell.
  e. In response, the T cell secretes cytokines, stimulates B cell proliferation, and attracts macrophages.
  f. Cytotoxic T cells recognize foreign antigens on tumor cells and cells whose surfaces indicate that they are infected by viruses.
  g. Memory T cells respond quickly to subsequent antigen exposure.
5. B cells and the humoral immune response
  a. Sometimes a B cell is activated when it encounters an antigen that fits its antigen receptors or more often a B cell is activated when stimulated by a helper T cell.
  b. An activated B cell proliferates (especially when stimulated by a T cell), enlarging its clone.
  c. Some activated B cells specialize into antibody-producing plasma cells.
  d. Antibodies react against the antigen-bearing agent that stimulated their production.
  e. An individual's diverse B cells defend against a very large number of pathogens.
6. Types of antibodies
  a. Antibodies are soluble proteins called immunoglobulins.
  b. The five major types of immunoglobulins are IgG, IgA, IgM, IgD, and IgE.
7. Antibody actions
  a. Antibodies directly attach to antigens, activate complement, or stimulate local tissue changes that are unfavorable to antigen-bearing agents.
  b. Direct attachment results in agglutination, precipitation, or neutralization.
  c. Activated proteins of complement attract phagocytes, alter cells so that they become more susceptible to phagocytosis, and rupture foreign cell membranes (lysis).
8. Immune responses
  a. The first reaction to an antigen is called a primary immune response.
    (1) During this response, antibodies are produced for several weeks.
    (2) Some T cells and B cells remain dormant as memory cells.
  b. A secondary immune response occurs rapidly as a result of memory cell response if the same antigen is encountered later.
9. Practical classification of immunity
  a. Naturally acquired immunity arises in the course of natural events, whereas artificially acquired immunity is the consequence of a medical procedure.
  b. Active immunity lasts much longer than passive immunity.
  c. A person who encounters a pathogen and has a primary immune response develops naturally acquired active immunity.
  d. A person who receives a vaccine containing a dead or weakened pathogen, or part of one, develops artificially acquired active immunity.
  e. A person who receives an injection of antibodies has artificially acquired passive immunity.
  f. When antibodies pass through a placental membrane from a pregnant woman to her fetus, the fetus develops naturally acquired passive immunity.

10. Allergic reactions
  a. Allergic reactions are excessive and misdirected immune responses that may damage tissue.
  b. Delayed-reaction allergy, which can occur in anyone and inflame the skin, results from repeated exposure to antigens.
  c. Immediate-reaction allergy is an inborn ability to overproduce IgE.
    (1) Allergic reactions result from mast cells bursting and releasing allergy mediators such as histamine.
    (2) The released chemicals cause allergy symptoms such as hives, hay fever, asthma, eczema, or gastric disturbances.
11. Transplantation and tissue rejection
  a. A transplant recipient's immune system may react against the donated tissue, an event termed a tissue rejection reaction.
  b. Matching donor and recipient tissues and using immunosuppressive drugs can minimize tissue rejection.
  c. Immunosuppressive drugs may increase susceptibility to infection.
12. Autoimmunity
  a. In autoimmune disorders, the immune system manufactures autoantibodies that attack a person's own body tissues.
  b. Autoimmune disorders may result from a previous viral infection, faulty T cell development, or reaction to a nonself antigen that resembles a self antigen.

CHAPTER ASSESSMENTS

14.1 Introduction
1. Explain the functions of the lymphatic system. (p. 377)

14.2 Lymphatic Pathways
2. Trace the general pathway of lymph from the interstitial spaces to the bloodstream. (p. 378)

14.3 Tissue Fluid and Lymph
3. Tissue fluid forms as a result of _____, whereas lymph forms due to increasing _____ in the tissue fluid. (p. 378)
4. Describe two functions of lymph. (p. 379)

14.4 Lymph Movement
5. Explain why physical exercise promotes lymphatic circulation. (p. 379)

14.5 Lymph Nodes
6. Describe the structure and functions of a lymph node. (p. 380)

14.6 Thymus and Spleen
7. Indicate the locations of the thymus and spleen. (p. 381)
8. Compare and contrast the functions of the thymus and spleen. (p. 381)

14.7 Body Defenses Against Infection
9. Defenses mechanisms that prevent the entry of many types of pathogens and destroy them if they enter provide _____ defense. Mechanisms that are very precise, targeting specific pathogens provide _____ defense. (p. 383)

14.8 Innate (Nonspecific) Defenses
10. Explain species resistance. (p. 384)
11. Identify the barriers that provide the body's first line of defense against infectious agents. (p. 384)
12. Describe how enzymatic actions function as defense mechanisms. (p. 384)

13. Define *interferon,* and explain its action. (p. 384)

14. _____ is a group of plasma proteins that when activated stimulate inflammation, attract phagocytes, and enhance phagocytosis. (p. 384)

15. _____ _____ _____ are specialized lymphocytes that secrete perforins to lyse cell membranes of virus-infected cells. (p. 384)

16. List the major effects of inflammation, and explain why each occurs. (p. 384)

17. Identify the major phagocytic cells in blood and other tissues. (p. 384)

18. Discuss why low-grade fever of short duration may be a desired response to infection. (p. 385)

14.9 Adaptive (Specific) Defenses, or Immunity

19. Review the origin of T cells and B cells. (p. 385)

20. Explain cellular immunity. (p. 385)

21. Explain humoral immunity. (p. 385)

22. Describe how T cells become activated. (p. 386)

23. Explain the function of memory cells. (p. 388)

24. Describe how B cells become activated. (p. 388)

25. Explain the function of plasma cells. (p. 388)

26. Match the major types of immunoglobulins with their function and/or where each is found. (p. 390)

(1) associated with allergic reactions	A. IgA
(2) important in B cell activation, on surfaces of most B cells	B. IgM
(3) activates complement, anti-A and anti-B in blood	C. IgG
(4) effective against bacteria, viruses, toxins in plasma and tissue fluids	D. IgD
(5) found in exocrine secretions, including breast milk	E. IgE

27. Explain two mechanisms by which antibodies directly attack antigens. (p. 390)

28. List the various effects of complement activation. (p. 391)

29. Contrast a primary and a secondary immune response. (p. 391)

30. Match the practical classifications of immunity with the examples. (p. 392)

(1) naturally acquired active immunity	A a breastfed newborn
(2) artificially acquired active immunity	B gamma globulin injection
(3) naturally acquired passive immunity	C vaccination
(4) artificially acquired passive immunity	D measles infection

31. List the major events leading to a delayed-reaction allergic response. (p. 393)

32. Describe how an immediate-reaction allergic response may occur. (p. 393)

33. Explain the relationship between tissue rejection and an immune response. (p. 393)

34. Explain the relationship between autoimmunity and an immune response. (p. 393)

INTEGRATIVE ASSESSMENTS/ CRITICAL THINKING

➡ OUTCOMES 14.2, 14.3, 14.4

1. Why is injecting a substance into the skin like injecting it into the lymphatic system?

➡ OUTCOME 14.5

2. How can the removal of enlarged lymph nodes for microscopic examination aid in diagnosing certain diseases?

➡ OUTCOME 14.9

3. The immune response is specific, diverse, and has memory. Give examples of each of these characteristics.

4. Some parents keep their preschoolers away from other children to prevent them from catching illnesses. How might these well-meaning parents actually be harming their children?

5. Why is a transplant consisting of fetal tissue less likely to provoke an immune rejection response than tissue from an adult?

6. People needing transplants outnumber available organs. Discuss the pros and cons of the following proposed rationing systems for determining who should receive transplants: (a) first come, first served; (b) people with the best tissue and blood type match; (c) patients whose need for an organ is caused by infection or disease, as opposed to those whose need for an organ was preventable, such as a lung destroyed by smoking; (d) the youngest people; (e) the wealthiest people; (f) the most important people.

7. A xenograft is tissue from a nonhuman animal used to replace a body part in a human. For example, pigs are being developed to provide cardiovascular spare parts because their hearts and blood vessels are similar to ours. To increase the likelihood of such a xenotransplant working, researchers genetically modify pigs to produce human antigens on their cell surfaces. How can this improve the chances of a human body not rejecting such a transplant?

8. Why does vaccination provide long-lasting protection against a disease, while gamma globulin (IgG) provides only short-term protection?

WEB CONNECTIONS

Visit the text website at **aris.mhhe.com** for additional quizzes, interactive learning exercises, and more.

AP R LYMPHATIC SYSTEM   ⬤

Anatomy & Physiology | REVEALED includes cadaver photos that allow you to peel away layers of the human body to reveal structures beneath the surface. This program also includes animations, radiologic imaging, audio pronunciation, and practice quizzing. To learn more visit **www.aprevealed.com**.

15

Digestive System and Nutrition

MICROBES ARE US. The human body is home to 10 to 100 million microorganisms, called microflora. Most live in the digestive tract, about 500 species in the mouth alone. To assess the population at the other end of the digestive tract, researchers analyzed DNA pieces in stool samples from a 28-year-old woman and a 37-year-old man, both of whom had not taken any medications over the past year that could have affected the microflora. By comparing the DNA pieces to those of known microorganisms, the researchers discovered that the "distal gut microbial community" had more than 6,800 species represented.

Researchers also tracked the formation and changing nature of the human gut microflora by classifying microbial DNA in a year's worth of babies' stool, collected daily from soiled diapers. The bacterial residents of the stool varied greatly from baby to baby at the onset, but by their first birthdays, the gut communities were more alike, and more closely resembled the microbial communities in adults.

The microorganisms that live in our large intestines are actually crucial to our health. They produce more than 80 types of enzymes that digest plant polysaccharides (such as pectin, xylan, and arabinose) that our bodies cannot break down, as well as easing processing of other sugars. Our "gut" residents also synthesize essential vitamins and amino acids, and break down certain toxins and drugs. The numbers and types of microorganisms that live in our intestines vary somewhat from person to person, and these differences may be one reason why

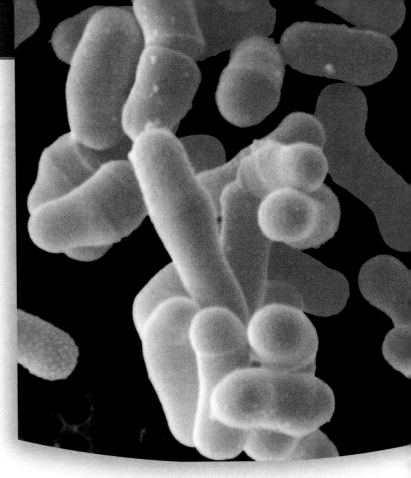

Several million microorganisms are normal residents of our digestive tracts. Tracking a year's worth of dirty diapers revealed that Bifidobacterium adolescentis, *pictured here, do not appear in the human intestines until several months of age. They aid digestion; produce certain acids; and inhibit growth of other bacteria and yeast that could cause disease.*

some people can eat a great deal and not gain weight, yet others gain weight easily. Studies show that an item of food may yield different numbers of calories in different people. One investigation of the energy in one kind of cookie found that even though the package listed 110 calories, the cookie yielded anywhere from 90 to 110 calories, depending upon who ate it.

Learning Outcomes   *After studying this chapter, you should be able to do the following:*

15.1 Introduction

1. Describe the general functions of the digestive system. (p. 402)

2. Name the major organs of the digestive system. (p. 402)

15.2 General Characteristics of the Alimentary Canal

3. Describe the structure of the wall of the alimentary canal. (p. 402)

4. Explain how the contents of the alimentary canal are mixed and moved. (p. 402)

15.3 Mouth

5. Describe the functions of the structures of the mouth. (p. 404)

6. Describe how different types of teeth are adapted for different functions, and list the parts of a tooth. (p. 408)

15.4–15.10 Salivary Glands–Large Intestine

7. Identify the function of each enzyme secreted by the digestive organs. (pp. 409–424)

8. Describe how digestive secretions are regulated. (pp. 409–424)

9. Describe the mechanisms of swallowing and defecating. (pp. 410 and 429)

10. Explain how the products of digestion are absorbed. (p. 424)

15.11 Nutrition and Nutrients

11. List the major sources of carbohydrates, lipids, and proteins. (p. 430)

12. Describe how cells utilize carbohydrates, lipids, and amino acids. (p. 430)

13. Identify the functions of each fat-soluble and water-soluble vitamin. (p. 433)

14. Identify the functions of each major mineral and trace element. (p. 435)

15. Describe an adequate diet. (p. 436)

Aids to Understanding Words   *(Appendix A on page 567 has a complete list of Aids to Understanding Words.)*

aliment- [food] *aliment*ary canal: Tubelike portion of the digestive system.

chym- [juice] *chym*e: Semifluid paste of food particles and gastric juice formed in the stomach.

decidu- [falling off] *decidu*ous teeth: Teeth that are shed during childhood.

gastr- [stomach] *gastr*ic gland: Portion of the stomach that secretes gastric juice.

hepat- [liver] *hepat*ic duct: Duct that carries bile from the liver to the common bile duct.

lingu- [tongue] *lingu*al tonsil: Mass of lymphatic tissue at the root of the tongue.

nutri- [nourish] *nutri*ent: Substance needed to nourish cells.

peri- [around] *peri*stalsis: Wavelike ring of contraction that moves material along the alimentary canal.

pyl- [gatekeeper] *pyl*oric sphincter: Muscle that serves as a valve between the stomach and small intestine.

vill- [hairy] *vill*i: Tiny projections of mucous membrane in the small intestine.

15.1 INTRODUCTION

Digestion (di-jest′yun) is the mechanical and chemical breakdown of foods and the absorption of the resulting nutrients by cells. *Mechanical digestion* breaks large pieces into smaller ones without altering their chemical composition. *Chemical digestion* breaks food into simpler chemicals. The organs of the **digestive system** carry out these processes.

The digestive system consists of the **alimentary canal** (al″i-men′tar-e kah-nal′), extending from the mouth to the anus, and several accessory organs, which secrete substances used in the process of digestion into the canal. The alimentary canal includes the mouth, pharynx, esophagus, stomach, small intestine, large intestine, rectum, and anus; the accessory organs include the salivary glands, liver, gallbladder, and pancreas (fig. 15.1; see reference plates 4, 5, and 6, pp. 26–28). Overall, the digestive system is a tube, open at both ends, that has a surface area of 186 square meters. It supplies nutrients for body cells.

15.2 GENERAL CHARACTERISTICS OF THE ALIMENTARY CANAL

The alimentary canal is a muscular tube about 8 meters long that passes through the body's thoracic and abdominopelvic cavities (fig. 15.2). The structure of its wall, how it moves food, and its innervation are similar throughout its length.

Structure of the Wall

The wall of the alimentary canal consists of four distinct layers that are developed to different degrees from region to region. Although the four-layered structure persists throughout the alimentary canal, certain regions are specialized for particular functions. Beginning with the innermost tissues, these layers are (fig. 15.3):

1. **Mucosa** (mu-ko′sah), or **mucous membrane** (mu′kus mem′brān) Surface epithelium, underlying connective tissue, and a small amount of smooth muscle form this layer. In some regions, the mucosa is folded, with tiny projections that extend into the passageway, or **lumen** (lu′men), of the digestive tube, which increases the absorptive surface area. The mucosa also has glands that are tubular invaginations into which the lining cells secrete mucus and digestive enzymes. The mucosa protects the tissues beneath it and carries on secretion and absorption.

2. **Submucosa** (sub″mu-ko′sah) The submucosa consists of considerable loose connective tissue as well as glands, blood vessels, lymphatic vessels, and nerves. Its vessels nourish surrounding tissues and carry away absorbed materials.

3. **Muscular layer** This layer, which produces movements of the tube, consists of two coats of smooth muscle tissue. The fibers of the inner coat encircle the tube. When these *circular fibers* contract, the tube's diameter decreases. The fibers of the outer muscular coat run lengthwise. When these *longitudinal fibers* contract, the tube shortens.

4. **Serosa** (sĕ′ro-sah), or **serous layer** (se′rus la′er) The *visceral peritoneum* of epithelium on the outside and the connective tissue beneath compose the serous layer, or outer covering, of the tube. The cells of the serosa protect underlying tissues and secrete serous fluid, which moistens and lubricates the tube's outer surface so that organs within the abdominal cavity slide freely against one another.

Movements of the Tube

The motor functions of the alimentary canal are of two basic types—*mixing movements* and *propelling movements*. Mixing occurs when smooth muscles in small segments of the tube contract rhythmically (fig. 15.4a). For example, when the stomach is full, waves of muscular contractions move along its walls from one end to the other. These waves mix food with digestive juices that the mucosa secretes. In the small intestine, **segmentation** aids mixing movements by alternately con-

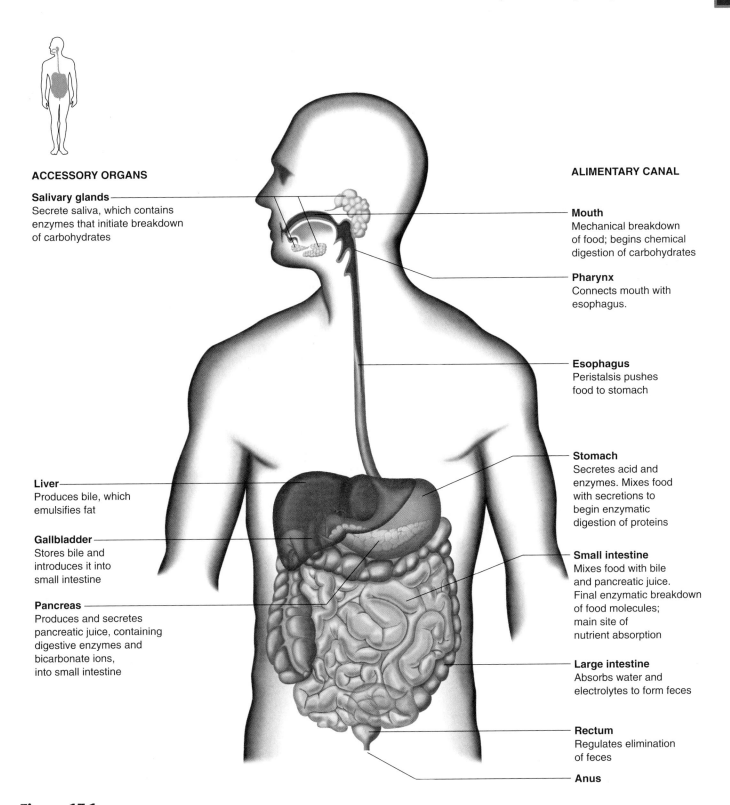

ACCESSORY ORGANS

Salivary glands
Secrete saliva, which contains
enzymes that initiate breakdown
of carbohydrates

Liver
Produces bile, which
emulsifies fat

Gallbladder
Stores bile and
introduces it into
small intestine

Pancreas
Produces and secretes
pancreatic juice, containing
digestive enzymes and
bicarbonate ions,
into small intestine

ALIMENTARY CANAL

Mouth
Mechanical breakdown
of food; begins chemical
digestion of carbohydrates

Pharynx
Connects mouth with
esophagus.

Esophagus
Peristalsis pushes
food to stomach

Stomach
Secretes acid and
enzymes. Mixes food
with secretions to
begin enzymatic
digestion of proteins

Small intestine
Mixes food with bile
and pancreatic juice.
Final enzymatic breakdown
of food molecules;
main site of
nutrient absorption

Large intestine
Absorbs water and
electrolytes to form feces

Rectum
Regulates elimination
of feces

Anus

Figure 15.1
Organs of the digestive system.

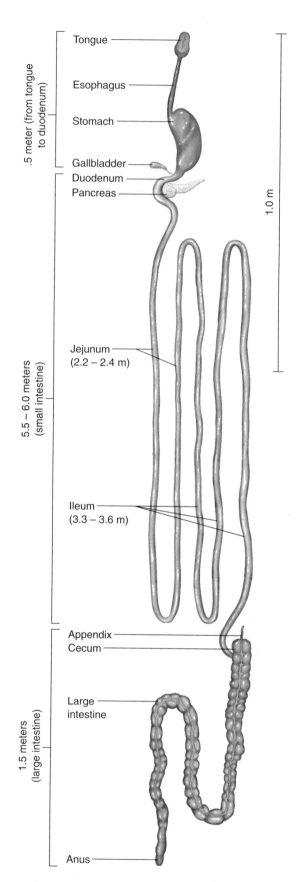

.5 meter (from tongue to duodenum)

- Tongue
- Esophagus
- Stomach
- Gallbladder
- Duodenum
- Pancreas

1.0 m

5.5 – 6.0 meters (small intestine)

- Jejunum (2.2 – 2.4 m)
- Ileum (3.3 – 3.6 m)
- Appendix
- Cecum
- Large intestine

1.5 meters (large intestine)

- Anus

Figure 15.2

The alimentary canal is a muscular tube about 8 meters long.

---

tracting and relaxing the smooth muscle in nonadjacent segments of the organ. Because segmentation does not follow a set pattern, materials are not propelled along the tract in one direction (fig. 15.4*b*).

Propelling movements include a wavelike motion called **peristalsis** (per″ĭ-stal′sis). When peristalsis occurs, a ring of contraction appears in the wall of the tube. At the same time, the muscular wall just ahead of the ring relaxes. As the peristaltic wave moves along, it pushes the tubular contents ahead of it (fig. 15.4*c*).

Check Your Recall

1. Which organs constitute the digestive system?
2. Describe the wall of the alimentary canal.
3. Name the two basic types of movements in the alimentary canal.

15.3 MOUTH

The **mouth** receives food and begins digestion by mechanically reducing the size of solid particles and mixing them with saliva. This action is called *mastication* (mas″tĭ-ka′shun). The lips, cheeks, tongue, and palate surround the mouth, which includes a chamber between the palate and tongue called the *oral cavity,* as well as a narrow space between the teeth, cheeks, and lips called the *vestibule* (fig. 15.5).

Cheeks and Lips

The **cheeks,** forming the lateral walls of the mouth, consist of outer layers of skin, pads of subcutaneous fat, muscles associated with expression and chewing, and inner linings of moist, stratified squamous epithelium. The **lips** are highly mobile structures that surround the mouth opening. They contain skeletal muscles and sensory receptors useful in judging the temperature and texture of foods. Blood vessels near lip surfaces impart a reddish hue.

Tongue

The **tongue** nearly fills the oral cavity when the mouth is closed. Mucous membrane covers the tongue, and a membranous fold called the **lingual frenulum** (ling′gwahl fren′u-lum) connects the midline of the tongue to the floor of the mouth.

The *body* of the tongue is mostly skeletal muscle. These muscles mix food particles with saliva during chewing and move food toward the pharynx during swallowing. The tongue also helps move food underneath the teeth for chewing. Rough projections called **papillae** on the tongue surface provide friction, which helps handle food. These papillae also bear taste buds (see chapter 10, p. 267).

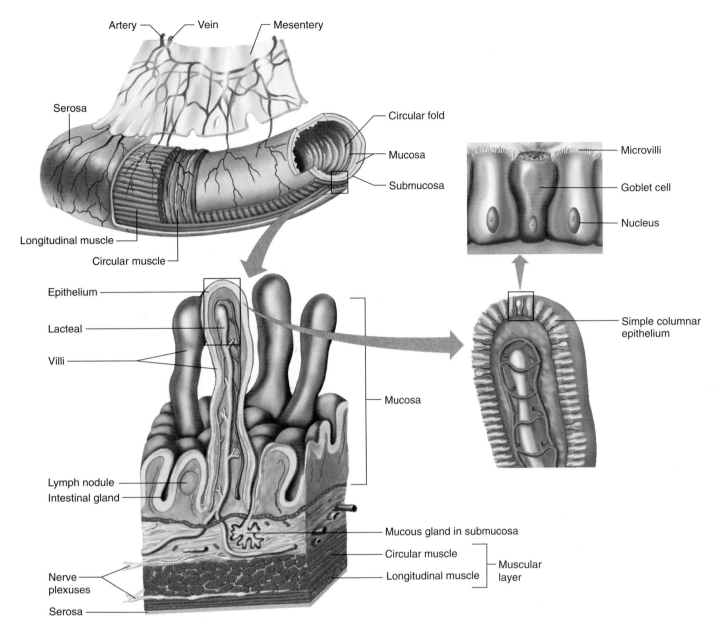

Figure 15.3
The wall of the small intestine, as in other portions of the alimentary canal, includes four layers: an inner mucosa, a submucosa, a muscular layer, and an outer serosa.

The posterior region, or *root,* of the tongue is anchored to the hyoid bone. It is covered with rounded masses of lymphatic tissue called **lingual tonsils** (ton′silz) (fig. 15.6).

Palate

The **palate** (pal′at) forms the roof of the oral cavity and consists of a bony anterior part (*hard palate*) and a muscular posterior part (*soft palate*). A muscular arch of the soft palate extends posteriorly and downward as a cone-shaped projection called the **uvula** (u′vu-lah).

In the back of the mouth, on either side of the tongue and closely associated with the palate, are masses

of lymphatic tissue called **palatine** (pal′ah-tīn) **tonsils** (see figs. 15.5 and 15.6). These structures lie beneath the epithelial lining of the mouth and, like other lymphatic tissues, help protect the body against infection.

The palatine tonsils are common sites of infection, and when inflamed, produce *tonsillitis*. Infected tonsils may swell so greatly that they block the passage through the pharynx and interfere with breathing and swallowing. Because the mucous membranes of the pharynx, auditory tubes, and middle ears are continuous, such an infection can travel from the throat into the middle ears (*otitis media*).

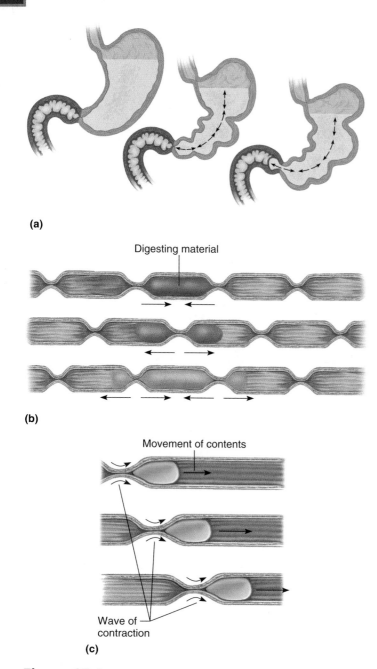

(a)

Digesting material

(b)

Movement of contents

Wave of contraction

(c)

Figure 15.4

Movements through the alimentary canal. (*a*) Mixing movements occur when small segments of the muscular wall of the stomach rhythmically contract. (*b*) Segmentation mixes contents of the small intestine. (*c*) Peristaltic waves move the contents along the canal.

When tonsillitis occurs repeatedly and does not respond to antibiotic treatment, the tonsils may be surgically removed. Such tonsillectomies are done less often today than they were a generation ago because the tonsils' role in immunity is now recognized.

Other masses of lymphatic tissue, called **pharyngeal** (fah-rin'je-al) **tonsils,** or *adenoids,* are on the posterior wall of the pharynx, above the border of the soft

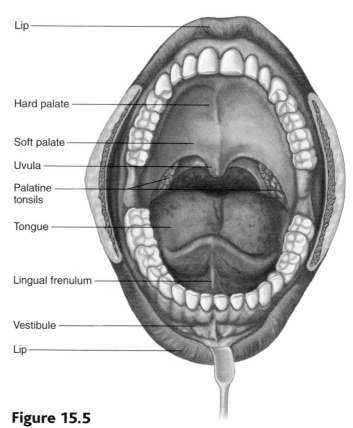

Lip

Hard palate

Soft palate

Uvula

Palatine tonsils

Tongue

Lingual frenulum

Vestibule

Lip

Figure 15.5

The mouth is adapted for ingesting food and beginning digestion, both mechanically and chemically.

palate (fig. 15.6). If the adenoids enlarge and block the passage between the pharynx and the nasal cavity, they also may be surgically removed.

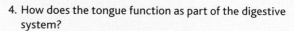

Check Your Recall

4. How does the tongue function as part of the digestive system?

5. Where are the tonsils located?

Teeth

Two different sets of **teeth** form during development. The members of the first set, the *primary teeth* (deciduous teeth), usually erupt through the gums at regular intervals between the ages of six months and two to four years (fig. 15.7). There are twenty deciduous teeth—ten in each jaw.

The primary teeth are usually shed in the same order they appeared. Before this happens, though, their roots are resorbed. Pressure from the developing *secondary teeth* (permanent teeth) then pushes the primary teeth out of their sockets. This secondary set consists of thirty-two teeth—sixteen in each jaw (fig. 15.8). The secondary teeth usually begin to appear at six years, but the set may not be complete until the third molars appear between seventeen and twenty-five years.

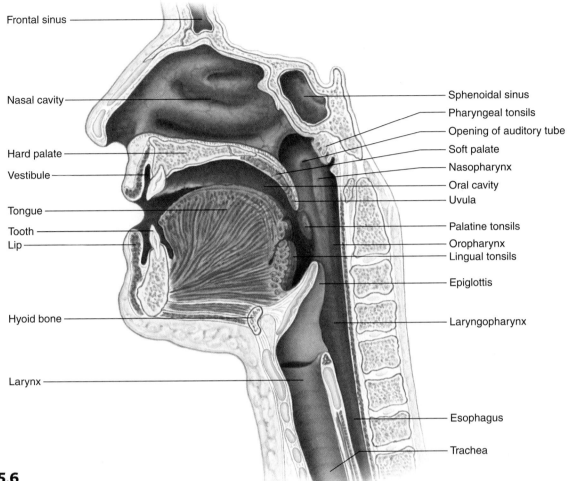

Frontal sinus

Nasal cavity

Hard palate

Vestibule

Tongue

Tooth

Lip

Hyoid bone

Larynx

Sphenoidal sinus

Pharyngeal tonsils

Opening of auditory tube

Soft palate

Nasopharynx

Oral cavity

Uvula

Palatine tonsils

Oropharynx

Lingual tonsils

Epiglottis

Laryngopharynx

Esophagus

Trachea

Figure 15.6

Sagittal section of the mouth, nasal cavity, and pharynx.

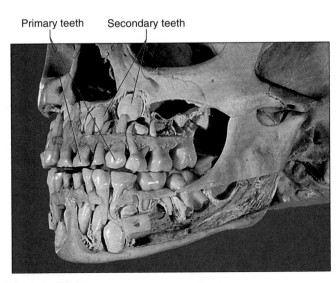

Primary teeth   Secondary teeth

Figure 15.7

This partially dissected child's skull reveals primary and developing secondary teeth in the maxilla and mandible.

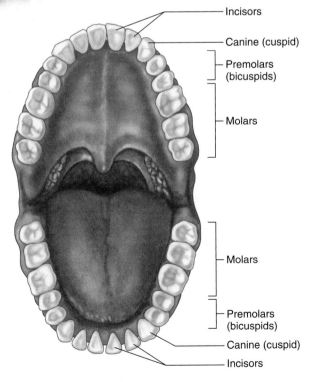

Incisors

Canine (cuspid)

Premolars (bicuspids)

Molars

Molars

Premolars (bicuspids)

Canine (cuspid)

Incisors

Figure 15.8

The secondary teeth of the upper and lower jaws.

Teeth begin mechanical digestion by breaking pieces of food into smaller pieces. This action increases the surface area of food particles, allowing digestive enzymes to react more effectively with the food molecules. Table 15.1 summarizes the number and kinds of teeth that appear during development and their functions.

Each tooth consists of two main portions—the *crown,* which projects beyond the gum (gingiva), and the *root,* which is anchored to the alveolar process of the jaw. Where these portions meet is called the *neck* of the tooth.

Glossy, white *enamel* covers the crown. Enamel mainly consists of calcium salts and is the hardest substance in the body. If enamel is damaged by abrasive action or injury, it is not replaced. Enamel also tends to wear away with age. The Topic of Interest on page 409 discusses tooth enamel destruction.

The bulk of a tooth beneath the enamel is *dentin,* a substance much like bone, but somewhat harder. Dentin surrounds the tooth's central cavity (pulp cavity), which contains blood vessels, nerves, and connective tissue, collectively called *pulp.* Blood vessels and nerves reach this cavity through tubular *root canals* extending into the root.

A thin layer of bonelike material called *cementum* encloses the root. A *periodontal ligament* surrounds the cementum. This ligament contains blood vessels and nerves as well as bundles of thick collagenous fibers which pass between the cementum and the bone of the alveolar process, firmly attaching the tooth to the jaw (fig. 15.9).

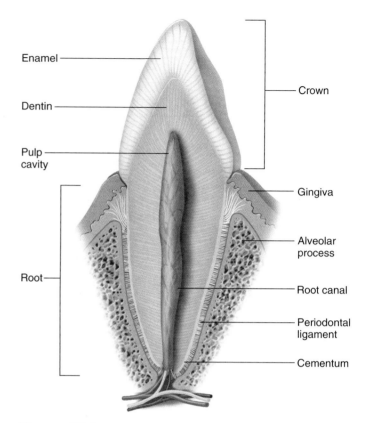

Figure 15.9
A section of a tooth.

Extracted primary and wisdom teeth may one day provide stem cells that can be used to regenerate tooth roots and supporting periodontal ligaments. The stem cells are in the pulp and a region called the apical papilla. Dental researchers hope that these stem cells may be cultured to yield replacement teeth for people who do not have enough jawbone to support dental implants.

Check Your Recall

6. How do primary teeth differ from secondary teeth?
7. Describe the structure of a tooth.
8. Explain how a tooth is attached to the bone of the jaw.

Table 15.1	Primary and Secondary Teeth			
	PRIMARY TEETH (DECIDUOUS)		**SECONDARY TEETH (PERMANENT)**	**FUNCTION**
Type	Number		Number	
Incisor				Bite off pieces of food
Central	4		4	
Lateral	4		4	
Canine (cuspid)	4		4	Grasp and tear food
Premolar (bicuspid)				Grind food particles
First			4	
Second			4	
Molar				Grind food particles
First	4		4	
Second	4		4	
Third			4	
Total	20		32	

Dental Caries

Sticky foods, such as caramel, lodge between the teeth and in the crevices of molars, feeding bacteria such as *Actinomyces, Streptococcus mutans,* and *Lactobacillus.* These microorganisms metabolize carbohydrates in the food, producing acid by-products that destroy tooth enamel and dentin. The bacteria also produce sticky substances that hold them in place.

If a person eats a candy bar but does not brush the teeth soon afterward, the acid-forming bacteria may decay tooth enamel, creating a condition called *dental caries.* Unless a dentist cleans and fills the resulting cavity that forms where enamel is destroyed, the damage will spread to the underlying dentin. As a result, the tooth becomes very sensitive.

Dental caries can be prevented in several ways:

1. Brush and floss teeth regularly.
2. Have regular dental exams and cleanings.
3. Drink fluoridated water or receive a fluoride treatment. Fluoride is incorporated into the enamel's chemical structure, strengthening it.
4. Have the dentist apply a sealant to children's and adolescents' teeth where crevices might hold onto decay-causing bacteria. The sealant is a coating that keeps acids from eating away at tooth enamel.

15.4 SALIVARY GLANDS

The **salivary** (sal′ĭ-ver-e) **glands** secrete saliva. This fluid moistens food particles, helps bind them, and begins the chemical digestion of carbohydrates. Saliva is also a solvent, dissolving foods so that they can be tasted, and it helps cleanse the mouth and teeth.

Salivary Secretions

A salivary gland has two types of secretory cells—*serous cells* and *mucous cells.* Proportions of these cells vary in the different types of salivary glands. Serous cells produce a watery fluid that includes the digestive enzyme **salivary amylase** (am′ĭ-lās). This enzyme splits starch and glycogen molecules into disaccharides—the first step in the chemical digestion of carbohydrates. Mucous cells secrete a thick liquid called **mucus,** which binds food particles and lubricates the food during swallowing.

When a person sees, smells, tastes, or even thinks about appealing food, parasympathetic nerve impulses elicit the secretion of a large volume of watery saliva. Conversely, food that looks, smells, or tastes unpleasant inhibits parasympathetic activity and less saliva is produced. Swallowing may become difficult.

Major Salivary Glands

Three pairs of major salivary glands—the parotid, submandibular, and sublingual glands—and many minor ones are associated with the mucous membranes of the tongue, palate, and cheeks (fig. 15.10). The **parotid glands** (pah-rot′id glandz) are the largest of the major salivary glands. Each gland lies anterior and somewhat inferior to each ear, between the skin of the cheek and the masseter muscle. The parotid glands secrete a clear, watery fluid that is rich in amylase.

The **submandibular** (sub″man-dib′u-lar) **glands** are located in the floor of the mouth on the inside surface of the lower jaw. The secretory cells of these glands are about equally serous and mucous. Consequently, the submandibular glands secrete a more viscous fluid than the parotid glands.

The **sublingual** (sub-ling′gwal) **glands,** the smallest of the major salivary glands, are on the floor of the mouth inferior to the tongue. Their secretory cells are primarily the mucous type, making their secretions thick and stringy.

Check Your Recall

9. What is the function of saliva?
10. What stimulates salivary glands to secrete saliva?
11. Where are the major salivary glands located?

15.5 PHARYNX AND ESOPHAGUS

The pharynx is a cavity posterior to the mouth from which the tubular esophagus leads to the stomach (see fig. 15.1). The pharynx and the esophagus do not digest food, but both are important passageways whose muscular walls function in swallowing.

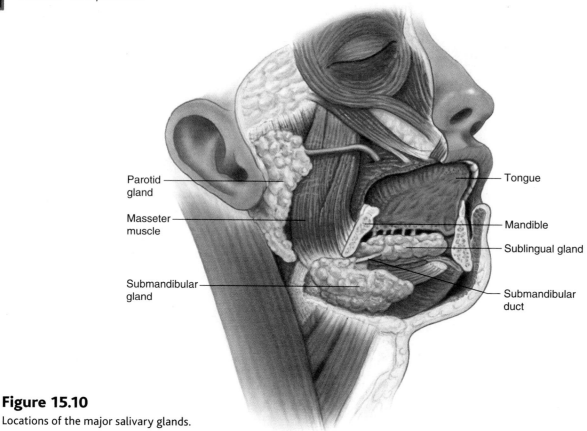

Figure 15.10
Locations of the major salivary glands.

Structure of the Pharynx

The **pharynx** (far'inks) connects the nasal and oral cavities with the larynx and esophagus (see fig. 15.6). It has three parts:

1. The **nasopharynx** (na"zo-far'inks) communicates with the nasal cavity and provides a passageway for air during breathing.
2. The **oropharynx** (o"ro-far'inks) is posterior to the soft palate and inferior to the nasopharynx. It is a passageway for food moving downward from the mouth and for air moving to and from the nasal cavity.
3. The **laryngopharynx** (lah-ring"go-far'inks), just inferior to the oropharynx, is a passageway to the esophagus.

Swallowing Mechanism

Swallowing has three stages. In the first stage, which is voluntary, food is chewed and mixed with saliva. Then the tongue rolls this mixture into a mass, or **bolus,** and forces it into the pharynx.

The second stage of swallowing begins as food stimulates sensory receptors around the pharyngeal opening. This triggers the swallowing reflex, which includes the following actions:

1. The soft palate (including the uvula) raises, preventing food from entering the nasal cavity.

2. The hyoid bone and the larynx are elevated. A flaplike structure attached to the larynx, called the *epiglottis* (ep"ĭ-glot'is), closes off the top of the larynx so that food is less likely to enter the trachea.
3. The tongue is pressed against the soft palate, sealing off the oral cavity from the pharynx.
4. The longitudinal muscles in the pharyngeal wall contract, pulling the pharynx upward toward the food.
5. Muscles in the lower portion of the pharynx relax, opening the esophagus.
6. A peristaltic wave begins in the pharyngeal muscles and forces food into the esophagus.

The swallowing reflex momentarily inhibits breathing. Then, during the third stage of swallowing, peristalsis transports the food in the esophagus to the stomach.

Computer simulation experiments show that each type of food requires an optimum range of number of chews to form a bolus. Eating raw carrots, for example, requires twenty to twenty-five chews.

Esophagus

The **esophagus** (ĕ-sof'ah-gus), a straight, collapsible tube about 25 centimeters long, is a food passageway from the pharynx to the stomach (see figs. 15.1 and 15.6). The esophagus begins at the base of the pharynx and descends posterior to the trachea, passing through the mediastinum. It penetrates the diaphragm through an opening, the *esophageal hiatus* (ĕ-sof″ah-je′al hi-a′tus), and is continuous with the stomach on the abdominal side of the diaphragm.

Mucous glands are scattered throughout the submucosa of the esophagus. Their secretions moisten and lubricate the tube's inner lining.

Just above where the esophagus joins the stomach, some circular smooth muscle fibers in the esophageal wall thicken, forming the **lower esophageal sphincter,** or cardiac sphincter (fig. 15.11). These fibers usually remain contracted, and they close the entrance to the stomach, preventing the stomach contents from regurgitating into the esophagus. When peristaltic waves reach the stomach, the muscle fibers temporarily relax and allow the swallowed food to enter.

In a *hiatal hernia,* a portion of the stomach protrudes through a weakened area of the diaphragm, through the esophageal hiatus, and into the thorax. As a result, regurgitation (reflux) of gastric juice into the esophagus may inflame the esophageal mucosa, causing heartburn, difficulty in swallowing, or ulceration and blood loss. In response to the destructive action of gastric juice, columnar epithelium may replace the squamous epithelium that normally lines the esophagus. This condition, called *Barrett's esophagus,* increases the risk of developing esophageal cancer.

Check Your Recall

12. Describe the regions of the pharynx.

13. List the major events that occur during swallowing.

14. What is the function of the esophagus?

15.6 STOMACH

The **stomach** is a J-shaped, pouchlike organ that hangs inferior to the diaphragm in the upper left portion of the abdominal cavity and has a capacity of about 1 liter or more (figs. 15.1 and 15.11; see reference plates 4 and 5, pp. 26–27). Thick folds (rugae) of mucosal and submucosal layers mark the stomach's inner lining and disappear when the stomach wall is distended. The stomach receives food from the esophagus, mixes the food with gastric juice, initiates protein digestion, carries on limited absorption, and moves food into the small intestine.

Parts of the Stomach

The stomach is divided into the cardiac, fundic, body, and pyloric regions (fig. 15.11). The *cardiac region* is a small area near the esophageal opening. The *fundic region,* which balloons superior to the cardiac portion,

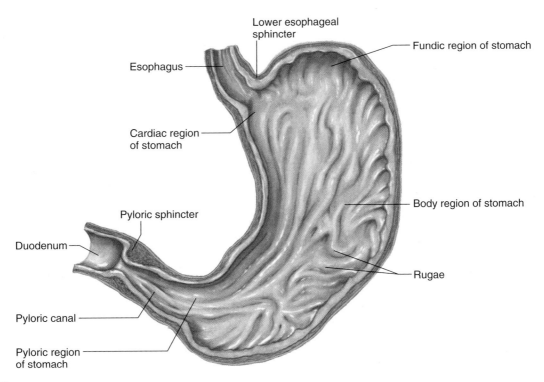

Figure 15.11

Major regions of the stomach and its associated structures.

is a temporary storage area. The dilated *body region,* which is the main part of the stomach, lies between the fundic and pyloric portions. The *pyloric region* narrows and becomes the *pyloric canal* as it approaches the small intestine.

At the end of the pyloric canal the muscular wall thickens, forming a powerful circular muscle, the **pyloric sphincter.** This muscle is a valve that controls gastric emptying.

Gastric Secretions

The mucous membrane that forms the inner lining of the stomach is thick. Its surface is studded with many small openings called *gastric pits* that are at the ends of tubular **gastric glands** (gas'trik glandz) (fig. 15.12).

Gastric glands generally have three types of secretory cells. *Mucous cells* are in the necks of the glands, near the openings of the gastric pits. *Chief cells* and *parietal cells* are in the deeper parts of the glands. The chief cells secrete digestive enzymes, and the parietal cells release hydrochloric acid. The products of the mucous cells, chief cells, and parietal cells together form **gastric juice** (gas'trik jo͞os).

Of the digestive enzymes in gastric juice, **pepsin** (pep'sin) is by far the most important. The chief cells secrete pepsin as the inactive enzyme precursor **pepsinogen** (pep-sin'o-jen). When pepsinogen contacts hydrochloric acid from the parietal cells, it breaks down rapidly, forming pepsin. Pepsin begins the digestion of nearly all types of dietary protein into polypeptides. This enzyme is most active in an acidic environment, which is provided by the hydrochloric acid in gastric juice.

The mucous cells of the gastric glands (*mucous neck cells*) and the mucous cells associated with the stomach's inner surface release a viscous, alkaline secretion which coats the inside of the stomach wall. This coating normally prevents the stomach from digesting itself.

Another component of gastric juice is **intrinsic factor** (in-trin'sik fak'tor), which the parietal cells secrete. Intrinsic factor helps the small intestine absorb vitamin B_{12}. Table 15.2 summarizes the components of gastric juice.

Check Your Recall

15. What are the secretions of the chief cells and parietal cells?
16. Which is the most important digestive enzyme in gastric juice?
17. Why doesn't the stomach digest itself?

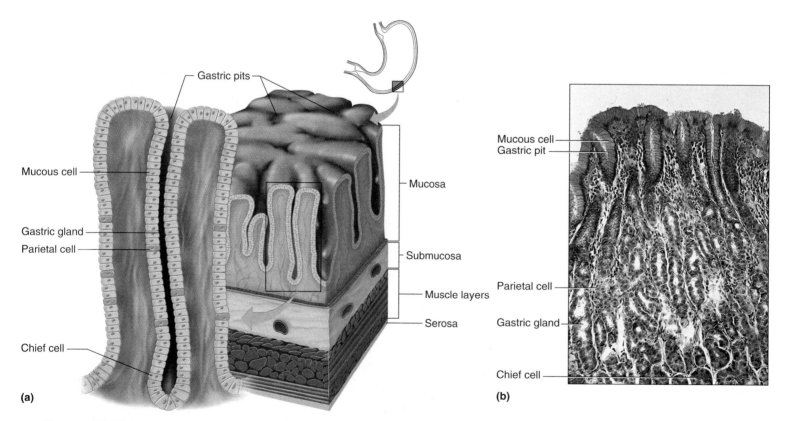

(a)

(b)

Mucous cell

Gastric pits

Mucosa

Gastric gland

Parietal cell

Submucosa

Muscle layers

Serosa

Chief cell

Mucous cell
Gastric pit

Parietal cell

Gastric gland

Chief cell

Figure 15.12

Lining of the stomach. (*a*) Gastric glands include mucous cells, parietal cells, and chief cells. (*b*) A light micrograph of cells associated with the gastric glands (50×).

Table 15.2	Major Components of Gastric Juice	
Component	**Source**	**Function**
Pepsinogen	Chief cells of the gastric glands	Inactive form of pepsin
Pepsin	Formed from pepsinogen in the presence of hydrochloric acid	A protein-splitting enzyme that digests nearly all types of dietary protein
Hydrochloric acid	Parietal cells of the gastric glands	Provides the acid environment needed for the conversion of pepsinogen into pepsin and for the action of pepsin
Mucus	Mucous cells	Provides a viscous, alkaline protective layer on the inside stomach wall
Intrinsic factor	Parietal cells of the gastric glands	Aids in vitamin B_{12} absorption

The 40 million cells that line the stomach's interior can secrete 2 to 3 quarts (about 2 to 3 liters) of gastric juice per day.

Gastrin stimulates cell growth in the mucosa of the stomach and intestines, except where gastrin is produced. This cell growth helps replace mucosal cells damaged by normal stomach function, disease, or medical treatments.

Regulation of Gastric Secretions

Gastric juice is produced continuously, but the rate varies considerably and is controlled both neurally and hormonally. When a person tastes, smells, or even sees appetizing food, or when food enters the stomach, parasympathetic impulses on the vagus nerves stimulate the release of acetylcholine (Ach) from nerve endings. This Ach stimulates gastric glands to secrete abundant gastric juice, which is rich in hydrochloric acid and pepsinogen. These parasympathetic impulses also stimulate certain stomach cells to release the peptide hormone **gastrin** (gas′trin), which increases the secretory activity of gastric glands (fig. 15.13).

As food moves into the small intestine, acid triggers sympathetic nerve impulses that inhibit gastric juice secretion. At the same time, proteins and fats in this region of the intestine cause the intestinal wall to release the peptide hormone **cholecystokinin** (ko″le-sis″to-ki′nin). This hormonal action decreases gastric motility as the small intestine fills with food.

An *ulcer* is an open sore in the skin or mucous membrane resulting from localized tissue breakdown. Gastric ulcers form in the stomach, and duodenal ulcers form in the region of the small intestine nearest the stomach.

For many years, gastric and duodenal ulcers were attributed to stress and treated with medications to decrease stomach acid secretion. In 1982, two Australian researchers boldly suggested that stomach infection by the bacterium *Helicobacter pylori* causes gastric ulcers. When the medical community did not believe them, one of the researchers swallowed some bacteria, calling it "swamp water," to demonstrate their effect—and soon developed stomach pain (gastritis). Still, it was twelve years before U.S. government physicians advised their colleagues to treat gastric ulcers as an infection in people with evidence of *Helicobacter pylori*. Today, a short course of antibiotics, often combined with acid-lowering drugs, can cure a gastric ulcer.

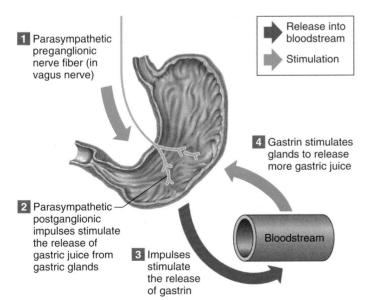

1 Parasympathetic preganglionic nerve fiber (in vagus nerve)

Release into bloodstream

Stimulation

4 Gastrin stimulates glands to release more gastric juice

2 Parasympathetic postganglionic impulses stimulate the release of gastric juice from gastric glands

3 Impulses stimulate the release of gastrin

Bloodstream

Figure 15.13

The secretion of gastric juice is regulated in part by parasympathetic nerve impulses that stimulate the release of gastric juice and gastrin.

Gastric Absorption

Gastric enzymes begin breaking down proteins, but the stomach wall is not well adapted to absorbing digestive products. The stomach absorbs only small volumes of water and certain salts as well as certain lipid-soluble drugs. Alcohol, which is not a nutrient, is absorbed both in the small intestine and the stomach. This is why the intoxicating effects of alcohol are felt soon after consuming alcoholic beverages.

Check Your Recall

18. What controls gastric juice secretion?
19. What is the function of cholecystokinin?
20. Which substances can the stomach absorb?

Mixing and Emptying Actions

Following a meal, the mixing movements of the stomach wall aid in producing a semifluid paste of food particles and gastric juice called **chyme** (kīm). Peristaltic waves push the chyme toward the pyloric region of the stomach, and as chyme accumulates near the pyloric sphincter, this muscle begins to relax. Stomach contractions push chyme a little at a time into the small intestine.

The rate at which the stomach empties depends on the fluidity of the chyme and the type of food present. Liquids usually pass through the stomach quite rapidly, but solids remain until they are well mixed with gastric juice. Fatty foods may remain in the stomach from three to six hours; foods high in proteins move through more quickly; carbohydrates usually pass through faster than either fats or proteins.

As chyme enters the duodenum (the first portion of the small intestine), accessory organs—the pancreas, liver, and gallbladder—add their secretions.

Vomiting results from a complex reflex that empties the stomach through the esophagus, pharynx, and mouth. Irritation or distension in the stomach or intestines can trigger vomiting. Sensory impulses travel from the site of stimulation to the *vomiting center* in the medulla oblongata, and several motor responses follow. These include taking a deep breath, raising the soft palate and thus closing the nasal cavity, closing the opening to the trachea (glottis), relaxing the circular muscle fibers at the base of the esophagus, contracting the diaphragm so it presses downward over the stomach, and contracting the abdominal wall muscles to increase pressure inside the abdominal cavity. The stomach, squeezed from all sides, forces its contents upward and out.

Check Your Recall

21. How is chyme produced?
22. What factors influence how quickly chyme leaves the stomach?

15.7 PANCREAS

The **pancreas** was discussed as an endocrine gland in chapter 11 (pp. 304–307). It also has an exocrine function—secretion of a digestive juice called **pancreatic juice** (pan″kre-at′ik jo͞os).

Structure of the Pancreas

The pancreas is closely associated with the small intestine. It extends horizontally across the posterior abdominal wall in the C-shaped curve of the duodenum (figs. 15.1 and 15.14).

The cells that produce pancreatic juice, called *pancreatic acinar* (a′sĭ-nar) *cells,* make up the bulk of the pancreas. These cells cluster around tiny tubes into which they release their secretions. The smaller tubes unite to form larger ones, which in turn give rise to a *pancreatic duct* extending the length of the pancreas. The pancreatic duct usually connects with the duodenum at the same place where the bile duct from the liver and gallbladder joins the duodenum, although other connections may be present (fig. 15.14). A *hepatopancreatic sphincter* controls the movement of pancreatic juices into the duodenum.

Pancreatic Juice

Pancreatic juice contains enzymes that digest carbohydrates, fats, nucleic acids, and proteins. The carbohydrate-digesting enzyme **pancreatic amylase** splits molecules of starch or glycogen into double sugars (disaccharides). The fat-digesting enzyme **pancreatic lipase** breaks triglyceride molecules into fatty acids and glycerol. Pancreatic juice also contains two **nucleases,** which are enzymes that break down nucleic acid molecules into nucleotides.

The protein-splitting (proteolytic) enzymes are **trypsin, chymotrypsin,** and **carboxypeptidase** (kar-bok″se-pep′tĭ-da-s). These enzymes split the bonds between particular combinations of amino acids in proteins. Because no single enzyme can split all the possible amino acid combinations, complete digestion of protein molecules requires several types of enzymes.

The proteolytic enzymes are stored within tiny structures in cells called *zymogen granules* (zi-mo′jen gran′ūlz). These enzymes, like gastric pepsin, are secreted in inactive forms. After the inactive forms of the proteolytic enzymes reach the small intestine, other enzymes activate them. For example, pancreatic cells release inactive *trypsinogen,* which is activated to trypsin when it contacts the enzyme **enterokinase** (en″tero-ki′na-s) secreted by the mucosa of the small intestine.

Painful *acute pancreatitis* results from blockage in the release of pancreatic juice. Trypsin, activated as pancreatic juice builds up, digests parts of the pancreas. Alcoholism, gallstones, certain infections, traumatic injuries, or the side effects of some drugs can cause pancreatitis.

Regulation of Pancreatic Secretion

The nervous and endocrine systems regulate release of pancreatic juice, as they do gastric and small intestinal secretions. For example, when parasympathetic

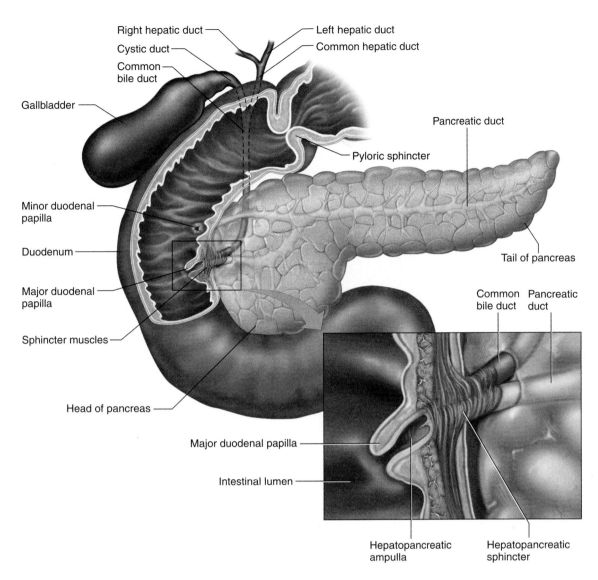

Figure 15.14
The pancreas is closely associated with the duodenum.

impulses stimulate gastric juice secretion, other parasympathetic impulses stimulate the pancreas to release digestive enzymes. Also, as acidic chyme enters the duodenum, the duodenal mucous membrane releases the peptide hormone **secretin** (se-kre′tin) into the bloodstream (fig. 15.15). This hormone stimulates secretion of pancreatic juice that has a high concentration of bicarbonate ions. These ions neutralize the acid of chyme and provide a favorable environment for digestive enzymes in the intestine.

Proteins and fats in chyme in the duodenum also stimulate the intestinal wall to release *cholecystokinin*. Like secretin, cholecystokinin travels via the bloodstream to the pancreas. Pancreatic juice secreted in response to cholecystokinin has a high concentration of digestive enzymes.

In cystic fibrosis, abnormal chloride channels in cells in various tissues draw water inward from interstitial spaces. This dries out secretions in the lungs and pancreas, leaving a very sticky mucus that impairs the functioning of these organs. When the pancreas is plugged with mucus, its secretions, containing digestive enzymes, cannot reach the duodenum. To maintain their body weight and prevent malnutrition, individuals with cystic fibrosis must take digestive enzyme supplements, usually as a powder mixed with a soft food such as applesauce.

Check Your Recall

23. List the enzymes in pancreatic juice.
24. What are the functions of the enzymes in pancreatic juice?
25. What regulates secretion of pancreatic juice?

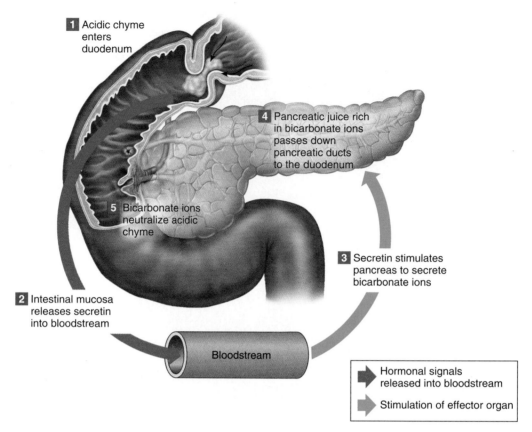

1 Acidic chyme enters duodenum

4 Pancreatic juice rich in bicarbonate ions passes down pancreatic ducts to the duodenum

5 Bicarbonate ions neutralize acidic chyme

3 Secretin stimulates pancreas to secrete bicarbonate ions

2 Intestinal mucosa releases secretin into bloodstream

Bloodstream

Hormonal signals released into bloodstream

Stimulation of effector organ

Figure 15.15

Acidic chyme entering the duodenum from the stomach stimulates the release of secretin, which, in turn, stimulates the release of pancreatic juice.

15.8 LIVER

The **liver** is in the upper right quadrant of the abdominal cavity, just inferior to the diaphragm. It is partially surrounded by the ribs, and extends from the level of the fifth intercostal space to the lower margin of the ribs. The reddish-brown liver is well supplied with blood vessels (see fig. 15.1 and reference plate 4, p. 26).

> The average adult liver is the heaviest organ in the body. It weighs about 3 pounds.

Liver Structure

A fibrous capsule encloses the liver, and connective tissue divides the organ into lobes, including a large *right lobe* and a smaller *left lobe* (fig. 15.16). Each lobe is separated into many tiny **hepatic lobules** (hĕ-pat′ik lob′ulz), which are the liver's functional units (fig. 15.17). A lobule consists of many hepatic cells radiating outward from a *central vein.* Vascular channels called **hepatic sinusoids** separate platelike groups of these cells from each other. Blood from the digestive tract, which is car-

ried in the *hepatic portal vein* (see chapter 13, p. 369), brings newly absorbed nutrients into the sinusoids and nourishes the hepatic cells (fig. 15.18).

Large phagocytic macrophages called *Kupffer cells* (koop′fer selz) are fixed to the inner linings of the hepatic sinusoids. They remove bacteria or other foreign particles that enter the blood in the portal vein through the intestinal wall. Blood passes from these sinusoids into the central veins of the hepatic lobules and exits the liver.

Within the hepatic lobules are many fine *bile canaliculi,* which carry secretions from hepatic cells to *bile ductules.* The ductules of neighboring lobules unite to form larger bile ducts which then converge to become the **hepatic ducts.** These ducts merge, in turn, to form the **common hepatic duct.**

Liver Functions

The liver carries on many important metabolic activities. Recall from chapter 11 (p. 306) that the liver plays a key role in carbohydrate metabolism by helping maintain concentration of blood glucose within the normal range. Hepatic cells responding to hormones such as insulin and glucagon lower the blood glucose level by polymerizing glucose to glycogen, and raise the blood

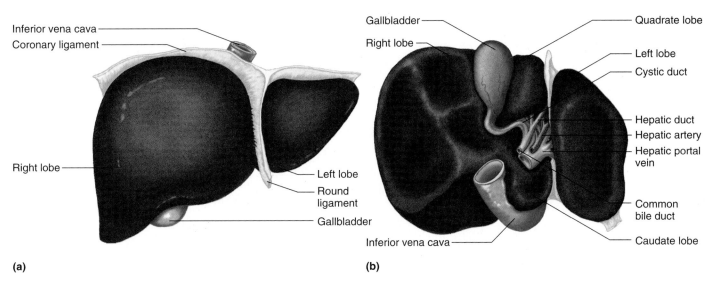

Inferior vena cava
Coronary ligament
Right lobe
Left lobe
Round ligament
Gallbladder

(a)

Gallbladder
Right lobe
Quadrate lobe
Left lobe
Cystic duct
Hepatic duct
Hepatic artery
Hepatic portal vein
Common bile duct
Inferior vena cava
Caudate lobe

(b)

Figure 15.16

Lobes of the liver, viewed (*a*) anteriorly and (*b*) inferiorly. Connective tissue divides the organ into the larger right lobe and smaller left lobe.

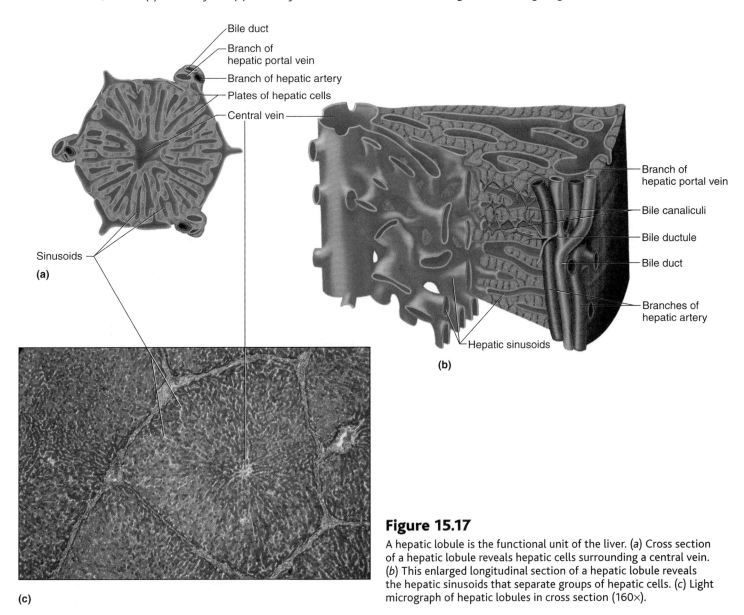

Bile duct
Branch of hepatic portal vein
Branch of hepatic artery
Plates of hepatic cells
Central vein
Sinusoids

(a)

Branch of hepatic portal vein
Bile canaliculi
Bile ductule
Bile duct
Branches of hepatic artery
Hepatic sinusoids

(b)

(c)

Figure 15.17

A hepatic lobule is the functional unit of the liver. (*a*) Cross section of a hepatic lobule reveals hepatic cells surrounding a central vein. (*b*) This enlarged longitudinal section of a hepatic lobule reveals the hepatic sinusoids that separate groups of hepatic cells. (*c*) Light micrograph of hepatic lobules in cross section (160×).

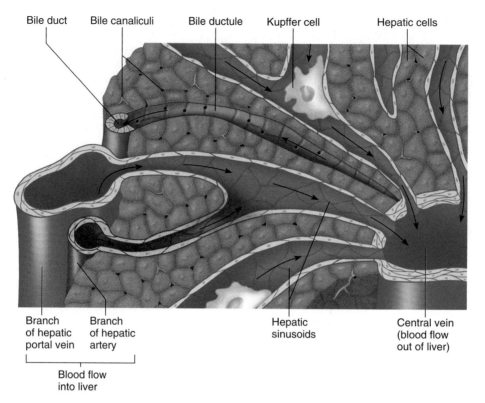

Bile duct | Bile canaliculi | Bile ductule | Kupffer cell | Hepatic cells

Branch of hepatic portal vein | Branch of hepatic artery | Hepatic sinusoids | Central vein (blood flow out of liver)

Blood flow into liver

Figure 15.18
The paths of blood and bile within a hepatic lobule.

glucose level by breaking down glycogen to glucose or by converting noncarbohydrates into glucose.

The liver's effects on lipid metabolism include oxidizing fatty acids at an especially high rate; synthesizing lipoproteins, phospholipids, and cholesterol; and converting portions of carbohydrate and protein molecules into fat molecules (see pp. 424–425). The blood transports fats synthesized in the liver to adipose tissue for storage.

The most vital liver functions concern protein metabolism. They include deaminating amino acids; forming urea (see p. 432); synthesizing plasma proteins, such as clotting factors (see chapter 12, p. 326); and converting certain amino acids to other amino acids.

The liver also stores many substances, including glycogen, iron, and vitamins A, D, and B_{12}. In addition, macrophages in the liver help destroy damaged red blood cells (see chapter 12, p. 322) and phagocytize foreign antigens. The liver also removes toxic substances such as alcohol and certain other drugs from blood (detoxification) and secretes bile.

Many of these liver functions are not directly related to the digestive system and are, as indicated above, discussed in other chapters. Bile secretion, however, is important to digestion and is explained next in this chapter. Table 15.3 summarizes the major functions of the liver. The Topic of Interest on page 419 discusses viral infections of the liver.

Check Your Recall

26. Locate the liver.
27. Describe a hepatic lobule.
28. Review liver functions.

Table 15.3	Major Functions of the Liver
General Function	**Specific Function**
Carbohydrate metabolism	Polymerizes glucose to glycogen; breaks down glycogen to glucose; converts noncarbohydrates to glucose
Lipid metabolism	Oxidizes fatty acids; synthesizes lipoproteins, phospholipids, and cholesterol; converts portions of carbohydrate and protein molecules into fats
Protein metabolism	Deaminates amino acids; forms urea; synthesizes plasma proteins; converts certain amino acids to other amino acids
Storage	Stores glycogen, iron, and vitamins A, D, and B_{12}
Blood filtering	Removes damaged red blood cells and foreign substances by phagocytosis
Detoxification	Removes toxins from blood
Secretion	Secretes bile

Topic of Interest

Hepatitis

Hepatitis is an inflammation of the liver. It has several causes, but the various types have similar symptoms.

For the first few days, hepatitis may resemble the flu, producing mild headache, low fever, fatigue, lack of appetite, nausea and vomiting, and sometimes stiff joints. By the end of the first week, more distinctive symptoms arise: a rash, pain in the upper right quadrant of the abdomen, dark and foamy urine, and pale feces. The skin and sclera of the eyes begin to turn yellow due to accumulating bile pigments (jaundice). Great fatigue may continue for two or three weeks, and then gradually the person begins to feel better.

This is hepatitis in its most common, least dangerous acute guise. About half a million people develop hepatitis in the United States each year, and 6,000 die. In a rare form called *fulminant hepatitis,* symptoms occur suddenly and severely, along with altered behavior and personality. Medical attention is necessary to prevent kidney or liver failure or coma. Hepatitis that persists for more than six months is termed chronic. As many as 300 million people worldwide are hepatitis carriers. They do not have symptoms but can infect others. Five percent of carriers eventually develop liver cancer.

Only rarely does hepatitis result from alcoholism, autoimmunity, or the use of certain drugs. Usually, one of several types of viruses cause hepatitis. Viral types are distinguished by the route of infection, surface features, and whether the viral genetic material is DNA or RNA. Hepatitis B virus has DNA; the others have RNA. The viral types are classified as follows:

Hepatitis A spreads by contact with food or objects contaminated with virus-containing feces, including diapers. The course of hepatitis A is short and mild.

Hepatitis B spreads by contact with virus-containing body fluids, such as blood, saliva, or semen. It may be transmitted by blood transfusions, hypodermic needles, or sexual activity.

Hepatitis C accounts for about half of all known cases of hepatitis. This virus is primarily transmitted in blood—by sharing razors or needles, from pregnant woman to fetus, or through blood transfusions or use of blood products. As many as 60% of individuals infected with the hepatitis C virus suffer chronic symptoms.

Hepatitis D occurs in people already infected with the hepatitis B virus. It is blood-borne and associated with blood transfusions and intravenous drug use. About 20% of individuals infected with this virus die from the infection.

Hepatitis E virus is usually transmitted in water contaminated with feces. It most often affects visitors to developing nations.

Hepatitis G is rare but seems to account for a significant percentage of cases of fulminant hepatitis. In people with healthy immune systems, it produces symptoms so mild that they may not even be noticed.

Because a virus usually causes hepatitis, antibiotic drugs, which are effective against bacteria, are not helpful. Usually, the person must just wait out the symptoms. Hepatitis C, however, sometimes responds to a form of interferon, an immune system biochemical given as a drug.

Composition of Bile

Bile (bīl) is a yellowish-green liquid continuously secreted from hepatic cells. In addition to water, bile contains *bile salts, bile pigments* (bilirubin and biliverdin), *cholesterol,* and *electrolytes.* Of these, bile salts are the most abundant and are the only bile substances that have a digestive function. Bile pigments are breakdown products of hemoglobin from red blood cells and are normally excreted in the bile (see chapter 12, p. 322).

Jaundice turns the skin and whites of the eyes yellow. The distinctive skin color reflects buildup of bile pigments. The condition can have several causes. In *obstructive jaundice,* bile ducts are blocked (as with gallstones or tumors). In *hepatocellular jaundice,* the liver is diseased (as in cirrhosis or hepatitis). In *hemolytic jaundice,* red blood cells are destroyed too rapidly (as with a blood transfusion from an incompatible blood group or a blood infection like malaria).

Gallbladder

The **gallbladder** (gawl′blad-er) is a pear-shaped sac in a depression on the liver's inferior surface. It connects to the **cystic duct** (sis′tik dukt), which in turn joins the common hepatic duct (figs. 15.1 and 15.19). The gallbladder is lined with epithelial cells and has a strong, muscular layer in its wall. The gallbladder stores bile between meals, reabsorbs water to concentrate bile, and contracts to release bile into the small intestine.

The common hepatic and cystic ducts join to form the *common bile duct.* It leads to the duodenum (the proximal part of the small intestine) (figs. 15.14 and 15.19), where the *hepatopancreatic sphincter* guards its exit. This sphincter normally remains contracted, so as bile collects in the common bile duct it backs up into the cystic duct. When this happens, bile flows into the gallbladder, where it is stored.

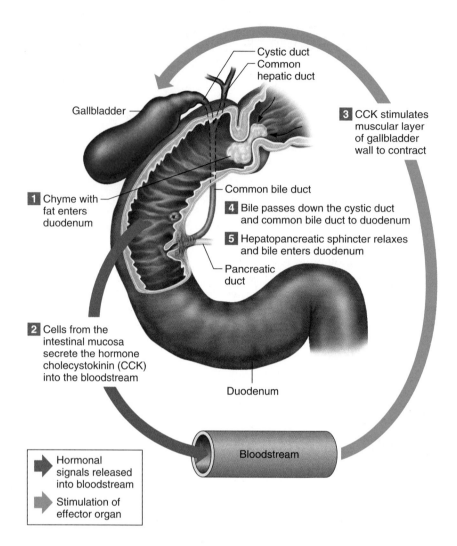

1 Chyme with fat enters duodenum

Gallbladder

Cystic duct
Common hepatic duct

3 CCK stimulates muscular layer of gallbladder wall to contract

Common bile duct

4 Bile passes down the cystic duct and common bile duct to duodenum

5 Hepatopancreatic sphincter relaxes and bile enters duodenum

Pancreatic duct

2 Cells from the intestinal mucosa secrete the hormone cholecystokinin (CCK) into the bloodstream

Duodenum

Bloodstream

Hormonal signals released into bloodstream

Stimulation of effector organ

Figure 15.19

Fatty chyme entering the duodenum stimulates the gallbladder to release bile.

Cholesterol in bile may precipitate and form crystals called *gallstones* under certain conditions (fig. 15.20). Gallstones entering the common bile duct may block bile flow into the small intestine and cause considerable pain. A surgical procedure called a *cholecystectomy* removes the gallbladder when gallstones are obstructive. The surgery can often be done with a laparoscope (small, lit probe) on an outpatient basis.

Regulation of Bile Release

Normally, bile does not enter the duodenum until *cholecystokinin* stimulates the gallbladder to contract. The intestinal mucosa releases this hormone in response to proteins and fats in the small intestine. (Recall its action to stimulate pancreatic enzyme secretion, p. 415.) The hepatopancreatic sphincter usually remains contracted until a peristaltic wave in the duodenal wall approaches it. Then the sphincter relaxes, and bile is squirted into the small intestine (see fig. 15.19). Table 15.4 summarizes the hormones that help control digestion.

Functions of Bile Salts

Bile salts aid digestive enzymes. Bile salts affect *fat globules* (clumped molecules of fats) much like a soap or detergent would affect them. That is, bile salts break fat globules into smaller droplets, an action called

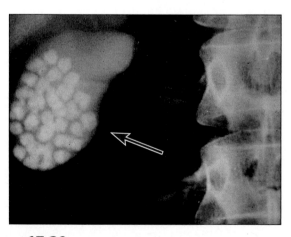

Figure 15.20

Radiograph of a gallbladder that contains gallstones (arrow).

Table 15.4	Hormones of the Digestive Tract	
Hormone	**Source**	**Function**
Gastrin	Gastric cells, in response to food	Causes gastric glands to increase their secretory activity
Cholecystokinin	Intestinal wall cells, in response to proteins and fats in the small intestine	Causes gastric glands to decrease their secretory activity and inhibits gastric motility; stimulates pancreas to secrete fluid with a high digestive enzyme concentration; stimulates gallbladder to contract and release bile
Secretin	Cells in the duodenal wall, in response to acidic chyme entering the small intestine	Stimulates pancreas to secrete fluid with a high bicarbonate ion concentration

emulsification (e-mul″sĭ-fĭ-ka′shun), that greatly increases the total surface area of the fatty substance. The tiny fat droplets then mix with water. Fat-splitting enzymes (lipases) can then digest fat molecules more effectively.

Bile salts also enhance absorption of fatty acids, cholesterol, and the fat-soluble vitamins A, D, E, and K. Lack of bile salts results in poor lipid absorption and vitamin deficiencies.

Check Your Recall

29. Explain how bile forms.
30. Describe the function of the gallbladder.
31. How is secretion of bile regulated?
32. How do bile salts function in digestion?

15.9 SMALL INTESTINE

The **small intestine** is a tubular organ that extends from the pyloric sphincter to the beginning of the large intestine. With its many loops and coils, it fills much of the abdominal cavity (see fig. 15.1 and reference plates 4 and 5, pp. 26–27).

The small intestine receives secretions from the pancreas and liver. It also completes digestion of the nutrients in chyme, absorbs the products of digestion, and transports the residues to the large intestine.

Parts of the Small Intestine

The small intestine consists of three portions: the duodenum, the jejunum, and the ileum (figs. 15.21 and 15.22). The **duodenum** (du″o-de′num), which is about 25 centimeters long and 5 centimeters in diameter, lies posterior to the parietal peritoneum and is the most fixed portion of the small intestine. It follows a C-shaped path as it passes anterior to the right kidney and the upper three lumbar vertebrae.

The remainder of the small intestine is mobile and lies free in the peritoneal cavity. The proximal two-fifths of

this portion is the **jejunum** (jě-joo′num), and the remainder is the **ileum** (il′e-um). A double-layered fold of peritoneal membrane called **mesentery** (mes′en-ter″e) suspends these portions from the posterior abdominal wall (figs. 15.21 and 15.23). The mesentery supports the blood vessels, nerves, and lymphatic vessels that supply the intestinal wall. The jejunum and ileum are not distinctly separate parts; however, the diameter of the jejunum is greater than that of the ileum, and its wall is thicker, more vascular, and more active.

A filmy, double fold of peritoneal membrane called the *greater omentum* drapes like an apron from the stomach over the transverse colon and the folds of the small intestine. If infections occur in the wall of the alimentary canal, cells from the omentum may adhere to the inflamed region, helping to wall off the area, thereby preventing spread of the infection to the peritoneal cavity (fig. 15.23).

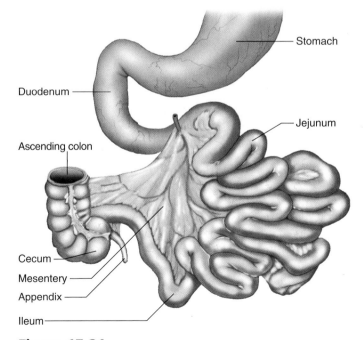

Figure 15.21

The three parts of the small intestine are the duodenum, the jejunum, and the ileum.

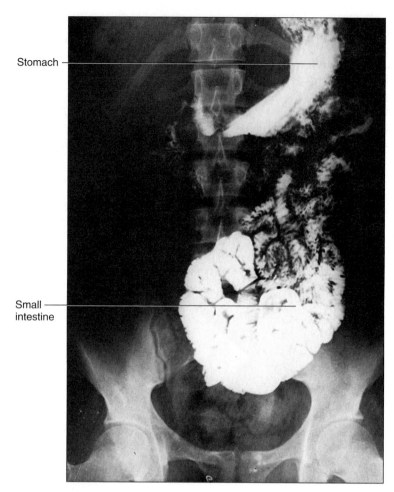

Figure 15.22

Radiograph showing a normal small intestine containing a radiopaque substance that the patient ingested.

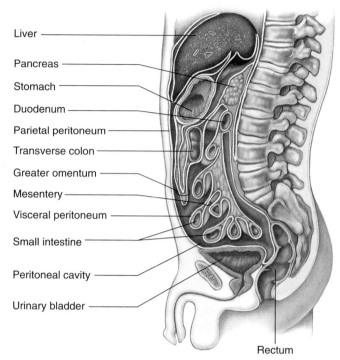

Liver
Pancreas
Stomach
Duodenum
Parietal peritoneum
Transverse colon
Greater omentum
Mesentery
Visceral peritoneum
Small intestine
Peritoneal cavity
Urinary bladder
Rectum

Figure 15.23

Mesentery formed by folds of the peritoneal membrane suspends portions of the small intestine from the posterior abdominal wall.

> The epithelial cells that form the lining of the small intestine are continually replaced. New cells form within the intestinal glands by mitosis and migrate outward onto the villus surface. When the migrating cells reach the tip of the villus, they are shed. This *cellular turnover* renews the small intestine's epithelial lining every three to six days. As a result, nearly one-quarter of the bulk of feces consists of dead epithelial cells from the small intestine.

Structure of the Small Intestine Wall

The inner wall of the small intestine throughout its length appears velvety due to many tiny projections of mucous membrane called **intestinal villi** (in-tes′tĭ-nal vil′i) (figs. 15.24 and 15.25; see fig. 15.3). These structures are most numerous in the duodenum and the proximal portion of the jejunum. They project into the lumen of the alimentary canal, contacting the intestinal contents. Villi greatly increase the surface area of the intestinal lining, aiding the absorption of digestive products.

Each villus consists of a layer of simple columnar epithelium and a core of connective tissue containing blood capillaries, a lymphatic capillary called a **lacteal,** and nerve fibers. Blood capillaries and lacteals carry away absorbed nutrients, and nerve fibers transmit impulses to stimulate or inhibit villus activities. Between the bases of adjacent villi are tubular **intestinal glands** that extend downward into the mucous membrane (figs. 15.24 and 15.25; see fig. 15.3).

Secretions of the Small Intestine

Mucus-secreting goblet cells are abundant throughout the mucosa of the small intestine. In addition, many specialized *mucus-secreting glands* in the submucosa in the proximal portion of the duodenum secrete a thick, alkaline mucus in response to certain stimuli.

The intestinal glands at the bases of the villi secrete large volumes of a watery fluid, which brings digestive products into the villi. The fluid has a nearly neutral pH (6.5–7.5), and it lacks digestive enzymes. However, the epithelial cells of the intestinal mucosa have digestive enzymes embedded in the membranes of their microvilli on their luminal surfaces. These enzymes break down food molecules just before absorption takes place. The enzymes include **peptidases,** which split peptides into their constituent amino acids; **sucrase, maltase,** and **lactase,** which split the double sugars (disaccharides) sucrose, maltose, and lactose into the simple sug-

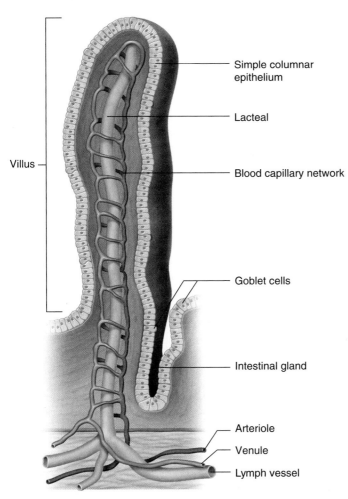

Figure 15.24

Structure of a single intestinal villus. The elongated shapes of intestinal villi dramatically increase the absorptive surface area of the small intestine.

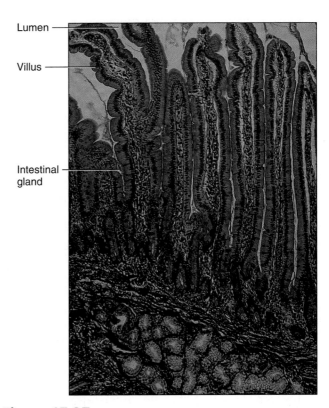

Figure 15.25

Light micrograph of intestinal villi from the wall of the duodenum (50×).

ars (monosaccharides) glucose, fructose, and galactose; and **intestinal lipase,** which splits fats into fatty acids and glycerol. Table 15.5 summarizes the sources and actions of the major digestive enzymes.

Regulation of Small Intestine Secretions

Goblet cells and intestinal glands secrete their products when chyme provides both mechanical and chemical stimulation. Distension of the intestinal wall activates the nerve plexuses within the wall and stimulates parasympathetic reflexes that also trigger release of small intestine secretions.

Check Your Recall

33. Describe the parts of the small intestine.
34. What is the function of an intestinal villus?
35. What is the function of the intestinal glands?
36. List the intestinal digestive enzymes.

Many adults do not produce sufficient lactase to adequately digest lactose, or milk sugar. In this condition, called *lactose intolerance,* lactose remains undigested, increasing the osmotic pressure of the intestinal contents and drawing water into the intestines. At the same time, intestinal bacteria metabolize undigested sugar, producing organic acids and gases. The overall result is bloating, intestinal cramps, and diarrhea. To avoid these symptoms, people with lactose intolerance can take lactase pills before eating dairy products. Infants with lactose intolerance can drink formula based on soybeans rather than milk.

Genetic evidence suggests that lactose intolerance may be the "normal" condition, with ability to digest lactose the result of a mutation that occurred recently in our evolutionary past and became advantageous when the advent of agriculture brought dairy foods to human populations. The trait of ability to digest lactose has increased in parallel to increased use of dairy foods at least three times in history, in different populations.

Absorption in the Small Intestine

Villi greatly increase the surface area of the intestinal mucosa, making the small intestine the most important absorbing organ of the alimentary canal. So effective

Table 15.5	Summary of the Major Digestive Enzymes	
Enzyme	**Source**	**Digestive Action**
Salivary enzyme		
Salivary amylase	Salivary glands	Begins carbohydrate digestion by breaking down starch and glycogen to disaccharides
Gastric enzyme		
Pepsin	Chief cells	Begins protein digestion
Pancreatic enzymes		
Pancreatic amylase	Pancreas	Breaks down starch and glycogen into disaccharides
Pancreatic lipase	Pancreas	Breaks down fats into fatty acids and glycerol
Proteolytic enzymes (a) Trypsin (b) Chymotrypsin (c) Carboxypeptidase	Pancreas	Breaks down proteins or partially digested proteins into peptides
Nucleases	Pancreas	Breaks down nucleic acids into nucleotides
Intestinal enzymes		
Peptidase	Mucosal cells	Breaks down peptides into amino acids
Sucrase, maltase, lactase	Mucosal cells	Breaks down disaccharides into monosaccharides
Intestinal lipase	Mucosal cells	Breaks down fats into fatty acids and glycerol
Enterokinase	Mucosal cells	Converts trypsinogen into trypsin

15.26, the resulting fatty acids and glycerol molecules diffuse into villi epithelial cells. The endoplasmic reticula of the cells use the fatty acids to resynthesize fat molecules similar to those previously digested. These fats are encased in protein to form *chylomicrons,* which make their way to the lacteals of the villi. Lymph in the lacteals and other lymphatic vessels carries chylomicrons to the bloodstream (see chapter 14, p. 377). Some fatty acids with very short carbon chains may be absorbed directly into the blood capillary of a villus without being changed back into fat.

Chylomicrons transport dietary fats to muscle and adipose cells. Similarly, VLDL (very low-density lipoprotein with a high concentration of triglycerides) molecules, produced in the liver, transport triglycerides synthesized from excess dietary carbohydrates. As VLDL molecules reach adipose cells, an enzyme, *lipoprotein lipase,* catalyzes reactions that unload their triglycerides, converting the VLDL to LDL (low-density lipoproteins). Because most of the triglycerides have been removed, LDL molecules have a higher cholesterol content than do the original VLDL molecules. Cells in the peripheral tissues remove LDL from plasma by receptor-mediated endocytosis, thus obtaining a supply of cholesterol (see chapter 3, p. 66).

While LDL delivers cholesterol to tissues, HDL (high-density lipoprotein with a high concentration of protein and a low concentration of lipids) removes cholesterol from tissues and delivers it to the liver. The liver produces the basic HDL framework and secretes the HDL molecules into the bloodstream. As they circulate, the HDL molecules pick up cholesterol from peripheral tissues and return to the liver, where they enter liver cells by receptor-mediated endocytosis. The liver disposes of the cholesterol it obtains in this manner by secreting it into bile or by using it to synthesize bile salts.

The intestine reabsorbs much of the cholesterol and bile salts in bile, which are then transported back to the liver, and the secretion-reabsorption cycle repeats. During each cycle, some of the cholesterol and bile salts escape reabsorption, reach the large intestine, and are eliminated with the feces.

In addition to the products of carbohydrate, protein, and fat digestion, the intestinal villi absorb electrolytes by active transport and water by osmosis. Table 15.6 summarizes the absorption process.

is the small intestine in absorbing digestive products, water, and electrolytes that very little absorbable material reaches its distal end.

Carbohydrate digestion begins in the mouth with the activity of salivary amylase, and enzymes from the intestinal mucosa and pancreas complete the process in the small intestine. Villi absorb the resulting monosaccharides, which enter blood capillaries. Simple sugars are absorbed by facilitated diffusion or active transport (see chapter 3, pp. 62 and 64).

Pepsin activity begins protein digestion in the stomach, and enzymes from the intestinal mucosa and the pancreas complete digestion in the small intestine. During this process, large protein molecules are broken down into amino acids, which are then actively transported into the villi and carried away by the blood.

Enzymes from the intestinal mucosa and pancreas digest fat molecules almost entirely. As shown in figure

Check Your Recall

37. Which substances resulting from digestion of carbohydrate, protein, and fat molecules does the small intestine absorb?

38. Describe how fatty acids are absorbed and transported.

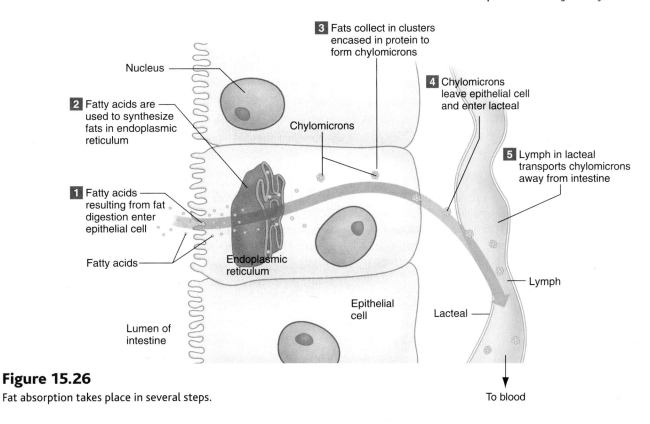

Figure 15.26

Fat absorption takes place in several steps.

Table 15.6	Intestinal Absorption of Nutrients	
Nutrient	**Absorption Mechanism**	**Means of Transport**
Monosaccharides	Facilitated diffusion and active transport	Blood in capillaries
Amino acids	Active transport	Blood in capillaries
Fatty acids and glycerol	Facilitated diffusion of glycerol; diffusion of fatty acids into cells	
	(a) Most fatty acids are resynthesized into fats and incorporated into chylomicrons for transport	Lymph in lacteals
	(b) Some fatty acids with relatively short carbon chains are transported without being changed back into fats	Blood in capillaries
Electrolytes	Diffusion and active transport	Blood in capillaries
Water	Osmosis	Blood in capillaries

In *malabsorption,* the small intestine digests, but does not absorb, some nutrients. Symptoms include diarrhea, weight loss, weakness, vitamin deficiencies, anemia, and bone demineralization. Causes of malabsorption include surgical removal of a portion of the small intestine, obstruction of lymphatic vessels due to a tumor, or interference with the production and release of bile as a result of liver disease.

Another cause of malabsorption is a reaction to *gluten,* found in certain grains, especially wheat and rye. This condition is called *celiac disease.* Microvilli are damaged, and in severe cases, villi may be destroyed. Both of these effects reduce the surface area of the small intestine, preventing absorption of some nutrients. Health food grocery stores sell gluten-free products, including pasta and baked goods.

Movements of the Small Intestine

Like the stomach, the small intestine carries on mixing movements and peristalsis. The major mixing movement is segmentation, in which small, ringlike contractions occur periodically, cutting chyme into segments and moving it back and forth. Segmentation also slows the movement of chyme through the small intestine.

Weak peristaltic waves propel chyme short distances through the small intestine. Consequently, chyme moves slowly through the small intestine, taking from three to ten hours to travel its length.

If the small intestine wall becomes overdistended or irritated, a strong *peristaltic rush* may pass along the organ's entire length. This movement sweeps the contents of the small intestine into the large intestine so quickly that water, nutrients, and electrolytes that would normally be absorbed are not. The result is *diarrhea,* characterized by more frequent defecation and watery stools. Prolonged diarrhea causes imbalances in water and electrolyte concentrations.

At the distal end of the small intestine, the **ileocecal** (il″e-o-se′kal) **sphincter** joins the small intestine's ileum to the large intestine's cecum (fig. 15.27). Normally, this sphincter remains constricted, preventing the contents of the small intestine from entering the large intestine, and the contents of the large intestine from backing up into the ileum. However, after a meal, a gastroileal reflex increases peristalsis in the ileum and relaxes the sphincter, forcing some of the contents of the small intestine into the cecum.

Check Your Recall

39. Describe the movements of the small intestine.
40. What stimulus relaxes the ileocecal sphincter?

15.10 LARGE INTESTINE

The **large intestine** is so named because its diameter is greater than that of the small intestine. This portion of the alimentary canal is about 1.5 meters long, and it begins in the lower right side of the abdominal cavity, where the ileum joins the cecum. From there, the large intestine

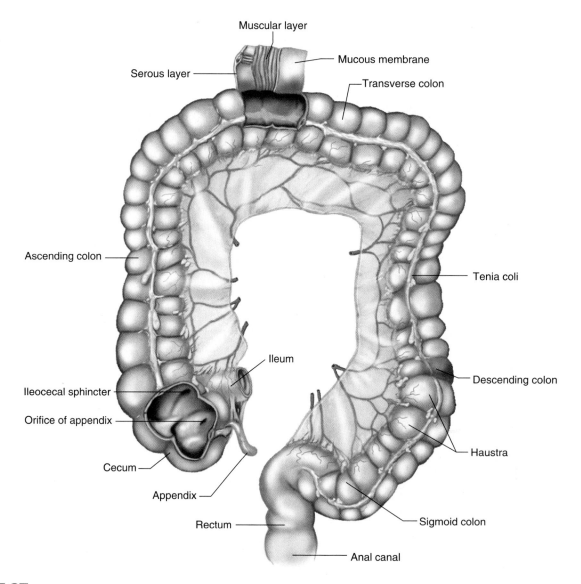

Figure 15.27

Parts of the large intestine (anterior view).

ascends on the right side, crosses obliquely to the left, and descends into the pelvis. At its distal end, it opens to the outside of the body as the anus (see fig. 15.1).

The large intestine absorbs water and electrolytes from chyme remaining in the alimentary canal. It also forms and stores feces.

Parts of the Large Intestine

The large intestine consists of the cecum, colon, rectum, and anal canal (figs. 15.27 and 15.28; see reference plates 4 and 5, pp. 26–27). The **cecum,** at the beginning of the large intestine, is a dilated, pouchlike structure that hangs slightly below the ileocecal opening. Projecting downward from it is a narrow tube with a closed end called the **appendix.** The human appendix has no known digestive function. However, it consists partly of lymphatic tissue.

In *appendicitis,* the appendix becomes inflamed and infected. Surgery is required to prevent the appendix from rupturing. If it does break open, the contents of the large intestine may enter the abdominal cavity and cause a serious infection of the peritoneum called *peritonitis.*

The **colon** is divided into four portions—the ascending, transverse, descending, and sigmoid colons. The **ascending colon** begins at the cecum and continues

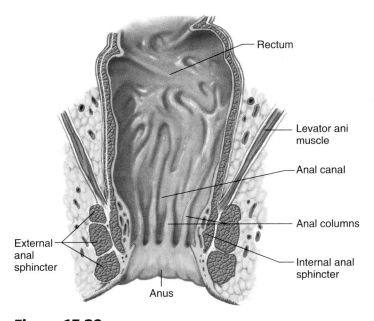

Figure 15.29

The rectum and the anal canal are at the distal end of the alimentary canal.

upward against the posterior abdominal wall to a point just inferior to the liver. There, it turns sharply to the left and becomes the **transverse colon.** The transverse colon is the longest and most movable part of the large intestine. It is suspended by a fold of peritoneum and sags in the middle below the stomach. As the transverse colon approaches the spleen, it turns abruptly downward and becomes the **descending colon.** At the brim of the pelvis, the descending colon makes an S-shaped curve called the **sigmoid colon** and then becomes the rectum.

The **rectum** lies next to the sacrum and generally follows its curvature. The peritoneum firmly attaches the rectum to the sacrum, and the rectum ends about 5 centimeters below the tip of the coccyx, where it becomes the anal canal (see fig. 15.27).

The last 2.5–4.0 centimeters of the large intestine form the **anal canal** (fig. 15.29). The mucous membrane in the canal is folded into six to eight longitudinal *anal columns.* At its distal end, the canal opens to the outside as the **anus.** Two sphincter muscles guard the anus—an *internal anal sphincter muscle,* composed of smooth muscle under involuntary control, and an *external anal sphincter muscle,* composed of skeletal muscle under voluntary control.

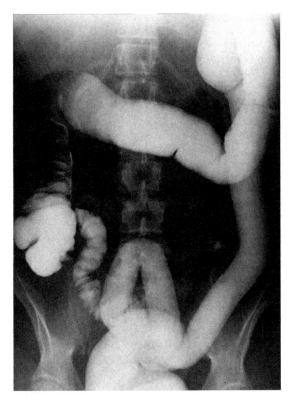

Figure 15.28

Radiograph of the large intestine containing a radiopaque substance that the patient ingested.

Check Your Recall

41. What is the general function of the large intestine?

42. Describe the parts of the large intestine.

Hemorrhoids are enlarged and inflamed branches of the rectal vein in the anal columns that cause intense itching, sharp pain, and sometimes bright red bleeding. The hemorrhoids may be internal or bulge out of the anus. Hemorrhoids can be caused by anything that puts prolonged pressure on the delicate rectal tissue, including obesity, pregnancy, constipation, diarrhea, and liver disease.

Eating more fiber-rich foods and drinking lots of water can usually prevent or cure hemorrhoids. Warm soaks in the tub, cold packs, and careful wiping of painful areas also help, as do external creams and ointments. Surgery—with a scalpel or a laser—can remove severe hemorrhoids.

Structure of the Large Intestine Wall

The wall of the large intestine is composed of the same types of tissues as other parts of the alimentary canal but has some unique features. The large intestine wall lacks the villi characteristic of the small intestine, and the layer of longitudinal muscle fibers is not uniformly distributed throughout the large intestine wall. Instead, the fibers form three distinct bands (teniae coli) that extend the entire length of the colon (see fig. 15.27). These bands exert tension lengthwise on the wall, creating a series of pouches (haustra).

It is recommended that people over 50 years of age have a colorectal cancer screening, performed with fiberoptic colonoscopy. Under sedation, a flexible lit tube is inserted into the rectum, and polyps and tumors identified and removed. Persons with a family history of colon cancer should be screened at an earlier age. Fiberoptic colonoscopy is invasive, uncomfortable, expensive, and takes about two hours. A newer procedure, computed tomographic colonography (popularly called a virtual colonoscopy), is beginning to become available. It requires the same preparatory bowel cleansing, but does not require sedation, and is faster and less expensive. However, if a lesion is detected, the more invasive approach must be used to remove the suspicious tissue.

Functions of the Large Intestine

Unlike the small intestine, which secretes digestive enzymes and absorbs the products of digestion, the large intestine has little or no digestive function. However, the mucous membrane that forms the large intestine's inner lining contains many tubular glands. Structurally, these glands are similar to those of the small intestine, but they are composed almost entirely of goblet cells (fig. 15.30). Consequently, mucus is the large intestine's only significant secretion.

Mucus secreted into the large intestine protects the intestinal wall against the abrasive action of the materials passing through it. Mucus also binds particles of fecal matter, and its alkalinity helps control the pH of the large intestine contents.

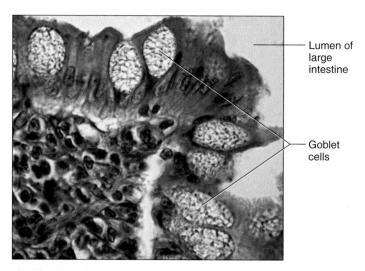

Figure 15.30
Light micrograph of the large intestinal mucosa (560×).

Chyme entering the large intestine contains materials that the small intestine did not digest or absorb. It also contains water, electrolytes, mucus, and bacteria. The large intestine normally absorbs water and electrolytes in the proximal half of the tube. Substances that remain in the tube become feces and are stored for a time in the distal portion of the large intestine.

The many bacteria that normally inhabit the large intestine, called *intestinal flora,* break down some of the molecules that escape the actions of human digestive enzymes. For instance, cellulose, a complex carbohydrate in food of plant origin, passes through the alimentary canal almost unchanged, but colon bacteria can break down cellulose and use it as an energy source. These bacteria, in turn, synthesize certain vitamins, such as K, B_{12}, thiamine, and riboflavin, which the intestinal mucosa absorbs. Bacterial actions in the large intestine may produce intestinal gas (flatus).

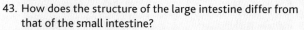
The colon is home to 100 trillion bacteria.

Check Your Recall
43. How does the structure of the large intestine differ from that of the small intestine?
44. What substances does the large intestine absorb?

Movements of the Large Intestine

The movements of the large intestine—mixing and peristalsis—are similar to those of the small intestine, although usually slower. Also, peristaltic waves of the large intestine happen only two or three times each day.

Topic of Interest

Inflammatory Bowel Disease

Inflammatory bowel disease is a group of disorders that includes ulcerative colitis and Crohn disease. The disorders differ by the site and the extent of inflammation and ulceration of the intestines. In the United States, about 100,000 people suffer the abdominal cramps and diarrhea of ulcerative colitis, and 500,000 individuals have similar symptoms of Crohn disease.

Ulcerative colitis affects the mucosa and submucosa of the distal large intestine and the rectum. In about 25% of cases, the disease extends no farther than the rectum. Bouts of bloody diarrhea and cramps may last for days or weeks, and may recur frequently or only rarely. The severe diarrhea leads to weight loss and electrolyte imbalances and may develop into colon cancer or affect other organs, including the skin, eyes, or liver. The inflamed and ulcerous tissue is continuous.

Crohn disease is more extensive than ulcerative colitis, extending into the small and large intestines and penetrating all tissue layers. In contrast to the uniformity of ulcerative colitis, affected portions of intestine in Crohn disease are interspersed with unaffected areas, producing a "cobblestone" effect after many years. The ileum and cecum are affected in about 40% of affected individuals, only the small intestine is involved in 30% of cases, and only the large intestine is involved in 25% of cases. Rarely, the disease affects more proximal structures of the gastrointestinal tract. The diarrhea is often not bloody, and complications such as cancer are rare.

Overall, about 20% of people with symptoms of irritable bowel disease seem to fall between the descriptions of ulcerative colitis and Crohn disease, and are classified as having "indeterminate colitis." Surgery is sometimes used to treat inflammatory bowel disease. Autoimmunity, infection, or a genetic predisposition may contribute to causing inflammatory bowel disease.

These waves produce *mass movements* in which a large section of the intestinal wall constricts vigorously, forcing the intestinal contents toward the rectum. Typically, mass movements follow a meal as a result of the gastrocolic reflex initiated in the small intestine. Irritations of the intestinal mucosa also can trigger such movements. For instance, a person with an inflamed colon (colitis) may experience frequent mass movements. The Topic of Interest above discusses inflammatory bowel disease.

A person can usually initiate a *defecation reflex* by holding a deep breath and contracting the abdominal wall muscles. This action increases internal abdominal pressure and forces feces into the rectum. As the rectum fills, its wall distends, triggering the defecation reflex that stimulates peristaltic waves in the descending colon. The internal anal sphincter relaxes. At the same time, other reflexes involving the sacral region of the spinal cord strengthen the peristaltic waves, lower the diaphragm, close the glottis, and contract the abdominal wall muscles. These actions further increase internal abdominal pressure and squeeze the rectum. The external anal sphincter is signaled to relax, and the feces are forced to the outside. Contracting the external anal sphincter allows voluntary inhibition of defecation.

Feces

Feces (fe'sēz) include materials that were not digested or absorbed, plus water, electrolytes, mucus, shed intestinal cells, and bacteria. Usually, feces are about 75% water, and their color derives from bile pigments altered by bacterial action. Feces' pungent odor results from a variety of compounds that bacteria produce.

Check Your Recall

45. How does peristalsis in the large intestine differ from peristalsis in the small intestine?

46. List the major events that occur during defecation.

47. Describe the composition of feces.

15.11 NUTRITION AND NUTRIENTS

Nutrition is the study of nutrients and how the body utilizes them. **Nutrients** (nu'tre-ents) include carbohydrates, lipids, proteins, vitamins, minerals, and water (see chapter 2, pp. 41–46, and chapter 4, p. 80). Carbohydrates, lipids, and proteins are called **macronutrients** because they are required in large amounts. They provide energy as well as other specific functions. Vitamins and minerals are required in much smaller amounts and are therefore called **micronutrients.** They do not directly provide energy, but make possible the biochemical reactions that extract energy from macronutrient molecules.

Macronutrients provide potential energy that can be expressed in calories, which are units of heat. A **calorie** (kal'o-re) is the amount of heat required to raise the temperature of a gram of water by 1° Celsius. The calorie used to measure food energy is 1,000 times greater.

This larger calorie (Cal) is technically a kilocalorie, but nutritional studies commonly refer to it simply as a calorie. As a result of cellular oxidation, 1 gram of carbohydrate or 1 gram of protein yields about 4 calories, but 1 gram of fat yields about 9 calories (twice as much chemical energy as carbohydrates or proteins).

Foods provide nutrients, and digestion breaks nutrients down to sizes that can be absorbed and transported in the bloodstream. Nutrients that human cells cannot synthesize, such as certain amino acids, are called **essential nutrients.**

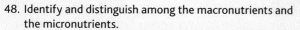

Check Your Recall

48. Identify and distinguish among the macronutrients and the micronutrients.
49. How is food energy measured?

Carbohydrates

Carbohydrates are organic compounds and include the sugars and starches. The energy held in their chemical bonds is used primarily to power cellular processes.

Carbohydrate Sources

Carbohydrates are ingested in a variety of forms, including starch from grains and vegetables; glycogen from meats; disaccharides from cane sugar, beet sugar, and molasses; and monosaccharides from honey and fruits (see chapter 2, p. 41). Digestion breaks down complex carbohydrates into monosaccharides, which are small enough to be absorbed.

Cellulose is a complex carbohydrate that is abundant in food—it gives celery its crunch and lettuce its crispness. Humans cannot digest cellulose, so the portion of it that is not broken down by intestinal flora passes through the alimentary canal largely unchanged. Thus, cellulose provides bulk (also called fiber or roughage) against which the muscular wall of the digestive system can push, facilitating the movement of food.

Carbohydrate Utilization

The monosaccharides absorbed from the digestive tract include *fructose, galactose,* and *glucose.* Liver enzymes catalyze reactions that convert fructose and galactose into glucose, which is the carbohydrate form most commonly oxidized for cellular fuel (see chapter 4, pp. 80–82).

Many cells obtain energy by oxidizing fatty acids. Some cells, however, such as neurons, require a continuous supply of glucose for survival. Even a temporary decrease in the glucose supply may seriously impair nervous system function. Consequently, the body requires a minimum of carbohydrates. If foods do not provide an adequate carbohydrate supply, the liver may convert some noncarbohydrates, such as amino acids from proteins, into glucose. Thus, the requirement for glucose has physiological priority over the requirement to synthesize proteins from available amino acids.

Carbohydrates have functions other than providing energy. Some excess glucose is polymerized to form *glycogen,* which is stored in the liver and muscles. When required to supply energy, glucose can be mobilized rapidly from glycogen. However, only a certain amount of glycogen can be stored, and excess glucose is usually converted into fat and stored in adipose tissue. To obtain energy, the body first metabolizes glucose, then glycogen stores, and finally fats and proteins.

Cells use carbohydrates as starting materials for the synthesis of such vital biochemicals as the five-carbon sugars *ribose* and *deoxyribose,* required for production of the nucleic acids RNA and DNA. Carbohydrates are also required to synthesize the disaccharide *lactose* (milk sugar) when the breasts are actively secreting milk.

Carbohydrate Requirements

Because carbohydrates provide the primary fuel source for cellular processes, the requirement for carbohydrates varies with individual energy expenditure. Physically active individuals require more fuel than sedentary ones. However, eating excess carbohydrates may promote obesity and increase cardiovascular disease risk factors. Because people vary greatly in their exercise habits and in how they respond to different carbohydrates, the optimal percentage of the diet that should be carbohydrates is highly individualized. It is estimated, however, that an intake of at least 125 to 175 grams daily is necessary to spare protein (that is, to avoid protein breakdown) and to avoid metabolic disorders resulting from excess fat utilization.

Check Your Recall

50. List several common sources of carbohydrates.
51. Explain the importance of cellulose in the diet.
52. Explain why the requirement for glucose has priority over protein synthesis.
53. Discuss why individual carbohydrate requirements may vary.

Lipids

Lipids are organic compounds that include fats, oils, and fatlike substances. They supply energy for cellular processes and for building structures such as cell membranes. Lipids include fats, phospholipids, and cholesterol. The most common dietary lipids are the fats called *triglycerides.* Recall from chapter 2 (p. 41) that a triglyceride molecule consists of a glycerol and three fatty acids.

Lipid Sources

Triglycerides are part of both plant- and animal-based foods. Saturated fats are found mainly in foods of animal origin, such as meats, eggs, milk, and lard, as well as in palm and coconut oils. Unsaturated fats are in seeds, nuts, and plant oils. Monounsaturated fats, such as those in olive, peanut, and canola oils, are the healthiest. Saturated fats in excess are a risk factor for cardiovascular disease.

Cholesterol is abundant in liver and egg yolk and, to a lesser extent, in whole milk, butter, cheese, and meats. It is not present in foods of plant origin.

Lipid Utilization

Many foods contain phospholipids, cholesterol, or triglycerides. Lipids serve a variety of physiological functions, but the main one is to supply energy. Before a triglyceride molecule can release energy, it must undergo hydrolysis (breakdown in the presence of water) as part of digestion, releasing the constituent fatty acids and glycerol. After being absorbed, these products are transported in lymph and blood to tissues. Figure 15.31 shows that some of the fatty acid portions can react to form molecules of acetyl coenzyme A by a series of reactions called **beta oxidation** (ba′tah ok″sĭ-da′shun). Excess acetyl coenzyme A can be converted into compounds called *ketone bodies,* such as acetone, which later may be changed back to acetyl coenzyme A. In either case, the resulting acetyl coenzyme A can be oxidized in the citric acid cycle. The glycerol portions of the triglyceride molecules can also enter metabolic pathways leading to the citric acid cycle, or they can be used to synthesize glucose. Fatty acid molecules released from fat hydrolysis can combine to form fat molecules and be stored in adipose tissue.

The liver can convert fatty acids from one form to another, but it cannot synthesize certain fatty acids, called **essential fatty acids.** *Linoleic acid,* for example, is required for phospholipid synthesis, which in turn is necessary for cell membrane formation and the transport of circulating lipids. Good sources of linoleic acid include corn, cottonseed, and soy oils. Another essential fatty acid is linolenic acid.

The liver regulates circulating lipids by using free fatty acids to synthesize triglycerides, phospholipids, and lipoproteins that may then be released into the bloodstream. Because lipids are less dense than proteins, as the proportion of lipids in a lipoprotein increases, the density of the particle decreases. Conversely, as the proportion of lipids decreases, the density increases. Lipoproteins are classified on the basis of their densities, which reflect their composition. *Very-low-density lipoproteins* (VLDL) have a relatively high concentration of triglycerides. *Low-density lipoproteins* (LDL) have a relatively high concentration of cholesterol and are the major cholesterol-carrying lipoproteins. *High-density lipoproteins* (HDL) have a relatively high concentration of protein and a lower concentration of lipids.

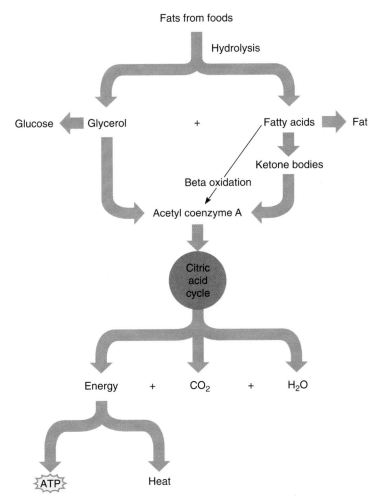

Figure 15.31
The body digests fats from foods into glycerol and fatty acids, which may enter catabolic pathways and be used as energy sources.

The liver controls the total amount of cholesterol in the body by synthesizing cholesterol and releasing it into the bloodstream, or by removing cholesterol from the bloodstream and excreting it into bile. The liver also uses cholesterol to produce bile salts. Cholesterol is not an energy source. It provides structural material for cell and organelle membranes and is a starting material for the synthesis of certain sex hormones and adrenal hormones.

Adipose tissue stores excess triglycerides. If the blood lipid concentration drops (in response to fasting, for example), some of these triglycerides are hydrolyzed into free fatty acids and glycerol, and then released into the bloodstream.

Lipid Requirements

The amounts and types of fats required for health vary with individuals' habits and goals. Because linoleic acid is an essential fatty acid, nutritionists recommend that formula-fed infants receive 3% of their energy intake in the form of linoleic acid to prevent deficiency conditions. Fat intake must be sufficient to carry fat-soluble vitamins. Lipids provide flavor to food, which is one reason why adhering to a very low-fat diet is difficult. The USDA and American Heart Association recommend that lipid intake not exceed 30% of calories.

> Beware the label "low-fat" on prepared foods such as ice cream. Often less fat means more sugar, so that the altered product has more calories than the original. This is one reason that years of following low-fat diets may have contributed to supersized Americans.

Check Your Recall

54. Which fatty acids are essential nutrients?
55. What is the liver's role in the utilization of lipids?
56. What are the functions of cholesterol?

Proteins

Proteins are polymers of amino acids with a wide variety of functions. Proteins include enzymes that control metabolic rates, clotting factors, the keratin of skin and hair, elastin and collagen of connective tissue, plasma proteins that regulate water balance, the muscle components actin and myosin, certain hormones, and the antibodies that protect against infection.

Proteins may also supply energy after digestion breaks them down into amino acids. The liberated amino acids are transported to the liver, where they undergo *deamination,* losing their nitrogen-containing portions (—NH$_2$ groups) (see fig. 2.17, p. 44). These —NH$_2$ groups then react to form the waste *urea* (u-re′ah), which is excreted in urine.

Depending upon the particular amino acids involved, the remaining deaminated portions are decomposed in one of several pathways (fig. 15.32). Some of these pathways lead to formation of acetyl coenzyme A, and others lead more directly to the steps of the citric acid cycle. As energy is released from the cycle, some of it is captured in ATP molecules. If energy is not required immediately, the deaminated portions of the amino acids may react to form glucose or fat molecules in other metabolic pathways.

Protein Sources

Foods rich in proteins include meats, fish, poultry, cheese, nuts, milk, eggs, and cereals. Legumes, including beans and peas, contain lesser amounts. The cells of an adult can synthesize all but eight required amino acids, and the cells of a child can produce all but ten. Amino acids that the body can synthesize are termed nonessential; those that it cannot synthesize are **essential amino acids.** This term refers only to dietary intake since all amino acids are required for normal protein synthesis. Table 15.7 lists the amino acids in foods and indicates those that are essential.

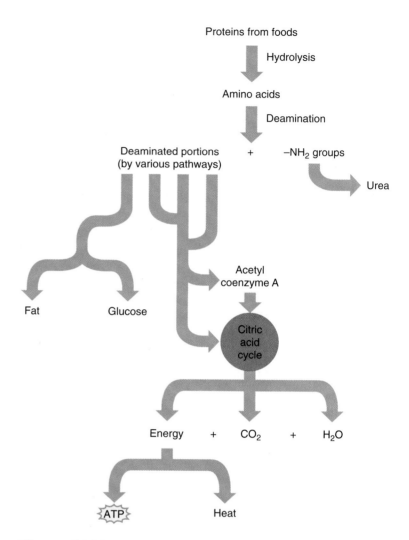

Figure 15.32

The body digests proteins from foods into amino acids but must deaminate these smaller molecules before they can be used as energy sources.

Table 15.7	Amino Acids in Foods
Alanine	Leucine (e)
Arginine (ch)	Lysine (e)
Asparagine	Methionine (e)
Aspartic acid	Phenylalanine (e)
Cysteine	Proline
Glutamic acid	Serine
Glutamine	Threonine (e)
Glycine	Tryptophan (e)
Histidine (ch)	Tyrosine
Isoleucine (e)	Valine (e)

Eight essential amino acids (e) cannot be synthesized by human cells and must be provided in the diet. Two additional amino acids (ch) are essential in growing children.

All twenty types of amino acids must be present in the body at the same time for growth and tissue repair to occur. If just one type of essential amino acid is missing from the diet, normal protein synthesis cannot take place, because many proteins include all twenty types of amino acids.

Proteins are classified as complete or incomplete on the basis of the amino acid types they provide. **Complete proteins,** such as those in milk, meats, and eggs, have adequate amounts of the essential amino acids. **Incomplete proteins,** such as *zein* in corn, which has too little of the essential amino acids tryptophan and lysine, are unable by themselves to maintain human tissues or to support normal growth and development. A protein called *gliadin* in wheat is an example of a **partially complete protein.** It does not have enough lysine to promote growth, but it has enough to maintain life.

> Many plant proteins have too little of one or more essential amino acids to provide adequate nutrition for a person. However, combining appropriate plant foods can provide an adequate diversity of dietary amino acids. For example, beans are low in methionine but have enough lysine. Rice lacks lysine but has enough methionine. A meal of beans and rice provides enough of both types of amino acids.

Protein Requirements

Proteins supply essential amino acids. They also provide nitrogen and other elements for the synthesis of nonessential amino acids and certain nonprotein nitrogenous substances. The amount of dietary protein individuals require varies according to body size, metabolic rate, and other factors, such as activity level. Bodybuilders, for example, require more protein to help heal small muscle tears that result from weight lifting.

For an average adult, nutritionists recommend a daily protein intake of about 0.8 grams per kilogram of body weight. Another way to estimate desirable protein intake is to divide weight in pounds by two. Most people should consume 60 to 150 grams of protein a day. For a pregnant woman, the recommendation increases by an additional 30 grams of protein per day. Similarly, a nursing mother requires an extra 20 grams of protein per day to maintain an adequate level of milk production.

Check Your Recall

57. List some examples of proteins associated with the body.
58. How does dietary protein provide energy?
59. Which foods are rich sources of proteins?
60. Why are some amino acids called essential?
61. Distinguish between complete and incomplete proteins.

Vitamins

Vitamins (vi'tah-minz) are organic compounds (other than carbohydrates, lipids, and proteins) that are required in small amounts for normal metabolism, but that cells cannot synthesize in adequate amounts. Thus, they are essential nutrients that must come from foods.

Vitamins are classified on the basis of solubility. Some are soluble in fats (or fat solvents) and others are soluble in water. *Fat-soluble vitamins* are vitamins A, D, E, and K; *water-soluble vitamins* are the B vitamins and vitamin C.

Fat-Soluble Vitamins

Because fat-soluble vitamins dissolve in fats, they associate with lipids and are influenced by the same factors that affect lipid absorption. For example, bile salts in the intestine promote absorption of these vitamins. Fat-soluble vitamins accumulate in various tissues, which is why excess intake can lead to overdose conditions. For example, too much beta carotene, a vitamin A precursor, can tinge the skin orange. Fat-soluble vitamins resist the effects of heat; therefore, cooking and food processing usually do not destroy them. Table 15.8 lists the fat-soluble vitamins and their characteristics, functions, sources, and recommended daily allowances (RDA) for adults.

Water-Soluble Vitamins

The water-soluble vitamins include the B vitamins and vitamin C. The **B vitamins** are several compounds that are essential for normal cellular metabolism. They help oxidize carbohydrates, lipids, and proteins. Because the B vitamins are often present together in foods, they are referred to as the *vitamin B complex.* Members of this group differ chemically and functionally. Cooking and food processing destroy some of them.

Vitamin C (ascorbic acid) is one of the least stable vitamins and is fairly widespread in plant foods. It is necessary for collagen production, the conversion of folacin to folinic acid, and the metabolism of certain amino acids. Vitamin C also promotes iron absorption and synthesis of certain hormones from cholesterol. Table 15.9 lists the water-soluble vitamins and their characteristics, functions, sources, and RDAs for adults.

> Sailors on English ships ate limes to protect them from scurvy (vitamin C deficiency). American ships carried cranberries for the same purpose.

Check Your Recall

62. What are vitamins?
63. What are the two major types of vitamins?
64. How can fat-soluble vitamins accumulate in the body?
65. List the fat-soluble and water-soluble vitamins.

Table 15.8	Fat-Soluble Vitamins		
Vitamin	**Characteristics**	**Functions**	**Sources and RDA* for Adults**
Vitamin A	Occurs in several forms; synthesized from carotenes; stored in liver; stable in heat, acids, and bases; unstable in light	Necessary for synthesis of visual pigments, mucoproteins, and mucopolysaccharides; for normal development of bones and teeth; and for maintenance of epithelial cells	Liver, fish, whole milk, butter, eggs, leafy green vegetables, yellow and orange vegetables and fruits; RDA = 4,000–5,000 IU†
Vitamin D	A group of steroids; resistant to heat, oxidation, acids, and bases; stored in liver, skin, brain, spleen, and bones	Promotes absorption of calcium and phosphorus; promotes development of teeth and bones	Produced in skin exposed to ultraviolet light; in milk, egg yolk, fish liver oils, fortified foods; RDA = 400 IU
Vitamin E	A group of compounds; resistant to heat and visible light; unstable in presence of oxygen and ultraviolet light; stored in muscles and adipose tissue	An antioxidant; prevents oxidation of vitamin A and polyunsaturated fatty acids; may help maintain stability of cell membranes	Oils from cereal seeds, salad oils, margarine, shortenings, fruits, nuts, and vegetables; RDA = 30 IU
Vitamin K	Occurs in several forms; resistant to heat, but destroyed by acids, bases, and light; stored in liver	Required for synthesis of prothrombin, which functions in blood clotting	Leafy green vegetables, egg yolk, pork liver, soy oil, tomatoes, cauliflower; RDA = 55–70 μg

*RDA = recommended daily allowance.

†IU = international unit.

Table 15.9	Water-Soluble Vitamins		
Vitamin	**Characteristics**	**Functions**	**Sources and RDA* for Adults**
Thiamine (vitamin B$_1$)	Destroyed by heat and oxygen, especially in alkaline environment	Part of coenzyme required to oxidize carbohydrates; coenzyme required for ribose synthesis	Lean meats, liver, eggs, whole-grain cereals, leafy green vegetables, legumes; RDA =1.5 mg
Riboflavin (vitamin B$_2$)	Stable to heat, acids, and oxidation; destroyed by bases and ultraviolet light	Part of enzymes and coenzymes required to oxidize glucose and fatty acids and for cellular growth	Meats, dairy products, leafy green vegetables, whole-grain cereals; RDA = 1.7 mg
Niacin (nicotinic acid) (vitamin B$_3$)	Stable to heat, acids, and bases; converted to niacinamide by cells; synthesized from tryptophan	Part of coenzymes required to oxidize glucose and to synthesize proteins, fats, and nucleic acids	Liver, lean meats, peanuts, legumes; RDA = 20 mg
Pantothenic acid (vitamin B$_5$)	Destroyed by heat, acids, and bases	Part of coenzyme A required to oxidize carbohydrates and fats	Meats, whole-grain cereals, legumes, milk, fruits, vegetables; RDA = 10 mg
Vitamin B$_6$	Group of three compounds; stable to heat and acids; destroyed by oxidation, bases, and ultraviolet light	Coenzyme required to synthesize proteins and certain amino acids, to convert tryptophan to niacin, to produce antibodies, and to synthesize nucleic acids	Liver, meats, bananas, avocados, beans, peanuts, whole-grain cereals, egg yolk; RDA = 2 mg
Cyanocobalamin (vitamin B$_{12}$)	Complex, cobalt-containing compound; stable to heat; inactivated by light, strong acids, and strong bases; absorption regulated by intrinsic factor from gastric glands; stored in liver	Part of coenzyme required to synthesize nucleic acids and to metabolize carbohydrates; plays role in myelin synthesis; needed for normal red blood cell production	Liver, meats, milk, cheese, eggs; RDA = 3–6 μg
Folacin (folic acid)	Occurs in several forms; destroyed by oxidation in acid environment or by heat in alkaline environment; stored in liver, where it is converted into folinic acid	Coenzyme required for metabolism of certain amino acids and for DNA synthesis; promotes red blood cell production	Liver, leafy green vegetables, whole-grain cereals, legumes; RDA = 0.4 mg
Biotin	Stable to heat, acids, and light; destroyed by oxidation and bases	Coenzyme required to metabolize amino acids and fatty acids, and to synthesize nucleic acids	Liver, egg yolk, nuts, legumes, mushrooms; RDA = 0.3 mg
Ascorbic acid (vitamin C)	Chemically similar to monosaccharides; stable in acids but destroyed by oxidation, heat, light, and bases	Required to produce collagen, to convert folacin to folinic acid, and to metabolize certain amino acids; promotes absorption of iron and synthesis of hormones from cholesterol	Citrus fruits, tomatoes, potatoes, leafy green vegetables; RDA = 60 mg

*RDA = recommended daily allowance.

Minerals

Dietary **minerals** (min′er-alz) are inorganic elements that are essential in human metabolism. Plants usually extract minerals from soil, and humans obtain minerals from plant foods or from animals that have eaten plants.

Characteristics of Minerals

Minerals are responsible for about 4% of body weight and are most concentrated in the bones and teeth. Minerals are usually incorporated into organic molecules. For example, phosphorus is found in phospholipids, iron in hemoglobin, and iodine in thyroxine. Some minerals are part of inorganic compounds, such as the calcium phosphate of bone. Other minerals are free ions, such as sodium, chloride, and calcium ions in blood.

Minerals compose parts of the structural materials in all body cells. They also constitute portions of enzyme molecules, contribute to the osmotic pressure of body fluids, and play vital roles in nerve impulse conduction, muscle fiber contraction, blood coagulation, and maintenance of the pH of body fluids.

Major Minerals

The minerals *calcium* and *phosphorus* account for nearly 75% by weight of the mineral elements in the body; thus, they are **major minerals.** Other major minerals, each of which accounts for 0.05% or more of the body weight, include potassium, sulfur, sodium, chlorine, and magnesium. Table 15.10 lists the distribution, functions, sources, and adult RDAs of major minerals.

The human body has enough phosphorus to make two thousand match tips.

Trace Elements

Trace elements are essential minerals found in minute amounts, each making up less than 0.005% of adult body weight. They include iron, manganese, copper, iodine, cobalt, zinc, fluorine, selenium, and chromium. Table 15.11 lists the distribution, functions, sources, and adult RDAs of the trace elements.

Table 15.10	Major Minerals		
Mineral	**Distribution**	**Functions**	**Sources and RDA* for Adults**
Calcium (Ca)	Mostly in the inorganic salts of bones and teeth	Structure of bones and teeth; essential for nerve impulse conduction, muscle fiber contraction, and blood coagulation; increases permeability of cell membranes; activates certain enzymes	Milk, milk products, leafy green vegetables; RDA = 800 mg
Phosphorus (P)	Mostly in the inorganic salts of bones and teeth	Structure of bones and teeth; component in nearly all metabolic reactions; constituent of nucleic acids, many proteins, some enzymes, and some vitamins; occurs in cell membrane, ATP, and phosphates of body fluids	Meats, cheese, nuts, whole-grain cereals, milk, legumes; RDA = 800 mg
Potassium (K)	Widely distributed; tends to be concentrated inside cells	Helps maintain intracellular osmotic pressure and regulate pH; promotes metabolism; required for nerve impulse conduction and muscle fiber contraction	Avocados, dried apricots, meats, nuts, potatoes, bananas; RDA = 2,500 mg
Sulfur (S)	Widely distributed; abundant in skin, hair, and nails	Essential part of various amino acids, thiamine, insulin, biotin, and mucopolysaccharides	Meats, milk, eggs, legumes; no RDA established
Sodium (Na)	Widely distributed; mostly in extracellular fluids and bound to inorganic salts of bone	Helps maintain osmotic pressure of extracellular fluids and regulate water movement; needed for conduction of nerve impulses and contraction of muscle fibers; aids in regulation of pH and in transport of substances across cell membranes	Table salt, cured ham, sauerkraut, cheese, graham crackers; RDA = 2,500 mg
Chlorine (Cl)	Closely associated with sodium (as chloride); most highly concentrated in cerebrospinal fluid and gastric juice	Helps maintain osmotic pressure of extracellular fluids, regulate pH, and maintain electrolyte balance; essential in formation of hydrochloric acid; aids transport of carbon dioxide by red blood cells	Same as for sodium; no RDA established
Magnesium (Mg)	Abundant in bones	Required in metabolic reactions that occur in mitochondria and are associated with ATP production; plays role in the breakdown of ATP to ADP	Milk, dairy products, legumes, nuts, leafy green vegetables; RDA = 300–350 mg

*RDA = recommended daily allowance.

There is enough iron in a human being to make a small nail.

Check Your Recall

66. What are minerals?

67. What are the major functions of minerals?

68. Distinguish between a major mineral and a trace element.

69. Name the major minerals and trace elements.

Adequate Diets

An adequate diet provides sufficient energy, essential fatty acids, essential amino acids, vitamins, and minerals to support optimal growth and to maintain and repair body tissues. Because individual requirements for nutrients vary greatly with age, sex, growth rate, physical activity, and level of stress, as well as with genetic and other environmental factors, designing a diet that is adequate for everyone is impossible. Diagrams called food guide pyramids are used to organize foods according to suggested relative amounts, often in serving sizes. Figure 15.33 depicts one such food pyramid, developed by the U.S. Department of Agriculture.

If the diet lacks essential nutrients or a person fails to use available foods to best advantage, **malnutrition** (mal″nu-trish′un) results. This condition may be due to either *undernutrition,* producing the symptoms of deficiency diseases, or to *overnutrition,* arising from excess nutrient intake.

A variety of factors can lead to malnutrition. For example, a deficiency condition may stem from lack of food or from poor-quality food. On the other hand, malnutrition may result from overeating or from taking too many vitamin supplements.

A measurement called the body mass index (BMI) is used to determine whether a person is of adequate weight, overweight, or obese. To calculate your BMI, divide your weight in kilograms (a kilogram equals 2.2 pounds) by your height in meters squared (one foot equals about .3 meters). Figure 15.34 interprets the BMI. The Topic of Interest on page 438 discusses the effects of undereating and overeating.

Table 15.11	Trace Elements		
Trace Element	**Distribution**	**Functions**	**Sources and RDA* for Adults**
Iron (Fe)	Primarily in blood; stored in liver, spleen, and bone marrow	Part of hemoglobin molecule; catalyzes vitamin A formation; incorporated into a number of enzymes	Liver, lean meats, dried apricots, raisins, enriched whole-grain cereals, legumes, molasses; RDA = 10–18 mg
Manganese (Mn)	Most concentrated in liver, kidneys, and pancreas	Activates enzymes required for fatty acid and cholesterol synthesis, urea formation, and normal functioning of the nervous system	Nuts, legumes, whole-grain cereals, leafy green vegetables, fruits; RDA = 2.5–5 mg
Copper (Cu)	Most highly concentrated in liver, heart, and brain	Essential for hemoglobin synthesis, bone development, melanin production, and myelin formation	Liver, oysters, crabmeat, nuts, whole-grain cereals, legumes; RDA = 2–3 mg
Iodine (I)	Concentrated in thyroid gland	Essential component for synthesis of thyroid hormones	Food content varies with soil content in different geographic regions; iodized table salt; RDA = 0.15 mg
Cobalt (Co)	Widely distributed	Component of cyanocobalamin; needed for synthesis of several enzymes	Liver, lean meats, milk; no RDA established
Zinc (Zn)	Most concentrated in liver, kidneys, and brain	Component of enzymes involved in digestion, respiration, bone metabolism, liver metabolism; necessary for normal wound healing and maintaining skin integrity	Meats, cereals, legumes, nuts, vegetables; RDA = 15 mg
Fluorine (F)	Primarily in bones and teeth	Component of tooth structure (enamel)	Fluoridated water; RDA = 1.5–4.0 mg
Selenium (Se)	Concentrated in liver and kidneys	Component of certain enzymes	Lean meats, fish, cereals; RDA = 0.05–2.00 mg
Chromium (Cr)	Widely distributed	Essential for use of carbohydrates	Liver, lean meats, wine; RDA = 0.05–2.00 mg

*RDA = recommended daily allowance.

Activity
Activity is represented by the steps and the person climbing them, as a reminder of the importance of daily physical activity.

Moderation
Moderation is represented by the narrowing of each food group from bottom to top. The wider base stands for foods with little or no solid fats or added sugars. These should be selected more often. The narrower top area stands for foods containing more added sugars and solid fats. The more active you are, the more of these foods can fit into your diet.

Personalization
Personalization is shown by the person on the steps and the slogan.

Proportionality
Proportionality is shown by the different widths of the food group bands. The widths suggest how much food a person should choose from each group. The widths are just a general guide, not exact proportions.

Variety
Variety is symbolized by the 6 color bands representing the 5 food groups of the Pyramid and oils. This illustrates that foods from all groups are needed each day for good health.

Gradual Improvement
Gradual improvement is encouraged by the slogan. It suggests that individuals can benefit from taking small steps to improve their diet and lifestyle each day.

MyPyramid.gov
STEPS TO A HEALTHIER YOU

U.S. Department of Agriculture
Center for Nutrition Policy
and Promotion
April 2005 CNPP-16

USDA is an equal opportunity provider and employer.

| GRAINS | VEGETABLES | FRUITS | OILS | MILK | MEAT & BEANS |

Figure 15.33
The USDA food pyramid symbolizes an individual approach to healthy eating and physical exercise. Using it requires entering information into the website at http://mypyramid.gov.

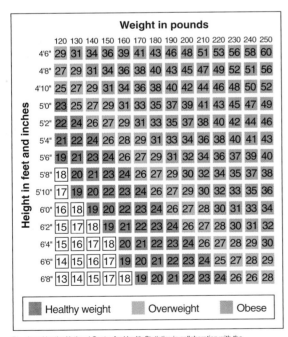

Figure 15.34
Body mass index (BMI). BMI equals weight/height^2, with weight measured in kilograms and height measured in meters. This chart provides a shortcut—the calculations have been done and converted to the English system of measurement. The uncolored squares indicate underweight.

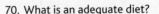

Check Your Recall
70. What is an adequate diet?
71. What factors influence individual requirements for nutrients?
72. What causes malnutrition?

Clinical Terms Related to the Digestive System and Nutrition

achalasia (ak″ah-la′ze-ah) Failure of the smooth muscle to relax at some junction in the digestive tube, such as between the esophagus and stomach.

achlorhydria (ah″klor-hi′dre-ah) Lack of hydrochloric acid in gastric secretions.

aphagia (ah-fa′je-ah) Inability to swallow.

cachexia (kah-kek′se-ah) State of chronic malnutrition and physical wasting.

cholecystitis (ko″le-sis-ti′tis) Inflammation of the gallbladder.

cholelithiasis (ko″le-lĭ-thi′ah-sis) Stones in the gallbladder.

cholestasis (ko″le-sta′sis) Blockage in bile flow from the gallbladder.

cirrhosis (sĭ-ro′sis) Condition in which liver cells degenerate and the surrounding connective tissues thicken.

diverticulitis (di″ver-tik″u-li′tis) Inflammation of small pouches (diverticula) that form in the lining and wall of the colon.

dumping syndrome (dum′ping sin′drōm) Symptoms, including diarrhea, that often occur following a gastrectomy.

Eating Extremes: Undereating and Overeating

In the United States, many people are underweight or overweight. Two-thirds of people over the age of 20 are overweight or obese, yet fashion models and celebrities grow ever more skeletal, wearing clothing much smaller than the average woman's dress size of 14.

Being thin is not a disease, but abnormal eating behavior is. Anorexia nervosa is self-imposed starvation. The sufferer perceives herself as overweight and eats barely enough to survive, losing as much as 25% of her body weight. She may hasten weight loss by vomiting, taking laxatives and diuretics, or intensely exercising. Anorexia leads to low blood pressure, slowed or irregular heartbeat, constipation, and constant chills. Menstruation stops as body fat level plunges. Hair becomes brittle, the skin dries out, and soft, pale, fine body hair called lanugo, normally seen only on a fetus, grows to preserve body heat. A person with anorexia may be hospitalized so that intravenous feeding can prevent sudden death from heart failure due to a mineral imbalance. Psychotherapy and nutritional counseling may help identify and remedy the underlying cause of the abnormal eating behavior. Despite these interventions, 15–21% of people with anorexia die from the disease.

In bulimia, a person eats large amounts of food, and then gets rid of the thousands of extra calories by vomiting, taking laxatives, or exercising frantically. The binge-and-purge cycle is very hard to break, even with psychotherapy and nutritional counseling.

Because body weight reflects energy balance, excess food means, ultimately, excess weight. Being overweight or obese raises the risk of developing hypertension, diabetes, stroke, gallstones, sleep apnea, and certain cancers. The body strains to support the extra weight—miles of blood vessels are required to nourish additional pounds. Usually, being overweight stems from overeating and inactivity. The average person in the United States today consumes 3,700 calories daily, compared to 3,100 in the 1960s. In many other nations people eat far less and exercise more.

Weight loss requires eating less and exercising more. A safe goal is to lose one pound of fat per week. A pound of fat contains 3,500 calories of energy. Shedding that pound requires an appropriate combination of calorie cutting and exercise—that is, eating 500 fewer calories per day or exercising off 500 calories each day, or some other combination. Certain weight-loss drugs inhibit the function of pancreatic lipase, preventing the digestion and absorption of about a third of dietary fat. For people with BMIs above 40, or above 35 in addition to an obesity-related disorder, bariatric surgery can lead to great weight loss. Two types of procedures are done. In laparoscopic adjustable gastric banding, a silicone band ties off part of the stomach, limiting the capacity of the organ to hold food. The band can be inflated or deflated in a doctor's office by adding or removing saline. The second type of bariatric surgery is gastric bypass, in which part of the stomach is stapled shut, forming a pouch that is surgically connected to the jejunum, bypassing the duodenum. Both procedures lead to decreased hunger, drastically reduced food intake, and some decrease in the absorption of nutrients. A special diet, liquid at first, must be followed. Many patients who have had bariatric surgery report improvement in or disappearance of type 2 diabetes, back pain, arthritis, varicose veins, sleep apnea, and hypertension.

---

dysentery (dis'en-ter"e) Intestinal infection by viruses, bacteria, or protozoans that causes diarrhea and cramps.

dyspepsia (dis-pep'se-ah) Indigestion; difficulty in digesting a meal.

dysphagia (dis-fa'je-ah) Difficulty in swallowing.

enteritis (en"tĕ-ri'tis) Inflammation of the intestine.

esophagitis (e-sof"ah-ji'tis) Inflammation of the esophagus.

gastrectomy (gas-trek'to-me) Partial or complete removal of the stomach.

gastrostomy (gas-tros'to-me) Creation of an opening in the stomach wall through which food and liquids can be administered when swallowing is not possible.

glossitis (glŏs-si'tis) Inflammation of the tongue.

hyperalimentation (hi"per-al"ĭ-men-ta'shun) Long-term intravenous nutrition.

ileitis (il"e-i'tis) Inflammation of the ileum.

pharyngitis (far"in-ji'tis) Inflammation of the pharynx.

polyphagia (pol"e-fa'je-ah) Overeating.

pyloric stenosis (pi-lor'ik stĕ-no'sis) Congenital obstruction at the pyloric sphincter due to an enlarged pyloric muscle.

pylorospasm (pi-lor'o-spazm) Spasm of the pyloric portion of the stomach or of the pyloric sphincter.

pyorrhea (pi"o-re'ah) Inflammation of the dental periosteum with pus formation.

stomatitis (sto"mah-ti'tis) Inflammation of the lining of the mouth.

Clinical Connection

Saliva is vital for tasting and processing food so that it can be swallowed and digested. It also keeps teeth healthy by washing away bacteria and plaque. In a condition called xerostomia, or "dry mouth," saliva production is insufficient. As a result, chewed food does not soften enough and is difficult to swallow. Even licking an envelope can be impossible for a person with this condition.

Digestive System

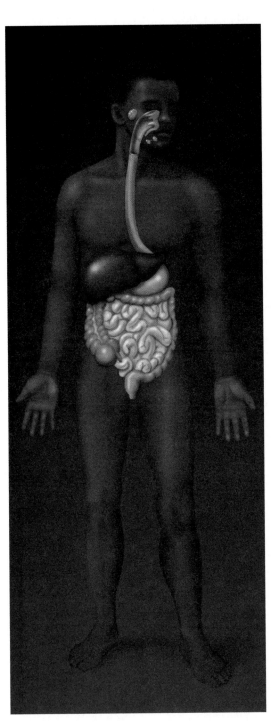

Integumentary System

Vitamin D activated in the skin plays a role in absorption of calcium from the digestive tract.

Cardiovascular System

The bloodstream carries absorbed nutrients to all body cells.

Skeletal System

Bones are important in mastication. Calcium absorption is necessary to maintain bone matrix.

Lymphatic System

The lymphatic system plays a major role in the absorption of fats.

Muscular System

Muscles are important in mastication, swallowing, and the mixing and moving of digestion products through the gastrointestinal tract.

Respiratory System

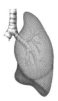

The digestive system and the respiratory system share common anatomical structures.

Nervous System

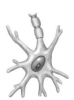

The nervous system can influence digestive system activity.

Urinary System

The kidneys and liver work together to activate vitamin D.

Endocrine System

Hormones can influence digestive system activity.

Reproductive System

In a woman, nutrition is essential for conception and normal development of an embryo and fetus.

The digestive system ingests, digests, and absorbs nutrients for use by all body cells.

Often, dry mouth is a side effect of a medication. Hundreds of medications can cause xerostomia, including drugs that treat depression, hypertension, cancer, and allergies. Radiation to the head or neck to treat cancer can cause xerostomia. Infection of the salivary glands or Sjögren's syndrome, a condition in which white blood cells attack salivary gland cells, can also cause dry mouth.

Sometimes identifying and avoiding a causative medication can relieve dry mouth. If this isn't possible, the Mayo Clinic Health Letter suggests the following strategies:

- Regularly sip water.
- Avoid mouth breathing.
- Suck on sugar-free hard candy or gum.
- Use room vaporizers to add moisture to the environment.

If these measures fail, saliva substitute sprays are available without a prescription, or a physician can prescribe a medication, such as pilocarpine, that increases the production of saliva.

SUMMARY OUTLINE

15.1 Introduction (p. 402)

Digestion mechanically and chemically breaks down foods and absorbs the products. The digestive system consists of an alimentary canal and several accessory organs.

15.2 General Characteristics of the Alimentary Canal (p. 402)

Regions of the alimentary canal perform specific functions.
1. Structure of the wall
   The wall consists of four layers—the mucosa, submucosa, muscular layer, and serosa.
2. Movements of the tube
   Motor functions include mixing and propelling movements.

15.3 Mouth (p. 404)

The mouth receives food and begins digestion.
1. Cheeks and lips
   a. Cheeks consist of outer layers of skin, pads of fat, muscles associated with expression and chewing, and inner linings of epithelium.
   b. Lips are highly mobile and have sensory receptors.
2. Tongue
   a. The tongue's rough surface handles food and has taste buds.
   b. Lingual tonsils are on the root of the tongue.
3. Palate
   a. The palate includes hard and soft portions.
   b. Palatine tonsils are located on either side of the tongue in the back of the mouth.
4. Teeth
   a. There are twenty primary and thirty-two secondary teeth.
   b. Teeth begin mechanical digestion by breaking food into smaller pieces, increasing the surface area exposed to digestive actions.
   c. Each tooth consists of a crown and root, and is composed of enamel, dentin, pulp, nerves, and blood vessels.
   d. A periodontal ligament attaches a tooth to the alveolar process.

15.4 Salivary Glands (p. 409)

Salivary glands secrete saliva, which moistens food, helps bind food particles, begins chemical digestion of carbohydrates, makes taste possible, and helps cleanse the mouth.
1. Salivary secretions
   Salivary glands include serous cells that secrete digestive enzymes and mucous cells that secrete mucus.
2. Major salivary glands
   a. The parotid glands secrete saliva rich in amylase.
   b. The submandibular glands produce viscous saliva.
   c. The sublingual glands primarily secrete mucus.

15.5 Pharynx and Esophagus (p. 409)

The pharynx and esophagus are important passageways.
1. Structure of the pharynx
   The pharynx is divided into a nasopharynx, oropharynx, and laryngopharynx.
2. Swallowing mechanism
   Swallowing occurs in three stages:
   a. Food is mixed with saliva and forced into the pharynx.
   b. Involuntary reflex actions move the food into the esophagus.
   c. Peristalsis transports food to the stomach.
3. Esophagus
   a. The esophagus passes through the diaphragm and joins the stomach.
   b. Circular muscle fibers at the distal end of the esophagus help prevent regurgitation of food from the stomach.

15.6 Stomach (p. 411)

The stomach receives food, mixes it with gastric juice, carries on a limited amount of absorption, and moves food into the small intestine.
1. Parts of the stomach
   a. The stomach is divided into cardiac, fundic, body, and pyloric regions.
   b. The pyloric sphincter is a valve between the stomach and small intestine.
2. Gastric secretions
   a. Gastric glands secrete gastric juice.
   b. Gastric juice contains pepsin (begins chemical digestion of proteins), hydrochloric acid, and intrinsic factor.
3. Regulation of gastric secretions
   a. Parasympathetic impulses and the hormone gastrin enhance gastric secretion.
   b. Food in the small intestine reflexly inhibits gastric secretions.
4. Gastric absorption
   The stomach wall may absorb a few substances, such as water and other small molecules.
5. Mixing and emptying actions
   a. Mixing movements help produce chyme. Peristaltic waves move chyme into the pyloric region.
   b. The muscular wall of the pyloric region regulates chyme movement into the small intestine.
   c. The rate of emptying depends on the fluidity of chyme and the type of food present.

15.7 Pancreas (p. 414)

1. Structure of the pancreas
   a. The pancreas produces pancreatic juice that is secreted into a pancreatic duct.
   b. The pancreatic duct leads to the duodenum.
2. Pancreatic juice
   a. Pancreatic juice contains enzymes that can split carbohydrates, fats, nucleic acids, and proteins.
   b. Pancreatic juice has a high bicarbonate ion concentration that helps neutralize chyme and causes intestinal contents to be alkaline.
3. Hormones regulate pancreatic secretion
   a. Secretin stimulates the release of pancreatic juice with a high bicarbonate ion concentration.
   b. Cholecystokinin stimulates the release of pancreatic juice with a high concentration of digestive enzymes.

15.8 Liver (p. 416)

1. Liver structure
   a. The lobes of the liver consist of hepatic lobules, the functional units of the gland.
   b. Bile canals carry bile from hepatic lobules to hepatic ducts.
2. Liver functions
   a. The liver metabolizes carbohydrates, lipids, and proteins; stores some substances; filters blood; destroys toxins; and secretes bile.
   b. Bile is the only liver secretion that directly affects digestion.
3. Composition of bile
   a. Bile contains bile salts, bile pigments, cholesterol, and electrolytes.
   b. Only the bile salts have digestive functions.
4. Gallbladder
   a. The gallbladder stores bile between meals.
   b. A sphincter muscle controls release of bile from the common bile duct.
5. Regulation of bile release
   a. Cholecystokinin from the small intestine stimulates bile release.
   b. The sphincter muscle at the base of the common bile duct relaxes as a peristaltic wave in the duodenal wall approaches.
6. Functions of bile salts
   Bile salts emulsify fats and aid in the absorption of fatty acids, cholesterol, and certain vitamins.

15.9 Small Intestine (p. 421)

The small intestine receives secretions from the pancreas and liver, completes nutrient digestion, absorbs the products of digestion, and transports the residues to the large intestine.

1. Parts of the small intestine
   The small intestine consists of the duodenum, jejunum, and ileum.
2. Structure of the small intestine wall
   a. The wall is lined with villi that greatly increase the surface area and aid in mixing and absorption.
   b. Intestinal glands are located between the villi.
3. Secretions of the small intestine
   a. Secretions include mucus and digestive enzymes.
   b. Digestive enzymes split molecules of sugars, proteins, and fats into simpler forms.
4. Regulation of small intestine secretions
   Secretion is stimulated by gastric juice, chyme, and reflexes stimulated by distension of the small intestine wall.

5. Absorption in the small intestine
   a. Enzymes on microvilli perform the final steps in digestion.
   b. Blood capillaries in the villi absorb monosaccharides, amino acids, fatty acids, and glycerol.
   c. Fat molecules with longer chains of carbon atoms enter the lacteals of the villi.
   d. Fatty acids with relatively short carbon chains enter blood capillaries of the villi.
6. Movements of the small intestine
   a. Movements include mixing by segmentation and peristalsis.
   b. The ileocecal sphincter controls movement of the intestinal contents from the small intestine into the large intestine.

15.10 Large Intestine (p. 426)

The large intestine reabsorbs water and electrolytes, and forms and stores feces.

1. Parts of the large intestine
   a. The large intestine consists of the cecum, colon, rectum, and anal canal.
   b. The colon is divided into ascending, transverse, descending, and sigmoid portions.
2. Structure of the large intestine wall
   a. The large intestine wall resembles the wall in other parts of the alimentary canal.
   b. The large intestine wall has a unique layer of longitudinal muscle fibers arranged in distinct bands.
3. Functions of the large intestine
   a. The large intestine has little or no digestive function.
   b. It secretes mucus.
   c. The large intestine absorbs water and electrolytes.
   d. The large intestine forms and stores feces.
4. Movements of the large intestine
   a. Movements are similar to those in the small intestine.
   b. Mass movements occur two to three times each day.
   c. A defecation reflex stimulates defecation.
5. Feces
   a. Feces consist largely of water, undigested material, electrolytes, mucus, and bacteria.
   b. The color of feces is due to bile pigments that have been altered by bacterial actions.

15.11 Nutrition and Nutrients (p. 429)

Nutrition is the study of nutrients and how the body utilizes them. The macronutrients (carbohydrates, lipids, and proteins) are required in large amounts. The micronutrients (vitamins and minerals) are not. Calories measure potential energy in foods.

1. Carbohydrates
   a. Carbohydrate sources
      (1) Starch, glycogen, disaccharides, and monosaccharides are carbohydrates.
      (2) Cellulose is a polysaccharide that human enzymes cannot digest.
   b. Carbohydrate utilization
      (1) Oxidation releases energy from glucose.
      (2) Excess glucose is stored as glycogen or converted to fat.
      (3) Carbohydrates supply energy and are also part of nucleic acids and milk.
   c. Carbohydrate requirements
      (1) Humans survive with a wide range of carbohydrate intakes.
      (2) Excess carbohydrates may lead to weight gain.

2. Lipids
   a. Lipid sources
      (1) Foods of plant and animal origin provide triglycerides.
      (2) Foods of animal origin provide dietary cholesterol.
   b. Lipid utilization
      (1) The liver and adipose tissue control triglyceride metabolism.
      (2) Linoleic acid and linolenic acid are essential fatty acids.
      (3) Lipids supply energy and are used to build membranes and steroid hormones.
   c. Lipid requirements
      (1) The amounts and types of fats required for health are highly individualized.
      (2) Fat intake must be sufficient to carry fat-soluble vitamins.
3. Proteins
   a. Protein sources
      (1) Meats, dairy products, cereals, and legumes provide most proteins.
      (2) Complete proteins contain adequate amounts of all the essential amino acids.
      (3) Incomplete proteins lack adequate amounts of one or more essential amino acids.
   b. Protein utilization
      Proteins serve as structural materials, function as enzymes, and provide energy.
   c. Protein requirements
      Proteins and amino acids must supply essential amino acids and nitrogen for the synthesis of nitrogen-containing molecules.
4. Vitamins
   a. Fat-soluble vitamins
      (1) These include vitamins A, D, E, and K.
      (2) They are carried in lipids and are influenced by the same factors that affect lipid absorption.
      (3) They resist the effects of heat; thus, cooking or food processing does not destroy them.
   b. Water-soluble vitamins
      (1) This group includes the B vitamins and vitamin C.
      (2) B vitamins make up a group (the vitamin B complex) and oxidize carbohydrates, lipids, and proteins.
      (3) Cooking or processing food destroys some water-soluble vitamins.
5. Minerals
   a. Characteristics of minerals
      (1) Most minerals are in the bones and teeth.
      (2) Minerals are usually incorporated into organic molecules; some occur in inorganic compounds or as free ions.
      (3) They serve as structural materials, function in enzymes, and play vital roles in metabolism.
   b. Major minerals include calcium, phosphorus, potassium, sulfur, sodium, chlorine, and magnesium.
   c. Trace elements include iron, manganese, copper, iodine, cobalt, zinc, fluorine, selenium, and chromium.
6. Adequate diets
   a. An adequate diet provides sufficient energy and essential nutrients to support optimal growth, maintenance, and repair of tissues.
   b. Individual requirements vary so greatly that designing a diet that is adequate for everyone is not possible. Food guide pyramids can help to personalize diets.
   c. Malnutrition is poor nutrition due to lack of food or failure to make the best use of available food.

CHAPTER ASSESSMENTS

◀15.1 Introduction

1. Functions of the digestive system include (p. 402)
   a. mechanical breakdown of foods.
   b. chemical breakdown of foods.
   c. breaking large pieces into smaller ones without altering their chemical composition.
   d. breaking food into simpler chemicals.
   e. all of the above
2. List the major parts of the alimentary canal, then separately list the accessory organs of the digestive system. (p. 402)

◀15.2 General Characteristics of the Alimentary Canal

3. Contrast the composition of the layers of the wall of the alimentary canal. (p. 402)
4. Distinguish between mixing and propelling movements. (p. 402)

◀15.3 Mouth

5. Discuss the functions of the mouth and its parts. (p. 404)
6. Distinguish between primary and secondary teeth. (p. 406)
7. Describe the structure of a tooth. (p. 408)
8. The teeth that are best adapted for grasping and tearing food are the (p. 408)
   a. incisors.   c. premolars.
   b. canines.   d. molars.

◀15.4–15.10 Salivary Glands–Large Intestine

9. Match the glands with the enzyme(s) they secrete. Enzymes may be used more than once. (pp. 409–424)

(1) salivary glands (serous cells)	A. peptidase
(2) stomach (chief cells)	B. amylase
(3) pancreas (acinar cells)	C. nuclease
(4) small intestine (mucosal cells)	D. lipase
	E. pepsin
	F. trypsin, chymotrypsin, carboxypeptidase
	G. sucrase, maltase, lactase

10. Match the enzyme(s) with their functions. (pp. 409–424)

(1) peptidase	A. Begins protein digestion
(2) amylase	B. Breaks fats into fatty acids and glycerol
(3) nuclease	C. Breaks down proteins or partially digested proteins into peptides
(4) lipase	D. Breaks down starch and glycogen into disaccharides
(5) pepsin	E. Breaks down peptides into amino acids
(6) trypsin, chymotrypsin, carboxypeptidase	F. Breaks down nucleic acids into nucleotides
(7) sucrase, maltase, lactase	G. Breaks down disaccharides into monosaccharides

11. List the steps in swallowing. (p. 410)
12. Explain the stimulus for and response of the parasympathetic nervous system in digestion. (p. 413)
13. Explain how hormones control the secretions and/or release of secretions from the stomach, pancreas, and gallbladder. (pp. 413, 415, 420)
14. List the steps in defecating. (p. 429)

15. Discuss absorption of amino acids, monosaccharides, glycerol, fatty acids, electrolytes, and water from substances in the small and large intestines. (pp. 424 and 428)

◄ 15.11 Nutrition and Nutrients

16. Identify dietary sources of carbohydrates, lipids, and proteins. (pp. 430–433)

17. Explain how cells utilize carbohydrates, lipids, and proteins for the normal functioning of the body. (pp. 430–433)

18. Match the vitamins with their general functions, and indicate if the vitamin is fat-soluble or water-soluble. Functions may be used more than once. (p. 434)

(1) vitamin A	A. Part of coenzyme A in oxidation of carbohydrates
(2) vitamin B_1 (thiamine)	B. Required for ribose synthesis
(3) vitamin B_2 (riboflavin)	C. Necessary for synthesis of visual pigments
(4) vitamin B_3 (niacin)	D. Required for synthesis of prothrombin
(5) vitamin B_5 (pantothenic acid)	E. Required to produce collagen
(6) vitamin B_6	F. Required to synthesize nucleic acids
(7) vitamin B_{12} (cyanocobalamin)	G. Promotes red blood cell production
(8) folacin	H. Plays a role in myelin synthesis
(9) biotin	I. Antioxidant, may help stabilize cell membranes
(10) vitamin C (ascorbic acid)	J. Promotes development of teeth and bones
(11) vitamin D	K. Required to produce antibodies
(12) vitamin E	L. Required for cellular growth
(13) vitamin K	M. Part of coenzymes to synthesize proteins, fats, and nucleic acids

19. Match the minerals/elements with their functions (functions may be used more than once), and indicate whether each is a major mineral or a trace element required for nutrition. Functions may be used more than once. (pp. 435–436)

(1) calcium	A. Essential for the use of carbohydrates
(2) chlorine	B. A component of certain enzymes
(3) chromium	C. A component of tooth structure (enamel)
(4) cobalt	D. A component of the structure of teeth and bones.
(5) copper	E. Helps maintain intracellular osmotic pressure
(6) fluorine	F. An essential part of certain amino acids
(7) iodine	G. Helps maintain extracellular fluid osmotic pressure
(8) iron	H. Necessary for normal wound healing
(9) magnesium	I. A component of cyanocobalamin
(10) manganese	J. Essential for the synthesis of thyroid hormones
(11) phosphorus	K. Required in metabolic reactions associated with ATP production
(12) potassium	L. A component of hemoglobin molecules
(13) selenium	M. Essential for hemoglobin synthesis and melanin production
(14) sodium	N. Required for cholesterol synthesis and urea formation
(15) sulfur	
(16) zinc	

20. Define *adequate diet*. (p. 436)

INTEGRATIVE ASSESSMENTS/ CRITICAL THINKING

➡ OUTCOMES 15.1, 15.2

1. How does mechanical digestion enhance chemical digestion?

➡ OUTCOME 15.6

2. How would removal of 95% of the stomach (subtotal gastrectomy) to treat severe ulcers or cancer affect digestion and absorption? How would the patient's eating habits have to be altered? Why? Do you think that people should have this type of surgery to treat life-threatening obesity?

➡ OUTCOMES 15.7, 15.8

3. Why might a person with inflammation of the gallbladder (cholecystitis) also develop inflammation of the pancreas (pancreatitis)?

➡ OUTCOMES 15.8, 15.11

4. Why does blood sugar concentration stay relatively stable in a person whose diet is low in carbohydrates?

➡ OUTCOME 15.11

5. How can people consume vastly different diets, yet all obtain adequate nourishment?

6. How can too little fat in the diet lead to a vitamin deficiency, even if a person takes vitamin supplements?

7. Examine the label information on the packages of a variety of dry breakfast cereals. Which types of cereals provide adequate sources of vitamins and minerals? Which major nutrients are lacking in these cereals?

WEB CONNECTIONS

Visit the text website at **aris.mhhe.com** for additional quizzes, interactive learning exercises, and more.

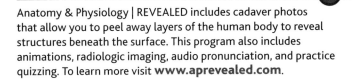 DIGESTIVE SYSTEM

Anatomy & Physiology | REVEALED includes cadaver photos that allow you to peel away layers of the human body to reveal structures beneath the surface. This program also includes animations, radiologic imaging, audio pronunciation, and practice quizzing. To learn more visit **www.aprevealed.com**.

16

Respiratory System

THE DANGERS OF SECONDHAND SMOKE. Evidence is mounting that exposure to environmental tobacco smoke (ETS)—also called secondhand smoke—is as dangerous as actually smoking. ETS has two sources: *sidestream smoke* comes from lit cigarettes, cigars, or pipes, and *mainstream smoke* is exhaled by smokers. The smoke contains more than 4,000 chemical compounds, many of which have been classified as irritants, carcinogens, mutagens, and systemic or developmental toxins. More than 60 carcinogens have been identified in tobacco smoke. Toxins include benzene, formaldehyde, vinyl chloride, ammonia, arsenic, and cyanide.

Altogether, exposure to ETS likely kills an estimated 53,000 nonsmokers in the U.S. each year. That includes:

- 46,000 heart disease deaths
- 3,400 lung cancer deaths

ETS is also responsible for 150,000 to 300,000 cases of lower respiratory infection (bronchitis or pneumonia) in children under 18 months of age; worsening of asthma in children; ear infections in children and low birth weight. Even short exposures are dangerous, to anyone. Just a half hour of breathing someone else's smoke can activate platelets, damage endothelium, decrease coronary artery blood flow, and decrease heart rate variability—all changes that set the stage for cardiovascular disease.

The danger of secondhand smoke, long debated, is now widely accepted, and new rules and regulations attempt to limit exposure.

The only way to decrease exposure to ETS is to eliminate it. Dividing a space into smoking and nonsmoking areas does not work, because the smoke lingers for hours—this is why airplanes, restaurants, and many workplaces have been made smoke-free. Even cleaning and ventilation does not eliminate smoke, as anyone who has ever stayed in a hotel room supposedly changed from smoking to nonsmoking can attest. The lingering of smoke on one's clothing is a warning sign of the lingering that occurs in the body of the nonsmoker as well. The *U.S. Surgeon General's Report on the Health Consequences of Involuntary Exposure to Tobacco Smoke* concluded that there is "no risk-free level of exposure to secondhand smoke."

Learning Outcomes   *After studying this chapter, you should be able to do the following:*

16.1 Introduction

1. Identify the general functions of the respiratory system. (p. 445)

16.2 Organs of the Respiratory System

2. Describe the locations of the organs of the respiratory system. (p. 445)

3. Describe the functions of each organ of the respiratory system. (p. 445)

16.3 Breathing Mechanism

4. Explain the mechanisms of inspiration and expiration. (p. 452)

5. Define each of the respiratory volumes and capacities. (p. 455)

16.4 Control of Breathing

6. Locate the respiratory areas in the brainstem and explain how they control breathing. (p. 457)

7. Discuss how various factors affect the respiratory areas. (p. 458)

16.5 Alveolar Gas Exchanges

8. Describe the structure and function of the respiratory membrane. (p. 460)

9. Explain how air and blood exchange gases. (p. 460)

16.6 Gas Transport

10. List the ways blood transports oxygen and carbon dioxide. (p. 461)

Aids to Understanding Words

(Appendix A on page 567 has a complete list of Aids to Understanding Words.)

alveol- [small cavity] *alveolus*: Microscopic air sac within a lung.

bronch- [windpipe] *bronchus*: Primary branch of the trachea.

cric- [ring] *cric*oid cartilage: Ring-shaped mass of cartilage at the base of the larynx.

epi- [upon] *epi*glottis: Flaplike structure that partially covers the opening into the larynx during swallowing.

hem- [blood] *hem*oglobin: Pigment in red blood cells that transports oxygen and carbon dioxide.

16.1 INTRODUCTION

Cells require oxygen to break down nutrients to release energy and produce ATP, and must excrete the carbon dioxide that results. Obtaining oxygen and removing carbon dioxide are the primary functions of the **respiratory system,** which includes tubes that remove particles from (filter) incoming air and transport air into and out of the lungs, as well as microscopic air sacs where gases are exchanged. The respiratory organs also entrap particles from incoming air, help control the temperature and water content of the air, produce vocal sounds, and participate in the sense of smell and the regulation of blood pH.

The entire process of gas exchange between the atmosphere and cells is called **respiration** (res″pĭ-ra′shun). The events of respiration include: (1) movement of air into and out of the lungs—commonly called breathing or *ventilation;* (2) gas exchange between the blood and the air in the lungs (external respiration); (3) gas transport in blood between the lungs and body cells; and (4) gas exchange between the blood and the cells (internal respiration). The process of oxygen utilization and carbon dioxide production at the cellular level is called *cellular respiration.*

16.2 ORGANS OF THE RESPIRATORY SYSTEM

The organs of the respiratory system can be divided into two groups, or tracts. Those in the *upper respiratory tract* include the nose, nasal cavity, paranasal sinuses, and pharynx. Those in the *lower respiratory tract* include the larynx, trachea, bronchial tree, and lungs (fig. 16.1; see reference plates 3, 4, 5, and 6, pp. 25–28).

Nose

Bone and cartilage support the **nose** internally. Its two *nostrils* are openings through which air can enter and leave the nasal cavity. Many internal hairs guard the nostrils, preventing entry of large particles carried in the air.

Nasal Cavity

The **nasal cavity** is a hollow space behind the nose (fig. 16.1). The **nasal septum,** composed of bone and cartilage, divides the nasal cavity into right and left portions. **Nasal conchae** are bones and bone processes that curl out from the lateral walls of the nasal cavity on each side, dividing the cavity into passageways (fig. 16.2). Nasal conchae also support the mucous membrane that lines the nasal cavity and help increase its surface area.

The mucous membrane has pseudostratified ciliated epithelium that is rich in mucus-secreting goblet

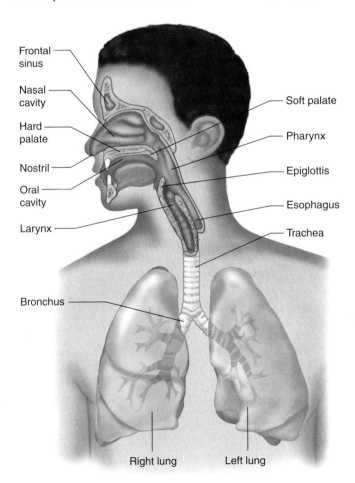

Figure 16.1

Organs and associated structures of the respiratory system.

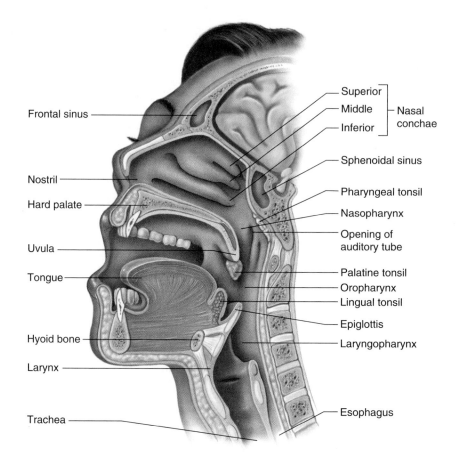

Figure 16.2

Major structures associated with the respiratory tract in the head and neck.

Labels (top to bottom, left): Frontal sinus, Nostril, Hard palate, Uvula, Tongue, Hyoid bone, Larynx, Trachea

Labels (right): Superior, Middle, Inferior — Nasal conchae; Sphenoidal sinus; Pharyngeal tonsil; Nasopharynx; Opening of auditory tube; Palatine tonsil; Oropharynx; Lingual tonsil; Epiglottis; Laryngopharynx; Esophagus

cells (see chapter 5, p. 98). It also includes an extensive network of blood vessels, and as air passes over the membrane, heat leaves the blood and warms the air, adjusting the air's temperature to that of the body. In addition, incoming air is moistened as water evaporates from the mucous lining. The sticky mucus that the mucous membrane secretes entraps dust and other small particles entering with the air.

> The nasal septum is usually straight at birth, although it can bend as the result of a birth injury. With age, the septum bends toward one side or the other. Such a *deviated septum* may obstruct the nasal cavity, making breathing difficult.

As the cilia of the epithelial lining move, they push a thin layer of mucus and entrapped particles toward the pharynx, where the mucus is swallowed (fig. 16.3). In the stomach, gastric juice destroys microorganisms in the mucus.

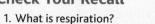

> A spore of the bacterium that causes anthrax is only half a micron wide. When spores are coated with powder to create a "bioweapon," they are still small enough to bypass the hairs and mucus in the nose, reaching the lungs, where they can cause inhalation anthrax. The bacteria release a toxin that causes death.

Paranasal Sinuses

Recall from chapter 7 (pp. 142–146) that the **paranasal sinuses** are air-filled spaces within the *maxillary, frontal, ethmoid,* and *sphenoid bones* of the skull and opening into the nasal cavity. Mucous membranes line the sinuses and are continuous with the lining of the nasal cavity. The paranasal sinuses reduce the weight of the skull and are resonant chambers that affect the quality of the voice.

> A painful sinus headache can result from blocked drainage caused by an infection or allergic reaction.

Check Your Recall

1. What is respiration?
2. Which organs constitute the respiratory system?
3. What are the functions of the mucous membrane that lines the nasal cavity?
4. Where are the paranasal sinuses located?
5. What are the functions of the paranasal sinuses?

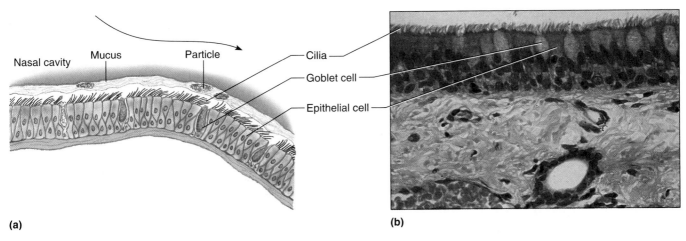

(a)

(b)

Figure 16.3

Mucus movement in the respiratory tract. (*a*) Cilia move mucus and trapped particles from the nasal cavity to the pharynx. (*b*) Micrograph of ciliated epithelium in the respiratory tract (275×).

Pharynx

The **pharynx,** or throat, is behind the oral cavity, the nasal cavity and the larynx (see fig. 16.1). It is a passageway for food traveling from the oral cavity to the esophagus and for air passing between the nasal cavity and the larynx. It also helps produce the sounds of speech. Chapter 15 (p. 410) describes the subdivisions of the pharynx—the nasopharynx, oropharynx, and laryngopharynx, which are shown on figure 16.2.

Larynx

The **larynx** (lar'inks) is an enlargement in the airway at the top of the trachea and below the pharynx. It conducts air in and out of the trachea and prevents foreign objects from entering the trachea. It also houses the *vocal cords.*

The larynx is composed of a framework of muscles and cartilages bound by elastic tissue. The largest of the cartilages are the *thyroid* ("Adam's apple"), *cricoid,* and *epiglottic cartilages* (fig. 16.4).

Inside the larynx, two pairs of horizontal *vocal folds* composed of muscle tissue and connective tissue with a covering of mucous membrane extend inward from the lateral walls. The upper folds are called *false vocal cords* because they do not produce sounds (fig. 16.5*a*). Muscle fibers within these folds help close the airway during swallowing.

The lower folds of muscle tissue and elastic fibers are the *true vocal cords.* Air forced between the vocal cords causes them to vibrate from side to side, which generates sound waves. Changing the shapes of the pharynx and oral cavity and using the tongue and lips transform these sound waves into words.

Contracting or relaxing muscles that alter the tension on the vocal cords controls the pitch (musical tone) of a sound. Increasing tension raises pitch, and decreasing tension lowers pitch. The intensity (loudness) of a sound reflects the force of air passing through the vocal folds. Stronger blasts of air produce louder sound; weaker blasts produce softer sound.

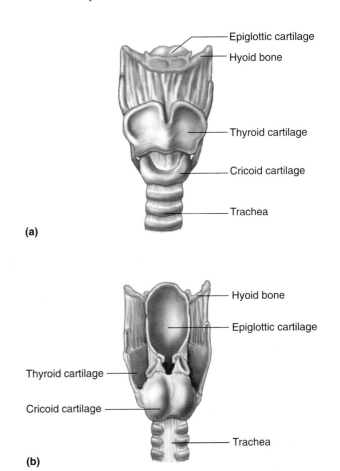

(a)

(b)

Figure 16.4

Larynx. (*a*) Anterior and (*b*) posterior views of the larynx.

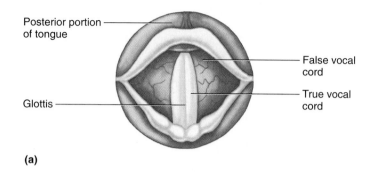

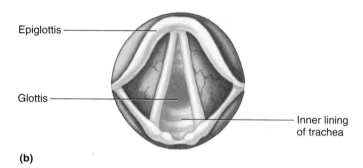

Posterior portion of tongue

False vocal cord

True vocal cord

Glottis

(a)

Epiglottis

Glottis

Inner lining of trachea

(b)

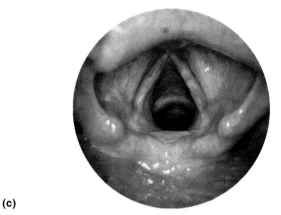

(c)

Figure 16.5

The vocal cords as viewed from above with the glottis (*a*) closed and (*b*) open. (*c*) Photograph of the glottis and vocal folds.

Damage to the nerves (recurrent laryngeal nerves) that supply the laryngeal muscles can alter the quality of a person's voice. These nerves pass through the neck as parts of the vagus nerves, and they can be injured by trauma or surgery to the neck or thorax. Nodules or other growths on the margins of the vocal folds that interfere with the free flow of air can also cause vocal problems. Surgery can remove such lesions.

During normal breathing, the vocal cords are relaxed and the opening between them, called the **glottis** (glot′is), is a triangular slit. However, when food or liquid is swallowed, muscles within the false vocal cords close the glottis, which prevents food or liquid from entering the trachea (fig. 16.5).

The epiglottic cartilage supports a flaplike structure called the **epiglottis.** This structure usually stands upright and allows air to enter the larynx. During swallowing, however, the larynx rises, and the epiglottis presses downward to partially cover the opening into the larynx. This helps prevent foods and liquids from entering the air passages (see chapter 15, p. 410).

Laryngitis—hoarseness or lack of voice—occurs when the mucous membrane of the larynx becomes inflamed and swollen due to an infection or an irritation from inhaled vapors, and prevents the vocal cords from vibrating as freely as before. Laryngitis is usually mild, but may be dangerous if swollen tissues obstruct the airway and interfere with breathing. Inserting a tube (endotracheal tube) into the trachea through the nose or mouth can restore the passageway until the inflammation subsides.

Check Your Recall

6. Describe the structure of the larynx.

7. How do the vocal cords produce sounds?

8. What is the function of the glottis? The epiglottis?

Trachea

The **trachea** (tra′ke-ah), or windpipe, is a flexible, cylindrical tube about 2.5 centimeters in diameter and 12.5 centimeters in length (fig. 16.6). It extends downward anterior to the esophagus and into the thoracic cavity, where it splits into right and left bronchi.

A ciliated mucous membrane with many goblet cells lines the trachea's inner wall. This membrane filters incoming air and moves entrapped particles upward into the pharynx, where the mucus can be swallowed.

Within the tracheal wall are about twenty C-shaped pieces of hyaline cartilage, one above the other. The open ends of these incomplete rings are directed posteriorly, and smooth muscle and connective tissues fill the gaps between the ends. These cartilaginous rings prevent the trachea from collapsing and blocking the airway. The soft tissues that complete the rings in the back allow the nearby esophagus to expand as food moves through it to the stomach.

Bronchial Tree

The **bronchial tree** (brong′ke-al tre) consists of branched airways leading from the trachea to the microscopic air sacs in the lungs (fig. 16.7). Its branches begin with the right and left **primary bronchi,** which arise from the trachea at the level of the fifth thoracic vertebra.

A short distance from its origin, each primary bronchus divides into secondary bronchi, which in turn

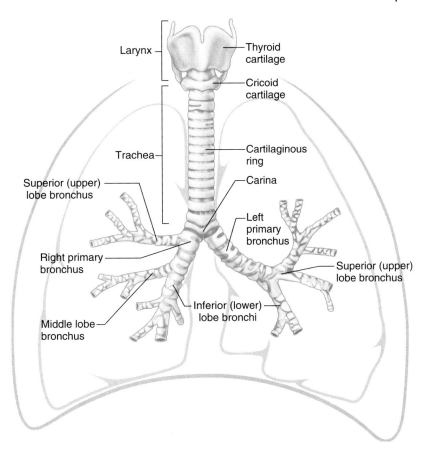

Figure 16.6

The trachea conducts air between the larynx and the bronchi.

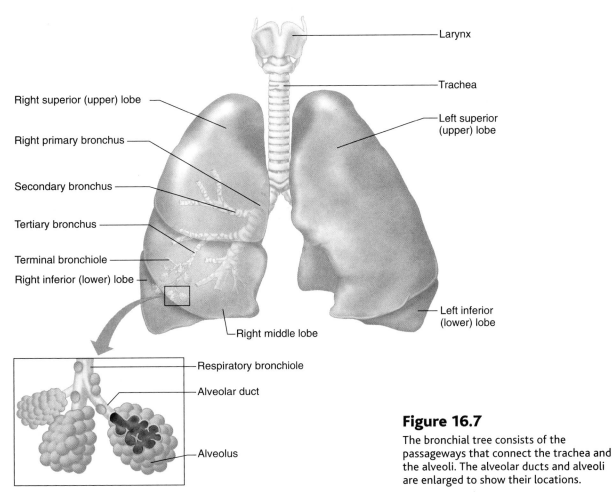

Figure 16.7

The bronchial tree consists of the passageways that connect the trachea and the alveoli. The alveolar ducts and alveoli are enlarged to show their locations.

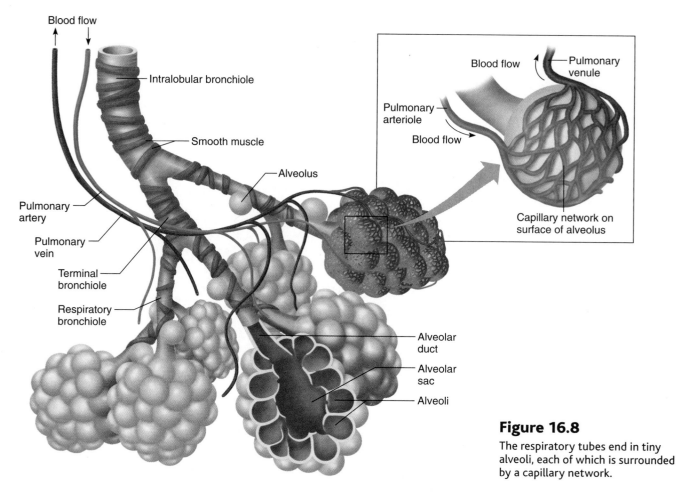

Figure 16.8

The respiratory tubes end in tiny alveoli, each of which is surrounded by a capillary network.

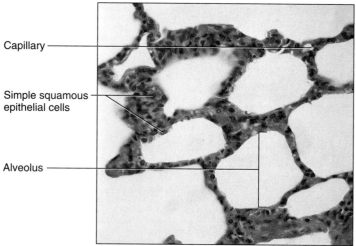

Figure 16.9

Light micrograph of alveoli (250×).

branch into tertiary bronchi, and then into finer and finer tubes. Among these smaller tubes are **bronchioles** that continue to divide, giving rise to terminal bronchioles, respiratory bronchioles, and finally to very thin tubes called **alveolar ducts.** These ducts lead to thin-walled outpouchings called **alveolar sacs.** Alveolar sacs lead to smaller microscopic air sacs called **alveoli**

(al-ve′o-li; singular, alveolus), which lie within capillary networks (figs. 16.8 and 16.9).

The structure of a bronchus is similar to that of the trachea, but the tubes that branch from it have less cartilage in their walls, and the bronchioles lack cartilage. As the cartilage diminishes, a layer of smooth muscle surrounding the tube becomes more prominent. This muscular layer persists even in the smallest bronchioles, but only a few muscle fibers are in the alveolar ducts.

The branches of the bronchial tree are air passages whose mucous membranes filter incoming air and distribute the air to alveoli throughout the lungs. (The Genetics Connection entitled Cystic Fibrosis on page 464 discusses what happens when the mucus formed is extremely thick.) The alveoli provide a large surface area of thin simple squamous epithelial cells through which gases can easily be exchanged. Oxygen diffuses from the alveoli into the blood in nearby capillaries, and carbon dioxide diffuses from the blood into the alveoli (fig. 16.10).

Combined, two adult lungs have about 300 million alveoli, providing a total surface area nearly half the size of a tennis court.

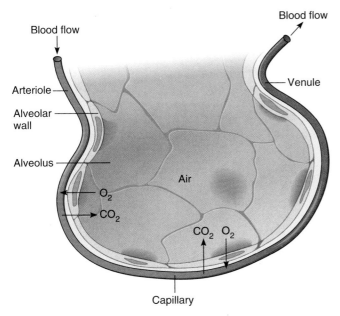

Figure 16.10

Oxygen (O_2) diffuses from air within the alveolus into the capillary, while carbon dioxide (CO_2) diffuses from blood within the capillary into the alveolus.

Check Your Recall

9. What is the function of the cartilaginous rings in the tracheal wall?

10. Describe the bronchial tree.

11. Predict the direction of diffusion of gases between alveoli and alveolar capillaries.

Lungs

The **lungs** are soft, spongy, cone-shaped organs in the thoracic cavity (see fig. 16.1 and reference plates 4 and 5, pp. 26–27). The mediastinum separates the right and left lungs medially, and the diaphragm and thoracic cage enclose them.

Each lung occupies most of the thoracic space on its side. A bronchus and some large blood vessels suspend each lung in the cavity. These tubular structures enter the lung on its medial surface. A layer of serous membrane, the **visceral pleura** (vis′er-al ploo′rah), firmly attaches to each lung surface and folds back to become the **parietal pleura** (pah-ri′ĕ-tal ploo′rah). The parietal pleura, in turn, forms part of the mediastinum and lines the inner wall of the thoracic cavity (fig. 16.11).

No significant space exists between the visceral and parietal pleurae, but the potential space between them is called the **pleural cavity** (ploo′ral kav′ĭ-te). It has a thin film of serous fluid that lubricates adjacent pleural surfaces, reducing friction as they move against one another during breathing. This fluid also helps hold the pleural membranes together, as explained in section 16.3.

The right lung is larger than the left one and is divided into three lobes. The left lung has two lobes (see figs. 16.1 and 16.7).

A major branch of the bronchial tree supplies each lobe. A lobe also has connections to blood and lymphatic vessels and lies within connective tissues. Thus, a lung includes air passages, alveoli, blood vessels, connective tissues, lymphatic vessels, and nerves. Table 16.1 summarizes the characteristics of the major parts of the respiratory system.

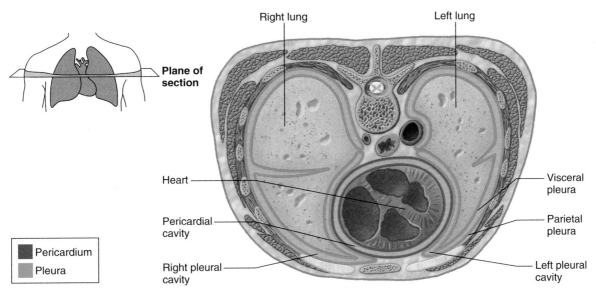

Figure 16.11

The potential spaces between the pleural membranes, called the left and right pleural cavities, are shown here as actual spaces.

Table 16.1	Parts of the Respiratory System	
Part	**Description**	**Function**
Nose	Part of face centered above mouth, in and below space between eyes	Nostrils provide entrance to nasal cavity; internal hairs begin to filter incoming air
Nasal cavity	Hollow space behind nose	Conducts air to pharynx; mucous lining filters, warms, and moistens incoming air
Paranasal sinuses	Hollow spaces in certain skull bones	Reduce weight of skull; serve as resonant chambers
Pharynx	Chamber behind nasal cavity, oral cavity, and larynx	Passageway for air moving from nasal cavity to larynx and for food moving from oral cavity to esophagus
Larynx	Enlargement at top of trachea	Passageway for air; prevents foreign objects from entering trachea; houses vocal cords
Trachea	Flexible tube that connects larynx with bronchial tree	Passageway for air; mucous lining continues to filter particles from incoming air
Bronchial tree	Branched tubes that lead from trachea to alveoli	Conducts air from trachea to alveoli; mucous lining continues to filter incoming air
Lungs	Soft, cone-shaped organs that occupy a large portion of the thoracic cavity	Contain air passages, alveoli, blood vessels, connective tissues, lymphatic vessels, and nerves of the lower respiratory tract

Check Your Recall

12. Where are the lungs located?
13. What is the function of serous fluid within the pleural cavity?
14. What kinds of structures make up a lung?

16.3 BREATHING MECHANISM

Breathing, or ventilation, is the movement of air from outside the body into and out of the bronchial tree and alveoli. The actions providing these air movements are termed **inspiration** (in″spĭ-ra′shun), or inhalation, and **expiration** (ek″spĭ-ra′shun), or exhalation.

Inspiration

Atmospheric pressure due to the weight of air is the force that moves air into the lungs. At sea level, this pressure is sufficient to support a column of mercury about 760 millimeters (mm) high in a tube. Thus, normal air pressure is equal to 760 mm of mercury (Hg).

Air pressure is exerted on all surfaces in contact with the air, and because people breathe air, the inside surfaces of their lungs also are subjected to pressure. The pressures on the inside of the lungs and alveoli and on the outside of the thoracic wall are about the same.

If the pressure inside the lungs and alveoli decreases, atmospheric pressure will push outside air into the airways. That is what happens during normal inspiration. Impulses carried on the phrenic nerves, which are associated with the cervical plexuses (see chapter 9, p. 248), stimulate muscle fibers in the dome-shaped *diaphragm*

below the lungs to contract. The diaphragm moves downward, the thoracic cavity enlarges, and the pressure within the alveoli falls to about 2 mm Hg below that of atmospheric pressure. In response, atmospheric pressure forces air into the airways (fig. 16.12).

Pressure and volume are related in an opposite or inverse way. For example, if we pull back on the plunger of a syringe, the volume inside the barrel increases, causing the air pressure inside to decrease. Outside air is then pushed into the syringe by atmospheric pressure. In contrast, if we push on the plunger of a syringe, the volume inside the syringe is reduced, but the pressure inside increases, forcing air out into the atmosphere. The movement of air into and out of the lungs occurs in much the same way.

While the diaphragm is contracting and moving downward, the *external (inspiratory) intercostal muscles* between the ribs may be stimulated to contract. This raises the ribs and elevates the sternum, enlarging the thoracic cavity even more. As a result, the pressure inside is reduced further, and the greater atmospheric pressure forces even more air into the airways.

Lung expansion in response to movements of the diaphragm and chest wall depends on movements of the pleural membranes. Any separation of the pleural membranes decreases pressure in the intrapleural space, holding these membranes together. In addition, only a thin film of serous fluid separates the parietal pleura on the inner wall of the thoracic cavity from the visceral pleura attached to the surface of the lungs. The water molecules in this fluid greatly attract the pleural membranes and each other, helping to adhere the moist surfaces of the pleural membranes, much as a wet coverslip sticks to a microscope slide. As a result of these

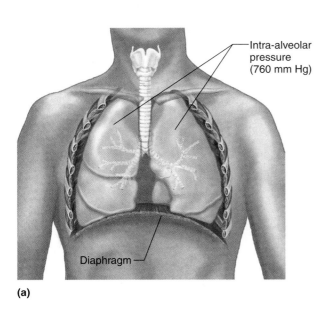

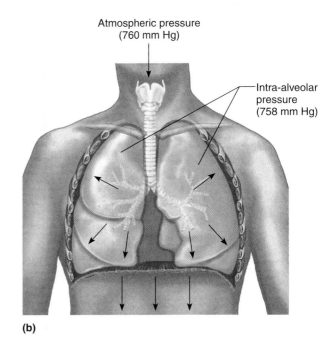

Figure 16.12

Normal inspiration. (*a*) Prior to inspiration, the intra-alveolar pressure is 760 mm Hg. (*b*) The intra-alveolar pressure decreases to about 758 mm Hg as the thoracic cavity enlarges, and atmospheric pressure forces air into the airways.

factors, when the external intercostal muscles move the thoracic wall upward and outward, the parietal pleura moves too, and the visceral pleura follows it. This helps expand the lung in all directions.

Although the moist pleural membranes play a role in expansion of the lungs, the moist inner surfaces of the alveoli have the opposite effect. Here, the attraction of water molecules creates a force called **surface tension** that makes it difficult to inflate the alveoli and may actually cause them to collapse. Certain alveolar cells, however, synthesize a mixture of lipids and proteins called **surfactant** (ser-fak′tant). It is secreted continuously into alveolar air spaces and reduces the alveoli's tendency to collapse, especially when lung volumes are low. Surfactant makes it easier for inspiratory efforts to inflate the alveoli.

Surfactant is particularly important in the minutes after birth, when the newborn's lungs inflate for the first time. Premature infants often suffer respiratory distress syndrome because they do not produce sufficient surfactant. To help many of these newborns survive, physicians inject synthetic surfactant into the tiny lungs through an endotracheal tube. A ventilator machine especially geared to an infant's size assists breathing.

If a person needs to take a deeper than normal breath, the diaphragm and external intercostal muscles contract more forcefully. Additional muscles, such as the pectoralis minor and the sternocleidomastoid, can also pull the thoracic cage farther upward and outward,

enlarging the thoracic cavity and decreasing internal pressure (fig. 16.13).

The first breath is the toughest. A newborn must use twenty times the energy to take the first breath as for subsequent breaths. This is because each of the millions of alveoli start out only partially inflated.

Expiration

The forces for expiration come from the *elastic recoil* of tissues and from surface tension. The lungs and thoracic wall contain considerable elastic tissue, which stretches with lung expansion during inspiration. Also, the diaphragm lowering compresses the abdominal organs beneath it. As the diaphragm and external intercostal muscles relax following inspiration, the elastic tissues cause the lungs and thoracic cage to recoil and return to their original shapes. Similarly, the abdominal organs spring back into their previous shapes, pushing the diaphragm upward (fig. 16.14*a*). At the same time, the surface tension that develops between the moist surfaces of the alveolar linings decreases the diameters of the alveoli. Each of these factors increases alveolar pressure about 1 mm Hg above atmospheric pressure, so that the air inside the lungs is forced out through respiratory passages. Thus, normal resting expiration is a passive process.

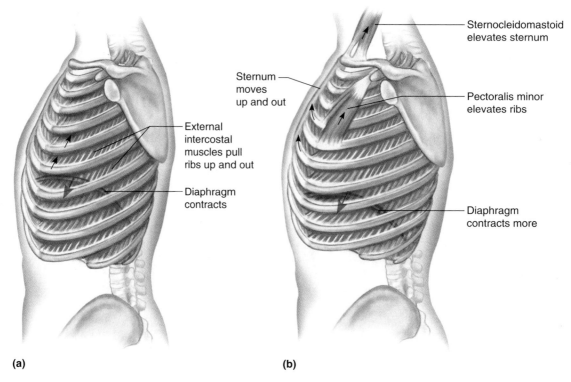

(a)   (b)

Figure 16.13

Maximal inspiration. (*a*) Shape of the thorax at the end of normal inspiration. (*b*) Shape of the thorax at the end of maximal inspiration, aided by contraction of the sternocleidomastoid and pectoralis minor muscles.

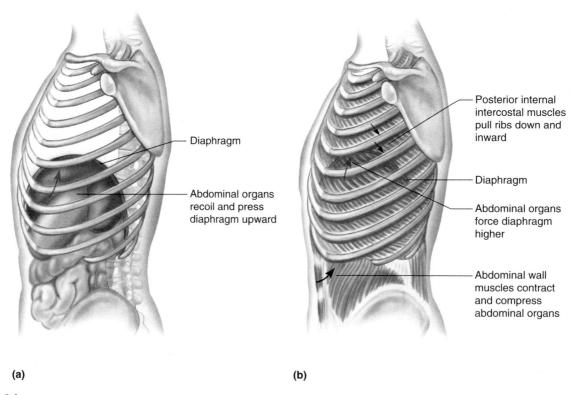

(a)   (b)

Figure 16.14

Expiration. (*a*) Normal resting expiration is due to elastic recoil of the lung tissues and the abdominal organs. (*b*) Contraction of the abdominal wall muscles and the posterior internal intercostal muscles aids maximal expiration.

Because low pressure and wet surfaces hold the visceral and parietal pleural membranes together, no actual space normally exists in the pleural cavity between them. A puncture in the thoracic wall, however, allows atmospheric air to enter the pleural cavity and create a real space between the membranes. This condition, called *pneumothorax,* may collapse the lung on the affected side because of the lung's elasticity. The condition of a collapsed lung is called *atelectasis.*

If a person needs to exhale more air than normal, the posterior *internal (expiratory) intercostal muscles* can be contracted (fig. 16.14*b*). These muscles pull the ribs and sternum downward and inward, increasing the pressure in the lungs. Also, the *abdominal wall muscles,* including the external and internal obliques, transversus abdominis, and rectus abdominis, can squeeze the abdominal organs inward (see fig. 8.19, p. 197). Thus, the abdominal wall muscles can increase pressure in the abdominal cavity and force the diaphragm still higher against the lungs. These actions squeeze additional air out of the lungs. The Topic of Interest on page 456 entitled Emphysema and Lung Cancer discusses problems in breathing associated with disease.

Air movements that occur in addition to breathing are called *nonrespiratory movements.* They are used to clear air passages, as in coughing and sneezing, or to express emotion, as in laughing and crying.

Nonrespiratory movements usually result from *reflexes,* although sometimes they are initiated voluntarily. A *cough,* for example, can be produced through conscious effort or may be triggered by a foreign object in an air passage.

Coughing involves taking a deep breath, closing the glottis, and forcing air upward from the lungs against the closure. Then the glottis is suddenly opened, and a blast of air is forced upward from the lower respiratory tract. Usually, this rapid rush of air removes the substance that triggered the reflex.

A *sneeze* is much like a cough, but it clears the upper respiratory passages rather than the lower ones. This reflex is usually initiated by a mild irritation in the lining of the nasal cavity, and in response, a blast of air is forced up through the glottis. This time, the air is directed into the nasal passages by depressing the uvula, thus closing the opening between the pharynx and the oral cavity. A sneeze can propel a particle out of the nose at 200 miles per hour.

Laughing involves taking a breath and releasing it in a series of short expirations. *Crying* consists of very similar movements, and sometimes it is necessary to note a person's facial expression in order to distinguish laughing from crying.

A *hiccup* is caused by sudden inspiration due to a spasmodic contraction of the diaphragm while the glottis is closed. Air striking the vocal folds causes the sound of the hiccup. We do not know the function, if any, of hiccups.

Yawning may aid respiration by providing an occasional deep breath. During normal, quiet breathing, not all of the alveoli are ventilated, and some blood may pass through the lungs without becoming well oxygenated. This low blood oxygen concentration triggers the yawn reflex, prompting a very deep breath that ventilates more of the alveoli.

Check Your Recall

15. Describe the events in inspiration.
16. How does expansion of the chest wall expand the lungs during inspiration?
17. What forces cause normal expiration?

Respiratory Air Volumes and Capacities

Different intensities in breathing move different volumes of air in or out of the lungs. *Spirometry* measures such air volumes, revealing four distinct **respiratory volumes** (re-spi′rah-to″re vol′ūmz).

One inspiration plus the following expiration is called a **respiratory cycle.** The volume of air that enters (or leaves) during a single respiratory cycle is termed the **tidal volume.** About 500 milliliters (mL) of air enter during a normal, resting inspiration. Approximately the same volume leaves during a normal, resting expiration. Thus, the **resting tidal volume** is about 500 mL (fig. 16.15).

During forced inspiration, air in addition to the resting tidal volume enters the lungs. This extra volume is the **inspiratory reserve volume** (complemental air), and at maximum, it equals about 3,000 mL.

During forced expiration, the lungs can expel up to about 1,100 mL of air beyond the resting tidal volume. This amount is called the **expiratory reserve volume** (supplemental air). However, even after the most forceful expiration, about 1,200 mL of air remains in the lungs. This is called the **residual volume.**

Because of the residual volume and the expiratory reserve volumes, newly inhaled air always mixes with air already in the lungs. This prevents the oxygen and carbon dioxide concentrations in the lungs from fluctuating greatly with each breath.

Combining two or more of the respiratory volumes yields four **respiratory capacities** (re-spi′rah-to″re kah-pas′ĭ-tēz). Combining the inspiratory reserve volume (3,000 mL) with the tidal volume (500 mL) and the expiratory reserve volume (1,100 mL) gives the **vital capacity** (4,600 mL). This is the maximum volume of air a person can exhale after taking the deepest breath possible.

The tidal volume (500 mL) plus the inspiratory reserve volume (3,000 mL) gives the **inspiratory capacity** (3,500 mL), which is the maximum volume of air a person can inhale following a resting expiration. Similarly, the expiratory reserve volume (1,100 mL) plus the residual volume (1,200 mL) equals the **functional residual capacity** (2,300 mL), which is the volume of air that remains in the lungs following a resting expiration.

The vital capacity plus the residual volume equals the **total lung capacity** (about 5,800 mL). This total varies with age, sex, and body size.

Some of the air that enters the respiratory tract during breathing does not reach the alveoli. This volume (about 150 mL) remains in the passageways of the trachea,

Topic of Interest

Emphysema and Lung Cancer

Emphysema is a progressive, degenerative disease that destroys alveolar walls. As a result, clusters of small air sacs merge to form larger chambers, which drastically decreases the surface area of the respiratory membrane and thereby reduces the volume of gases that can be exchanged through the membrane. Alveolar walls lose some of their elasticity, and capillary networks associated with the alveoli diminish (fig. 16A).

Loss of tissue elasticity in the lungs makes it increasingly difficult for a person with emphysema to force air out, because normal expiration requires the passive elastic recoil of inflated tissues. Consequently, the person must exert abnormal muscular effort to exhale.

Emphysema may develop in response to prolonged exposure to respiratory irritants, such as those in tobacco smoke and polluted air. The disease may also result from an inherited enzyme deficiency.

Lung cancer, like other cancers, is the uncontrolled division of abnormal cells that rob normal cells of nutrients and oxygen, eventually crowding them out. Some cancerous growths in the lungs result secondarily from cancer cells that have spread (metastasized) from other parts of the body, such as the breasts, intestines, liver, or kidneys. Cancers that begin in the lungs are called *primary pulmonary cancers.* These may arise from epithelia, connective tissue, or blood cells. The most common form originates from epithelium in a bronchiole (fig. 16B) and is called *bronchogenic carcinoma.* This type of cancer is a response to irritation, such as prolonged exposure to tobacco smoke. Susceptibility to primary pulmonary cancers may be inherited.

Cancer cells divide to form tumor masses that obstruct air passages and reduce gas exchange. Bronchogenic carcinoma can spread quickly to the circulation, establishing secondary cancers in the lymph nodes, liver, bones, brain, or kidneys. Lung cancer is treated with surgery, ionizing radiation, and drugs, but the survival rate is low.

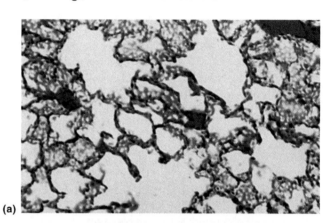
(a)

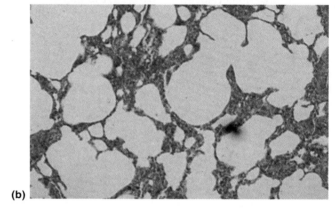
(b)

Figure 16A

Comparison of lung tissues. (*a*) Normal lung tissue. (*b*) As emphysema develops, alveoli merge, forming larger chambers (100×).

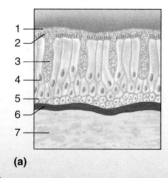
(a)

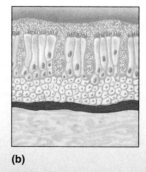
(b)

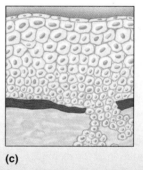
(c)

Figure 16B

About 95% of lung cancers start in the lining (epithelium) of a bronchiole. (*a*) The normal lining shows (*4*) columnar cells with (*2*) hairlike cilia, (*3*) goblet cells that secrete (*1*) mucus, and (*5*) basal cells from which new columnar cells arise. (*6*) A basement membrane separates the epithelial cells from (*7*) the underlying connective tissue. (*b*) In the first stage of lung cancer, the basal cells divide repeatedly. The goblet cells secrete excess mucus, and the cilia are less efficient in moving the heavy mucus secretion. (*c*) Continued division of basal cells displaces the columnar and goblet cells. The basal cells penetrate the basement membrane and invade the deeper connective tissue.

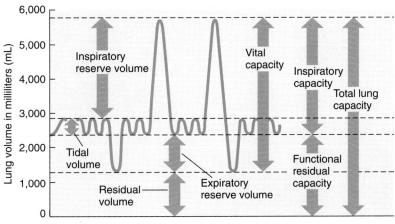

Figure 16.15

Respiratory volumes and capacities.

bronchi, and bronchioles. Because gas is not exchanged through the walls of these passages, this air is said to occupy *anatomic dead space*. Table 16.2 summarizes the respiratory air volumes and capacities.

An instrument called a *spirometer* measures respiratory air volumes, except residual volume, which requires a special technique. Such measurements are used to evaluate the courses of emphysema, pneumonia, and lung cancer, conditions in which functional lung tissue is lost. These measurements may also track the progress of diseases such as bronchial asthma that obstruct air passages.

Check Your Recall

18. What is tidal volume?
19. Distinguish between inspiratory and expiratory reserve volumes.
20. How is vital capacity determined?
21. How is total lung capacity calculated?

16.4 CONTROL OF BREATHING

Normal breathing is a rhythmic, involuntary act that continues even when a person is unconscious. The respiratory muscles, however, are also under voluntary control. (Take a deep breath and consider this!)

Respiratory Areas

Groups of neurons in the brainstem comprise the **respiratory areas,** which control both inspiration and expiration. The components of the respiratory areas are widely scattered throughout the pons and medulla oblongata. Two parts of the respiratory areas are of special interest: the respiratory center of the medulla and the respiratory group of the pons (fig. 16.16).

The **medullary respiratory center** includes two bilateral groups of neurons that extend throughout the length of the medulla oblongata. They are called the dorsal respiratory group and the ventral respiratory group.

The *dorsal respiratory group* is most important in stimulating the muscles of inspiration, primarily the

Table 16.2	Respiratory Air Volumes and Capacities	
Name	Volume*	Description
Tidal volume (TV)	500 mL	Volume moved in or out of lungs during respiratory cycle
Inspiratory reserve volume (IRV)	3,000 mL	Volume that can be inhaled during forced breathing in addition to tidal volume
Expiratory reserve volume (ERV)	1,100 mL	Volume that can be exhaled during forced breathing in addition to tidal volume
Residual volume (RV)	1,200 mL	Volume that remains in lungs even after maximal expiration
Inspiratory capacity (IC)	3,500 mL	Maximum volume of air that can be inhaled following exhalation of tidal volume: IC = TV + IRV
Functional residual capacity (FRC)	2,300 mL	Volume of air that remains in the lungs following exhalation of tidal volume: FRC = ERV + RV
Vital capacity (VC)	4,600 mL	Maximum volume of air that can be exhaled after taking the deepest breath possible: VC = TV + IRV + ERV
Total lung capacity (TLC)	5,800 mL	Total volume of air that the lungs can hold: TLC = VC + RV

*Values are typical for a tall, young adult.

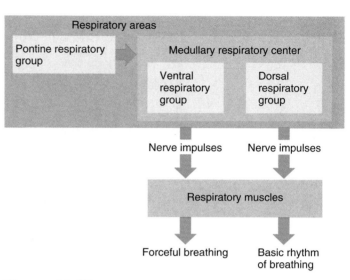

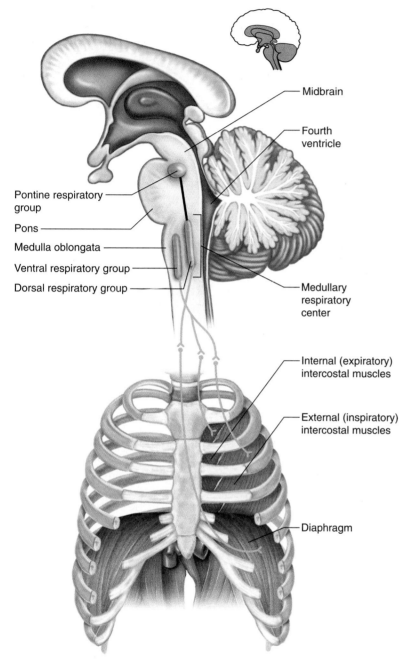

Figure 16.16

The respiratory areas are located in the pons and the medulla oblongata.

diaphragm. The more impulses they send, the more forceful the muscle contractions, and the greater the inspiration. When they stop sending impulses, the inspiratory muscles relax, and expiration occurs passively.

The *ventral respiratory group* is comprised of neurons that control other respiratory muscles, primarily the intercostals and abdominals. During more forceful breathing, some of these neurons increase inspiratory efforts, while others increase the force of expiration.

Neurons in another part of the brainstem, the pons, compose the pontine respiratory group (formerly the

Figure 16.17

The medullary respiratory center and the pontine respiratory group control breathing.

pneumotaxic center). These neurons make connections with the medullary rhythmicity center, and they may contribute to the basic rhythm of breathing (fig. 16.17).

Check Your Recall

22. Where are the respiratory areas?

23. Describe how the respiratory areas maintain a normal breathing pattern.

24. Explain how the breathing pattern may change.

Factors Affecting Breathing

The respiratory areas affect breathing rate and depth, and so do certain chemicals in body fluids, the degree to which lung tissues stretch, a person's emotional state, and level of physical activity (Topic of Interest, Exercise and Breathing, p. 459). For example, *chemosensitive areas* (central chemoreceptors), located in the ventral portion of the medulla oblongata near the origins of the vagus nerves, sense changes in the cerebrospinal fluid (CSF) levels of carbon dioxide and hydrogen ions. If either of these levels rises, the central chemoreceptors signal the respiratory areas, and respiratory rate and tidal volume increase. As a result of the increased ventilation, more carbon dioxide is exhaled, the blood and CSF levels of these chemicals fall, and breathing rate decreases.

Adding carbon dioxide to air can stimulate the rate and depth of breathing. Ordinary air is about 0.04% carbon dioxide. Inhaling air containing 4% carbon dioxide usually doubles the breathing rate.

Exercise and Breathing

Moderate to heavy physical exercise greatly increases the volume of oxygen the skeletal muscles use. For example, a young man at rest utilizes about 250 mL of oxygen per minute, but maximal exercise may require 3,600 mL of oxygen per minute.

As oxygen utilization increases, the volume of carbon dioxide produced also increases. Because decreased blood oxygen and increased blood carbon dioxide concentrations stimulate the respiratory areas, exercise increases breathing rate. Studies reveal, however, that blood oxygen and carbon dioxide concentrations usually do not change during exercise. This reflects the respiratory system's effectiveness in obtaining oxygen and releasing carbon dioxide to the outside.

The cerebral cortex and sensory structures called *proprioceptors* that are associated with muscles and joints cause much of the increased breathing rate during vigorous exercise. Specifically, whenever the cerebral cortex signals skeletal muscles to contract, it also transmits stimulating impulses to the respiratory areas. Muscular movements stimulate proprioceptors, triggering a *joint reflex* that sends impulses to the respiratory center, increasing breathing rate.

When breathing rate increases during exercise, increased blood flow is also required to power skeletal muscles. Thus, physical exercise taxes both the cardiovascular and respiratory systems. If either of these systems fails to keep pace with cellular demands, the person begins to feel out of breath. This feeling usually reflects an inability of the heart and blood vessels to move enough blood between the lungs and cells, rather than the respiratory system's inability to provide enough air.

Low blood oxygen has little direct effect on the central chemoreceptors associated with the respiratory areas. Instead, *peripheral chemoreceptors* in specialized structures called the *carotid bodies* and the *aortic bodies* sense changes in blood oxygen levels. Peripheral chemoreceptors are in the walls of certain large arteries (the carotid arteries and the aorta) in the neck and thorax (fig. 16.18). Stimulated peripheral chemoreceptors transmit impulses to the respiratory areas, increasing the breathing rate. However, blood oxygen levels must be very low to trigger this mechanism. Thus, oxygen plays only a minor role in the control of normal respiration.

An *inflation reflex* helps regulate the depth of breathing. This reflex occurs when stretched lung tissues stimulate stretch receptors in the visceral pleura, bronchioles, and alveoli. The sensory impulses of this reflex travel via the vagus nerves to the pontine respiratory group and shorten the duration of inspiratory movements. This action prevents overinflation of the lungs during forceful breathing.

Emotional upset can alter the normal breathing pattern. Fear and pain typically increase the breathing rate. Conscious control of breathing is also possible because the respiratory muscles are voluntary.

A person can voluntarily stop breathing for a very short time. If breathing stops, blood levels of carbon dioxide and hydrogen ions rise, and oxygen levels fall. These changes (primarily the increased carbon dioxide) stimulate the chemoreceptors, and soon the urge to inhale overpowers the desire to hold the breath.

A person can increase breath-holding time by breathing rapidly and deeply in advance. This action,

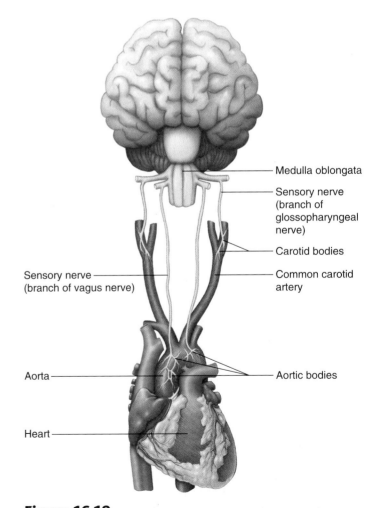

Figure 16.18

Decreased blood oxygen concentration stimulates peripheral chemoreceptors in the carotid and aortic bodies.

called **hyperventilation** (hi″per-ven″tĭ-la′shun), lowers the blood carbon dioxide level. Following hyperventilation, it takes longer for carbon dioxide to rise to a level that produces an overwhelming effect on the respiratory areas. (*Note:* Prolonging breath-holding in this way can cause abnormally low blood oxygen levels. Hyperventilation should never be used to help hold the breath while swimming because the person may lose consciousness underwater and drown.)

Interference with the oxygen supply to the brain causes fainting. A person who is emotionally upset may hyperventilate, become dizzy, and lose consciousness. This condition is due to a lowered carbon dioxide concentration followed by a rise in pH (respiratory alkalosis), a localized vasoconstriction of cerebral arterioles, and resulting decreased blood flow to nearby brain cells.

Check Your Recall

25. Which chemical factors affect breathing?
26. Describe the inflation reflex.
27. How does hyperventilation decrease the respiratory rate?

16.5 ALVEOLAR GAS EXCHANGES

The parts of the respiratory system discussed so far conduct air in and out of air passages. The alveoli carry on the vital process of exchanging gases between the air and the blood.

Alveoli

Alveoli are microscopic air sacs clustered at the distal ends of the narrowest respiratory tubes, the alveolar ducts (see fig. 16.8). Each alveolus consists of a tiny space within a thin wall that separates it from adjacent alveoli.

Respiratory Membrane

The wall of an alveolus consists of an inner lining of simple squamous epithelium. In close association with an alveolus is a dense network of capillaries, which are also lined with simple squamous epithelial cells. Thin, fused basement membranes separate the layers of these flattened cells, and in the spaces between the cells are elastic and collagenous fibers that support the alveolar wall. At least two thicknesses of epithelial cells and a layer of fused basement membranes separate the air in an alveolus from the blood in a capillary (fig. 16.19). These layers comprise the **respiratory membrane** (re-spi′rah-to″re mem′brān) across which blood and alveolar air exchange gases.

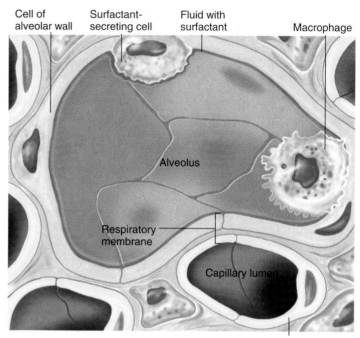

Figure 16.19

The respiratory membrane consists of the wall of the alveolus and the wall of the capillary.

If all of the capillaries that surround the alveoli were unwound and laid end to end, they would extend for about 620 miles.

Diffusion Across the Respiratory Membrane

Recall from chapter 3 (p. 61) that molecules diffuse from regions where they are in higher concentration toward regions where they are in lower concentration. For gases, it is more useful to think of diffusion occurring from regions of higher pressure toward regions of lower pressure. The pressure of a gas determines the rate at which it diffuses from one region to another.

Measured by volume, ordinary air is about 78% nitrogen, 21% oxygen, and 0.04% carbon dioxide. Air also has traces of other gases that have little or no physiological importance.

In a mixture of gases such as air, each gas accounts for a portion of the total pressure the mixture produces. The amount of pressure each gas contributes is called the **partial pressure** (par′shal presh′ur) of that gas and is proportional to its concentration. For example, because air is 21% oxygen, oxygen accounts for 21% of the atmospheric pressure (21% of 760 mm Hg), or 160 mm Hg. Thus, the partial pressure of oxygen, symbolized P_{O_2}, in atmospheric air is 160 mm Hg. Similarly, the partial pressure of carbon dioxide (P_{CO_2}) in air is 0.3 mm Hg.

Gas molecules from the air may enter, or dissolve in, a liquid. This is what happens when carbon dioxide is added to a carbonated beverage, or when inspired gases dissolve in the blood in the alveolar capillaries.

When a mixture of gases dissolves in blood, the resulting concentration of each gas is proportional to its partial pressure. Each gas diffuses between blood and its surroundings from areas of higher partial pressure to areas of lower partial pressure until the partial pressures in the two regions reach equilibrium. For example, the P_{CO_2} in capillary blood is 45 mm Hg, but the P_{CO_2} in alveolar air is 40 mm Hg. Because of the difference in these partial pressures, carbon dioxide diffuses from blood, where its partial pressure is higher, across the respiratory membrane and into alveolar air (fig. 16.20). When blood leaves the lungs, its P_{CO_2} is 40 mm Hg, which is the same as the P_{CO_2} of alveolar air. Similarly, the P_{O_2} of capillary blood is 40 mm Hg, but that of alveolar air is 104 mm Hg. Thus, oxygen diffuses from alveolar air into blood, and blood leaves the lungs with a P_{O_2} of 104 mm Hg. (Because of the large volume of air always in the lungs, as long as breathing continues, alveolar P_{O_2} stays relatively constant at 104 mm Hg.)

A number of factors affect diffusion across the respiratory membrane. More surface area, shorter distance, greater solubility of gases, and a steeper partial pressure gradient all favor increased diffusion. Thus, diseases that harm the respiratory membrane, such as pneumonia, or diseases that reduce the surface area for diffusion, such as emphysema, require increased P_{O_2} for treatment.

The respiratory membrane is normally so thin that certain soluble chemicals other than carbon dioxide may diffuse into alveolar air and be exhaled. This is why breath analysis can reveal alcohol in the blood or acetone on the breath of a person who has untreated diabetes mellitus. Breath analysis may also detect substances associated with kidney failure, certain digestive disturbances, and liver disease.

Check Your Recall

28. Describe the structure of the respiratory membrane.

29. What is the partial pressure of a gas?

30. What force moves oxygen and carbon dioxide across the respiratory membrane?

16.6 GAS TRANSPORT

Blood transports oxygen and carbon dioxide between the lungs and the cells. As these gases enter blood, they dissolve in the liquid portion (plasma) or combine chemically with blood components.

Oxygen Transport

Almost all the oxygen (over 98%) that blood transports binds the iron-containing protein **hemoglobin** (he″mo-glo′bin) in red blood cells. The remainder of the oxygen dissolves in plasma.

In the lungs, where the P_{O_2} is relatively high, oxygen dissolves in blood and combines rapidly with the iron atoms of hemoglobin, forming **oxyhemoglobin** (ok″sĭ-he″mo-glo′bin) (fig. 16.21*a*). The chemical bonds between oxygen and hemoglobin molecules are unstable, and as the P_{O_2} decreases, oxyhemoglobin molecules release oxygen, which diffuses into nearby cells that have depleted their oxygen supplies in cellular respiration (fig. 16.21*b*).

Several other factors affect how much oxygen oxyhemoglobin releases. More oxygen is released as the blood concentration of carbon dioxide increases, as blood becomes more acidic, or as blood temperature increases.

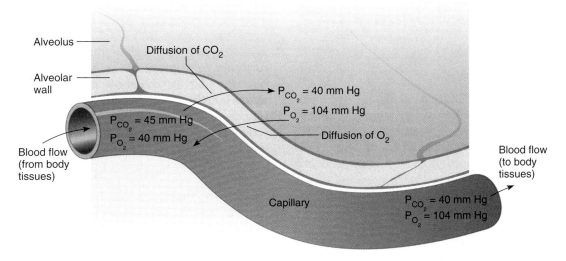

Figure 16.20

Gases are exchanged between alveolar air and capillary blood because of differences in partial pressures.

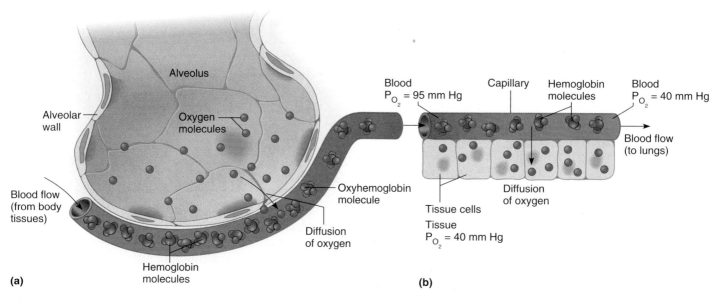

Figure 16.21

Blood transports oxygen. (*a*) Oxygen molecules, entering blood from the alveolus, bond to hemoglobin, forming oxyhemoglobin. (*b*) Near cells, oxyhemoglobin releases oxygen. Note that much oxygen is still bound to hemoglobin at the P_{O_2} of systemic venous blood.

This explains why more oxygen is released to skeletal muscles during physical exercise. The increased muscular activity and oxygen utilization increase carbon dioxide concentration, decrease pH, and raise temperature. Less active cells receive proportionately less oxygen.

A deficiency of O_2 reaching the tissues is called **hypoxia.** It may occur because of decreased arterial P_{O_2} (*hypoxemia*), diminished ability of the blood to transport O_2 (anemic hypoxia), inadequate blood flow (ischemic hypoxia), or a defect at the cellular level (histotoxic hypoxia), such as in cyanide poisoning.

Check Your Recall

31. How is oxygen transported from the lungs to cells?
32. What stimulates blood to release oxygen to tissues?

Carbon Dioxide Transport

Blood flowing through capillaries gains carbon dioxide because tissues have a relatively high P_{CO_2}. Blood transports carbon dioxide to the lungs in one of three forms: as carbon dioxide dissolved in plasma, as part of a compound formed by bonding to hemoglobin, or in the form of a bicarbonate ion (fig. 16.22).

The amount of carbon dioxide that dissolves in plasma is determined by its partial pressure. The higher the P_{CO_2} of the tissues, the more carbon dioxide will go into solution. However, only about 7% of the carbon dioxide that blood transports is in this form.

Unlike oxygen, which binds the iron atoms (part of the "heme" portion) of hemoglobin molecules, carbon dioxide bonds with the amino groups ($-NH_2$) of the "globin" or protein portion of these molecules. Consequently, oxygen and carbon dioxide do not compete for binding sites, and a hemoglobin molecule can transport both gases at the same time.

Carbon dioxide loosely bonds with hemoglobin, forming **carbaminohemoglobin** (kar-bam″ĭ-no he″mo-glo′bin). This molecule decomposes readily in regions of low P_{CO_2}, releasing its carbon dioxide. Transporting carbon dioxide this way is theoretically quite effective, but carbaminohemoglobin forms slowly. Only about 23% of the carbon dioxide that blood transports is in this form.

The most important carbon dioxide transport mechanism forms **bicarbonate ions** (HCO_3^-). Carbon dioxide reacts with water to form carbonic acid (H_2CO_3):

$$CO_2 + H_2O \rightarrow H_2CO_3$$

This reaction occurs slowly in plasma, but much of the carbon dioxide diffuses into red blood cells. These cells have the enzyme **carbonic anhydrase** (kar-bon′ik an-hi′drās), which speeds the reaction between carbon dioxide and water. The resulting carbonic acid then dissociates, releasing hydrogen ions (H^+) and bicarbonate ions (HCO_3^-):

$$H_2CO_3 \rightarrow H^+ + HCO_3^-$$

Most of the hydrogen ions bind hemoglobin molecules quickly, and thus do not accumulate and greatly change blood pH. The bicarbonate ions diffuse out of red blood cells and enter the plasma. Nearly 70% of the carbon dioxide that blood transports is in this form.

When blood passes through the capillaries of the lungs, its dissolved carbon dioxide diffuses into alve-

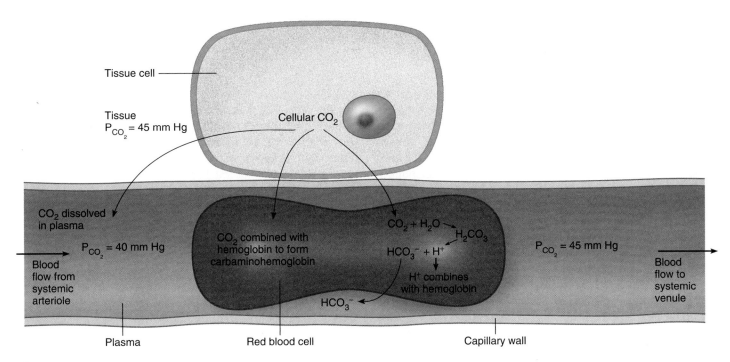

Figure 16.22
Carbon dioxide produced by cells is transported in the blood plasma in a dissolved state, bound to hemoglobin, or in the form of bicarbonate ions (HCO_3^-).

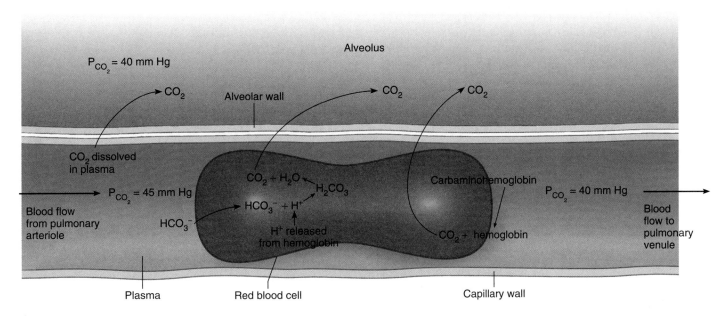

Figure 16.23
In the lungs, carbon dioxide diffuses from the blood into the alveoli.

oli in response to the relatively low P_{CO_2} of alveolar air (fig. 16.23). At the same time, hydrogen ions and bicarbonate ions in red blood cells recombine to form carbonic acid, and under the influence of carbonic anhydrase, the carbonic acid quickly breaks down to yield carbon dioxide and water:

$$H^+ + HCO_3^- \rightarrow H_2CO_3 \rightarrow CO_2 + H_2O$$

Carbaminohemoglobin also releases its carbon dioxide, and carbon dioxide continues to diffuse out of the blood until the P_{CO_2} of the blood and that of alveolar air are in equilibrium. Table 16.3 summarizes transport of blood gases.

Genetics Connection

Cystic Fibrosis

"Woe to that child which when kissed on the forehead tastes salty. He is bewitched and soon must die." So went a seventeenth-century British saying about a child with cystic fibrosis (CF). Salty skin is still one of the first symptoms that parents of a child with CF notice. The disease, inherited from two carrier parents, affects about 30,000 people in the United States and 70,000 worldwide. It isn't known how many people have mild forms of the disease, merely with symptoms of frequent respiratory infection.

In 1938, physicians first described CF as a defect in channels leading from certain glands. This causes formation of extremely thick, sticky mucus, which encourages infections by microorganisms not otherwise common in the lungs. A clogged pancreas prevents digestive juices from reaching the intestines and thus impairs absorption of nutrients. Affected children are subject to frequent respiratory infections, and in the past were often diagnosed with "failure to thrive."

In the 1930s, life expectancy for a child with CF was five years, but by 1960 it became possible to treat the symptoms. Antibiotics control the respiratory infections, and daily "postural drainage" exercises shake the stifling mucus free from the lungs—a parent catches the expelled mucus in a jar. Digestive enzymes mixed into soft foods enhance nutrient absorption, helping the person gain weight.

In 1989, geneticists discovered the gene and protein defect behind CF, allowing development of more targeted treatments. The gene encodes a protein called the "cystic fibrosis transmembrane regulator," or CFTR for short. It is an ion channel that controls chloride transport out of cells, but also controls flow of water, ATP, and sodium in ways that are still not well understood. In most people with CF, the chloride channel is missing one amino acid, and is so deformed that it never reaches the cell's surface to function. With chloride ions unable to leave the cell, water moves in by osmosis, drying out the mucus. Organs become clogged.

By the early 1990s, new drugs to treat CF became available. Some drugs allowed more sodium to enter or chloride to leave the cells lining the lungs. A natural enzyme, deoxyribonuclease, used as a drug degrades the DNA that accumulates in infected lungs as white blood cells cause inflammation. Several experimental gene therapies can introduce functional CFTR genes into affected cells, but so far these treatments work only for short times or restore function to only a small part of the respiratory tract.

Even though CF cannot yet be cured, the various treatments have been so successful that life expectancy has increased tenfold since 1940. Researchers in the United Kingdom tracked individuals born with CF from 1968 to 1995, consulting a national registry and death certificates. They project that children born with CF since 2000 can expect to live, on average, at least 50 years.

Table 16.3	Gases Transported in Blood	
Gas	**Reaction Involved**	**Substance Transported**
Oxygen	1–2% dissolves in plasma; 98–99% combines with iron atoms of hemoglobin molecules	Oxyhemoglobin
Carbon dioxide	About 7% dissolves in plasma	Carbon dioxide
	About 23% combines with amino groups of hemoglobin molecules	Carbaminohemoglobin
	About 70% reacts with water to form carbonic acid; the carbonic acid then dissociates to release hydrogen ions and bicarbonate ions	Bicarbonate ions

The normal percentage of hemoglobin molecules that bind carbon monoxide (CO) in people who do not smoke is 2%. In people with CO poisoning, levels may exceed 20%. The increased CO binding prevents oxygen from binding, starving tissues of oxygen and causing symptoms of chest pain, shortness of breath, fatigue, confusion, an irregular pulse, and abnormal heart rhythm.

Check Your Recall

33. Describe three forms in which blood can transport carbon dioxide from cells to the lungs.

34. How can hemoglobin carry oxygen and carbon dioxide at the same time?

35. How is carbon dioxide released from blood into the lungs?

Respiratory System

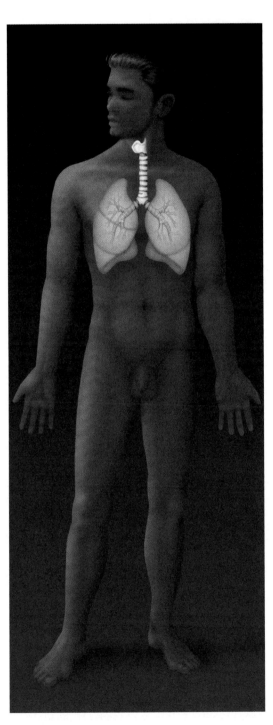

Integumentary System

Stimulation of skin receptors may alter respiratory rate.

Skeletal System

Bones provide attachments for muscles involved in breathing.

Muscular System

The respiratory system eliminates carbon dioxide produced by exercising muscles.

Nervous System

The brain controls the respiratory system. The respiratory system helps control pH of the internal environment.

Endocrine System

Hormonelike substances control the production of red blood cells that transport oxygen and carbon dioxide.

Cardiovascular System

As the heart pumps blood through the lungs, the lungs oxygenate the blood and excrete carbon dioxide.

Lymphatic System

Cells of the immune system patrol the lungs and defend against infection.

Digestive System

The digestive system and respiratory system share openings to the outside.

Urinary System

The kidneys and the respiratory system work together to maintain blood pH. The kidneys compensate for water lost through breathing.

Reproductive System

Respiration increases during sexual activity. Fetal gas exchange begins before birth.

The respiratory system provides oxygen for the internal environment and excretes carbon dioxide.

Clinical Terms Related to the Respiratory System

anoxia (ah-nok′se-ah) Absence or deficiency of oxygen within tissues.

apnea (ap-ne′ah) Temporary cessation of breathing.

asphyxia (as-fik′se-ah) Oxygen deficiency and excess carbon dioxide in blood and tissues.

atelectasis (at″e-lek′tah-sis) Collapse of a lung or part of a lung.

bradypnea (brad″e-ne′ah) Abnormally slow breathing.

bronchiolectasis (brong″ke-o-lek′tah-sis) Chronic dilation of the bronchioles.

bronchitis (brong-ki′tis) Inflammation of the bronchial lining.

Cheyne-Stokes respiration (chān stōks res″pĭ-ra′shun) Irregular breathing consisting of a series of shallow breaths that increase in depth and rate, followed by breaths that decrease in depth and rate.

dyspnea (disp′ne-ah) Difficulty breathing.

eupnea (up-ne′ah) Normal breathing.

hemothorax (he″mo-tho′raks) Blood in the pleural cavity.

hypercapnia (hi″per-kap′ne-ah) Excess carbon dioxide in the blood.

hyperoxia (hi″per-ok′se-ah) Excess oxygen in the blood.

hyperpnea (hi″perp-ne′ah) Increase in the depth and rate of breathing.

hyperventilation (hi″per-ven″tĭ-la′shun) Prolonged, rapid, and deep breathing.

hypoxemia (hi″pok-se′me-ah) Deficiency in blood oxygenation.

hypoxia (hi-pok′se-ah) Diminished availability of oxygen in tissues.

lobar pneumonia (lo′ber nu-mo′ne-ah) Pneumonia that affects an entire lung lobe.

pleurisy (ploo′rĭ-se) Inflammation of the pleural membranes.

pneumoconiosis (nu″mo-ko″ne-o′sis) Accumulation of particles from the environment in the lungs and the reaction of tissues to them.

pneumothorax (nu″mo-tho′raks) Entrance of air into the space between the pleural membranes, followed by lung collapse.

rhinitis (ri-ni′tis) Inflammation of the nasal cavity lining.

sinusitis (si″nŭ-si′tis) Inflammation of the sinus cavity lining.

tachypnea (tak″ip-ne′ah) Rapid, shallow breathing.

tracheotomy (tra″ke-ot′o-me) Incision in the trachea for exploration or for removal of a foreign object.

Clinical Connection

Emphysema can make routine tasks such as shopping or walking a dog impossible. One treatment is to remove the affected portions of the lungs. Supposedly, the remaining lung tissue regains some elastic recoil, the diaphragm returns to a normal shape, and lung function improves. The technique, called "lung volume reduction surgery," was developed in the 1970s but was not widely performed until the 1990s. It is helpful for only some patients, and may actually harm certain others.

Of the 700 lung volume reduction surgeries performed in the mid-1990s, many patients never left the intensive care unit afterwards. So the National Institutes of Health and the Health Care Financing Administration called for controlled clinical trials of the procedure, comparing it to standard medical management for emphysema. Results from the National Emphysema Treatment Trial indicate that the procedure is most likely to help people with severe disease in the upper lobes who can withstand surgery. For patients with emphysema widespread throughout the lungs, very low forced expiratory volume, or hampered ability to get rid of carbon monoxide—about one in eight individuals requesting the procedure—lung volume reduction surgery causes more harm than good. The procedure is valuable only if physicians carefully screen candidates.

SUMMARY OUTLINE

16.1 Introduction (p. 445)
The respiratory system includes tubes that remove particles from incoming air and transport air to and from the lungs and the air sacs where gases are exchanged. Respiration is the entire process of gas exchange between the atmosphere and body cells.

16.2 Organs of the Respiratory System (p. 445)
The organs of the respiratory system can be divided into two groups. The upper respiratory tract includes the nose, nasal cavity, paranasal sinuses, and pharynx; the lower respiratory tract includes the larynx, trachea, bronchial tree, and lungs.

1. Nose
   a. Bone and cartilage support the nose.
   b. The nostrils are openings for air.
2. Nasal cavity
   a. Nasal conchae divide the nasal cavity into passageways and help increase the surface area of the mucous membrane.
   b. The mucous membrane filters, warms, and moistens incoming air.
   c. Ciliary action carries particles trapped in mucus to the pharynx, where they are swallowed.
3. Paranasal sinuses
   a. The paranasal sinuses are spaces in the bones of the skull that open into the nasal cavity.
   b. Mucous membrane lines the sinuses.
4. Pharynx
   a. The pharynx is behind the nasal cavity, oral cavity, and larynx.
   b. It is a passageway for air and food.
5. Larynx
   a. The larynx conducts air and helps prevent foreign objects from entering the trachea.
   b. It is composed of muscles and cartilages and is lined with mucous membrane.
   c. The larynx contains the vocal cords, which vibrate from side to side and produce sounds when air passes between them.
   d. The glottis and epiglottis help prevent foods and liquids from entering the trachea.
6. Trachea
   a. The trachea extends into the thoracic cavity anterior to the esophagus.
   b. It divides into right and left bronchi.

7. Bronchial tree
   a. The bronchial tree consists of branched air passages that lead from the trachea to the air sacs.
   b. Alveoli are at the distal ends of the narrowest tubes, the alveolar ducts.
8. Lungs
   a. The mediastinum separates the left and right lungs, and the diaphragm and thoracic cage enclose them.
   b. The visceral pleura attaches to the surface of the lungs. The parietal pleura lines the thoracic cavity.
   c. Each lobe of the lungs is composed of alveoli, blood vessels, and supporting tissues.

16.3 Breathing Mechanism (p. 452)

Changes in the size of the thoracic cavity accompany inspiration and expiration.

1. Inspiration
   a. Atmospheric pressure forces air into the lungs.
   b. Inspiration occurs when the pressure inside alveoli decreases.
   c. Pressure within alveoli decreases when the diaphragm moves downward and the thoracic cage moves upward and outward.
   d. Surface tension aids lung expansion.
2. Expiration
   a. Elastic recoil of tissues and surface tension within alveoli provide the forces of expiration.
   b. Thoracic and abdominal wall muscles aid expiration.
3. Respiratory air volumes and capacities
   a. One inspiration followed by one expiration is a respiratory cycle.
   b. The amount of air that moves in (or out) during a single respiratory cycle is the tidal volume.
   c. Additional air that can be inhaled is the inspiratory reserve volume. Additional air that can be exhaled is the expiratory reserve volume.
   d. Residual volume remains in the lungs after a maximal expiration.
   e. The vital capacity is the maximum amount of air a person can exhale after taking the deepest breath possible.
   f. The inspiratory capacity is the maximum volume of air a person can inhale following exhalation of the tidal volume.
   g. The functional residual capacity is the volume of air that remains in the lungs after a person exhales the tidal volume.
   h. The total lung capacity equals the vital capacity plus the residual volume.

16.4 Control of Breathing (p. 457)

Normal breathing is rhythmic and involuntary.

1. Respiratory areas
   a. The respiratory areas are in the brainstem and include portions of the medulla oblongata and pons.
   b. The medullary respiratory center includes two groups of neurons.
      (1) The dorsal respiratory group controls the basic rhythm of breathing.
      (2) The ventral respiratory group increases inspiratory and expiratory movements during forceful breathing.
   c. The pontine respiratory group contributes to the breathing rate.
2. Factors affecting breathing
   a. Chemicals, stretching of lung tissues, emotional state, and exercise affect breathing.
   b. Chemosensitive areas (central chemoreceptors) are associated with the respiratory center.

(1) Blood levels of carbon dioxide and hydrogen ions influence the central chemoreceptors.
(2) Stimulation of these receptors increases breathing rate.
   c. Peripheral chemoreceptors are in the walls of certain large arteries.
      (1) These chemoreceptors sense low oxygen levels.
      (2) When oxygen levels are low, breathing rate increases.
   d. Overstretching lung tissues triggers an inflation reflex.
      (1) This reflex shortens the duration of inspiratory movements.
      (2) The inflation reflex prevents overinflation of the lungs during forceful breathing.
   e. Hyperventilation decreases blood carbon dioxide levels, but *this is very dangerous when done before swimming underwater.*

16.5 Alveolar Gas Exchanges (p. 460)

Gas exchange between air and blood occurs in alveoli.

1. Alveoli
   Alveoli are tiny air sacs clustered at the distal ends of alveolar ducts.
2. Respiratory membrane
   a. This membrane consists of alveolar and capillary walls.
   b. Blood and alveolar air exchange gases across this membrane.
3. Diffusion across the respiratory membrane
   a. The partial pressure of a gas is proportional to the concentration of that gas in a mixture or the concentration dissolved in a liquid.
   b. Gases diffuse from regions of higher partial pressure toward regions of lower partial pressure.
   c. Carbon dioxide diffuses from blood into alveolar air. Oxygen diffuses from alveolar air into blood.

16.6 Gas Transport (p. 461)

Blood transports gases between the lungs and cells.

1. Oxygen transport
   a. Blood mainly transports oxygen in combination with hemoglobin molecules.
   b. The resulting oxyhemoglobin is unstable and releases its oxygen in regions where the P_{O_2} is low.
   c. More oxygen is released as the plasma P_{CO_2} increases, as blood becomes more acidic, and as blood temperature increases.
2. Carbon dioxide transport
   a. Carbon dioxide may be carried in solution, bound to hemoglobin, or as a bicarbonate ion.
   b. Most carbon dioxide is transported in the form of bicarbonate ions.
   c. The enzyme carbonic anhydrase speeds the reaction between carbon dioxide and water to form carbonic acid.
   d. Carbonic acid dissociates to release hydrogen ions and bicarbonate ions.

CHAPTER ASSESSMENTS

16.1 Introduction

1. List the general functions of the respiratory system. (p. 445)

16.2 Organs of the Respiratory System

2. Which one of the following is the beginning of the lower respiratory tract? (p. 445)

   a. nostril   c. pharynx

   b. nasal cavity   d. larynx

3. Explain how the nose and nasal cavity filter the incoming air. (p. 445)

4. Identify the locations of the major paranasal sinuses. (p. 446)

5. Match the following structures with their descriptions: (p. 447–451)

   (1) true vocal cords   A. Serous membrane on lungs

   (2) false vocal cords   B. Contains the vocal cords

   (3) larynx   C. Vibrate to make sound

   (4) visceral pleura   D. Air sacs

   (5) alveoli   E. Muscular folds

6. Name and describe the locations of the larger cartilages of the larynx. (p. 447)

16.3 Breathing Mechanism

7. Explain how inspiration and expiration depend on pressure changes. (p. 452)

8. Compare the muscles used in a resting inspiration with those in a forced inspiration. (p. 452)

9. Define *surface tension* and explain how it aids breathing. (p. 453)

10. Define *surfactant* and explain its function. (p. 453)

11. Compare the muscles used (if any) in a resting expiration with those in a forced expiration. (p. 453)

12. Distinguish between the vital capacity and the total lung capacity. (p. 455)

16.4 Control of Breathing

13. Describe the location of the respiratory areas and name the major components. (p. 457)

14. Which one of the following is most important in forceful breathing? (p. 457)

    a. dorsal respiratory group

    b. ventral respiratory group

    c. pontine respiratory group

15. Describe the function of the chemoreceptors in the carotid and aortic bodies. (p. 459)

16. Describe the inflation reflex. (p. 459)

17. Hyperventilation is which one of the following? (p. 459)

    a. any increase in breathing

    b. an increase in breathing that brings in oxygen too quickly

    c. an increase in breathing that eliminates carbon dioxide too quickly

    d. an increase in breathing that has no effect on blood gases

16.5 Alveolar Gas Exchanges

18. Define *respiratory membrane* and explain its function. (p. 460)

19. Explain the relationship between the partial pressure of a gas and diffusion of that gas. (p. 461)

20. Summarize the exchange of oxygen and carbon dioxide across the respiratory membrane. (p. 461)

16.6 Gas Transport

21. Identify how blood transports oxygen. (p. 461)

22. List three factors that increase the release of oxygen from hemoglobin. (p. 461)

23. Identify the three ways blood transports carbon dioxide. (p. 462)

INTEGRATIVE ASSESSMENTS/ CRITICAL THINKING

➡ OUTCOME 16.2

1. Why does breathing through the mouth dry out the throat?

➡ OUTCOMES 16.2, 16.3

2. It is below 0°F outside, but the dedicated runner bundles up and hits the road anyway. "You're crazy," shouts a neighbor. "Your lungs will freeze." Why is the well-meaning neighbor wrong?

➡ OUTCOMES 16.2, 16.3, 16.4

3. Emphysema reduces the lungs' capacity to recoil elastically. Which respiratory air volumes does emphysema affect?

➡ OUTCOMES 16.2, 16.3, 16.4, 16.5

4. George Washington went for a walk in the freezing rain on a bleak December day in 1799. The next day, he had trouble breathing and swallowing. A doctor suggested cutting a hole in the president's throat so he could breathe, but other doctors voted him down, instead bleeding the patient, plastering his throat with bran and honey, and placing beetles on his legs to produce blisters. Soon, Washington's voice became muffled, his breathing was more labored, and he grew restless. For a short time, he seemed euphoric; then he died. Washington had epiglottitis, in which the epiglottis swells to ten times its normal size. How does this diagnosis explain his symptoms? Which suggested treatment might have worked?

➡ OUTCOMES 16.3, 16.4, 16.5

5. When a woman is very close to delivering a baby, she may hyperventilate. Breathing into a paper bag regulates her breathing. How does this action return her breathing to normal?

➡ OUTCOMES 16.3, 16.4, 16.6

6. Why can you not commit suicide by holding your breath?

➡ OUTCOMES 16.3, 16.5, 16.6

7. What changes would you expect in the relative concentrations of blood oxygen in a patient who breathes rapidly and deeply for a prolonged time? Why?

➡ OUTCOMES 16.4, 16.5

8. If a person has stopped breathing and is receiving pulmonary resuscitation, would it be better to administer pure oxygen or a mixture of oxygen and carbon dioxide? Why?

➡ OUTCOMES 16.5, 16.6

9. Why were the finishing times of endurance events rather slow at the 1968 Olympics, held in 2,200-meter-high Mexico City?

WEB CONNECTIONS

Visit the text website at **aris.mhhe.com** for additional quizzes, interactive learning exercises, and more.

AP R RESPIRATORY SYSTEM ●

Anatomy & Physiology | REVEALED includes cadaver photos that allow you to peel away layers of the human body to reveal structures beneath the surface. This program also includes animations, radiologic imaging, audio pronunciation, and practice quizzing. To learn more visit **www.aprevealed.com**.

17

Urinary System

HEMOLYTIC UREMIC SYNDROME. In late summer of 2006, spinach went from being a health food to being a source of food poisoning. Within a few weeks 199 people became infected with *Escherichia coli* strain 0157:H7 from eating raw spinach. Among them was a 2-year-old, named Kyle, whose mother had mixed spinach into a fruit smoothie. Others had eaten spinach in sandwiches or salads. As the number of poisoned people grew, many commented on the irony of becoming ill from such a healthy food.

E. coli is a common bacterium that can cause severe illness when it produces a poison called shigatoxin. Food poisoning from toxin-producing *E. coli* begins with sharp abdominal pain and bloody diarrhea. Intensifying pain sends many victims to hospitals. For 16% of them, the condition worsens to hemolytic uremic syndrome (HUS), which develops when the bloodstream transports the toxin to the kidneys, where the toxin destroys the microscopic capillaries that normally filter proteins and blood cells from forming urine. With the capillaries compromised, proteins and blood cells, as well as damaged kidney cells, appear in the urine. HUS causes acute kidney failure. Often blood clots around the sites of the damaged kidney cells, and new cells can form—after weeks of hospitalization, the person recovers, although kidney damage may be permanent. For those not so lucky, such as 2-year-old Kyle, HUS is deadly.

Past *E. coli* outbreaks were associated with eating undercooked hamburger, drinking unprocessed apple cider, or exposure to the bacteria at petting zoos or country fairs. All of these infection routes came from a single type of source—manure. Epidemiologists used DNA

Spinach tainted with Escherichia coli *strain 0157:H7 caused food poisoning in at least 199 people in late summer, 2006. Several died of hemolytic uremic syndrome.*

tests to trace the spinach that had sickened people in 22 states to a single facility in California that had processed spinach contaminated with runoff from a nearby cattle ranch. Tainted cider was made from apples that had dropped and touched manure. People who became ill after visiting petting zoos and county fairs had ingested the bacteria from not washing their hands after close contact with the animals and their excrement. Most cases of *E. coli* poisoning, however, are caused by toxin-tainted hamburger. The illness was first recognized in 1982 and was attributed to practices that began at about that time—raising cattle on huge ranches, then mixing together the meat from thousands of animals. More than 400 outbreaks have occurred since then.

Learning Outcomes   *After studying this chapter, you should be able to do the following:*

17.1 Introduction

1. List the general functions of the organs of the urinary system. (p. 470)

17.2 Kidneys

2. Describe the locations and structure of the kidneys. (p. 470)

3. List the functions of the kidneys. (p. 471)

4. Trace the pathway of blood through the major vessels within a kidney. (p. 471)

5. Describe a nephron, and explain the functions of its major parts. (p. 472)

17.3 Urine Formation

6. Explain how glomerular filtrate is produced, and describe its composition. (p. 474)

7. Explain the factors that affect the rate of glomerular filtration and how this rate is regulated. (p. 477)

8. Discuss the role of tubular reabsorption in urine formation. (p. 479)

9. Define tubular secretion, and explain its role in urine formation. (p. 481)

17.4 Urine Elimination

10. Describe the structure of the ureters, urinary bladder, and urethra. (p. 483)

11. Explain the process and control of micturition. (p. 485)

Aids to Understanding Words   *(Appendix A on page 567 has a complete list of Aids to Understanding Words.)*

calyc- [small cup] major *calyc*es: Cuplike divisions of the renal pelvis.

cort- [covering] renal *cort*ex: Shell of tissues surrounding the inner kidney.

detrus- [to force away] *detrus*or muscle: Muscle within the bladder wall that expels urine.

glom- [little ball] *glom*erulus: Cluster of capillaries within a renal corpuscle.

mict- [to pass urine] *mict*urition: Process of expelling urine from the bladder.

nephr- [pertaining to the kidney] *nephr*on: Functional unit of a kidney.

papill- [nipple] renal *papill*ae: Small elevations that project into a renal calyx.

trigon- [triangle] *trigon*e: Triangular area on the internal floor of the urinary bladder.

17.1 INTRODUCTION

Cells produce a variety of wastes that are toxic if they accumulate. Body fluids, such as blood and lymph, carry wastes from the tissues that produce them, while other structures remove wastes from the blood and transport them to the outside. The respiratory system removes carbon dioxide from the blood, and the *urinary system* removes certain salts and nitrogenous wastes. The urinary system also helps maintain the normal concentrations of water and electrolytes in body fluids, regulates the pH and volume of body fluids, and helps control red blood cell production and blood pressure.

The urinary system consists of a pair of kidneys, which remove substances from blood, form urine, and help regulate certain metabolic processes; a pair of tubular ureters, which transport urine from the kidneys; a saclike urinary bladder, which stores urine; and a tubular urethra, which conveys urine to the outside of the body. Figure 17.1 and reference plate 6 (p. 28) show these organs.

17.2 KIDNEYS

A **kidney** is a reddish-brown, bean-shaped organ with a smooth surface. An adult kidney is about 12 centimeters long, 6 centimeters wide, and 3 centimeters thick, and is enclosed in a tough, fibrous capsule (fig. 17.2).

Location of the Kidneys

The kidneys lie on either side of the vertebral column in a depression high on the posterior wall of the abdominal cavity. The upper and lower borders of the kidneys are generally at the levels of the twelfth thoracic and third lumbar vertebrae, respectively. The left kidney is usually 1.5–2.0 centimeters higher than the right one.

The kidneys are positioned **retroperitoneally** (ret″ro-per″ĭ-to-ne′ale), which means they are behind the parietal peritoneum and against the deep muscles of the back. Connective tissue and masses of adipose tissue surround the kidneys and hold them in position (see fig. 1.11, p. 11).

Kidney Structure

The lateral surface of each kidney is convex, but its medial side is deeply concave. The resulting medial depression leads into a hollow chamber called the **renal sinus.** The entrance to this sinus is the *hilum,* and through it pass blood vessels, nerves, lymphatic vessels, and the ureter (see fig. 17.1).

The superior end of the ureter expands to form a funnel-shaped sac called the **renal pelvis** (re′nal pel′vis) inside the renal sinus. The pelvis is subdivided into two or three tubes, called *major calyces* (singular, *calyx*), and these in turn are subdivided into several *minor calyces* (fig. 17.2a).

A series of small elevations called *renal papillae* project into the renal sinus from its wall. Tiny openings that lead into a minor calyx pierce each projection.

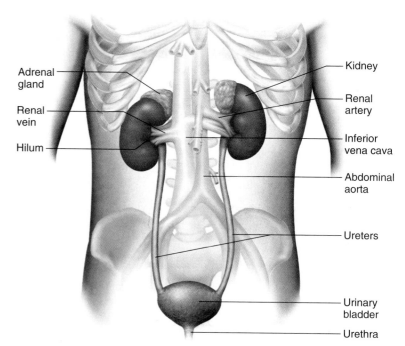

Figure 17.1

The urinary system includes the kidneys, ureters, urinary bladder, and urethra. Note the relationship of these structures to the major blood vessels.

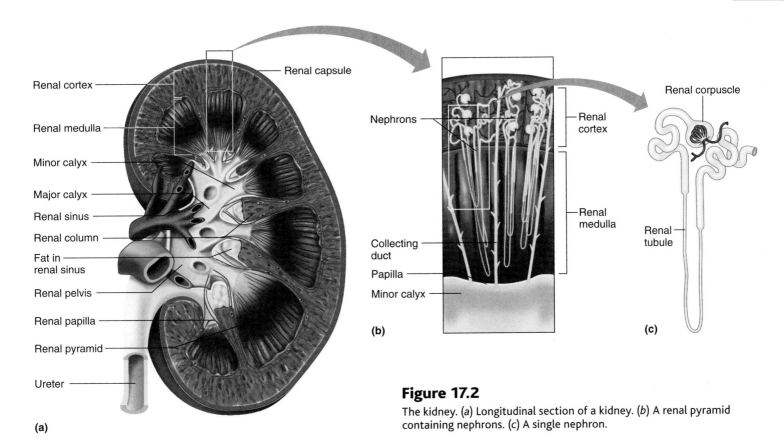

Figure 17.2

The kidney. (*a*) Longitudinal section of a kidney. (*b*) A renal pyramid containing nephrons. (*c*) A single nephron.

Each kidney has two distinct regions—an inner medulla and an outer cortex. The **renal medulla** (re′nal mĕ-dul′ah) is composed of conical masses of tissue called *renal pyramids* and appears striated. The **renal cortex** (re′nal kor′teks) forms a shell around the medulla and dips into the medulla between renal pyramids, forming *renal columns*. The granular appearance of the cortex is due to the random organization of tiny tubules associated with the **nephrons** (nef′ronz), which are the kidney's functional units (fig. 17.2*b,c*).

Check Your Recall

1. Where are the kidneys located?
2. Describe kidney structure.
3. Name the kidney's functional unit.

Kidney Functions

The primary function of the kidneys is to help maintain homeostasis by regulating the composition, volume, and the pH of the extracellular fluid. They accomplish this by removing metabolic wastes from the blood and diluting them with water and electrolytes to form urine, which they then excrete.

The kidneys have several other important functions:

- Secreting the hormone erythropoietin (see chapter 12, p. 320) to help control the rate of red blood cell production.
- Playing a role in the activation of vitamin D.
- Helping to maintain blood volume and blood pressure by secreting the enzyme renin.

Renal Blood Vessels

The **renal arteries,** which arise from the abdominal aorta, supply blood to the kidneys. These arteries transport a large volume of blood. When a person is at rest, the renal arteries usually carry 15–30% of the total cardiac output into the kidneys.

A renal artery enters a kidney through the hilum and gives off several branches, called *interlobar arteries,* which pass between the renal pyramids. At the junction between the medulla and the cortex, the interlobar arteries branch, forming a series of incomplete arches, the *arcuate arteries,* which in turn give rise to *interlobular arteries.* The final branches of the interlobular arteries, called **afferent arterioles** (af′er-ent ar-te′re-ōlz), lead to the nephrons (figs. 17.3 and 17.4).

Venous blood returns through a series of vessels that correspond generally to arterial pathways. The **renal vein** then joins the inferior vena cava as it courses through the abdominal cavity (see fig. 17.1).

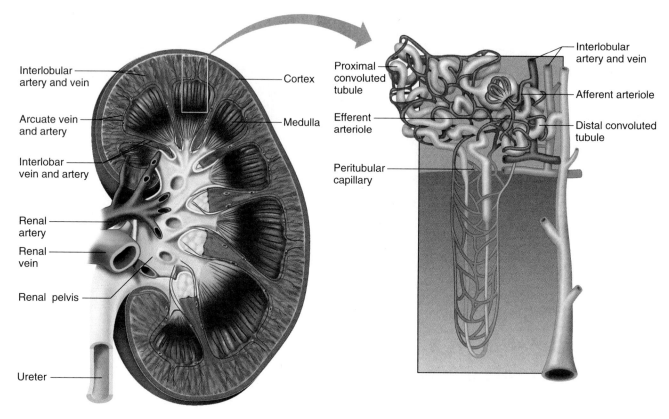

Figure 17.3

Main branches of the renal artery and renal vein.

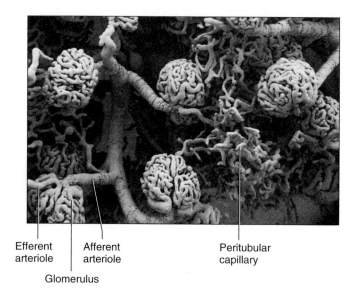

Figure 17.4

Scanning electron micrograph of a cast of the renal blood vessels associated with glomeruli (200×). From *Tissues and Organs: A Text-Atlas of Scanning Electron Microscopy,* by R. G. Kessel and R. H. Kardon, © 1979 W. H. Freeman and Company, all rights reserved.

A kidney transplant can help patients with end-stage renal disease. This procedure requires a kidney from a living or recently deceased donor whose tissues are antigenically similar (histocompatible) to those of the recipient. A surgeon places the kidney in the depression on the medial surface of the right or left ilium (iliac fossa). The surgeon then connects the renal artery and vein of the donor kidney to the recipient's iliac artery and vein, respectively, and the ureter of the donor kidney to the dome of the recipient's urinary bladder.

Nephrons

Nephron Structure

A kidney contains about 1 million nephrons. Each nephron consists of a **renal corpuscle** (re'nal kor'pusl) and a **renal tubule** (re'nal tu'būl) (see fig. 17.2*c*). Fluid flows through renal tubules on its way out of the body.

A renal corpuscle is composed of a tangled cluster of blood capillaries called a **glomerulus** (glo-mer'u-lus). Glomerular capillaries filter fluid, the first step in urine formation. A thin-walled, saclike structure called a **glomerular capsule** (glo-mer'u-lar kap'sūl) surrounds the glomerulus (fig. 17.5). The glomerular capsule, an expansion at the proximal end of a renal tubule, receives the fluid filtered at the glomerulus. The renal tubule leads away from the glomerular capsule and coils into a portion called the *proximal convoluted tubule.*

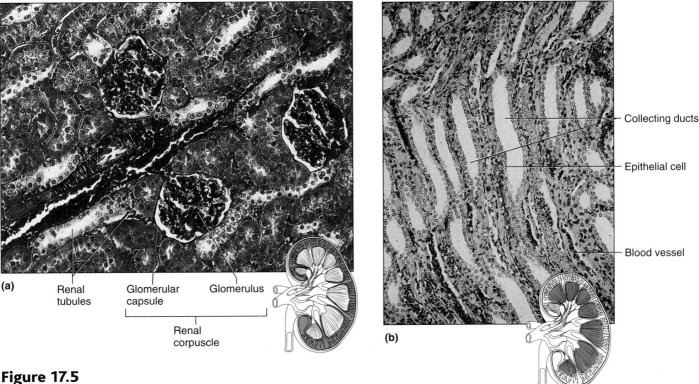

(a) Renal tubules   Glomerular capsule   Glomerulus

Renal corpuscle

Collecting ducts

Epithelial cell

Blood vessel

(b)

Figure 17.5

Microscopic view of the kidney. (*a*) Light micrograph of a section of the human renal cortex (220×). (*b*) Light micrograph of the renal medulla (80×).

Where the proximal convoluted tubule dips toward the renal pelvis, it becomes the *descending limb of the nephron loop* (loop of Henle). The tubule then curves back toward its renal corpuscle and forms the *ascending limb of the nephron loop.* The ascending limb returns to the region of the renal corpuscle, where it coils tightly again and is called the *distal convoluted tubule.*

Distal convoluted tubules from several nephrons merge in the renal cortex to form a *collecting duct* (technically not part of the nephron), which in turn passes into the renal medulla and enlarges as other distal convoluted tubules join it. The resulting tube empties into a minor calyx through an opening in a renal papilla. Figure 17.6 summarizes the structure of a nephron and its associated blood vessels.

Check Your Recall

4. List the general functions of the kidneys.
5. Trace the blood supply to the nephron.
6. Name the parts of a nephron.

Blood Supply of a Nephron

The cluster of capillaries that forms a glomerulus arises from an afferent arteriole. After passing through the glomerular capillaries, blood (minus any filtered fluid)

enters an **efferent arteriole** (ef′er-ent ar-te′re-ōl), whose diameter is smaller than that of the afferent vessel (see fig. 17.4). This is instead of entering a venule, the usual circulatory route. The efferent arteriole resists blood flow to some extent, which backs up blood into the glomerulus, increasing pressure in the glomerular capillary.

The efferent arteriole branches into a complex, freely interconnecting network of capillaries, called the **peritubular capillary** (per″ĭ-tu′bu-lar kap′ĭ-ler″e) system, that surrounds the renal tubule (see figs. 17.4 and 17.6). Blood in the peritubular capillary system is under low pressure. After flowing through the capillary network, the blood rejoins blood from other branches of the peritubular capillary system and enters the venous system of the kidney.

Juxtaglomerular Apparatus

Near its origin, the distal convoluted tubule passes between and contacts afferent and efferent arterioles. At the point of contact, the epithelial cells of the distal tubule are quite narrow and densely packed. These cells comprise a structure called the *macula densa.*

Close by, in the walls of the arterioles near their attachments to the glomerulus, are some enlarged smooth muscle cells called *juxtaglomerular cells.* With cells of the macula densa, they constitute the **juxtaglomerular apparatus** (juks″tah-glo-mer′u-lar

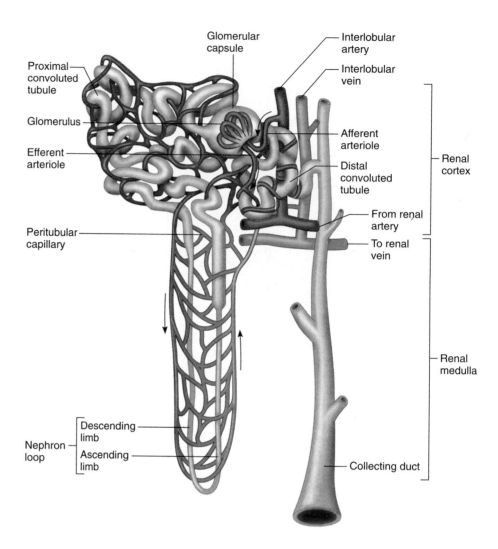

Figure 17.6

Structure of a nephron and associated blood vessels. Arrows indicate the direction of blood flow.

ap"ah-ra'tus), or juxtaglomerular complex (fig. 17.7). Its function in the control of renin secretion is described later in this chapter on page 478.

Check Your Recall

7. Describe the system of blood vessels associated with a nephron.

8. What structures form the juxtaglomerular apparatus?

17.3 URINE FORMATION

Urine formation begins with filtration of plasma by the glomerular capillaries, a process called **glomerular filtration.** Recall from chapter 13 (p. 356) that the force of blood pressure causes filtration to occur at capillaries throughout the body, but most of this fluid is reabsorbed into the bloodstream by the colloid osmotic pressure of the plasma (fig. 17.8*a*). Nephrons take this to another level, using two capillaries working in series. The first capillary bed is specialized only to filter, and instead of forming interstitial fluid, the filtered fluid (filtrate) moves

into the renal tubule, where much of it is destined to become urine (fig. 17.8*b*).

Glomerular filtration produces 180 liters of fluid, more than four times the total body water, every 24 hours. However, glomerular filtration could not continue for very long unless most of this filtered fluid were returned to the internal environment. Thus, in addition to filtration, two other processes contribute to urine formation. **Tubular reabsorption** (tu'bu-lar re-absorp'shun) moves substances from the tubular fluid back into the blood within the peritubular capillary. **Tubular secretion** (tu'bu-lar se-kre'shun), the reverse process, moves substances from the blood within the peritubular capillary into the renal tubule (fig. 17.8*b*). In tubular reabsorption, the kidney selectively reclaims just the right amounts of substances, such as water, electrolytes, and glucose, that the body requires. Waste and substances that are in excess exit the body. In tubular secretion, by contrast, some substances that the body must excrete, such as hydrogen ions and certain toxins, are removed even faster than through filtration alone.

The final product of these three processes is **urine.** The following relationship determines the amount of any given substance excreted in the urine:

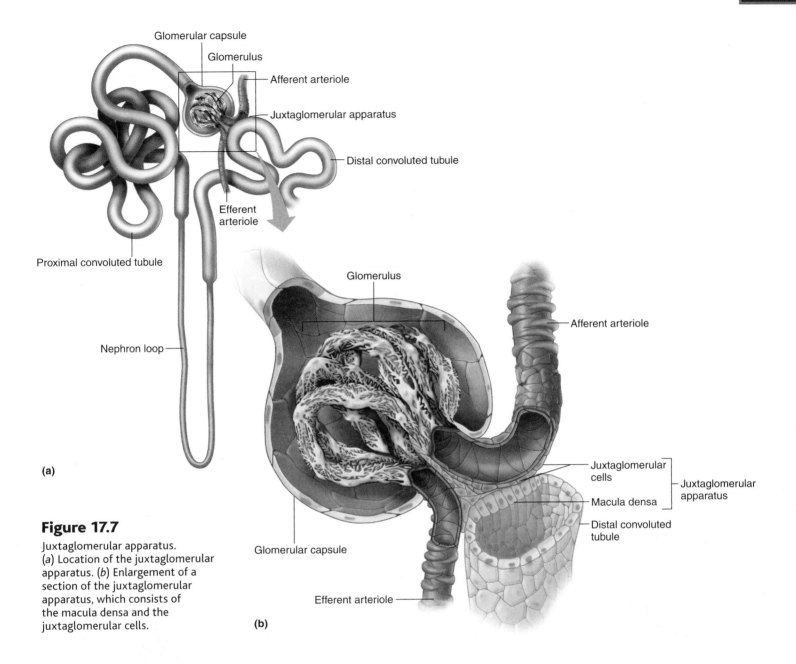

Figure 17.7

Juxtaglomerular apparatus.
(*a*) Location of the juxtaglomerular apparatus. (*b*) Enlargement of a section of the juxtaglomerular apparatus, which consists of the macula densa and the juxtaglomerular cells.

> Amount filtered at the glomerulus
> − Amount reabsorbed by the tubule
> + Amount secreted by the tubule
> _____
> = Amount excreted in the urine

As the kidneys selectively excrete waste products and excess materials in the urine, they contribute to homeostasis by maintaining the composition of the internal environment.

Glomerular Filtration

Urine formation begins when water and certain dissolved substances are filtered out of glomerular capillaries and into glomerular capsules (fig. 17.9*a*). This filtration is similar to filtration at the arteriolar ends of other capillaries. However, many tiny openings (fenestrae) in glomerular capillary walls make glomerular capillaries much more permeable than capillaries in other tissues, even though cells called *podocytes* cover these capillaries and help make them impermeable to plasma proteins (fig. 17.9*b*).

The glomerular capsule receives the resulting **glomerular filtrate,** which is similar in composition to the filtrate that becomes tissue fluid elsewhere in the body. That is, glomerular filtrate is mostly water and the same components as blood plasma, except for the large protein molecules. Table 17.1 shows the relative concentrations of some substances in plasma, glomerular filtrate, and urine.

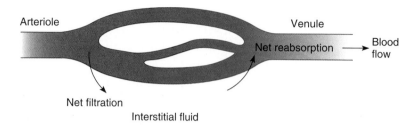

(a) In most systemic capillaries, filtration predominates at the arteriolar end and osmotic reabsorption predominates at the venular end.

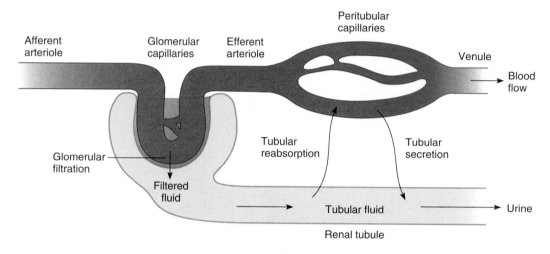

(b) In the kidneys, the glomerular capillaries are specialized for filtration. The renal tubule is specialized to control movements of substances back into the blood of the peritubular capillaries (tubular reabsorption) or from the blood into the renal tubule (tubular secretion).

Figure 17.8

Capillaries in the kidneys are highly specialized to perform processes that occur in capillaries elsewhere as well.

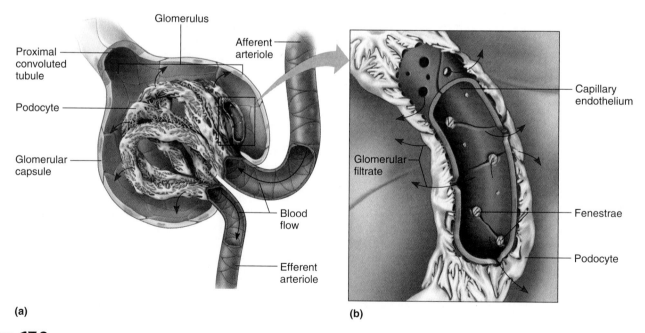

(a)   **(b)**

Figure 17.9

Glomerular filtration. (*a*) The first step in urine formation is filtration of substances out of glomerular capillaries and into the glomerular capsule. (*b*) Glomerular filtrate passes through the fenestrae of the capillary endothelium.

Filtration Pressure

As in other capillaries, the hydrostatic pressure of blood forces substances through the glomerular capillary wall. (Recall that glomerular capillary pressure is high compared to that of other capillaries.) The osmotic pressure of plasma in the glomerulus and the hydrostatic pressure inside the glomerular capsule also influence this movement. An increase in either of these pressures opposes movement out of the capillary and thus reduces filtration. The net pressure forcing substances out of the glomerulus is the **net filtration pressure,** and it is normally always positive, favoring filtration at the glomerulus (fig. 17.10).

> If arterial blood pressure plummets, as can occur during *shock,* glomerular hydrostatic pressure may fall below the level required for filtration. At the same time, epithelial cells of the renal tubules may not receive sufficient nutrients to maintain their high metabolic rates. As a result, cells die (tubular necrosis), impairing renal functions. Such changes can cause renal failure.

Filtration Rate

The glomerular filtration rate is directly proportional to net filtration pressure. Consequently, factors that affect glomerular hydrostatic pressure, glomerular plasma osmotic pressure, or hydrostatic pressure in the glomerular capsule also affect filtration rate. For example, any change in the diameters of the afferent and efferent arterioles changes glomerular hydrostatic pressure, also altering the glomerular filtration rate.

The afferent arteriole, which delivers blood to the glomerulus, may constrict in response to sympathetic nerve impulses. Blood flow diminishes, filtration pressure decreases, and filtration rate drops. On the other hand, if the efferent arteriole (which transports blood from the glomerulus) constricts, blood backs up into the glomerulus, net filtration pressure increases, and filtration rate rises. Vasodilation of these vessels causes opposite effects.

In capillaries, the plasma colloid osmotic pressure that attracts water inward (see chapter 12, p. 326) opposes the blood pressure that forces water and dissolved substances outward. During filtration through the capillary wall, proteins remaining in the plasma raise colloid osmotic pressure within the glomerular capillary. As this pressure rises, filtration decreases. Conversely, conditions that decrease plasma colloid osmotic pressure, such as a decrease in plasma protein concentration, increase the filtration rate.

> In *glomerulonephritis,* the glomerular capillaries are inflamed and become more permeable to proteins, which appear in the glomerular filtrate and in urine (proteinuria). At the same time, the protein concentration in blood plasma decreases (hypoproteinemia), and this decreases plasma colloid osmotic pressure. As a result, less tissue fluid moves into the capillaries, and edema develops.

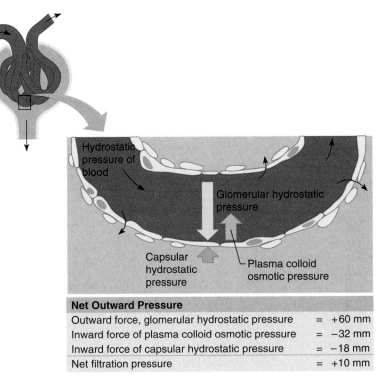

Net Outward Pressure	
Outward force, glomerular hydrostatic pressure	= +60 mm
Inward force of plasma colloid osmotic pressure	= −32 mm
Inward force of capsular hydrostatic pressure	= −18 mm
Net filtration pressure	= +10 mm

Figure 17.10

Normally the glomerular net filtration pressure is positive, causing filtration. The forces involved include the hydrostatic and osmotic pressure of the plasma and the hydrostatic pressure of the fluid in the glomerular capsule.

Table 17.1	Relative Concentrations of Substances in the Plasma, Glomerular Filtrate, and Urine

CONCENTRATIONS (mEq/L)			
Substance	Plasma	Glomerular Filtrate	Urine
Sodium (Na⁺)	142	142	128
Potassium (K⁺)	5	5	60
Calcium (Ca⁺²)	4	4	5
Magnesium (Mg⁺²)	3	3	15
Chloride (Cl⁻)	103	103	134
Bicarbonate (HCO₃⁻)	27	27	14
Sulfate (SO₄⁻²)	1	1	33
Phosphate (PO₄⁻³)	2	2	40

CONCENTRATIONS (mg/100 mL)			
Substance	Plasma	Glomerular Filtrate	Urine
Glucose	100	100	0
Urea	26	26	1,820
Uric acid	4	4	53

Note: mEq/L = milliequivalents per liter.

The hydrostatic pressure in the glomerular capsule sometimes changes because of an obstruction, such as a stone in a ureter or an enlarged prostate gland pressing on the urethra. If this occurs, fluids back up into renal tubules and raise the hydrostatic pressure in the glomerular capsule. Because any increase in capsular pressure opposes glomerular filtration, the filtration rate may decrease significantly.

At rest, the kidneys receive about 25% of the cardiac output, and about 20% of the blood plasma is filtered as it flows through the glomerular capillary. This means that in an average adult, the glomerular filtration rate for the nephrons of both kidneys is about 125 milliliters per minute, or 180,000 milliliters (180 liters, or nearly 45 gallons) in 24 hours. Only a small fraction is excreted as urine. Instead, most of the fluid that passes through the renal tubules is reabsorbed and reenters the plasma.

Check Your Recall

9. Which processes form urine?
10. Which forces affect net filtration pressure?
11. Which factors influence the rate of glomerular filtration?

Regulation of Filtration Rate

The glomerular filtration rate is usually relatively constant. To help maintain homeostasis, however, the glomerular filtration rate may increase when body fluids are in excess and decrease when the body must conserve fluid.

Sympathetic nervous system reflexes that respond to changes in blood pressure and blood volume can alter the glomerular filtration rate. If blood pressure or volume drops sufficiently, afferent arterioles vasoconstrict, decreasing the glomerular filtration rate. This helps ensure that less urine forms when the body must conserve water. Conversely, vasodilation of afferent arterioles increases the glomerular filtration rate to counter increased blood volume or blood pressure.

Another mechanism to control filtration rate involves the enzyme *renin.* Juxtaglomerular cells secrete renin in response to three types of stimuli: (1) whenever special cells in the afferent arteriole sense a drop in blood pressure; (2) in response to sympathetic stimulation; and (3) when the macula densa (see fig. 17.7) senses decreased quantities of chloride, potassium, and sodium ions reaching the distal tubule. Once in the bloodstream, renin reacts with the plasma protein *angiotensinogen* to form *angiotensin I.* A second enzyme (*angiotensin-converting enzyme,* or ACE) in the lungs and in plasma quickly converts angiotensin I to *angiotensin II.*

Angiotensin II carries out a number of actions that help maintain sodium balance, water balance, and

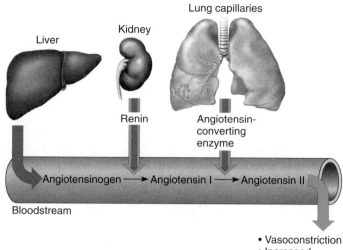

Figure 17.11

The formation of angiotensin II in the bloodstream involves several organs and results in multiple actions that conserve sodium and water.

blood pressure (fig. 17.11). Angiotensin II vasoconstricts the efferent arteriole, which causes blood to back up into the glomerulus, raising glomerular capillary hydrostatic pressure. This important action helps minimize the decrease in glomerular filtration rate when systemic blood pressure is low. Angiotensin II has a major effect on the kidneys by stimulating secretion of the adrenal hormone aldosterone, which stimulates tubular reabsorption of sodium.

The heart secretes another hormone, atrial natriuretic peptide (ANP), when blood volume increases. ANP increases sodium excretion by a number of mechanisms, including increasing the glomerular filtration rate.

> Elevated blood pressure (hypertension) is sometimes associated with excessive release of renin, followed by increased formation of the vasoconstrictor angiotensin II. Patients with this form of high blood pressure often take a drug called an *angiotensin-converting enzyme inhibitor.* These "ACE inhibitors" prevent the formation of angiotensin II by inhibiting the action of the enzyme that converts angiotensin I into angiotensin II.

Check Your Recall

12. What is the function of the macula densa?
13. How does renin help regulate filtration rate?

Tubular Reabsorption

Comparing the composition of glomerular filtrate entering the renal tubule with that of urine leaving the tubule reveals that the fluid changes as it passes through the tubule (see table 17.1). For example, glucose is present in glomerular filtrate but absent in urine. In contrast, urea and uric acid are much more concentrated in urine than in glomerular filtrate. Such changes in fluid composition are largely the result of tubular reabsorption, the process by which filtered substances are returned to the bloodstream. In this process, substances are transported out of the tubular fluid, through the epithelium of the renal tubule, and into the interstitial fluid. These substances then diffuse into the peritubular capillaries (fig. 17.12a).

Tubular reabsorption returns substances to the internal environment. The term *tubular* is used because the epithelial cells that make up the renal tubules control this process. In tubular reabsorption, substances must first cross the cell membrane facing the inside of the tubule and then cross the cell membrane facing the interstitial fluid.

The basic rules for movements across cell membranes apply to tubular reabsorption. Substances moving down a concentration gradient must be lipid-soluble, or there must be a carrier or channel for that substance in the renal tubular cells. Active transport, requiring ATP, may move substances uphill against a concentration gradient.

Peritubular capillary blood is under relatively low pressure because it has already passed through two arterioles. Also, the walls of the peritubular capillaries are more permeable than other capillaries. Finally, because fluid is lost through glomerular filtration, the plasma protein concentration in the peritubular capillaries is relatively high. All of these factors enhance the rate of fluid reabsorption from the renal tubule.

Tubular reabsorption occurs throughout the renal tubule, but most of it takes place in the proximal convoluted portion. The epithelial cells here have many microscopic projections called *microvilli* that form a "brush border" on their free surfaces. These tiny extensions greatly increase the surface area exposed to glomerular filtrate and enhance reabsorption.

Segments of the renal tubule are adapted to reabsorb specific substances, using particular modes of transport. Active transport, for example, reabsorbs glucose through the walls of the proximal convoluted tubule. Water is then reabsorbed by osmosis through the epithelium of the proximal convoluted tubule. However, portions of the distal convoluted tubule and collecting duct may be almost impermeable to water, a characteristic important in the regulation of urine concentration and volume, as described later in this chapter on pages 481–482.

Active transport utilizes carrier molecules in cell membranes (see chapter 3, p. 64). These carriers transport certain molecules across the membrane, release them, and then repeat the process. Such a mechanism,

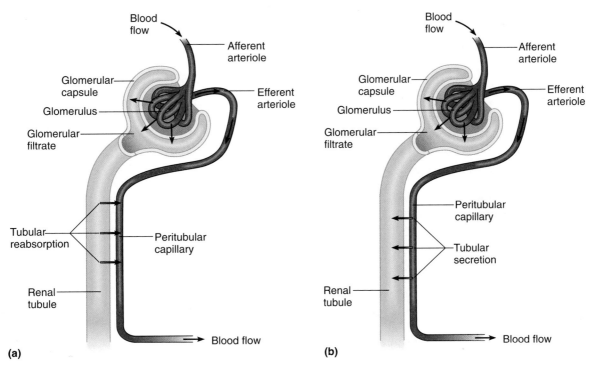

(a)

(b)

Figure 17.12

Two processes in addition to glomerular filtration contribute to form urine. (*a*) Tubular reabsorption transports substances from the glomerular filtrate into the blood within the peritubular capillary. (*b*) Tubular secretion transports substances from the blood within the peritubular capillary into the renal tubule.

however, has a *limited transport capacity;* that is, it can only transport a certain number of molecules in a given time because the number of carriers is limited.

Usually, carrier molecules are able to transport all of the glucose in glomerular filtrate. But when the plasma glucose concentration increases to a critical level, called the *renal plasma threshold,* more glucose molecules are in the filtrate than can be actively transported. As a result, some glucose remains in the tubular fluid and is excreted in urine.

> Glucose in urine, called *glucosuria* (or *glycosuria*), may occur following intravenous administration of glucose. It may also occur in a patient with diabetes mellitus whose blood glucose concentration rises abnormally (see chapter 11, p. 308).

Amino acids enter the glomerular filtrate and are reabsorbed in the proximal convoluted tubule. Three different active transport mechanisms reabsorb different groups of amino acids whose members have similar structures. Normally, only a trace of amino acids remains in urine.

Glomerular filtrate is nearly free of protein except for traces of albumin, a small protein that is taken up by endocytosis through the brush border of epithelial cells lining the proximal convoluted tubule. Once they are inside an epithelial cell, these proteins are broken down to amino acids, which then move into the blood of the peritubular capillary.

The epithelium of the proximal convoluted tubule reabsorbs other substances, including creatine; lactic, citric, uric, and ascorbic (vitamin C) acids; and phosphate, sulfate, calcium, potassium, and sodium ions. Active transport mechanisms with limited transport capacities reabsorb these chemicals. However, these substances usually do not appear in urine until glomerular filtrate concentration exceeds a particular substance's threshold.

Sodium and Water Reabsorption

Substances that remain in the renal tubule become more concentrated as water is reabsorbed from the filtrate. Water reabsorption occurs passively by osmosis, primarily in the proximal convoluted tubule, and is closely associated with the active reabsorption of sodium ions. It increases if sodium reabsorption increases, and decreases if sodium reabsorption decreases (fig. 17.13).

Active transport (the sodium pump) reabsorbs about 70% of sodium ions in the proximal segment of the renal tubule. As these positively charged ions (Na^+) move through the tubular wall, negatively charged ions, including chloride ions (Cl^-), phosphate ions (PO_4^{-3}), and bicarbonate ions (HCO_3^-), accompany them. These negatively charged ions move because of the electrochemical attraction between particles of opposite

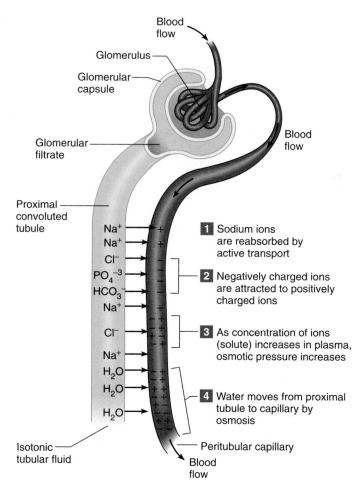

Figure 17.13

In the proximal portion of the renal tubule, osmosis reabsorbs water in response to active transport reabsorbing sodium and other solutes.

charge, a form of *passive transport* because it does not require direct expenditure of cellular energy.

As active transport moves more sodium ions out of the proximal tubule, along with (passively) various negatively charged ions, the concentration of solutes within the peritubular blood increases. Since water moves across cell membranes from regions of lesser solute concentration toward regions of greater solute concentration, water moves by osmosis from the renal tubule into the peritubular capillary. Movement of solutes and water into the peritubular capillary greatly reduces the fluid volume within the renal tubule. The end of the proximal convoluted tubule is in osmotic equilibrium, and the remaining tubular fluid is isotonic.

Active transport continues to reabsorb sodium ions as the tubular fluid moves through the nephron loop, the distal convoluted tubule, and the collecting duct. Water is absorbed passively by osmosis in various segments of the renal tubule. As a result, almost all the sodium ions and water that enter the renal tubule as part of the glomerular filtrate are reabsorbed before urine is excreted.

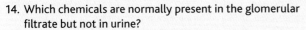
Tubular Secretion

Although 20% of the plasma flowing through the kidneys is filtered in the glomeruli, approximately 80% escapes filtration and continues on through the peritubular capillaries. In tubular secretion, certain substances move from the plasma of blood in the peritubular capillary into the fluid of the renal tubule. As a result, the amount of a particular chemical excreted in the urine may exceed the amount filtered from the plasma in the glomerulus (fig. 17.12*b*). As in the case of tubular reabsorption, the term *tubular* refers to control by the epithelial cells that make up the renal tubules.

Active transport mechanisms similar to those that function in reabsorption secrete some substances. Secretory mechanisms, however, transport substances in the opposite direction. For example, the epithelium of the proximal convoluted segment actively secretes certain organic compounds, including penicillin, creatinine, and histamine, into the tubular fluid.

Hydrogen ions are also actively secreted throughout the entire renal tubule. Secretion of hydrogen ions is important in regulating the pH of body fluids, as chapter 18 (p. 501) explains.

Most potassium ions in the glomerular filtrate are actively reabsorbed in the proximal convoluted tubule, but some may be secreted in the distal segment and collecting duct. During this process, active reabsorption of sodium ions from the tubular fluid results in a negative electrical charge within the tubule. Because positively charged potassium ions (K^+) and hydrogen ions (H^+) are attracted to negatively charged regions, these ions move passively through the tubular epithelium and enter the tubular fluid (fig. 17.14). Potassium ions are also secreted by active processes.

To summarize, urine forms as a result of the following:

- Glomerular filtration of materials from blood plasma.
- Reabsorption of substances, including glucose; water; creatine; amino acids; lactic, citric, and uric acids; and phosphate, sulfate, calcium, potassium, and sodium ions.
- Secretion of substances, including penicillin, histamine, phenobarbital, hydrogen ions, ammonia, and potassium ions.

Regulation of Urine Concentration and Volume

The hormones aldosterone and ADH (antidiuretic hormone) may stimulate additional reabsorption of sodium and water, respectively. The changes in sodium and

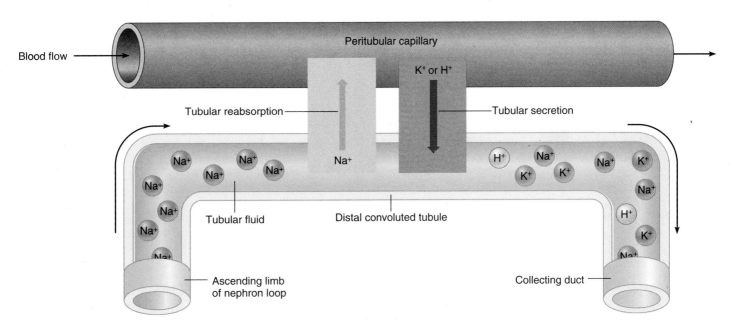

Figure 17.14

In the distal convoluted tubule, potassium ions (and hydrogen ions) may be passively secreted in response to the active reabsorption of sodium ions.

water excretion in response to these hormones are the final adjustments the kidney makes to maintain a constant internal environment.

As discussed in chapter 11 (p. 304), the adrenal glands secrete aldosterone in response to changes in the blood concentrations of sodium and potassium ions. Aldosterone stimulates the distal convoluted tubule to reabsorb sodium and secrete potassium. Angiotensin II is another important stimulator of aldosterone secretion.

Neurons in the hypothalamus produce ADH, which the posterior pituitary releases in response to a decreasing water concentration in blood or a decrease in blood volume. When ADH reaches the kidney, it increases the water permeability of the epithelial linings of the distal convoluted tubule and collecting duct, and water moves rapidly out of these segments by osmosis—that is, water is reabsorbed. Urine volume falls, and soluble wastes and other substances become more concentrated, which minimizes loss of body fluids when dehydration is likely.

If body fluids have excess water, ADH secretion decreases. As blood levels of ADH drop, the epithelial linings of the distal segment and collecting duct become less permeable to water, less water is reabsorbed, and urine is more dilute, excreting the excess water. Table 17.2 summarizes the role of ADH in urine production.

Urea and Uric Acid Excretion

Urea (u-re′ah) is a by-product of amino acid catabolism. Consequently, its plasma concentration reflects the amount of protein in the diet. Urea enters the renal tubule by filtration. About 80% of it is reabsorbed, and the remainder is excreted in urine.

Uric acid is a product of the metabolism of certain organic bases in nucleic acids. Active transport reabsorbs all the uric acid normally present in glomerular filtrate, but a small amount is secreted into the renal tubule and is excreted in urine. Table 17.3 summarizes some specific functions of the nephron segments and the collecting duct.

Table 17.2	**Role of ADH in Regulating Urine Concentration and Volume**
1.	Concentration of water in blood decreases.
2.	Increase in osmotic pressure of body fluids stimulates osmoreceptors in hypothalamus of brain.
3.	Hypothalamus signals posterior pituitary to release ADH.
4.	Blood carries ADH to kidneys.
5.	ADH causes distal convoluted tubules and collecting ducts to increase water reabsorption by osmosis.
6.	Urine becomes concentrated, and urine volume decreases.

Excess uric acid may precipitate in the plasma and be deposited as crystals in joints, causing inflammation and extreme pain, particularly in the digits. King Charles I of Spain (also known as Holy Roman Emperor Charles V) ruled a vast empire from 1516 until his abdication in 1556 due to what his physicians diagnosed as gout. In 2006, Spanish researchers confirmed the diagnosis by detecting uric acid deposits in the terminal joint of a finger that, for reasons unknown, had been preserved in a small box apart from the rest of the king. Today, gout can be treated with drugs that inhibit uric acid reabsorption, increasing its excretion. Gout is usually inherited; it was once thought to be caused by gluttony because only wealthy people could afford to eat meat, which worsens the condition.

Check Your Recall

20. How does the hypothalamus regulate urine concentration and volume?

21. Explain how urea and uric acid are excreted.

Urine Composition

Urine composition reflects the volumes of water and amounts of solutes that the kidneys must eliminate from the body or retain in the internal environment to maintain homeostasis. Urine composition differs considerably from time to time because of variations in dietary intake and physical activity. About 95% water, urine usually contains urea and uric acid, but may also have a trace of amino acids and a variety of electrolytes, whose concentrations vary directly with amounts in the diet (see table 17.1).

The volume of urine produced is usually between 0.6 and 2.5 liters per day, depending on fluid intake, environmental temperature and relative humidity of the surrounding air, as well as the person's emotional condition, respiratory rate, and body temperature. Urine output of 50–60 milliliters per hour is normal; output of less than 30 milliliters per hour may indicate kidney failure.

Glucose, proteins, hemoglobin, ketones, and blood cells are not normally in urine, but circumstances may explain their presence. Glucose in urine may follow a large intake of carbohydrates, proteins may appear following vigorous physical exercise, and ketones may appear after a prolonged fast. Pregnant women may have glucose in their urine as birth nears.

Table 17.3	Functions of Nephron Components
Part	**Function**
Renal corpuscle	
Glomerulus	Filtration of water and dissolved substances from plasma
Glomerular capsule	Receives glomerular filtrate
Renal tubule	
Proximal convoluted tubule	Reabsorption of glucose; amino acids; creatine; lactic, uric, citric, and ascorbic acids; phosphate, sulfate, calcium, potassium, and sodium ions by active transport
	Reabsorption of water by osmosis
	Reabsorption of chloride ions and other negatively charged ions by electrochemical attraction
	Active secretion of substances such as penicillin, histamine, creatinine, and hydrogen ions
Descending limb of nephron loop	Reabsorption of water by osmosis
Ascending limb of nephron loop	Reabsorption of sodium, potassium, and chloride ions by active transport
Distal convoluted tubule	Reabsorption of sodium ions by active transport
	Reabsorption of water by osmosis
	Secretion of hydrogen and potassium ions both actively and passively by electrochemical attraction
Collecting duct	Reabsorption of water by osmosis

Note: Although the collecting duct is not anatomically part of the nephron, it is included here because of its functional importance.

Check Your Recall

22. List the normal constituents of urine.
23. What factors affect urine volume?

17.4 URINE ELIMINATION

After urine forms in the nephrons, it passes from the collecting ducts through openings in the renal papillae and enters the calyces of the kidney (see fig. 17.2). From there, it passes through the renal pelvis, and a ureter conveys it to the urinary bladder (see fig. 17.1 and reference plate 6, p. 28). The urethra passes urine to the outside.

Ureters

Each **ureter** (u-re′ter) is a tube about 25 centimeters long that begins as the funnel-shaped renal pelvis. It descends behind the parietal peritoneum and runs parallel to the vertebral column. In the pelvic cavity, each ureter courses forward and medially, joining the urinary bladder from underneath.

The ureter wall has three layers. The inner layer, or *mucous coat,* is continuous with the linings of the renal tubules and the urinary bladder. The middle layer, or *muscular coat,* consists largely of smooth muscle fibers. The outer layer, or *fibrous coat,* is connective tissue (fig. 17.15).

The muscular walls of the ureters propel the urine. Muscular peristaltic waves, originating in the renal pelvis, force urine along the length of the ureter. When a peristaltic wave reaches the urinary bladder, a jet of urine spurts into the bladder. A flaplike fold of mucous membrane covers the opening through which urine enters the bladder. This fold acts as a valve, allowing urine to enter the bladder from the ureter but preventing it from backing up.

Check Your Recall

24. Describe the structure of a ureter.
25. How is urine moved from the renal pelvis to the urinary bladder?
26. What prevents urine from backing up from the urinary bladder into the ureters?

Urinary Bladder

The **urinary bladder** is a hollow, distensible, muscular organ that stores urine and forces it into the urethra (see fig. 17.1 and reference plate 6, p. 28). It is in the pelvic cavity, behind the symphysis pubis and beneath the parietal peritoneum.

The pressure of surrounding organs alters the shape of the somewhat spherical bladder. When empty, the

Kidney Stones

Kidney stones, which are usually composed of uric acid, calcium oxalate, calcium phosphate, or magnesium phosphate, can form in the collecting ducts and renal pelvis (fig. 17A). Such a stone passing into a ureter causes sudden, severe pain that begins in the region of the kidney and radiates into the abdomen, pelvis, and lower limbs. It may also cause nausea and vomiting, and blood in the urine.

About 60% of kidney stones pass from the body on their own. Other stones were once removed surgically but are now shattered with intense sound waves. In this procedure, called *extracorporeal shock-wave lithotripsy (ESWL)*, the patient is placed in a stainless steel tub filled with water. A spark-gap electrode produces shock waves underwater, and a reflector concentrates and focuses the shock-wave energy on the stones. The resulting sandlike fragments then leave in urine.

The tendency to form kidney stones is inherited, particularly the stones that contain calcium, which account for more than half of all cases. Eating calcium-rich foods does not increase the risk, but taking calcium supplements can. People who have calcium oxalate stones can reduce the risk of recurrence by avoiding specific foods: chocolate, coffee, wheat bran, cola, strawberries, spinach, nuts, and tea. Other causes of kidney stones include excess vitamin D, blockage of the urinary tract, or a complication of a urinary tract infection.

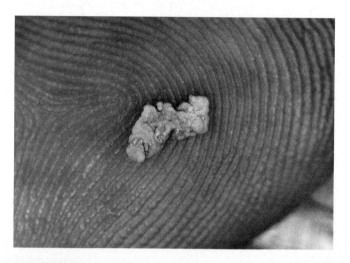

Figure 17A
This kidney stone is small, held against this fingertip, but it is large enough to cause severe pain.

It is very helpful for a physician to analyze the composition of the stones, because certain drugs can prevent recurrence. Stones can be obtained during surgery or by the person, using a special collection device. The best advice to steer clear of kidney stones, however, is simple: drink a lot of water.

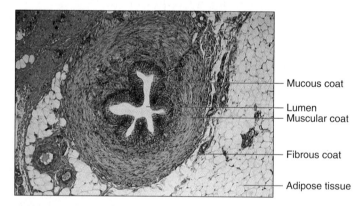

Mucous coat

Lumen
Muscular coat

Fibrous coat

Adipose tissue

Figure 17.15
Cross section of a ureter (75×).

inner wall of the bladder forms many folds, but as the bladder fills with urine, the wall becomes smoother. At the same time, the superior surface of the bladder expands upward into a dome.

The internal floor of the bladder includes a triangular area called the *trigone*, which has an opening at each of its three angles (fig. 17.16*a*). Posteriorly, at the base of the trigone, the openings are those of the ureters. Anteriorly, at the apex of the trigone, a short, funnel-shaped extension called the *neck* of the bladder, contains the opening into the urethra.

The wall of the urinary bladder has four layers. The inner layer, or *mucous coat*, includes several thicknesses of transitional epithelial cells. The thickness of this tissue changes as the bladder expands and contracts. During distension, the tissue may be only two or three cells thick; during contraction, it may be five or six cells thick (see chapter 5, p. 98).

The second layer of the bladder wall is the *submucous coat*. It consists of connective tissue and has many elastic fibers. The third layer of the bladder wall, or *muscular coat,* is composed primarily of coarse bundles of smooth muscle fibers. These bundles are interlaced in all directions and at all depths, and together they comprise the **detrusor muscle** (de-truz'or mus'l). The portion of the detrusor muscle that surrounds the neck of the bladder forms an *internal urethral sphincter.* Sustained contraction of this muscle prevents the bladder from emptying until pressure within the blad-

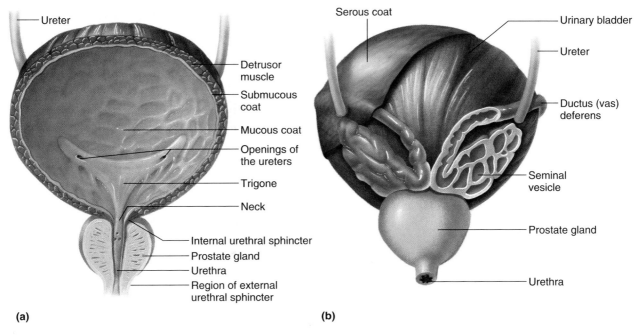

Figure 17.16

A urinary bladder in a male. (*a*) Longitudinal section. (*b*) Posterior view.

der increases to a certain level. The detrusor muscle is innervated with parasympathetic nerve fibers that function in the micturition reflex, discussed next.

The outer layer of the bladder wall, or *serous coat,* consists of the parietal peritoneum. This layer is only on the bladder's upper surface. Elsewhere, the outer coat is connective tissue.

> Because the linings of the ureters and the urinary bladder are continuous, infectious agents, such as bacteria, may ascend from the urinary bladder into the ureters. Inflammation of the urinary bladder, called *cystitis,* is more common in women than in men because the female urethral pathway is shorter. Inflammation of the ureter is called *ureteritis.*

Check Your Recall

27. Describe the trigone of the urinary bladder.
28. Describe the structure of the bladder wall.
29. What kind of nerve fibers supply the detrusor muscle?

Micturition

Micturition (mik″tu-rish′un), or urination, is the process that expels urine from the urinary bladder. In micturition, the detrusor muscle contracts, as do muscles in the abdominal wall and pelvic floor. At the same time, muscles in the thoracic wall and diaphragm do not contract. Micturition also requires relaxation of the

external urethral sphincter. This muscle, which is part of the urogenital diaphragm described in chapter 8 (p. 200), surrounds the urethra about 3 centimeters from the bladder and is composed of voluntary skeletal muscle tissue.

Distension of the bladder wall as it fills with urine stimulates stretch receptors, triggering the micturition reflex. The *micturition reflex center* is in the spinal cord. When sensory impulses from the stretch receptors signal the reflex center, parasympathetic motor impulses travel to the detrusor muscle, which contracts rhythmically in response. A sensation of urgency accompanies this action.

The urinary bladder may hold as much as 600 milliliters of urine before stimulating pain receptors, but the urge to urinate usually begins when it contains about 150 milliliters. As urine volume increases to 300 milliliters or more, the sensation of fullness intensifies, and contractions of the bladder wall become more powerful. When these contractions are strong enough to force the internal urethral sphincter open, another reflex signals the external urethral sphincter to relax, and the bladder can empty.

Because the external urethral sphincter is composed of skeletal muscle, it is under conscious control, and is typically contracted until a person decides to urinate. Nerve centers in the brainstem and cerebral cortex that can partially inhibit the micturition reflex aid this control. When a person decides to urinate, the external urethral sphincter relaxes, and the micturition reflex is no longer inhibited. Nerve centers within the pons and the hypothalamus of the brain heighten the micturition reflex.

Then the detrusor muscle contracts, and urine is excreted through the urethra. Within a few moments, the neurons of the micturition reflex undergo adaptation, the detrusor muscle relaxes, and the bladder begins to fill with urine again.

Damage to the spinal cord above the sacral region destroys voluntary control of urination. However, if the micturition reflex center and its sensory and motor fibers are uninjured, micturition may continue to occur reflexly. In this case, the bladder collects urine until its walls stretch enough to trigger a micturition reflex, and the detrusor muscle contracts in response. This condition is called an *automatic bladder.*

Urethra

The **urethra** (u-re'thrah) is a tube that conveys urine from the urinary bladder to the outside (see fig. 17.1 and reference plate 7, p. 29). Its wall is lined with mucous membrane and has a thick layer of smooth muscle tissue, whose fibers are generally directed longitudinally. The urethral wall also has abundant mucous glands, called *urethral glands,* which secrete mucus into the urethral canal (fig. 17.17).

In a female, the urethra is about 4 centimeters long. Its opening, the *external urethral orifice* (urinary meatus) is anterior to the vaginal opening and posterior to the clitoris. In a male, the urethra functions as part of both the urinary system and the reproductive system and extends from the bladder to the tip of the penis.

Check Your Recall

30. Describe micturition.
31. How is it possible to consciously inhibit the micturition reflex?
32. Describe the structure of the urethra.

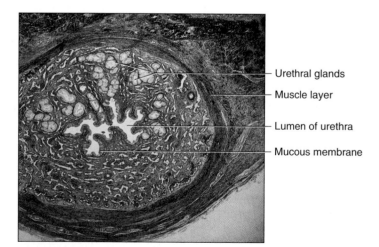

- Urethral glands
- Muscle layer
- Lumen of urethra
- Mucous membrane

Figure 17.17
Cross section through the urethra (10×).

Clinical Terms Related to the Urinary System

anuria (ah-nu're-ah) Absence of urine due to failure of kidney function or to an obstruction in a urinary pathway.
bacteriuria (bak-te"re-u're-ah) Bacteria in urine.
cystectomy (sis-tek'to-me) Surgical removal of the urinary bladder.
cystitis (sis-ti'tis) Inflammation of the urinary bladder.
cystoscope (sis'to-skōp) Instrument used to visually examine the interior of the urinary bladder.
cystotomy (sis-tot'o-me) Incision of the urinary bladder wall.
diuresis (di"u-re'sis) Increased urine excretion.
diuretic (di"u-ret'ik) Substance that increases urine production.
dysuria (dis-u're-ah) Painful or difficult urination.
hematuria (hem"ah-tu're-ah) Blood in urine.
incontinence (in-kon'ti-nens) Inability to control urination and/or defecation reflexes.
nephrectomy (ně-frek'to-me) Surgical removal of a kidney.
nephrolithiasis (nef"ro-lĭ-thi'ah-sis) Kidney stones.
nephroptosis (nef"rop-to'sis) Movable or displaced kidney.
oliguria (ol"ĭ-gu're-ah) Scanty urine output.
polyuria (pol"ě-u're-ah) Excess urine output.
pyelolithotomy (pi"ě-lo-lĭ-thot'o-me) Removal of a stone from the renal pelvis.
pyelonephritis (pi"ě-lo-ne-fri'tis) Inflammation of the renal pelvis.
pyelotomy (pi"ě-lot'o-me) Incision into the renal pelvis.
pyuria (pi-u're-ah) Pus (white blood cells) in urine.
uremia (u-re'me-ah) Accumulation in blood of substances ordinarily excreted in urine.
ureteritis (u-re"ter-i'tis) Inflammation of the ureter.
urethritis (u"re-thri'tis) Inflammation of the urethra.

Clinical Connection

A twenty-five-year-old man arrived at the hospital in acute renal failure following three days of malaise. He claimed to have taken only cough medicine, and then slept thirty-six hours. A urinalysis revealed blood and protein, but it was the renal biopsy that alarmed the emergency room physician—the renal tubules were damaged, with telltale crystals of calcium oxalate obstructing their lumens. The physician knew that these crystals result from drinking ethylene glycol, a component of antifreeze, but the patient denied such activity. He was sent home. A week later, he returned with new symptoms—seizures and other signs of neurological impairment. Over the next two months, with aggressive treatment, his kidney function returned to normal. It was then that he admitted to having drunk the colorless, odorless, sweet-tasting antifreeze, in a suicide attempt. Because the dose that he took was not large enough to damage his kidneys to the extent seen, physicians suspected that an ingredient in the cough medicine increased the toxicity of the antifreeze.

Urinary System

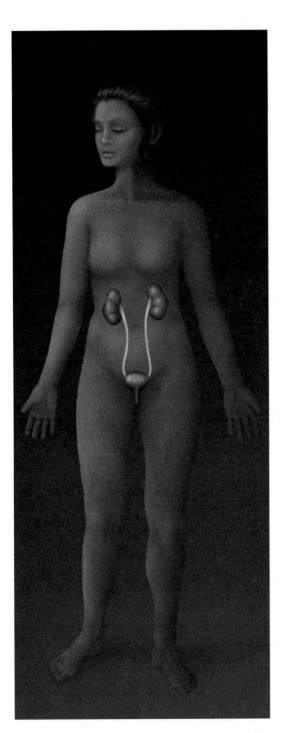

Integumentary System

The urinary system compensates for water loss due to sweating. The kidneys and skin both play a role in vitamin D production.

Skeletal System

The kidneys and bone tissue work together to control plasma calcium levels.

Muscular System

Muscle tissue controls urine elimination from the bladder.

Nervous System

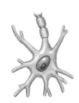

The nervous system influences urine production and elimination.

Endocrine System

The endocrine system influences urine production.

Cardiovascular System

The urinary system controls blood volume. Blood volume and blood pressure play a role in determining water and solute excretion.

Lymphatic System

The kidneys control extracellular fluid volume and composition (including lymph).

Digestive System

The kidneys compensate for fluids lost by the digestive system.

Respiratory System

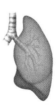

The kidneys and the lungs work together to control the pH of the internal environment.

Reproductive System

The urinary system in males shares organs with the reproductive system. The kidneys compensate for fluids lost from the male and female reproductive systems.

The urinary system controls the composition of the internal environment.

SUMMARY OUTLINE

17.1 Introduction (p. 470)

The urinary system consists of the kidneys, ureters, urinary bladder, and urethra.

17.2 Kidneys (p. 470)

1. Location of the kidneys
   a. The kidneys are high on the posterior wall of the abdominal cavity.
   b. They are behind the parietal peritoneum.
2. Kidney structure
   a. A kidney has a hollow renal sinus.
   b. The ureter expands into the renal pelvis.
   c. Renal papillae project into the renal sinus.
   d. Each kidney is subdivided into a medulla and a cortex.
3. Kidney functions
   a. The kidneys maintain homeostasis by removing metabolic wastes from blood and excreting them.
   b. They also help regulate red blood cell production; blood volume and blood pressure; and the volume, composition, and pH of body fluids.
4. Renal blood vessels
   a. Arterial blood flows through the renal artery, interlobar arteries, arcuate arteries, interlobular arteries, and afferent arterioles to the nephrons.
   b. Venous blood returns through a series of vessels that correspond to the arterial pathways.
5. Nephrons
   a. Nephron structure
      (1) A nephron is the functional unit of the kidney.
      (2) It consists of a renal corpuscle and a renal tubule.
         (a) The corpuscle consists of a glomerulus and a glomerular capsule.
         (b) Segments of the renal tubule include the proximal convoluted tubule, nephron loop (ascending and descending limbs), and distal convoluted tubule, which empties into a collecting duct.
      (3) The collecting duct empties into the minor calyx of the renal pelvis.
   b. Blood supply of a nephron
      (1) The glomerular capillary receives blood from the afferent arteriole and passes it to the efferent arteriole.
      (2) The efferent arteriole gives rise to the peritubular capillary system, which surrounds the renal tubule.
   c. Juxtaglomerular apparatus
      (1) The juxtaglomerular apparatus is at the point of contact between the distal convoluted tubule and the afferent and efferent arterioles.
      (2) It consists of the macula densa and juxtaglomerular cells.

17.3 Urine Formation (p. 474)

Nephrons remove wastes from blood and regulate water and electrolyte concentrations. Urine is the end product.

1. Glomerular filtration
   a. Urine formation begins when water and dissolved materials filter out of glomerular capillaries.
   b. Glomerular capillaries are much more permeable than the capillaries in other tissues.
   c. The composition of the filtrate is similar to that of tissue fluid.
2. Filtration pressure
   a. Filtration is due mainly to hydrostatic pressure inside glomerular capillaries.
   b. The osmotic pressure of plasma and the hydrostatic pressure in the glomerular capsule also affect filtration.
   c. Filtration pressure is the net force moving material out of the glomerulus and into the glomerular capsule.
3. Filtration rate
   a. Rate of filtration varies with filtration pressure.
   b. Filtration pressure changes with the diameters of the afferent and efferent arterioles.
   c. As colloid osmotic pressure in the glomerulus increases, filtration rate decreases.
   d. As hydrostatic pressure in a glomerular capsule increases, filtration rate decreases.
   e. The kidneys produce about 125 milliliters of glomerular fluid per minute, most of which is reabsorbed.
4. Regulation of filtration rate
   a. Glomerular filtration rate remains relatively constant, but may increase or decrease as required.
   b. Increased sympathetic nerve activity can decrease glomerular filtration rate.
   c. When the macula densa senses decreased amounts of chloride, potassium, and sodium ions in the distal tubule, it causes juxtaglomerular cells to release renin.
   d. This triggers a series of changes leading to vasoconstriction of afferent and efferent arterioles, which may affect glomerular filtration rate, and aldosterone secretion, which stimulates tubular sodium reabsorption.
5. Tubular reabsorption
   a. Substances are selectively reabsorbed from glomerular filtrate.
   b. The peritubular capillary's permeability adapts it for reabsorption.
   c. Most reabsorption occurs in the proximal tubule, where epithelial cells have microvilli.
   d. Different modes of transport reabsorb various substances in particular segments of the renal tubule.
      (1) Active transport reabsorbs glucose and amino acids.
      (2) Osmosis reabsorbs water.
   e. Active transport mechanisms have limited transport capacities.
6. Sodium and water reabsorption
   a. Substances that remain in the filtrate are concentrated as water is reabsorbed.
   b. Active transport reabsorbs sodium ions.
   c. As positively charged sodium ions move out of the filtrate, negatively charged ions follow them.
   d. Water is passively reabsorbed by osmosis.
7. Tubular secretion
   a. Secretion transports substances from plasma to the tubular fluid.
   b. Various organic compounds are secreted actively.
   c. Potassium and hydrogen ions are secreted both actively and passively.
8. Regulation of urine concentration and volume
   a. Most sodium is reabsorbed before urine is excreted.
   b. Antidiuretic hormone increases the permeability of the distal convoluted tubule and collecting duct, promoting water reabsorption.

9. Urea and uric acid excretion
    a. Diffusion passively reabsorbs urea. About 50% of the urea is excreted in urine.
    b. Active transport reabsorbs uric acid. Some uric acid is secreted into the renal tubule.
10. Urine composition
    a. Urine is about 95% water, and it also usually contains urea and uric acid.
    b. Urine contains varying amounts of electrolytes and may contain a trace of amino acids.
    c. Urine volume varies with fluid intake and with certain environmental factors.

17.4 Urine Elimination (p. 483)

1. Ureters
    a. The ureter extends from the kidney to the urinary bladder.
    b. Peristaltic waves in the ureter force urine to the urinary bladder.
2. Urinary bladder
    a. The urinary bladder stores urine and forces it through the urethra during micturition.
    b. The openings for the ureters and urethra are at the three angles of the trigone.
    c. A portion of the detrusor muscle forms an internal urethral sphincter.
3. Micturition
    a. Micturition expels urine.
    b. Micturition contracts the detrusor muscle and relaxes the external urethral sphincter.
    c. Micturition reflex
        (1) Distension stimulates stretch receptors in the bladder wall.
        (2) The micturition reflex center in the spinal cord sends parasympathetic motor impulses to the detrusor muscle.
        (3) As the bladder fills, its internal pressure increases, forcing the internal urethral sphincter open.
        (4) A second reflex relaxes the external urethral sphincter unless voluntary control maintains its contraction.
        (5) Nerve centers in the cerebral cortex and brainstem aid control of urination.
4. Urethra
    The urethra conveys urine from the urinary bladder to the outside.

CHAPTER ASSESSMENTS

17.1 Introduction

1. Name and identify the general functions of the organs of the urinary system. (p. 470)

17.2 Kidneys

2. Explain why the kidneys are said to be retroperitoneal. (p. 470)
3. Describe the external and internal structure of a kidney. (p. 470)
4. Identify the functions of the kidneys. (p. 471)
5. List in correct order the vessels through which blood passes as it travels from the renal artery to the renal vein. (pp. 471 and 473)
6. Distinguish between a renal corpuscle and a renal tubule. (p. 472)
7. Name in correct order the parts of the nephron through which fluid passes from the glomerulus to the collecting duct. (p. 472)
8. Describe the location and structure of the juxtaglomerular apparatus. (p. 473)

17.3 Urine Formation

9. Define *filtration pressure*. (p. 477)
10. Which one of the following is abundant in blood plasma, but only in small amounts in glomerular filtrate? (p. 477)
    a. sodium ions
    b. water
    c. glucose
    d. protein
    e. potassium ions
11. Explain how the diameters of the afferent and efferent arterioles affect the rate of glomerular filtration. (p. 477)
12. Explain how changes in the osmotic pressure of blood plasma affect the glomerular filtration rate. (p. 477)
13. Explain how the hydrostatic pressure of a glomerular capsule affects the rate of glomerular filtration. (p. 478)
14. Describe two mechanisms by which the body regulates filtration rate. (p. 478)
15. Discuss how tubular reabsorption is selective. (p. 479)
16. Explain how the peritubular capillary is adapted for reabsorption. (p. 479)
17. Explain how epithelial cells of the proximal convoluted tubule are adapted for reabsorption. (p. 479)
18. Explain why active transport mechanisms have limited transport capacities. (p. 480)
19. Define *renal plasma threshold*. (p. 480)
20. Explain how amino acids and proteins are reabsorbed. (p. 480)
21. Describe the effect of sodium reabsorption on the reabsorption of negatively charged ions. (p. 480)
22. Explain how sodium reabsorption affects water reabsorption. (p. 480)
23. Explain how potassium ions may be secreted passively. (p. 481)
24. The major action of of ADH in the kidneys is to (p. 481)
    a. increase water absorption by the proximal convoluted tubule.
    b. increase glomerular filtration rate
    c. increase water reabsorption by the collectiong duct
    d. increase potassium's excretion
25. Compare the processes that reabsorb urea and uric acid. (p. 482)
26. List the common constituents of urine and their sources. (p. 482)
27. Identify some of the factors that affect the volume of urine produced daily. (p. 482)

17.4 Urine Elimination

28. Describe the structure and function of a ureter. (p. 483)
29. Explain how the muscular wall of the ureter helps move urine. (p. 483)
30. Describe the structure and location of the urinary bladder. (p. 483)
31. Define *detrusor muscle*. (p. 484)
32. Distinguish between the internal and external urethral sphincters. (p. 484)
33. Describe the micturition reflex. (p. 485)
34. Which of the following involves skeletal muscle? (p. 485).
    a. contraction of the internal urethral sphincter
    b. contraction of the external urethral sphincter
    c. ureteral peristalsis
    d. detrusor muscle contraction

INTEGRATIVE ASSESSMENTS/ CRITICAL THINKING

OUTCOMES 17.1, 17.2, 17.3

1. Imagine you are adrift at sea. Why will you dehydrate more quickly if you drink seawater instead of fresh water to quench your thirst?

2. Why are people following high-protein diets advised to drink large volumes of water?

OUTCOMES 17.2, 17.3

3. Why may protein in the urine be a sign of kidney damage? What structures in the kidney are probably affected?

4. An infant is born with narrowed renal arteries. What effect will this condition have on urine volume?

5. If blood pressure plummets in a patient in shock as a result of a severe injury, how would you expect urine volume to change? Why?

OUTCOMES 17.2, 17.4

6. Why do urinary tract infections frequently accompany sexually transmitted diseases?

OUTCOMES 17.2, 17.3, 17.4

7. If a patient who has had major abdominal surgery receives intravenous fluids equal to the blood volume lost during surgery, would you expect urine volume to be greater or less than normal? Why?

WEB CONNECTIONS

Visit the text website at **aris.mhhe.com** for additional quizzes, interactive learning exercises, and more

AP R URINARY SYSTEM

Anatomy & Physiology | REVEALED includes cadaver photos that allow you to peel away layers of the human body to reveal structures beneath the surface. This program also includes animations, radiologic imaging, audio pronunciations, and practice quizzing. To learn more visit **www.aprevealed.com**.

18

Water, Electrolyte, and Acid-Base Balance

Korey Stringer was an offensive tackle for the Minnesota Vikings who died of heatstroke.

HEATSTROKE CAN BE DEADLY. August 2, 2001, was another 90° high-humidity day at training camp for the Minnesota Vikings in Mankato. The day before, offensive tackle Korey Stringer hadn't been able to participate in afternoon practice, citing exhaustion—but he vowed to make it the next morning. He did, but still felt unwell. After vomiting three times, he walked over to an air-conditioned shelter, dizzy and breathing heavily. Trainers recognized the signs of heat exhaustion and took Stringer to a nearby medical facility, but it was too late. On arrival, Stringer's body temperature was a life-threatening 108°F, and he soon lost consciousness. He died at 1:50 the next morning.

Korey Stringer had heatstroke, which occurs rapidly when the body is exposed to a heat index (heat considering humidity) of more than 105°F and body temperature rises above 106°F. On that August day, the heat index was 110°F. Under these conditions, evaporation of sweat is less efficient at cooling the body and the organs begin to fail. The situation is worse if the individual is heavy or if the body is covered. Stringer weighed 335 pounds and was exercising in full football gear.

During the heat wave of 2001 in the weeks following Stringer's death, several athletes in their teens also succumbed to heatstroke. More than 300 people die in the United States each year from this preventable condition, most of them either elderly people or infants, who may have poor temperature control. Heatstroke is unpredictable because people have different limits. After Stringer's death, many players remembered feeling dizzy or chilled when the weather was hot, but continuing to exercise anyway. Athletic trainers typically weigh players twice a day and suspect heatstroke if an athlete suddenly loses 6 to 8 pounds. After Stringer's death, sports medicine specialists advised the National Football League to shorten or change the time of practices when heat and humidity become dangerous, to enforce water breaks, and to allow players at least a week to adjust to a different climate before wearing full gear.

Following is a list of the symptoms of heatstroke:

Headache
Dizziness
Exhaustion
Profuse sweating, which then stops
Dry, hot, and red skin
Pulse elevated as high as 180 beats per minute
Increased respiratory rate
Disorientation
Losing consciousness or having a seizure
Rapid rise in body temperature

Treatment for heatstroke must be fast. First, call 911 for assistance. Check the person's airway, and if he or she is not breathing, administer cardiopulmonary resuscitation (CPR). Move the victim to a cooler environment, and take off the clothing. If possible, place ice on the armpits, groin, and neck. Even more effective is a water-soaked sheet wrapped around the entire body, for rapid cooling. Do not give the person anything to drink.

Learning Outcomes   *After studying this chapter, you should be able to do the following:*

18.1 Introduction
1. Explain water and electrolyte balance. (p. 492)

18.2 Distribution of Body Fluids
2. Explain body fluid distribution in compartments. (p. 492)

18.3 Water Balance
3. List the routes by which water enters and leaves the body, and explain how water intake and output are regulated. (p. 494)

18.4 Electrolyte Balance

4.   Explain how electrolytes enter and leave the body, and describe how the intake and output of electrolytes are regulated. (p. 495)

18.5 Acid-Base Balance

5.   List the major sources of hydrogen ions in the body. (p. 498)

6.   Distinguish between strong and weak acids and bases. (p. 499)

7.   Explain how chemical buffer systems, the respiratory center, and the kidneys minimize changing pH values of the body fluids. (p. 500)

18.6 Acid-Base Imbalances

8.   Describe the causes and consequences of elevation or decrease in body fluid pH. (p. 502)

Aids to Understanding Words   *(Appendix A on page 567 has a complete list of Aids to Understanding Words.)*

de- [separation from] *de*hydration: Removal of water from the cells or body fluids.

extra- [outside] *extra*cellular fluid: Fluid outside the body cells.

im- [not] *im*balance: Condition in which factors are not in equilibrium.

intra- [within] *intra*cellular fluid: Fluid within body cells.

neutr- [neither one nor the other] *neutr*al: Solution that is neither acidic nor basic.

18.1 INTRODUCTION

Two types of substances that are important in maintaining homeostasis in the body are water and **electrolytes,** which are molecules that release ions in water. To maintain homeostasis, the quantities of water and electrolytes must be in balance; that is, the amounts entering the body must equal the amounts leaving it. Thus, the body requires mechanisms to (1) replace lost water and electrolytes, and (2) excrete any excess.

Because electrolytes are dissolved in the water of body fluids, water balance and electrolyte balance are interdependent. Consequently, anything that alters electrolyte concentrations necessarily alters the water concentration by either adding or removing solutes. Likewise, anything that changes the water concentration changes electrolyte concentrations by either concentrating or diluting them.

> A human being is 60% water. Losing one-tenth of that volume can be fatal.

18.2 DISTRIBUTION OF BODY FLUIDS

Body fluids are not uniformly distributed throughout tissues but are in regions, or *compartments,* of different volumes that contain fluids of varying compositions. The movement of water and electrolytes between these compartments is regulated to stabilize both their distribution and the composition of body fluids.

Fluid Compartments

The body of an average adult female is about 52% water by weight, and that of an average male is about 63% water. These proportions differ because females gener-

ally have more adipose tissue, which has little water, than do males. Water in the body (about 40 liters), together with its dissolved electrolytes, is distributed into two major compartments—an intracellular fluid compartment and an extracellular fluid compartment.

The **intracellular** (in"trah-sel'u-lar) **fluid compartment** includes all the water and electrolytes that cell membranes enclose. In other words, intracellular fluid is fluid within cells, and in an adult, it represents about 63% by volume of total body water.

The **extracellular** (ek"strah-sel'u-lar) **fluid compartment** includes all the fluid outside cells—within the tissue spaces (interstitial fluid), blood vessels (plasma), and lymphatic vessels (lymph). Epithelial layers separate a specialized fraction of extracellular fluid from other extracellular fluids. This **transcellular** (trans-sel'u-lar) **fluid** includes *cerebrospinal fluid* of the central nervous system, *aqueous* and *vitreous humors* of the eyes, *synovial fluid* of the joints, *serous fluid* in body cavities, and fluid *secretions* of the exocrine glands. The fluids of the extracellular compartment constitute about 37% by volume of total body water (fig. 18.1).

Body Fluid Composition

Most *extracellular fluids* are chemically similar with high concentrations of sodium, chloride, and bicarbonate ions. These fluids include a greater concentration of calcium ions and lesser concentrations of potassium, magnesium, phosphate, and sulfate ions than does intracellular fluid. The blood plasma fraction of extracellular fluid has considerably more protein than does either interstitial fluid or lymph.

Intracellular fluid has high concentrations of potassium, magnesium, and phosphate ions. It includes a greater concentration of sulfate ions and lesser concentrations of sodium, chloride, and bicarbonate ions than do extracellular fluids. Intracellular fluid also has a greater

Total body water

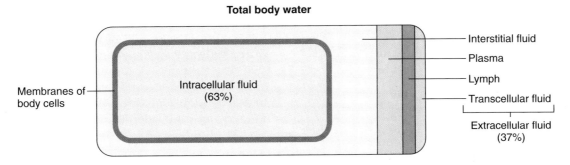

Figure 18.1

Cell membranes separate fluid in the intracellular compartment from fluid in the extracellular compartment. Approximately two-thirds of the water in the body is inside cells.

protein concentration than does plasma. Figure 18.2 shows these relative concentrations.

Check Your Recall

1. How are water balance and electrolyte balance interdependent?

2. Describe the normal distribution of water in the body.

3. Which electrolytes are in higher concentrations in extracellular fluids? In intracellular fluid?

4. How does protein concentration vary in different body fluids?

Movement of Fluid Between Compartments

Two major factors regulate the movement of water and electrolytes from one fluid compartment to another: *hydrostatic pressure* and *osmotic pressure* (fig. 18.3). Recall from chapter 13 (pp. 356–357) that fluid leaves the plasma at the arteriolar ends of capillaries and enters the interstitial spaces because of the net outward force of hydrostatic pressure (blood pressure). Fluid returns to the plasma from the interstitial spaces at the venular ends of capillaries because of the net inward force of *colloid osmotic pressure* due to the plasma proteins. Likewise, as mentioned in chapter 14 (p. 379), the hydrostatic pressure that develops within interstitial spaces forces the fluid therein into lymph capillaries. Lymph circulation returns interstitial fluid to the plasma.

Pressures similarly control fluid movement between the intracellular and extracellular compartments. Because hydrostatic pressure within the cells and surrounding interstitial fluid is ordinarily equal and stable, a change in osmotic pressure is the likely cause of any net fluid movement.

The sodium ion concentration in extracellular fluids is especially high. A decrease in this concentration causes net movement of water from the extracellular

compartment into the intracellular compartment by osmosis. As a consequence, the cells swell. Conversely, if the sodium ion concentration in interstitial fluid increases, the net movement of water is outward from the intracellular compartment, and cells shrink as they lose water.

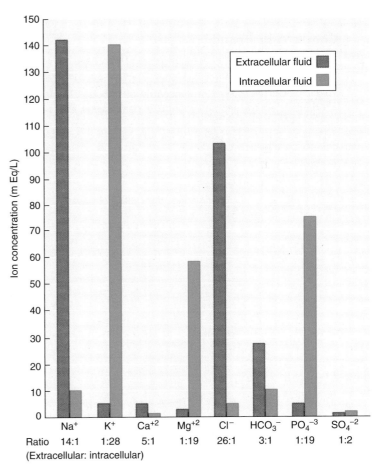

Figure 18.2

Extracellular fluids have relatively high concentrations of sodium (Na^+), calcium (Ca^{+2}), chloride (Cl^-), and bicarbonate (HCO_3^-) ions. Intracellular fluid has relatively high concentrations of potassium (K^+), magnesium (Mg^{+2}), phosphate (PO_4^{-3}), and sulfate (SO_4^{-2}) ions.

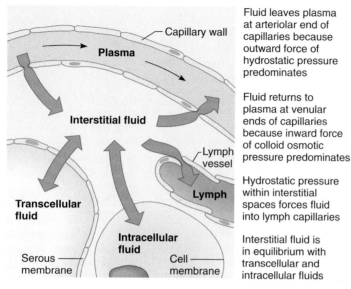

Capillary wall

Plasma

Interstitial fluid

Lymph vessel

Lymph

Transcellular fluid

Intracellular fluid

Serous membrane

Cell membrane

Fluid leaves plasma at arteriolar end of capillaries because outward force of hydrostatic pressure predominates

Fluid returns to plasma at venular ends of capillaries because inward force of colloid osmotic pressure predominates

Hydrostatic pressure within interstitial spaces forces fluid into lymph capillaries

Interstitial fluid is in equilibrium with transcellular and intracellular fluids

Figure 18.3

Net movements of fluids between compartments result from differences in hydrostatic and osmotic pressures.

Check Your Recall

5. Which factors control the movement of water and electrolytes from one fluid compartment to another?

6. How does the sodium ion concentration of body fluids affect the net movement of water between the compartments?

18.3 WATER BALANCE

Water balance exists when total water intake equals total water output. Homeostatic mechanisms maintain water balance. The Topic of Interest entitled Water Balance Disorders on pages 496–497 discusses dehydration, edema, and water intoxication.

Water Intake

The volume of water gained each day varies from individual to individual. An average adult living in a moderate environment takes in about 2,500 milliliters. Of this volume, drinking water or beverages supply about 60%, while moist foods provide another 30%. The remaining 10% is a by-product of the oxidative metabolism (chapter 4, p. 80) of nutrients and is called **water of metabolism** (fig. 18.4*a*).

> The kangaroo rat is a desert rodent that does not have to drink water. It survives on the water of metabolism alone.

Regulation of Water Intake

The primary regulator of water intake is thirst. The intense feeling of thirst derives from the effect of osmotic pressure of extracellular fluids on a *thirst center* in the hypothalamus. As the body loses water, the osmotic pressure of extracellular fluids increases. This

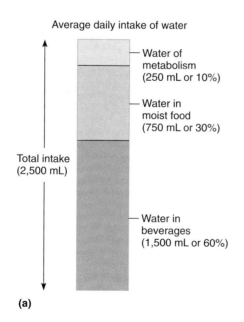

Average daily intake of water

Water of metabolism (250 mL or 10%)

Water in moist food (750 mL or 30%)

Total intake (2,500 mL)

Water in beverages (1,500 mL or 60%)

(a)

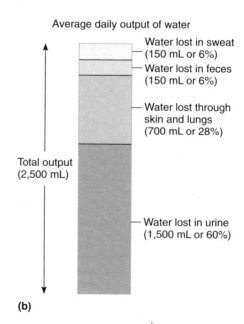

Average daily output of water

Water lost in sweat (150 mL or 6%)

Water lost in feces (150 mL or 6%)

Water lost through skin and lungs (700 mL or 28%)

Total output (2,500 mL)

Water lost in urine (1,500 mL or 60%)

(b)

Figure 18.4

Water balance. (*a*) Major sources of body water. (*b*) Routes by which the body loses water. Urine production is most important in the regulation of water balance.

stimulates *osmoreceptors* in the thirst center, which cause the person to feel thirsty and to seek water.

Thirst is a homeostatic mechanism, normally triggered whenever total body water decreases by as little as 1%. Drinking distends the stomach wall, triggering nerve impulses that inhibit the thirst mechanism. In this way, drinking stops even before the swallowed water is absorbed, preventing the person from drinking too much.

> **Check Your Recall**
> 7. What is water balance?
> 8. Where is the thirst center located?
> 9. What stimulates fluid intake? What inhibits it?

Water Output

Water normally enters the body only through the mouth, but it can be lost by a variety of routes. These include obvious losses in urine, feces, and sweat (sensible perspiration), as well as less obvious losses, such as evaporation of water from the skin (insensible perspiration) and from the lungs during breathing.

If an average adult takes in 2,500 milliliters of water each day, then 2,500 milliliters must be eliminated to maintain water balance. Of this volume, perhaps 60% is lost in urine, 6% in feces, and 6% in sweat. About 28% is lost by evaporation from the skin and lungs (fig. 18.4*b*). These percentages vary with environmental temperature and relative humidity, as well as with physical exercise.

Regulation of Water Output

The primary means of regulating water output is urine production. The distal convoluted tubules of the nephrons and collecting ducts are most important in regulating the volume of water excreted in the urine. The epithelial linings of these segments of the renal tubule remain relatively impermeable to water unless antidiuretic hormone (ADH) is present. ADH increases the permeability of the distal convoluted tubule and collecting duct, thereby increasing water reabsorption and reducing urine production. In the absence of ADH, less water is reabsorbed, and more urine is produced (see chapter 17, p. 482).

> *Diuretics* are substances that promote urine production. A number of familiar chemicals, such as caffeine in coffee and tea, have diuretic effects, as do a variety of drugs used to reduce the volume of body fluids.
>
> Diuretics produce their effects in different ways. Some, such as alcohol and certain narcotic drugs, promote urine formation by inhibiting ADH release. Other diuretics, such as caffeine, inhibit the reabsorption of sodium ions or other solutes in portions of the renal tubules. When less sodium is reabsorbed, less water is also reabsorbed, therefore resulting in a larger urine volume.

> **Check Your Recall**
> 10. By what routes does the body lose water?
> 11. What role do renal tubules play in water balance regulation?

18.4 ELECTROLYTE BALANCE

Electrolyte balance exists when the quantities of electrolytes the body gains equal those lost. Homeostatic mechanisms maintain electrolyte balance.

Electrolyte Intake

The electrolytes of greatest importance to cellular functions dissociate to release sodium, potassium, calcium, magnesium, chloride, sulfate, phosphate, bicarbonate, and hydrogen ions. Foods provide most of these electrolytes, but drinking water and other beverages are also sources. Some electrolytes are by-products of metabolic reactions.

Regulation of Electrolyte Intake

Ordinarily, responding to hunger and thirst provides sufficient electrolytes. A severe electrolyte deficiency may produce a *salt craving,* a strong desire to eat salty foods.

Electrolyte Output

The body loses some electrolytes by perspiring, with more lost in sweat on warmer days and during strenuous exercise. Varying amounts of electrolytes are lost in the feces. The greatest electrolyte output occurs as a result of kidney function and urine production. The kidneys alter electrolyte output to maintain balance.

> **Check Your Recall**
> 12. Which electrolytes are of greatest importance to cellular functions?
> 13. Which mechanisms ordinarily regulate electrolyte intake?
> 14. By what routes does the body lose electrolytes?

Regulation of Electrolyte Output

Precise concentrations of positively charged ions, such as sodium (Na^+), potassium (K^+), and calcium (Ca^{+2}), are required for nerve impulse conduction, muscle fiber contraction, and maintenance of cell membrane potential. *Sodium ions* account for nearly 90% of positively charged ions in extracellular fluids. The kidneys and the hormone aldosterone regulate these ions. Aldosterone,

Water Balance Disorders

Among the more common disorders that reflect an imbalance in the concentration of body fluids are dehydration, water intoxication, and edema.

Dehydration

In *dehydration,* water output exceeds water intake. Dehydration may follow excess sweating or prolonged water deprivation while water is still lost from the body. The extracellular fluid becomes increasingly more concentrated, and water leaves cells by osmosis (fig. 18A). Dehydration may also accompany prolonged vomiting or diarrhea that depletes body fluids.

During dehydration, the skin and mucous membranes of the mouth feel dry, and body weight drops. Hyperthermia may develop as the body's temperature-regulating mechanism becomes less effective due to lack of water for sweat.

Infants' kidneys are less able to conserve water than those of adults, so infants are more likely to become dehydrated. Elderly people are also especially susceptible to developing water imbalances because the sensitivity of their thirst mechanism decreases with age, and physical disabilities may make it difficult for them to obtain adequate fluids.

The treatment for dehydration is to replace the lost water and electrolytes. If only water is replaced, the extracellular fluid becomes more dilute than normal, producing a condition called water intoxication.

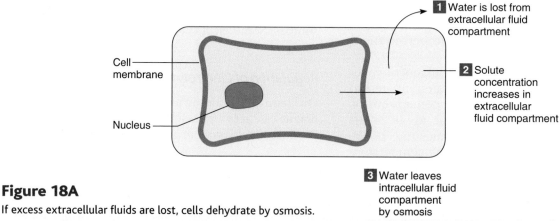

1 Water is lost from extracellular fluid compartment

2 Solute concentration increases in extracellular fluid compartment

3 Water leaves intracellular fluid compartment by osmosis

Figure 18A

If excess extracellular fluids are lost, cells dehydrate by osmosis.

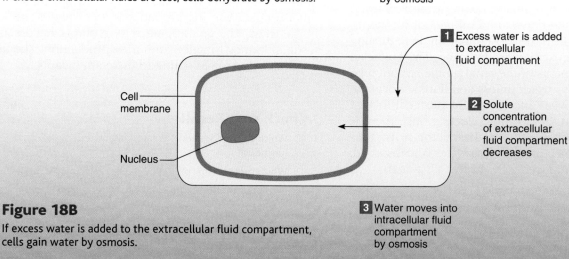

1 Excess water is added to extracellular fluid compartment

2 Solute concentration of extracellular fluid compartment decreases

3 Water moves into intracellular fluid compartment by osmosis

Figure 18B

If excess water is added to the extracellular fluid compartment, cells gain water by osmosis.

Water Intoxication

Babies rushed to emergency rooms because they are having seizures sometimes have drunk too much water, a rare condition called *water intoxication*. This can occur when a baby under six months of age is given several bottles of water a day or very dilute infant formula. The hungry infant drinks the water, and soon its tissues swell with the excess fluid. When the serum sodium level drops, the eyes begin to flutter, and a seizure occurs. As extracellular fluid becomes hypotonic, water enters the cells rapidly by osmosis (fig. 18B). Coma resulting from swelling brain tissues may follow unless water intake is restricted and hypertonic salt solutions given. Usually, recovery is complete within a few days. Water intoxication caused the death of a fraternity member forced to drink gallons of water as an initiation test.

Edema

Edema is an abnormal accumulation of extracellular fluid within the interstitial spaces. Several factors cause edema, including decrease in the plasma protein concentration (hypoproteinemia), obstruction of lymphatic vessels, increased venous pressure, and increased capillary permeability.

Hypoproteinemia may result from liver disease that hinders plasma protein synthesis; kidney disease (glomerulonephritis) that damages glomerular capillaries, allowing proteins to enter urine; or starvation, in which amino acid intake is insufficient to support synthesis of plasma proteins. In each case, the plasma protein concentration decreases, which decreases plasma colloid osmotic pressure, reducing the normal return of tissue fluid to the venular ends of capillaries. Consequently, tissue fluid accumulates in the interstitial spaces.

Edema may result from *lymphatic obstructions* due to surgery or parasitic infections of lymphatic vessels, as discussed in chapter 14 (p. 380). Back pressure develops in the lymphatic vessels, interfering with the normal movement of tissue fluid into them. At the same time, proteins that the lymphatic circulation ordinarily removes accumulate in interstitial spaces, raising the osmotic pressure of interstitial fluid. This effect draws still more fluid into the interstitial spaces.

If blood outflow from the liver into the inferior vena cava is blocked, venous pressure within the liver and portal blood vessels increases greatly. As a result, fluid with a high protein concentration is exuded from the surfaces of the liver and intestine into the peritoneal cavity. This increases the osmotic pressure of the abdominal fluid, which in turn attracts more water into the peritoneal cavity by osmosis. This condition, called *ascites,* distends the abdomen and is quite painful.

Edema may also result from increased capillary permeability accompanying *inflammation*. Recall that inflammation is a response to tissue damage and usually releases chemicals such as histamine from damaged cells. Histamine causes vasodilation and increased capillary permeability, so that excess fluid is filtered out of capillaries and enters interstitial spaces. Table 18A summarizes the factors that can cause edema.

Table 18A	Factors Associated with Edema	
Factor	**Cause**	**Effect**
Low plasma protein concentration	Liver disease and failure to synthesize proteins; kidney disease and loss of proteins in urine; lack of proteins in diet due to starvation	Plasma colloid osmotic pressure decreases; less fluid enters venular ends of capillaries by osmosis
Obstruction of lymphatic vessels	Surgical removal of portions of lymphatic pathways; certain parasitic infections	Back pressure in lymphatic vessels interferes with movement of fluid from interstitial spaces into lymph capillaries
Increased venous pressure	Venous obstructions or faulty venous valves	Back pressure in veins interferes with reabsorption of fluid from interstitial spaces into venular ends of capillaries
Inflammation	Tissue damage	Capillaries become abnormally permeable; fluid leaks from plasma into interstitial spaces

which the adrenal cortex secretes, increases sodium ion reabsorption in the distal convoluted tubules of the kidneys' nephrons and collecting ducts.

Aldosterone also regulates potassium ions. A rising potassium ion concentration directly stimulates the adrenal cortex to secrete aldosterone. This hormone enhances tubular reabsorption of sodium ions, and at the same time, causes tubular secretion of potassium ions (fig. 18.5). Conditions caused by an imbalance of these ions are discussed in the Topic of Interest entitled Sodium and Potassium Imbalances on page 499.

Recall from chapter 11 (p. 301) that the calcium ion concentration dropping below normal directly stimulates the parathyroid glands to secrete parathyroid hormone. This hormone returns the concentration of calcium in extracellular fluids toward normal.

Generally, the regulatory mechanisms that control positively charged ions secondarily control the concentrations of negatively charged ions. For example, renal tubules passively reabsorb chloride ions (Cl^-), the most abundant negatively charged ions in extracellular fluids, in response to active tubular reabsorption of sodium ions. That is, negatively charged chloride ions are electrically attracted to positively charged sodium ions and accompany them as they are reabsorbed (see chapter 17, p. 480).

Active transport mechanisms with limited transport capacities partially regulate some negatively charged ions, such as phosphate ions (PO_4^{-3}) and sulfate ions (SO_4^{-2}). Thus, if extracellular phosphate ion concentration is low, renal tubules reabsorb phosphate ions.

On the other hand, if the renal plasma threshold is exceeded, excess phosphate is excreted in urine.

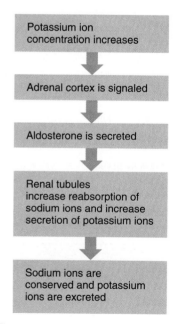

Figure 18.5

If potassium ion concentration increases, the kidneys conserve sodium ions and excrete potassium ions.

Check Your Recall

15. How does aldosterone regulate sodium and potassium ion concentration?

16. How is calcium regulated?

17. What mechanism regulates the concentrations of most negatively charged ions?

18.5 ACID-BASE BALANCE

Recall from chapter 2 (pp. 38–39) that electrolytes that dissociate in water and release hydrogen ions are called **acids** and electrolytes that release ions that combine with hydrogen ions are called **bases.** Maintenance of homeostasis depends on controlling the concentrations of acids and bases within body fluids.

Sources of Hydrogen Ions

Most of the hydrogen ions in body fluids originate as by-products of metabolic processes, although the digestive tract may directly absorb small quantities. The major metabolic sources of hydrogen ions include the following (fig. 18.6):

1. **Aerobic respiration of glucose** This process produces carbon dioxide and water. Carbon dioxide diffuses out of cells and reacts with the water in extracellular fluids to form *carbonic acid,* which then ionizes to release hydrogen ions and bicarbonate ions:

$$H_2CO_3 \rightarrow H^+ + HCO_3^-$$

2. **Anaerobic respiration of glucose** Anaerobically metabolized glucose produces *lactic acid,* which adds hydrogen ions to body fluids.

3. **Incomplete oxidation of fatty acids** This process produces *acidic ketone bodies,* which increase hydrogen ion concentration.

4. **Oxidation of sulfur-containing amino acids** This process yields *sulfuric acid* (H_2SO_4), which ionizes to release hydrogen ions.

5. **Breakdown (hydrolysis) of phosphoproteins and nucleic acids** Phosphoproteins and nucleic acids include phosphorus. Their oxidation produces *phosphoric acid* (H_3PO_4), which ionizes to release hydrogen ions.

The acids resulting from metabolism vary in strength. Thus, their effects on the hydrogen ion concentration of body fluids vary.

Check Your Recall

18. Distinguish between an acid and a base.

19. What are the major sources of hydrogen ions in the body?

Topic of Interest

Sodium and Potassium Imbalances

Extracellular fluids usually have high sodium ion concentrations, and intracellular fluid usually has a high potassium ion concentration. Renal regulation of sodium is closely related to that of potassium because secretion (and excretion) of potassium accompanies active reabsorption of sodium (under the influence of aldosterone). Therefore, conditions resulting from sodium ion imbalance often also involve potassium ion imbalance. Such disorders include:

1. *Low sodium concentration (hyponatremia)* Possible causes of sodium deficiencies include prolonged sweating, vomiting, or diarrhea; renal disease in which sodium is inadequately reabsorbed; adrenal cortex disorders in which aldosterone secretion is insufficient to promote sodium reabsorption (Addison disease); and drinking too much water. One possible effect of hyponatremia is the development of hypotonic extracellular fluid that promotes water movement into cells by osmosis, producing symptoms of water intoxication.

2. *High sodium concentration (hypernatremia)* Possible causes of elevated sodium concentration include excess water loss by evaporation (despite decreased sweating), as may occur during high fever, and increased water loss accompanying diabetes insipidus. In one form of

diabetes insipidus, ADH secretion is insufficient for renal tubules to maintain water conservation. Hypernatremia may disturb the central nervous system, causing confusion, stupor, and coma.

3. *Low potassium concentration (hypokalemia)* Possible causes of potassium deficiency include the release of excess aldosterone by the adrenal cortex (Cushing syndrome), which increases renal excretion of potassium; use of diuretic drugs that promote potassium excretion; kidney disease; and prolonged vomiting or diarrhea. Possible effects of hypokalemia include muscular weakness or paralysis, respiratory difficulty, and severe cardiac disturbances, such as atrial or ventricular arrhythmias.

4. *High potassium concentration (hyperkalemia)* Possible causes of elevated potassium concentration include renal disease, which decreases potassium excretion; use of drugs that promote renal conservation of potassium; the release of insufficient aldosterone by the adrenal cortex (Addison disease); and a shift of potassium from intracellular to extracellular fluid, a change that accompanies an increase in plasma hydrogen ion concentration (acidosis). Possible effects of hyperkalemia include paralysis of the skeletal muscles and severe cardiac disturbances, such as cardiac arrest.

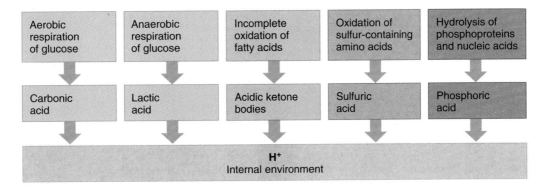

Figure 18.6

Some of the metabolic processes that provide hydrogen ions.

Strengths of Acids and Bases

Acids that dissociate to release hydrogen ions more completely are *strong acids,* and those that dissociate to release hydrogen ions less completely are *weak acids.* For example, the hydrochloric acid (HCl) of gastric juice is a strong acid, but the carbonic acid (H_2CO_3) produced when carbon dioxide reacts with water is weak.

Bases release ions, such as hydroxide ions (OH^-), which can combine with hydrogen ions and thereby lower their own concentration. Thus, sodium hydroxide (NaOH), which releases hydroxide ions, and sodium bicarbonate ($NaHCO_3$), which releases bicarbonate ions (HCO_3^-), are bases. Strong bases dissociate to release more OH^- or its equivalent than do weak bases. Often, the negative ions themselves are called bases. For example, HCO_3^- acting as a base combines with H^+ from the strong acid HCl to form the weak acid carbonic acid (H_2CO_3).

Regulation of Hydrogen Ion Concentration

Acid-base buffer systems, the respiratory center in the brainstem, and the nephrons in the kidneys regulate hydrogen ion concentration in body fluids. The pH scale is used to measure hydrogen ion concentration.

Acid-Base Buffer Systems

Acid-base buffer systems, in all body fluids, consist of chemicals that combine with excess acids or bases. More specifically, the chemical components of a buffer system can combine with strong acids, which release many hydrogen ions, converting them into weak acids, which release fewer hydrogen ions. Likewise, these buffers can combine with strong bases to convert them into weak bases. Such actions help minimize pH changes in body fluids. The three most important acid-base buffer systems in body fluids are:

1. **Bicarbonate buffer system** The bicarbonate buffer system, which is present in both intracellular and extracellular fluids, uses the bicarbonate ion (HCO_3^-), acting as a weak base, and carbonic acid (H_2CO_3), acting as a weak acid. In the presence of excess hydrogen ions, bicarbonate ions combine with hydrogen ions to form carbonic acid, thus minimizing any increase in the hydrogen ion concentration of the body fluids:

$$H^+ + HCO_3^- \rightarrow H_2CO_3$$

On the other hand, if conditions are basic or alkaline, carbonic acid dissociates to release bicarbonate ion and hydrogen ion:

$$H_2CO_3 \rightarrow H^+ + HCO_3^-$$

It is important to remember that even though this reaction releases bicarbonate ion, the increase of free hydrogen ions at equilibrium is what minimizes the shift toward a more alkaline pH.

2. **Phosphate buffer system** The phosphate acid-base buffer system also operates in both intracellular and extracellular body fluids. It is particularly important in the control of hydrogen ion concentrations in the tubular fluid of the nephrons and in urine. This buffer system consists of two phosphate ions—monohydrogen phosphate (HPO_4^{-2}) and dihydrogen phosphate ($H_2PO_4^-$). Under acidic conditions, monohydrogen phosphate ions react with hydrogen ions to produce dihydrogen phosphate:

$$H^+ + HPO_4^{-2} \rightarrow H_2PO_4^-$$

Under alkaline conditions, dihydrogen phosphate ions release hydrogen ions:

$$H_2PO_4^- \rightarrow H^+ + HPO_4^{-2}$$

3. **Protein buffer system** The protein acid-base buffer system consists of the plasma proteins, such as albumins, and certain proteins within cells, including the hemoglobin of red blood cells. As described in chapter 2 (p. 44), proteins are chains of amino acids. Some of these amino acids have freely exposed amino groups (–NH_2). When the solution pH falls, these amino groups can accept hydrogen ions:

$$-NH_2 + H^+ \rightarrow -NH_3^+$$

Some amino acids of a protein also have freely exposed *carboxyl groups* (–COOH). When the solution pH rises, these carboxyl groups can ionize, releasing hydrogen ions:

$$-COOH \rightarrow -COO^- + H^+$$

Thus, protein molecules can function as bases by accepting hydrogen ions into their amino groups or as acids by releasing hydrogen ions from their carboxyl groups. This special property allows protein molecules to operate as an acid-base buffer system, minimizing changes in pH.

Table 18.1 summarizes the actions of the three major buffer systems.

Neurons are particularly sensitive to changes in the pH of body fluids. If the interstitial fluid becomes more alkaline than normal (alkalosis), neurons become more excitable, and seizures may result. Conversely, acidic conditions (acidosis) depress neuron activity, reducing the level of consciousness.

Table 18.1	Chemical Acid-Base Buffer System	
Buffer System	**Constituents**	**Actions**
Bicarbonate system	Bicarbonate ion (HCO_3^-)	Combines with a hydrogen ion under acidic conditions
	Carbonic acid (H_2CO_3)	Releases a hydrogen ion under alkaline conditions
Phosphate system	Monohydrogen phosphate (HPO_4^{-2})	Combines with a hydrogen ion under acidic conditions
	Dihydrogen phosphate ($H_2PO_4^-$)	Releases a hydrogen ion under alkaline conditions
Protein system (and amino acids)	–NH_2 group of an amino acid or protein	Combines with a hydrogen ion in the presence of excess acid
	–COOH group of an amino acid or protein	Releases a hydrogen ion in the presence of excess base

The Medullary Respiratory Center

The medullary respiratory center in the brainstem helps regulate the hydrogen ion concentration in body fluids by controlling the rate and depth of breathing (see chapter 16, pp. 457–458). Figure 18.7 traces this process. Specifically, if cells increase carbon dioxide production, as during physical exercise, carbonic acid production increases. As carbonic acid dissociates, the concentration of hydrogen ions increases, and the pH of the internal environment drops. Such an increasing concentration of carbon dioxide in the central nervous system and the subsequent increase in hydrogen ion concentration in the cerebrospinal fluid stimulate chemoreceptors in the medulla oblongata.

The respiratory center responds by increasing the depth and rate of breathing, so the lungs excrete more carbon dioxide. This returns the hydrogen ion concentration in body fluids toward normal because the released carbon dioxide comes from carbonic acid:

$$H_2CO_3 \rightarrow CO_2 + H_2O$$

Conversely, if cells are less active, production of carbon dioxide and hydrogen ions in body fluids remains relatively low. As a result, breathing rate and depth stay closer to resting levels.

The Kidneys

Nephrons help regulate the hydrogen ion concentration of body fluids by excreting hydrogen ions in urine. Recall from chapter 17 (p. 481) that epithelial cells lining certain segments of the renal tubules secrete hydrogen ions into the tubular fluid.

Rates of Regulation

The various regulators of hydrogen ion concentration operate at different rates. Acid-base buffers can convert strong acids or bases into weak acids or bases almost immediately. For this reason, these chemical buffer systems are sometimes called the body's *first line of defense* against shifts in pH.

Physiological buffer systems, such as the respiratory and renal mechanisms, function more slowly and constitute the body's *second line of defense* against shifts in pH. The respiratory mechanism may require several minutes to begin resisting a change in pH, and the renal mechanism may require one to three days to regulate a changing hydrogen ion concentration. Figure 18.8 compares the actions of chemical buffers and physiological buffers.

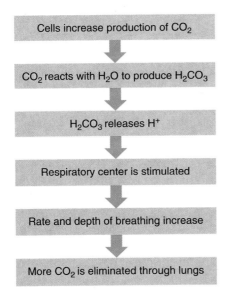

Figure 18.7

An increase in carbon dioxide production increases carbon dioxide elimination.

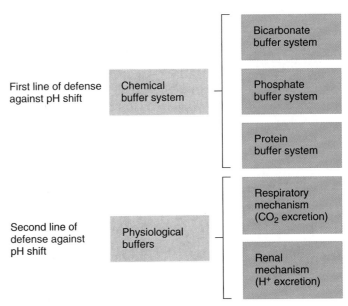

Figure 18.8

Chemical buffers act rapidly, whereas physiological buffers may require several minutes to several days to begin resisting a change in pH.

18.6 ACID-BASE IMBALANCES

Chemical and physiological buffer systems generally maintain the hydrogen ion concentration of body fluids within very narrow pH ranges. The pH of arterial blood is normally 7.35–7.45. Abnormal conditions may disturb the acid-base balance. A pH value below 7.35 produces *acidosis*. A pH above 7.45 produces *alkalosis*. Such shifts in the pH of body fluids can be life-threatening. A person usually cannot survive if the pH of body fluids drops to 6.8 or rises to 8.0 for longer than a few hours (fig. 18.9).

Acidosis results from an accumulation of acids or loss of bases, either of which increases the hydrogen ion concentration of body fluids. Conversely, alkalosis results from a loss of acids or an accumulation of bases accompanied by a decrease in hydrogen ion concentration (fig. 18.10).

Acidosis

The two major types of acidosis are respiratory acidosis and metabolic acidosis. Factors that increase carbon dioxide concentration, also increasing the concentration of carbonic acid (the respiratory acid), cause respiratory acidosis. Metabolic acidosis is due to accumulation of any other acids in the body fluids or to loss of bases, including bicarbonate ions.

Respiratory acidosis may be due to hindered pulmonary ventilation, which increases carbon dioxide concentration. This may result from the following conditions:

1. Injury to the respiratory center of the brainstem, decreasing rate and depth of breathing.
2. Obstruction of air passages that interferes with air movement into alveoli.
3. Diseases that decrease gas exchange, such as pneumonia, or that reduce the surface area of the respiratory membrane, such as emphysema.

Figure 18.11 summarizes the factors that can lead to respiratory acidosis. Any of these conditions can increase the level of carbonic acid and hydrogen ions in body

fluids, lowering pH. Chemical buffers, such as hemoglobin, may resist this shift in pH. At the same time, rising concentrations of carbon dioxide and hydrogen ions stimulate the respiratory center, increasing the breathing rate and depth and thereby lowering the carbon dioxide concentration. Also, the kidneys may begin to excrete more hydrogen ions. Eventually, these chemical and physiological buffers return the pH of the body fluids to normal. The acidosis is thus *compensated*.

The symptoms of respiratory acidosis result from depression of central nervous system function. They include drowsiness, disorientation, stupor, labored breathing, and

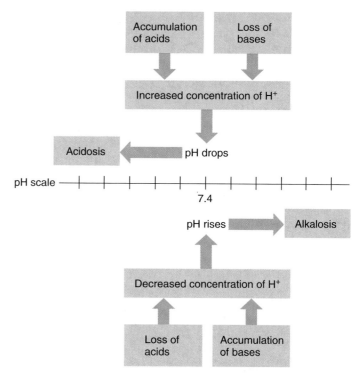

Figure 18.10

Acidosis results from accumulation of acids or loss of bases. Alkalosis results from loss of acids or accumulation of bases.

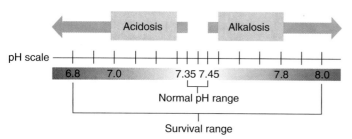

Figure 18.9

If the pH of arterial blood drops to 6.8 or rises to 8.0 for more than a few hours, the person usually cannot survive.

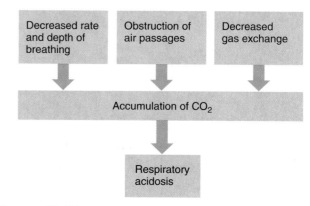

Figure 18.11

Some of the factors that lead to respiratory acidosis.

cyanosis. In *uncompensated acidosis,* the person may become comatose and die.

Metabolic acidosis is due to accumulation of non-respiratory acids or loss of bases. Factors that may lead to this condition include the following:

1. Kidney disease that reduces glomerular filtration so that the kidneys fail to excrete acids produced in metabolism (uremic acidosis).

2. In diabetes mellitus, metabolic reactions convert some fatty acids into ketone bodies, such as *acetoacetic acid, beta-hydroxybutyric acid,* and *acetone.* Normally, these molecules are rare, and cells oxidize them as energy sources. However, if fats are being utilized too quickly, as in diabetes mellitus, ketone bodies may accumulate faster than they can be oxidized and be excreted in urine (ketonuria). The lungs may excrete acetone, which is volatile and imparts a fruity odor to the breath. Acetoacetic acid and beta-hydroxybutyric acid may accumulate and lower pH (ketoacidosis), and also combine with bicarbonate ions in the renal tubules. As a result, excess bicarbonate ions are excreted in the urine, interfering with the function of the bicarbonate acid-base buffer system.

3. Prolonged vomiting with loss of the alkaline contents of the upper intestine and the stomach contents. (Losing only the stomach contents produces metabolic alkalosis.) Vomiting can empty not only the stomach, but also the first foot or so of the intestine.

4. Prolonged diarrhea with loss of excess alkaline intestinal secretions (especially in infants).

Figure 18.12 summarizes the factors leading to metabolic acidosis. In each case, pH is lowered. Countering this lower pH are chemical buffer systems, which accept excess hydrogen ions; the respiratory center, which increases breathing rate and depth; and the kidneys, which excrete more hydrogen ions.

Alkalosis

The two major types of alkalosis are respiratory alkalosis and metabolic alkalosis. Respiratory alkalosis results from excessive loss of carbon dioxide and consequent loss of carbonic acid. Metabolic alkalosis is due to excessive loss of hydrogen ions or gain of bases.

Respiratory alkalosis develops as a result of *hyperventilation* (described in chapter 16, pp. 459–460), in which too much carbon dioxide is lost, consequently decreasing carbonic acid and hydrogen ion concentrations. Hyperventilation may occur in response to anxiety or may accompany fever or poisoning from salicylates, such as aspirin. At high altitudes, hyperventilation may be a response to low oxygen partial pressure. Musicians can hyperventilate when providing the large volume of air needed to play sustained passages on wind instruments. In each case, rapid, deep breathing depletes carbon dioxide, and the pH of body fluids increases. Figure 18.13 illustrates the factors leading to respiratory alkalosis.

Chemical buffers, such as hemoglobin, that release hydrogen ions resist the increase in pH. The lower concentrations of carbon dioxide and hydrogen ions decrease stimulation of the respiratory center. This inhibits the hyperventilation, thus reducing further carbon dioxide loss. At the same time, the kidneys excrete fewer hydrogen ions, and the urine becomes alkaline as bases are excreted.

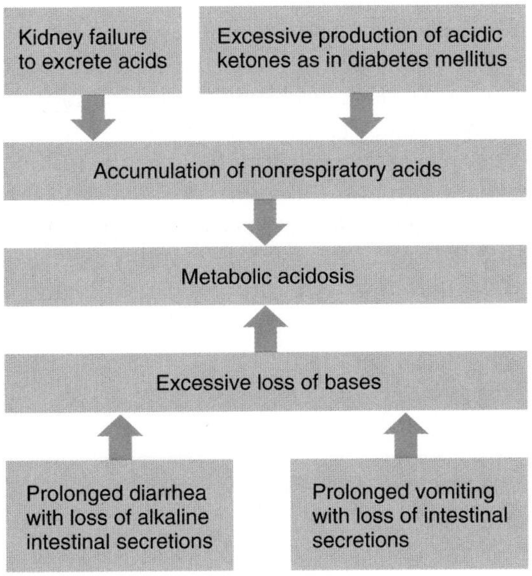

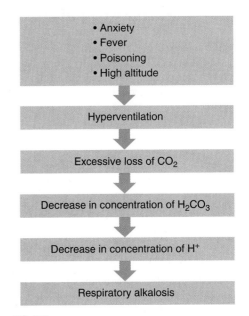

Figure 18.12

Some of the factors that lead to metabolic acidosis.

Figure 18.13

Some of the factors that lead to respiratory alkalosis.

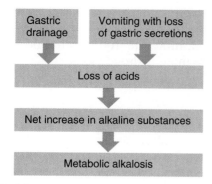

Figure 18.14
Some of the factors that lead to metabolic alkalosis.

glucosuria (glu″ko-su′re-ah) Excess sugar in urine.
hyperglycemia (hi″per-gli-se′me-ah) Abnormally high blood sugar level.
hyperkalemia (hi″per-kah-le′me-ah) Excess potassium in the blood.
hypernatremia (hi″per-na-tre′me-ah) Excess sodium in the blood.
hyperuricemia (hi″per-u″rĭ-se′me-ah) Excess uric acid in the blood.
hypoglycemia (hi″po-gli-se′me-ah) Abnormally low blood sugar level.
ketonuria (ke″to-nu′re-ah) Ketone bodies in the urine.
ketosis (ke″to′sis) Acidosis due to excess ketone bodies in body fluids.
proteinuria (pro″te-ĭ-nu′re-ah) Protein in the urine.
uremia (u-re′me-ah) Toxic condition resulting from nitrogenous wastes in the blood.

Symptoms of respiratory alkalosis include lightheadedness, agitation, dizziness, and tingling sensations. In severe cases, peripheral nerves may spontaneously trigger impulses, and muscles may respond with tetanic contractions (see chapter 8, p. 188).

Metabolic alkalosis results from a great loss of hydrogen ions or from a gain in bases, both of which increase blood pH (alkalemia). This condition may follow gastric drainage (lavage), prolonged vomiting of stomach contents, or use of certain diuretic drugs. Because gastric juice is very acidic, its loss leaves body fluids more basic. Metabolic alkalosis may also develop from ingesting too much antacid, such as sodium bicarbonate. Symptoms of metabolic alkalosis include decreased breathing rate and depth, which in turn increases the blood carbon dioxide concentration. Figure 18.14 illustrates the factors leading to metabolic alkalosis.

Clinical Terms Related to Water and Electrolyte Balance

acetonemia (as″ĕ-to-ne′me-ah) Excess acetone in blood.
acetonuria (as″ĕ-to-nu′re-ah) Excess acetone in urine.
albuminuria (al-bu″mĭ-nu′re-ah) Albumin in urine.
anasarca (an″ah-sar′kah) Widespread accumulation of tissue fluid.
antacid (ant-as′id) Substance that neutralizes an acid.
anuria (ah-nu′re-ah) Absence of urine excretion.
azotemia (az″o-te′me-ah) Accumulation of nitrogenous wastes in blood.
diuresis (di″u-re′sis) Increased urine production.

Clinical Connection

From 10% to 20% of people who have cancer develop excess calcium ions in the blood, called hypercalcemia. Normally, only about 1% of the body's calcium is in the bloodstream, but in these cancer patients, the bones lose more than this. Hypercalcemia is difficult to diagnose because symptoms are similar to those of cancer—fatigue, weakness, lack of appetite, nausea, vomiting, constipation, thirst, and clouded thinking.

Two mechanisms underlie the calcium imbalance. The cancer itself may invade the bone, releasing stored calcium. Cancer cells also secrete parathyroid hormone-related proteins that stimulate osteoclasts to destroy bone tissue, releasing calcium. Hypercalcemia worsens if a person is bedbound, which hastens calcium loss from bone, and if the person is dehydrated. The high blood calcium may also impair the ability of the kidneys to retain water, producing copious urine (polyuria), which exacerbates the dehydration.

Hypercalcemia most often develops in people who have breast or lung cancer, but is also associated with cancers of the head, neck, and kidney, and the blood cancer multiple myeloma. Treatment is to provide fluids and to give drugs that restore calcium balance, including pamidronate disodium and the more recently approved zoledronic acid. Treating cancer-associated hypercalcemia can extend life, improve quality of life, and make the person feel well enough to undergo further cancer treatment.

SUMMARY OUTLINE

18.1 Introduction (p. 492)

Maintenance of water and electrolyte balance requires that equal quantities of these substances enter and leave the body. Altering the water balance affects the electrolyte balance.

18.2 Distribution of Body Fluids (p. 492)

1. Fluid compartments
   a. The intracellular fluid compartment includes the fluids and electrolytes that are enclosed by cell membranes.
   b. The extracellular fluid compartment includes all the fluids and electrolytes outside the cell membranes.

2. Body fluid composition
   a. Extracellular fluids have high concentrations of sodium, chloride, and bicarbonate ions. These fluids include a greater concentration of calcium ions with less potassium, magnesium, phosphate, and sulfate ions than does intracellular fluid. Plasma contains more protein than does either interstitial fluid or lymph.
   b. Intracellular fluid contains high concentrations of potassium, magnesium, and phosphate ions. It also has a greater concentration of sulfate ions and lesser concentrations of sodium, chloride, calcium, and bicarbonate ions than does extracellular fluid.
3. Movement of fluid between compartments
   a. Hydrostatic and osmotic pressure regulate fluid movements.
      (1) Hydrostatic pressure forces fluid out of plasma, and colloid osmotic pressure returns fluid to plasma.
      (2) Hydrostatic pressure drives fluid into lymph vessels.
      (3) Osmotic pressure regulates fluid movement in and out of cells.
   b. Sodium ion concentrations are especially important in regulating fluid movement.

18.3 Water Balance (p. 494)

1. Water intake
   a. Most water comes from consuming liquids or moist foods.
   b. Oxidative metabolism produces some water.
2. Regulation of water intake
   a. Thirst is the primary regulator of water intake.
   b. Drinking and the resulting stomach distension inhibit thirst.
3. Water output
   Water is lost in urine, feces, and sweat, and by evaporation from the skin and lungs.
4. Regulation of water output
   The distal convoluted tubules of the nephrons and collecting ducts regulate water output.

18.4 Electrolyte Balance (p. 495)

1. Electrolyte intake
   a. The electrolytes of greatest importance to cellular functions in body fluids dissociate to release ions of sodium, potassium, calcium, magnesium, chloride, sulfate, phosphate, bicarbonate, and hydrogen.
   b. These ions are obtained in foods and beverages or as by-products of metabolic processes.
2. Regulation of electrolyte intake
   a. Food and drink usually provide sufficient electrolytes.
   b. A severe electrolyte deficiency may produce a salt craving.
3. Electrolyte output
   a. Electrolytes are lost through perspiration, feces, and urine.
   b. Quantities lost vary with temperature and physical exercise.
   c. Most electrolytes are lost through the kidneys.
4. Regulation of electrolyte output
   a. Concentrations of sodium, potassium, and calcium ions in body fluids are particularly important.
   b. The adrenal cortex secretes aldosterone to regulate sodium and potassium ions.
   c. Parathyroid hormone regulates calcium ions.
   d. The mechanisms that control positively charged ions secondarily regulate negatively charged ions.

18.5 Acid-Base Balance (p. 498)

Acids are electrolytes that dissociate to release hydrogen ions. Bases release ions that combine with hydrogen ions. Body fluid pH must remain within a certain range.

1. Sources of hydrogen ions
   a. Aerobic respiration of glucose produces carbonic acid.
   b. Anaerobic respiration of glucose produces lactic acid.
   c. Incomplete oxidation of fatty acids releases acidic ketone bodies.
   d. Oxidation of sulfur-containing amino acids produces sulfuric acid.
   e. Hydrolysis of phosphoproteins and nucleic acids produces phosphoric acid.
2. Strengths of acids and bases
   a. Acids vary in the extent to which they dissociate to release ions.
      (1) Strong acids, such as hydrochloric acid, dissociate more completely.
      (2) Weak acids, such as carbonic acid, dissociate less completely.
   b. Bases also vary in strength.
3. Regulation of hydrogen ion concentration
   a. Acid-base buffer systems
      (1) Buffer systems convert strong acids into weaker acids or strong bases into weaker bases.
      (2) They include the bicarbonate buffer system, phosphate buffer system, and protein buffer system.
      (3) Buffer systems minimize pH changes.
   b. The respiratory center controls the rate and depth of breathing to regulate pH.
   c. Kidney nephrons excrete hydrogen ions to regulate pH.
   d. Chemical buffers act more rapidly. Physiological buffers act less rapidly.

18.6 Acid-Base Imbalances (p. 502)

1. Acidosis
   a. Respiratory acidosis results from increases in the concentrations of carbon dioxide and carbonic acid.
   b. Metabolic acidosis results from accumulation of other acids or loss of bases.
2. Alkalosis
   a. Respiratory alkalosis results from loss of carbon dioxide and carbonic acid.
   b. Metabolic alkalosis results from loss of hydrogen ions or gain of bases.

CHAPTER ASSESSMENTS

◀ **18.1 Introduction**

1. Describe how water balance and electrolyte balance are interdependent. (p. 492)

◀ **18.2 Distribution of Body Fluids**

2. All of the water and electrolytes enclosed by cell membranes constitute the: (p. 492)

   a. transcellular fluid   d. lymph
   b. intracellular fluid   e. plasma
   c. extracellular fluid

3. Contrast the composition of intracellular fluid and extracellular fluid. (p. 492)

4. Describe the control of fluid movement between body fluid compartments. (p. 493)

◀ **18.3 Water Balance**

5. Prepare a list of sources of normal water gain and loss to illustrate how water intake equals water output. (p. 494)

6. Define *water of metabolism*. (p. 494)

7. Explain how water intake is regulated. (p. 494)

8. Explain how nephrons regulate water output. (p. 495)

18.4 Electrolyte Balance

9. The electrolytes of greatest importance to cellular functions in body fluids include: (p. 495)

   a. sodium   d. chloride

   b. potassium   e. all of the above

   c. calcium

10. Explain how electrolyte intake is regulated. (p. 495)

11. List the routes by which electrolytes leave the body. (p. 495)

12. Explain how the adrenal cortex regulates electrolyte output. (p. 495)

13. Describe the role of the parathyroid glands in regulating electrolyte balance. (p. 498)

18.5 Acid-Base Balance

14. List five sources of hydrogen ions in body fluids, and name an acid that originates from each source. (p. 498)

15. _____ _____ dissociate to release hydrogen ions more completely. An example is hydrochloric acid. (p. 499)

16. _____ _____ dissociate to release fewer hydroxide ions. (p. 499)

17. Describe how the bicarbonate and phosphate buffer systems resist pH changes. (p. 500)

18. Explain why a protein has acidic as well as basic properties. (p. 500)

19. Discuss how the respiratory system and the kidneys function in the regulation of the acid-base balance. (p. 501)

18.6 Acid-Base Imbalances

20. Distinguish between respiratory and metabolic acid-base imbalances. (p. 502)

21. Explain how the body compensates for acid-base imbalances. (p. 502)

INTEGRATED ASSESSMENTS/ CRITICAL THINKING

OUTCOME 18.1

1. Radiation therapy may damage the mucosa of the stomach and intestines. What effect might this have on the patient's electrolyte balance?

OUTCOME 18.2

2. If the right ventricle of a patient's heart is failing, increasing venous pressure, what changes might occur in the patient's extracellular fluid compartments?

OUTCOMES 18.3, 18.4

3. A thirty-eight-year-old woman contracted *Mycoplasma* pneumonia and ran a temperature of 104°F for five days. Even though the woman drank great volumes of liquid, her blood pressure dropped to 70/50, indicating dehydration. Should the woman receive intravenous hypertonic glucose or normal isotonic saline? Why?

4. Some time ago, several newborn infants died after sodium chloride was accidentally substituted for sugar in their formula. What symptoms would this produce? Why are infants more prone to the hazard of excess salt intake than adults?

OUTCOMES 18.5, 18.6

5. After eating an undercooked hamburger, a twenty-five-year-old male developed diarrhea due to infection with a strain of *Escherichia coli* that produces a shigatoxin. How would this affect his blood pH, urine pH, and respiratory rate?

6. A student hyperventilates and is disoriented just before an exam. Is this student likely to be experiencing acidosis or alkalosis? How will the body compensate in an effort to maintain homeostasis?

7. A ten-year-old female is rescued from a swimming pool after several minutes of floundering in the water. What is (are) the cause(s) of the girl's acidosis? What treatment(s) will bring the body back to homeostasis?

WEB CONNECTIONS

Visit the text website at **aris.mhhe.com** for additional quizzes, interactive learning exercises, and more.

AP R URINARY SYSTEM

Anatomy & Physiology | REVEALED includes cadaver photos that allow you to peel away layers of the human body to reveal structures beneath the surface. This program also includes animations, radiologic imaging, audio pronunciations, and practice quizzing. To learn more visit **www.aprevealed.com**.

19
Reproductive Systems

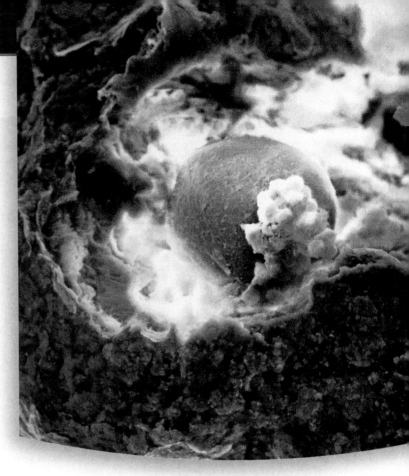

SELLING EGGS. The ad in the student newspaper seemed too good to be true—the fee for donating a few eggs would pay nearly a semester's tuition. Intrigued, the young woman submitted a health history, had a checkup, and a month later received a call. A young couple struggling with infertility sought an egg donor. They'd chosen Sherrie because, with her strawberry-blond hair, she looked just like Linda, the woman whose cancer had left her unable to conceive. The donor eggs would be fertilized in a laboratory dish (*in vitro*) with sperm from Linda's partner, and then implanted in Linda's uterus.

For two weeks Sherrie injected herself in the thigh with a drug that acts like gonadotropin-releasing hormone, suppressing release of an egg from an ovary (ovulation). When daily hormone checks indicated that her endocrine system was in sync with Linda's, Sherrie began giving herself shots twice a day at the back of the hip. This second drug mimicked follicle-stimulating hormone, and it caused several ovarian follicles to mature. Finally, injections of luteinizing hormone brought the eggs to full maturity. Then, at a health-care facility, Sherrie was given pain medication and light sedation. A needle was inserted through her vaginal wall to remove a dozen of the most mature eggs as they swelled to the surface of her ovary. This is the least invasive way to retrieve eggs.

Two *in vitro* fertilized ova divided a few times, forming an early embryo, and were then implanted into Linda's uterus. The rest were frozen, for possible later use. The preparation and procedure weren't too painful. Sherrie had felt a dull aching the last day, and felt bloated for a few days after the egg retrieval, but she did not experience

Ads in student newspapers attempt to entice young women to donate eggs. The protocol is very uncomfortable and takes several weeks. Pay is high, and is often based on SAT scores, college grades, and even the reputation of a particular university.

bleeding, infection, cramping, or mood swings. Nor did she develop a complication in which too many eggs mature, causing fluid to leak from blood vessels and accumulate in the abdomen. About 6% of egg donors develop the syndrome, which can cause infertility, kidney failure, and even death. Future risks, however, are uncertain, because eggs haven't yet been collected long enough to know the consequences. Case reports point to ovary scarring and possibly even cancer. Another side effect that Sherrie had not fully considered was how she would feel afterward. Although she was happy to have helped the couple and to have paid her tuition, she feared she would always wonder about the twins her eggs had become.

Learning Outcomes   *After studying this chapter, you should be able to do the following:*

19.1 Introduction

1. State the general functions of the male and female reproductive systems. (p. 508)

19.2 Organs of the Male Reproductive System

2. Describe the general functions of each part of the male reproductive system. (p. 508)

3. Outline the process of spermatogenesis. (p. 509)

4. Describe semen production and exit from the body. (p. 513)

19.3 Hormonal Control of Male Reproductive Functions

5. Explain how hormones control the activities of male reproductive organs and the development of male secondary sex characteristics. (p. 515)

19.4 Organs of the Female Reproductive System

6. Describe the general functions of each part of the female reproductive system. (p. 518)

7. Outline the process of oogenesis. (p. 519)

19.5 Hormonal Control of Female Reproductive Functions

8. Explain how hormones control the activities of female reproductive organs and the development of female secondary sex characteristics. (p. 524)

9. Describe the major events that occur during a reproductive cycle. (p. 524)

19.6 Mammary Glands

10. Review the structure of the mammary glands. (p. 526)

Aids to Understanding Words   *(Appendix A on page 567 has a complete list of Aids to Understanding Words.)*

andr- [man] *andr*ogens: Male sex hormones.

ejacul- [to shoot forth] *ejacul*ation: Process of expelling semen from the male reproductive tract.

fimb- [fringe] *fimb*riae: Irregular extensions on the margin of the infundibulum of the uterine tube.

follic- [small bag] *follic*le: Ovarian structure that contains an egg.

genesis- [origin] spermato*genesis*: Formation of sperm cells.

germ- [to bud or sprout] *germ*inal epithelium: Tissue that gives rise to sex cells by special cell division.

labi- [lip] *labi*a minora: Flattened, longitudinal folds that extend along the margins of the female vestibule.

mens- [month] *mens*es: Monthly flow of blood from the female reproductive tract.

mons- [an eminence] *mons* pubis: Rounded elevation overlying the symphysis pubis in a female.

puber- [adult] *puber*ty: Time when a person becomes able to reproduce.

19.1 INTRODUCTION

The male and female reproductive systems are a connected series of organs and glands that produce and nurture sex cells and transport them to sites of fertilization. Male sex cells are **sperm.** Female sex cells are eggs, or **oocytes** (o-o-sitz), which in Latin means "egg cells."

Sex cells have one set of genetic instructions, carried on 23 chromosomes, compared to two sets on 46 chromosomes in other cells. When sex cells join at fertilization, the amount of genetic information held in 46 chromosomes is restored. Some of the reproductive organs secrete hormones vital to the development and maintenance of secondary sex characteristics and the regulation of reproductive physiology.

19.2 ORGANS OF THE MALE REPRODUCTIVE SYSTEM

Organs of the male reproductive system produce and maintain male sex cells, or *sperm cells;* transport these cells and supporting fluids to the outside; and secrete male sex hormones. A male's *primary sex organs* (gonads) are the two testes in which sperm cells and male sex hormones form. The *accessory sex organs* of the male reproductive system are the internal and external reproductive organs (fig. 19.1; reference plates 3 and 4, pp. 25–26).

Testes

The **testes** (tes'tēz; sing., *testis*) are ovoid structures about 5 centimeters in length and 3 centimeters in diameter. Both testes are within the cavity of the saclike *scrotum*.

Structure of the Testes

A tough, white, fibrous capsule encloses each testis. Along the capsule's posterior border, the connective tissue thickens and extends into the testis, forming thin septa that divide the testis into about 250 *lobules*.

Each lobule contains one to four highly coiled, convoluted **seminiferous tubules** (se″mĭ-nif′er-us too′būlz), each approximately 70 centimeters long uncoiled. These tubules course posteriorly and unite to form a complex network of channels. The channels give rise to several ducts that join a tube called the *epididymis*. The epididymis coils on the outer surface of the testis and continues to become the *ductus deferens* (fig. 19.2*a*).

A specialized stratified epithelium with **spermatogenic** (sper″mah-to-jen′ik) **cells,** which give rise to sperm cells, lines the seminiferous tubules. Other specialized cells, called **interstitial cells** (cells of Leydig), lie in the spaces between the seminiferous tubules (fig. 19.2*b,c*). Interstitial cells produce and secrete male sex hormones.

The epithelial cells of the seminiferous tubules can give rise to *testicular cancer,* a common cancer in young men. In most cases, the first sign is a painless testis enlargement or a scrotal mass that attaches to a testis.

If a biopsy (tissue sample) reveals cancer cells, surgery can remove the affected testis (orchiectomy). Radiation and/or chemotherapy are very successful in preventing the cancer from recurring.

Check Your Recall

1. Describe the structure of a testis.

2. Where in the testes are the sperm cells produced?

3. Which cells produce male sex hormones?

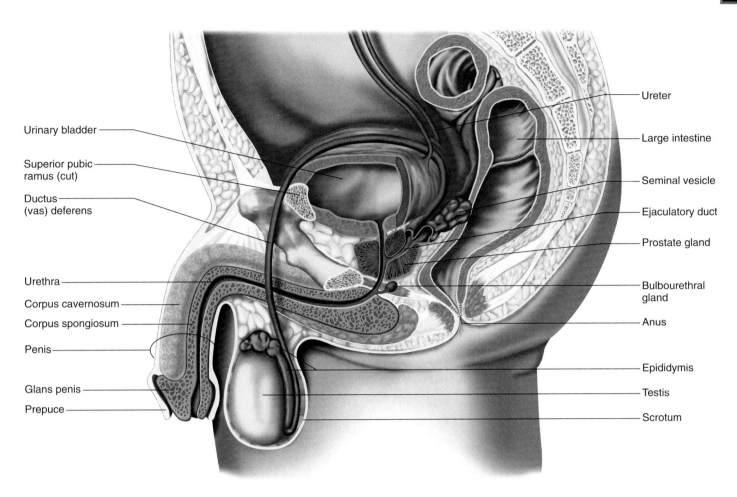

Figure 19.1

Male reproductive organs (sagittal view). The paired testes are the primary sex organs, and the other structures, both internal and external, are accessory sex organs.

Formation of Sperm Cells

The epithelium of the seminiferous tubules consists of supporting cells (sustentacular cells, or Sertoli cells) and spermatogenic cells. Supporting cells provide a scaffolding for the spermatogenic cells, and also nourish and regulate them.

Males produce sperm cells continually, starting a puberty. Sperm cells collect in the lumen of each seminiferous tubule, and then pass to the epididymis, where they accumulate and mature.

A mature sperm cell is a tiny, tadpole-shaped structure about 0.06 millimeters long. It consists of a flattened head, a cylindrical midpiece (body), and an elongated tail (fig. 19.3; see fig. 3.9*b*, p. 59).

The oval *head* of a sperm cell is composed primarily of a nucleus and contains highly compacted chromatin consisting of 23 chromosomes. A small protrusion at its anterior end, called the *acrosome,* contains enzymes that help the sperm cell penetrate an egg cell during fertilization. (Chapter 20, pp. 539–540, describes this process.)

The *midpiece* of a sperm cell has a central, filamentous core and many mitochondria in a spiral. The *tail* (flagellum) consists of several microtubules enclosed in an extension of the cell membrane. The mitochondria provide ATP for the tail's lashing movement, which propels the sperm cell through fluid.

Spermatogenesis

Sperm cells form in a process called **spermatogenesis** (sper″mah-to-jen′ĕ-sis). In the male embryo, spermatogenic cells are undifferentiated and are also called **spermatogonia** (sper″mah-to-go′ne-ah) (fig. 19.4). Each spermatogonium contains 46 chromosomes in its nucleus, the usual number for human cells. Beginning during embryonic development, hormones stimulate spermatogonia to undergo mitosis (see chapter 3, p. 68). Each cell division gives rise to two daughter cells, one a new type A spermatogonium which maintains the supply of undifferentiated cells, the other a type B spermatogonium which enlarges to become a *primary spermatocyte.* Supporting cells help sustain the developing sperm cells.

At puberty, the primary spermatocytes then reproduce by a special type of cell division called **meiosis** (mi-o′sis). Meiosis includes two successive divisions,

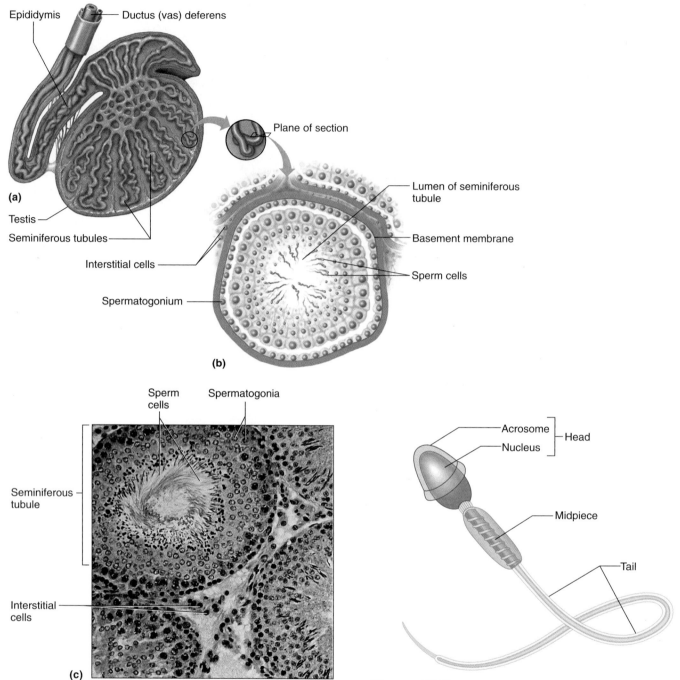

(a)

Epididymis — Ductus (vas) deferens

Plane of section

Testis

Seminiferous tubules

Interstitial cells

Spermatogonium

Lumen of seminiferous tubule

Basement membrane

Sperm cells

(b)

Sperm cells   Spermatogonia

Seminiferous tubule

Interstitial cells

(c)

Figure 19.2

Structure of the testes. (*a*) Sagittal section of the testis. (*b*) Cross section of the seminiferous tubule. (*c*) Light micrograph of seminiferous tubules (200×).

Acrosome ⎱ Head
Nucleus ⎰

Midpiece

Tail

Figure 19.3

A mature sperm cell has three distinctive parts: a tail that provides motility, a midpiece rich in mitochondria, and a head that houses the nucleus. The acrosome at the anterior end of the sperm cell contains enzymes that enable the sperm cell to penetrate the egg cell.

called the *first* and *second meiotic divisions*. The first meiotic division (meiosis I) separates homologous chromosome pairs. Homologous pairs are the same, gene for gene. They may not be identical, however, because a gene may have variants, and the chromosome that comes from the person's mother may carry a different variant for the corresponding gene from the father's homologous chromosome. Before meiosis I, each homologous chromosome is replicated, so it consists of two complete DNA strands called *chromatids*. The chromatids of a replicated chromosome attach at regions called *centromeres*. Each chromatid has the complete genetic information associated with that chromosome.

Each of the cells that undergoes the second meiotic division (meiosis II) begins with one member of each homologous pair, a condition termed **haploid** (hap′loyd).

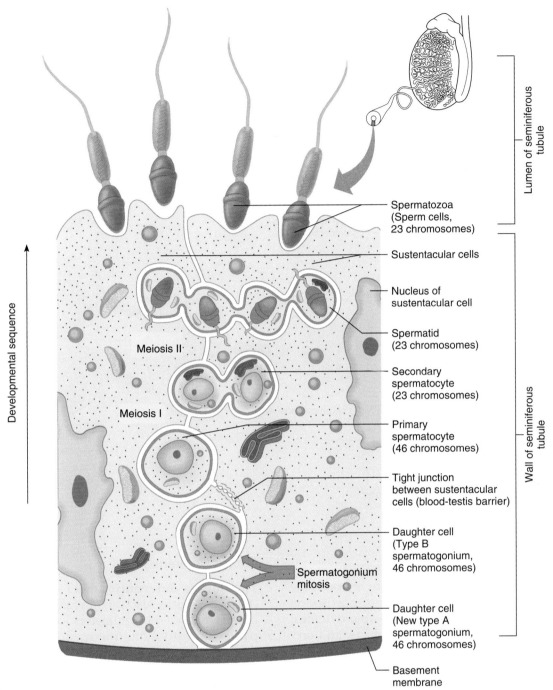

Lumen of seminiferous tubule

Spermatozoa (Sperm cells, 23 chromosomes)

Sustentacular cells

Nucleus of sustentacular cell

Spermatid (23 chromosomes)

Secondary spermatocyte (23 chromosomes)

Primary spermatocyte (46 chromosomes)

Tight junction between sustentacular cells (blood-testis barrier)

Daughter cell (Type B spermatogonium, 46 chromosomes)

Daughter cell (New type A spermatogonium, 46 chromosomes)

Basement membrane

Developmental sequence

Wall of seminiferous tubule

Meiosis II

Meiosis I

Spermatogonium mitosis

Figure 19.4

Spermatogonia (type A) continue the germ cell line. Spermatogonia (type B) give rise to primary spermatocytes by mitosis; the spermatocytes, in turn, give rise to sperm cells by meiosis. Note that as the cells approach the lumen, they mature.

That is, a haploid cell has one set of chromosomes. This second division separates the chromatids, producing cells that are still haploid, but whose chromosomes are no longer in the replicated form. After meiosis II, each of the chromatids is an independent chromosome.

During spermatogenesis, each primary spermatocyte divides to form two *secondary spermatocytes.* Each of these cells, in turn, divides to form two *spermatids,* which mature into sperm cells. Consequently, for each primary spermatocyte that undergoes meiosis,

four sperm cells, with 23 chromosomes in each of their nuclei, form. Figure 19.5 depicts spermatogenesis.

Check Your Recall

4. Explain the function of supporting cells in the seminiferous tubules.

5. Describe the structure of a sperm cell.

6. Review the events of spermatogenesis.

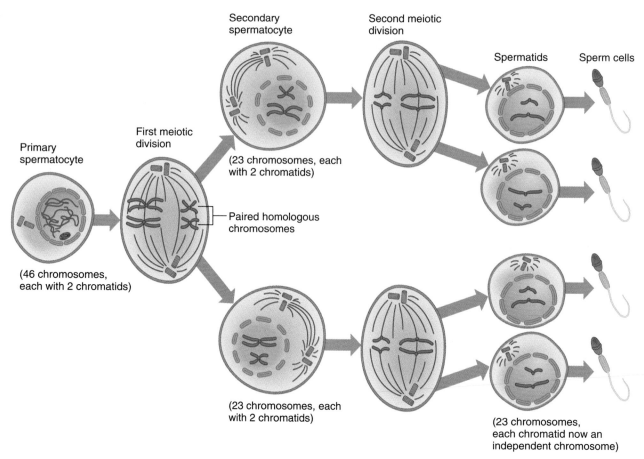

Figure 19.5

In spermatogenesis, two successive meiotic divisions.

Male Internal Accessory Organs

The internal accessory organs of the male reproductive system are specialized to nurture and transport sperm cells. These structures include the two epididymides, two ductus deferentia, two ejaculatory ducts, and ure-thra, as well as the two seminal vesicles, prostate gland, and two bulbourethral glands.

Epididymides

The **epididymides** (ep″ĭ-di-dy′mides; sing., *epididymis*) are tightly coiled, threadlike tubes about 6 meters long (see figs. 19.1 and 19.2). Each epididymis is connected to ducts within a testis. The epididymis emerges from the top of the testis, descends along the posterior surface of the testis, and then courses upward to become the duc-tus deferens.

Immature sperm cells reaching the epididymis are nonmotile. As rhythmic peristaltic contractions help move these cells through the epididymis, the cells mature. After this aging process, sperm cells have the potential to move independently and fertilize egg cells, but they usually do not "swim" until after ejaculation.

Ductus Deferentia

The **ductus deferentia** (duk′tus def″er-en′sha; sing., *ductus deferens*), also called vasa deferentia, are muscu-lar tubes about 45 centimeters long (see fig. 19.1). Each passes upward along the medial side of a testis and through a passage in the lower abdominal wall (ingui-nal canal), enters the pelvic cavity, and ends behind the urinary bladder. Just outside the prostate gland, the duc-tus deferens unites with the duct of a seminal vesicle to form an **ejaculatory duct,** which passes through the prostate gland and empties into the urethra.

Seminal Vesicles

The **seminal vesicles** are convoluted, saclike structures about 5 centimeters long. Each attaches to the ductus deferens near the base of the urinary bladder (see fig. 19.1). The glandular tissue lining the inner wall of a seminal vesicle secretes a slightly alkaline fluid. This fluid helps regulate the pH of the tubular contents as sperm cells travel to the outside. Seminal vesicle secre-tions also contain *fructose,* a monosaccharide that pro-vides energy to sperm cells, and *prostaglandins* (see

Topic of Interest

Prostate Enlargement

The prostate gland is small in boys, begins to grow in early adolescence, and reaches adult size several years later. Usually, the gland does not grow again until age fifty, when in about half of all men it enlarges enough to press on the urethra. This produces a feeling of pressure because the bladder cannot empty completely and the man feels the need to urinate frequently. Retained urine can lead to infection and inflammation, bladder stones, or kidney disease.

Risk factors for prostate enlargement include a fatty diet, having had a vasectomy, occupational exposure to batteries or to the metal cadmium, and inheriting a particular gene that also causes breast cancer. The enlargement may be benign or cancerous. Because prostate cancer is highly treatable if detected early, men should have their prostates examined regularly.

Diagnostic tests for prostate cancer include a rectal exam, visualization of the prostate, urethra, and urinary bladder with a device that is inserted through the penis, called a cytoscope; as well as a blood test to detect prostate-specific antigen (PSA), a cell surface protein normally found on prostate cells. Elevated PSA levels can indicate an enlarged prostate. Ultrasound then may provide further information on whether a benign or cancerous growth is present. Table 19A summarizes some treatments for an enlarged prostate.

Table 19A	Some Medical Treatments for an Enlarged Prostate Gland
Surgical removal of prostate	
Radiation	
Drug (Proscar, or finasteride) to block testosterone's growth-stimulating effect on prostate	
Alpha blocker drugs, which relax muscles near the prostate, relieving pressure	
Microwave energy delivered through a probe inserted into urethra or rectum	
Balloon inserted into urethra and inflated with liquid	
Tumor frozen with liquid nitrogen delivered by probe through skin	
Device (stent) inserted between lobes of the prostate to relieve pressure on the urethra	

chapter 11, p. 294), which stimulate muscular contractions within the female reproductive organs, aiding the movement of sperm cells toward the egg cell.

Check Your Recall

7. Describe the structure of the epididymis.
8. Trace the path of the ductus deferens.
9. What is the function of a seminal vesicle?

Prostate Gland

The **prostate** (pros'tāt) **gland** is a chestnut-shaped structure about 4 centimeters across and 3 centimeters thick that surrounds the proximal portion of the urethra, just inferior to the urinary bladder (see fig. 19.1). It is enclosed in connective tissue and composed of many branched tubular glands, whose ducts open into the urethra.

The prostate gland secretes a thin, milky fluid with an alkaline pH. This secretion neutralizes the fluid containing sperm cells, which is acidic from accumulation of metabolic wastes that stored sperm cells produce. Prostatic fluid also enhances the motility of sperm cells and helps neutralize the acidic secretions of the vagina.

The Topic of Interest above discusses prostate enlargement, which is common among older men.

Bulbourethral Glands

The two **bulbourethral** (bul"bo-u-re'thral) **glands** (Cowper's glands) are each about a centimeter in diameter and are inferior to the prostate gland surrounded by muscle fibers of the external urethral sphincter (see fig. 19.1). Bulbourethral glands have many tubes whose epithelial linings secrete a mucuslike fluid in response to sexual stimulation. This fluid lubricates the end of the penis in preparation for sexual intercourse. However, females secrete most of the lubricating fluid for sexual intercourse.

Semen

Semen (se'men) is the fluid the male urethra conveys to the outside during ejaculation. It consists of sperm cells from the testes and secretions of the seminal vesicles, prostate gland, and bulbourethral glands. Semen is slightly alkaline (pH about 7.5), and it includes prostaglandins and nutrients.

The volume of semen released at one time varies from 2 to 5 milliliters. The average number of sperm cells in the fluid is about 120 million per milliliter.

Sperm cells are nonmotile while in the ducts of the testis and epididymis, but begin to swim as they mix with accessory gland secretions. Sperm cells cannot naturally fertilize an egg cell, however, until they enter the female reproductive tract. Acquiring the ability to fertilize an egg cell is called *capacitation,* and it reflects weakening of the sperm cells' acrosomal membranes.

Check Your Recall

10. Where is the prostate gland located?
11. What are the functions of the prostate gland's secretion?
12. What are the components of semen?

Male External Reproductive Organs

The male external reproductive organs are the scrotum, which encloses the testes, and the penis. The urethra passes through the penis.

Scrotum

The **scrotum** is a pouch of skin and subcutaneous tissue that hangs from the lower abdominal region posterior to the penis (see fig. 19.1). A medial septum subdivides the scrotum into two chambers, each of which encloses a testis. Each chamber also contains a serous membrane, which covers the testis and helps ensure that it moves smoothly within the scrotum. The scrotum protects and helps regulate the temperature of the testes, factors that are important to sex cell production.

Exposure to cold stimulates the smooth muscle in the wall of the scrotum to contract, the scrotal skin to wrinkle, and the testes to move closer to the pelvic cavity, where they can absorb heat. Exposure to warmth stimulates the smooth muscle to relax and the scrotum to hang loosely, providing an environment 3°C (about 5°F) below body temperature, which is more conducive to sperm cell production and survival.

Penis

The **penis** is a cylindrical organ that conveys urine and semen through the urethra to the outside (see fig. 19.1). During erection, it enlarges and stiffens, enabling it to be inserted into the vagina during sexual intercourse.

The *body,* or shaft, of the penis has three columns of erectile tissue—a pair of dorsally located *corpora cavernosa* and a single, ventral *corpus spongiosum.* A tough capsule of dense connective tissue surrounds each column. Skin, a thin layer of subcutaneous tissue, and a layer of connective tissue enclose the penis.

The corpus spongiosum, through which the urethra extends, enlarges at its distal end to form a sensitive, cone-shaped **glans penis.** The glans covers the ends of the corpora cavernosa and bears the urethral opening (exter-

nal urethral orifice). The skin of the glans is very thin and hairless, and has sensory receptors for sexual stimulation. A loose fold of skin called the *prepuce* (foreskin) originates just posterior to the glans and extends anteriorly to cover the glans as a sheath. The prepuce is sometimes removed by a surgical procedure called *circumcision.*

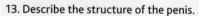

Check Your Recall

13. Describe the structure of the penis.
14. What is circumcision?

Erection, Orgasm, and Ejaculation

During sexual stimulation, parasympathetic nerve impulses from the sacral portion of the spinal cord release the vasodilator nitric oxide (NO), dilating the arteries leading into the penis. At the same time, the increasing pressure of arterial blood entering the vascular spaces of erectile tissue compresses the veins of the penis, reducing the flow of venous blood away from the penis. Consequently, blood accumulates in erectile tissues, and the penis swells and elongates, producing an **erection.**

The culmination of sexual stimulation is **orgasm** (or'gazm), a pleasurable feeling of physiological and psychological release. Emission and ejaculation accompany male orgasm.

Emission (e-mish'un) is the movement of sperm cells from the testes and secretions from the prostate gland and seminal vesicles into the urethra, where they mix to form semen. Emission occurs in response to sympathetic nerve impulses from the spinal cord, which stimulate peristaltic contractions in smooth muscle in the walls of the testicular ducts, epididymides, ductus deferentia, and ejaculatory ducts. At the same time, other sympathetic impulses stimulate rhythmic contractions of the seminal vesicles and prostate gland.

As the urethra fills with semen, sensory impulses are stimulated and pass into the sacral portion of the spinal cord. In response, motor impulses are transmitted from the cord to certain skeletal muscles at the base of the penile erectile columns, contracting them rhythmically. This increases the pressure within the erectile tissues and helps force semen through the urethra to the outside, a process called **ejaculation** (e-jak"u-la'shun).

The sequence of events during emission and ejaculation is coordinated so that fluid from the bulbourethral glands is expelled first. This is followed by the release of fluid from the prostate gland, the passage of sperm cells, and finally the ejection of fluid from the seminal vesicles.

Immediately after ejaculation, sympathetic impulses constrict the arteries that supply the erectile tissue, reducing blood inflow. Smooth muscles in the walls of the vascular spaces partially contract again, and the veins of the penis carry excess blood out of these spaces. The penis gradually returns to its flaccid state. Table 19.1 summarizes the functions of the male reproductive organs. The Topic of Interest on page 517 discusses male infertility.

> Spontaneous emissions and ejaculations commonly occur in adolescent males during sleep and thus are called *nocturnal emissions.* Changes in hormonal concentrations that accompany adolescent development and sexual maturation cause these emissions.

Check Your Recall

15. What controls blood flow into penile erectile tissues?
16. Distinguish among orgasm, emission, and ejaculation.
17. Review the events associated with emission and ejaculation.

19.3 HORMONAL CONTROL OF MALE REPRODUCTIVE FUNCTIONS

The hypothalamus, anterior pituitary gland, and testes secrete hormones that control male reproductive functions. These hormones initiate and maintain sperm cell production and oversee the development and maintenance of male secondary sex characteristics.

Hypothalamic and Pituitary Hormones

Prior to ten years of age, the male body is reproductively immature. It is childlike, with spermatogenic cells undifferentiated. Then, a series of changes leads to development of a reproductively functional adult. The hypothalamus controls many of these changes.

Recall from chapter 11 (pp. 295 and 299) that the hypothalamus secretes gonadotropin-releasing hormone (GnRH), which enters blood vessels leading to the anterior pituitary gland. In response, the anterior pituitary secretes the **gonadotropins** (go-nad"o-trōp'inz) *luteinizing hormone (LH)* and *follicle-stimulating hormone (FSH).* LH, which in males is sometimes called interstitial cell stimulating hormone (ICSH), promotes development of testicular interstitial cells, and they in turn secrete male sex hormones. FSH stimulates the supporting cells of the seminiferous tubules to respond to the effects of the male sex hormone *testosterone.* Then, in the presence of FSH and testosterone, these supporting cells stimulate spermatogenic cells to undergo spermatogenesis, giving rise to sperm cells (fig. 19.6). The supporting cells also secrete a hormone called *inhibin,* which inhibits the anterior pituitary gland by negative feedback. This action prevents oversecretion of FSH.

Male Sex Hormones

Male sex hormones are termed **androgens** (an'drojenz). Testicular interstitial cells produce most androgens, but the adrenal cortex synthesizes small amounts (see chapter 11, p. 304). **Testosterone** (tes-tos'tĕ-rōn) is the most important androgen. It loosely attaches to plasma proteins for secretion and transport in blood.

Testosterone secretion begins during fetal development and continues for several weeks following birth; then it nearly ceases during childhood. Between the ages of thirteen and fifteen, a young man's androgen production usually increases rapidly. This phase in development, when an individual becomes reproductively functional, is **puberty** (pu'ber-te). After puberty, testosterone secretion continues throughout the life of a male.

Actions of Testosterone

During puberty, testosterone stimulates enlargement of the testes and accessory organs of the reproductive system, as well as development of *male secondary sex characteristics,* which are special features associated with the adult male body. Secondary sex characteristics in the male include:

1. Increased growth of body hair, particularly on the face, chest, axillary region, and pubic region. Sometimes, hair growth on the scalp slows.

Table 19.1	Functions of the Male Reproductive Organs
Organ	**Function**
Testis	
Seminiferous tubules	Produce sperm cells
Interstitial cells	Produce and secrete male sex hormones
Epididymis	Stores sperm cells undergoing maturation; conveys sperm cells to ductus deferens
Ductus deferens	Conveys sperm cells to ejaculatory duct
Seminal vesicle	Secretes an alkaline fluid containing nutrients and prostaglandins that helps neutralize the acidic components of semen
Prostate gland	Secretes an alkaline fluid that helps neutralize semen's acidity and enhances sperm cell motility
Bulbourethral gland	Secretes fluid that lubricates end of penis
Scrotum	Encloses, protects, and regulates temperature of testes
Penis	Conveys semen into vagina during sexual intercourse; glans penis is richly supplied with sensory nerve endings associated with feelings of pleasure during sexual stimulation

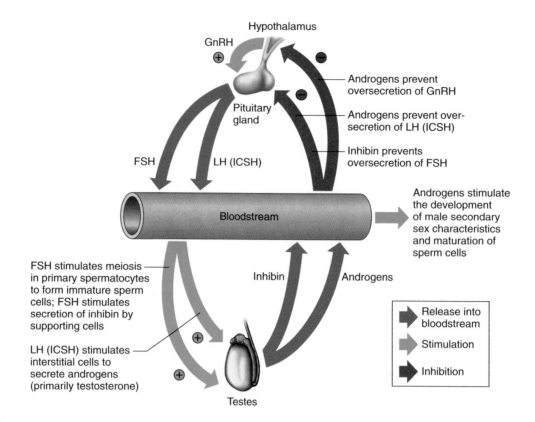

Figure 19.6

The hypothalamus controls maturation of sperm cells and development of male secondary sex characteristics. Negative feedback among the hypothalamus, the anterior lobe of the pituitary gland, and the testes controls the concentration of testosterone in the male body.

2. Enlargement of the larynx and thickening of the vocal folds, with lowering of the pitch of the voice.
3. Thickening of the skin.
4. Increased muscular growth, broadening of the shoulders, and narrowing of the waist.
5. Thickening and strengthening of the bones.

Testosterone also increases the rate of cellular metabolism and red blood cell production. This is why the average number of red blood cells in a microliter of blood is usually greater in males than in females. Testosterone stimulates sexual activity by affecting certain portions of the brain.

Regulation of Male Sex Hormones

The extent to which male secondary sex characteristics develop is directly related to the amount of testosterone that interstitial cells secrete. A negative feedback system involving the hypothalamus regulates testosterone output (fig. 19.6).

An increasing blood testosterone concentration inhibits the hypothalamus, and hypothalamic stimulation of the anterior pituitary gland by GnRH decreases. As the pituitary's secretion of LH falls in response, testosterone release from the interstitial cells decreases.

As the blood testosterone concentration drops, the hypothalamus becomes less inhibited, and it once again stimulates the anterior pituitary to release LH. Increasing LH secretion then causes interstitial cells to release more testosterone, and the blood testosterone concentration increases.

Testosterone level decreases somewhat during and after the *male climacteric,* a decline in sexual function associated with aging. At any given age, the testosterone concentration in the male body is regulated to remain relatively constant.

Check Your Recall

18. Which hormone initiates the changes associated with male sexual maturity?
19. Describe several male secondary sex characteristics.
20. List the functions of testosterone.
21. Explain how the secretion of male sex hormones is regulated.

19.4 ORGANS OF THE FEMALE REPRODUCTIVE SYSTEM

The organs of the female reproductive system produce and maintain the female sex cells, the egg cells (or oocytes); transport these cells to the site of fertilization;

Male Infertility

Male infertility—the inability of sperm cells to fertilize an egg cell—has several causes. If, during fetal development, the testes do not descend into the scrotum, the higher temperature of the abdominal cavity or inguinal canal destroys any sperm cells developing in the seminiferous tubules, causing infertility. Certain diseases, such as mumps, may inflame the testes (orchitis) and cause infertility by destroying cells in the seminiferous tubules.

Both the quality and the abundance of sperm cells are essential factors in a man's ability to father a child. If a sperm head is misshapen, if a sperm cell cannot swim, or if sperm cells are too few, completing the journey to the egg cell may be impossible.

Computer-aided sperm analysis (CASA) is standardizing and expanding criteria for normalcy in human semen and the sperm it contains. For this analysis, a man abstains from intercourse for two to three days and then provides a sperm sample, which must be examined within the hour. The man also provides information about his reproductive history and possible exposure to toxins. The sperm sample is placed on a slide under a microscope, and a video camera sends an image to a videocassette recorder, which projects a live or digitized image. The camera also sends the image to a com-

puter, which traces sperm trajectories and displays them on a monitor. Figure 19A shows a CASA of normal sperm cells, depicting different swimming patterns as they travel. Table 19B lists the components of a semen analysis. CASA analyses indicate that sperm movements differ so greatly from man to man that each individual may have his own sperm "motility signature."

Table 19B	Semen Analysis
Characteristic	**Normal Value**
Volume	2–5 milliliters/ejaculate
Sperm cell density	60–150 million cells/milliliter
Percent motile sperm	> 40%
Motile sperm cell density	> 24 million/milliliter
Average velocity of sperm	> 20 micrometers/second
Motility of sperm	> 8 micrometers/second
Percent normal sperm morphology	> 80%
White blood cells	Occasional or absent

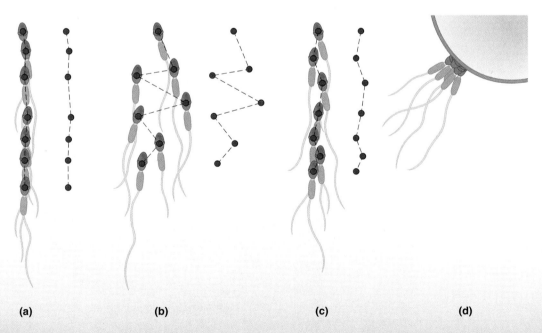

(a)   (b)   (c)   (d)

Figure 19A

A computer tracks sperm cell movements. In semen, sperm cells swim in a straight line (a), but as they are activated by biochemicals normally found in the woman's body, their trajectories widen (b). The sperm cells in (c) are in the mucus of a woman's cervix, and the sperm cells in (d) are attempting to digest through the structures surrounding an egg cell.

provide a favorable environment for a developing off-spring; move the offspring to the outside; and produce female sex hormones. A female's *primary sex organs* (gonads) are the two ovaries, which produce the female sex cells and sex hormones. The *accessory sex organs* of the female reproductive system are the internal and external reproductive organs (fig. 19.7; reference plates 5 and 6, pp. 27–28).

Ovaries

The two **ovaries** are solid, ovoid structures, each about 3.5 centimeters long, 2 centimeters wide, and 1 centimeter thick. The ovaries lie in shallow depressions in the lateral wall of the pelvic cavity (fig. 19.7).

Ovary Structure

Ovarian tissues are subdivided into two indistinct regions—an inner *medulla* and an outer *cortex*. The ovarian medulla is composed of loose connective tissue and has many blood vessels, lymphatic vessels, and nerve fibers. The ovarian cortex consists of more compact tissue and has a granular appearance due to tiny masses of cells called *ovarian follicles.*

A layer of cuboidal epithelium covers the ovary's free surface. Just beneath this epithelium is a layer of dense connective tissue.

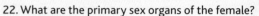

Check Your Recall
22. What are the primary sex organs of the female?
23. Describe the structure of an ovary.

Primordial Follicles

During prenatal (before birth) development of a female, small groups of cells in the outer region of the ovarian cortex form several million **primordial follicles.** Each of these structures consists of a single, large cell, called a *primary oocyte,* which is closely surrounded by epithelial cells called *follicular cells.*

Early in development, primary oocytes begin to undergo meiosis, but the process soon halts and does not continue until the individual reaches puberty. Once the primordial follicles appear, no new ones form. Instead, the number of oocytes in the ovary steadily declines as many degenerate. Of the several million oocytes that

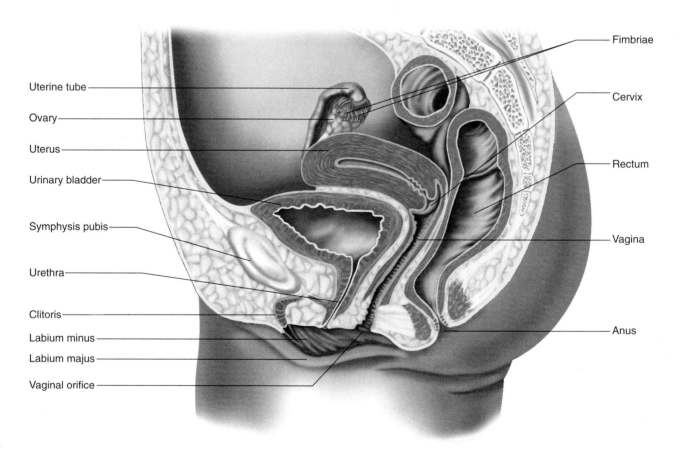

Figure 19.7

Female reproductive organs (sagittal view). The paired ovaries are the primary sex organs, and the other structures, both internal and external, are accessory sex organs.

formed in the embryo, only a million or so remain at birth, and perhaps 400,000 are present at puberty. The ovary releases fewer than 400 or 500 oocytes during a female's reproductive life.

Oogenesis

Oogenesis (o″o-jen′ĕ-sis) is the process of egg cell formation. Beginning at puberty, some primary oocytes are stimulated to continue meiosis. As in the case of sperm cells, the resulting cells have one-half as many chromosomes (23) in their nuclei as their parent cells.

When a primary oocyte divides, the distribution of the cytoplasm is unequal. One of the resulting cells, called a *secondary oocyte* (egg cell), is large, and the other, called the *first polar body,* is small (fig. 19.8). This uneven cytoplasm distribution differs from that of spermatogenesis.

The large secondary oocyte can be fertilized by a sperm cell. Upon fertilization, the secondary oocyte divides unequally to produce a tiny *second polar body* and a large fertilized egg cell, also called a **zygote** (zi′gōt).

The polar bodies have no further function and soon degenerate. Their role in reproduction allows for production of an egg cell that has the massive amounts of cytoplasm and the abundant organelles required to carry the zygote through the first few cell divisions, yet with the right number of chromosomes.

> The largest cell in the human body is the egg cell. The smallest cell is the sperm.

Check Your Recall

24. Describe the major events of oogenesis.
25. What is the function of polar body formation?

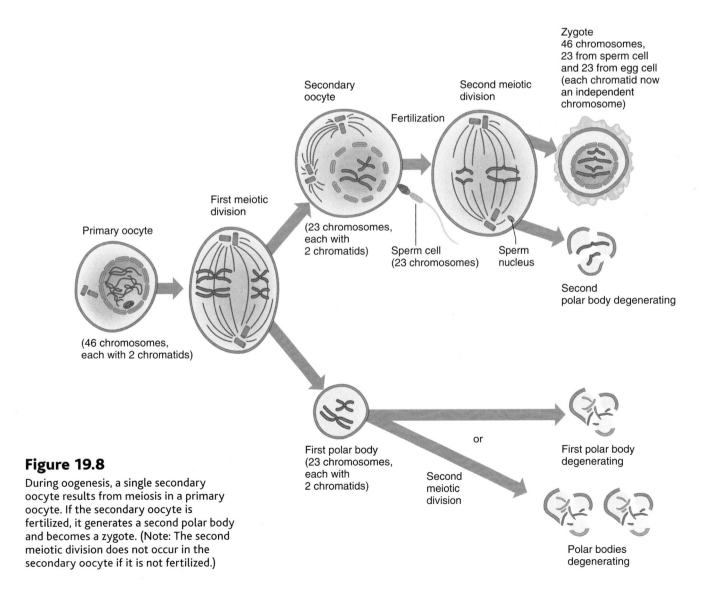

Figure 19.8

During oogenesis, a single secondary oocyte results from meiosis in a primary oocyte. If the secondary oocyte is fertilized, it generates a second polar body and becomes a zygote. (Note: The second meiotic division does not occur in the secondary oocyte if it is not fertilized.)

Primary oocyte

(46 chromosomes, each with 2 chromatids)

First meiotic division

Secondary oocyte

(23 chromosomes, each with 2 chromatids)

Fertilization

Sperm cell (23 chromosomes)

First polar body (23 chromosomes, each with 2 chromatids)

Second meiotic division

Second meiotic division

Sperm nucleus

Zygote 46 chromosomes, 23 from sperm cell and 23 from egg cell (each chromatid now an independent chromosome)

Second polar body degenerating

or

First polar body degenerating

Polar bodies degenerating

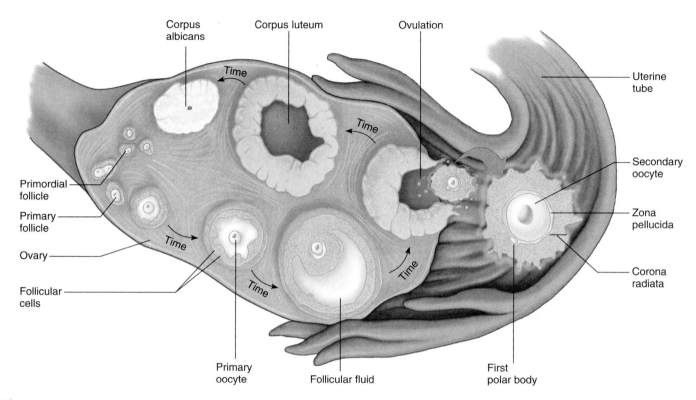

Figure 19.9

In an ovary, as a follicle matures, a developing primary oocyte enlarges and becomes surrounded by follicular cells and fluid. Eventually, the mature follicle ruptures, releasing the secondary oocyte and layers of surrounding follicular cells.

Follicle Maturation

At puberty, the anterior pituitary gland secretes increased amounts of FSH, and the ovaries enlarge in response. At the same time, some of the primordial follicles mature into **primary follicles** (pri'ma-re fol'ĭ-klz). Figure 19.9 traces the maturation of a follicle in an ovary.

During maturation, a primary oocyte enlarges, and surrounding follicular cells proliferate by mitosis. These follicular cells organize into layers, and soon a cavity (*antrum*) appears in the cellular mass. A clear *follicular fluid* fills the cavity and bathes the primary oocyte. The enlarging fluid-filled cavity presses the primary oocyte to one side.

In time, the mature follicle reaches a diameter of 10 millimeters or more and bulges outward on the ovary surface like a blister. The secondary oocyte within the mature follicle is a large, spherical cell, surrounded by a layer of glycoprotein called the zona pellucida and attached to a mantle of follicular cells (corona radiata) (fig. 19.10). Processes from the follicular cells extend through the zona pellucida and supply the secondary oocyte with nutrients.

As many as twenty primary follicles may begin maturing at any one time, but one follicle usually outgrows the others. Typically, only the dominant follicle fully develops, and the other follicles degenerate.

Ovulation

As a follicle matures, its primary oocyte undergoes oogenesis, giving rise to a secondary oocyte and a first polar body. The process called **ovulation** (o″vu-la'shun) releases the secondary oocyte and first polar body with

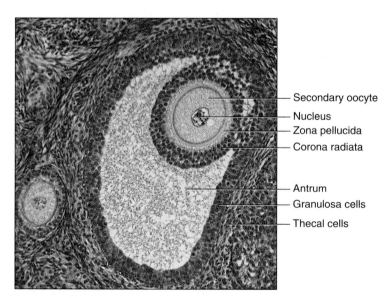

Figure 19.10

Light micrograph of a maturing follicle (200×).

one or two surrounding layers of follicular cells from the mature follicle.

Hormones from the anterior pituitary gland trigger ovulation, rapidly swelling the mature follicle and weakening its wall. Eventually, the wall ruptures, and follicular fluid and the secondary oocyte ooze from the ovary's surface (see fig. 19.9).

After ovulation, the secondary oocyte and surrounding follicular cells are usually propelled to the opening of a nearby uterine tube (fig. 19.11). If the oocyte is not fertilized within a short time, it degenerates.

Check Your Recall

26. What changes occur in a follicle and its oocyte during maturation?
27. What causes ovulation?
28. What happens to an oocyte following ovulation?

Female Internal Accessory Organs

The internal accessory organs of the female reproductive system include a pair of uterine tubes, a uterus, and a vagina.

Uterine Tubes

The **uterine tubes** (fallopian tubes or oviducts) open near the ovaries (fig. 19.11). Each tube is about 10 centimeters long and passes medially to the uterus, penetrates its wall, and opens into the uterine cavity.

Near each ovary, a uterine tube expands, forming a funnel-shaped *infundibulum* (in"fun-dib'u-lum), which partially encircles the ovary. Fingerlike extensions called *fimbriae* (fim'bre) fringe the infundibulum margin. Although the infundibulum generally does not touch the ovary, one of the larger extensions connects directly to the ovary.

Simple columnar epithelial cells, some *ciliated,* line the uterine tube. The epithelium secretes mucus, and the cilia beat toward the uterus. These actions help draw the secondary oocyte and expelled follicular fluid into the infundibulum following ovulation. Ciliary action and peristaltic contractions of the uterine tube's muscular layer help transport the secondary oocyte down the uterine tube. Fertilization may occur in the uterine tube.

Uterus

If the secondary oocyte is fertilized in the uterine tube, becoming a zygote, the **uterus** receives the developing embryo and sustains its development. The uterus is a hollow, muscular organ shaped somewhat like an inverted pear.

The size of the uterus changes greatly during pregnancy. In its nonpregnant, adult state, the uterus is about 7 centimeters long, 5 centimeters wide (at its broadest point), and 2.5 centimeters in diameter. It is located medially within the anterior portion of the pelvic cavity, superior to the vagina, and usually bends forward over the urinary bladder.

The upper two-thirds, or *body,* of the uterus has a dome-shaped top (fig. 19.11). The uterine tubes enter the top of the uterus at its broadest part. The lower third

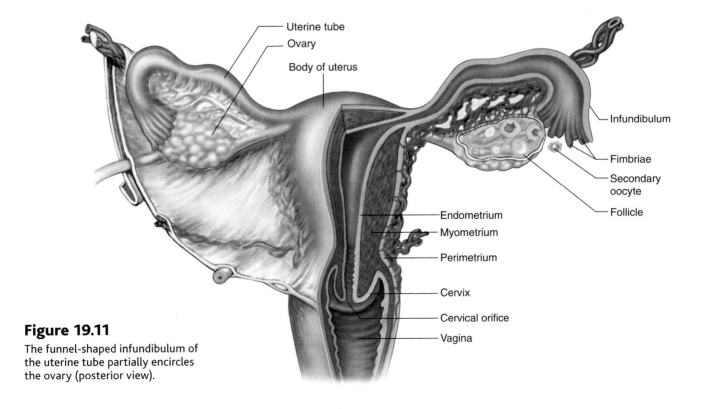

Figure 19.11
The funnel-shaped infundibulum of the uterine tube partially encircles the ovary (posterior view).

of the uterus is called the **cervix.** This tubular part extends downward into the upper portion of the vagina. The cervix surrounds the opening called the *cervical orifice,* through which the uterus opens to the vagina

The uterine wall is thick and has three layers (figs. 19.11 and 19.12). The **endometrium** (en"do-me're-um), the inner mucosal layer, is covered with columnar epithelium and contains abundant tubular glands. The **myometrium** (mi"o-me're-um), a thick, middle, muscular layer, consists largely of bundles of smooth muscle fibers. During the monthly female reproductive cycles and during pregnancy, the endometrium and myometrium change extensively. The **perimetrium** (per-ĭ-me're-um) is an outer serosal layer that covers the body of the uterus and part of the cervix.

During pregnancy, the uterus expands to 500 times its normal size.

A procedure called the *Pap (Papanicolaou) smear test* can usually detect cancer of the cervix. A sample of cervical tissue is smeared on a glass slide, stained, and sent to a laboratory, where computer image recognition software scans for cancer cells. Cervical cancer detected and treated early has a high cure rate. A vaccine is available that protects against strains of the human papilloma virus that cause cervical cancer.

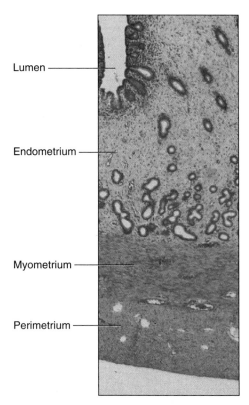

Lumen

Endometrium

Myometrium

Perimetrium

Figure 19.12
Light micrograph of the uterine wall (10.5×).

Vagina

The **vagina** is a fibromuscular tube, about 9 centimeters long, extending from the uterus to the outside (see fig. 19.7). It conveys uterine secretions, receives the erect penis during sexual intercourse, and provides an open channel for the offspring during birth.

The vagina extends upward and back into the pelvic cavity. It is posterior to the urinary bladder and urethra, anterior to the rectum, and attached to these structures by connective tissues.

A thin membrane of connective tissue and stratified squamous epithelium called the **hymen** partially covers the *vaginal orifice.* A central opening of varying size allows uterine and vaginal secretions to pass to the outside.

The vaginal wall has three layers. The inner *mucosal layer* is stratified squamous epithelium. This layer lacks mucous glands; the mucus in the lumen of the vagina comes from uterine glands and from vestibular glands at the mouth of the vagina.

The middle *muscular layer* consists mainly of smooth muscle fibers. A thin band of striated muscle at the lower end of the vagina helps close the vaginal opening. Another voluntary muscle (bulbospongiosus) is primarily responsible for closing this orifice.

The outer *fibrous layer* consists of dense connective tissue interlaced with elastic fibers. It attaches the vagina to surrounding organs.

Check Your Recall

29. How is a secondary oocyte moved along a uterine tube?

30. Describe the structure of the uterus.

31. Describe the structure of the vagina.

Female External Reproductive Organs

The external accessory organs of the female reproductive system include the labia majora, labia minora, clitoris, and vestibular glands. These structures surround the openings of the urethra and vagina, and compose the **vulva** (see fig. 19.7).

Labia Majora

The **labia majora** (singular, *labium majus*) enclose and protect the other external reproductive organs. They correspond to the scrotum in males and are composed of rounded folds of adipose tissue and a thin layer of smooth muscle, covered by skin.

The labia majora lie close together. A cleft that includes the urethral and vaginal openings separates the labia longitudinally. At their anterior ends, the labia merge to form a medial, rounded elevation of adipose tissue called the *mons pubis,* which overlies the symphysis pubis.

Labia Minora

The **labia minora** (singular, *labium minus*) are flattened, longitudinal folds between the labia majora (see fig. 19.7). They are composed of connective tissue richly supplied with blood vessels, giving a pinkish appearance. Posteriorly, the labia minora merge with the labia majora, while anteriorly, they converge to form a hood-like covering around the clitoris.

Clitoris

The **clitoris** (klit′o-ris) is a small projection at the anterior end of the vulva between the labia minora (see fig. 19.7). It is usually about 2 centimeters long and 0.5 centimeters in diameter, including a portion embedded in surrounding tissues. The clitoris corresponds to the penis in males and has a similar structure. It is composed of two columns of erectile tissue called *corpora cavernosa.* At its anterior end, a small mass of erectile tissue forms a glans, which is richly supplied with sensory nerve fibers.

Vestibule

The labia minora enclose a space called the **vestibule.** The vagina opens into the posterior portion of the vestibule, and the urethra opens in the midline, just anterior to the vagina and about 2.5 centimeters posterior to the glans of the clitoris.

A pair of **vestibular glands,** corresponding to the bulbourethral glands in males, lie one on either side of the vaginal opening. Beneath the mucosa of the vestibule on either side is a mass of vascular erectile tissue called the *vestibular bulb.*

Check Your Recall

32. What is the male counterpart of the labia majora? Of the clitoris?

33. Which structures are in the vestibule?

Erection, Lubrication, and Orgasm

Erectile tissues in the clitoris and around the vaginal entrance respond to sexual stimulation. Following such stimulation, parasympathetic nerve impulses from the sacral portion of the spinal cord release the vasodilator nitric oxide (NO), causing the arteries associated with the erectile tissues to dilate. As a result, blood inflow increases, and the erectile tissues swell. At the same time, the vagina expands and elongates.

If sexual stimulation is sufficiently intense, parasympathetic impulses stimulate the vestibular glands to secrete mucus into the vestibule. This moistens and lubricates the tissues surrounding the vestibule and the lower end of the vagina, facilitating insertion of the penis into the vagina.

The clitoris is supplied with abundant sensory nerve fibers, which are especially sensitive to local stimulation. The culmination of such stimulation is orgasm.

Just prior to orgasm, the tissues of the outer third of the vagina engorge with blood and swell. This increases the friction on the penis during intercourse. Orgasm initiates a series of reflexes involving the sacral and lumbar portions of the spinal cord. In response to these reflexes, the muscles of the perineum and the walls of the uterus and uterine tubes contract rhythmically. These contractions help transport sperm cells through the female reproductive tract toward the upper ends of the uterine tubes. Table 19.2 summarizes the functions of the female reproductive organs.

The clitoris has nearly 8,000 nerve fibers, the densest collection of any body part. The corresponding part of the penis has only 4,000 nerve fibers.

Check Your Recall

34. What events result from parasympathetic stimulation of the female reproductive organs?

35. What changes occur in the vagina just prior to and during orgasm?

Table 19.2	Functions of the Female Reproductive Organs
Organ	**Function**
Ovary	Produces oocytes and female sex hormones
Uterine tube	Conveys secondary oocyte toward uterus; site of fertilization; conveys developing embryo to uterus
Uterus	Protects and sustains embryo during pregnancy
Vagina	Conveys uterine secretions to outside of body; receives erect penis during sexual intercourse; provides open channel for offspring during birth process
Labia majora	Enclose and protect other external reproductive organs
Labia minora	Form margins of vestibule; protect openings of vagina and urethra
Clitoris	Produces feelings of pleasure during sexual stimulation due to abundant sensory nerve endings in glans
Vestibule	Space between labia minora that contains vaginal and urethral openings
Vestibular glands	Secrete fluid that moistens and lubricates vestibule

19.5 HORMONAL CONTROL OF FEMALE REPRODUCTIVE FUNCTIONS

The hypothalamus, anterior pituitary gland, and ovaries secrete hormones that control maturation of female sex cells, the development and maintenance of female secondary sex characteristics, and changes that occur during the monthly reproductive cycle.

Female Sex Hormones

The female body is reproductively immature until about age ten years. Then the hypothalamus begins to secrete increasing amounts of GnRH, which in turn stimulates the anterior pituitary to release the gonadotropins FSH and LH. These hormones play primary roles in controlling female sex cell maturation and in producing female sex hormones.

Several tissues, including the ovaries, the adrenal cortices, and the placenta (during pregnancy), secrete female sex hormones belonging to two major groups—**estrogens** (es′tro-jenz) and **progesterone** (pro-jes′tĭ-rōn). *Estradiol* is the most abundant of the estrogens, which also include *estrone* and *estriol.*

The ovaries are the primary source of estrogens (in a nonpregnant female). At puberty, under the influence of the anterior pituitary, the ovaries secrete increasing amounts of estrogens. Estrogens stimulate enlargement of accessory organs, including the vagina, uterus, uterine tubes, ovaries, and external reproductive structures. Estrogens also develop and maintain the *female secondary sex characteristics,* which include:

1. Development of the breasts and the ductile system of the mammary glands in the breasts.
2. Increased deposition of adipose tissue in the subcutaneous layer generally and in the breasts, thighs, and buttocks particularly.
3. Increased vascularization of the skin.

The ovaries are also the primary source of progesterone (in a nonpregnant female). This hormone promotes changes in the uterus during the female reproductive cycle, affects the mammary glands, and helps regulate the secretion of gonadotropins from the anterior pituitary.

Androgen (male sex hormone) concentrations produce certain other changes in females at puberty. For example, increased hair growth in the pubic and axillary regions is due to androgen secreted by the adrenal cortices. Conversely, development of the female skeletal configuration, which includes narrow shoulders and broad hips, is a response to a low androgen concentration.

Check Your Recall

36. What stimulates sexual maturation in a female?
37. What is the function of estrogens?
38. What is the function of androgen in a female?

Female Reproductive Cycle

The female reproductive cycle consists of regular, recurring changes in the uterine lining, which culminate in menstrual bleeding (menses). Such cycles usually begin around age thirteen and continue into middle age, then cease.

Elite female athletes may have disturbed reproductive cycles, ranging from diminished menstrual flow (oligomenorrhea) to complete stoppage (amenorrhea). The more active an athlete, the more likely it is that she will have menstrual irregularities, and this may impair her ability to conceive. The culprit in infertility appears to be too little body fat. The diminished fat reserves decreases secretion of the hormone leptin, which lowers secretion of gonadotropin-releasing hormone from the hypothalamus, which in turn lowers estrogen levels. The infertility apparently results from too little estrogen. Adipose tissue itself also contains some estrogen, a small supply made even smaller in the elite athlete.

A female's first reproductive cycle, called **menarche** (mĕ-nar′ke), occurs after the ovaries and other organs of the reproductive control system have matured and begun responding to certain hormones. Then, hypothalamic secretion of GnRH stimulates the anterior pituitary to release threshold levels of FSH and LH. FSH stimulates maturation of an ovarian follicle. The follicular cells produce increasing amounts of estrogens and some progesterone. LH stimulates certain ovarian cells to secrete precursor molecules (such as testosterone), which are also used to produce estrogens.

In a young female, estrogens stimulate the development of secondary sex characteristics. Estrogens secreted during subsequent reproductive cycles continue the development and maintenance of these characteristics.

An increasing concentration of estrogens during the first week or so of a reproductive cycle changes the uterine lining, thickening the glandular endometrium (proliferative phase). This is shown in the diagram of the reproductive cycle in figure 19.13. Meanwhile, the developing follicle completes maturation, and by around the fourteenth day of the cycle, the follicle appears on the ovary surface as a blisterlike bulge. Within the follicle, the follicular cells, which surround and connect the sec-

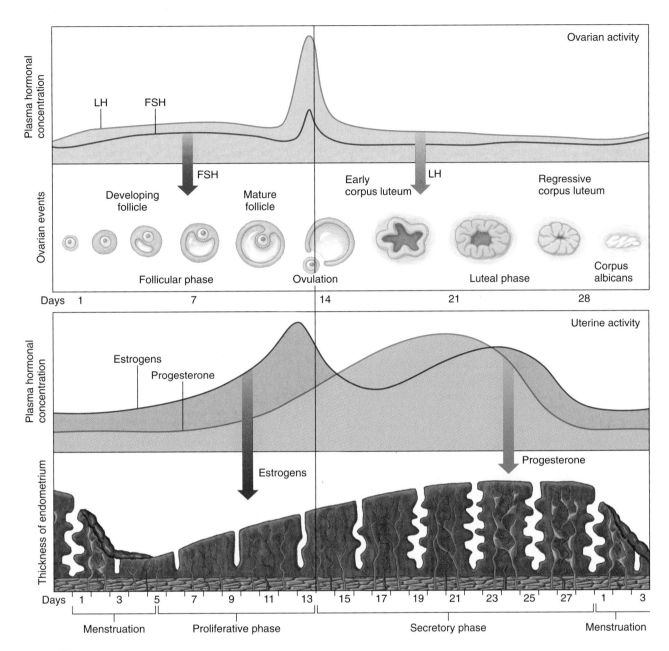

Figure 19.13

Major events in the female reproductive cycle.

ondary oocyte to the inner wall, loosen. Follicular fluid accumulates rapidly.

While the follicle matures, it secretes estrogens that inhibit the anterior pituitary's release of LH, but allow LH to be stored in the gland. Estrogens also make anterior pituitary cells more sensitive to the action of GnRH, which the hypothalamus releases in rhythmic pulses about ninety minutes apart.

Near the fourteenth day of follicular development, anterior pituitary cells finally release the stored LH in response to the GnRH pulses. The resulting surge in LH concentra-

tion, which lasts about thirty-six hours, weakens and ruptures the bulging follicular wall, which sends the secondary oocyte and follicular fluid from the ovary (ovulation).

Following ovulation, the remnants of the follicle within the ovary change rapidly. The space containing the follicular fluid fills with blood, which soon clots. Under the influence of LH, follicular cells enlarge to form a temporary glandular structure called a **corpus luteum** (kor'pus loot'e-um) ("yellow body").

Follicular cells secrete progesterone during the first part of the reproductive cycle. Corpus luteum cells secrete abundant progesterone and estrogens during the last half of the cycle. Consequently, as a corpus luteum forms, the blood progesterone concentration increases sharply.

Progesterone causes the endometrium to become more vascular and glandular. It also stimulates uterine glands to secrete more glycogen and lipids (secretory phase). As a result, endometrial tissues fill with fluids containing nutrients and electrolytes, providing a favorable environment for embryo development.

High levels of estrogens and progesterone inhibit the anterior pituitary's release of LH and FSH. Consequently, no other follicles are stimulated to develop when the corpus luteum is active. However, if a sperm cell does not fertilize the egg cell released at ovulation, the corpus luteum begins to degenerate on about the twenty-fourth day of the cycle. Eventually, connective tissue replaces it. The remnant of such a corpus luteum is called a *corpus albicans.*

When the corpus luteum ceases to function, concentrations of estrogens and progesterone decline rapidly, and blood vessels in the endometrium constrict. This reduces the supply of oxygen and nutrients to the thickened uterine lining, and these lining tissues soon disintegrate and slough off. At the same time, blood leaves damaged capillaries, creating a flow of blood and cellular debris that passes through the vagina as the *menstrual flow* (menses). This flow usually begins about the twenty-eighth day of the cycle and continues for three to five days while the concentrations of estrogens are low. The beginning of the menstrual flow marks the end of a reproductive cycle and the beginning of a new cycle. Table 19.3 summarizes the reproductive cycle.

Low blood concentrations of estrogens and progesterone at the beginning of the reproductive cycle mean that the hypothalamus and anterior pituitary are no longer inhibited. Consequently, FSH and LH concentrations soon increase, stimulating a new follicle to mature. As this follicle secretes estrogens, the uterine lining undergoes repair, and the endometrium begins to thicken again.

Menopause

After puberty, reproductive cycles continue at regular intervals into the late forties or early fifties, when they usually become increasingly irregular. Then, within a few months or years, the cycles cease altogether. This period in life is called **menopause** (men'o-pawz), or the female climacteric.

Aging of the ovaries causes menopause. After about thirty-five years of cycling, few primary follicles remain to respond to pituitary gonadotropins. Consequently, the follicles no longer mature, ovulation does not occur, and the blood concentration of estrogens plummets.

Reduced concentrations of estrogens and lack of progesterone may change the female secondary sex characteristics. The breasts, vagina, uterus, and uterine tubes may shrink, and the pubic and axillary hair may thin.

Table 19.3	Major Events in a Reproductive Cycle

1. Anterior pituitary gland secretes follicle-stimulating hormone (FSH) and luteinizing hormone (LH).

2. FSH stimulates maturation of a follicle.

3. Follicular cells produce and secrete estrogens.

   a. Estrogens maintain secondary sex characteristics.

   b. Estrogens cause endometrium to thicken.

4. Anterior pituitary releases a surge of LH, which stimulates ovulation.

5. Follicular cells become corpus luteum cells, which secrete estrogens and progesterone.

   a. Estrogens continue to stimulate uterine wall development.

   b. Progesterone stimulates the endometrium to become more glandular and vascular.

   c. Estrogens and progesterone inhibit anterior pituitary from secreting LH and FSH.

6. If the egg cell is not fertilized, the corpus luteum degenerates and no longer secretes estrogens and progesterone.

7. As concentrations of estrogens and progesterone decline, blood vessels in the endometrium constrict.

8. Uterine lining disintegrates and sloughs off, producing menstrual flow.

9. Anterior pituitary is no longer inhibited and again secretes FSH and LH.

10. The reproductive cycle repeats.

Check Your Recall

39. Trace the events of the female reproductive cycle.

40. What causes menstrual flow?

41. What are some changes that may occur at menopause?

19.6 MAMMARY GLANDS

The **mammary glands** are accessory organs of the female reproductive system that are specialized to secrete milk following pregnancy. They are in the subcutaneous tissue of the anterior thorax within elevations called *breasts* (fig. 19.14*a*; reference plate 1, p. 23). The breasts overlie the *pectoralis major* muscles and extend from the second to the sixth ribs and from the sternum to the axillae.

A *nipple* is located near the tip of each breast at about the level of the fourth intercostal space. A circular area of pigmented skin, called the *areola,* surrounds each nipple (fig. 19.14*b*).

A mammary gland is composed of fifteen to twenty lobes. Each lobe contains glands (alveolar glands) and an alveolar duct that leads to a lactiferous duct, which in turn leads to the nipple and opens to the outside

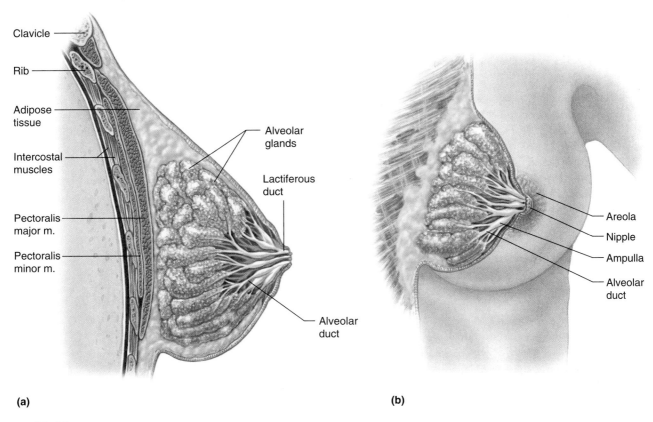

Figure 19.14
Structure of the female breast and mammary glands. (*a*) Sagittal section. (*b*) Anterior view. (*m.* stands for *muscle.*)

(fig. 19.14). Dense connective and adipose tissues separate the lobes. These tissues also support the glands and attach them to the fascia of the underlying pectoral muscles. Other connective tissue, which forms dense strands called *suspensory ligaments,* extends inward from the dermis of the breast to the fascia, helping to support the breast's weight.

The mammary glands of males and females are similar. As children reach puberty, the glands in males do not develop, whereas in females, ovarian hormones stimulate development of the glands. The alveolar glands and ducts enlarge, and fat forms deposits around and within the breasts. Chapter 20 (pp. 555–556) describes the hormonal mechanism that stimulates mammary glands to produce and secrete milk. The Topic of Interest entitled Treating Breast Cancer on pages 528–529 discusses diagnosis, treatment, and prevention of breast cancer.

Check Your Recall

42. Describe the structure of a mammary gland.
43. How does ovarian hormone secretion change the mammary glands?

19.7 BIRTH CONTROL

Birth control is the voluntary regulation of the number of offspring produced and the time at which they are conceived. This control requires a method of **contraception** (kon″trah-sep′shun) to avoid fertilization of an egg cell following sexual intercourse (coitus) or to prevent the hollow ball of cells (a blastocyst) that will develop into an embryo from implanting in the uterine wall. Several methods of contraception are available, and they have varying degrees of effectiveness.

Coitus Interruptus

Coitus interruptus is withdrawal of the penis from the vagina before ejaculation, which prevents entry of sperm cells into the female reproductive tract. This method can still result in pregnancy because a male may find it difficult to withdraw just prior to ejaculation. Also, some semen containing sperm cells may reach the vagina before ejaculation occurs.

Rhythm Method

The *rhythm method* (also called timed coitus or natural family planning) requires abstinence from sexual

Treating Breast Cancer

One in eight women will develop breast cancer at some point in her life (table 19C). About 1% of breast cancer cases are in men. Breast cancer is really several illnesses. As medical research reveals the cellular and molecular characteristics that distinguish subtypes of the disease, treatments old and new are being increasingly tailored to individuals. This "rational" approach may delay progression of the disease and increase the survival rate.

Warning Signs

Changes that could signal breast cancer include a small area of thickened tissue, a dimple, a change in contour, or a nipple that is flattened, points in an unusual direction, or produces a discharge. A woman can note these changes by performing a monthly "breast self-exam," in which she lies flat on her back with the arm raised behind her head and systematically feels all parts of each breast. But sometimes breast cancer gives no warning at all—early signs of fatigue and feeling ill may not occur until the disease has spread beyond the breast.

After finding a lump, the next step is a physical exam, in which a health-care provider palpates the breast and does a mammogram, an X-ray scan that can pinpoint the location and approximate extent of abnormal tissue (fig. 19B). An ultrasound scan can distinguish between a cyst (a fluid-filled sac of glandular tissue) and a tumor (a solid mass). If an area is suspicious, a thin needle is used to take a biopsy (sample)

Table 19C	Breast Cancer Risk		
By Age	**Odds**	**By Age**	**Odds**
25	1 in 19,608	60	1 in 24
30	1 in 2,525	65	1 in 17
35	1 in 622	70	1 in 14
40	1 in 217	75	1 in 11
45	1 in 93	80	1 in 10
50	1 in 50	85	1 in 9
55	1 in 33	95 or older	1 in 8

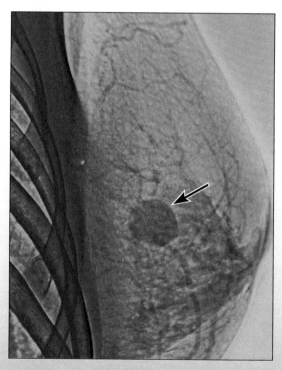

Figure 19B
Mammogram of a breast with a tumor (arrow).

intercourse 2 days before and 1 day after ovulation. The rhythm method results in a relatively high rate of pregnancy because accurately identifying infertile times to have intercourse is difficult. Another disadvantage of the rhythm method is that it restricts spontaneity in sexual activity.

Mechanical Barriers

Mechanical barrier contraceptives prevent sperm cells from entering the female reproductive tract during sexual intercourse. The *male condom* consists of a thin latex or natural membrane sheath placed over the erect

penis before intercourse to prevent semen from entering the vagina upon ejaculation (fig. 19.15*a*). A *female condom* resembles a small plastic bag. A woman inserts it into her vagina prior to intercourse. The device blocks sperm cells from reaching the cervix.

Some men feel that a condom decreases the sensitivity of the penis during intercourse, and its use may interrupt spontaneity. However, condoms are inexpensive and may also prevent transmission of some sexually transmitted diseases.

Another mechanical barrier is the *diaphragm*, a cup-shaped device with a flexible ring forming the rim.

of the tissue, whose cells will be scrutinized for the telltale characteristics of cancer.

Eighty percent of the time, a breast lump is a sign of fibrocystic breast disease, which is benign (noncancerous). The lump may be a cyst or a solid, fibrous mass of connective tissue called a fibroadenoma.

Surgery, Radiation, and Chemotherapies

If biopsied breast cells are cancerous, treatment usually begins with surgery. A lumpectomy removes a small tumor and some surrounding tissue; a simple mastectomy removes a breast; and a modified radical mastectomy removes the breast and surrounding lymph nodes, but preserves the pectoral muscles. A few lymph nodes are typically examined, and if cancer cells are detected, further surgery is performed.

After surgery, most breast cancers are treated with radiation and combinations of chemotherapeutic drugs, plus sometimes newer drugs that are targeted to certain types of breast cancer. Standard chemotherapies kill all rapidly dividing cells, and those used for breast cancer include fluorouracil, doxorubicin, cyclophosphamide, methotrexate, and paclitaxol. New protocols that provide more frequent, lower doses can temper some of the side effects of these powerful drugs.

Newer treatments developed for specific subtypes of breast cancer are easier to tolerate and can be extremely effective. Three types of drugs keep signals (estrogen and growth factors) from stimulating cancer cells to divide:

1. Selective estrogen receptor modulators (SERMs), such as tamoxifen and raloxifene, block estrogen receptors. About half of all people with breast cancer have receptors for estrogen on their cancer cells.
2. Aromatase inhibitors block an enzyme that is required for tissues other than those of the ovaries to synthesize estrogens. These drugs are used in women who are past menopause, whose ovaries no longer synthesize estrogen. They are prescribed after a five-year course of a SERM.

3. Trastuzumab can help people whose cancer cells bear too many receptors that bind a particular growth factor. It is a monoclonal antibody, which is based on an immune system protein. Trastuzumab blocks the growth factor from signaling cell division. Marketed as Herceptin, this drug treats a particularly aggressive form of the disease that strikes younger women.

Genetic tests performed on cancer cells can provide an estimate of the risk that the disease will recur and a prediction of whether or not a particular drug treatment is likely to be effective.

Prevention Strategies

Health-care providers advise women to have baseline mammograms by the age of forty and yearly mammograms after that, or beginning at age fifty, depending upon medical and family histories. Although a mammogram can detect a tumor up to two years before it can be felt, it can also miss some tumors. Thus, breast self-exam is also important in early detection.

Genetic tests can identify women who have inherited certain variants of genes—such as *BRCA1, BRCA2, p53,* and *HER-2/neu*—that place them at very high risk for developing breast cancer. Some women at high risk for these rare cancers have their breasts removed to prevent the disease. In one family, a genetic test told a woman whose two sisters and mother had breast cancer that she had escaped their fate, and she canceled surgery. Yet her young cousin, who thought she was free of the gene because it was inherited through her father, found by genetic testing that she would likely develop breast cancer. A subsequent mammogram revealed that the disease had already begun.

Only 5% to 10% of all breast cancers arise from an inherited tendency. Much current research seeks to identify the environmental triggers that cause the majority of cases.

A woman inserts the diaphragm into the vagina so that it covers the cervix, preventing sperm cells from entering the uterus (fig. 19.15*b*). To be effective, a diaphragm must be fitted for size by a physician, inserted properly, and used with a spermicide applied to the diaphragm surface adjacent to the cervix and to the rim of the device. It must be left in position for several hours following sexual intercourse. A diaphragm can be inserted up to six hours prior to sexual contact.

Similar to but smaller than the diaphragm is the *cervical cap*, which adheres to the cervix by suction.

A woman inserts it with her fingers before intercourse. For centuries, different societies have used cervical caps made of such varied substances as beeswax, lemon halves, paper, and opium poppy fibers.

Chemical Barriers

Chemical barrier contraceptives include creams, foams, and jellies with spermicidal properties (fig. 19.15*c*). These chemicals create an unfavorable environment in the vagina for sperm cells.

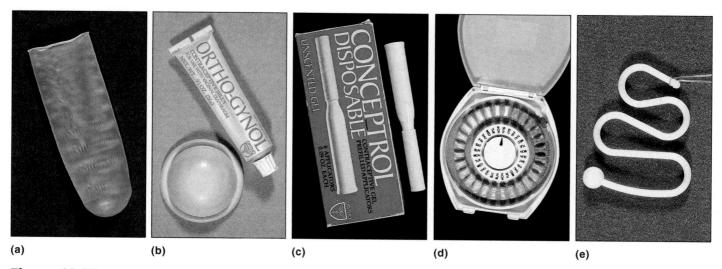

(a) (b) (c) (d) (e)

Figure 19.15

Devices and chemicals used for birth control include (a) male condom, (b) diaphragm, (c) spermicidal gel, (d) oral contraceptive, and (e) IUD.

Chemical barrier contraceptives are fairly easy to use but have a high failure rate when used alone. They are most effective with a condom or diaphragm.

Combined Hormone Contraceptives

Combined hormone contraceptives deliver estrogen and progestin. Various methods are used to administer the hormones, but all work on the same principle with about the same efficacy. A monthly injection of Lunelle is one such method. A small, flexible chemical ring (Nuvaring) may be inserted deep into the vagina once a month, remaining in place three out of four weeks. A plastic patch (Orthro Evra) impregnated with the hormones may be applied to the skin on the buttocks, stomach, arm, or upper torso once a week for three out of four weeks. The most commonly used method to deliver the hormones is orally in pill form.

An *oral contraceptive,* or birth control pill, contains synthetic estrogen-like and progesterone-like chemicals (fig. 19.15*d*). In women, these drugs disrupt the normal pattern of gonadotropin (FSH and LH) secretion preventing follicle maturation and the LH surge that triggers ovulation. They also interfere with buildup of the uterine lining necessary for implantation.

Oral contraceptives, if used correctly, prevent pregnancy nearly 100% of the time. However, they may cause nausea, retention of body fluids, increased skin pigmentation, and breast tenderness. Also, some women, particularly those over age thirty-five who smoke, may develop intravascular blood clots, liver disorders, or high blood pressure when using certain types of oral contraceptives.

Injectable Contraception

An intramuscular injection of Depo-Provera (medroxyprogesterone acetate) protects against pregnancy for three months by preventing the maturation and release of a secondary oocyte. It also alters the uterine lining, making it less hospitable for a developing embryo. Use of Depo Provera requires a doctor's care because potential side effects make it risky for women with certain medical conditions.

Intrauterine Devices

An *intrauterine device (IUD)* is a small, solid object that a physician places within the uterine cavity (fig. 19.15*e*). An IUD interferes with implantation of a blastocyst, perhaps by inflaming the uterine tissues.

The uterus may spontaneously expel the IUD, or the IUD may produce abdominal pain or excessive menstrual bleeding. It may also injure the uterus or produce other serious health problems. A physician should regularly check IUD placement.

Surgical Methods

Surgical methods of contraception sterilize the male or female. In the male, a physician performs a *vasectomy,* removing a small section of each ductus (vas) deferens near the epididymis and tying (ligating) the cut ends of the ducts. This is a vasectomy, an operation that produces few side effects, although it may cause some pain for a week or two.

After a vasectomy, sperm cells cannot leave the epididymis; thus, they are excluded from the semen. However, sperm cells may already be present in portions of the ducts distal to the cuts. Consequently, the sperm count may not reach zero for several weeks.

The corresponding procedure in the female is *tubal ligation.* The uterine tubes are cut and ligated so that sperm cells cannot reach an egg cell.

Neither a vasectomy nor a tubal ligation changes hormonal concentrations or sex drives. These procedures, shown in figure 19.16, are the most reliable forms of contraception. Reversing them requires microsurgery.

Check Your Recall

44. What factors make the rhythm method less reliable than some other methods of contraception?

45. Describe two methods of contraception that use mechanical barriers.

46. How do combined hormone oral and injectable contraceptives prevent pregnancy?

47. How does an IUD prevent pregnancy?

19.8 SEXUALLY TRANSMITTED DISEASES

The recognized **sexually transmitted diseases (STDs)** are often called "silent infections" because the early stages may not produce symptoms, especially in women. Table 19.4 describes some prevalent STDs. By the time symptoms appear, it is often too late to prevent complications or spread of the infection to sexual partners. Many STDs have similar symptoms, some of which are also seen in diseases or allergies that are not sexually related. A physician should be consulted if one or a combination of the following symptoms appears:

1. Burning sensation during urination.
2. Pain in the lower abdomen.

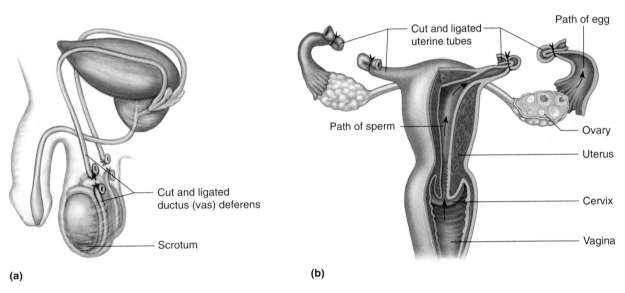

(a) **(b)**

Figure 19.16

Surgical methods of birth control. (*a*) In a vasectomy, each ductus (vas) deferens is cut and ligated. (*b*) In a tubal ligation, each uterine tube is cut and ligated.

Table 19.4	Some Sexually Transmitted Diseases		
Disease	**Cause**	**Symptoms**	**Treatment**
Acquired immune deficiency syndrome	Human immunodeficiency virus	Fever, weakness, infections, cancer	Drugs to treat or delay symptoms; no cure
Chlamydia infection	*Chlamydia trachomatis* bacteria	Painful urination and intercourse, mucous discharge from penis or vagina	Antibiotics
Genital herpes	Herpes simplex 2 virus	Genital sores, fever	Antiviral drug (acyclovir)
Genital warts	Human papilloma virus	Warts on genitals	Chemical or surgical removal
Gonorrhea	*Neisseria gonorrhoeae* bacteria	In women, usually none; in men, painful urination	Antibiotics
Syphilis	*Treponema pallidum* bacteria	Initial chancre sore usually on genitals or mouth; rash 6 months later; several years with no symptoms as infection spreads; finally damage to heart, liver, nerves, brain	Antibiotics

Reproductive Systems

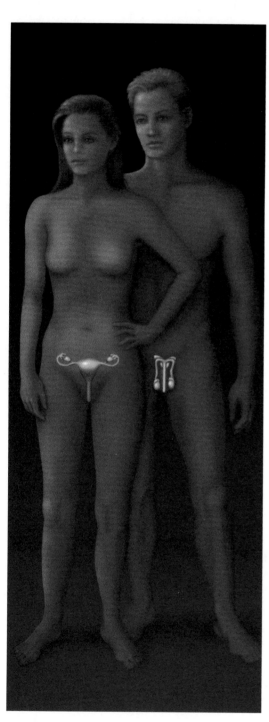

Integumentary System

Skin sensory receptors play a role in sexual pleasure.

Skeletal System

Bones can be a temporary source of calcium during lactation.

Muscular System

Skeletal, cardiac, and smooth muscles all play a role in reproductive processes and sexual activity.

Nervous System

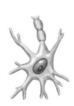

The nervous system plays a major role in sexual activity and sexual pleasure.

Endocrine System

Hormones control the production of eggs in the female and sperm in the male.

Cardiovascular System

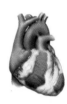

Blood pressure is necessary for the normal function of erectile tissue in the male and female.

Lymphatic System

Special mechanisms inhibit the female immune system from attacking sperm as foreign invaders.

Digestive System

Proper nutrition is essential for the formation of normal gametes.

Respiratory System

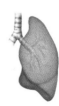

Breathing provides oxygen that assists in the production of ATP needed for egg and sperm development.

Urinary System

Male urinary and reproductive systems share common structures. Kidneys help compensate for fluid loss from the reproductive systems.

Gamete production, fertilization, fetal development, and childbirth are essential for survival of the species.

3. Fever or swollen glands in the neck.

4. Discharge from the vagina or penis.

5. Pain, itching, or inflammation in the genital or anal area.

6. Pain during intercourse.

7. Sores, blisters, bumps, or a rash anywhere on the body, particularly the mouth or genitals.

8. Itchy, runny eyes.

One possible complication of the STDs gonorrhea and chlamydia is **pelvic inflammatory disease,** in which bacteria enter the vagina and spread throughout the reproductive organs. The disease begins with intermittent cramps, followed by sudden fever, chills, weakness, and severe cramps. Hospitalization and intravenous antibiotics can stop the infection. The uterus and uterine tubes are often scarred, resulting in infertility and increased risk of ectopic pregnancy, in which the embryo develops within a uterine tube.

Acquired immune deficiency syndrome (AIDS) is an STD that destroys the immune system. Infections and often cancer, diseases that the immune system usually conquers, overrun the body. The AIDS virus (human immunodeficiency virus, or HIV) passes from one person to another in body fluids such as semen and blood. Unprotected intercourse and using a needle containing contaminated blood are the most frequent routes of transmission in the United States. The Topic of Interest in chapter 14 (p. 389) explores HIV infection further.

Check Your Recall

48. Why are sexually transmitted diseases often called "silent infections"?

49. What are some common symptoms of sexually transmitted diseases?

Clinical Terms Related to the Reproductive Systems

amenorrhea (a-men″o-re′ah) Absence of menstrual flow, usually due to a disturbance in hormonal concentrations.

conization (ko″nĭ-za′shun) Surgical removal of a cone of tissue from the cervix for examination.

curettage (ku″rĕ-tahzh′) Surgical procedure in which the cervix is dilated and the endometrium of the uterus is scraped. It is commonly called D and C, for dilatation (dilation) and curettage.

dysmenorrhea (dis″men-ŏ-re′ah) Painful menstruation.

endometriosis (en″do-me″tre-o′sis) Tissue similar to the inner lining of the uterus in the pelvic cavity.

endometritis (en″do-mĕ-tri′tis) Inflammation of the uterine lining.

epididymitis (ep″ĭ-did″ĭ-mi′tis) Inflammation of the epididymis.

hematometra (hem″ah-to-me′trah) Accumulation of menstrual blood within the uterine cavity.

hysterectomy (his″tĕ-rek′to-me) Surgical removal of the uterus.

mastitis (mas′ti′tis) Inflammation of a mammary gland.

oophorectomy (o″of-o-rek′to-me) Surgical removal of an ovary.

oophoritis (o″of-o-ri′tis) Inflammation of an ovary.

orchiectomy (or″ke-ĕk′to-me) Surgical removal of a testis.

orchitis (or-ki′tis) Inflammation of a testis.

prostatectomy (pros″tah-tek′to-me) Surgical removal of a portion or all of the prostate gland.

prostatitis (pros″tah-ti′tis) Inflammation of the prostate gland.

salpingectomy (sal″pin-jek′to-me) Surgical removal of a uterine tube.

vaginitis (vaj″ĭ-ni′tis) Inflammation of the vaginal lining.

varicocele (var′ĭ-ko-sēl″) Distension of the veins within the spermatic cord.

Clinical Connection

Bruce Reimer was born in 1965. At age eight months, most of his penis was accidentally burned off during a circumcision procedure. Physicians and psychologists advised the parents to "reassign" the child's gender as female. At twenty-two months of age, corrective surgery created Brenda from Bruce. But Brenda continually fought attempts to being raised as a girl and at age fourteen threatened suicide unless allowed to live as a male. He took the name David Reimer, and eventually married, adopted his wife's children, and was a young grandfather, but committed suicide in May 2004.

Apparently, surgery could not silence David's XY chromosome constitution—that of a male. Since his case, several studies of infants born with very small penises and reared as girls overwhelmingly confirm Reimer's experience that nature has a greater effect on gender identity than nurture.

In the past, physicians based the decision to remove a small or damaged penis and reassign sex as female on size. If a newborn's stretched organ exceeded an inch, he was deemed a he. If the protrusion was under 3/8 of an inch, she was deemed a she. Organs that fell in between were shortened into a clitoris during the first week of life, and girlhood officially began. Today, such decisions rest more on an individual's chromosomal sex, and in some cases, surgery is delayed until a person can decide for him- or herself.

SUMMARY OUTLINE

19.1 Introduction (p. 508)

Reproductive organs produce sex cells and sex hormones, sustain these cells, or transport them.

19.2 Organs of the Male Reproductive System (p. 508)

The primary male sex organs are the testes, which produce sperm cells and male sex hormones. Accessory organs include the internal and external reproductive organs.

1. Testes
   a. Structure of the testes
      (1) The testes are composed of lobules separated by connective tissue and filled with seminiferous tubules.
      (2) The seminiferous tubules contain undifferentiated cells that give rise to sperm cells.
      (3) The interstitial cells produce male sex hormones.
   b. Formation of sperm cells
      (1) The epithelium lining the seminiferous tubules includes supporting cells and spermatogenic cells.
         (a) Supporting cells support and nourish spermatogenic cells.
         (b) Spermatogenic cells give rise to sperm cells.
      (2) A sperm cell consists of a head, midpiece, and tail.
   c. Spermatogenesis
      (1) Spermatogonia give rise to sperm cells.
      (2) Meiosis reduces the number of chromosomes in sperm cells by one-half (from 46 to 23).
      (3) Spermatogenesis produces four sperm cells from each primary spermatocyte.
2. Male internal accessory organs
   a. Epididymides
      (1) Each epididymis is a tightly coiled tube that leads into a ductus deferens.
      (2) They store and nourish immature sperm cells and promote their maturation.
   b. Ductus deferentia
      (1) Each ductus deferens is a muscular tube.
      (2) They pass through the inguinal canal, enter the pelvic cavity, course medially and end behind the urinary bladder.
      (3) They fuse with the ducts from seminal vesicles to form the ejaculatory ducts.
   c. Seminal vesicles
      (1) Each seminal vesicle is a saclike structure attached to a ductus deferens.
      (2) They secrete an alkaline fluid that contains nutrients, such as fructose, and prostaglandins.
   d. Prostate gland
      (1) The prostate gland surrounds the urethra just inferior to the urinary bladder.
      (2) It secretes a thin, milky fluid that neutralizes the pH of semen and the acidic secretions of the vagina.
   e. Bulbourethral glands
      (1) The bulbourethral glands are two small structures inferior to the prostate gland.
      (2) They secrete a fluid that lubricates the penis in preparation for sexual intercourse.
   f. Semen
      (1) Semen consists of sperm cells and secretions of the seminal vesicles, prostate gland, and bulbourethral glands.
      (2) This fluid is slightly alkaline and contains nutrients and prostaglandins.
      (3) Sperm cells in semen begin to swim, but are unable to fertilize egg cells until they are activated in the female reproductive tract.
3. Male external reproductive organs
   a. Scrotum
      The scrotum is a pouch of skin and subcutaneous tissue that encloses the testes for protection and temperature regulation.
   b. Penis
      (1) The penis is specialized to become erect for insertion into the vagina during sexual intercourse.
      (2) Its body is composed of three columns of erectile tissue.
4. Erection, orgasm, and ejaculation
   a. During erection, the vascular spaces within the erectile tissue engorge with blood.
   b. Orgasm is the culmination of sexual stimulation. Emission and ejaculation accompany male orgasm.
   c. Semen moves along the reproductive tract as smooth muscle in the walls of the tubular structures contracts by reflex.

19.3 Hormonal Control of Male Reproductive Functions (p. 515)

1. Hypothalamic and pituitary hormones
   a. The male body remains reproductively immature until the hypothalamus releases gonadotropin-releasing hormone (GnRH), which stimulates the anterior pituitary gland to release gonadotropins.
   b. Follicle-stimulating hormone (FSH) stimulates spermatogenesis.
   c. Luteinizing hormone (LH), sometimes known in males as interstitial cell stimulating hormone (ICSH), stimulates interstitial cells to produce male sex hormones.
2. Male sex hormones
   a. Male sex hormones are called androgens, with testosterone the most important.
   b. Androgen production increases rapidly at puberty.
   c. Actions of testosterone
      (1) Testosterone stimulates development of the male reproductive organs.
      (2) It also develops and maintains male secondary sex characteristics.
   d. Regulation of male sex hormones
      (1) A negative feedback mechanism regulates testosterone concentration.
         (a) A rising testosterone concentration inhibits the hypothalamus and reduces the anterior pituitary's secretion of gonadotropins.
         (b) As testosterone concentration falls, the hypothalamus signals the anterior pituitary to secrete gonadotropins.
      (2) The testosterone concentration remains relatively stable from day to day.

19.4 Organs of the Female Reproductive System (p. 516)

The primary female sex organs are the ovaries, which produce female sex cells and sex hormones. Accessory organs are internal and external.

1. Ovaries
   a. Ovary structure
      (1) Each ovary is subdivided into a medulla and a cortex.
      (2) The medulla is composed of connective tissue, blood vessels, lymphatic vessels, and nerves.
      (3) The cortex contains ovarian follicles and is covered by cuboidal epithelium.

b. Primordial follicles
   (1) During prenatal development, groups of cells in the ovarian cortex form millions of primordial follicles.
   (2) Each primordial follicle contains a primary oocyte and a layer of follicular cells.
   (3) The primary oocyte begins meiosis, but the process halts until puberty.
   (4) The number of primary oocytes steadily declines throughout a female's life.
c. Oogenesis
   (1) Beginning at puberty, some primary oocytes are stimulated to continue meiosis.
   (2) When a primary oocyte undergoes meiosis, it gives rise to a secondary oocyte in which the original chromosome number is reduced by one-half (from 46 to 23).
   (3) Fertilization of a secondary oocyte produces a zygote.
d. Follicle maturation
   (1) At puberty, FSH initiates follicle maturation.
   (2) During maturation of the follicle, the primary oocyte undergoes meiosis giving rise to a secondary oocyte and first polar body, the follicular cells multiply, and a fluid-filled cavity forms.
   (3) Usually, only one follicle at a time fully develops.
e. Ovulation
   (1) Ovulation is the release of a secondary oocyte from an ovary.
   (2) A rupturing follicle releases the secondary oocyte.
   (3) After ovulation, the secondary oocyte is drawn into the opening of the uterine tube.
2. Female internal accessory organs
a. Uterine tubes
   (1) The end of each uterine tube expands, and its margin bears irregular extensions.
   (2) Ciliated cells that line the tube and peristaltic contractions in the wall of the tube help transport the secondary oocyte down the uterine tube. Fertilization may occur.
b. Uterus
   (1) The uterus receives the embryo and sustains it during development.
   (2) The uterine wall includes the endometrium, myometrium, and perimetrium.
c. Vagina
   (1) The vagina receives the erect penis, conveys uterine secretions to the outside, and provides an open channel for the fetus during birth.
   (2) Its wall consists of mucosal, muscular, and fibrous layers.
3. Female external reproductive organs
a. Labia majora
   (1) The labia majora are rounded folds of adipose tissue and skin.
   (2) The upper ends form a rounded elevation over the symphysis pubis.
b. Labia minora
   (1) The labia minora are flattened, longitudinal folds between the labia majora.
   (2) They are well supplied with blood vessels.
c. Clitoris
   (1) The clitoris is a small projection at the anterior end of the vulva. It corresponds to the male penis.
   (2) It is composed of two columns of erectile tissue.
d. Vestibule
   (1) The vestibule is the space between the labia minora.
   (2) The vestibular glands secrete mucus into the vestibule during sexual stimulation.

4. Erection, lubrication, and orgasm
a. During periods of sexual stimulation, the erectile tissues of the clitoris and vestibular bulbs engorge with blood and swell.
b. The vestibular glands secrete mucus into the vestibule and vagina.
c. During orgasm, the muscles of the perineum, uterine wall, and uterine tubes contract rhythmically.

19.5 Hormonal Control of Female Reproductive Functions (p. 524)

The hypothalamus, anterior pituitary gland, and ovaries secrete hormones that control sex cell maturation, the development and maintenance of female secondary sex characteristics, and changes that occur during the monthly reproductive cycle.

1. Female sex hormones
a. A female body remains reproductively immature until about ten years of age, when gonadotropin secretion increases.
b. The most important female sex hormones are estrogens and progesterone.
   (1) Estrogens develop and maintain most female secondary sex characteristics.
   (2) Progesterone prepares the uterus for pregnancy.
2. Female reproductive cycle
a. FSH from the anterior pituitary initiates a reproductive cycle by stimulating follicle maturation.
b. Maturing follicular cells secrete estrogens, which maintain the secondary sex characteristics and thicken the uterine lining.
c. Secretion of a relatively large amount of LH by the anterior pituitary triggers ovulation.
d. Following ovulation, follicular cells give rise to the corpus luteum.
   (1) The corpus luteum secretes progesterone, which causes the endometrium to become more vascular and glandular.
   (2) If a secondary oocyte is not fertilized, the corpus luteum begins to degenerate.
   (3) As concentrations of estrogens and progesterone decline, the uterine lining disintegrates, causing menstrual flow.
e. During this cycle, estrogens and progesterone inhibit the release of LH and FSH. As concentrations of estrogens and progesterone decline, the anterior pituitary secretes FSH and LH again, stimulating a new reproductive cycle.
3. Menopause
a. Menopause is termination of reproductive cycles due to aging of the ovaries.
b. Reduced concentrations of estrogens and lack of progesterone may cause regressive changes in female secondary sex characteristics.

19.6 Mammary Glands (p. 526)

1. The mammary glands are in the subcutaneous tissue of the anterior thorax.
2. They are composed of lobes that contain glands and a duct.
3. Dense connective and adipose tissues separate the lobes.
4. Ovarian hormones stimulate female breast development.
a. Alveolar glands and ducts enlarge.
b. Fat is deposited around and within the breasts.

19.7 Birth Control (p. 527)

Birth control is voluntary regulation of how many offspring are produced and when they are conceived. It usually involves some method of contraception.

1. Coitus interruptus is withdrawal of the penis from the vagina before ejaculation.
2. The rhythm method is abstinence from sexual intercourse for several days before and after ovulation.
3. Mechanical barriers
   a. Males and females can use condoms.
   b. Females can also use diaphragms and cervical caps.
4. Chemical barriers
   Spermicidal creams, foams, and jellies provide an unfavorable environment in the vagina for sperm survival.
5. Combined hormone contraceptives
   a. A monthly injection, a flexible ring inserted deep into the vagina, or a plastic patch can deliver estrogen and progestin to prevent pregnancy.
   b. Birth control pills contain synthetic estrogen-like and progesterone-like substances that disrupt a female's normal pattern of gonadotropin secretion preventing follicle maturation, ovulation, and the normal buildup of the uterine lining.
6. Injectable contraception
   Intramuscular injection with Depo-Provera every three months acts similarly to oral contraceptives to prevent pregnancy.
7. Intrauterine devices (IUDs)
   An IUD is a solid object inserted in the uterine cavity that prevents pregnancy by interfering with implantation of a blastocyst.
8. Surgical methods
   Vasectomies in males and tubal ligations in females are surgical sterilization procedures.

19.8 Sexually Transmitted Diseases (p. 531)

1. Sexually transmitted diseases (STDs) are passed during sexual contact and may go undetected for years.
2. The twenty recognized STDs share similar symptoms.

CHAPTER ASSESSMENTS

19.1 Introduction

1. General functions of the male and female reproductive systems include: (p. 508)
   a. producing sex cells
   b. nurturing sex cells
   c. transporting sex cells to sites of fertilization
   d. secreting hormones
   e. all of the above

19.2 Organs of the Male Reproductive System

2. List the organs (both primary and accessory) of the male reproductive system, and explain how each organ's structure affects the organ's function. (pp. 508–514)
3. List the major steps in spermatogenesis. (p. 509)
4. Trace the path of sperm cells from their site of formation to the outside. Indicate composition and when and where secretions are added to produce semen. (pp. 509–513)
5. Contrast emission and ejaculation. (p. 514)

19.3 Hormonal Control of Male Reproductive Functions

6. Describe the role of gonadotropin-releasing hormone (GnRH) in the control of male reproductive functions. (p. 515)

7. Discuss the actions of testosterone. (p. 515)

19.4 Organs of the Female Reproductive System

8. List the organs (both primary and accessory) of the female reproductive system, and explain how each organ's structure affects the organ's function. (pp. 518–523)
9. List the major steps in oogenesis. (p. 519)
10. Describe how a follicle matures. (p. 520)
11. Define *ovulation*. (p. 520)

19.5 Hormonal Control of Female Reproductive Functions

12. Describe the role of gonadotropin-releasing hormone (GnRH) in the control of female reproductive functions. (p. 524)
13. Discuss the actions of estrogens. (p. 524)
14. Summarize the major events in the reproductive cycle. (p. 524)

19.6 Mammary Glands

15. Describe the structure of a mammary gland. (p. 526)

19.7 Birth Control

16. Match the birth control method with its description. (pp. 527–531)

    (1) withdrawal
    (2) rhythm method
    (3) condom
    (4) spermicide (foam, gel)
    (5) estrogen/progesterone
    (6) IUD
    (7) vasectomy
    (8) tubal ligation

    A. Kills sperm (not very effective when used alone)
    B. Keeps sperm out of vagina or from entering cervix (additionally, may help prevent disease)
    C. Prevents implantation
    D. No intercourse during fertile times (ineffective)
    E. Penis removed from vagina before ejaculation
    F. Sperm cells never reach penis (very effective)
    G. Prevents follicle maturation and ovulation
    H. Oocytes never reach uterus (very effective)

19.8 Sexually Transmitted Diseases

17. Common symptoms of sexually transmitted diseases include: (p. 531)
    a. a burning sensation during urination
    b. discharge from vagina or penis
    c. pain during intercourse
    d. sores, blisters, or rash on genitals
    e. all of the above

18. If left untreated, a complication of the sexually transmitted diseases gonorrhea and chlamydia is _____. (p. 533)

INTEGRATED ASSESSMENTS/ CRITICAL THINKING

OUTCOMES 19.1, 19.2, 19.4

1. Why must the chromosome number be halved in sperm cells and secondary oocytes?

OUTCOME 19.2

2. Some men are unable to become fathers because their spermatids do not mature into sperm. Injection of their spermatids into their partners' secondary oocytes sometimes

results in conception. A few men have fathered healthy babies this way. Why would this procedure work with spermatids, but not with primary spermatocytes?

OUTCOME 19.4

3. Sometimes a sperm cell fertilizes a polar body rather than a secondary oocyte. An embryo does not develop, and the fertilized polar body degenerates. Why is a polar body unable to support development of an embryo?

OUTCOMES 19.2, 19.4

4. How are the human male and female reproductive tracts similar? How are the structures of the testis and ovary similar?

5. What changes, if any, would a male who has had one testis removed experience? A female who has had one ovary removed?

OUTCOME 19.7

6. Understanding the causes of infertility can be valuable in developing new birth control methods. Cite a type of contraceptive based on each of the following causes of infertility: (a) failure to ovulate due to a hormonal imbalance; (b) a large fibroid tumor that disturbs the uterine lining; (c) endometrial tissue blocking uterine tubes; (d) low sperm count (too few sperm per ejaculate).

7. Does a tubal ligation cause a woman to enter menopause prematurely? Why or why not?

WEB CONNECTIONS

Visit the text website at **aris.mhhe.com** for additional quizzes, interactive learning exercises, and more.

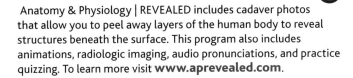 REPRODUCTIVE SYSTEM

Anatomy & Physiology | REVEALED includes cadaver photos that allow you to peel away layers of the human body to reveal structures beneath the surface. This program also includes animations, radiologic imaging, audio pronunciations, and practice quizzing. To learn more visit **www.aprevealed.com**.

20

Pregnancy, Growth, Development, and Genetics

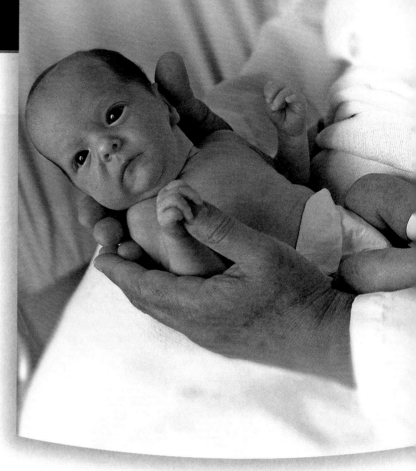

Experiments suggest that even the smallest premature babies can feel pain. In the future, they may be routinely given painkillers.

DO PREMATURE BABIES FEEL PAIN? Tyler was born at 26 weeks' gestation, just over two weeks past the time at which he would probably not have survived. Life in the neonatal intensive care unit is not easy. He is hooked up to monitors, and various procedures are performed, in total, at least a dozen times a day. He breathes and feeds with the help of tubes. Tyler is probably quite uncomfortable, but he has no way of communicating that.

For many years, health-care practitioners who monitor and treat the youngest of the young have assumed that the grimaces and squirms of their charges do not necessarily indicate discomfort or pain. The premature baby's brain, it was thought, was not mature enough to overlay noxious stimuli with the experience necessary to form what we think of as pain. So treating pain in "preemies" has largely been a case-by-case challenge. If facial expression, breathing pattern, and vital signs suggest distress, pain medication might be given—but such drugs have never been tested in newborns, let alone in those who arrived well before their due dates.

What was needed in assessing preemie pain was a way to look at the brain as it responds to a noxious stimulus. A form of neuroimaging called near-infrared spectroscopy (NIRS) has done that. Rather than having to transport the child to a large scanner, as is done with MRI or

PET imaging, NIRS tracks changes in oxygen concentration in the brain cribside, with gentle sensors. In a study to assess response to pain, researchers used NIRS when a vein was punctured or a heel was pricked to sample blood for routine diagnostic testing. When this happened, the part of the cerebral cortex that processes sensation into perception was activated. Touch alone did not have this effect. It appears, then, that the cerebral cortex of a premature infant can indeed register pain. As a result of these and other studies, a government task force of clinicians, scientists, and bioethicists is currently developing clinical trial procedures to assess the effects of various painkillers on premature babies.

Learning Outcomes   *After studying this chapter, you should be able to do the following:*

Aids to Understanding Words *(Appendix A on page 567 has a complete list of Aids to Understanding Words.)*

allant- [sausage] *allant*ois: Tubelike structure extending from the yolk sac into the connecting stalk of the embryo.

chorio- [skin] *chorio*n: Outermost membrane surrounding the fetus and its membranes.

cleav- [to divide] *cleav*age: Period of development when the zygote divides into smaller and smaller cells.

hetero- [other, different] *hetero*zygous: Condition in which the members of a gene pair are different.

hom- [same, common] *hom*ozygous: Condition in which the members of a gene pair are the same.

lacun- [pool] *lacun*a: Space between the chorionic villi that fills with maternal blood.

morul- [mulberry] *morul*a: Embryonic structure consisting of a solid ball of about sixteen cells that resembles a mulberry.

nat- [to be born] pre*nat*al: Period of development before birth.

troph- [well fed] *troph*oblast: Cellular layer that surrounds the inner cell mass and helps nourish it.

umbil- [navel] *umbil*ical cord: Structure attached to the fetal naval (umbilicus) that connects the fetus to the placenta.

20.1 INTRODUCTION

A sperm cell and a secondary oocyte unite, forming a zygote, and the journey of prenatal development begins. Following thirty-eight weeks of cell division, growth, and specialization into distinctive tissues and organs, a new human being enters the world.

Before birth, an individual grows and develops. **Growth** is an increase in size. In humans and other many-celled organisms, growth entails an increase in cell numbers, followed by enlargement of the newly formed cells. **Development,** which includes growth, is the continuous process by which an individual changes from one life phase to another. These life phases include a **prenatal period** (pre-na′tal pe′re-od), which begins with fertilization and ends at birth, and a **postnatal** (pōst-na′tal) **period,** which begins at birth and ends at death.

20.2 PREGNANCY

The union of an egg cell and a sperm cell is called **fertilization** (fer″tĭ-lĭ-za′shun), or conception, which typically occurs in a uterine tube. **Pregnancy** (preg′nan-se) begins when the developing offspring nestles into the lining of the uterus. Pregnancy consists of three periods called trimesters, each about three months long.

Transport of Sex Cells

Each month, a female of reproductive age usually ovulates a secondary oocyte, unless she is anovulatory, as discussed in the Topic of Interest on page 543. The released egg cell then usually enters a uterine tube. During sexual intercourse, the male deposits semen containing sperm cells in the vagina near the cervix. To reach the secondary oocyte, the sperm cells must move upward through the uterus and uterine tube. Prostaglandins in the semen stimulate lashing of sperm tails and muscular contractions within the walls of the uterus and uterine tube, which help sperm cells move.

Also, high concentrations of estrogens during the first part of the menstrual cycle stimulate the uterus and cervix to secrete a thin, watery fluid that promotes sperm transport and survival. Conversely, during the latter part of the cycle, when progesterone concentration is high, the female reproductive tract secretes a viscous fluid that hampers sperm transport and survival. These changes in the penetrability of the cervical mucus increase the chance that sperm will reach the egg cell when a woman is most fertile.

Sperm cells reach the upper portions of the uterine tube within an hour following sexual intercourse. Many sperm cells may reach the secondary oocyte, but only one actually fertilizes it (fig. 20.1; see fig. 20.4a).

Fertilization

A sperm cell first invades the follicular cells that adhere to the secondary oocyte's surface (corona radiata) then

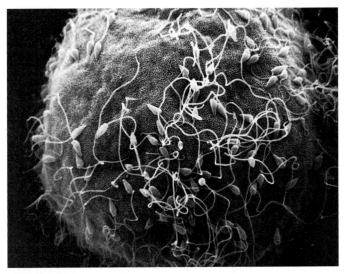

Figure 20.1

Scanning electron micrograph of sperm cells on the surface of a secondary oocyte (1,100×). Only one sperm cell actually fertilizes a secondary oocyte.

binds the *zona pellucida* surrounding the oocyte's cell membrane. The acrosome of the sperm cell releases enzymes (including hyaluronidase) that aid penetration of the sperm head by digesting proteins in the zona pellucida (fig. 20.2). However, at least several hundred sperm cells must be present to produce enough enzymes to enable one to penetrate. This is why males with very low sperm counts are said to be subfertile.

The head portion of one sperm cell enters the oocyte, leaving the mitochondria-rich middle section and tail outside. This action triggers lysosome-like vesicles just beneath the oocyte cell membrane to release enzymes that harden the zona pellucida. As a result, additional sperm heads cannot enter. Entry of more than one sperm cell would be a developmental disaster, because the zygote would have too much genetic material.

Figure 20.3 shows what happens once a sperm cell head enters the secondary oocyte's cytoplasm. The secondary oocyte then divides to form a large cell, whose nucleus contains the female's genetic contribution, and a tiny second polar body, which is later expelled. Meiosis is completed. The approaching nuclei from the two sex cells are called pronuclei, until they meet and merge. Next, the pronuclei unite. Their nuclear membranes fall apart, and their chromosomes mingle, completing fertilization. The Topic of Interest on page 544 describes fertilization accomplished in the laboratory.

Because each sex cell provides 23 chromosomes, the product of fertilization is a cell with 46 chromosomes—the usual number in a human body cell (somatic cell). This cell is the **zygote** (zi′gōt), and it is the first cell of the future offspring.

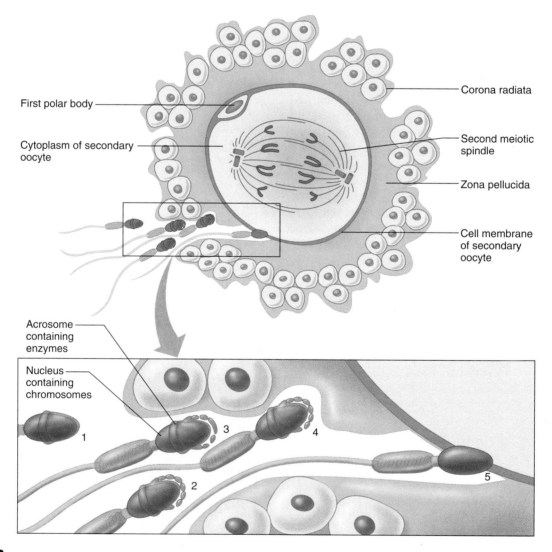

First polar body

Cytoplasm of secondary oocyte

Corona radiata

Second meiotic spindle

Zona pellucida

Cell membrane of secondary oocyte

Acrosome containing enzymes

Nucleus containing chromosomes

1

2

3

4

5

Figure 20.2

Steps in fertilization: (*1*) The sperm cell reaches the corona radiata surrounding the secondary oocyte. (*2*) and (*3*) The acrosome of the sperm cell releases a protein-digesting enzyme. (*4*) The sperm cell penetrates the zona pellucida surrounding the oocyte. (*5*) The sperm cell's membrane fuses with the oocyte's cell membrane.

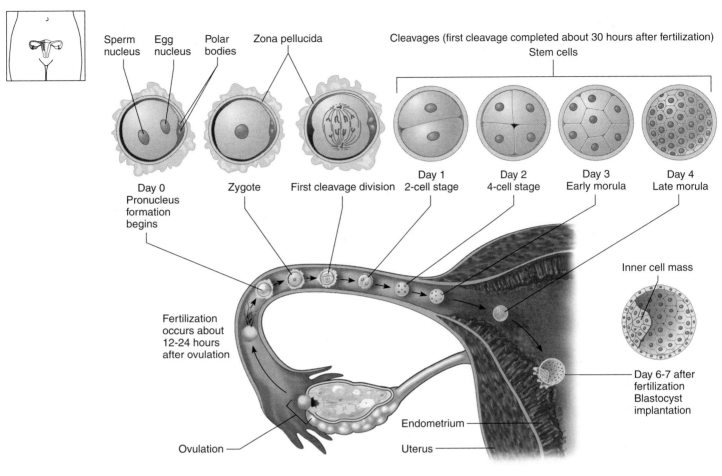

Figure 20.3
Stages of early human prenatal development.

Check Your Recall

1. What factors aid the movements of the secondary oocyte and sperm cells through the female reproductive tract?

2. Where in the female reproductive system does fertilization normally take place?

3. List the events of fertilization.

20.3 PRENATAL PERIOD

Early Embryonic Development

About 30 hours after forming, the zygote undergoes *mitosis,* giving rise to two new cells (blastomeres) (fig. 20.4*b*). These cells, in turn, divide into four cells, which divide into eight cells, and so forth. The divisions occur rapidly, with little time for growth. Thus, each division yields smaller cells. This phase of early rapid cell division is termed **cleavage** (klēv′ij) (see fig. 20.3).

During cleavage, the tiny mass of cells moves through the uterine tube to the uterine cavity. The trip takes about three days, and by then the structure consists of a solid ball of about sixteen cells. The ball is called a *morula,* (mor′u-lah) (figs. 20.3 and 20.4*c*).

The morula remains free within the uterine cavity for about three days. During this stage, the zona pellucida of the original secondary oocyte degenerates. Then the morula hollows out, forming a *blastocyst,*

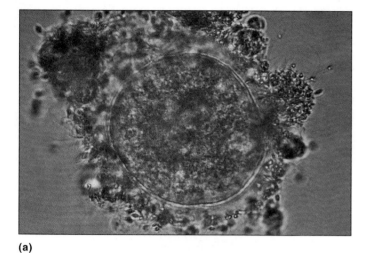

(a)

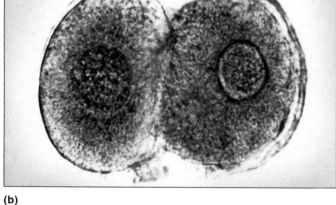

(b)

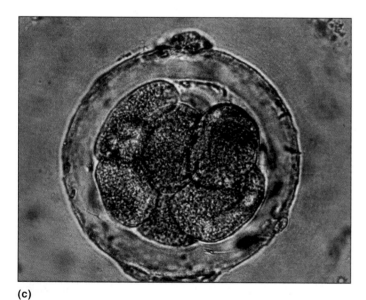

(c)

Figure 20.4

Light micrographs of (*a*) a human secondary oocyte surrounded by follicular cells and sperm cells (250×), (*b*) the two-cell stage (600×), and (*c*) a morula (500×).

which begins to attach to the endometrium. By the end of the first week of development, the blastocyst superficially implants in the endometrium (fig. 20.5). Up until this point, the cells that will become the developing offspring are pluripotent stem cells (see fig. 20.3), which means that they can give rise to several specialized types of cells, as well as yield additional stem cells.

Occasionally, the developing mass of cells that is the embryo implants in tissues outside the uterus, such as those of the uterine tube, an ovary, the cervix, or an organ in the abdominal cavity. The result is an *ectopic pregnancy*. If a fertilized egg implants in the uterine tube, it is specifically called a *tubal pregnancy*. The tube usually ruptures as the embryo enlarges and causes severe pain and heavy vaginal bleeding, threatening the pregnant woman and the embryo. Treatment is prompt surgical removal of the embryo and repair or removal of the damaged uterine tube.

About the time of implantation, certain cells on the inner face of the blastocyst organize into a group, called the inner cell mass, that will give rise to the body of the developing offspring. This marks the beginning of

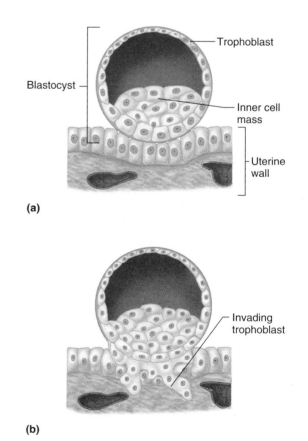

Figure 20.5

About the sixth day of prenatal development, the blastocyst (*a*) contacts the uterine wall and (*b*) begins to implant. The trophoblast, which will help form the placenta, secretes hCG, a hormone that maintains the pregnancy.

Female Infertility

Infertility is the inability to conceive after a year of trying. In 90% of cases, infertility has a physical cause, and 60% of the time, the abnormality is in the female's reproductive system.

A common cause of female infertility is insufficient secretion (hyposecretion) of gonadotropic hormones by the anterior pituitary, preventing ovulation (anovulation). Testing the urine for *pregnanediol,* a product of progesterone metabolism, detects an anovulatory cycle. Because progesterone concentration normally rises after ovulation, no increase in pregnanediol in the urine during the latter part of the menstrual cycle suggests lack of ovulation.

Fertility specialists can treat anovulation due to hyposecretion of gonadotropic hormones by administering human chorionic gonadotropin (hCG) obtained from human placentas. Another ovulation-stimulating biochemical, human menopausal gonadotropin (hMG), contains luteinizing hormone (LH) and follicle-stimulating hormone (FSH) and is collected from the urine of postmenopausal women. However, either hCG or hMG may overstimulate the ovaries and cause many follicles to release secondary oocytes simultaneously, which may result in multiple births if several are fertilized.

Another cause of female infertility is *endometriosis,* in which small pieces of the inner uterine lining (endometrium) move up through the uterine tubes during menstruation and implant in the abdominal cavity. Here, the tissue changes in a similar way to the uterine lining during the reproductive cycle. The abnormally located tissue breaks down at the end of the cycle but cannot be expelled. Instead, it remains in the abdominal cavity, irritating the lining (peritoneum) and causing considerable pain. This tissue also stimulates formation of fibrous tissue (fibrosis), which may encase the ovary, preventing ovulation or obstructing the uterine tubes.

Sexually transmitted diseases (STDs), such as gonorrhea, can cause female infertility. These infections can inflame and obstruct the uterine tubes or stimulate production of viscous mucus that plugs the cervix and prevents sperm entry.

Women become infertile if their ovaries must be removed, which may be part of cancer treatment. Secondary oocytes can be removed before cancer treatment and frozen, then later thawed, fertilized, and implanted into the woman's uterus (see Topic of Interest on page 544). However, many oocytes do not survive freezing. Despite two decades of attempts, only 100 babies have been born using frozen oocytes. The probability of pregnancy occurring using a frozen oocyte and current technology is about 3%.

Finding the right treatment for a particular patient requires determining the infertility's cause. Table 20A describes diagnostic tests for female infertility.

Table 20A	Tests to Assess Female Infertility
Test	**What It Checks**
Hormone levels	Whether ovulation occurs
Ultrasound	Placement and appearance of reproductive organs and structures
Postcoital test	Cervix examined soon after unprotected intercourse to see if mucus is thin enough to allow sperm through
Endometrial biopsy	Small piece of uterine lining sampled and viewed under microscope to see if it can support an embryo
Hysterosalpingogram	Dye injected into uterine tube and followed with scanner shows if tube is clear or blocked
Laparoscopy	Small, lit optical device inserted near navel to detect scar tissue blocking tubes, which ultrasound may miss

the **embryonic stage** of development. The offspring is termed an **embryo** (em'bre-o) until the end of the eighth week, when the basic structural form of the human body is recognizable. After the eighth week and until birth, the offspring is called a **fetus** (fe'tus). Rudiments of all organs are present by the end of embryonic development. These organs and other structures enlarge and specialize during fetal development.

During the embryonic stage, the cells surrounding the embryo, with cells of the endometrium, form a complex vascular structure called the **placenta** (plah-sen'tah). This organ attaches the embryo to the uterine wall and exchanges nutrients, gases, and wastes between maternal blood and the embryo's blood. The placenta also secretes hormones.

Check Your Recall

4. What is cleavage?

5. What is implantation?

6. How do an embryo and a fetus differ?

Quintuplets

Michele and Ray L'Esperance wanted children badly, but Michele's uterine tubes had been removed due to scarring. A procedure called *in vitro fertilization* (IVF) enabled the couple not only to have children, but to become the parents of the first "test tube" quintuplets.

At Beaumont Hospital in Royal Oak, Michigan, Michele first received human menopausal gonadotropin to stimulate development of ovarian follicles. When an ultrasound scan showed that the follicles had grown to a certain diameter, she received human chorionic gonadotropin to induce ovulation. Then Michele's physician used an optical instrument called a laparoscope to examine the interior of her abdomen and take the largest secondary oocytes from an ovary. The secondary oocytes were incubated at 37°C in a medium buffered at pH 7.4. When the secondary oocytes matured, they were mixed in a laboratory dish with Ray's sperm cells, which had been washed to remove inhibitory factors. Secretions from Michele's reproductive tract were added to activate the sperm.

Next, fertilized secondary oocytes were selected and incubated in a special medium for about sixty hours. At this stage, five balls of eight to sixteen cells each were transferred through Michele's cervix and into her uterus to increase the chances that one or two would complete development. (Today, fewer fertilized secondary oocytes are transferred because of medical problems associated with multiple births.) The L'Esperances beat the odds—they had healthy quintuplets, born 1 minute apart. Today, the quints are in college (fig. 20A)!

Success rates for IVF vary from clinic to clinic, ranging from 0% to 40%, with the average about 17%. Pregnancy via IVF is expensive, costing thousands of dollars. Table 20B describes other assisted reproductive technologies.

Figure 20A

The L'Esperance quintuplets. *In vitro* fertilization worked for Michele and Ray L'Esperance. Five fertilized eggs implanted in Michele's uterus are now Raymond, Alexandria, Danielle, Erica, and Veronica. But many couples are disappointed with the high failure rate of the technology.

Table 20B	Assisted Reproductive Technologies	
Technology	**Procedure**	**Condition It Treats**
Intrauterine insemination	Donated sperm cells are placed into the cervix or uterus.	Male infertility—lack of sperm cells or low sperm count
Surrogate mother	A secondary oocyte fertilized *in vitro* is implanted in a woman other than its donor. The surrogate, or "gestational mother," gives the newborn to the "genetic mother" and her partner, the sperm donor.	Female infertility—lack of a uterus
Gamete intrafallopian transfer (GIFT)	Secondary oocytes are removed from a woman's ovary, then placed along with donated sperm cells into a uterine tube	Female infertility—bypasses blocked uterine tube.
Zygote intrafallopian transfer (ZIFT)	A secondary oocyte fertilized *in vitro* is placed in a uterine tube. It travels to the uterus on its own.	Female infertility—bypasses blocked uterine tube
Embryo adoption	A woman is artificially inseminated with sperm cells from a man whose partner cannot ovulate. If the woman conceives, the morula is flushed from her uterus and implanted in the uterus of the sperm donor's partner.	Female infertility—a woman has nonfunctional ovaries, but a healthy uterus
Intracytoplasmic sperm injection	Sperm or spermatids are injected into secondary oocytes *in vitro*.	Low sperm count; sperm that cannot mature past spermatid stage; men with paralysis who cannot ejaculate.

Hormonal Changes During Pregnancy

During a typical reproductive cycle, the corpus luteum degenerates about two weeks after ovulation. Consequently, concentrations of estrogens and progesterone decline rapidly, the uterine lining breaks down, and the endometrium sloughs away as menstrual flow. If this occurs following implantation, the embryo is lost in a spontaneous abortion.

The hormone **human chorionic gonadotropin (hCG)** normally prevents spontaneous abortion. Cells from the outer blastocyst form a layer called the trophoblast, which surrounds the developing embryo and later helps form the placenta (fig. 20.5). The trophoblast secretes hCG. This hormone, similar in function to luteinizing hormone (LH), maintains the corpus luteum, which continues secreting estrogens and progesterone, stimulating the uterine wall to grow and develop. At the same time, hCG inhibits the anterior pituitary's release of follicle-stimulating hormone (FSH) and LH, halting the normal reproductive cycle.

Secretion of hCG continues at a high level for about two months, and then declines by the end of four months. Detecting this hormone in urine or blood is the basis of pregnancy tests. The corpus luteum persists throughout pregnancy, but its function as a hormone source becomes less important after the first three months (first trimester), when the placenta secretes sufficient estrogens and progesterone (fig. 20.6).

Placental estrogens and *placental progesterone* maintain the uterine wall during the second and third trimesters of pregnancy. The placenta also secretes a hormone called **placental lactogen** that, with placental estrogens and progesterone, stimulates breast development and prepares the mammary glands for milk secretion. Placental progesterone and a polypeptide hormone called *relaxin* from the corpus luteum inhibit the smooth muscles in the myometrium, suppressing uterine contractions until the birth process begins.

Very early in pregnancy, while vast hormonal changes sweep a woman's body and the embryo rapidly increases in size and complexity, the woman may not yet realize what is happening. Early signs of pregnancy resemble those of approaching menstruation, such as bloating and irritable mood. As the pregnancy continues, the woman's blood volume increases by one-third, and her bones may weaken if she does not receive adequate dietary calcium. Muscle spasms may occur in response to rapid weight gain. In the later months, the fetus pushing against the woman's internal organs can produce heartburn, shortness of breath, and frequent urination. Fetal movements become noticeable by the fourth or fifth month, first as slight flutterings, then as jabs, kicks, and squirming movements.

The high concentration of placental estrogens during pregnancy enlarges the vagina and external reproductive organs. Also, relaxin relaxes the ligaments joining the symphysis pubis and sacroiliac joints during the last week of pregnancy, allowing greater movement at these joints and aiding the passage of the fetus through the birth canal.

Other hormonal changes of pregnancy include increased adrenal secretion of aldosterone, which promotes renal reabsorption of sodium and leads to fluid retention. The parathyroid glands secrete parathyroid hormone, which helps maintain a high concentration of maternal blood calcium (see chapter 11, p. 301). Table 20.1 summarizes the hormonal changes of pregnancy.

Check Your Recall

7. Which hormone normally prevents spontaneous abortion?
8. What is the source of the hormones that sustain the uterine wall during pregnancy?
9. What other hormonal changes occur during pregnancy?

Embryonic Stage

The embryonic stage extends through the eighth week of prenatal development. During this time, the placenta forms, the main internal organs develop, and the major external body structures appear.

Early in the embryonic stage, the inner cell mass organizes into a flattened **embryonic disc** with two distinct layers—an outer *ectoderm* and an inner *endoderm*. A short time later, the ectoderm and endoderm fold, and a third layer of cells, the *mesoderm*, forms between

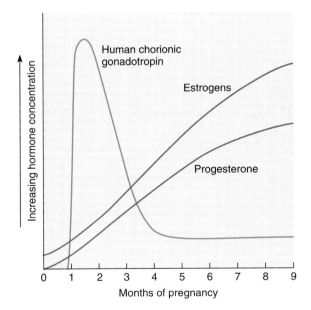

Figure 20.6

The relative concentrations of three hormones in maternal blood change during pregnancy.

Table 20.1	Hormonal Changes During Pregnancy

1. Following implantation, cells of the embryo begin to secrete human chorionic gonadotropin (hGC).
2. Human chorionic gonadotropin maintains the corpus luteum, which continues to secrete estrogens and progesterone.
3. The developing placenta secretes abundant estrogens and progesterone.
4. Placental estrogens and progesterone:
    a. stimulate the uterine lining to continue development.
    b. maintain the uterine lining.
    c. inhibit the anterior pituitary's secretion of follicle-stimulating hormone (FSH) and luteinizing hormone (LH).
    d. stimulate development of mammary glands.
    e. inhibit uterine contractions (progesterone).
    f. enlarge the reproductive organs (estrogens).
5. Relaxin from the corpus luteum also inhibits uterine contractions and relaxes the pelvic ligaments.
6. The placenta secretes placental lactogen that stimulates breast development.
7. Aldosterone from the adrenal cortex promotes renal reabsorption of sodium.
8. Parathyroid hormone from the parathyroid glands helps maintain a high concentration of maternal blood calcium.

Table 20.2	Stages and Events of Early Human Embryonic Development		

Stage	Time Period	Principal Events
Zygote	12–24 hours following ovulation	Secondary oocyte is fertilized, meiosis is completed; zygote has 46 chromosomes and is genetically distinct
Cleavage	30 hours to third day	Mitosis increases cell number
Morula	Third to fourth day	Solid ball of cells
Blastocyst	Fifth day through second week	Hollowed ball forms trophoblast (outside) and inner cell mass, which implants and flattens to form embryonic disc
Gastrula	End of second week	Primary germ layers form

three primary germ layers, is called a **gastrula** (gas′troo-lah). Table 20.2 summarizes the stages of early embryonic development.

Ectodermal cells give rise to the nervous system, portions of special sensory organs, the epidermis, hair, nails, glands of the skin, and linings of the mouth and anal canal. Mesodermal cells form all types of muscle tissue, bone tissue, bone marrow, blood, blood vessels, lymphatic vessels, connective tissues, internal reproductive organs, kidneys, and the epithelial linings of the body cavities. Endodermal cells produce the epithelial linings of the digestive tract, respiratory tract, urinary bladder, and urethra.

them. All organs form from these three cell layers, called the **primary germ layers** (pri′mar-e jerm la′erz) (fig. 20.7). A *connecting stalk* attaches the embryonic disc to the developing placenta. The two-week embryo, with its

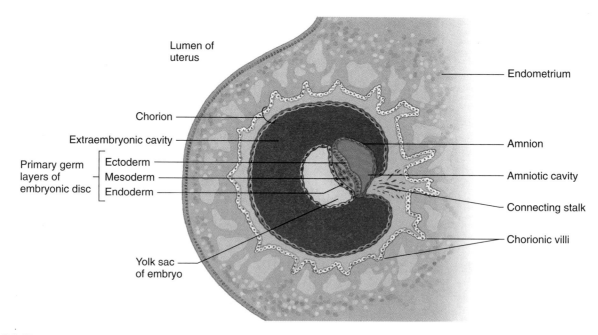

Figure 20.7

Early in the embryonic stage of development, the three primary germ layers form. The ectoderm and the endoderm form first, and then the mesoderm forms between them.

As the embryo implants in the uterus, proteolytic enzymes from the trophoblast break down endometrial tissue, providing nutrients for the developing embryo. A second layer of cells begins to line the trophoblast, and together these two layers form a structure called the **chorion** (ko're-on). Soon, slender projections grow out from the trophoblast, including the new cell layer, eroding their way into the surrounding endometrium by continuing to secrete proteolytic enzymes. These projections become increasingly intricate and form the highly branched **chorionic villi,** which are well established by the end of the fourth week (fig. 20.7).

As the chorionic villi develop, embryonic blood vessels form within them and are continuous with those passing through the connecting stalk to the body of the embryo. At the same time, irregular spaces called **lacunae** form around and between the villi. These spaces fill with maternal blood that escapes from eroded endometrial blood vessels.

Also, during the fourth week of development, the flat embryonic disc becomes cylindrical. By the end of week four, the head and jaws develop, the heart beats and forces blood through the blood vessels, and tiny buds form, which will give rise to the upper and lower limbs (fig. 20.8).

During the fifth through the seventh weeks, as figure 20.8 shows, the head grows rapidly and becomes rounded and erect. The face, with developing eyes, nose, and mouth, becomes more humanlike. The upper and lower limbs elongate, and fingers and toes are sculpted from plate-like precursors of the hands and feet. Apoptosis (programmed cell death) carves the digits from the webbing (fig. 20.9). By the end of the seventh week, all the main internal organs are present, and as these structures enlarge, the body takes on a more humanlike appearance.

Until about the end of the eighth week, the chorionic villi cover the entire surface of the former trophoblast. However, as the embryo and the chorion surrounding it enlarge, only those villi that remain in contact with the endometrium endure. The others degenerate, and the portions of the chorion to which they were attached become smooth. Thus, the region of the chorion still in contact with the uterine wall is restricted to a disc-shaped area that becomes the placenta.

A thin **placental membrane** separates embryonic blood within the capillary of a chorionic villus from maternal blood in a lacuna. Across this membrane, which is composed of the epithelium of the chorionic villus and the epithelial wall of the capillary inside the villus, maternal and embryonic blood exchange substances (fig. 20.10). Oxygen and nutrients diffuse from the maternal blood into the embryo's blood, and carbon dioxide and other wastes diffuse from the embryo's blood into the

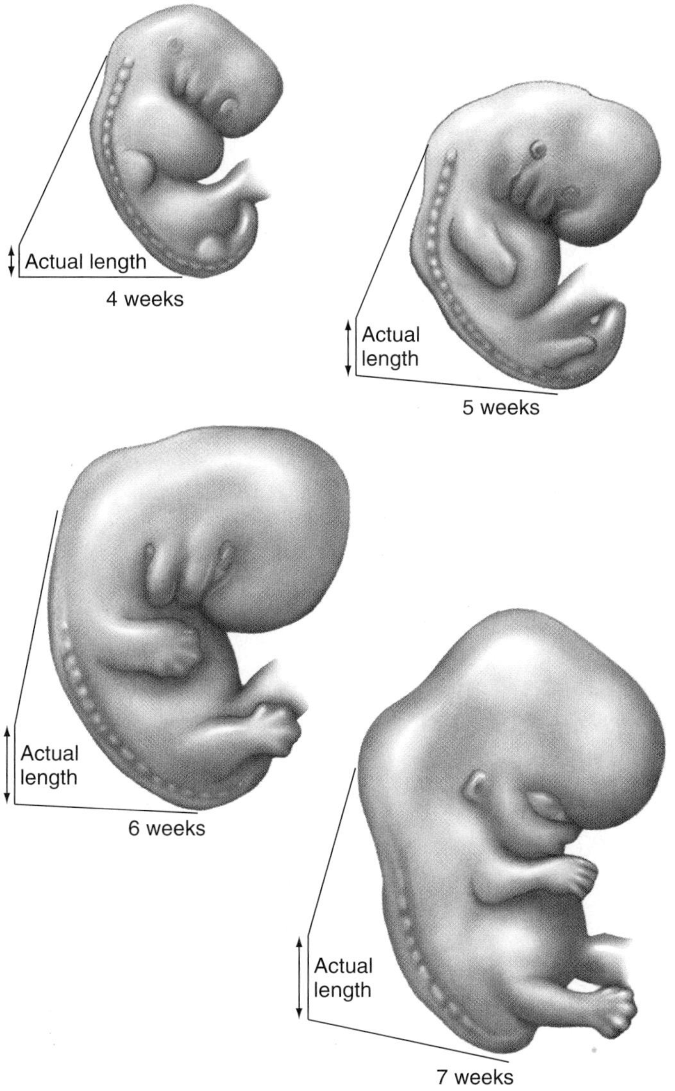

Actual length
4 weeks

Actual length
5 weeks

Actual length
6 weeks

Actual length
7 weeks

Figure 20.8

In the fifth through the seventh weeks of development, the embryonic body and face develop a humanlike appearance.

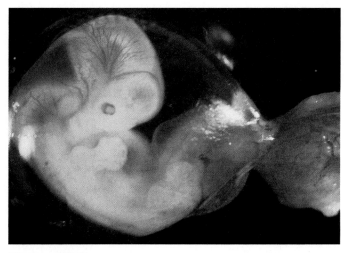

Figure 20.9

Human embryo after about six weeks of development (6.5×).

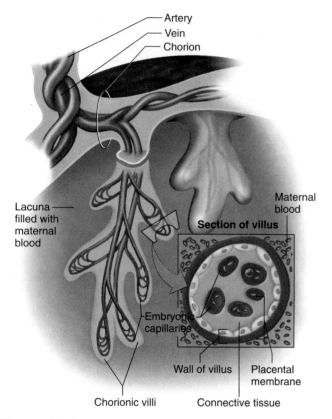

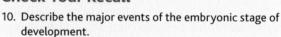

Figure 20.10

The placental membrane consists of the epithelial wall of an embryonic capillary and the epithelial wall of a chorionic villus, as illustrated in the section of the villus (lower part of the figure).

maternal blood. Various substances also cross the placental membrane by active transport and pinocytosis.

If a pregnant woman takes an addictive substance, her newborn may suffer from withdrawal symptoms when amounts of the chemical it is accustomed to receiving suddenly plummet. Newborn addiction occurs with certain drugs of abuse, with some drugs used to treat anxiety, and even with very large doses of vitamin C. Although vitamin C is not addictive, if a fetus is accustomed to megadoses, the sudden drop in vitamin C level after birth may bring on symptoms of deficiency.

Check Your Recall

10. Describe the major events of the embryonic stage of development.
11. Which tissues and structures develop from ectoderm? From mesoderm? From endoderm?
12. Describe how the placenta forms.
13. How are substanced exchanged between the embryo's blood and the maternal blood?

The embryonic portion of the placenta is the chorion and its villi; the maternal portion is the area of the uterine wall (decidua basalis) where the villi attach (fig. 20.11). The fully formed placenta is a reddish-brown disc about 20 centimeters long and 2.5 centimeters thick, and weighs about 0.5 kilogram.

Also during the embryonic stage, another membrane, called the **amnion** (am′ne-on), develops around

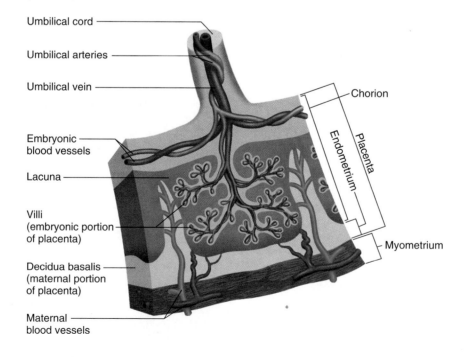

Figure 20.11

The placenta consists of an embryonic portion and a maternal portion.

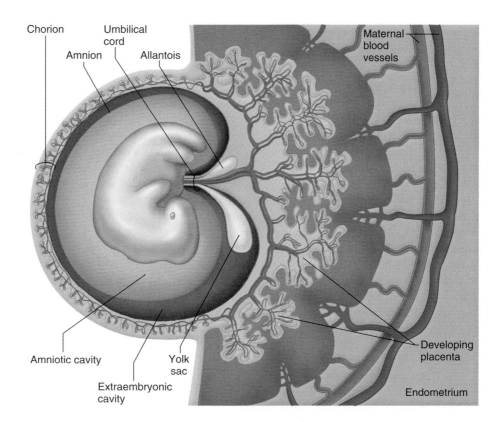

Figure 20.12

As the amnion develops, it surrounds the embryo, and the umbilical cord begins to form from structures in the connecting stalk.

the embryo. It appears during the second week. Its margin attaches around the edge of the embryonic disc, and **amniotic fluid** fills the space between the amnion and the embryonic disc.

As the embryo becomes more cylindrical, the amnion margins fold, enclosing the embryo in the amnion and amniotic fluid. The amnion envelops the tissues on the underside of the embryo, by which the embryo attaches to the chorion and the developing placenta. In this manner, the **umbilical cord** (um-bil′ĭ-kal kord) forms (fig. 20.12).

The umbilical cord contains three blood vessels—two *umbilical arteries* and one *umbilical vein*—that transport blood between the embryo and the placenta (see fig. 20.11). The umbilical cord suspends the embryo in the *amniotic cavity*. The amniotic fluid allows the embryo to grow freely without compression from surrounding tissues and also protects the embryo from jarring movements of the woman's body.

Two other extra-embryonic membranes form during development—the yolk sac and the allantois (fig. 20.12). The **yolk sac** forms during the second week and attaches to the underside of the embryonic disc. It forms blood cells in the early stages of development and gives rise to the cells that later become sex cells. The **allantois** (ah-lan′to-is) forms during the third week as a tube extending from the early yolk sac into the connecting stalk of the embryo. It, too, forms blood cells and gives rise to the umbilical arteries and vein.

By the beginning of the eighth week, the embryo is usually 30 millimeters long and weighs less than 5 grams. It is recognizable as human (fig. 20.13).

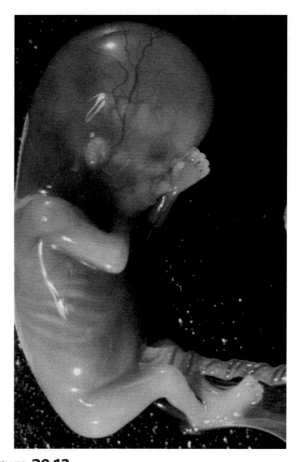

Figure 20.13

By the beginning of the eighth week of development, the embryonic body is recognizable as human (6×).

Some Causes of Birth Defects

Thalidomide

The idea that the placenta always protects the embryo and fetus from harmful substances was tragically disproven between 1957 and 1961, when more than 10,000 children in Europe were born with flippers for limbs. Doctors soon determined that the teratogen (an agent that causes a birth defect) was the mild tranquilizer *thalidomide,* which all of the mothers of deformed infants had taken early in pregnancy, at the time when limbs form. The United States was spared a thalidomide disaster because an astute government physician noted the drug's adverse effects on monkeys used in experiments and halted testing. Thalidomide is used today to treat leprosy, AIDS, and certain blood and bone marrow cancers, but is never prescribed for pregnant women.

Rubella

Also in the early 1960s, another teratogen, a virus, was sweeping the United States. A *rubella* (German measles) epidemic caused 20,000 birth defects and 30,000 stillbirths. Successful vaccination programs have since greatly lowered the incidence of "congenital rubella syndrome" in many countries.

Alcohol

A pregnant woman who has as few as one or two alcoholic drinks a day, or perhaps a large amount at a crucial time in prenatal development, risks *fetal alcohol syndrome (FAS)* in her unborn child. Animal studies show that even small amounts of alcohol can alter fetal brain chemistry. Until tests become available that can identify women who are genetically predisposed to developing the syndrome, it is best to avoid drinking alcohol entirely when pregnant or when trying to become pregnant.

A child with FAS has a small head, misshapen eyes, and a flat face and nose (fig. 20B). He or she grows slowly before and after birth. Intellect is impaired, ranging from minor learning disabilities to mental retardation. Teens and young adults with FAS are short and have small heads. Many remain at an early grade-school level of intellectual development, and they often lack social and communication skills.

In the United States today, FAS is the third most common cause of mental retardation in newborns. One to three of every 1,000 infants has the syndrome, and more than 40,000 affected children are born each year.

Cigarettes

Chemicals in cigarette smoke stress a fetus. Carbon monoxide crosses the placenta and plugs sites on the fetus's hemoglobin molecules that bind oxygen. Other chemicals in smoke prevent nutrients from reaching the fetus. Studies comparing the placentas of smokers and nonsmokers show that smoke-exposed placentas lack important growth factors. The result of these assaults is poor growth before and after birth. Cigarette smoking during pregnancy is linked to spontaneous abortion, stillbirth, prematurity, and low birth weight.

Nutrients and Malnutrition

Certain nutrients in large amounts, particularly vitamins, act in the body as drugs. The acne medication *isotretinoin* (Accutane) is a derivative of vitamin A that causes spontaneous abortions and defects of the heart, nervous system, and face. A vitamin A-based drug used to treat psoriasis, as well as excesses of vitamin A itself, also cause birth defects because some forms of the vitamin are stored in body fat for up to three years after ingestion.

Environmental factors that cause congenetial malformations by interfering with prenatal growth and/or development are called **teratogens.** The Topic of Interest on this page and page 551 discusses some teratogens.

Check Your Recall

14. Which blood vessels are in the umbilical cord?
15. What is the function of amniotic fluid?
16. What is the function of the yolk sac?

Fetal Stage

The **fetal stage** begins at the end of the eighth week of development and lasts until birth. Growth is rapid, and body proportions change considerably. At the beginning of the fetal stage, the head is disproportionately large, and the lower limbs are short. Gradually, the proportions become more like those of a child.

During the third month, body lengthening accelerates, but head growth slows. The upper limbs reach the relative length they will maintain throughout develop-

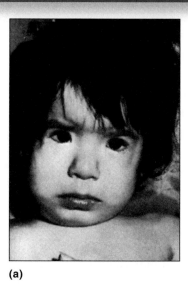 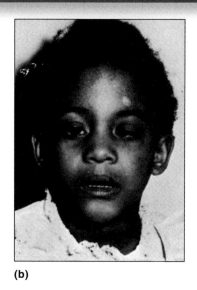 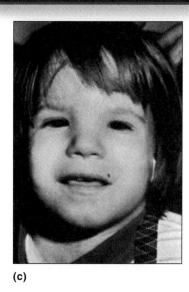

(a)   (b)   (c)

Figure 20B

Fetal alcohol syndrome. Some children whose mothers drank alcohol during pregnancy have characteristic flat faces that are strikingly similar in different races. Women who drink excessively while pregnant have a 30–45% chance of having a child affected to some degree by prenatal exposure to alcohol. However, only 6% of exposed offspring have full-blown fetal alcohol syndrome.

Malnutrition during pregnancy threatens the fetus. Obstetric records of pregnant women before, during, and after World War II link inadequate nutrition early in pregnancy to an increase in spontaneous abortions. The aborted fetuses had very little brain tissue. Poor nutrition affects placenta development. More recent studies reveal that malnutrition or starvation before birth causes shifts in metabolism to make the most of calories from food. This protective action, however, sets the stage for developing obesity and associated disorders, such as type 2 diabetes and cardiovascular disease, in adulthood.

Occupational Hazards

The workplace can be a source of teratogens. Women who work with textile dyes, lead, certain photographic chemicals, semiconductor materials, mercury, and cadmium have increased rates of spontaneous abortion and delivering children with birth defects. Men whose jobs expose them to sustained heat, such as smelter workers, glass manufacturers, and bakers, may produce sperm that can fertilize a secondary oocyte but possibly lead to spontaneous abortion or a birth defect. A virus or a toxic chemical carried in semen may also cause a birth defect.

ment, and ossification centers appear in most bones. By the twelfth week, the external reproductive organs are distinguishable as male or female.

In the fourth month, the body grows rapidly and reaches a length of up to 20 centimeters. The lower limbs lengthen considerably, and the skeleton continues to ossify. A four-month-old fetus will startle and turn away from a bright light flashed on a pregnant woman's belly, and may also react to sudden loud noises.

In the fifth month, growth slows. The lower limbs reach their final relative proportions. Skeletal muscles

About 5 in every 1,000 pregnant women have a form of "morning sickness" that is very severe and is termed hyperemesis gravidarum. It occurs between the 5th and 22nd weeks of the pregnancy. In the worst cases, a woman may vomit up to 20 times a day, and lose 5% of her pre-pregnancy weight. For many years the condition was thought to be psychological in origin, due perhaps to a fear of pregnancy or motherhood, but today the most likely recognized cause is excess hCG or excessive response to the hormone. Hospitalization is required to control hydration and electrolyte balance.

contract, and the pregnant woman may feel fetal movements. Hair begins to grow on the head. Fine, downy hair and a cheesy mixture of dead epidermal cells and sebum from the sebaceous glands cover the skin.

During the sixth month, the fetus gains substantial weight. Eyebrows and eyelashes grow. The skin is quite wrinkled and translucent, and blood vessels in the skin give the fetus a reddish appearance.

In the seventh month, the skin becomes smoother as fat is deposited in subcutaneous tissues. The eyelids, which fused during the third month, reopen. At the end of this month, the fetus is about 40 centimeters long.

In the final trimester, fetal brain cells rapidly form networks, as organs specialize and grow. Fat continues to be laid down beneath the skin. In the male, the testes descend from regions near the developing kidneys, through the inguinal canal, and into the scrotum. The digestive and respiratory systems mature last, which is why some premature infants have difficulty digesting milk and breathing.

At the end of the ninth month (on average, 266 days), the fetus is *full-term*. It is about 50 centimeters long and weighs 2.7–3.6 kilograms. The skin has lost its downy hair, but sebum and dead epidermal cells still coat it. Hair usually covers the scalp. The fingers and toes have well-developed nails. The skull bones are largely ossified. As figure 20.14 shows, the fetus is usually positioned upside down, with its head toward the cervix.

Check Your Recall

17. What major changes occur during the fetal stage of development?

18. Describe a full-term fetus.

Fetal Blood and Circulation

Throughout fetal development, the maternal blood supplies oxygen and nutrients and carries away wastes. These substances diffuse between maternal and fetal blood through the placental membrane, and umbilical blood vessels carry them to and from the fetus.

The fetal blood and cardiovascular system must adapt to intrauterine existence. The concentration of oxygen-carrying hemoglobin in fetal blood is about 50% greater than in maternal blood, and fetal hemoglobin has a greater attraction for oxygen than does adult hemoglobin. At a particular oxygen partial pressure, fetal hemoglobin can carry 20–30% more oxygen than can adult hemoglobin. Different genes encode the protein subunits of hemoglobin in embryos, fetuses, and individuals after birth.

Figure 20.15 shows the path of blood in the fetal cardiovascular system. The umbilical vein transports blood rich in oxygen and nutrients from the placenta to the fetus. This vein enters the body and extends along

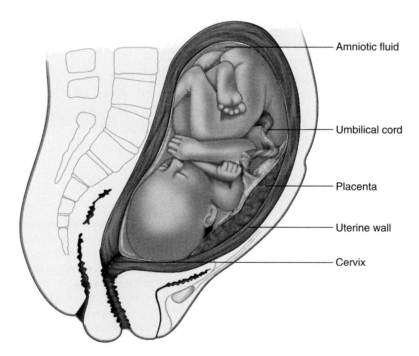

Amniotic fluid

Umbilical cord

Placenta

Uterine wall

Cervix

Figure 20.14

A full-term fetus is usually positioned with its head near the cervix.

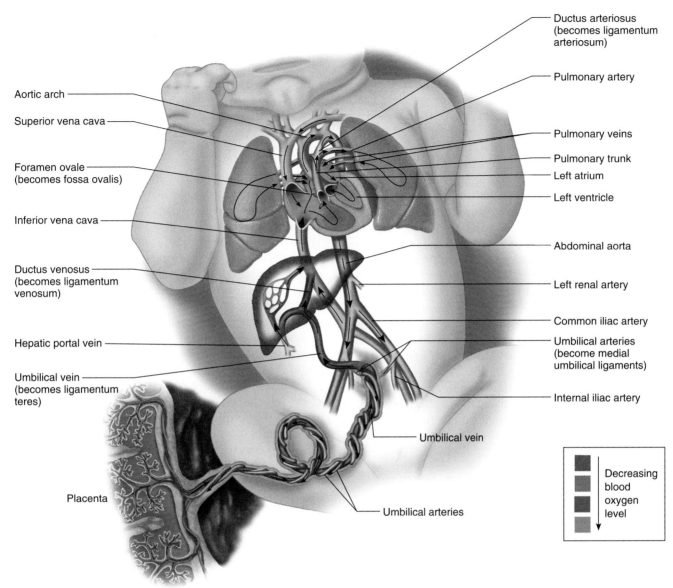

Figure 20.15
The general pattern of fetal circulation.

the anterior abdominal wall to the liver. About half the blood it carries passes into the liver, and the rest enters a vessel called the **ductus venosus** (duk′tus ve′no-sus), which bypasses the liver.

The ductus venosus extends a short distance and joins the inferior vena cava. There, oxygenated blood from the placenta mixes with deoxygenated blood from the lower parts of the fetal body. This mixture continues through the inferior vena cava to the right atrium.

In an adult heart, blood from the right atrium enters the right ventricle and is pumped through the pulmonary trunk and arteries to the lungs (see chapter 13, p. 343). The fetal lungs, however, are nonfunctional, and blood largely bypasses them. Much of the blood from the inferior vena cava that enters the fetal right atrium is shunted directly into the left atrium through an open-

ing in the atrial septum called the **foramen ovale** (fo-ra′man ova′le). Blood passes through the foramen ovale because blood pressure is somewhat greater in the right atrium than in the left atrium. Furthermore, a small valve on the left side of the atrial septum overlies the foramen ovale and helps prevent blood from moving in the reverse direction.

The rest of the fetal blood entering the right atrium, including a large proportion of the deoxygenated blood entering from the superior vena cava, passes into the right ventricle and out through the pulmonary trunk. Only a small volume of blood enters the pulmonary circuit because the lungs are collapsed, and their blood vessels have a high resistance to blood flow. However, enough blood does reach lung tissues to sustain them.

Most of the blood in the pulmonary trunk bypasses the lungs by entering a fetal vessel called the **ductus arteriosus** (duk′tus ar-te″re-o′sus), which connects the pulmonary trunk to the descending portion of the aortic arch. As a result of this connection, blood with a relatively low oxygen concentration, which is returning to the heart through the superior vena cava, bypasses the lungs. At the same time, it is prevented from entering the portion of the aorta that branches to the heart and brain.

The more highly oxygenated blood that enters the left atrium through the foramen ovale mixes with a small amount of deoxygenated blood returning from the pulmonary veins. This mixture moves into the left ventricle and is pumped into the aorta. Some of it reaches the myocardium through the coronary arteries, and some reaches the brain tissues through the carotid arteries.

Blood carried by the descending aorta includes the less oxygenated blood from the ductus arteriosus. Some of the blood is carried into the branches of the aorta that lead to the lower regions of the body. The rest passes into the umbilical arteries, which branch from the internal iliac arteries and lead to the placenta. There, the blood is reoxygenated (fig. 20.15).

Table 20.3 summarizes the major features of fetal circulation. At birth, the fetal cardiovascular system must adjust when the placenta ceases to function and the newborn begins to breathe.

> The umbilical cord usually contains two arteries and one vein. A small percentage of newborns have only one umbilical artery. Since this condition is often associated with other cardiovascular disorders, the vessels within the severed cord are routinely counted following birth. Some inherited conditions are also associated with an abnormal number of umbilical cord vessels.

Table 20.3	Fetal Cardiovascular Adaptations
Adaptation	**Function**
Fetal blood	Hemoglobin has greater oxygen-carrying capacity than adult hemoglobin
Umbilical vein	Carries nutrient-rich oxygenated blood from placenta to fetus
Ductus venosus	Conducts about half the blood from the umbilical vein directly to the inferior vena cava, bypassing the liver
Foramen ovale	Conveys much blood entering right atrium from inferior vena cava, through the atrial septum, and into the left atrium, bypassing the lungs
Ductus arteriosus	Conducts some blood from the pulmonary trunk to the aorta, bypassing the lungs
Umbilical arteries	Carry blood from the internal iliac arteries to placenta

Check Your Recall

19. Which umbilical vessel carries oxygenated blood to the fetus?
20. What is the function of the ductus venosus?
21. How does fetal circulation allow blood to bypass the lungs?

Birth Process

Pregnancy usually continues for thirty-eight weeks from conception. Pregnancy ends with the *birth process.* A period of rapid changes and intense physical demands on the pregnant woman begins hours or days before the birth. The work of giving birth is appropriately called labor.

A declining progesterone concentration plays a major role in initiating birth. During pregnancy, progesterone suppresses uterine contractions. As the placenta ages, the progesterone concentration in the uterus declines, which stimulates synthesis of a prostaglandin that promotes uterine contractions. At the same time, the cervix begins to thin and then open. Changes in the cervix may begin a week or two before other signs of labor appear.

Another stimulant of the birth process is stretching of the uterine and vaginal tissues late in pregnancy. This initiates nerve impulses to the hypothalamus, which in turn signals the posterior pituitary gland to release the hormone **oxytocin** (see chapter 11, p. 298). Oxytocin stimulates powerful uterine contractions. Combined with the greater excitability of the myometrium due to the decline in progesterone secretion, stimulation by oxytocin aids later stages of labor.

During labor, rhythmic muscular contractions begin at the top of the uterus and extend down its length. Since the fetus is usually positioned head downward, labor contractions force the head against the cervix (fig. 20.16). This action stretches the cervix, which elicits a reflex that stimulates still stronger labor contractions. Thus, there is a *positive feedback system* in which uterine contractions produce more intense uterine contractions. At the same time, continuing cervix dilation reflexly stimulates the posterior pituitary to increase oxytocin release. As labor continues, positive feedback stimulates abdominal wall muscles to contract, which also helps force the fetus through the cervix and vagina to the outside.

> An infant passing through the birth canal can tear the delicate tissues between the vulva and anus (perineum). To avoid a ragged tear, a physician makes an *episiotomy,* a clean cut in the perineal tissues.

Following birth of the fetus, the placenta separates from the uterine wall, and uterine contractions expel it through the birth canal. This expulsion, termed the *afterbirth,* is accompanied by bleeding, because the separation damages vascular tissues. However, oxytocin stimulates

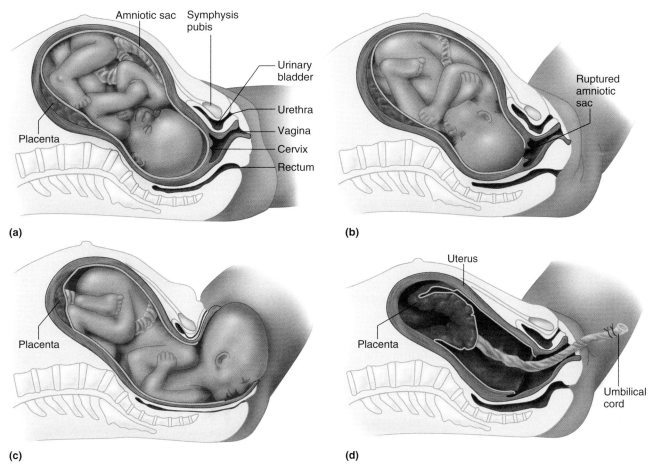

Figure 20.16

Stages in birth (*a*) Fetal position before labor, (*b*) dilation of the cervix, (*c*) expulsion of the fetus, (*d*) expulsion of the placenta.

continued uterine contraction, which compresses the bleeding vessels and minimizes blood loss. Breast-feeding also contributes to returning the uterus to its original, pre-pregnancy size, as the suckling of the newborn stimulates the release of oxytocin from the posterior pituitary.

Check Your Recall

22. Describe the role of progesterone in initiating labor.
23. Explain how dilation of the cervix affects labor.

20.4 POSTNATAL PERIOD

Following birth, both mother and newborn experience physiological and structural changes.

Milk Production and Secretion

During pregnancy, placental estrogens and progester-one stimulate further development of the mammary glands. Estrogens cause the ductile systems to extend and branch, and deposit abundant fat around them. Pro-gesterone stimulates development of the alveolar glands at the ends of the ducts. Placental lactogen also pro-motes these growth changes.

Hormonal activity doubles breast size during preg-nancy, and the mammary glands become capable of secreting milk. However, milk is not secreted because placental progesterone inhibits milk production, and placental lactogen blocks the action of *prolactin* (see chapter 11, p. 296).

Following childbirth and the expulsion of the pla-centa, maternal blood concentrations of placental hor-mones decline rapidly. In two or three days, prolactin, which is no longer inhibited, stimulates the mammary glands to secrete milk. Meanwhile, the glands secrete a thin, watery fluid called *colostrum* that has more pro-tein, but less carbohydrate and fat, than milk. Colostrum contains antibodies from the mother's immune system that protect the newborn from certain infections.

Milk ejection requires contraction of specialized *myo-epithelial cells* surrounding the alveolar glands (fig. 20.17). Suckling or mechanical stimulation of sensory receptors in the nipple or areola elicits the reflex action that con-trols this process. Impulses from these receptors go to

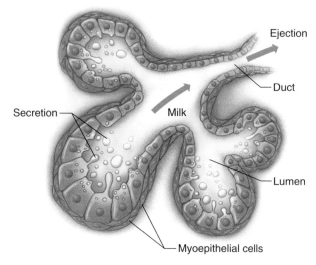

Figure 20.17

Myoepthelial cells contract, releasing milk from an alveolar gland.

the hypothalamus, which signals the posterior pituitary gland to release oxytocin. Oxytocin travels in the bloodstream to the breasts and stimulates myoepithelial cells to contract. As a result, milk is ejected into a suckling infant's mouth in about 30 seconds.

As long as milk is removed from the breasts, release of prolactin and oxytocin continues, and the mammary glands produce milk. If milk is not removed regularly, the hypothalamus inhibits prolactin secretion, and within about one week, the mammary glands stop producing milk.

Human milk is the best possible food for human babies. The milk of other animals contains different proportions of nutrients.

Human milk is 4.5% fat and 1.1% protein, which is suitable for providing the fatty insulation that enables neurons in the developing brain to communicate effectively. In contrast, cow's milk has 3.5% fat and 3.1% protein, because it is more important for survival for the calf to bulk up its muscles than to develop its brain. Milk of the gray seal has a whopping 53% fat, but this animal uses the fat for insulation in its frigid environment.

Check Your Recall

24. How does pregnancy affect the mammary glands?
25. What stimulates the mammary glands to produce milk?
26. What causes milk ejection?

Neonatal Period

The **neonatal** (ne″o-na′tal) **period** begins abruptly at birth and extends to the end of the first four weeks. At birth, the newborn must make quick physiologi-

cal adjustments to become self-reliant. It must respire, obtain and digest nutrients, excrete wastes, and regulate body temperature.

The largest newborn of recent times was a 24-pound 4-ounce baby boy born in Turkey.

A newborn's most immediate need is to obtain oxygen and excrete carbon dioxide. The first breath must be particularly forceful because the lungs are collapsed, and their small airways considerably resistant to air movement. Also, surface tension tends to hold the moist membranes of the lungs together. However, the lungs of a full-term fetus continuously secrete *surfactant* (see chapter 16, p. 453), which reduces surface tension. After the first powerful breath begins to expand the lungs, breathing eases.

Premature infants' survival chances increase directly with age and weight, and parallel the increasing maturity of the lungs. The ability of the aveoli to exchange gases and the presence of surfactant to reduce alveolar surface tension are important. A baby born at 25 weeks has a 50% chance of survival; at 24 weeks, 39% survival; and at 23 weeks, 17% survival. The smallest and earliest premature birth to survive in recent times was Amilla Taylor, born at 21 weeks, 6 days, and weighing slightly under 10 ounces.

The newborn has a high metabolic rate, and its immature liver may be unable to supply enough glucose to support metabolism. Consequently, the newborn typically utilizes stored fat for energy.

A newborn's kidneys are usually unable to produce concentrated urine, so they excrete a dilute fluid. For this reason, the newborn may become dehydrated and develop a water and electrolyte imbalance. Also, some of the newborn's homeostatic control mechanisms may not function adequately, such as the temperature-regulating system.

When the placenta ceases to function and breathing begins, the newborn's cardiovascular system also changes. Following birth, the umbilical vessels constrict. The umbilical arteries close first, and if the umbilical cord is not clamped or severed for a minute or so, blood continues to flow from the placenta to the newborn through the umbilical vein, adding to the newborn's blood volume. Similarly, the ductus venosus constricts shortly after birth and appears in the adult as a fibrous cord (ligamentum venosum) superficially embedded in the wall of the liver.

The foramen ovale closes as a result of blood pressure changes in the right and left atria as fetal vessels constrict. As blood ceases to flow from the umbilical vein into the inferior vena cava, the blood pressure in the right atrium falls. Also, as the lungs expand with the first breathing movements, resistance to blood flow through the pulmonary circuit decreases, more blood enters the left atrium through the pulmonary veins, and blood pressure in the left atrium increases.

As the blood pressure in the left atrium rises and that in the right atrium falls, the valve on the left side of the atrial septum closes the foramen ovale. In most individuals, this valve gradually fuses with the tissues along the margin of the foramen. In an adult, a depression called the *fossa ovalis* marks the site of the previous opening.

The ductus arteriosus, like the other fetal vessels, constricts after birth. After the ductus arteriosus closes, blood can no longer bypass the lungs by moving from the pulmonary trunk directly into the aorta. In an adult, a cord called the *ligamentum arteriosum* represents the ductus arteriosus.

In *patent ductus arteriosus (PDA)*, the ductus arteriosus fails to close completely. After birth, the metabolic rate and oxygen consumption in neonatal tissues increase, in large part to maintain body temperature. If the ductus arteriosus remains open, the neonate's blood oxygen concentration may be too low to adequately supply tissues, including the myocardium. If PDA is not corrected surgically, the heart may fail, even though the myocardium is normal.

Changes in the newborn's cardiovascular system are gradual. Constriction of the ductus arteriosus may be functionally complete within 15 minutes, but the permanent closure of the foramen ovale may take up to a year.

Fetal hemoglobin production falls after birth, and by the time an infant is four months old, most of the circulating hemoglobin is the adult type. Figure 20.18 illustrates cardiovascular changes in the newborn.

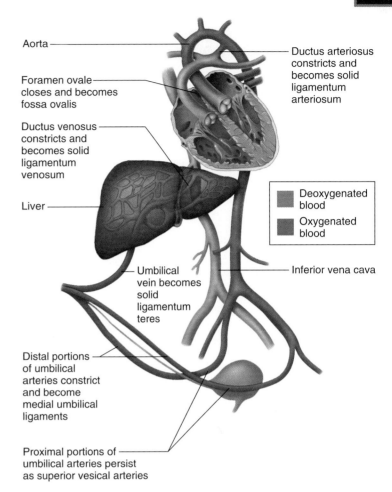

Figure 20.18

Major changes in the newborn's cardiovascular system.

Table 20.4 summarizes the major events during the neonatal period as well as those of the later stages of human development. Table 20.5 outlines aging-related changes.

Table 20.4	Stages in Postnatal Development	
Stage	**Time Period**	**Major Events**
Neonatal period	Birth to end of fourth week	Newborn begins to respire, eat, digest nutrients, excrete wastes, regulate body temperature, and make cardiovascular adjustments
Infancy	End of fourth week to one year	Growth rate is high; teeth begin to erupt; muscular and nervous systems mature so that coordinated activities are possible; communication begins
Childhood	One year to puberty	Growth rate is high; deciduous teeth erupt and are then replaced by permanent teeth; high degree of muscular control is achieved; bladder and bowel controls are established; intellectual abilities mature
Adolescence	Puberty to adulthood	Person becomes reproductively functional and emotionally more mature; growth spurts occur in skeletal and muscular systems; high levels of motor skills are developed; intellectual abilities increase
Adulthood	Adolescence to old age	Person remains relatively unchanged anatomically and physiologically; degenerative changes begin to occur
Senescence	Old age to death	Degenerative changes continue; body becomes less able to cope with demands; death usually results from mechanical disturbances in the cardiovascular system or from disease processes that affect vital organs

Table 20.5	Aging-Related Changes
Organ System	**Aging-Related Changes**
Integumentary system	Degenerative loss of collagenous and elastic fibers in dermis; decreased production of pigment in hair follicles; reduced activity of sweat and sebaceous glands Skin thins, wrinkles, and dries out; hair turns gray and then white
Skeletal system	Degenerative loss of bone matrix Bones become thinner, less dense, and more likely to fracture; stature may shorten due to compression of intervertebral discs and vertebrae
Muscular system	Loss of skeletal muscle fibers; degenerative changes in neuromuscular junctions; loss of muscular strength
Nervous system	Degenerative changes in neurons; loss of dendrites and synaptic connections; accumulation of lipofuscin in neurons; decreases in sensation Decreasing efficiency in processing and recalling information; decreasing ability to communicate; diminished sense of smell and taste; loss of elasticity of lenses and consequent loss of ability to accommodate for close vision
Endocrine system	Reduced hormonal secretions; decreased metabolic rate; reduced ability to cope with stress; reduced ability to maintain homeostasis
Cardiovascular system	Degenerative changes in cardiac muscle; decrease in lumen diameters of arteries and arterioles; decreased cardiac output; increased resistance to blood flow; increased blood pressure
Lymphatic system	Decrease in efficiency of immune system; increased incidence of infections and neoplastic diseases; increased incidence of autoimmune diseases
Digestive system	Decreased motility in gastrointestinal tract; reduced secretion of digestive juices; reduced efficiency of digestion
Respiratory system	Degenerative loss of elastic fibers in lungs; fewer alveoli; reduced vital capacity; increase in dead air space; reduced ability to clear airways by coughing
Urinary system	Degenerative changes in kidneys; fewer functional nephrons; reductions in filtration rate, tubular secretion, and reabsorption
Reproductive systems	
Male	Reduced secretion of sex hormones; enlargement of prostate gland; decrease in sexual energy
Female	Degenerative changes in ovaries; decrease in secretion of sex hormones; menopause; regression of secondary sex characteristics

Check Your Recall

27. Why must a newborn's first breath be particularly forceful?

28. What does a newborn use for energy during its first few days?

29. How do the kidneys of a newborn differ from those of an adult?

30. What changes occur in the newborn's cardiovascular system?

20.5 GENETICS

The newborn enters the world, and the elated parents look for family resemblances. Does she have her father's nose, or her grandmother's curly hair?

As the child grows, a unique mix of traits emerges. Inherited traits are determined by DNA sequences that comprise genes, which instruct cells to synthesize particular proteins, as discussed in section 4.6 (pp. 83–85). When a gene's DNA sequence changes, or *mutates,* illness may result. The Genetics Connections in Chapters 3 (p. 54), 8 (p. 191), 12 (p. 333), 14 (pp. 394–395), and 16 (p. 464) describe illnesses that arise from mutations.

The field of **genetics** (jĕ-net′iks) investigates how genes confer specific characteristics that affect health or contribute to our natural variation, and how genes are passed from generation to generation. Fetal chromosome checks, as discussed in the Genetics Connection (pp. 562–563), provide clues to an individual's future health. The environment influences how most genes are expressed. For example, a person who inherits gene variants that confer susceptibility to lung cancer may not develop the illness if he or she breathes clean air and never smokes.

Chromosomes and Genes Come in Pairs

Charts called karyotypes display by size the 23 chromosome pairs of a human somatic cell (fig. 20.19). Pairs 1 through 22 are **autosomes** (aw′to-somz), which do not carry genes that determine sex. The other two chromosomes, the X and the Y, determine sex and are called **sex chromosomes.** Females have two X chromosomes, and males have one X and one Y.

Each chromosome except the tiny Y includes hundreds of genes. Somatic cells have two copies of each autosome, and therefore two copies of each gene. Gene copies can be identical or slightly different in DNA sequence. Such variant forms of a gene are called **alleles** (ah-léels). An individual who has two identical

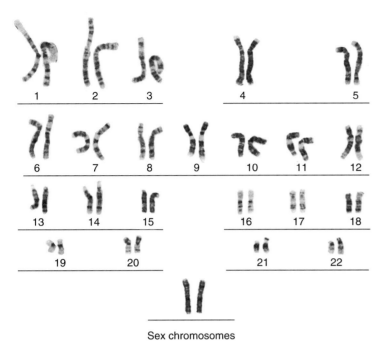

Figure 20.19

The human karyotype. Each somatic cell contains 23 chromosome pairs, or 46 chromosomes. This cell is from a female—it has two X chromosomes.

alleles of a gene is **homozygous** (ho″mo-zi′gus) for that gene. A person with two different alleles is **heterozygous** (het′er-o-zi′gus) for it. A heterozygote is also called a *carrier*.

The combination of alleles, for one gene or many, constitutes a person's **genotype** (jē′no-tīp). The appearance, health condition, or other characteristics associated with a particular genotype is the **phenotype** (fe′no-tīp). An allele is *wild-type* (indicated with a plus sign) if its associated phenotype is either normal function or the most common expression in a particular population. An allele that differs from wild-type has undergone a mutation, which may lead to a *mutant* (abnormal or unusual) phenotype.

Check Your Recall

31. Distinguish between autosomes and sex chromosomes.

32. How does a homozygote differ from a heterozygote?

33. Distinguish between genotype and phenotype.

Modes of Inheritance

We can predict the probability that a certain inherited trait will occur in the offspring of two individuals by considering how genes and chromosomes are distributed in meiosis, and the combinations in which they can unite at fertilization. Patterns of inheritance through families are called *modes of inheritance.*

Dominant and Recessive Inheritance

A **dominant** allele masks expression of a **recessive** allele. Dominant alleles are usually indicated with a capital letter. An allele that causes a trait or disease can be recessive or dominant, and inherited in either an autosomal or an X-linked manner. Y-linked conditions are extremely rare because that chromosome has very few genes.

The following generalizations describe modes of inheritance:

1. An autosomal condition affects both sexes. X-linked characteristics affect males much more than females. Y-linked traits are only passed from father to son.

2. A person inherits an autosomal recessive condition from two healthy carrier parents. Recessive conditions can "skip" generations.

3. A person who inherits a dominant condition has at least one affected parent. Therefore, generations are not skipped.

Three major modes of inheritance are autosomal recessive, autosomal dominant, and X-linked recessive. Cystic fibrosis illustrates *autosomal recessive* inheritance, in which two recessive alleles, one from each parent, transmit a trait. Receiving two disease-causing alleles impairs chloride channels in cells lining the pancreas, respiratory tract, intestines, and testes. Half of a heterozygous man's sperm have the disease-causing allele, as do half of the woman's secondary oocytes. Because sperm and oocytes combine at random, each offspring has a 25% chance of inheriting two wild-type alleles, a 50% chance of inheriting a disease-causing allele from either parent and being a carrier, and a 25% chance of inheriting a disease-causing allele from each parent.

Figure 20.20 illustrates two ways to depict the possible offspring of two carriers of cystic fibrosis. A **Punnett square** symbolizes the logic used to deduce the probabilities of inheriting particular genotypes in offspring. Each box records an allele combination at fertilization. A **pedigree** shows family members, how they are related, and their genotypes. Males are squares, females are circles, and the symbols for carriers are half-filled in while those for affected individuals are all filled in. Geneticists use Punnett squares and pedigrees to predict the outcome in all modes of inheritance.

Only one disease-causing allele is necessary to inherit an *autosomal dominant* condition. Huntington disease, which is inherited in an autosomal dominant manner, is characterized by loss of coordination, uncontrollable dancelike movements, cognitive impairment, and personality changes, typically beginning gradually near age 40. An affected person has an affected parent. Autosomal dominant disorders tend to begin in adulthood; autosomal recessive disorders usually have an early onset.

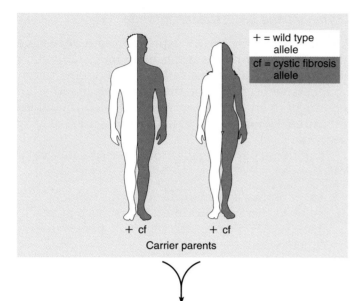

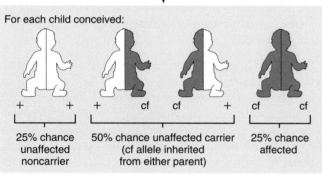

(a)

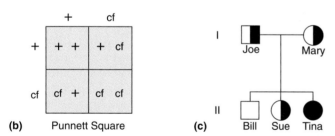

(b)   Punnett Square

(c)

Figure 20.20

Inheritance of cystic fibrosis from carrier parents illustrates autosomal recessive inheritance. (*a*) Each child has a 25% chance of being unaffected and not a carrier, a 50% chance of being an unaffected carrier, and a 25% chance of being affected. Sexes are affected with equal frequency. A Punnett square (*b*) and a pedigree (*c*) are other ways of depicting this information. Symbols in the pedigree with both black and white indicate unaffected carriers (heterozygotes).

A third major mode of inheritance is *X-linked recessive.* For a female, X-linked inheritance is like autosomal recessive inheritance, because she has two X chromosomes. That is, she can be a heterozygote or a homozygote. For a male, however, recessive alleles on the lone X chromosome are always expressed. A male

with an X-linked condition inherits it from a mother who is either a carrier or affected; he does not inherit an X chromosome from his father (or he wouldn't be male). Colorblindness and the blood-clotting disorder hemophilia A are X-linked recessive conditions.

Check Your Recall

34. Distinguish between dominant and recessive alleles.

35. How do Punnett squares and pedigrees depict gene transmission?

36. Compare the three modes of inheritance.

Multifactorial Traits

Most inherited traits and disorders are influenced to some extent by environmental factors, such as nutrition, physical activity, and exposure to toxins and pathogens. Environmental influences are particularly noticeable for traits determined by more than one gene, termed *polygenic.* Usually, several genes contribute to differing degrees toward molding the overall phenotype of a polygenic trait. Such a trait is said to be "continuously varying," which means that there are many degrees of its expression. Height, skin color, and intelligence are polygenic traits that show great variation.

When individuals with a polygenic trait are categorized into classes and the frequencies of the classes plotted as a bar graph, a bell-shaped curve emerges. The curves are strikingly similar for different polygenic traits. Figure 20.21 vividly shows the effect of the environment in the bell curve for height—as nutrition improved during the time span between the two photos, the heights of the tallest people increased.

Traits molded by one or more genes plus the environment are termed *multifactorial,* or complex. Height and skin color are multifactorial as well as polygenic, because they are influenced by environmental factors—nutrition and sun exposure, respectively. Most of the more common illnesses, including heart disease, diabetes mellitus, hypertension, and cancers, are multifactorial, as are most polygenic traits. An exception may be eye color, on which the environment has little, if any, impact.

Check Your Recall

37. What is a polygenic trait?

38. Why are different polygenic traits described by similar bell curves?

39. What is a multifactorial trait?

(a)

(b)

Figure 20.21

Previous editions of this (and other) textbooks have used the photograph in (a) to illustrate the continuously varying nature of height. In the photo, taken around 1920, 175 cadets at the Connecticut Agricultural College lined up by height. In 1997, Professor Linda Strausbaugh asked her genetics students at the school, today the University of Connecticut at Storrs, to recreate the scene (b). They did, and confirmed the continuously varying nature of human height. But they also demonstrated how height increased during the twentieth century. Improved nutrition has definitely played a role in expressing genetic potential for height. The tallest people in the old photograph (a) are 5'9" tall, whereas the tallest people in the more recent photograph (b) are 6'5" tall.

Clinical Terms Related to Pregnancy, Growth, Development, and Genetics

abruptio placentae (ab-rup'she-o plah-cen'tā) Premature separation of the placenta from the uterine wall.

dizygotic twins (di"zi-got'ik twinz) Twins resulting from two sperm cells fertilizing two egg cells.

hydatidiform mole (hi"dah-tid'ĭ-form mōl) Abnormal pregnancy resulting from fertilization of a polar body instead of a secondary oocyte; a mass of cysts.

hydramnios (hi-dram'ne-os) Excess amniotic fluid.

intrauterine transfusion (in"trah-u'ter-in trans-fu'zhun) Transfusion administered by injecting blood into the fetal peritoneal cavity before birth.

lochia (lo'ke-ah) Vaginal discharge following childbirth.

meconium (mĕ-ko'ne-um) Anal discharge from the digestive tract of a full-term fetus.

monozygotic twins (mon"o-zi-got'ik twinz) Twins resulting from one sperm cell fertilizing one egg cell, which then splits.

perinatology (per"ĭ-na-tol'o-je) Branch of medicine concerned with the fetus after the age of viability and with the newborn for the first four weeks after birth.

postpartum (pōst-par'tum) Occurring after birth.

teratology (ter"ah-tol'o-je) Study of substances that disrupt development, causing congenital malformations.

trimester (tri-mes'ter) Each third of the total period of pregnancy.

ultrasonography (ul"trah-son-og'rah-fe) Technique used to visualize the size and position of fetal structures from patterns of deflected ultrasonic waves.

Genetics Connection

Fetal Chromosome Checks

Chromosomes provide clues to health. A chromosome number other than 46 usually signals a serious medical condition, as do chromosomes that have missing or extra material. Sampling cells from a fetus and preparing charts of the chromosomes can detect these conditions.

Ultrasound, in which sound waves bounced off a fetus are converted into an image, can detect large-scale structural anomalies that are part of certain chromosomal syndromes. Also, blood tests performed on a pregnant woman at fifteen weeks detect levels of maternal serum markers (alpha fetoprotein, human chorionic gonadotropin, a form of estrogen, and certain other biochemicals). Abnormal levels sometimes reflect an extra chromosome. Doctors follow up questionable ultrasound or blood test results or a family history of abnormal chromosomes with one of the following procedures that examines fetal chromosomes.

Chorionic Villus Sampling

Chorionic villus sampling (CVS) (fig. 20C*a*) examines the chromosomes in chorionic villus cells, which are genetically identical to fetal cells because they are derived from the same fertilized egg. Rarely, the test causes spontaneous abortion. Thus, only women who have previously had a child with a detectable chromosome abnormality usually have the test. CVS is performed at the tenth week of gestation.

Amniocentesis

Amniocentesis is performed after the fourteenth week of gestation. A physician uses ultrasound to guide a needle into the amniotic sac and withdraws about 5 milliliters of fluid (fig. 20C*b*). Fetal fibroblasts in the fluid are cultured and their chromosomes checked. It takes about a week to grow these cells.

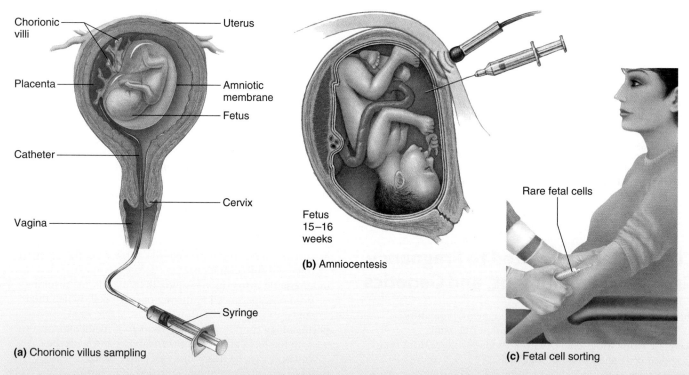

(a) Chorionic villus sampling

(b) Amniocentesis

(c) Fetal cell sorting

Figure 20C

Three ways to check a fetus's chromosomes. (a) Chorionic villus sampling (CVS) removes cells of the chorionic villi, whose chromosomes match those of the fetus. (b) In amniocentesis, a needle is inserted into the uterus to collect a sample of amniotic fluid, which contains fetal cells. (c) Fetal cell sorting separates fetal cells in the woman's circulation, but is still experimental. For all three types of tests, fetal chromosomes are stained and examined, and a genetic counselor interprets the results for patients. Additional tests are required for specific genetic disorders.

Until recently, amniocentesis carried a risk of about 0.5% of being followed by spontaneous abortion. But in the thirty or so years since those statistics were collected, the procedure has become much safer. The current risk of amniocentesis causing a miscarriage is about 1 in 1,600. Previously, only women over age 35, when the risk of conceiving a fetus with abnormal chromosomes is about 0.5% (risk increases with age), and women who have already had a child with detectable chromosomal abnormality, were offered amniocentesis. Many more women are likely to be offered the test if the new study results are confirmed.

Fetal Cell Sorting

Fetal cell sorting separates rare fetal cells from a pregnant woman's bloodstream (fig. 20Cc). A device called a fluorescence-activated cell sorter can pull out the fetal cells. The technique is safer than CVS or amniocentesis because the fetus and its membranes are not touched.

Fetal cell sorting traces its roots to 1957, when an autopsy on a pregnant woman revealed cells from a very early embryo lodged in a blood vessel in her lung. Researchers realized that the cells were from an embryo because of the Y chromosomes, which only male cells have. Fetal cells enter the maternal circulation in up to 70% of all pregnancies, and may remain for decades in the woman's body.

Preimplantation Genetic Diagnosis (PGD)

If a couple has a family history of a chromosomal or single gene condition that could affect their offspring, a procedure called preimplantation genetic diagnosis (PGD) can allow them to select early embryos that have not inherited the condition (fig. 20D). After secondary oocytes are fertilized *in vitro* (in glassware) and allowed to divide to the 8-celled stage, one cell from each of several embryos is removed and tested for the disease-causing gene variant or combination. If the genes or chromosomes are unaffected, a tested 7-celled embryo is implanted in the woman, and if all goes well, development ensues. Embryos destined to develop the family's disorder are discarded or used for research. PGD has enabled hundreds of children to be born free of their lethal legacies. A famous case was Adam Nash, who was selected as an embryo because his bone marrow matched that of his sister Molly, who was dying of an inherited anemia. Adam saved Molly's life.

PGD was originally developed to help families with known genetic conditions, but it is increasingly being used in IVF procedures to ensure that a chromosomally correct embryo is implanted. A bioethical issue has arisen over the traits considered in selecting embryos. In the United Kingdom, for example, inherited cancer susceptibility is an approved indication for having PGD. Bioethicists object because these cancers do not begin until adulthood, not everyone who inherits the susceptibility actually develops the cancer, and the cancers may be treatable. Another controversy concerning PGD is its use to select gender.

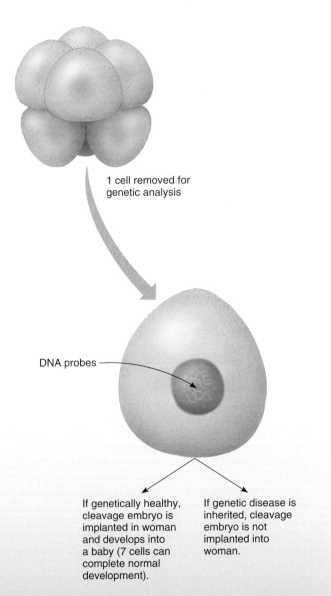

1 cell removed for genetic analysis

DNA probes

If genetically healthy, cleavage embryo is implanted in woman and develops into a baby (7 cells can complete normal development).

If genetic disease is inherited, cleavage embryo is not implanted into woman.

Figure 20D

Preimplantation genetic diagnosis probes disease-causing genes in an 8-celled cleavage embryo.

Clinical Connection

Preeclampsia, also called "toxemia of pregnancy," is a complication that produces dangerously high blood pressure in a pregnant woman. Evidence suggests that preeclampsia may be passed on through the male. For many years, obstetricians routinely asked their patients if their mothers had had preeclampsia, because it has a tendency to occur in women whose mothers were affected. However, a study of 1.7 million pregnancies in Norway revealed that if a man's first wife had preeclampsia, his second wife had double the average risk of developing the condition too. Another study on 298 men and 237 women in Utah found that women whose mothers-in-law had experienced preeclampsia when pregnant with the women's husbands faced approximately twice the risk of developing the condition themselves. A gene from the male likely affects the placenta in a way that elevates the pregnant woman's blood pressure. Researchers recently discovered that the placentas of women with preeclampsia have an excess of two substances (soluble endoglin and a tyrosine kinase) and a deficiency of another (placental growth factor).

SUMMARY OUTLINE

20.1 Introduction (p. 539)

Growth is an increase in size. Development is the process of changing from one life phase to another.

20.2 Pregnancy (p. 539)

Pregnancy is the presence of a developing offspring in the uterus.

1. Transport of sex cells
   a. A male deposits semen in the vagina during sexual intercourse.
   b. A sperm cell lashes its tail to move, and is aided by muscular contractions in the female reproductive tract.
2. Fertilization
   a. An enzyme helps a sperm cell penetrate the zona pellucida.
   b. When a sperm cell head penetrates a secondary oocyte's cell membrane, changes in the membrane and the zona pellucida prevent entry of additional sperm cells.
   c. Completion of meiosis forming the second polar body.
   d. Fusion of the pronuclei completes fertilization.
   e. The product of fertilization is a zygote with 46 chromosomes.

20.3 Prenatal Period (p. 541)

1. Early embryonic development
   a. Cells undergo mitosis, giving rise to smaller and smaller cells during cleavage.
   b. The developing offspring moves down the uterine tube to the uterus, where it implants in the endometrium.
   c. Once implanted, the offspring is called an embryo through the eighth week of development. Thereafter, it is a fetus.
   d. Eventually, embryonic and maternal cells form a placenta.
2. Hormonal changes during pregnancy
   a. Embryonic cells produce human chorionic gonadotropin (hCG), which maintains the corpus luteum.
   b. Placental tissue produces high concentrations of estrogens and progesterone.
      (1) Estrogens and progesterone maintain the uterine wall and inhibit secretion of follicle-stimulating hormone (FSH) and luteinizing hormone (LH).
      (2) Progesterone and relaxin inhibit contraction of uterine muscles.
      (3) Estrogens enlarge the vagina.
      (4) Relaxin helps relax the ligaments of the pelvic joints.
   c. Placental lactogen stimulates development of the breasts and mammary glands.
   d. During pregnancy, increased aldosterone secretion promotes retention of sodium and body fluid. Increased secretion of parathyroid hormone helps maintain a high concentration of maternal blood calcium.
3. Embryonic stage
   a. The embryonic stage extends from the beginning of the second week through the eighth week of development.
   b. During this stage, the placenta and main internal and external body structures develop.
   c. The cells of the inner cell mass organize into primary germ layers.
   d. The embryonic disc becomes cylindrical and attaches to the developing placenta.
   e. The placental membrane consists of the epithelium of the chorionic villi and the epithelium of the capillaries inside the villi.
      (1) Oxygen and nutrients diffuse from maternal blood across the placental membrane and into fetal blood.
      (2) Carbon dioxide and other wastes diffuse from fetal blood across the placental membrane and into maternal blood.
   f. A fluid-filled amnion develops around the embryo.
   g. The umbilical cord forms as the amnion envelops the tissues attached to the underside of the embryo.
   h. The yolk sac forms on the underside of the embryonic disc.
   i. The allantois extends from the yolk sac into the connecting stalk.
   j. By the beginning of the eighth week, the embryo is recognizable as human.
4. Fetal stage
   a. The fetal stage extends from the end of the eighth week of development until birth.
   b. Existing structures grow and mature. Only a few new parts appear.
   c. The fetus is full-term at the end of thirty-eight weeks.
5. Fetal blood and circulation
   a. Umbilical vessels carry blood between the placenta and the fetus.
   b. Fetal blood carries a greater concentration of oxygen than does maternal blood because the concentration of oxygen-carrying hemoglobin is greater in fetal blood and fetal hemoglobin has greater affinity for oxygen.
   c. Blood enters the fetus through the umbilical vein and partially bypasses the liver through the ductus venosus.
   d. Blood enters the right atrium and partially bypasses the lungs through the foramen ovale.

e. Blood entering the pulmonary trunk partially bypasses the lungs through the ductus arteriosus.

f. Blood enters the umbilical arteries from the internal iliac arteries.

6. Birth process
   a. During pregnancy, placental progesterone inhibits uterine contractions.
   b. A variety of factors promote birth.
      (1) A decreasing progesterone concentration and the release of a prostaglandin initiate the birth process.
      (2) The posterior pituitary gland releases oxytocin.
      (3) Oxytocin stimulates uterine muscles to contract, and labor begins.
   c. Following birth, placental tissues are expelled.

20.4 Postnatal Period (p. 555)

1. Milk production and secretion
   a. Following childbirth, concentrations of placental hormones decline, the action of prolactin is no longer blocked, and the mammary glands begin to secrete milk.
   b. A reflex response to mechanical stimulation of the nipple stimulates the posterior pituitary to release oxytocin, which causes the alveolar ducts to eject milk.
2. Neonatal period
   a. The neonatal period extends from birth to the end of the first four weeks.
   b. The newborn must begin to respire, obtain nutrients, excrete wastes, and regulate body temperature.
   c. The first breath must be powerful to expand the lungs.
   d. The liver is immature and unable to supply sufficient glucose, so the newborn depends primarily on stored fat for energy.
   e. A newborn's immature kidneys cannot concentrate urine well.
   f. A newborn's homeostatic mechanisms may function imperfectly, and body temperature may be unstable.
   g. The cardiovascular system changes when placental circulation ceases.
      (1) Umbilical vessels constrict.
      (2) The ductus venosus constricts.
      (3) A valve closes the foramen ovale as blood pressure in the right atrium falls and pressure in the left atrium rises.
      (4) The ductus arteriosus constricts.

20.5 Genetics (p. 558)

1. Chromosomes and genes come in pairs
   a. Karyotypes are charts that display the two copies of each of the 22 autosomes, which do not determine sex, and the sex chromosomes (X and Y), which do.
   b. A person with two identical variants, or alleles, for a gene is homozygous. A person with two different alleles is heterozygous.
   c. The combination of alleles is the genotype; their expression as a trait is the phenotype.
2. Modes of inheritance
   a. A dominant allele masks the expression of a recessive allele.
   b. Modes of inheritance include autosomal recessive, in which an affected individual inherits an illness from heterozygous or affected parents; autosomal dominant, in which one affected parent passes on the condition; and X-linked recessive, which affects males more than females because males lack a second X chromosome to mask disease-causing recessive alleles.
   c. Punnett squares and pedigrees depict gene transmission in families.

3. Multifactorial traits
   a. Traits determined by more than one gene are polygenic.
   b. The continuously varying nature of polygenic traits can be depicted in bell-shaped curves.
   c. One or more genes and environmental influences cause a multifactorial trait.

CHAPTER ASSESSMENTS

20.1 Introduction

1. _____ is an increase in the size of the individual, whereas _____ is the contiuous process by which an individual changes from one life phase to another. (p. 539)

2. _____ is the period of development from fertilization to birth, whereas _____ is the period of development from birth to death. (p. 539)

20.2 Pregnancy

3. Define *pregnancy*. (p. 539)

4. Describe how sperm cells move within the female reproductive tract. (p. 539)

5. Summarize the events occurring after the sperm cell head enters the oocyte's cytoplasm. (p. 540)

20.3 Prenatal Period

6. Explain how cleavage forms a blastocyst. (p. 541)

7. Contrast an embryo and a fetus. (p. 543)

8. List hormones associated with pregnancy, and describe the function of each. (p. 545)

9. Ectodermal cells of the developing embryo give rise to: (p. 546)
   a. bone tissue
   b. the kidneys
   c. the lining of the urethra
   d. the epidermis
   e. connective tissues

10. Describe the composition and formation of the placenta. (p. 547)

11. List the function(s) of the placenta. (p. 547)

12. Distinguish between the chorion and the amnion. (p. 547)

13. Explain the function of the amniotic fluid. (p. 549)

14. Describe the formation of the umbilical cord. (p. 549)

15. List the major changes of the fetal stage of development. (p. 550)

16. Describe a full-term fetus. (p. 552)

17. List the structures through which blood passes as it travels from the placenta to the fetus and back to the placenta. (p. 552)

18. Explain positive feedback and the role of hormones in the expulsion of the fetus and the afterbirth. (p. 554)

20.4 Postnatal Period

19. Explain why a newborn's first breath must be particularly forceful. (p. 556)

20. Relate the difficulties of the fetus in maintaining water/electrolyte and body temperature homeostasis. (p. 556)

21. Describe the changes in the newborn's cardiovascular system. (p. 556)

20.5 Genetics

22. Distinguish between autosomes and sex chromosomes. (p. 558)

23. An individual with two different alleles is _____. (p. 559)

24. Match the mode of inheritance with its description. (p. 559)

(1) autosomal recessive   A. Inherited by male from carrier mothers

(2) autosomal dominant   B. Inherited from one affected parent

(3) X-linked recessive   C. Inherited from two carrier (unaffected) parents

25. Traits that are polygenic and affected by the environment are termed _____. Name two such traits. (p. 560)

INTEGRATIVE ASSESSMENTS/ CRITICAL THINKING

➡ OUTCOME 20.2

1. Why can twins resulting from a single fertilized egg exchange blood or receive organ transplants from each other without rejection, while twins resulting from two fertilized eggs sometimes cannot?

➡ OUTCOME 20.3

2. What technology would enable a fetus born in the fourth month to survive in a laboratory setting? (This is not yet possible.)

3. Toxins usually cause more severe medical problems if exposure is during the first eight weeks of pregnancy rather than during the later weeks. Why?

➡ OUTCOMES 20.2, 20.3, 20.5

4. What kinds of studies and information are required to determine whether a man's exposure to a potential teratogen can cause birth defects years later? How would such analysis differ if a woman were exposed?

➡ OUTCOMES 20.3, 20.4

5. What symptoms may appear if a newborn's ductus arteriosus fails to close?

➡ OUTCOME 20.5

6. Bob and Joan know from a blood test that they are each heterozygous (carriers) for the autosomal recessive gene that causes sickle cell disease. If their first three children are healthy, what is the probability that their fourth child will have the disease?

WEB CONNECTIONS

Visit the text website at **aris.mhhe.com** for additional quizzes, interactive learning exercises, and more.

AP R REPRODUCTIVE SYSTEM

Anatomy & Physiology | REVEALED includes cadaver photos that allow you to peel away layers of the human body to reveal structures beneath the surface. This program also includes animations, radiologic imaging, audio pronunciations, and practice quizzing. To learn more visit **www.aprevealed.com**.

AIDS TO UNDERSTANDING WORDS

acetabul-, vinegar cup: *acetabul*um
adip-, fat: *adip*ose tissue
agglutin-, to glue together: *agglutin*ation
aliment-, food: *aliment*ary canal
allant-, sausage: *allant*ois
alveol-, small cavity: *alveol*us
an-, without: *an*aerobic respiration
ana-, up: *ana*bolism
andr-, man: *andr*ogens
append-, to hang something: *append*icular
ax-, axis: *ax*ial skeleton, *ax*on
bil-, bile: *bil*irubin
-blast, bud: osteo*blast*
brady-, slow: *brady*cardia
bronch-, windpipe: *bronch*us
calat-, something inserted: inter*calat*ed disc
calyc-, small cup: major *calyc*es
cardi-, heart: peri*cardi*um
carp-, wrist: *carp*als
cata-, down: *cata*bolism
chondr-, cartilage: *chondr*ocyte
chorio-, skin: *chorio*n
choroid, skinlike: *choroid* plexus
chym-, juice: *chym*e
-clast, break: osteo*clast*
cleav-, to divide: *cleav*age
cochlea, snail: *cochlea*
condyl-, knob: *condyl*e
corac-, a crow's beak: *corac*oid process
cort-, covering: renal *cort*ex
cran-, helmet: *cran*ial
cribr-, sieve: *cribr*iform plate
cric-, ring: *cric*oid cartilage
-crin, to secrete: endo*crin*e
crist-, crest: *crist*a galli
cut-, skin: sub*cut*aneous
cyt-, cell: *cyt*oplasm
de-, separation from: *de*hydration
decidu-, falling off: *decidu*ous teeth
dendr-, tree: *dendr*ite
derm-, skin: *derm*is
detrus-, to force away: *detrus*or muscle
di-, two: *di*saccharide
diastol-, dilation: *diastol*ic pressure
diuret-, to pass urine: *diuret*ic
dors-, back: *dors*al
ejacul-, to shoot forth: *ejacul*ation
embol-, stopper: *embol*us
endo-, within: *endo*plasmic reticulum, *endo*crine gland
epi-, upon: *epi*thelial tissue, *epi*dermis, *epi*glottis
erg-, work: syn*erg*ist
erythr-, red: *erythr*ocyte
exo-, outside: *exo*crine gland
extra-, outside: *extra*cellular fluid
fimb-, fringe: *fimb*riae

follic-, small bag: hair *follic*le, ovarian *follic*le
fov-, pit: *fov*ea capitis
funi-, small cord or fiber: *funi*culus
gangli-, a swelling: *gangli*on
gastr-, stomach: *gastr*ic gland
-gen, to be produced: aller*gen*
-genesis, origin: spermato*genesis*
germ-, to bud or sprout: *germ*inal epithelium
glen-, joint socket: *glen*oid cavity
-glia, glue: neuro*glia*
glom-, little ball: *glom*erulus
glyc-, sweet: *glyc*ogen
-gram, something written: electrocardio*gram*
hema-, blood: *hema*tocrit
hemo-, blood: *hemo*globin
hepat-, liver: *hepat*ic duct
hetero-, other, different: *hetero*zygous
hom-, same, common: *hom*ozygous
homeo-, same: *homeo*stasis
humor-, fluid: *humor*al immunity
hyper-, above, more, over: *hyper*tonic, *hyper*trophy, *hyper*thyroidism
hypo-, below: *hypo*tonic, *hypo*thyroidism
im-, not: *im*balance
immun-, free: *immun*ity
inflamm-, set on fire: *inflamm*ation
inter-, among, between: *inter*phase, *inter*calated disc, *inter*vertebral disc
intra-, inside, within: *intra*membranous bone, *intra*cellular fluid
iris, rainbow: *iris*
iso-, equal: *iso*tonic
kerat-, horn: *kerat*in
labi-, lip: *labi*a minora
labyrinth, maze: *labyrinth*
lacri-, tears: *lacri*mal gland
lacun-, pool: *lacun*a
laten-, hidden: *laten*t period
-lemm, rind or peel: neuri*lemm*a
leuko-, white: *leuko*cyte
lingu-, tongue: *lingu*al tonsil
lip-, fat: *lip*ids
-logy, study of: physio*logy*
-lyt, dissolvable: electro*lyt*e
macr-, large: *macr*ophage
macula, spot: *macula* lutea
meat-, passage: auditory *meat*us
melan-, black: *melan*in
mening-, membrane: *mening*es
mens-, month: *mens*es
meta-, change: *meta*bolism
mict-, to pass urine: *mict*urition
mit-, thread: *mit*osis
mono-, one: *mono*saccharide
mons-, mountain: *mons* pubis
morul-, mulberry: *morul*a

moto-, moving: *moto*r neuron
mut-, change: *mut*ation
myo-, muscle: *myo*fibril
nat-, to be born: pre*nat*al
nephr-, pertaining to the kidney: *nephr*on
neutr-, neither one nor the other: *neutr*al
nod-, knot: *nod*ule
nutri-, nourish: *nutri*ent
odont-, tooth: *odont*oid process
olfact-, to smell: *olfact*ory
-osis, abnormal condition: leukocyt*osis*
os-, bone: *os*seous tissue
papill-, nipple: *papill*ary muscle, renal *papill*ae
para-, beside: *para*thyroid glands
pariet-, wall: *pariet*al membrane
path-, disease: *path*ogen
pelv-, basin: *pelv*ic cavity
peri-, around: *peri*cardial membrane, *peri*pheral nervous system, *peri*stalsis
phag-, to eat: *phag*ocytosis
pino-, to drink: *pino*cytosis
pleur-, rib: *pleur*al membrane
plex-, interweaving: choroid *plex*us
-poie, make, produce: hemato*poie*sis, erythro*poie*tin
poly-, many: *poly*unsaturated
pseudo-, false: *pseudo*stratified epithelium
puber-, adult: *puber*ty
pyl-, gatekeeper: *pyl*oric sphincter
sacchar-, sugar: mono*sacchar*ide
sarco-, flesh: *sarco*plasm
scler-, hard: *scler*a
seb-, grease: *seb*aceous gland
sens-, feeling: *sens*ory neuron
-som, body: ribo*som*e
squam-, scale: *squam*ous epithelium
-stasis, standing still, halt: homeo*stasis*, hemo*stasis*
strat-, layer: *strat*ified
syn-, together: *syn*thesis, *syn*ergist, *syn*apse, *syn*ctium
systol-, contraction: *systol*ic pressure
tachy-, rapid: *tachy*cardia
tetan-, stiff: *tetan*ic contraction
thromb-, clot: *thromb*ocyte
toc-, birth: oxy*toc*in
-tomy, cutting: ana*tomy*
trigon-, triangle: *trigon*e
-troph, well fed: muscular hyper*troph*y, *troph*oblast
-tropic, influencing: adrenocortico*tropic*
tympan-, drum: *tympan*ic membrane
umbil-, navel: *umbil*ical cord
ventr-, belly or stomach: *ventr*icle
vill-, hairy: *vill*i
vitre-, glass: *vitre*ous humor
-zym, ferment: en*zym*e

PERIODIC TABLE OF ELEMENTS

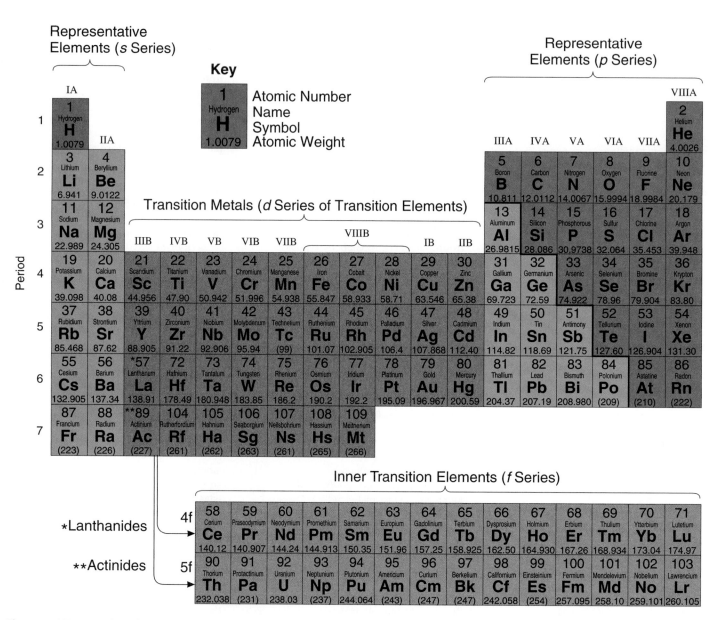

Representative Elements (*s* Series)

Representative Elements (*p* Series)

Key

1	Atomic Number
Hydrogen	Name
H	Symbol
1.0079	Atomic Weight

Transition Metals (*d* Series of Transition Elements)

	IA																	VIIIA
1	1 Hydrogen **H** 1.0079	IIA											IIIA	IVA	VA	VIA	VIIA	2 Helium **He** 4.0026
2	3 Lithium **Li** 6.941	4 Beryllium **Be** 9.0122											5 Boron **B** 10.811	6 Carbon **C** 12.0112	7 Nitrogen **N** 14.0067	8 Oxygen **O** 15.9994	9 Fluorine **F** 18.9984	10 Neon **Ne** 20.179
3	11 Sodium **Na** 22.989	12 Magnesium **Mg** 24.305	IIIB	IVB	VB	VIB	VIIB	VIIIB			IB	IIB	13 Aluminum **Al** 26.9815	14 Silicon **Si** 28.086	15 Phosphorous **P** 30.9738	16 Sulfur **S** 32.064	17 Chlorine **Cl** 35.453	18 Argon **Ar** 39.948
4	19 Potassium **K** 39.098	20 Calcium **Ca** 40.08	21 Scandium **Sc** 44.956	22 Titanium **Ti** 47.90	23 Vanadium **V** 50.942	24 Chromium **Cr** 51.996	25 Manganese **Mn** 54.938	26 Iron **Fe** 55.847	27 Cobalt **Co** 58.933	28 Nickel **Ni** 58.71	29 Copper **Cu** 63.546	30 Zinc **Zn** 65.38	31 Gallium **Ga** 69.723	32 Germanium **Ge** 72.59	33 Arsenic **As** 74.922	34 Selenium **Se** 78.96	35 Bromine **Br** 79.904	36 Krypton **Kr** 83.80
5	37 Rubidium **Rb** 85.468	38 Strontium **Sr** 87.62	39 Yttrium **Y** 88.905	40 Zirconium **Zr** 91.22	41 Niobium **Nb** 92.906	42 Molybdenum **Mo** 95.94	43 Technetium **Tc** (99)	44 Ruthenium **Ru** 101.07	45 Rhodium **Rh** 102.905	46 Palladium **Pd** 106.4	47 Silver **Ag** 107.868	48 Cadmium **Cd** 112.40	49 Indium **In** 114.82	50 Tin **Sn** 118.69	51 Antimony **Sb** 121.75	52 Tellurium **Te** 127.60	53 Iodine **I** 126.904	54 Xenon **Xe** 131.30
6	55 Cesium **Cs** 132.905	56 Barium **Ba** 137.34	*57 Lanthanum **La** 138.91	72 Hafnium **Hf** 178.49	73 Tantalum **Ta** 180.948	74 Tungsten **W** 183.85	75 Rhenium **Re** 186.2	76 Osmium **Os** 190.2	77 Iridium **Ir** 192.2	78 Platinum **Pt** 195.09	79 Gold **Au** 196.967	80 Mercury **Hg** 200.59	81 Thallium **Tl** 204.37	82 Lead **Pb** 207.19	83 Bismuth **Bi** 208.980	84 Polonium **Po** (209)	85 Astatine **At** (210)	86 Radon **Rn** (222)
7	87 Francium **Fr** (223)	88 Radium **Ra** (226)	**89 Actinium **Ac** (227)	104 Rutherfordium **Rf** (261)	105 Hahnium **Ha** (262)	106 Seaborgium **Sg** (263)	107 Neilsbohrium **Ns** (261)	108 Hassium **Hs** (265)	109 Meitnerium **Mt** (266)									

Period

Inner Transition Elements (*f* Series)

*Lanthanides	4f	58 Cerium **Ce** 140.12	59 Praseodymium **Pr** 140.907	60 Neodymium **Nd** 144.24	61 Promethium **Pm** 144.913	62 Samarium **Sm** 150.35	63 Europium **Eu** 151.96	64 Gadolinium **Gd** 157.25	65 Terbium **Tb** 158.925	66 Dysprosium **Dy** 162.50	67 Holmium **Ho** 164.930	68 Erbium **Er** 167.26	69 Thulium **Tm** 168.934	70 Ytterbium **Yb** 173.04	71 Lutetium **Lu** 174.97
**Actinides	5f	90 Thorium **Th** 232.038	91 Protactinium **Pa** (231)	92 Uranium **U** 238.03	93 Neptunium **Np** (237)	94 Plutonium **Pu** 244.064	95 Americium **Am** (243)	96 Curium **Cm** (247)	97 Berkelium **Bk** (247)	98 Californium **Cf** 242.058	99 Einsteinium **Es** (254)	100 Fermium **Fm** 257.095	101 Mendelevium **Md** 258.10	102 Nobelium **No** 259.101	103 Lawrencium **Lr** 260.105

Elements 110 to 114 have been reported in experiments, but have not yet been confirmed.

CHANGES OCCURRING IN THE HEART DURING A CARDIAC CYCLE

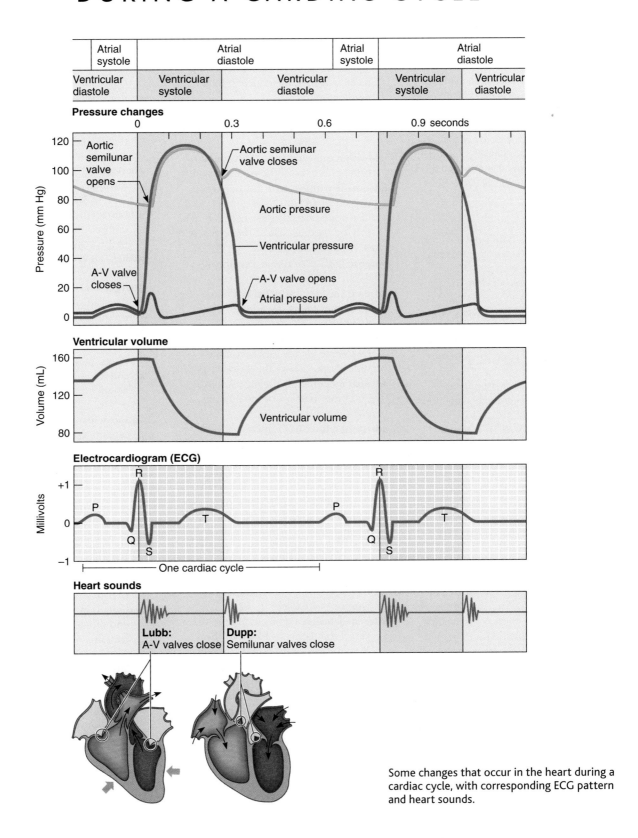

Some changes that occur in the heart during a cardiac cycle, with corresponding ECG pattern and heart sounds.

Glossary

A phonetic guide to pronunciation follows each glossary word. Any unmarked vowel that ends a syllable or stands alone as a syllable has the long sound. Thus, the word *play* is phonetically spelled *pla*. Any unmarked vowel followed by a consonant has the short sound. The word *tough*, for instance, is phonetically spelled *tuf*. If a long vowel appears in the middle of a syllable (followed by a consonant), it is marked with a macron (ˉ), the sign for a long vowel. Thus, the word *plate* is phonetically spelled *plāt*. Similarly, if a vowel stands alone or ends a syllable, but has a short sound, it is marked with a breve (˘).

A

abdominal cavity (ab-dom′ĭ-nal kav′ĭ-te) The space between the diaphragm and the pelvis. p. 8

abdominopelvic cavity (ab-dom″ĭ-no-pel′vik kav′ĭ-te) The space between the diaphragm and the lower portion of the trunk of the body. p. 8

abduction (ab-duk′shun) Movement of a body part away from the midline. p. 166

accessory organ (ak-ses′o-re or′gan) Organ that supplements the functions of other organs. p. 276

accommodation (ah-kom″o-da′shun) Adjustment of the lens of the eye for close or distant vision. p. 279

acetylcholine (as″ĕ-til-ko′lēn) A type of neurotransmitter, which is a biochemical secreted at axon ends of many neurons; transmits nerve messages across synapses. p. 181

acetylcholinesterase (as″ĕ-til-ko″lin-es′ter-a-s) An enzyme that catalyzes breakdown of acetylcholine. p. 183

acid (as′id) A substance that ionizes in water to release hydrogen ions. p. 38

acidosis (as″ĭ-do′sis) Increase in acidity of body fluids below pH 7.35. p. 502

ACTH Adrenocorticotropic hormone. p. 298

actin (ak′tin) A protein that forms filaments that slide between filaments of the protein myosin, contracting muscle fibers. p. 177

action potential (ak′shun po-ten′shal) The sequence of electrical changes that occurs in a portion of a nerve cell membrane that is exposed to a stimulus that exceeds the membrane's threshold. p. 223

active site (ak′tiv sīt) The part of an enzyme molecule that binds a substrate. p. 79

active transport (ak′tiv trans′port) Process that requires energy to move a substance across a cell membrane, usually against the concentration gradient. p. 64

adaptive immunity (a-dap′tiv ĭ-mu′nĭ-te) Specific defenses carried out by T and B cells. p. 383

adduction (ah-duk′shun) Movement of body part toward the midline. p. 166

adenosine triphosphate (ah-den′o-sēn tri-fos′fāt) (ATP) Organic molecule that stores and releases energy, which may be used in cellular processes. p. 80

ADH Antidiuretic hormone. p. 293

adipose (ad′ĭ-pōs) **tissue** Fat-storing tissue. p. 185

adrenal cortex (ah-dre′nal kor′teks) The outer portion of the adrenal gland. p. 302

adrenal glands (ah-dre′nal glandz) Endocrine glands located on the superior portion of the kidneys. p. 302

adrenal medulla (ah-dre′nal me-dul′ah) The inner portion of the adrenal gland. p. 302

adrenergic fiber (ad″ren-er′jik fi′ber) Axon that secretes norepinephrine at its terminal. p. 251

adrenocorticotropic hormone (ad-re″no-kor″te-ko-trōp′ik hor′mōn) **(ACTH)** Hormone that the anterior pituitary secretes to stimulate activity in the adrenal cortex. p. 298

aerobic respiration (a″er-o′bik res″pĭ-ra′shun) The complete, energy-releasing, breakdown of glucose to carbon dioxide and water, in the presence of oxygen. p. 82

afferent arteriole (af′er-ent ar-te′re-ōl) Vessel that conveys blood to the glomerulus of each kidney nephron in a kidney. p. 471

agglutination (ah-gloo″tĭ-na′shun) Clumping of blood cells in response to a reaction between an antibody and an antigen. p. 331

agonist (ago′-nist) Muscle responsible for a body movement. Also called a prime mover. p. 192

agranulocyte (a-gran′u-lo-sīt) A nongranular leukocyte. p. 323

albumins (al-bu′minz) Plasma proteins that help regulate osmotic concentration of the blood. p. 325

aldosterone (al dos′ter-ōn″) A hormone that the adrenal cortex secretes, which regulates sodium and potassium ion concentrations and fluid volume. p. 304

alimentary canal (al″i-men′tar-e kah-nal′) Tubular portion of the digestive tract that leads from the mouth to the anus. p. 402

alkalosis (al″kah-lo′sis) Increase in alkalinity of body fluids above pH 7.45. p. 502

allantois (ah-lan′to-is) A structure in the embryo that forms umbilical cord blood vessels. p. 549

alleles (ah-le′elz) Variant forms of a gene. p. 558

allergens (al′er-jenz) Foreign substances that can provoke an allergic reaction. p. 392

alveolar ducts (al-ve′o-lar duktz′) Fine tubes that carry air to air sacs of the lungs. p. 450

alveoli (al-ve′o-li) (singular, alveolus) An air sac of a lung; saclike structure. p. 450

amino acids (ah-me′no as′idz) Small organic compounds that include an amino group (–NH₂) and a carboxyl group (–COOH); structural units of a protein molecule. p. 44

amnion (am′ne-on) An extraembryonic membrane that encircles a developing fetus and contains amniotic fluid. p. 548

amniotic fluid (am″ne-ot′ik floo′id) Fluid within the amniotic cavity that surrounds the developing fetus. p. 549

ampulla (am-pul′ah) An expansion at the end of each semicircular canal that contains a crista ampullaris. p. 274

anabolism (ah-nab′o-lizm) Synthesis of larger molecules from smaller ones; anabolic metabolism. p. 77

anaphase (an′ah-fāz) Stage in mitosis when duplicate chromosomes move to opposite poles of cell. p. 68

anatomy (ah-nat′o-me) Branch of science dealing with the form and structure of body parts. p. 3

androgens (an′dro-jenz) Male sex hormones such as testosterone. p. 515

antagonists (an-tag´o-nists) A muscle that opposes a prime mover. p. 192

antebrachial (an″te-bra´ke-al) Pertaining to the forearm. p. 16

antecubital (an″te-ku´bĭ-tal) The region in front of the elbow joint. p. 16

anterior (an-te´re-or) Pertaining to the front. p. 14

anterior pituitary (pĭ-tu´ĭ-tār″e) The front lobe of the pituitary gland. p. 295

antibodies (an″tĭ-bod″ēz) Proteins (immunoglobulins) that B cells of the immune system produce in response to nonself antigens; antibodies react with the antigens. p. 332

anticodon (an″tĭ-ko´don) Three contiguous nucleotides of a transfer RNA molecule that are complementary to a specific mRNA codon. p. 87

antidiuretic hormone (an″tĭ-di´u-ret´ik hor´mōn) (ADH) Hormone of the posterior pituitary lobe that enhances water conservation in the kidneys. p. 293

antigens (an´tĭ-jenz) Chemicals that stimulate B cells to produce antibodies. p. 331

aorta (a-or´tah) Major systemic artery that receives blood from left ventricle. p. 344

aortic sinus (a-or´tik si´nus) Swelling in the aortic wall behind each cusp of the semilunar valve. p. 362

aortic valve (a-or´tik valv) Flaplike structures in the wall of the aorta near its origin that prevent blood from returning to the left ventricle of heart. p. 344

apocrine gland (ap´o-krin gland) A type of gland whose secretions contain parts of secretory cells. p. 101

aponeuroses (ap″o-nu-ro´sēz) Sheets of connective tissue that attach muscles. p. 177

apoptosis (ayp-o-toe´sis) Programmed cell death. p. 72

appendicular (ap″en-dik´u-lar) Pertaining to upper or lower limbs. p. 8

aqueous humor (a´kwe-us hu´mor) Watery fluid that fills the anterior cavity of the eye. p. 280

arachnoid mater (ah-rak´noid ma´ter) Delicate, weblike middle layer of meninges. p. 231

areolar (ah-re´o-lar) **tissue** Delicate, thin, loose connective tissue that is widespread in the body. p. 105

arrector pili muscle (ah-rek´tor pil´i mus´l) Smooth muscle in the skin associated with a hair follicle. p. 123

arterioles (ar-te´re-ōlz) Small branches of an artery that communicate with a capillary network. p. 353

arteries (ar´ter-ēz) Blood vessels that transport blood away from the heart. p. 353

articular cartilage (ar-tik´u-lar kar´tĭ-lij) Hyaline cartilage that covers ends of bones in synovial joints. p. 131

ascending tract (ah-send´ing trakt) Group of nerve fibers in the spinal cord that transmits sensory impulses upward to the brain. p. 233

association area (ah-so″se-a´shun a´re-ah) Region of the cerebral cortex controlling memory, reasoning, judgment, and emotions. p. 237

astrocyte (as´tro-sīt) A type of neuroglial cell that connects neurons to blood vessels. p. 214

atmospheric pressure (at″mos-fēr´ik presh´ur) Pressure exerted by the weight of air; about 760 millimeters of mercury at sea level. p. 452

atoms (at´omz) Smallest particles of an element that have the properties of that element. p. 3

atomic number (ah-tom´ik num´ber) Number of protons in an atom of an element. p. 32

atomic weight (ah-tom´ik wāt) The number of protons and neutrons in an atom. p. 32

ATP Adenosine triphosphate. p. 80

ATPase Enzyme that releases energy stored in the terminal phosphate bonds of ATP molecules. p. 181

atria (a´tre-ah) (singular, atrium) Chambers of the heart that receive blood from the veins. p. 342

atrioventricular bundle (a″tre-o-ven-trik´u-lar bun´dl) (A-V bundle) Group of specialized fibers that conducts impulses from the atrioventricular node to the ventricular muscle of heart. p. 349

atrioventricular node (a″tre-o-ven-trik´u-lar nōd) (A-V node) Specialized mass of cardiac muscle fibers in the interatrial septum of heart; transmits cardiac impulses from sinoatrial node to the A-V bundle. p. 349

auditory ossicles (aw´di-to″re os´i-klz) Bones of the middle ear. p. 269

auditory tube (aw´di-to″re tūb) The tube that connects the middle ear cavity to the pharynx; eustachian tube. p. 270

auricle (aw´ri-kl) An earlike structure; the portion of the heart that forms the wall of an atrium. p. 269

autonomic nervous system (aw″to-nom´ik ner´vus sis´tem) Part of nervous system that controls the viscera. p. 214

autosome (aw´to-some) A chromosome that does not include a gene that determines sex. p. 000

axial (ak´se-al) Pertaining to the head, neck, and trunk. p. 8

axillary (ak´sĭ-ler″e) Pertaining to the armpit. p. 16

axon (ak´son) A nerve fiber. It conducts a nerve impulse away from the neuron cell body. p. 212

B

B cell (sel) Lymphocyte that produces and secretes antibodies that bind and destroy foreign antigens; B lymphocyte. p. 385

baroreceptor (bar″o-re-sep´tor) Sensory receptor in a blood vessel wall that responds to changes in pressure. p. 360

basal nucleus (ba´sal nu´kle-us) Mass of gray matter deep within a cerebral hemisphere of the brain. p. 238

base (bās) Substance that ionizes in water, releasing hydroxide ions (OH⁻) or other ions that combine with hydrogen ions. p. 38

basement membrane A layer of extracellular matrix that anchors epithelial tissue to underlying connective tissue. p. 95

basophils (ba´so-filz) White blood cells containing cytoplasmic granules that stain with basic dye. p. 323

beta oxidation (ba´tah ok″sĭ-da´shun) Chemical process that breaks fatty acids down into molecules of acetyl coenzyme A. p. 431

bile (bīl) Fluid secreted by the liver and stored in the gallbladder. p. 419

bilirubin (bil″ĭ-roo´bin) A bile pigment produced from hemoglobin breakdown. p. 322

biliverdin (bil″ĭ-ver´din) A bile pigment produced from hemoglobin breakdown. p. 322

blastocyst (blas´to-sist) An early stage of prenatal development in which the embryo is a hollow ball of cells. p. 541

blood (blud) A connective tissue consisting of formed elements (white blood cells, red blood cells, and platelets) in a liquid extracellular matrix (plasma). p. 109

bone (bōn) A hard, mineralized connective tissue that forms the skeleton. p. 108

brachial (bra´ke-al) Pertaining to the arm. p. 16

brainstem (brānstem) Portion of the brain that includes the midbrain, pons, and medulla oblongata. p. 000

bronchial tree (brong´ke-al tre) The bronchi and their branches that carry air from the trachea to the alveoli of the lungs. p. 448

bronchiole (brong′ke-ōl) A small branch of a bronchus within the lung. p. 450

bronchus (brong′kus) (singular, bronchus) A branch of the trachea that leads to the lung. p. 448

buccal (buk′al) Pertaining to the mouth and inner lining of the cheeks. p. 16

buffer (buf′er) A substance that can react with a strong acid or base to form a weaker acid or base and thus resist change in pH. p. 39

bulbourethral glands (bul″bo-u-re′thral glandz) Glands that secrete viscous fluid into the male urethra during sexual excitement; Cowper's glands. p. 513

bursae (ber′se) (singular, *bursa*) Saclike, fluid-filled structures, lined with synovial membrane, near a joint. p. 163

C

calcitonin (kal″sĭ-to′nin) Hormone the thyroid gland secretes that helps regulate blood calcium concentration. p. 300

calorie (kal′o-re) A unit used to measure heat energy and the energy content of foods. p. 429

canaliculus (kan″ah-lik′u-lus) Microscopic canal that connects lacunae of bone tissue. p. 108

capacitation (kah-pas″i-ta′shun) Activation of sperm cell to fertilize an oocyte. p. 514

capillaries (kap′ĭ-lar″ēz) Small blood vessels that connect arterioles and venules. p. 355

carbohydrates (kar″bo-hi′drātz) Organic compounds consisting of carbon, hydrogen, and oxygen, in a 1:2:1 ratio. p. 41

carbonic anhydrase (kar-bon′ik an-hi′drās) Enzyme that catalyzes the reaction between carbon dioxide and water to form carbonic acid. p. 462

carbaminohemoglobin (kar-bam″ĭ-no-he″mo-glo′bin) Compound formed by the union of carbon dioxide and hemoglobin. p. 462

carboxypeptidase (kar-bok″se-pep′tĭ-dās) A protein-splitting enzyme in pancreatic juice. p. 414

cardiac conduction system (kar′de-ak kon-duk′shun sis′tem) System of specialized cardiac muscle fibers that conducts cardiac impulses from the S-A node into the myocardium. p. 349

cardiac cycle (kar′de-ak si′kl) A series of myocardial contractions and relaxations that constitutes a complete heartbeat. p. 346

cardiac muscle tissue (kar′de-ak mus′l tish′u) Specialized muscle tissue found only in the heart. p. 111

cardiac output (kar′de-ak owt′poot) The volume of blood per minute that the heart pumps (multiply stroke volume in milliliters by heart rate in beats per minute). p. 359

cardiovascular (kahr″de-o-vas′ku-lur) Pertaining to the heart and blood vessels. p. 12

carpals (kar′pals) Bones of the wrist. p. 16

cartilage (kar′ti-lij) Type of connective tissue with cells in lacunae separated by a semisolid extracellular matrix. p. 106

cartilaginous (kar-tĭ-laj′ĭ-nus) **joints** Two or more bones joined by cartilage. p. 162

catabolism (kă-tab′o-lizm) Breakdown of large molecules into smaller ones; catabolic metabolism. p. 77

catalyst (kat′ah-list) A chemical that increases the rate of a chemical reaction but is not permanently altered by the reaction. p. 38

celiac (se′le-ak) Pertaining to the abdomen. p. 16

cell (sel) The structural and functional unit of life. p. 3

cell body (sel bod′e) Portion of nerve cell that includes a cytoplasmic mass and a nucleus, and from which nerve fibers extend. p. 212

cell membrane (sel mem′brān) The selectively permeable outer boundary of a cell consisting of a phospholipid bilayer embedded with proteins; plasma membrane or cytoplasmic membrane. p. 52

cellular immune response (sel′u-lar i-mūn′ ri-spons′) The attack of T cells and their secreted products on foreign cells. p. 385

cellular respiration (sel′u-lar res″pĭ-ra′shun) Cellular process that releases energy from organic compounds. p. 80

cellulose (sel′u-lōs) Polysaccharide abundant in plant tissues that human digestive enzymes cannot break down. p. 430

cementum (se-men′tum) Bonelike material that fastens the root of a tooth into its bony socket. p. 408

central canal (sen′tral kah-nal′) Tiny channel in bone tissue that houses a blood vessel, p. 108; tube in spinal cord that is continuous with brain ventricles and contains cerebrospinal fluid. p. 232

central nervous system (sen′tral ner′vus sis′tem) **(CNS)** The brain and spinal cord. p. 212

centriole (sen′tre-ōl) A cellular organelle built of microtubules that organizes the mitotic spindle. p. 57

centromeres (sen′tro-mērz) Portion of a chromosome to which spindle fibers attach during mitosis. p. 510

centrosome (sen′tro-sōm) Cellular organelle consisting of two centrioles. p. 57

cephalic (sě-fal′ik) Pertaining to the head. p. 16

cerebellar cortex (ser″ě-bel′ar kor′teks) The outer layer of the cerebellum. p. 236

cerebellum (ser″ě-bel′um) Part of the brain that coordinates skeletal muscle movement. p. 234

cerebral cortex (ser′ě-bral kor′teks) The outer layer of the cerebrum. p. 236

cerebral hemispheres (ser′ě-bral hem′ĭ-sfērz) Large, paired structures that constitute the cerebrum. p. 235

cerebrospinal fluid (ser″ě-bro-spi′nal floo′id) Fluid in the ventricles of the brain, subarachnoid space of the meninges, and the central canal of the spinal cord. p. 231

cerebrum (ser′ě-brum) Part of the brain that occupies the upper part of the cranial cavity and provides higher mental functions. p. 234

cervical (ser′vĭ-kal) Pertaining to the neck. p. 17

cervix (ser′viks) Narrow, inferior end of the uterus that leads into the vagina. p. 522

chemical bond (kem′ĭkel bond) Attractive force between atoms. p. 32

chemistry (kem′is-trē) The study of matter p. 31

chemoreceptors (ke″mo-re-sep′torz) A receptor that is stimulated by the binding of certain chemicals. p. 261

chief cells (chēf selz) Cells of gastric gland that secrete digestive enzymes. p. 412

cholecystokinin (ko″le-sis″to-ki′nin) Hormone the small intestine secretes that stimulates release of pancreatic juice from the pancreas and bile from the gallbladder. p. 413

cholesterol (ko-les′ter-ol) A lipid produced by body cells and used to synthesize steroid hormones. p. 431

cholinergic fiber (ko″lin-er′jik fi′ber) Axon that secretes acetylcholine at its terminal. p. 251

chondrocytes (kon′dro-sītz) Cartilage cells. p. 106

chordae tendineae (kor′de ten′dĭ-ne) Fibrous strings attached to the cusps of the tricuspid and mitral valves in the heart. p. 343

chorion (ko′re-on) Extraembryonic membrane that forms the outermost covering around a fetus and contributes to formation of the placenta. p. 547

chorionic villi (ko″re-on′ik vil′i) Projections that extend from the outer surface of the chorion and help attach an embryo to the uterine wall. p. 547

choroid coat (ko′roid kōt) The vascular, pigmented middle layer of the wall of the eye. p. 279

choroid plexus (ko′roid plek′sus) Mass of specialized capillaries that secretes cerebrospinal fluid into a brain ventricle. p. 239

chromatids (kro′mah-tidz) One longitudinal half of a replicated chromosome. p. 510

chromatin (kro′mah-tin) DNA and complexed protein that condenses to form chromosomes during mitosis. p. 59

chromatophilic substance (kro″mah-to-fil′ik sub′stans) **(Nissl bodies)** Membranous sacs in the cytoplasm of nerve cells that have ribosomes attached to their surfaces. p. 214

chromosomes (kro′mo-sōmz) Rodlike structures that condense from chromatin in a cell's nucleus during mitosis. p. 59

chylomicron (ki″lo-mi′kron) A microscopic droplet of fat in the blood following fat digestion. p. 422

chyme (kīm) Semifluid mass of partially digested food that passes from the stomach to the small intestine. p. 414

chymotrypsin (ki″mo-trip′sin) A protein-splitting enzyme in pancreatic juice. p. 414

cilia (sil′e-ah) Microscopic, hairlike processes on the exposed surfaces of certain epithelial cells. p. 58

ciliary body (sil′e-er″e bod′e) Structure associated with the choroid layer of the eye that secretes aqueous humor and houses the ciliary muscle. p. 279

circadian rhythm (ser″kah-de′an rithm) Pattern of repeated behavior associated with cycles of night and day. p. 307

cisternae (sis-ter′ne) Enlarged portions of sarcoplasmic reticulum near the actin and myosin filaments of a muscle fiber. p. 180

citric acid cycle (sit′rik as′id si′kl) Series of chemical reactions that oxidizes certain molecules, releasing energy; Krebs cycle. p. 80

cleavage (klēv′ij) Early successive divisions of the zygote into a ball of progressively smaller cells. p. 541

clitoris (klit′o-ris) Small, erectile organ in the anterior portion of the vulva; corresponding to the penis. p. 523

clone (klōn) A group of cells that originate from a single cell and are therefore genetically identical. p. 386

CNS Central nervous system. p. 212

coagulation (ko-ag″u-la′shun) Blood clotting. p. 329

cochlea (kok′le-ah) Portion of inner ear that has hearing receptors. p. 270

codon (ko′don) Set of three contiguous nucleotides of a messenger RNA molecule that specifies a particular amino acid. p. 87

coenzyme (ko-en′zīm) A nonprotein organic molecule required for the activity of a particular enzyme. p. 80

cofactor (ko′fak-tor) A small molecule or ion that must combine with an enzyme for activity. p. 80

collagenous (kol′ah-jen-us) **fibers** White fibers consisting of the protein collagen, common in connective tissues, including bone matrix. p. 104

common hepatic duct (kom′mon hepat′ik dukt) Tube that transports bile from the liver to the common bile duct. p. 416

compact bone (kom′-pakt bōn) Dense tissue in which cells are organized in osteons (Haversian systems) with no spaces. p. 131

complement (kom′ple-ment) A group of enzymes activated when an antibody binds an antigen; enhances reaction against foreign substances. p. 384

complete protein (kom-plēt′ pro′te-in) A protein that contains adequate amounts of the essential amino acids to maintain body tissues and to promote normal growth and development. p. 433

compound (kom′pownd) A substance composed of two or more chemically bonded elements. p. 36

cone (kōn) Color receptor in the retina of the eye. p. 282

conformation (kon″for-ma′shen) Three-dimensional shape of a molecule p. 44

conjunctiva (kon″junk-ti′vah) Membranous covering on the anterior surface of the eye and lining the eyelids. p. 276

connective tissue (kŏ-nek′tiv tish′u) A basic type of tissue that consists of cells within an extracellular matrix. Connective tissues include bone, cartilage, and blood. p. 102

convergence (kon-ver′jens) Nerve impulses arriving at the same neuron. p. 227

cornea (kor′ne-ah) Transparent anterior portion of the outer layer of the eye wall. p. 278

coronal (kor′o-nul) A plane that divides a structure into anterior and posterior portions. p. 15

coronary artery (kor′o-na″re ar′ter-e) An artery that supplies blood to the wall of the heart. p. 346

coronary sinus (kor′o-na″re si′nus) A large vessel on the posterior surface of the heart into which cardiac veins drain. p. 346

corpus callosum (kor′pus kah-lo′sum) A mass of white matter in the brain composed of nerve fibers connecting the right and left cerebral hemispheres. p. 235

corpus luteum (kor′pus loot′e-um) Structure that forms from the tissues of a ruptured ovarian follicle and secretes female hormones. p. 525

cortisol (kor′tĭ-sol) A glucocorticoid that the adrenal cortex secretes. p. 304

costal (kos′tal) Pertaining to the ribs. p. 17

covalent bond (ko′va-lent bond) Chemical bond formed by electron sharing between atoms. p. 35

cranial nerve (kra′ne-al nerv) Nerve that arises from the brain or brainstem. p. 243

creatine phosphate (kre′ah-tin fos′fāt) A biochemical that stores energy in muscle tissue. p. 183

crista ampullaris (kris′tah am-pul′ar-is) Sensory organ in a semicircular canal that functions in the sense of dynamic equilibrium. p. 274

cubital (ku′bi-tal) Pertaining to the elbow. p. 17

cutaneous membrane (ku-ta′ne-us mem′brane) Pertaining to the skin. p. 110

cyclic AMP A second messenger molecule in a signal transduction pathway. p. 293

cystic duct (sis′tik dukt) The tube that connects the gallbladder to the common bile duct. p. 419

cytocrine secretion (si′to-krin se-kre′shun) Transfer of melanin granules from melanocytes into epithelial cells. p. 119

cytokinesis (si″to-kĭ-ne′sis) Division of the cytoplasm. p. 68

cytoplasm (si′to-plazm) The gel-like contents of a cell (cytosol) and organelles, excluding the nucleus, enclosed by the cell membrane. p. 52

cytoskeleton (si″to-skel′e-ten) A cell's framework of protein filaments and tubules. p. 55

D

deamination (de-am″ĭ-na′shun) Removing amino groups (–NH₂) from amino acids. p. 432

decomposition (de″-kom-po-zish′un) The breakdown of molecules into simpler compounds. p. 37

dehydration synthesis (de″hi-dra′shun sin′thĕ-sis) Anabolic process that joins small molecules by releasing the equivalent of a water molecule; synthesis. p. 77

dendrite (den′drīt) Process of a neuron that receives input from other neurons. p. 212

dense connective tissue (dens kō-nek′tiv tish′u) A connective tissue with many collagenous fibers, a fine network of elastin fibers, and sparse fibroblasts. p. 106

dentin (den′tin) Bonelike substance that forms the bulk of a tooth. p. 408

deoxyhemoglobin (de-ok″se-he″mo-glo′bin) Hemoglobin to which oxygen is not bound. p. 319

deoxyribonucleic acid (de-ok′si-ri″bo-nu-kle″ik as′id) **(DNA)** The genetic material; a double-stranded polymer of nucleotides, each containing a phosphate group, a nitrogenous base (adenine, thymine, guanine, or cytosine), and the sugar deoxyribose. p. 83

depolarization (de-po″lar-ĭ-za′shun) The loss of an electrical charge on the surface of a cell membrane. p. 222

dermis (der′mis) The thick layer of the skin beneath the epidermis. p. 117

descending tract (de-send′ing trakt) Group of nerve fibers that carries nerve impulses downward from the brain through the spinal cord. p. 233

detrusor muscle (de-trūz′or mus′l) Muscular wall of the urinary bladder. p. 484

diapedesis (di″ah-pĕ-de′sis) Squeezing of leukocytes between the cells of blood vessel walls. p. 324

diaphragm (di′ah-fram) A sheetlike structure largely composed of skeletal muscle and connective tissue that separates the thoracic and abdominal cavities, p. 8; a caplike contraceptive device inserted in the vagina. p. 452

diaphysis (di-af′ĭ-sis) The shaft of a long bone. p. 131

diastole (di-as′to-le) Phase of the cardiac cycle when a heart chamber wall relaxes. p. 346

diastolic pressure (di-a-stol′ik presh′ur) Lowest arterial blood pressure reached during diastolic phase of the cardiac cycle. p. 358

diencephalon (di″en-sef′ah-lon) Part of the brain in the region of the third ventricle that includes the thalamus and hypothalamus. p. 234

differentiation (dif″er-en″she-a′shun) Cell specialization. p. 70

diffusion (dĭ-fu′zhun) Random movement of molecules from a region of higher concentration toward one of lower concentration. p. 61

digestion (di-jest′yun) Breaking down of large nutrient molecules into molecules small enough to be absorbed; hydrolysis. p. 402

dipeptide (di-pep′tīd) A molecule composed of two amino acids. p. 78

disaccharides (di-sak′ah-rīdz) Sugars produced by the union of two monosaccharides. p. 41

distal (dis′tal) Further from a point of attachment; opposite of *proximal*. p. 15

divergence (di-ver′jens) Spreading apart. p. 227

DNA Deoxyribonucleic acid. p. 46

dominant (dom′eh-nant) A gene variant (allele) that masks the expression of another. p. 559

dorsal (dors′al) Pertaining to the back surface of a body part. p. 17

dorsal root (dor′sal root) The sensory branch of a spinal nerve by which it joins the spinal cord. p. 248

ductus arteriosus (duk′tus ar-te″re-o′sus) Blood vessel that connects the pulmonary artery and the aorta in a fetus. p. 554

ductus deferentia (duk′tus def″er-en′sha) Tubes that lead from the epididymides to the urethra in the male reproductive tract. p. 508

ductus venosus (duk′tus ven-o′sus) Blood vessel that connects the umbilical vein and the inferior vena cava in a fetus. p. 553

dura mater (du′rah ma′ter) Tough outer layer of the meninges. p. 230

dynamic equilibrium (di-nam′ik e″kwĭ-lib′re-um) Maintenance of balance when the head and body are suddenly moved or rotated. p. 273

E

eardrum Tympanic membrane; a thin membrane that covers the auditory canal and separates the external ear from the middle ear. p. 269

eccrine (ek′rin) **glands** Sweat glands that maintain body temperature. p. 124

ECG Electrocardiogram. p. 350

ectoderm (ek′to-derm) The outermost primary germ layer of the embryo. p. 545

edema (ĕ-de′mah) Fluid accumulation within tissue spaces. p. 64

effectors (e-fek′torz) Muscles or glands that effect changes in the body. p. 214

efferent arteriole (ef′er-ent ar-te′re-ōl) The vessel that conducts blood away from the glomerulus of each kidney nephron. p. 473

ejaculation (e-jak″u-la′shun) Discharge of sperm-containing semen from the male urethra. p. 514

elastic cartilage (ē-las-tik kar′-ti-lij) A flexible type of cartilage with many elastic fibers. p. 106

elastic fibers (e-las′tic fiberz) Stretchy yellow connective tissue fibers consisting of the protein elastin. p. 104

electrocardiogram (e-lek″tro-kar′de-o-gram″) **(ECG)** A recording of the electrical activity associated with the heartbeat. p. 350

electrolyte (e-lek′tro-līt) A substance that ionizes in water solution. p. 38

electrolyte balance (e-lek′tro-līt bal′ans) Condition when electrolytes entering the body equal those leaving it. p. 495

electrons (e-lek′tronz) Small, negatively charged particles that encircle the nucleus of an atom. p. 32

electron transport chain (e-lektron tranz′port chān) Series of metabolic reactions that capture the energy in the bonds of nutrient molecules as ATP. p. 80

elements (el′ĕ-mentz) Pure chemical substances with only one type of atom. p. 31

embolus (em′bo-lus) A blood clot or gas bubble that obstructs a blood vessel. p. 331

embryo (em′bre-o) A prenatal stage of development after primary germ layers form and rudiments of all organs are present. p. 543

emission (e-mish'un) The movement of sperm cells from the ductus deferentia into the ejaculatory ducts and urethra. p. 514

emulsification (e-mul"sĭ-fĭ-ka'shun) Breaking up of fat globules into smaller droplets by the action of bile salts. p. 421

enamel (e-nam'el) Hard covering on the exposed surface of a tooth. p. 408

endocardium (en"do-kar'de-um) Inner lining of the heart chambers. p. 342

endochondral bones (en"do-kon'dral bōnz) Bones that originate as hyaline cartilage and are subsequently replaced by bone tissue. p. 134

endocrine (en'do-krin) **glands** Glands that secrete hormones into the blood. p. 101

endocytosis (en"do-si-to'sis) Process by which a cell membrane envelops a substance and draws it into the cell in a vesicle. p. 65

endoderm (en'do-derm) The innermost primary germ layer in the embryo. p. 545

endolymph (en'do-limf) Fluid in the membranous labyrinth of the inner ear. p. 270

endometrium (en"do-me'tre-um) The inner lining of the uterus. p. 522

endomysium (en"do-mis'e-um) The sheath of connective tissue surrounding each skeletal muscle fiber. p. 177

endoplasmic reticulum (en'do-plaz mik rĕ-tik'u-lum) Organelle composed of a network of connected membranous tubules and vesicles. p. 55

endosteum (en-dos'te-um) Tissue lining the medullary cavity in a bone. p. 132

endothelium (en"do-the'le-um) The layer of epithelial cells that forms the inner lining of blood vessels and heart chambers. p. 353

energy (en'er-je) An ability to move something and thus do work. p. 80

enzymes (en'zīmz) Proteins that catalyze specific biochemical reactions. p. 77

eosinophils (e"o-sin'o-filz) White blood cells containing cytoplasmic granules that stain with acidic dye. p. 323

ependymal cells (ĕ-pen'dĭ-mal selz) Neuroglial cells that line the ventricles of the brain. p. 214

epicardium (ep"ĭ-kar'de-um) The visceral portion of the pericardium on the surface of heart. p. 342

epidermis (ep"i-der'mis) The outer epithelial layer of the skin. p. 117

epididymides (ep"ĭ-did'ĭ-mī dēz) Highly coiled tubules that lead from

the seminiferous tubules of the testes to the ductus deferentia. p. 508

epidural space (ep"ĭ-du'ral spās) The space between the dural sheath of the spinal cord and the bone of the vertebral canal. p. 230

epigastric region (ep"ĭ-gas'trik re'jun) The upper middle portion of the abdomen. p. 16

epiglottis (ep"ĭ-glot'is) Flaplike, cartilaginous structure at the back of the tongue near the entrance to the trachea. p. 448

epimysium (ep"i-mis'e-um) The outer sheath of connective tissue surrounding a skeletal muscle. p. 177

epinephrine (ep"ĭ-nef'rin) A hormone the adrenal medulla secretes during times of stress. p. 302

epiphyseal plate (ep"ĭ-fiz'e-al plāt) Cartilaginous layer between the epiphysis and diaphysis of a long bone that grows, lengthening the bone. p. 134

epiphysis (ĕ-pif'ĭ-sis) The end of a long bone. p. 131

epithelial tissues (ep"ĭ-the'le-al tish'uz) Tissues that cover all external and line all internal body surfaces. Varieties are classified by cell shape (squamous, cuboidal, or columnar) and number of layers (simple, stratified, or pseudostratified). p. 95

equilibrium (e"kwĭ-lib're-um) A state of balance between opposing forces. p. 61

erythrocytes (ĕ-rith'ro-sītz) Red blood cells. p. 319

erythropoietin (ĕ-rith"ro-poi'ĕ-tin) A kidney hormone that promotes red blood cell formation. p. 320

esophagus (ĕ-sof'ah-gus) Tubular portion of the digestive tract that leads from the pharynx to the stomach. p. 410

essential amino acids (ĕ-sen'shal ah-me'no as'idz) Amino acids required for health that body cells cannot synthesize in adequate amounts. p. 432

essential fatty acids (ĕ-sen'shal fat'e as'idz) Fatty acids required for health that body cells cannot synthesize in adequate amounts. p. 431

estrogens (es'tro-jenz) A group of hormones that stimulates the development of female secondary sex characteristics. p. 524

eumelanin (u-mel'ah-nin) A brownish-black pigment. p. 123

eversion (e-ver'zhun) Outward turning movement of the foot so that the sole faces laterally. p. 166

exchange reaction (eks-chānj re-ak'shun) A chemical reaction in which parts of two kinds of molecules trade positions. p. 38

exocrine (ek'so-krin) **glands** Glands that secrete products into ducts or onto body surfaces. p. 101

exocytosis (ek"so-si-tōsis) Transport of substances out of a cell in membrane-bounded vesicles. p. 65

expiration (ek"spĭ-ra'shun) Expulsion of air from the lungs. p. 452

extension (ek-sten'shun) Movement increasing the angle between parts at a joint. p. 165

extracellular fluid compartment (eks"trah-sel'u-lar floo'id cum-partment) Fluid outside cells; includes plasma and interstitial fluid. p. 492

extracellular matrix (eks"trah-sel'u-lar ma'triks) Molecules that fill spaces between cells, consisting mostly of protein fiber networks. p. 102

extrapyramidal tract (ek"strah-pĭ-ram'ĭ-dal trakt) Nerve tracts, other than the corticospinal tracts, that transmit impulses from the cerebral cortex to the spinal cord. p. 233

F

facilitated diffusion (fah-sil"ĭ-tāt'ed dĭ-fu'zhun) Diffusion in which carrier molecules transport substances across membranes from a region of higher concentration to a region of lower concentration. p. 62

facilitation (fah-sil"ĭ-ta'shun) Subthreshold stimulation of a neuron that increases responsiveness to further stimulation. p. 227

fascia (fash'e-ah) A sheet of fibrous connective tissue that encloses a muscle. p. 177

fat (fat) Adipose tissue; an organic molecule that consists of glycerol and fatty acids. p. 41

fatty acids (fat'e as'idz) Building blocks of fat molecules. p. 41

feces (fe'sēz) Material expelled from the digestive tract during defecation. p. 429

fertilization (fer"tĭ-lĭ-za'shun) The union of a secondary oocyte and a sperm cell. p. 539

fetus (fe'tus) A prenatal human after eight weeks of development. p. 543

fibrin (fi'brin) Insoluble, fibrous protein formed from fibrinogen during blood coagulation. p. 330

fibrinogen (fi-brin'o-jen) Plasma protein converted into fibrin during blood coagulation. p. 326

fibroblasts (fi'bro-blasts) Cells that produce fibers in connective tissues. p. 102

fibrocartilage (fi'bro kar'tĭ-lij) A tough connective tissue with many collagenous fibers. p. 106

fibrous joints (fi'brus joints) Two or more bones joined by fibrous connective tissue. p. 162

filtration (fil-tra'shun) Movement of small molecules through a membrane by hydrostatic pressure, while large molecules are restricted. p. 63

fissure (fish'ur) A narrow cleft that separates parts, such as the lobes of the cerebrum. p. 236

flagellum (fla-jel'um) Motile tail-like cellular structure such as on a sperm cell. p. 58

flexion (flek'shun) Bending at a joint that decreases the angle between bones. p. 165

follicle-stimulating hormone (fol'ĭ-kl stim'u-la"ting hor'mōn) **(FSH)** A hormone that the anterior pituitary secretes to stimulate follicular development in a female or sperm cell production in a male. p. 298

follicular cells (fŏ-lik'u-lar selz) Ovarian cells that surround a developing oocyte and secrete female sex hormones. p. 000

fontanel (fon"tah-nel') Membranous region between certain cranial bones in the skull of a fetus or infant. p. 147

foramen magnum (fo-ra'men mag'num) Opening in the occipital bone of the skull through which the spinal cord passes. p. 142

foramen ovale (fo-ra'men o-val'e) Opening in the interatrial septum of the fetal heart. p. 553

fovea centralis (fo've-ah sen-tral'is) Region of the retina, consisting of densely packed cones, that provides the greatest visual acuity. p. 281

free nerve endings (frē nerv end-ingz) Receptors abundant in epithelium, associated with sensing temperature and pain. p. 262

frontal (frun'tal) Pertaining to the forehead. p. 17

FSH Follicle-stimulating hormone. p. 298

functional syncytium (funk'shun-al sin-sish'e-um) Merging cells that function as a unit; those of the heart are joined electrically. p. 348

G

gallbladder (gawl'blad-er) Saclike organ associated with the liver that stores and concentrates bile. p. 419

ganglia (gang'glē-ah) (singular, *ganglion*) Masses of neuron cell bodies, usually outside the central nervous system. p. 218

gastric glands (gas'trik glandz) Glands in the stomach wall that secrete gastric juice. p. 412

gastric juice (gas'trik jōōs) Secretion of the gastric glands in the stomach. p. 412

gastrin (gas'trin) Hormone that the stomach lining secretes to stimulate gastric juice secretion. p. 443

gene (jēn) Part of a DNA molecule that encodes information to synthesize a protein, a control sequence, or tRNA or rRNA; the unit of inheritance. p. 84

genetic code (jĕ-net'ik kōd) Correspondence between DNA base triplets and particular amino acids. p. 85

genome (je'nōm) The complete set of genetic instructions in a cell. p. 84

genotype (jē-no-tīp) The alleles of a particular individual. p. 000

GH Growth hormone. p. 296

gland (gland) A group of cells that secrete a product. p. 101

globin (glo'bin) The protein part of a hemoglobin molecule. p. 322

globulins (glob'u-linz) Types of proteins in blood plasma. p. 326

glomerular capsule (glo-mer'u-lar kap'sūl) Double-walled enclosure of the glomerulus of a nephron; Bowman's capsule. p. 472

glomerulus (glo-mer'u-lus) A capillary tuft in the glomerular capsule of a nephron. p. 472

glottis (glot'is) Slitlike opening between the true vocal folds or vocal cords. p. 448

glucagon (gloo'kah-gon) Hormone secreted by the pancreatic islets that releases stored glucose. p. 306

glucocorticoid (gloo"ko-kor'tĭ-koid) Any of several hormones that the adrenal cortex secretes that affects carbohydrate, fat, and protein metabolism. p. 304

glucose (gloo'kōs) A monosaccharide in blood that is the primary source of cellular energy. p. 37

gluteal (gloo'te-al) Pertaining to the buttocks. p. 17

glycerol (glis'er-ol) An organic compound that is a building block for fat molecules. p. 41

glycogen (gli'ko-jen) A polysaccharide that stores glucose in the liver and muscles. p. 430

glycolysis (gli-kol'ĭ-sis) The energy-releasing breakdown of glucose to produce 2 pyruvic acid molecules. p. 80

goblet cell (gob'let sel) An epithelial cell specialized to secrete mucus in the respiratory tract and intestines. p. 97

Golgi apparatus (gol'je ap"ah-ra'tus) An organelle that prepares and modifies cellular products for secretion. p. 55

gonadotropins (go-nad"o-trōp'inz) Hormones that stimulate activity in the gonads (testes and ovaries). p. 515

granulocytes (gran'u-lo-sītz) Leukocytes with granules in the cytoplasm. p. 323

growth hormone (grōth hor'mōn) **(GH)** A hormone of the anterior pituitary that promotes growth of the organism; somatotropin. p. 296

gyrus (ji'rus) Elevations on a surface that form as the structure infolds; convolution. p. 235

H

hair follicle (hār fol'i-kl) Tubelike depression in the skin in which a hair develops. p. 122

haploid (hap'loyd) A cell with one copy of each chromosome. p. 510

hapten (hap'ten) A small molecule that combines with a larger one, forming an antigen. p. 385

hematocrit (he-mat'o-krit) The volume percentage of red blood cells in a sample of whole blood. p. 318

hematopoiesis (he"-mă-to-poi-e'sis) The production of blood cells from dividing stem and progenitor cells. p. 138

heme (hēm) The iron-containing portion of a hemoglobin molecule. p. 322

hemoglobin (he"mo-glo'bin) Oxygen-carrying pigment in red blood cells. p. 138

hemostasis (he"mo-sta'sis) The stoppage of bleeding. p. 328

hepatic lobule (hĕ-pat'ik lob'ul) A functional unit of the liver. p. 416

heterozygous (het'er-o-zi'gus) The presence of two different alleles (variants) of a gene in an individual. p. 559

holocrine (ho'lo-krin) **glands** Glands whose secretion contains entire secretory cells. p. 101

homeostasis (ho"me-o-sta'sis) A state of balance in which the body's internal environment remains in the normal range. p. 5

homozygous (ho'mo-zi'gus) The presence of identical alleles (variants) of a gene in an individual. p. 559

hormones (hor'mōnz) Substances secreted by an endocrine gland and transported in the blood. p. 290

humoral immune response (hu′mor-al i-mūn′ ri-spons′) Circulating antibodies' destruction of cells bearing foreign (nonself) antigens. p. 385

hyaline cartilage (hī-ah-lin kar′ti-lij) The most abundant type of cartilage, with many thin collagenous fibers. p. 106

hydrogen bond (hi′dro-jen bond) A weak chemical attraction between a hydrogen atom and an atom of oxygen or nitrogen. p. 36

hydrolysis (hi-drol′ĭ-sis) Enzymatically adding a water molecule to split a molecule into smaller portions. p. 78

hydrostatic pressure (hy″dro-stat′ik presh′ur) Pressure exerted by fluids, such as blood pressure. p. 5

hymen (hi′men) A membranous fold of tissue that partially covers the vaginal opening. p. 522

hyperplasia (hi′per-pla′ze-ah) Excess cell division in a tissue. p. 70

hypertonic (hi″per-ton′ik) A solution with a greater osmotic pressure than the solution (usually body fluids) with which it is compared. p. 83

hyperventilation (hi″per-ven″tĭ-la′shun) Abnormally deep and prolonged breathing. p. 460

hypochondriac region (hi′po-kon′dre-ak re′jun) The portion of the abdomen on either side of the upper middle or epigastric region. p. 16

hypogastric region (hi′po-gas′trik re′jun) The lower middle portion of the abdomen. p. 16

hypothalamus (hi′po-thal′ah-mus) Part of the brain below the thalamus and forming the floor of the third ventricle. p. 240

hypotonic (hi′po-ton′ik) A solution with a lower osmotic pressure than the solution (usually body fluids) with which it is compared. p. 63

I

iliac region (il′e-ak re′jun) Part of the abdomen on either side of the lower middle or hypogastric region. p. 16

ilium (il′e-um) One of the bones making up the hipbone. p. 157

immunity (ĭ-mu′nĭ-te) Resistance to the effects of specific disease-causing agents. p. 325

immunoglobulins (im″u-no-glob′u-linz) Globular plasma proteins that function as antibodies. p. 385

incomplete protein (in″kom-plēt′ pro′te-in) A protein with inadequate amounts of essential amino acids. p. 433

inferior (in-fer′e-or) Below something else; pertaining to the lower surface of a part. p. 14

inflammation (in″flah-ma′shun) A tissue response to stress that dilates blood vessels and accumulates fluid in the affected region. p. 384

inguinal (ing′gwĭ-nal) Pertaining to the groin region. p. 17

innate defenses (in′ate dē-fens-ez) Nonspecific immune defenses that block entry of or destroy pathogens. p. 383

inorganic (in″or-gan′ik) Chemical substances that do not include carbon and hydrogen. p. 39

insertion (in-ser′shun) The end of a muscle attached to a movable part. p. 192

inspiration (in″spĭ-ra′shun) Breathing in; inhalation. p. 452

insula (in′su-lah) A cerebral lobe deep within the lateral sulcus. p. 236

insulin (in′su-lin) A hormone secreted by beta cells in pancreatic islets that stimulates cells to take up glucose. p. 306

integumentary (in-teg-u-men′tar-e) **system** The skin and its accessory structures. p. 12

intercalated disc (in-ter″kah-lāt′ed disk) Membranous boundary between cardiac muscle cells. p. 111

internal environment (in-ter′nĕl en-vi-ruhment) The fluid surrounding body cells. p. 5

interneuron (in″ter-nu′ron) A neuron between a sensory neuron and a motor neuron; internuncial or association neuron. p. 218

interphase (in′ter-fāz) Period between cell divisions when a cell metabolizes and prepares to divide. p. 67

interstitial cell (in″ter-stish′al sel) A hormone-secreting cell between seminiferous tubules of the testis. p. 508

intervertebral disc (in″ter-ver′tĕ-bral disk) A layer of fibrocartilage between bodies of adjacent vertebrae. p. 140

intestinal glands (in-tes′tĭ-nal glandz) Tubular glands at the base of villi in the intestinal wall. p. 422

intestinal villi (in-tes′tĭ-nal vil′ī) (singular, *villus*) Fingerlike extensions of the small intestinal lining. p. 422

intracellular fluid compartment (in″trah-sel′u-lar floo′id cum-part′ment) Fluid in cells. p. 492

intramembranous bones (in″trah-mem′brah-nus bōnz) Bones that form from membranelike layers of primitive connective tissue. p. 133

intrinsic factor (in-trin′sik fak′tor) A substance produced by the gastric glands that promotes absorption of vitamin B12. p. 412

inversion (in-ver′zhun) Inward turning movement of the foot so that the sole faces medially. p. 166

ions (i′onz) Electrically charged atoms or molecules. p. 33

ionic bond (i-on′ik bond) A chemical attraction between two ions by transfer of electrons. p. 35

iris (i′ris) Colored, muscular portion of the eye that surrounds the pupil and regulates its size. p. 280

isotonic (i″so-ton′ik) A solution with the same osmotic pressure as the solution (usually body fluids) with which it is compared. p. 63

isotopes (i′so-tōpz) Atoms that have the same number of protons as other atoms of the same element but a different number of neutrons. p. 33

J

joint The union of two or more bones; articulation. p. 161

juxtaglomerular apparatus (juks″tah-glo-mer′u-lar ap″ah-ra′tus) Structure in the arteriolar walls near the glomerulus that regulates renal blood flow. p. 473

K

keratin (ker′ah-tin) Protein in epidermis, hair, and nails. p. 98

keratinization (ker″ah-tin′i-za′shun) The process by which cells form fibrils of keratin and harden. p. 118

ketone bodies (ke′tōn bod′ēz) Compounds produced in fat catabolism. p. 431

Kupffer cells (koop′fer selz) Large, fixed phagocytes in the liver that remove bacterial cells from the blood. p. 370

L

labyrinth (lab′ĭ-rinth) The system of connecting tubes in the inner ear, including the cochlea, vestibule, and semicircular canals. p. 270

lacrimal gland (lak′rĭ-mal gland) Tear-secreting gland. p. 276

lactase (lak′tās) Enzyme that catalyzes the breakdown of lactose into glucose and galactose. p. 422

lacteals (lak′te-alz) Lymphatic capillaries associated with villi of the small intestine. p. 377

lactic acid (lak'tik as'id) An organic compound formed from pyruvic acid during anaerobic respiration. p. 184

lacunae (lah-ku'nā) Hollow cavities. p. 132

lamellae (lah-mel'ā) Layers of matrix surrounding the central canal of an osteon. p. 108

laryngopharynx (lah-ring″go-far'ingks) The lower portion of the pharynx extending posterior to the larynx. p. 410

larynx (lar'inks) Structure between the pharynx and trachea that houses the vocal cords. p. 447

latent period (la'tent pe're-od) Time between application of a stimulus and the beginning of a response in a muscle fiber. p. 187

lateral (lat'er-al) Pertaining to the side. p. 15

leukocytes (lu'ko-sītz) White blood cells. p. 323

levers (lev'erz) Simple mechanical devices consisting of a rod, fulcrum, weight, and a source of energy that is applied to some point on the rod. p. 138

ligaments (lig'ah-mentz) Cords or sheets of connective tissue binding two or more bones at a joint. p. 104

limbic system (lim'bik sis'tem) A group of connected structures in the brain that produces emotional feelings. p. 240

lingual frenulum (ling'gwahl fren'u-lum) Pertaining to the tongue. p. 404

lipase (li'pās) A fat-digesting enzyme. p. 424

lipids (lip'idz) Fats, oils, or fatlike compounds that usually include fatty acids. p. 41

lumbar (lum'bar) Pertaining to the region of the loins. p. 16

lumen (lu'men) Space inside a tubular structure, such as a blood vessel or intestine. p. 402

luteinizing hormone (lu'te-in-īz″ing hor'mōn) (**LH; ICSH** in males) A hormone that the anterior pituitary secretes that controls formation of the corpus luteum in females and testosterone secretion in males. p. 298

lymph (limf) Fluid carried in lymphatic vessels. p. 378

lymphatic pathway (lim-fat'ik path'wa) A pattern of connected vessels that transport lymph. p. 377

lymph node (limf nōd) A mass of lymphoid tissue. p. 378

lymphocytes (lim'fo-sītz) Type of white blood cells that provide immunity; B cell or T cell. p. 323

lysosome (li'so-sōm) Organelle that contains digestive enzymes. p. 56

M

macromolecules (mak-rō mol'ĕ-kūlz) Large molecules, such as proteins or nucleic acids. p. 3

macronutrients (mak-rō nu'tree-entz) The nutrients (carbohydrates, lipids, and proteins) that are required in large amounts. p. 429

macrophages (mak'ro-fājez) Large phagocytic cells. p. 104

macula lutea (mak'u-lah lu'te-ah) A yellowish depression in the retina where acute vision arises. p. 281

malnutrition (mal″nu-trish'un) Symptoms resulting from lack of specific nutrients. p. 436

mammary (mam'er-e) Pertaining to the breast. p. 17

marrow (mar'o) Connective tissue in spaces in bones that includes blood-forming stem and progenitor cells. p. 132

mast cells (mast selz) Cells to which antibodies, formed in response to allergens, attach, bursting the cells and releasing allergy mediators, which cause symptoms. p. 104

matter (mat'er) Anything that has weight and occupies space. p. 31

mechanoreceptors (mek″ah-no-re-sep'torz) Sensory receptors that sense mechanical stimulation, such as changes in pressure or tension. p. 261

medial (me'de-al) Toward or near the midline. p. 14

mediastinum (me″de-as-ti'num) Tissues and organs of the thoracic cavity that form a septum between the lungs. p. 8

medulla oblongata (mĕ-dul'ah ob″long-gah'tah) Part of the brainstem between the pons and the spinal cord. p. 241

medullary cavity (med'u-lār″e kav'ĭ-te) Cavity containing red or yellow marrow within the diaphysis of a long bone. p. 132

megakaryocytes (meg″ah-kar'e-o-sītz) Large cells in red bone marrow that shatter, giving rise to blood platelets. p. 323

meiosis (mi-o'sis) A form of cell division that halves the genetic material, resulting in an oocyte or four sperm cells. p. 68

melanin (mel'ah-nin) Dark pigment normally found in skin and hair. p. 119

melanocyte (mel'ah-no-sīt) Melanin-producing cell. p. 119

melatonin (mel″ah-to'nin) A hormone that the pineal gland secretes that controls daily rhythms. p. 307

memory cells (mem'o-re selz) B cells or T cells produced in the primary immune response that respond rapidly if the same antigen is encountered again. p. 388

menarche (mĕ-nar'kē) A female's first reproductive cycle p. 524

meninges (mĕ-nin'jēz) (singular, *meninx*) Membranes that cover the brain and spinal cord. p. 230

menisci (mĕ-nis'ke) (singular, *meniscus*) Fibrocartilages that separate the articulating surfaces of bones in the knees. p. 163

menopause (men'o-pawz) Termination of a woman's reproductive cycle. p. 526

merocrine (mer'o-krin) **gland** A gland whose cells secrete a fluid without losing cytoplasm. p. 101

mesentery (mes'en-ter″e) A fold of peritoneal membrane that attaches an abdominal organ to the abdominal wall. p. 421

mesoderm (mez'o-derm) The middle primary germ layer. p. 545

messenger RNA (mes'in-jer) RNA that transmits information for a protein's amino acid sequence from the nucleus to the cytoplasm. p. 86

metabolism (mĕ-tab'o-lizm) All of the chemical reactions in cells that break down or build up substances. p. 4

metacarpal (met″ah-kar'pal) A bone between the wrist and finger bones. p. 161

metaphase (met'ah-fāz) Stage in mitosis when chromosomes align in the middle of the cell. p. 68

metatarsal (met″ah-tar'sal) A bone between the ankle and the toe bones. p. 161

microfilament (mi″kro-fil'ah-ment) A tiny rod of actin protein in the cytoplasm that provides structural support or movement. p. 57

microglial cell (mi-krog'le-al sel) A neuroglial cell that supports neurons and phagocytizes. p. 214

micronutrients (mi-kro nu'tree-entz) Nutrients (vitamins and minerals) required in small amounts. p. 429

microtubule (mi″kro-tu'būl) A minute, hollow rod constructed of many molecules of the protein tubulin. p. 57

microvilli (mi″kro-vil'i) Tiny, cylindrical processes that extend from some epithelial cells, increasing membrane surface area. p. 479

micturition (mik″tu-rish'un) Urination. p. 485

midbrain (mid'brān) A small region of the brainstem between the diencephalon and the pons. p. 241

minerals (min′er-alz) Inorganic element essential in human metabolism. p. 435

mineralocorticoid (min″er-al-o-kor′tĭ-koid) Any of a group of hormones that the adrenal cortex secretes that affects electrolyte concentrations in body fluids. p. 304

mitochondria (mi″to-kon′dre-ah) (singular, *mitochondrion*) Organelles housing enzymes that catalyze the reactions of aerobic respiration. p. 58

mitosis (mi-to′sis) A form of cell division that produces two somatic cells with identical chromosome numbers as the original somatic cell. p. 68

mitral valve (mi′trul valv) Heart valve between the left atrium and the left ventricle; bicuspid valve. p. 343

mixed nerve (mikst nerv) Nerve that includes both sensory and motor neuron fibers. p. 228

molecular formula (mo-lek′u-lar fōr′mu-lah) An abbreviation for the number of atoms of each element in a compound. p. 37

molecule (mol′ĕ-kūl) A particle composed of two or more joined atoms. p. 3

monocyte (mon′o-sīt) A type of white blood cell that is a phagocyte. p. 323

monosaccharides (mon″o-sak′ah-rīdz) Simple sugars, such as glucose or fructose. p. 41

motor area (mo′tor a′re-ah) The region of the brain from which impulses to muscles or glands originate. p. 237

motor end plate Specialized part of a muscle fiber membrane at a neuromuscular junction. p. 180

motor nerve (mo′tor nerv) A nerve that consists of motor neuron axons. p. 228

motor neuron (mo′tor nu′ron) A neuron that transmits impulses from the central nervous system to an effector. p. 180

motor speech area (mo′tor spēch ār′e-ah) Region of the frontal lobe that coordinates complex muscular actions of mouth, tongue, and larynx, making speech possible; Broca's area. p. 237

motor unit (mo′tor u′nit) A motor neuron and its associated muscle fibers. p. 180

mucosa (mu-ko′sah) Innermost layer of the alimentary canal. p. 402

mucous cells (mu′kus selz) Glandular cells that secrete mucus. p. 101

mucous membrane (mu′kus mem-brān) A membrane that lines cavities and tubes that open to the outside of the body. p. 110

mucus (mu′kus) Fluid secretion of the mucous cells. p. 101

muscle impulse (mus′el im′puls) Impulse that travels along the sarcolemma to the transverse tubules. p. 181

muscle tissue (mus′el tish′u) Contractile tissue consisting of filaments of actin and myosin, which slide past each other, shortening cells. p. 110

myelin (mi′ĕ-lin) Fatty material that forms a sheathlike covering around some axons. p. 214

myelin sheaths (mi′ĕ-lin shēthz) Fatty layers formed from certain neuroglia that surround axons, providing insulation. p. 215

myocardium (mi″o-kar′de-um) Muscle tissue of the heart. p. 342

myofibrils (mi″o-fi′brilz) Contractile fibers in muscle cells. p. 177

myoglobin (mi″o-glo′bin) A pigmented protein in muscle that carries oxygen. p. 184

myometrium (mi″o-me′tre-um) The layer of smooth muscle tissue in the uterine wall. p. 522

myosin (mi′o-sin) A protein that, with actin, forms the filaments that contract muscle fibers. p. 177

N

nasal cavity (na′zal kav′ĭ-te) Space in the nose. p. 8

nasal concha (na′zal kong′kah) Shell-like bone extending out from the wall of the nasal cavity; a turbinate bone. p. 445

nasal septum (na′zal sep′tum) A wall of bone and cartilage that separates the nasal cavity into two parts. p. 445

nasopharynx (na″zo-far′inks) Part of the pharynx in the posterior part of the nasal cavity. p. 410

negative feedback (neg′ah-tiv fēd′bak) A mechanism activated by an imbalance that corrects the imbalance. p. 6

neonatal (ne″o-na′tal) The first four weeks of life. p. 556

nephrons (nef′ronz) Functional units of a kidney, consisting of renal corpuscles and renal tubules. p. 471

nerve (nerv) A bundle of axons in the peripheral nervous system. p. 212

nerve impulse (nerv im′puls) Depolarization and repolarization along an axon. p. 212

nervous tissue (ner′vus tish′u) Neurons and neuroglia. p. 111

neurilemma (nu″rĭ-lem′ah) Sheath formed from Schwann cells on the exterior of some axons. p. 215

neurofibrils (nu″ro-fi′brilz) Fine, cytoplasmic threads that extend from the cell bodies into the processes of neurons. p. 214

neuroglial cells (nu-rog′le-ahl selz) Specialized cells of the nervous system that produce myelin, communicate between cells, maintain the ionic environment, and nurture the differentiation of neurons. p. 111

neuromuscular junction (nu″ro-mus′ku-lar jungk′shun) Synapse between a motor neuron and a skeletal muscle fiber. p. 180

neurons (nu′ronz) Nerve cells. p. 111

neurotransmitters (nu″ro-trans′mit-erz) Chemicals that axons secrete on effectors (muscles or glands) or other neurons. p. 220

neutral (nu′tral) A chemical that is neither acidic nor alkaline; a chemical that is pH7. p. 39

neutrons (nu′tronz) Electrically neutral subatomic particles. p. 32

neutrophil (nu′tro-fil) A type of phagocytic leukocyte. p. 323

nonelectrolyte (non″e-lek′tro-līt) A substance that does not dissociate into ions when dissolved in water. p. 40

nonprotein nitrogenous substance (non-pro′te-in ni-troj′ĕ-nus sub′stans) A substance, such as urea or uric acid, that contains nitrogen but is not a protein. p. 327

norepinephrine (nor″ep-ĭ-nef′rin) A neurotransmitter released from the terminals of some nerve cells. p. 302

nuclease (nu′kle-ās) An enzyme that catalyzes decomposition of nucleic acids. p. 32

nuclei (nu′kle-ī) (singular, *nucleus*) The dense cores of atoms, composed of protons and neutrons, p. 32. Cellular organelles enclosed by double-layered selective membranes that contain the genetic materials, p. 52. Masses of interneuron cell bodies in the central nervous system. p. 218.

nucleic acid (nu-kle′ik as′id) A molecule that is a polymer of nucleotides; RNA or DNA. p. 46

nucleoli (nu-kle′o-lī) (singular, *nucleolus*) Small structures in the cell nucleus that contain RNA and proteins. p. 59

nucleotide (nu′kle-o-tīd″) A building block of a nucleic acid molecule, consisting of a sugar, a nitrogenous base, and a phosphate group. p. 46

nutrient (nu′tre-ent) A chemical that the body requires from the environment. p. 429

O

occipital (ok-sip′ĭ-tal) Pertaining to the lower, back portion of the head. p. 142

olfactory (ol-fak′to-re) Pertaining to the sense of smell. p. 266

olfactory nerves (ol-fak′to-re nervz) The first pair of cranial nerves, which conduct impulses associated with the sense of smell. p. 243

oligodendrocyte (ol″ĭ-go-den′dro-sīt) A type of neuroglial cell that forms myelin. p. 214

oocytes (o′o-sīts) Cells formed by oogenesis. Egg cells. p. 508

oogenesis (o″o-jen′ĕ-sis) Formation of an oocyte (egg cell). p. 519

optic chiasma (op′tik ki-az′mah) X-shaped structure on the underside of the brain formed by optic nerve fibers (axons) that partially cross over. p. 240

optic disc (op′tik disk) Region in the retina where nerve fibers (axons) exit, becoming part of the optic nerve. p. 281

oral (o′ral) Pertaining to the mouth. p. 8

organ (or′gan) A structure consisting of a group of tissues with a specialized function. p. 3

organelle (or″gah-nel′) A structure in a cell that has a specialized function. p. 3

organic (or-gan′ik) Carbon-containing molecules. p. 39

organism (or′gah-nizm) An individual living thing. p. 3

organ system (or′gan sis′tem) A group of organs coordinated to carry on a specialized function. p. 3

orgasm (or′gazm) The culmination of sexual excitement. p. 514

origin (or′ĭ-jin) The end of a muscle that attaches to a relatively immovable part. p. 192

oropharynx (o″ro-far′inks) Part of the pharynx in the posterior part of the oral cavity. p. 410

osmosis (oz-mo′sis) Movement of water through a semipermeable membrane from an area of greater water concentration to an area of lesser water concentration. p. 63

osmotic pressure (oz-mot′ik presh′ur) The amount of pressure needed to stop osmosis; a solution's potential pressure caused by nondiffusible (impermeant) solute particles in the solution. p. 493

ossification (os″ĭ-fĭ-ka′shun) The formation of bone tissue. p. 134

osteoblasts (os′te-o-blastz″) Bone-forming cells. p. 134

osteoclasts (os′te-o-klastz″) Cells that erode bone. p. 134

osteocytes (os′te-o-sīts) Mature bone cells. p. 108

osteon (os′te-on) A cylinder-shaped unit including bone cells that surround a central canal; Haversian system. p. 108

oval window Opening between the stapes and the inner ear. p. 269

ovaries (o′vah-rēz) Paired female reproductive organs that produce oocytes. p. 518

ovulation (o″vu-la′shun) The release of an oocyte from a mature ovarian follicle. p. 520

oxidation (ok″sĭ-da′shun) Process by which oxygen combines with another chemical; the removal of hydrogen or the loss of electrons; opposite of reduction. p. 80

oxygen debt (ok′sĭ-jen det) The amount of oxygen required after physical exercise to convert accumulated lactic acid to glucose. p. 184

oxyhemoglobin (ok″sĭ-he″mo-glo′bin) A hemoglobin molecule that has bound an oxygen atom. p. 461

oxytocin (ok″sĭ-to′sin) A hormone released from the posterior pituitary that contracts smooth muscles in the uterus and mammary glands. p. 298

P

pacemaker (pās′māk-er) Mass of specialized cardiac muscle tissue that controls the rhythm of the heartbeat; sinoatrial node. p. 349

pain receptors (pān re″sep′torz) Sensory nerve endings associated with pain. p. 261

palate (pal′at) The roof of the mouth. p. 405

palatine (pal′ah-tīn) Pertaining to the palate. p. 405

palmar (pahl′mar) Pertaining to the palm of the hand. p. 17

pancreas (pan′kre-as) Glandular organ in the abdominal cavity that secretes hormones and digestive enzymes. p. 414

pancreatic juice (pan″kre-at′ik jōōs) Digestive secretions of the pancreas. p. 414

papillae (pah-pil′ā) Tiny, nipplelike projections. p. 404

papillary muscle (pap′ĭ-ler″e mus′l) Muscle that extends inward from the ventricular walls of the heart and to which the chordae tendineae attach. p. 343

paranasal sinuses (par″ah-na′zal si-nusez) Air-filled cavities in a cranial or facial bone lined with mucous membrane and connected to the nasal cavity. p. 446

parasympathetic division (par″ah-sim″ pah-thet′ik de-vijh′in) Part of the autonomic nervous system that arises from the brain and sacral region of the spinal cord. p. 249

parathyroid glands (par″ah-thi′roid glandz) Four small endocrine glands embedded in the posterior portion of the thyroid gland. p. 301

parathyroid hormone (par″ah-thi′roid hor′mōn) **(PTH)** Hormone that the parathyroid glands secrete that regulates the levels of blood calcium and phosphate ions. p. 301

parietal (pah-ri′ĕ-tal) Pertaining to the wall of a cavity. p. 10

parietal cells (pah-ri′ĕ-tal selz) Cells of a gastric gland that secrete hydrochloric acid and intrinsic factor. p. 412

parietal pericardium (pah-ri′ĕ-tal per″ĭ-kar′de-um) Membrane that lines the inner wall of the pericardial cavity. p. 10

parietal peritoneum (pah-ri′ĕ-tal per″ĭ-to-ne′-um) Membrane that lines the inner wall of the peritoneal cavity. p. 10

parietal pleura (pah-ri′ĕ-tal ploo′rah) Membrane that lines the inner wall of the thoracic cavity. p. 451

parotid glands (pah-rot′id glandz) Large salivary glands on the sides of the face just in front and below the ears. p. 409

partial pressure (par′shal presh′ur) The pressure one gas produces in a mixture of gases. p. 460

pathogen (path′o-jen) A disease-causing agent. p. 383

pectoral (pek′tor-al) Pertaining to the chest. p. 17

pectoral girdle (pek′tor-al ger′dl) Part of the skeleton that supports and attaches the upper limbs. p. 141

pedigree (ped′eh-gree) A chart that displays how members of a family are related and which hereditary traits or disorders they have. p. 559

pelvic cavity (pel′vik kav′ĭ-te) The space between the hipbones that encloses the terminal portion of the large intestine, the urinary bladder, and the internal reproductive organs. p. 8

pelvic girdle (pel′vik ger′dl) Part of the skeleton to which the lower limbs attach. p. 141

pelvis (pel′vis) Bony ring formed by the sacrum and hipbones. p. 157

penis (pe′nis) Male external reproductive organ through which the urethra passes. p. 514

pepsin (pep′sin) Protein-splitting enzyme that the gastric glands secrete. p. 412

pepsinogen (pep-sin′o-jen) Inactive form of pepsin. p. 412

pericardial cavity (per″ĭ-kar′de-al kav′ĭ-te) The space between the visceral and parietal pericardial membranes. p. 10

pericardium (per″ĭ-kar′de-um) Serous membrane that surrounds the heart. p. 341

perichondrium (per″ĭ-kon′dre-um) Layer of fibrous connective tissue that encloses cartilaginous structures. p. 106

perilymph (per′ĭ-limf) Fluid in the space between the membranous and osseous labyrinths of the inner ear. p. 270

perimetrium (per-ĭ-me′tre-um) The outer serosal layer of the uterine wall. p. 522

perimysium (per″i-mis′e-um) Sheath of connective tissue that encloses a bundle of skeletal muscle fibers. p. 177

periodontal ligament (per″e-o-don′tal lig′ah-ment) Fibrous membrane around a tooth that attaches it to the jawbone. p. 408

periosteum (per″e-os′te-um) Fibrous connective tissue covering on the surface of a bone. p. 131

peripheral nervous system (pĕ-rif′er-al ner′vus sis′tem) **(PNS)** Part of the nervous system outside the central nervous system. p. 212

peripheral resistance (pĕ-rif′er-al re-zis′tans) Resistance to blood flow due to friction between the blood and the walls of the blood vessels. p. 350

peristalsis (per″ĭ-stal′sis) Rhythmic waves of muscular contraction in the walls of certain tubular organs. p. 404

peritoneal cavity (per″ĭ-to-ne′al kav′ĭ-te) The space between the visceral and parietal peritoneal membranes. p. 10

peritubular capillary (per″ĭ-tu′bu-lar kap′ĭ-ler″e) Capillary that surrounds a renal tubule and functions in tubular reabsorption and tubular secretion during urine formation. p. 473

peroxisomes (pĕ-roks′ĭ-sōmz) Membranous sacs abundant in kidney and liver cells that contain enzymes that catalyze a variety of biochemical reactions. p. 56

pH (pH) The negative logarithm of the hydrogen ion concentration used to indicate the acidic or alkaline condition of a solution; values range from 0 to 14. p. 39

phagocytosis (fag″o-si-to′sis) Process by which a cell engulfs and digests solids. p. 65

phalanx (fa′langks) (plural, *phalanges*) Bone of a finger or toe. p. 161

pharynx (far′inks) Part of the digestive tube posterior to the nasal and oral cavities, as well as the larynx. p. 447

phenotype (fe′no-tīp) The expression of a gene variant or gene combination. p. 559

pheomelanin (fe″o-mel′ah-nīn) A reddish-yellow pigment p. 123

phospholipid (fos″fo-lip′id) A lipid that includes two fatty acid molecules and a phosphate group bound to a glycerol molecule. p. 42

photoreceptors (fo″to-re-sep′torz) Sensory receptors sensitive to light energy; rods and cones of the eyes. p. 261

physiology (fiz″e-ol′o-je) The study of body functions. p. 3

pia mater (pi′ah ma′ter) Inner layer of meninges that encloses the brain and spinal cord. p. 231

pineal gland (pin′e-al gland) A small structure in the central brain that secretes the hormone melatonin, which controls certain biological rhythms. p. 240

pinocytosis (pi″no-si-to′sis) Process by which a cell engulfs droplets of fluid from its surroundings. p. 65

pituitary gland (pĭ-tu′ĭ-tār′e gland) Endocrine gland attached to the base of the brain consisting of anterior and posterior lobes. p. 295

placenta (plah-sen′tah) An organ that attaches the fetus to the uterine wall, providing for delivery of nutrients to and removal of wastes from the fetus. p. 308

plantar (plan′tar) Pertaining to the sole of the foot. p. 17

plasma (plaz′mah) Fluid portion of circulating blood. p. 318

plasma cells (plaz′mah selz) Antibody-producing cells that form when activated B cells proliferate. p. 385

plasma proteins (plaz′mah pro′te-inz) Proteins dissolved in blood plasma. p. 325

platelets (plāt′letz) Cytoplasmic fragments formed in red bone marrow that help blood clot. p. 325

pleural cavity (ploo′ral kav′ĭ-te) Potential space between pleural membranes. p. 10

pleural membranes (ploo′ral mem′brānz) Serous membranes that enclose the lungs and line the chest wall. p. 10

plexus (plek′sus) A network of interlaced nerves or blood vessels. p. 248

polar (pō′lar) A molecule in which charge distribution is uneven. p. 35

polarization (po″lar-ĭ-za′shun) An electrical charge on a cell membrane surface due to an unequal distribution of positive and negative ions on either side of the membrane. p. 220

polysaccharides (pol″e-sak′ah-rīdz) Carbohydrates composed of many bonded monosaccharides. p. 41

polyunsaturated (pol″e-un-sach′ĕ-ra-ted) **fatty acids** Fatty acids with more than one double carbon bond. p. 42

pons (ponz) Part of the brainstem above the medulla oblongata and below the midbrain. p. 241

popliteal (pop″lĭ-te′al) Pertaining to the region behind the knee. p. 17

positive feedback system (poz′ĭ-tiv fēd′bak sis′tem) Process by which changes cause additional similar changes, producing unstable conditions. p. 7

posterior (pos-tēr′e-or) Toward the back; the opposite of *anterior*. p. 14

posterior pituitary (pos-tēr′e-or pĭ-tu′ĭ-tār″e) The lobe of the pituitary gland that secretes oxytocin and antidiuretic hormone (vasopressin). p. 240

postganglionic fiber (pōst″gang-gle-on′ik fi′ber) Autonomic nerve fiber on the distal side of a ganglion. p. 249

postnatal (pōst-na′tal) After birth. p. 000

preganglionic fiber (pre″gang-gle-on′ik fi′ber) Autonomic nerve fiber on the proximal side of a ganglion. p. 249

pregnancy (preg′nan-se) The condition in which a female has a developing offspring in her uterus. p. 539

prenatal (pre-na′tal) Before birth. p. 000

primary follicle (pri′ma-re fol′ĭ-kl) Primordial follicle that begins to mature in response to hormonal changes in a female. p. 520

primary germ layers (pri′mar-e jerm la′erz) Three layers of cells in the embryo that develop into specific tissues and organs; ectoderm, mesoderm, and endoderm. p. 546

primary sex organs (pri′ma-re seks or′ganz) Sex-cell-producing parts; testes in males and ovaries in females. p. 508

prime mover (prīm moov′er) Muscle responsible for a particular body movement. Also called an agonist. p. 192

progenitor cell (pro-jen′ĭ-tor sel) A daughter cell of a stem cell that is partially specialized. p. 70

progesterone (pro-jes′tĭ-rōn) A female hormone secreted by the corpus luteum and placenta. p. 524

projection (pro-jek′shun) Process by which the brain causes a sensation to seem to come from the region of the body being stimulated. p. 261

prolactin (pro-lak′tin) **(PRL)** A hormone secreted by the anterior pituitary that stimulates milk production in the mammary glands. p. 296

pronation (pro-na′shun) Turning the palm of the hand downward while the forearm is parallel to the ground. p. 166

prophase (pro′fāz) Stage of mitosis when chromosomes become visible. p. 68

prostaglandins (pros″tah-glan′dins) A group of compounds with powerful, hormonelike effects. p. 294

prostate gland (pros′tāt gland) Gland surrounding the male urethra below the urinary bladder that adds its secretion to semen just prior to ejaculation. p. 513

proteins (pro′te-inz) Nitrogen-containing organic compounds consisting of amino acids. p. 43

prothrombin (pro-throm′bin) Plasma protein that functions in blood clotting. p. 330

protons (pro′tonz) Positively charged particles in an atomic nucleus. p. 32

protraction (pro-trak′shun) A forward movement of a body part. p. 166

proximal (prok′sĭ-mal) Closer to the point of attachment; opposite of *distal*. p. 15

pseudostratified epithelium (soo″do-strat′ĭ-fīd ep″ĭ-the′ lē-um) A single layer of cells appearing as more than one layer because the nuclei are at different positions in the cells. p. 98

PTH Parathyroid hormone. p. 301

puberty (pu′ber-te) Stage of development in which the reproductive organs become functional. p. 515

pulmonary circuit (pul′mo-ner″e ser′kit) System of blood vessels that carries blood between the heart and the lungs. p. 340

pulse (puls) The surge of blood felt through the walls of arteries in response to the contraction of the heart ventricles. p. 358

Punnett square (pun-it sqware) A diagram used to follow parental gene variant contributions to offspring. p. 559

pupil (pu′pil) Opening in iris through which light enters the eye. p. 280

Purkinje fibers (pur-kin′je fi′berz) Specialized cardiac muscle fibers that conduct cardiac impulses from the A-V bundle into the ventricular walls. p. 349

pyramidal cell (pĭ-ram′ĭ-dal sel) A large, pyramid-shaped neuron in the cerebral cortex. p. 237

pyruvic acid (pi-roo′vik as′id) An intermediate product of carbohydrate oxidation. p. 184

R

radioactive (ra″de-o-ak′tiv) An atom that releases energy at a constant rate. p. 33

rate-limiting enzyme (rāt lim′i-ting en′zīm) An enzyme, usually present in small amounts, that controls the rate of a metabolic pathway by regulating one of its steps. p. 82

receptors (re-sep′torz) Specialized cells that provide information about the environment. Also, cell surface structures that bind particular molecules, called ligands, thereby transmitting a signal to inside the cell. p. 6

recessive (re-sess′iv) A gene variant whose expression is masked by another. p. 559

recruitment (re-kroot′ment) Increase in the number of motor units that are activated as stimulation intensity increases. p. 188

red marrow (red mar′o) Blood-cell-forming tissue in spaces within bones. p. 138

referred pain (re-ferd′ pān) Pain that feels as if it is originating from a part other than the site being stimulated. p. 263

reflex (re′fleks) A rapid, automatic response to a stimulus. p. 228

reflex arc (re′fleks ark) A nerve pathway, consisting of a sensory neuron, interneuron, and motor neuron, that forms the structural and functional bases for a reflex. p. 228

refraction (re-frak′shun) A bending of light as it passes between media of different densities. p. 282

relaxin (re-lak′sin) Hormone from the corpus luteum that inhibits uterine contractions during pregnancy. p. 545

renal corpuscle (re′nal kor′pusl) Part of a nephron that consists of a glomerulus and a glomerular capsule. p. 472

renal cortex (re′nal kor′teks) The outer portion of a kidney. p. 471

renal medulla (re′nal mĕ-dul′ah) The inner portion of a kidney. p. 471

renal pelvis (re′nal pel′vis) The cavity in a kidney that channels urine to the ureter. p. 470

renal tubule (re′nal tu′būl) Part of a nephron that extends from the renal corpuscle to the collecting duct. p. 472

renin (re′nin) Enzyme that kidneys release that maintains blood pressure and blood volume. p. 478

replication (rep″lĭ-ka′shun) Production of an exact copy of a DNA molecule. p. 84

respiration (res″pĭ-ra′shun) Cellular process that releases energy from nutrients; breathing. p. 445

respiratory capacities (re-spi′rah-to″re kah-pas′ĭ-tēz) The sum of two or more respiratory volumes. p. 455

respiratory membrane (re-spi′rah-to″re mem′brān) Membrane composed of a capillary wall, an alveolar wall, and their basement membranes through which blood and inspired air exchange gases. p. 460

resting potential (res′ting po-ten′shal) The difference in electrical charge between the inside and the outside of an undisturbed nerve cell membrane. p. 221

reticular (rĕ-tik′u-lar) **fibers** Thin collagenous fibers. p. 104

reticular formation (rĕ-tik′u-lar for-ma′ shun) A complex network of nerve fibers in the brainstem that arouses the cerebrum. p. 242

retina (ret′ĭ-nah) Inner layer of the eye wall that includes the visual receptors. p. 281

retraction (rĕ-trak′shun) Movement of a part toward the back. p. 166

retroperitoneal (ret″ro-per″i-to-ne′al) Behind the peritoneum. p. 470

reversible reaction (re-ver′sĭ-bl re-ak′shun) Chemical reaction in which the products react, reforming the reactants. p. 38

rhodopsin (ro-dop′sin) Light-sensitive pigment in the rods of the retina; visual purple. p. 283

ribonucleic acid (ri″bo-nu-kle′ik as′id) **(RNA)** Nucleic acid whose nucleotides each include the sugar ribose, a phosphate group, and a nitrogenous base (adenine, uracil, guanine, or cytosine). p. 46

ribosome (ri′bo-sōm) Organelle composed of RNA and protein that is a structural support for protein synthesis and includes RNA molecules that function as enzymes. p. 55

RNA Ribonucleic acid. p. 46

rods (rodz) Receptors that provide colorless (black and white) vision. p. 282

rotation (ro-ta′shun) Movement turning a body part on its longitudinal axis. p. 166

round window (rownd win′do) A membrane-covered opening between the inner ear and the middle ear. p. 270

S

saccule (sak′ūl) An enlarged region of the membranous labyrinth of the inner ear. p. 273

sagittal (saj′ĭ-tal) A plane or section that divides a structure into right and left portions. p. 15

salivary amylase (sal′i-ver-e am′i-lās) An enzyme that hydrolyzes (digests) starch in the mouth. p. 409

salt (salt) A compound composed of oppositely-charged ions. p. 40

S-A node (nōd) Sinoatrial node. p. 349

sarcomeres (sar′ko-mērz) The structural and functional units of a myofibril. p. 177

sarcoplasmic reticulum (sar″ko-plaz′mik rĕ-tik′u-lum) Membranous network of channels and tubules of a muscle fiber, corresponding to the endoplasmic reticulum of other cells. p. 179

saturated fatty acids (sat′u-rāt″ed fat′e as′idz) Fatty acid molecules that include maximal hydrogens and therefore have no double carbon bonds. p. 41

Schwann cell (shwahn sel) A type of neuroglial cell that surrounds an axon of a peripheral neuron, forming the neurilemmal sheath and myelin. p. 214

sclera (skle′rah) White, fibrous outer layer of the eyeball. p. 278

scrotum (skro′tum) A pouch of skin that encloses the testes. p. 508

sebaceous glands (se-ba′shus glandz) Skin glands that secrete sebum. p. 123

sebum (se′bum) Oily secretion of sebaceous glands. p. 123

secretin (se-kre′tin) Hormone secreted in the small intestine that stimulates the pancreas to release pancreatic juice. p. 415

selectively permeable (se-lek′tiv-le per′me-ah-bl) A membrane that allows some molecules through but not others; semipermeable. p. 53

semen (se′men) Fluid containing sperm cells and secretions discharged from the male reproductive tract at ejaculation. p. 513

semicircular canals (sem″ĭ-ser′ku-lar kah-nalz′) Tubular structures of the inner ear that house receptors providing the sense of dynamic equilibrium. p. 270

seminiferous tubules (sem″ĭ-nif′er-us tu′būlz) Tubules in the testes where sperm cells form. p. 508

sensation (sen-sa′shun) A feeling resulting from the brain's interpretation of sensory nerve impulses. p. 261

sensory adaptation (sen′so-re ad″ap-ta′shun) Sensory receptors becoming unresponsive or inhibition along the CNS pathways leading to sensory regions of the cerebral cortex. p. 261

sensory area (sen′so-re a′re-ah) Part of the cerebral cortex that receives and interprets sensory nerve impulses. p. 237

sensory nerve (sen′so-re nerv) A nerve composed of sensory nerve fibers. p. 228

sensory neurons (sen′so-re nu′ronz) Neurons that transmit impulses from receptors to the central nervous system. p. 218

sensory receptors (sen′so-re re″sep′torz) Specialized structures associated with the peripheral ends of sensory neurons specific to detecting a particular sensation and triggering nerve impulses in response, which are transmitted to the central nervous system. p. 213

sensory speech area (sen′so-re spēch ār′e-ah) Region of the parietal lobe, near the temporal lobe and just posterior to the lateral sulcus, that is necessary for understanding written and spoken language; Wernicke's area. p. 237

serosa (sĕ′ro-sah) Outer covering of the alimentary canal. p. 402

serotonin (se″ro-to′nin) A vasoconstrictor released from blood platelets when blood vessels break, controlling bleeding. Also a neurotransmitter. pp. 329

serous cell (ser′us sel) Glandular cell that secretes a watery fluid (serous fluid) with high enzyme content. p. 101

serous membrane (ser′us mem′brān) Membrane that lines a cavity that does not open to the outside of the body. p. 110

set point A component of a homeostatic mechanism that establishes the range that is optimal for a particular measurement. p. 6

sex chromosome (seks crō-mo-some) A chromosome that includes a gene that determines sex. p. 558

simple sugar (sim′pl shoog′ar) Monosaccharide. p. 41

sinoatrial node (si″no-a′tre-al nōd) **(S-A node)** Specialized tissue in the wall of the right atrium that initiates cardiac cycles; pacemaker. p. 349

skeletal muscle tissue (skel′ĭ-tal mus′l tish′u) Type of voluntary muscle tissue in muscles attached to bones. p. 110

smooth muscle tissue (smooth mus′l tish′u) Type of involuntary muscle tissue in the walls of hollow viscera. p. 110

solute (sol′ūt) Chemical dissolved in a solution. p. 40

solvent (sol′vent) The liquid portion of a solution in which a solute is dissolved. p. 40

somatic nervous system (so-mat′ik ner′vus sis′tem) Motor pathways of the peripheral nervous system that lead to the skin and skeletal muscles. p. 214

special sense (spesh′al sens) Sense that stems from receptors associated with specialized sensory organs. p. 261

spermatids (sper′mah-tidz) Cells that represent an intermediate stage in sperm cell formation. p. 511

spermatogenesis (sper″mah-to-jen′ĕ-sis) Sperm cell production. p. 509

spermatogonia (sper″mah-to-go′ne-ah) Undifferentiated spermatogenic cells in the wall of a seminiferous tubule. p. 509

sphincter (sfingk′ter) A circular muscle that closes an opening or the lumen of a tubule. p. 395

spinal cord (spi′nal kord) Part of the central nervous system extending from the brainstem through the vertebral canal. p. 232

spinal nerve (spi′nal nerv) Nerve that arises from the spinal cord. p. 232

spleen (splēn) A large organ in the upper left region of the abdomen that serves as a blood reservoir. p. 383

spongy bone (spun′jē bōn) Bone that consists of bars and plates separated by irregular spaces; cancellous bone. p. 131

static equilibrium (stat′ik e″kwĭ-lib′re-um) The maintenance of balance when the head and body are motionless. p. 273

stem cell (stem sel) An undifferentiated cell that can divide to yield two daughter stem cells, or a stem cell and a progenitor cell. p. 70

steroid (ste′roid) A type of organic molecule including complex rings of carbon and hydrogen atoms. p. 42

stomach (stum′ak) Digestive organ between the esophagus and small intestine that stores food. p. 411

stratum basale (strat'tum ba'sal-e) The deepest layer of the epidermis, where cells divide; stratum germinativum. p. 118

stratum corneum (stra'tum kor'ne-um) Outer, horny layer of the epidermis. p. 118

stressor (stres'or) A factor capable of stimulating a stress response. p. 309

stroke volume (strōk vol'ūm) The volume of blood that each ventricle discharges in a heartbeat. p. 359

structural formula (struk'cher-al for'mu-lah) A representation of the way atoms bond to form a molecule, using symbols for each element and lines to indicate chemical bonds. p. 37

subarachnoid space (sub"ah-rak'noid spās) The space in the meninges between the arachnoid mater and the pia mater. p. 231

subcutaneous (sub"ku-ta'ne-us) **layer** The layer beneath the skin. p. 117

sublingual (sub-ling'gwal) Beneath the tongue. p. 409

submucosa (sub"mu-ko'sah) The layer of the alimentary canal underneath the mucosa. p. 402

substrate (sub'strāt) The target of enzyme action. p. 79

sucrase (su'krās) Digestive enzyme that catalyzes the breakdown of sucrose. p. 422

sulcus (sul'kus) (plural, *sulci*) A shallow groove, such as that between gyri on the brain surface. p. 236

summation (sum-ma'shun) Increased force of contraction by a skeletal muscle fiber when a twitch occurs before the previous twitch relaxes. p. 188

superior (soo-pe're-or) Structure above another structure. p. 14

surface tension (sur'fis ten'shun) The force that adheres moist membranes due to the attraction of water molecules. p. 453

surfactant (ser-fak'tant) Substance produced by the lungs that reduces the surface tension in alveoli. p. 453

sweat (swet) **glands** Exocrine glands in skin that secrete a mixture of water, salt, urea, and other bodily wastes. p. 124

sympathetic division (sim"pah-thet'ik de-vij'in) Part of the autonomic nervous system that arises from the thoracic and lumbar regions of the spinal cord. p. 249

synapse (sin'aps) The functional connection between the axon of a neuron and the dendrite or cell body of another neuron or the membrane of another cell type. p. 219

synaptic cleft (sĭ-nap'tik kleft) The space between two cells forming a synapse. p. 219

synaptic knob (sĭ-nap'tik nob) Tiny enlargement at the end of an axon that secretes a neurotransmitter. p. 220

synergists (sin'er-jists) Muscles that assist the action of a prime mover. p. 192

synovial (sĭ-no've-al) **joints** Freely movable joints. p. 162

synovial membrane (sĭ-no've-al mem'brān) Membrane that forms the inner lining of the capsule of a freely movable joint. p. 110

synthesis (sin'thē-sis) Building large molecules from smaller ones that join. p. 37

systemic circuit (sis-tem'ik ser'kit) The vessels that conduct blood between the heart and all body tissues except the lungs. p. 340

systole (sis'to-le) Phase of the cardiac cycle when a heart chamber wall contracts. p. 346

systolic pressure (sis-tol'ik presh'ur) Peak arterial blood pressure reached during the systolic phase of the cardiac cycle. p. 358

T

target cells (tar'get selz) Cells with specific receptors on which hormones exert their effect. p. 290

tarsals (tahr'sulz) The ankle bones. p. 17

taste bud (tāst bud) Organ including receptors associated with the sense of taste. p. 267

T cell (sel) A type of lymphocyte that interacts directly with antigens, producing the cellular immune response. p. 383

telophase (tel'o-fāz) Stage in mitosis when newly formed cells separate. p. 68

tendons (ten'donz) Cordlike or bandlike masses of white fibrous connective tissue that connect muscles to bones. p. 104

testes (tes'tēz) (singular, *testis*) Primary male reproductive organs; sperm-cell-producing organs. p. 298

testosterone (tes-tos'tĕ-rōn) Male sex hormone secreted by the interstitial cells of the testes. p. 515

tetanic (tĕ-tan'ik) **contraction** Continuous, forceful muscular contraction without relaxation. p. 188

thalamus (thal'ah-mus) A mass of gray matter at the base of the cerebrum in the wall of the third ventricle. p. 240

thermoreceptors (ther"mo-re-sep'torz) A sensory receptor sensitive to temperature changes; heat and cold receptors. p. 261

thoracic cavity (tho-ras'ik kav'ĭ-te) The space above the diaphragm in the chest. p. 8

threshold stimulus (thresh'old stim'u-lus) Stimulation level that must be exceeded to elicit a nerve impulse or a muscle contraction. p. 187

thrombus (throm'bus) A blood clot that remains where it forms in a blood vessel. p. 331

thymosins (thi'mo-sinz) A group of peptides secreted from the thymus that increases production of certain types of white blood cells. p. 308

thymus (thi'mus) A glandular organ in the mediastinum behind the sternum and between the lungs. p. 307

thyroid gland (thi'roid gland) Endocrine gland located just below the larynx and in front of the trachea that secretes thyroid hormones. p. 299

thyroid-stimulating hormone (thi-roid stim-ū-lay-ting hor-mone) **(TSH)** A hormone secreted from the anterior pituitary that controls secretion from the thyroid gland. p. 296

thyroxine (thi-rok'sin) A type of thyroid hormone. p. 299

tissues (tish'uz) Groups of similar cells that perform a specialized function. p. 3

trachea (tra'ke-ah) Tubular organ that leads from the larynx to the bronchi. p. 448

transcellular fluid (trans"sel'u-lar floo'id) A portion of the extracellular fluid, including the fluid within special body cavities. p. 492

transcription (trans-krip'shun) Manufacturing a complementary RNA from DNA. p. 86

transfer RNA (trans'fer) RNA molecule that carries an amino acid to a ribosome in protein synthesis. p. 87

translation (trans-la'shun) Assembly of an amino acid chain according to the sequence of base triplets in an mRNA molecule. p. 87

transverse (tranz-vers') A plane that divides a structure into superior and inferior portions. p. 15

transverse tubules (trans-vers' tu'būlz) Membranous channels that extend inward from a muscle fiber membrane. p. 180

tricuspid valve (tri-kus′pid valv) Heart valve between the right atrium and the right ventricle. p. 343

triglycerides (tri-glis′er-īdz) Lipids composed of three fatty acids and a glycerol molecule. p. 41

triiodothyronine (tri″i-o″do-thi′ro-nēn) A type of thyroid hormone. p. 299

trypsin (trip′sin) An enzyme in pancreatic juice that breaks down protein molecules. p. 414

twitch (twich) A brief contraction of a muscle fiber followed by relaxation. p. 187

U

umbilical cord (um-bil′ĭ-kal kord) Cordlike structure, containing one vein and two arteries, that connects the fetus to the placenta. p. 549

umbilical region (um-bil′ĭ-kal re′jun) The central portion of the abdomen. p. 16

unsaturated fatty acid (un-sat′u-rāt″ed fat′e as′id) Fatty acid molecule with one or more double carbon bonds. p. 41

urea (u-re′ah) A nonprotein nitrogenous substance resulting from protein metabolism. p. 482

ureter (u-re′ter) A muscular tube that carries urine from the kidney to the urinary bladder. p. 483

urethra (u-re′thrah) Tube leading from the urinary bladder to the outside of body. p. 486

urine (u′rin) Wastes and excess water removed from the blood and excreted by the kidneys into the ureters, to the urinary bladder, and out of the body through the urethra. p. 470

uterine tube (u′ter-in tūb) Tube that extends from the uterus on each side toward an ovary and transports sex cells; oviduct or fallopian tube. p. 521

uterus (u′ter-us) Hollow, muscular organ in the female pelvis where a fetus develops. p. 521

utricle (u′trĭ-kl) An enlarged portion of the membranous labyrinth of the inner ear. p. 273

uvula (u′vu-lah) A fleshy portion of the soft palate that extends down above the root of the tongue. p. 405

V

vaccine (vak′sēn) A substance that includes antigens that stimulate an immune response against a particular pathogen. p. 392

vagina (vah-ji′nah) Tubular organ that leads from the uterus to the vestibule of the female reproductive tract. p. 522

vasoconstriction (vas″o-kon-strik′shun) A decrease in the diameter of a blood vessel. p. 353

vasodilation (vas″o-di-la′shun) An increase in the diameter of a blood vessel. p. 355

veins (vānz) Vessels that carry blood toward the heart. p. 357

vena cava (ve′nah kav′ah) One of two large veins (superior and inferior) that convey deoxygenated blood to the right atrium of the heart. p. 368

ventral root (ven′tral root) Motor branch of a spinal nerve by which it connects with the spinal cord. p. 248

ventricles (ven′trĭ-klz) Cavities, such as brain ventricles filled with cerebrospinal fluid, or heart ventricles that contain blood. pp. 238, 342

venules (ven′ūlz) Vessels that carry blood from capillaries to veins. p. 357

vertebral (ver′te-bral) Pertaining to the spinal column. p. 8

vesicles (ves′ĭ-klz) Membranous cytoplasmic sacs formed by infoldings of the cell membrane. p. 58

viscera (vis′er-ah) Organs in body cavities, especially in the abdomen. p. 8

visceral pericardium (vis′er-al per″ĭ-kar′de-um) Membrane that covers the surface of the heart. p. 10

visceral peritoneum (vis′er-al per″ĭ-to-ne′-um) Membrane that covers the abdominal organs. p. 10

visceral pleura (vis′er-al ploo′rah) Membrane that covers the surfaces of the lungs. p. 451

viscosity (vis-kos′ĭ-te) The tendency for a fluid to resist flowing due to the internal friction of its molecules. p. 359

vitamins (vi′tah-minz) Organic compounds required for normal metabolism that the body cannot synthesize in adequate amounts and must therefore be obtained in the diet. p. 433

vitreous humor (vit′re-us hu′mor) Fluid between the lens and the retina of the eye. p. 281

vocal cords (vo′kal kordz) Folds of tissue of the larynx that vibrate and produce sounds. p. 447

vulva (vul′vah) The external female reproductive parts that surround the vaginal opening. p. 522

W

water balance (wot′er bal′ans) Equivalence of the volume of water entering the body with the volume leaving it. p. 494

water of metabolism (wot′er uv mĕ-tab′o-lizm) Water produced as a by-product of metabolism. p. 494

Y

yellow marrow (yel′o mar′o) Fat storage tissue in certain bone cavities. p. 138

Z

zygote (zi′gōt) Cell produced when an oocyte and a sperm fuse; fertilized ovum. p. 540

zymogen granules (zi-mo′jen gran′ūlz) Cellular structures that store inactive forms of enzymes in a cell. p. 414

Photo Credits

Application Index

Tables

Subject Index